ULLMANN'S

Modeling and Simulation

1807–2007 Knowledge for Generations

Each generation has its unique needs and aspirations. When Charles Wiley first opened his small printing shop in lower Manhattan in 1807, it was a generation of boundless potential searching for an identity. And we were there, helping to define a new American literary tradition. Over half a century later, in the midst of the Second Industrial Revolution, it was a generation focused on building the future. Once again, we were there, supplying the critical scientific, technical, and engineering knowledge that helped frame the world. Throughout the 20th Century, and into the new millennium, nations began to reach out beyond their own borders and a new international community was born. Wiley was there, expanding its operations around the world to enable a global exchange of ideas, opinions, and know-how.

For 200 years, Wiley has been an integral part of each generation's journey, enabling the flow of information and understanding necessary to meet their needs and fulfill their aspirations. Today, bold new technologies are changing the way we live and learn. Wiley will be there, providing you the must-have knowledge you need to imagine new worlds, new possibilities, and new opportunities.

Generations come and go, but you can always count on Wiley to provide you the knowledge you need, when and where you need it!

William J. Pesce
President and Chief Executive Officer

Peter Booth Wiley
Chairman of the Board

ULLMANN'S

Modeling and Simulation

WILEY-VCH Verlag GmbH & Co. KGaA

Library of Congress Card No.:
applied for

British Library Cataloguing-in-Publication Data:
A catalogue record for this book is available from the British Library.

Bibliographic information published by The Deutsche Nationalbibliothek
The Deutsche Nationalibliothek lists this publication in the Deutsche Nationalbibliografie; detailed bibliographic data is available in the Internet at http://dnb.d-nb.de

Typesetting: Steingraeber Satztechnik GmbH, Ladenburg

Printing: Strauss GmbH, Mörlenbach

Binding: Litges & Dopf Buchbinderei GmbH, Heppenheim

Cover Design: Grafik-Design Schulz, Fußgönheim

Wiley Bicentennial Logo Richard J. Pacifico

Printed in the Federal Republic of Germany
Printed on acid-free paper

ISBN: 978-3-527-31605-2

Preface

The major tasks facing engineers and scientists in industrial chemistry are the operation and optimization of existing processes and the design of new or improved ones. Modeling and simulation involves working with what is usually a simplified description of a system. The increasing pressure on development time and product quality in industrial process development stimulates the application of numerical methods in addition to traditional experimental ones. In many cases the numerical tools can save time and costs if they are used complementary to experimental methods.

Based on the latest online edition of Ullmann's Encyclopedia of Industrial Chemistry, and containing articles yet unpublished in print, this ready reference meets the need for a comprehensive survey of the mathematical fundamentals, complementary computational approaches as well as the application of modeling and simulation in chemistry and engineering.

Each of the detailed and carefully edited articles is written by renowned international experts. Readers benefit from the rigorous and cross-indexed nature of the parent reference, and will find both broad introductory information as well as in-depth details of significance to industrial and academic environments.

We are convinced that this handbook is a convenient source of information, tailor-made for engineers, scientists and computational chemists.

The Publisher

Contents

Symbols and Units IX
Conversion Factors XI
Abbreviations XII
Country Codes XVII
Periodic Table of Elements XVIII

Mathematics in Chemical Engineering .. 3
Model Reactors and Their Design Equations 151
Mathematical Modeling 203
Molecular Modeling 307
Molecular Dynamics Simulation 323
Computational Fluid Dynamics 341
Design of Experiments 363
Microreactors – Modeling and Simulation 401

Author Index 443

Subject Index 447

Ullmann's Modeling and Simulation

ISBN: 978-3-527-31605-2

Symbols and Units

Symbols and units agree with SI standards (for conversion factors see page XI). The following list gives the most important symbols used in the encyclopedia. Articles with many specific units and symbols have a similar list as front matter.

Symbol	Unit	Physical Quantity
a_{B}		activity of substance B
A_{r}		relative atomic mass (atomic weight)
A	m^2	area
c_{B}	mol/m^3, mol/L (M)	concentration of substance B
C	C/V	electric capacity
c_p, c_v	$J\,kg^{-1}K^{-1}$	specific heat capacity
d	cm, m	diameter
d		relative density ($\varrho/\varrho_{\mathrm{water}}$)
D	m^2/s	diffusion coefficient
D	Gy (= J/kg)	absorbed dose
e	C	elementary charge
E	J	energy
E	V/m	electric field strength
E	V	electromotive force
E_{A}	J	activation energy
f		activity coefficient
F	C/mol	Faraday constant
F	N	force
g	m/s^2	acceleration due to gravity
G	J	Gibbs free energy
h	m	height
$\hbar$	$W \cdot s^2$	Planck constant
H	J	enthalpy
I	A	electric current
I	cd	luminous intensity
k	(variable)	rate constant of a chemical reaction
k	J/K	Boltzmann constant
K	(variable)	equilibrium constant
l	m	length
m	g, kg, t	mass
M_{r}		relative molecular mass (molecular weight)
n_{D}^{20}		refractive index (sodium D-line, 20 °C)
n	mol	amount of substance
N_{A}	mol^{-1}	Avogadro constant ($6.023 \times 10^{23}\ mol^{-1}$)
p	Pa, bar *	pressure
Q	J	quantity of heat
r	m	radius
R	$J\,K^{-1}mol^{-1}$	gas constant
R	Ω	electric resistance
S	J/K	entropy
t	s, min, h, d, month, a	time

Symbols and Units (Continued from p. IX)

Symbol	Unit	Physical Quantity
t	°C	temperature
T	K	absolute temperature
u	m/s	velocity
U	V	electric potential
U	J	internal energy
V	m^3, L, mL, µL	volume
w		mass fraction
W	J	work
x_B		mole fraction of substance B
Z		proton number, atomic number
α		cubic expansion coefficient
α	$W\,m^{-2}K^{-1}$	heat-transfer coefficient (heat-transfer number)
α		degree of dissociation of electrolyte
$[\alpha]$	10^{-2}deg cm^2g^{-1}	specific rotation
η	Pa · s	dynamic viscosity
θ	°C	temperature
$\varkappa$		c_p/c_v
λ	$W\,m^{-1}K^{-1}$	thermal conductivity
λ	nm, m	wavelength
μ		chemical potential
ν	Hz, s^{-1}	frequency
ν	m^2/s	kinematic viscosity (η/ϱ)
π	Pa	osmotic pressure
ϱ	g/cm^3	density
σ	N/m	surface tension
τ	Pa (N/m^2)	shear stress
φ		volume fraction
χ	Pa^{-1} (m^2/N)	compressibility

* The official unit of pressure is the pascal (Pa).

Conversion Factors

SI unit	Non-SI unit	From SI to non-SI multiply by
Mass		
kg	pound (avoirdupois)	2.205
kg	ton (long)	9.842×10^{-4}
kg	ton (short)	1.102×10^{-3}
Volume		
m^3	cubic inch	6.102×10^{4}
m^3	cubic foot	35.315
m^3	gallon (U.S., liquid)	2.642×10^{2}
m^3	gallon (Imperial)	2.200×10^{2}
Temperature		
°C	°F	$°C \times 1.8 + 32$
Force		
N	dyne	1.0×10^{5}
Energy, Work		
J	Btu (int.)	9.480×10^{-4}
J	cal (int.)	2.389×10^{-1}
J	eV	6.242×10^{18}
J	erg	1.0×10^{7}
J	$kW \cdot h$	2.778×10^{-7}
J	$kp \cdot m$	1.020×10^{-1}
Pressure		
MPa	at	10.20
MPa	atm	9.869
MPa	bar	10
kPa	mbar	10
kPa	mm Hg	7.502
kPa	psi	0.145
kPa	torr	7.502

Powers of Ten

E (exa)	10^{18}	d (deci)	10^{-1}
P (peta)	10^{15}	c (centi)	10^{-2}
T (tera)	10^{12}	m (milli)	10^{-3}
G (giga)	10^{9}	μ (micro)	10^{-6}
M (mega)	10^{6}	n (nano)	10^{-9}
k (kilo)	10^{3}	p (pico)	10^{-12}
h (hecto)	10^{2}	f (femto)	10^{-15}
da (deca)	10	a (atto)	10^{-18}

Abbreviations

The following is a list of the abbreviations used in the text. Common terms, the names of publications and institutions, and legal agreements are included along with their full identities. Other abbreviations will be defined wherever they first occur in an article. For further abbreviations, see page IX, Symbols and Units; page XVI, Frequently Cited Companies (Abbreviations), and page XVII, Country Codes in patent references. The names of periodical publications are abbreviated exactly as done by Chemical Abstracts Service.

abs. absolute
a.c. alternating current
ACGIH American Conference of Governmental Industrial Hygienists
ACS American Chemical Society
ADI acceptable daily intake
ADN accord européen relatif au transport international des marchandises dangereuses par voie de navigation interieure (European agreement concerning the international transportation of dangerous goods by inland waterways)
ADNR ADN par le Rhin (regulation concerning the transportation of dangerous goods on the Rhine and all national waterways of the countries concerned)
ADP adenosine 5′-diphosphate
ADR accord européen relatif au transport international des marchandises dangereuses par route (European agreement concerning the international transportation of dangerous goods by road)
AEC Atomic Energy Commission (United States)
a.i. Active ingredient
AIChE American Institute of Chemical Engineers
AIME American Institute of Mining, Metallurgical, and Petroleum Engineers
ANSI American National Standards Institute
AMP adenosine 5′-monophosphate
APhA American Pharmaceutical Association
API American Petroleum Institute
ASTM American Society for Testing and Materials
ATP adenosine 5′-triphosphate
BAM Bundesanstalt für Materialprüfung (Federal Republic of Germany)
BAT Biologischer Arbeitsstoff-Toleranz-Wert (biological tolerance value for a working material, established by MAK Commission, see MAK)
Beilstein Beilstein's Handbook of Organic Chemistry, Springer, Berlin – Heidelberg – New York
BET Brunauer – Emmett – Teller
BGA Bundesgesundheitsamt (Federal Republic of Germany)
BGBl. Bundesgesetzblatt (Federal Republic of Germany)
BIOS British Intelligence Objectives Subcommitee Report (see also FIAT)
BOD biological oxygen demand
bp boiling point
B.P. British Pharmacopeia
BS British Standard
ca. circa
calcd. calculated
CAS Chemical Abstracts Service
cat. catalyst, catalyzed
CEN Comité Européen de Normalisation
cf. compare
CFR Code of Federal Regulations (United States)
cfu colony forming units
Chap. chapter
ChemG Chemikaliengesetz (Federal Republic of Germany)
C.I. Colour Index
CIOS Combined Intelligence Objectives Subcommitee Report (see also FIAT)
CNS central nervous system
Co. Company
COD chemical oxygen demand
conc. concentrated
const. constant
Corp. Corporation
crit. critical

CTFA The Cosmetic, Toiletry and Fragrance Association (United States)
DAB 9 Deutsches Arzneibuch, 9th ed., Deutscher Apotheker-Verlag, Stuttgart 1986
d.c. direct current
decomp. decompose, decomposition
DFG Deutsche Forschungsgemeinschaft (German Science Foundation)
dil. dilute, diluted
DIN Deutsche Industrie Norm (Federal Republic of Germany)
DMF dimethylformamide
DNA deoxyribonucleic acid
DOE Department of Energy (United States)
DOT Department of Transportation – Materials Transportation Bureau (United States)
DTA differential thermal analysis
EC effective concentration
EC European Community
ed. editor, edition, edited
e.g. for example
emf electromotive force
EmS Emergency Schedule
EN European Standard (European Community)
EPA Environmental Protection Agency (United States)
EPR electron paramagnetic resonance
Eq. equation
ESCA electron spectroscopy for chemical analysis
esp. especially
ESR electron spin resonance
Et ethyl substituent ($-C_2H_5$)
et al. and others
etc. et cetera
EVO Eisenbahnverkehrsordnung (Federal Republic of Germany)
exp (...) $e^{(\cdots)}$, mathematical exponent
FAO Food and Agriculture Organization (United Nations)
FDA Food and Drug Administration (United States)
FD & C Food, Drug and Cosmetic Act (United States)
FHSA Federal Hazardous Substances Act (United States)
FIAT Field Information Agency, Technical (United States reports on the chemical industry in Germany, 1945)
Fig. figure
fp freezing point
Friedländer P. Friedländer, Fortschritte der Teerfarbenfabrikation und verwandter Industriezweige, Vol. 1 – 25, Springer, Berlin 1888 – 1942
FT Fourier transform
(g) gas, gaseous
GC gas chromatography
GefStoffV Gefahrstoffverordnung (regulations in the Federal Republic of Germany concerning hazardous substances)
GGVE Verordnung in der Bundesrepublik Deutschland über die Beförderung gefährlicher Güter mit der Eisenbahn (regulation in the Federal Republic of Germany concerning the transportation of dangerous goods by rail)
GGVS Verordnung in der Bundesrepublik Deutschland über die Beförderung gefährlicher Güter auf der Straße (regulation in the Federal Republic of Germany concerning the transportation of dangerous goods by road)
GGVSee Verordnung in der Bundesrepublik Deutschland über die Beförderung gefährlicher Güter mit Seeschiffen (regulation in the Federal Republic of Germany concerning the transportation of dangerous goods by sea-going vessels)
GLC gas-liquid chromatography
Gmelin Gmelin's Handbook of Inorganic Chemistry, 8th ed., Springer, Berlin – Heidelberg – New York
GRAS generally recognized as safe
Hal halogen substituent (−F, −Cl, −Br, −I)
Houben-Weyl Methoden der organischen Chemie, 4th ed., Georg Thieme Verlag, Stuttgart
HPLC high performance liquid chromatography
IAEA International Atomic Energy Agency
IARC International Agency for Research on Cancer, Lyon, France

IATA-DGR International Air Transport Association, Dangerous Goods Regulations
ICAO International Civil Aviation Organization
i.e. that is
i.m. intramuscular
IMDG International Maritime Dangerous Goods Code
IMO Inter-Governmental Maritime Consultive Organization (in the past: IMCO)
Inst. Institute
i.p. intraperitoneal
IR infrared
ISO International Organization for Standardization
IUPAC International Union of Pure and Applied Chemistry
i.v. intravenous
Kirk-Othmer Encyclopedia of Chemical Technology, 3rd ed., J. Wiley & Sons, New York – Chichester – Brisbane – Toronto 1978 – 1984; 4th ed., J. Wiley & Sons, New York – Chichester – Brisbane – Toronto 1991 – 1998
(l) liquid
Landolt-Börnstein Zahlenwerte u. Funktionen aus Physik, Chemie, Astronomie, Geophysik u. Technik, Springer, Heidelberg 1950 – 1980; Zahlenwerte und Funktionen aus Naturwissenschaften und Technik, Neue Serie, Springer, Heidelberg, since 1961
LC_{50} lethal concentration for 50 % of the test animals
LCLo lowest published lethal concentration
LD_{50} lethal dose for 50 % of the test animals
LDLo lowest published lethal dose
ln logarithm (base e)
LNG liquefied natural gas
log logarithm (base 10)
LPG liquefied petroleum gas
M mol/L
M metal (in chemical formulas)
MAK Maximale Arbeitsplatz-Konzentration (maximum concentration at the workplace in the Federal Republic of Germany); cf. Deutsche Forschungsgemeinschaft (ed.): Maximale Arbeitsplatz-konzentrationen (MAK) und Biologische Arbeitsstoff-Toleranz-Werte (BAT), WILEY-VCH Verlag, Weinheim (published annually)
max. maximum
MCA Manufacturing Chemists Association (United States)
Me methyl substituent ($-CH_3$)
Methodicum Chimicum Methodicum Chimicum, Georg Thieme Verlag, Stuttgart
MFAG Medical First Aid Guide for Use in Accidents Involving Dangerous Goods
MIK maximale Immissionskonzentration (maximum immission concentration)
min. minimum
mp melting point
MS mass spectrum, mass spectrometry
NAS National Academy of Sciences (United States)
NASA National Aeronautics and Space Administration (United States)
NBS National Bureau of Standards (United States)
NCTC National Collection of Type Cultures (United States)
NIH National Institutes of Health (United States)
NIOSH National Institute for Occupational Safety and Health (United States)
NMR nuclear magnetic resonance
no. number
NOEL no observed effect level
NRC Nuclear Regulatory Commission (United States)
NRDC National Research Development Corporation (United States)
NSC National Service Center (United States)
NSF National Science Foundation (United States)
NTSB National Transportation Safety Board (United States)
OECD Organization for Economic Cooperation and Development
OSHA Occupational Safety and Health Administration (United States)

p., pp. page, pages
Patty G. D. Clayton, F. E. Clayton (eds.): Patty's Industrial Hygiene and Toxicology, 3rd ed., Wiley Interscience, New York
PB report Publication Board Report (U.S. Department of Commerce, Scientific and Industrial Reports)
PEL permitted exposure limit
Ph phenyl substituent ($-C_6H_5$)
Ph. Eur. European Pharmacopoeia, 2nd. ed., Council of Europe, Strasbourg 1981
phr part per hundred rubber (resin)
PNS peripheral nervous system
ppm parts per million
q. v. which see (quod vide)
ref. refer, reference
resp. respectively
R_f retention factor (TLC)
R. H. relative humidity
RID règlement international concernant le transport des marchandises dangereuses par chemin de fer (international convention concerning the transportation of dangerous goods by rail)
RNA ribonucleic acid
R phrase (R-Satz) risk phrase according to ChemG and GefStoffV (Federal Republic of Germany)
rpm revolutions per minute
RTECS Registry of Toxic Effects of Chemical Substances, edited by the National Institute of Occupational Safety and Health (United States)
(s) solid
SAE Society of Automotive Engineers (United States)
s.c. subcutaneous
SI International System of Units
SIMS secondary ion mass spectrometry
S phrase (S-Satz) safety phrase according to ChemG and GefStoffV (Federal Republic of Germany)
STEL Short Term Exposure Limit (see TLV)
STP standard temperature and pressure (0° C, 101.325 kPa)
T_g glass transition temperature
TA Luft Technische Anleitung zur Reinhaltung der Luft (clean air regulation in Federal Republic of Germany)
TA Lärm Technische Anleitung zum Schutz gegen Lärm (low noise regulation in Federal Republic of Germany)
TDLo lowest published toxic dose
THF tetrahydrofuran
TLC thin layer chromatography
TLV Threshold Limit Value (TWA and STEL); published annually by the American Conference of Governmental Industrial Hygienists (ACGIH), Cincinnati, Ohio
TOD total oxygen demand
TRK Technische Richtkonzentration (lowest technically feasible level)
TSCA Toxic Substances Control Act (United States)
TÜV Technischer Überwachungsverein (Technical Control Board of the Federal Republic of Germany)
TWA Time Weighted Average
UBA Umweltbundesamt (Federal Environmental Agency)
Ullmann Ullmann's Encyclopedia of Industrial Chemistry, 6th ed., Wiley-VCH, Weinheim, 2002, 5th ed., VCH Verlagsgesellschaft, Weinheim, 1985 – 1996; Ullmanns Encyklopädie der Technischen Chemie, 4th ed., Verlag Chemie, Weinheim 1972 – 1984
USAEC United States Atomic Energy Commission
USAN United States Adopted Names
USD United States Dispensatory
USDA United States Department of Agriculture
U.S.P. United States Pharmacopeia
UV ultraviolet
UVV Unfallverhütungsvorschriften der Berufsgenossenschaft (workplace safety regulations in the Federal Republic of Germany)
VbF Verordnung in der Bundesrepublik Deutschland über die Errichtung und den Betrieb von Anlagen zur Lagerung, Abfüllung und Beförderung brennbarer Flüssigkeiten (regulation in the Federal Republic of Germany

	concerning the construction and operation of plants for storage, filling, and transportation of flammable liquids; classification according to the flash point of liquids, in accordance with the classification in the United States)
VDE	Verband Deutscher Elektroingenieure (Federal Republic of Germany)
VDI	Verein Deutscher Ingenieure (Federal Republic of Germany)
vol	volume
vol.	volume (of a series of books)
vs.	versus
WGK	Wassergefährdungsklasse (water hazard class)
WHO	World Health Organization (United Nations)
Winnacker-Küchler	Chemische Technologie, 4th ed., Carl Hanser Verlag, München, 1982-1986; Winnacker-Küchler, Chemische Technik: Prozesse und Produkte, Wiley-VCH, Weinheim, from 2003
wt	weight
$	U.S. dollar, unless otherwise stated

Frequently Cited Companies (Abbreviations)

Air Products	Air Products and Chemicals
Akzo	Algemene Koninklijke Zout Organon
Alcoa	Aluminum Company of America
Allied	Allied Corporation
Amer. Cyanamid	American Cyanamid Company
BASF	BASF Aktiengesellschaft
Bayer	Bayer AG
BP	British Petroleum Company
Celanese	Celanese Corporation
Daicel	Daicel Chemical Industries
Dainippon	Dainippon Ink and Chemicals Inc.
Dow Chemical	The Dow Chemical Company
DSM	Dutch Staats Mijnen
Du Pont	E.I. du Pont de Nemours & Company
Exxon	Exxon Corporation
FMC	Food Machinery & Chemical Corporation
GAF	General Aniline & Film Corporation
W.R. Grace	W.R. Grace & Company
Hoechst	Hoechst Aktiengesellschaft
IBM	International Business Machines Corporation
ICI	Imperial Chemical Industries
IFP	Institut Français du Pétrole
INCO	International Nickel Company
3M	Minnesota Mining and Manufacturing Company
Mitsubishi Chemical	Mitsubishi Chemical Industries
Monsanto	Monsanto Company
Nippon Shokubai	Nippon Shokubai Kagaku Kogyo
PCUK	Pechiney Ugine Kuhlmann
PPG	Pittsburg Plate Glass Industries
Searle	G.D. Searle & Company
SKF	Smith Kline & French Laboratories
SNAM	Societá Nazionale Metandotti
Sohio	Standard Oil of Ohio
Stauffer	Stauffer Chemical Company
Sumitomo	Sumitomo Chemical Company
Toray	Toray Industries Inc.
UCB	Union Chimique Belge
Union Carbide	Union Carbide Corporation
UOP	Universal Oil Products Company
VEBA	Vereinigte Elektrizitäts- und Bergwerks-AG
Wacker	Wacker Chemie GmbH

Country Codes

The following list contains a selection of standard country codes used in the patent references.

AT	Austria	ID	Indonesia
AU	Australia	IL	Israel
BE	Belgium	IT	Italy
BG	Bulgaria	JP	Japan *
BR	Brazil	LU	Luxembourg
CA	Canada	MA	Morocco
CH	Switzerland	NL	Netherlands *
DE	Federal Republic of Germany (and Germany before 1949) *	NO	Norway
		NZ	New Zealand
DK	Denmark	PL	Poland
ES	Spain	PT	Portugal
FI	Finland	SE	Sweden
FR	France	US	United States of America
GB	United Kingdom	ZA	South Africa
GR	Greece	EP	European Patent Office *
HU	Hungary	WO	World Intellectual Property Organization

* For Europe, Federal Republic of Germany, Japan, and the Netherlands, the type of patent is specified: EP (patent), EP-A (application), DE (patent), DE-OS (Offenlegungsschrift), DE-AS (Auslegeschrift), JP (patent), JP-Kokai (Kokai tokkyo koho), NL (patent), and NL-A (application).

Periodic Table of Elements

element symbol, atomic number, and relative atomic mass (atomic weight)

1A "European" group designation and old IUPAC recommendation
1 group designation to 1986 IUPAC proposal
IA "American" group designation, also used by the Chemical Abstracts Service until the end of 1986

1A **1** IA	2A **2** IIA	3A **3** IIIB	4A **4** IVB	5A **5** VB	6A **6** VIB	7A **7** VIIB	8 **8** VIII	8 **9** VIII	8 **10** VIII	1B **11** IB	2B **12** IIB	3B **13** IIIA	4B **14** IVA	5B **15** VA	6B **16** VIA	7B **17** VIA	0 **18** VIIIA
1 **H** 1.0079																	2 **He** 4.0026
3 **Li** 6.941	4 **Be** 9.0122											5 **B** 10.811	6 **C** 12.011	7 **N** 14.007	8 **O** 15.999	9 **F** 18.998	10 **Ne** 20.180
11 **Na** 22.990	12 **Mg** 24.305											13 **Al** 26.982	14 **Si** 28.086	15 **P** 30.974	16 **S** 32.066	17 **Cl** 35.453	18 **Ar** 39.948
19 **K** 39.098	20 **Ca** 40.078	21 **Sc** 44.956	22 **Ti** 47.867	23 **V** 50.942	24 **Cr** 51.996	25 **Mn** 54.938	26 **Fe** 55.845	27 **Co** 58.933	28 **Ni** 58.693	29 **Cu** 63.546	30 **Zn** 65.409	31 **Ga** 69.723	32 **Ge** 72.61	33 **As** 74.922	34 **Se** 78.96	35 **Br** 79.904	36 **Kr** 83.80
37 **Rb** 85.468	38 **Sr** 87.62	39 **Y** 88.906	40 **Zr** 91.224	41 **Nb** 92.906	42 **Mo** 95.94	43 **Tc*** 98.906	44 **Ru** 101.07	45 **Rh** 102.91	46 **Pd** 106.42	47 **Ag** 107.87	48 **Cd** 112.41	49 **In** 114.82	50 **Sn** 118.71	51 **Sb** 121.76	52 **Te** 127.60	53 **I** 126.90	54 **Xe** 131.29
55 **Cs** 132.91	56 **Ba** 137.33		72 **Hf** 178.49	73 **Ta** 180.95	74 **W** 183.84	75 **Re** 186.21	76 **Os** 190.23	77 **Ir** 192.22	78 **Pt** 195.08	79 **Au** 196.97	80 **Hg** 200.59	81 **Tl** 204.38	82 **Pb** 207.2	83 **Bi** 208.98	84 **Po*** 208.98	85 **At*** 209.99	86 **Rn*** 222.02
87 **Fr*** 223.02	88 **Ra*** 226.03		104 **Rf*** 261.11	105 **Db*** 262.11	106 **Sg**	107 **Bh**	108 **Hs**	109 **Mt**	110 **Uun**[a]	111 **Uuu**[a]	112 **Uub**[a]		114 **Uuq**[a]		116 **Uuh**[a]		

[a] provisional IUPAC symbol

57 **La** 138.91	58 **Ce** 140.12	59 **Pr** 140.91	60 **Nd** 144.24	61 **Pm*** 146.92	62 **Sm** 150.36	63 **Eu** 151.97	64 **Gd** 157.25	65 **Tb** 158.93	66 **Dy** 162.50	67 **Ho** 164.93	68 **Er** 167.26	69 **Tm** 168.93	70 **Yb** 173.04	71 **Lu** 174.97
89 **Ac*** 227.03	90 **Th*** 232.04	91 **Pa*** 231.04	92 **U*** 238.03	93 **Np*** 237.05	94 **Pu*** 244.06	95 **Am*** 243.06	96 **Cm*** 247.07	97 **Bk*** 247.07	98 **Cf*** 251.08	99 **Es*** 252.08	100 **Fm*** 257.10	101 **Md*** 258.10	102 **No*** 259.10	103 **Lr*** 260.11

* radioactive element; mass of most important isotope given.

Modeling and Simulation

Mathematics in Chemical Engineering

BRUCE A. FINLAYSON, Department of Chemical Engineering, University of Washington, Seattle, Washington, United States (Chap. 1, 2, 3, 4, 5, 6, 7, 8, 9, 11 and 12)

LORENZ T. BIEGLER, Carnegie Mellon University, Pittsburgh, Pennsylvania, United States (Chap. 10)

IGNACIO E. GROSSMANN, Carnegie Mellon University, Pittsburgh, Pennsylvania, United States (Chap. 10)

1. **Solution of Equations** 5
1.1. **Matrix Properties** 5
1.2. **Linear Algebraic Equations** 7
1.3. **Nonlinear Algebraic Equations** . . . 10
1.4. **Linear Difference Equations** 12
1.5. **Eigenvalues** 12
2. **Approximation and Integration** . . . 13
2.1. **Introduction** 13
2.2. **Global Polynomial Approximation** . 13
2.3. **Piecewise Approximation** 15
2.4. **Quadrature** 18
2.5. **Least Squares** 20
2.6. **Fourier Transforms of Discrete Data** 21
2.7. **Two-Dimensional Interpolation and Quadrature** 22
3. **Complex Variables** 22
3.1. **Introduction to the Complex Plane** . 22
3.2. **Elementary Functions** 23
3.3. **Analytic Functions of a Complex Variable** 25
3.4. **Integration in the Complex Plane** . . 26
3.5. **Other Results** 28
4. **Integral Transforms** 29
4.1. **Fourier Transforms** 29
4.2. **Laplace Transforms** 33
4.3. **Solution of Partial Differential Equations by Using Transforms** . . . 37
5. **Vector Analysis** 40
6. **Ordinary Differential Equations as Initial Value Problems** 49
6.1. **Solution by Quadrature** 50
6.2. **Explicit Methods** 50
6.3. **Implicit Methods** 54
6.4. **Stiffness** 55
6.5. **Differential – Algebraic Systems** . . . 56
6.6. **Computer Software** 57
6.7. **Stability, Bifurcations, Limit Cycles** 58
6.8. **Sensitivity Analysis** 60
6.9. **Molecular Dynamics** 61
7. **Ordinary Differential Equations as Boundary Value Problems** 61
7.1. **Solution by Quadrature** 62
7.2. **Initial Value Methods** 62
7.3. **Finite Difference Method** 63
7.4. **Orthogonal Collocation** 65
7.5. **Orthogonal Collocation on Finite Elements** 69
7.6. **Galerkin Finite Element Method** . . 71
7.7. **Cubic B-Splines** 73
7.8. **Adaptive Mesh Strategies** 73
7.9. **Comparison** 73
7.10. **Singular Problems and Infinite Domains** 74
8. **Partial Differential Equations** 75
8.1. **Classification of Equations** 75
8.2. **Hyperbolic Equations** 77
8.3. **Parabolic Equations in One Dimension** 79
8.4. **Elliptic Equations** 84
8.5. **Parabolic Equations in Two or Three Dimensions** 86
8.6. **Special Methods for Fluid Mechanics** 86
8.7. **Computer Software** 88
9. **Integral Equations** 89
9.1. **Classification** 89
9.2. **Numerical Methods for Volterra Equations of the Second Kind** 91
9.3. **Numerical Methods for Fredholm, Urysohn, and Hammerstein Equations of the Second Kind** 91
9.4. **Numerical Methods for Eigenvalue Problems** 92
9.5. **Green's Functions** 92
9.6. **Boundary Integral Equations and Boundary Element Method** 94
10. **Optimization** 95
10.1. **Introduction** 95
10.2. **Gradient-Based Nonlinear Programming** 96
10.3. **Optimization Methods without Derivatives** 105
10.4. **Global Optimization** 106
10.5. **Mixed Integer Programming** 110
10.6. **Dynamic Optimization** 121

Ullmann's Modeling and Simulation

ISBN: 978-3-527-31605-2

10.7. **Development of Optimization Models** . . . 124
11. **Probability and Statistics** . . . 125
11.1. **Concepts** . . . 125
11.2. **Sampling and Statistical Decisions** . 129
11.3. **Error Analysis in Experiments** . . . 132
11.4. **Factorial Design of Experiments and Analysis of Variance** . . . 133
12. **Multivariable Calculus Applied to Thermodynamics** . . . 135
12.1. **State Functions** . . . 135
12.2. **Applications to Thermodynamics** . . 136
12.3. **Partial Derivatives of All Thermodynamic Functions** . . . 137
13. **References** . . . 138

Symbols

Variables

a	1, 2, or 3 for planar, cylindrical, spherical geometry
a_i	acceleration of i-th particle in molecular dynamics
A	cross-sectional area of reactor; Helmholtz free energy in thermodynamics
Bi	Biot number
Bim	Biot number for mass
c	concentration
C_p	heat capacity at constant pressure
C_s	heat capacity of solid
C_v	heat capacity at constant volume
Co	Courant number
D	diffusion coefficient
Da	Damköhler number
D_e	effective diffusivity in porous catalyst
E	efficiency of tray in distillation column
F	molar flow rate into a chemical reactor
G	Gibbs free energy in thermodynamics
h_p	heat transfer coefficient
H	enthalpy in thermodynamics
J	mass flux
k	thermal conductivity; reaction rate constant
k_e	effective thermal conductivity of porous catalyst
k_g	mass transfer coefficient
K	chemical equilibrium constant
L	thickness of slab; liquid flow rate in distillation column; length of pipe for flow in pipe
m_i	mass of i-th particle in molecular dynamics
M	holdup on tray of distillation column
n	power-law exponent in viscosity formula for polymers
p	pressure
Pe	Peclet number
q	heat flux
Q	volumetric flow rate; rate of heat generation for heat transfer problems
r	radial position in cylinder
$\boldsymbol{r}_i$	position of i-th particle in molecular dynamics
R	radius of reactor or catalyst pellet
Re	Reynolds number
S	entropy in thermodynamics
Sh	Sherwood number
t	time
T	temperature
$\boldsymbol{u}$	velocity
U	internal energyy in thermodynamics
$\boldsymbol{v}_i$	velocity of i-th particle in molecular dynamics
V	volume of chemical reactor; vapor flow rate in distillation column; potential energy in molecular dynamics; specific volume in thermodynamics
x	position
z	position from inlet of reactor

Greek symbols

a	thermal diffusivity
δ	Kronecker delta
Δ	sampling rate
ε	porosity of catalyst pellet
η	viscosity in fluid flow; effectiveness factor for reaction in a catalyst pellet
η_0	zero-shear rate viscosity
ϕ	Thiele modulus
φ	void fraction of packed bed
λ	time constant in polymer flow
ρ	density
ρ_s	density of solid
τ	shear stress
μ	viscosity

Special symbols

| | subject to
: mapping. For example, $h: R^n \to R^m$, states that functions h map real numbers into m real numbers. There are m functions h written in terms of n variables
$\in$ member of
$\to$ maps into

1. Solution of Equations

Mathematical models of chemical engineering systems can take many forms: they can be sets of algebraic equations, differential equations, and/or integral equations. Mass and energy balances of chemical processes typically lead to large sets of *algebraic* equations:

$$a_{11}x_1 + a_{12}x_2 = b_1$$
$$a_{21}x_1 + a_{22}x_2 = b_2$$

Mass balances of stirred tank reactors may lead to ordinary *differential* equations:

$$\frac{\mathrm{d}y}{\mathrm{d}t} = f\left[y(t)\right]$$

Radiative heat transfer may lead to *integral equations:*

$$y(x) = g(x) + \lambda \int_0^1 K(x,s) f(s)\,\mathrm{d}s$$

Even when the model is a differential equation or integral equation, the most basic step in the algorithm is the solution of sets of algebraic equations. The solution of sets of algebraic equations is the focus of Chapter 1.

A single linear equation is easy to solve for either x or y:

$$y = ax + b$$

If the equation is nonlinear,

$$f(x) = 0$$

it may be more difficult to find the x satisfying this equation. These problems are compounded when there are more unknowns, leading to simultaneous equations. If the unknowns appear in a linear fashion, then an important consideration is the structure of the matrix representing the equations; special methods are presented here for special structures. They are useful because they increase the speed of solution. If the unknowns appear in a nonlinear fashion, the problem is much more difficult. Iterative techniques must be used (i.e., make a guess of the solution and try to improve the guess). An important question then is whether such an iterative scheme converges. Other important types of equations are linear difference equations and eigenvalue problems, which are also discussed.

1.1. Matrix Properties

A matrix is a set of real or complex numbers arranged in a rectangular array.

$$\boldsymbol{A} = \begin{bmatrix} a_{11} & a_{12} & \dots & a_{1n} \\ a_{21} & a_{22} & \dots & a_{2n} \\ \vdots & \vdots & \vdots & \vdots \\ a_{m1} & a_{m2} & \dots & a_{mn} \end{bmatrix}$$

The numbers a_{ij} are the elements of the matrix $\boldsymbol{A}$, or $(A)_{ij} = a_{ij}$. The transpose of $\boldsymbol{A}$ is $(\boldsymbol{A}^T) = a_{ji}$.

The determinant of a square matrix $\boldsymbol{A}$ is

$$\boldsymbol{A} = \begin{vmatrix} a_{11} & a_{12} & \dots & a_{1n} \\ a_{21} & a_{22} & \dots & a_{2n} \\ \vdots & \vdots & \vdots & \vdots \\ a_{n1} & a_{n2} & \dots & a_{nn} \end{vmatrix}$$

If the i-th row and j-th column are deleted, a new determinant is formed having $n-1$ rows and columns. This determinant is called the minor of a_{ij} denoted as $\boldsymbol{M}_{ij}$. The cofactor A'_{ij} of the element a_{ij} is the signed minor of a_{ij} determined by

$$A'_{ij} = (-1)^{i+j} M_{ij}$$

The value of $|\boldsymbol{A}|$ is given by

$$|\boldsymbol{A}| = \sum_{j=1}^{n} a_{ij} A'_{ij} \text{ or } \sum_{i=1}^{n} a_{ij} A'_{ij}$$

where the elements a_{ij} must be taken from a single row or column of $\boldsymbol{A}$.

If all determinants formed by striking out whole rows or whole columns of order greater than r are zero, but there is at least one determinant of order r which is not zero, the matrix has rank r.

The value of a determinant is not changed if the rows and columns are interchanged. If the elements of one row (or one column) of a determinant are all zero, the value of the determinant is zero. If the elements of one row or column are multiplied by the same constant, the determinant is the previous value times that constant. If two adjacent rows (or columns) are interchanged, the value of the new determinant is the negative of the value of the original determinant. If two rows (or columns) are identical, the determinant is zero. The value of a determinant is not changed if one row (or column) is multiplied by a constant and added to another row (or column).

A matrix is *symmetric* if

$$a_{ij}=a_{ji}$$

and it is *positive definite* if

$$\boldsymbol{x}^T\boldsymbol{A}\boldsymbol{x}=\sum_{i=1}^{n}\sum_{j=1}^{n}a_{ij}x_ix_j\geq 0$$

for all $\boldsymbol{x}$ and the equality holds only if $\boldsymbol{x}=0$.

If the elements of $\boldsymbol{A}$ are complex numbers, $\boldsymbol{A}^*$ denotes the complex conjugate in which $(A^*)_{ij}=a^*_{ij}$. If $\boldsymbol{A}=\boldsymbol{A}^*$ the matrix is Hermitian.

The *inverse of a matrix* can also be used to solve sets of linear equations. The inverse is a matrix such that when $\boldsymbol{A}$ is multiplied by its inverse the result is the identity matrix, a matrix with 1.0 along the main diagonal and zero elsewhere.

$$\boldsymbol{A}\boldsymbol{A}^{-1}=\boldsymbol{I}$$

If $\boldsymbol{A}^T=\boldsymbol{A}^{-1}$ the matrix is orthogonal.

Matrices are added and subtracted element by element.

$$\boldsymbol{A}+\boldsymbol{B} \text{ is } a_{ij}+b_{ij}$$

Two matrices $\boldsymbol{A}$ and $\boldsymbol{B}$ are equal if $a_{ij}=b_{ij}$. Special relations are

$$(\boldsymbol{A}\boldsymbol{B})^{-1}=\boldsymbol{B}^{-1}\boldsymbol{A}^{-1},\quad (\boldsymbol{A}\boldsymbol{B})^T=\boldsymbol{B}^T\boldsymbol{A}^T$$
$$(\boldsymbol{A}^{-1})^T=(\boldsymbol{A}^T)^{-1},\quad (\boldsymbol{A}\boldsymbol{B}\boldsymbol{C})^{-1}=\boldsymbol{C}^{-1}\boldsymbol{B}^{-1}\boldsymbol{A}^{-1}$$

A diagonal matrix is zero except for elements along the diagonal.

$$a_{ij}=\begin{cases} a_{ii}, & i=j \\ 0, & i\neq j \end{cases}$$

A tridiagonal matrix is zero except for elements along the diagonal and one element to the right and left of the diagonal.

$$a_{ij}=\begin{cases} 0 \text{ if } j<i-1 \\ a_{ij} \text{ otherwise} \\ 0 \text{ if } j>i+1 \end{cases}$$

Block diagonal and pentadiagonal matrices also arise, especially when solving partial differential equations in two- and three-dimensions.

QR Factorization of a Matrix. If A is an $m \times n$ matrix with $m \geq n$, there exists an $m \times m$ unitary matrix $\boldsymbol{Q}=[\boldsymbol{q}_1, \boldsymbol{q}_2, \ldots \boldsymbol{q}_m]$ and an $m \times n$ right-triangular matrix $\boldsymbol{R}$ such that $\boldsymbol{A}=\boldsymbol{Q}\boldsymbol{R}$. The QR factorization is frequently used in the actual computations when the other transformations are unstable.

Singular Value Decomposition. If $\boldsymbol{A}$ is an $m \times n$ matrix with $m \geq n$ and rank $k \leq n$, consider the two following matrices.

$$\boldsymbol{A}\boldsymbol{A}^* \text{ and } \boldsymbol{A}^*\boldsymbol{A}$$

An $m \times m$ unitary matrix $\boldsymbol{U}$ is formed from the eigenvectors $\boldsymbol{u}_i$ of the first matrix.

$$\boldsymbol{U}=[\boldsymbol{u}_1,\boldsymbol{u}_2,\ldots\boldsymbol{u}_m]$$

An $n \times n$ unitary matrix $\boldsymbol{V}$ is formed from the eigenvectors $\boldsymbol{v}_i$ of the second matrix.

$$\boldsymbol{V}=[\boldsymbol{v}_1,\boldsymbol{v}_2,\ldots,\boldsymbol{v}_n]$$

Then the matrix $\boldsymbol{A}$ can be decomposed into

$$\boldsymbol{A}=\boldsymbol{U}\Sigma\boldsymbol{V}^*$$

where Σ is a $k \times k$ diagonal matrix with diagonal elements $d_{ii}=\sigma_i>0$ for $1\leq i\leq k$. The eigenvalues of $\Sigma^*\Sigma$ are σ_i^2. The vectors $\boldsymbol{u}_i$ for $k+1\leq i\leq m$ and $\boldsymbol{v}_i$ for $k+1\leq i\leq n$ are eigenvectors associated with the eigenvalue zero; the eigenvalues for $1\leq i\leq k$ are σ_i^2. The values of σ_i are called the singular values of the matrix $\boldsymbol{A}$. If $\boldsymbol{A}$ is real, then $\boldsymbol{U}$ and $\boldsymbol{V}$ are real, and hence orthogonal matrices. The value of the singular value decomposition comes when a process is represented by a linear transformation and the elements of $\boldsymbol{A}$, a_{ij}, are the contribution to an output i for a particular variable as input variable j. The input may be the size of a disturbance, and the output is the gain [1]. If the rank is less than n, not all the variables are independent and they cannot all be controlled. Furthermore, if the singular values are widely separated, the process is sensitive to small changes in the elements of the matrix and the process will be difficult to control.

1.2. Linear Algebraic Equations

Consider the $n \times n$ linear system

$$a_{11}\, x_1 + a_{12}\, x_2 + \ldots + a_{1n}\, x_n = f_1$$
$$a_{21}\, x_1 + a_{22}\, x_2 + \ldots + a_{2n}\, x_n = f_2$$
$$\ldots$$
$$a_{n1}\, x_1 + a_{n2}\, x_2 + \ldots + a_{nn}\, x_n = f_n$$

In this equation $a_{11}, \ldots, a_{nn}$ are known parameters, $f_1, \ldots, f_n$ are known, and the unknowns are $x_1, \ldots, x_n$. The values of all unknowns that satisfy every equation must be found. This set of equations can be represented as follows:

$$\sum_{j=1}^{n} a_{ij}\, x_j = f_j \text{ or } \boldsymbol{A}\boldsymbol{x} = \boldsymbol{f}$$

The most efficient method for solving a set of linear algebraic equations is to perform a lower – upper (LU) decomposition of the corresponding matrix $\boldsymbol{A}$. This decomposition is essentially a Gaussian elimination, arranged for maximum efficiency [2, 3].

The LU decomposition writes the matrix as

$$\boldsymbol{A} = \boldsymbol{L}\boldsymbol{U}$$

The $\boldsymbol{U}$ is upper triangular; it has zero elements below the main diagonal and possibly nonzero values along the main diagonal and above it (see Fig. 1). The $\boldsymbol{L}$ is lower triangular. It has the value 1 in each element of the main diagonal, nonzero values below the diagonal, and zero values above the diagonal (see Fig. 1). The original problem can be solved in two steps:

$$\boldsymbol{L}\boldsymbol{y} = \boldsymbol{f}, \boldsymbol{U}\boldsymbol{x} = \boldsymbol{y} \text{ solves } \boldsymbol{A}\boldsymbol{x} = \boldsymbol{L}\boldsymbol{U}\boldsymbol{x} = \boldsymbol{f}$$

Each of these steps is straightforward because the matrices are upper triangular or lower triangular.

When f is changed, the last steps can be done without recomputing the LU decomposition. Thus, multiple right-hand sides can be computed efficiently. The number of multiplications and divisions necessary to solve for m right-hand sides is:

$$\text{Operation count} = \frac{1}{3} n^3 - \frac{1}{3} n + m\, n^2$$

The *determinant* is given by the product of the diagonal elements of $\boldsymbol{U}$. This should be calculated as the LU decomposition is performed.

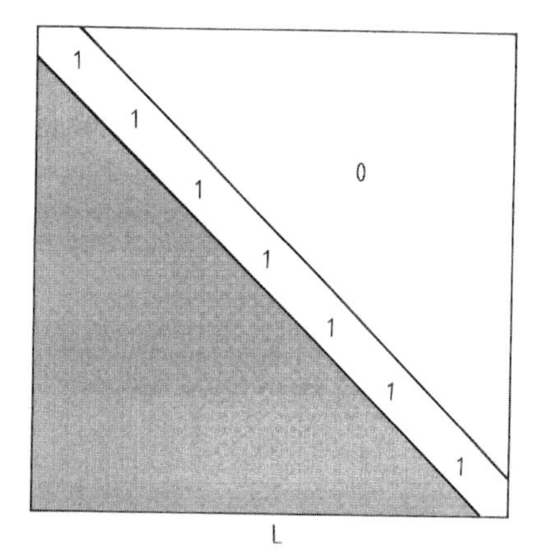

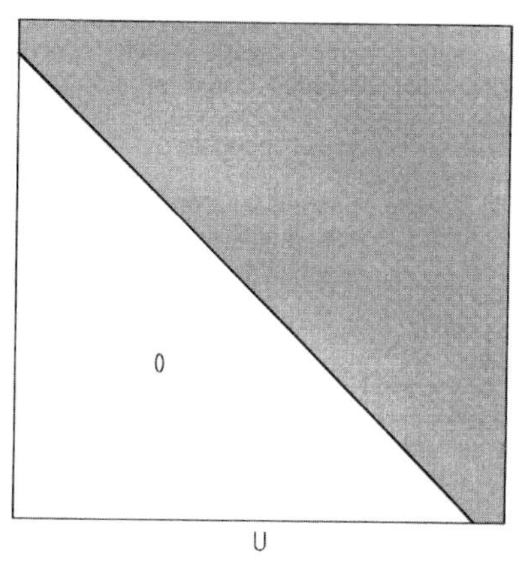

Figure 1. Structure of $\boldsymbol{L}$ and $\boldsymbol{U}$ matrices

If the value of the determinant is a very large or very small number, it can be divided or multiplied by 10 to retain accuracy in the computer; the scale factor is then accumulated separately. The condition number κ can be defined in terms of the singular value decomposition as the ratio of the largest d_{ii} to the smallest d_{ii} (see above). It can also be expressed in terms of the norm of the matrix:

$$\kappa(A) = \|\boldsymbol{A}\| \|\boldsymbol{A}^{-1}\|$$

where the norm is defined as

$$\|\boldsymbol{A}\| \equiv \sup_{x \neq 0} \frac{\|\boldsymbol{A}\boldsymbol{x}\|}{\|\boldsymbol{x}\|} = \max_k \sum_{j=1}^{n} |a_{jk}|$$

If this number is infinite, the set of equations is singular. If the number is too large, the matrix is said to be ill-conditioned. Calculation of the condition number can be lengthy so another criterion is also useful. Compute the ratio of the largest to the smallest pivot and make judgments on the ill-conditioning based on that.

When a matrix is ill-conditioned the LU decomposition must be performed by using pivoting (or the singular value decomposition described above). With pivoting, the order of the elimination is rearranged. At each stage, one looks for the largest element (in magnitude); the next stages if the elimination are on the row and column containing that largest element. The largest element can be obtained from only the diagonal entries (partial pivoting) or from all the remaining entries. If the matrix is nonsingular, Gaussian elimination (or LU decomposition) could fail if a zero value were to occur along the diagonal and were to be a pivot. With full pivoting, however, the Gaussian elimination (or LU decomposition) cannot fail because the matrix is nonsingular.

The *Cholesky decomposition* can be used for real, symmetric, positive definite matrices. This algorithm saves on storage (divide by about 2) and reduces the number of multiplications (divide by 2), but adds n square roots.

The linear equations are solved by

$$\boldsymbol{x} = \boldsymbol{A}^{-1}\boldsymbol{f}$$

Generally, the inverse is not used in this way because it requires three times more operations than solving with an LU decomposition. However, if an inverse is desired, it is calculated most efficiently by using the LU decomposition and then solving

$$\boldsymbol{A}\boldsymbol{x}^{(i)} = \boldsymbol{b}^{(i)}$$

$$b_j^{(i)} = \begin{cases} 0 & j \neq i \\ 1 & j = i \end{cases}$$

Then set

$$\boldsymbol{A}^{-1} = \left(\boldsymbol{x}^{(1)}|\boldsymbol{x}^{(2)}|\boldsymbol{x}^{(3)}|\cdots|\boldsymbol{x}^{(n)}\right)$$

Solutions of Special Matrices. Special matrices can be handled even more efficiently. A *tridiagonal matrix* is one with nonzero entries along the main diagonal, and one diagonal above and below the main one (see Fig. 2). The corresponding set of equations can then be written as

$$a_i x_{i-1} + b_i x_i + c_i x_{i+1} = d_i$$

The LU decomposition algorithm for solving this set is

$b_1' = b_1$
for $k = 2, n$ do
$a_k' = \frac{a_k}{b_{k-1}'}, \; b_k' = b_k - \frac{a_k}{b_{k-1}'} c_{k-1}$
enddo
$d_1' = d_1$
for $k = 2, n$ do
$d_k' = d_k - a_k' d_{k-1}'$
enddo
$x_n = d_n' / b_n'$
for $k = n-1, 1$do
$x_k = \frac{d_k' - c_k x_{k+1}}{b_k'}$
enddo

The number of multiplications and divisions for a problem with n unknowns and m right-hand sides is

$$\text{Operation count} = 2\,(n-1) + m\,(3\,n-2)$$

If

$$|b_i| > |a_i| + |c_i|$$

no pivoting is necessary. For solving two-point boundary value problems and partial differential equations this is often the case.

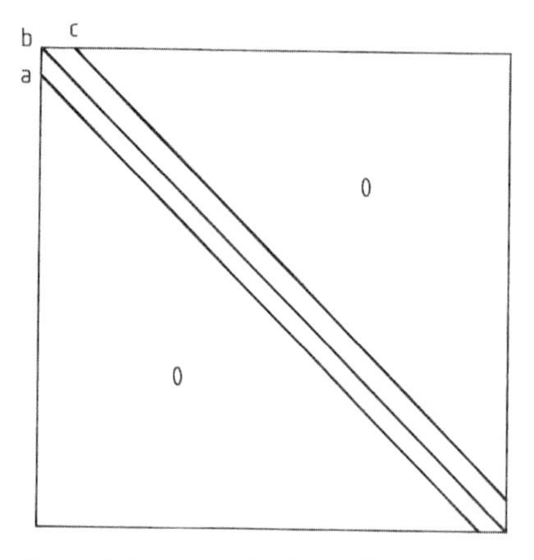

Figure 2. Structure of tridiagonal matrices

Sparse matrices are ones in which the majority of elements are zero. If the zero entries occur in special patterns, efficient techniques can be used to exploit the structure, as was done above for tridiagonal matrices, block tridiagonal matrices, arrow matrices, etc. These structures typically arise from numerical analysis applied to solve differential equations. Other problems, such as modeling chemical processes, lead to sparse matrices but without such a neatly defined structure—just a lot of zeros in the matrix. For matrices such as these, special techniques must be employed: efficient codes are available [4]. These codes usually employ a symbolic factorization, which must be repeated only once for each structure of the matrix. Then an LU factorization is performed, followed by a solution step using the triangular matrices. The symbolic factorization step has significant overhead, but this is rendered small and insignificant if matrices with exactly the same structure are to be used over and over [5].

The efficiency of a technique for solving sets of linear equations obviously depends greatly on the arrangement of the equations and unknowns because an efficient arrangement can reduce the bandwidth, for example. Techniques for renumbering the equations and unknowns arising from elliptic partial differential equations are available for finite difference methods [6] and for finite element methods [7].

Solutions with Iterative Methods. Sets of linear equations can also be solved by using iterative methods; these methods have a rich historical background. Some of them are discussed in Chapter 8 and include Jacobi, Gauss – Seidel, and overrelaxation methods. As the speed of computers increases, direct methods become preferable for the general case, but for large three-dimensional problems iterative methods are often used.

The *conjugate gradient method* is an iterative method that can solve a set of n linear equations in n iterations. The method primarily requires multiplying a matrix by a vector, which can be done very efficiently on parallel computers: for sparse matrices this is a viable method. The original method was devised by Hestenes and Stiefel [8]; however, more recent implementations use a *preconditioned* conjugate gradient method because it converges faster, provided a good "preconditioner" can be found. The system of n linear equations

$$\boldsymbol{A}\boldsymbol{x}=\boldsymbol{f}$$

where $\boldsymbol{A}$ is symmetric and positive definite, is to be solved. A preconditioning matrix $\boldsymbol{M}$ is defined in such a way that the problem

$$\boldsymbol{M}\boldsymbol{t}=\boldsymbol{r}$$

is easy to solve exactly ($\boldsymbol{M}$ might be diagonal, for example). Then the preconditioned conjugate gradient method is

Guess $\boldsymbol{x}_0$

Calculate $\boldsymbol{r}_0=\boldsymbol{f}-\boldsymbol{A}\boldsymbol{x}_0$

Solve $\boldsymbol{M}\,\boldsymbol{t}_0=\boldsymbol{r}_0$, and set $\boldsymbol{p}_0=\boldsymbol{t}_0$

for $k=1,n$ (or until convergence)

$$a_k=\frac{\boldsymbol{r}_k^{\mathrm{T}}\boldsymbol{t}_k}{\boldsymbol{p}_k^{\mathrm{T}}\boldsymbol{A}\boldsymbol{p}_k}$$

$$\boldsymbol{x}_{k+1}=\boldsymbol{x}_k+a_k\boldsymbol{p}_k$$

$$\boldsymbol{r}_{k+1}=\boldsymbol{r}_k-a_k\boldsymbol{A}\boldsymbol{p}_k$$

Solve $\boldsymbol{M}\boldsymbol{t}_{k+1}=\boldsymbol{r}_{k+1}$

$$b_k=\frac{\boldsymbol{r}_{k+1}^{\mathrm{T}}\boldsymbol{t}_{k+1}}{\boldsymbol{r}_k^{\mathrm{T}}\boldsymbol{t}_k}$$

$$\boldsymbol{p}_{k+1}=\boldsymbol{t}_{k+1}+b_k\boldsymbol{p}_k$$

test for convergence

enddo

Note that the entire algorithm involves only matrix multiplications. The generalized minimal residual method (GMRES) is an iterative method that can be used for nonsymmetric systems and is based on a modified Gram – Schmidt orthonormalization. Additional information, including software for a variety of methods, is available [9 – 13].

In dimensional analysis if the dimensions of each physical variable P_j (there are n of them) are expressed in terms of fundamental measurement units m_j (such as time, length, mass; there are m of them):

$$[P_j]=m_1^{\alpha_{1j}}m_2^{\alpha_{2j}}\cdots m_m^{\alpha_{mj}}$$

then a matrix can be formed from the α_{ij}. If the rank of this matrix is r, $n-r$ independent dimensionless groups govern that phenomenon. In chemical reaction engineering the chemical reaction stoichiometry can be written as

$$\sum_{i=1}^{n}\alpha_{ij}\,C_i=0,\ j=1,2,\ldots,m$$

where there are n species and m reactions. Then if a matrix is formed from the coefficients α_{ij}, which is an $n\times m$ matrix, and the rank of the matrix is r, there are r independent chemical reactions. The other $n-r$ reactions can be deduced from those r reactions.

1.3. Nonlinear Algebraic Equations

Consider a single nonlinear equation in one unknown,

$$f(x) = 0$$

In Microsoft Excel, roots are found by using Goal Seek or Solver. Assign one cell to be x, put the equation for $f(x)$ in another cell, and let Goal Seek or Solver find the value of x making the equation cell zero. In MATLAB, the process is similar except that a function (m-file) is defined and the command fzero('f',x0) provides the solution x, starting from the initial guess x_0.

Iterative methods applied to single equations include the *successive substitution* method

$$x^{k+1} = x^k + \beta f\left(x^k\right) \equiv g\left(x^k\right)$$

and the Newton – Raphson method.

$$x^{k+1} = x^k - \frac{f(x^k)}{\mathrm{d}f/\mathrm{d}x\,(x^k)}$$

The former method converges if the derivative of $g(x)$ is bounded [3]. The latter method

$$\left|\frac{\mathrm{d}g}{\mathrm{d}x}(x)\right| \leq \mu \text{ for } |x-\alpha| < h$$

is based on a Taylor series of the equation about the k-th iterate:

$$f\left(x^{k+1}\right) = f\left(x^k\right) + \frac{\mathrm{d}f}{\mathrm{d}x}\Big|_{x^k}\left(x^{k+1}-x^k\right) + \frac{\mathrm{d}^2 f}{\mathrm{d}x^2}\Big|_{x^k}\frac{1}{2}\left(x^{k+1}-x^k\right)^2 + \cdots$$

The second and higher-order terms are neglected and $f(x^{k+1}) = 0$ to obtain the method.

$$\left|\frac{\mathrm{d}f}{\mathrm{d}x}\right|_{x^0} > 0,\ \left|x^1 - x^0\right| = \left|\frac{f(x^0)}{\mathrm{d}f/\mathrm{d}x\,(x^0)}\right|_{x^0} \leq b,$$

$$\text{and } \left|\frac{\mathrm{d}^2 f}{\mathrm{d}x^2}\right| \leq c$$

Convergence of the Newton–Raphson method depends on the properties of the first and second derivative of $f(x)$ [3, 14]. In practice the method may not converge unless the initial guess is good, or it may converge for some parameters and not others. Unfortunately, when the method is nonconvergent the results look as though a mistake occurred in the computer programming; distinguishing between these situations is difficult, so careful programming and testing are required. If the method converges the difference between successive iterates is something like 0.1, 0.01, 0.0001, 10^{-8}. The error (when it is known) goes the same way; the method is said to be quadratically convergent when it converges. If the derivative is difficult to calculate a numerical approximation may be used.

$$\frac{\mathrm{d}f}{\mathrm{d}x}\Big|_{x^k} = \frac{f\left(x^k + \varepsilon\right) - f\left(x^k\right)}{\varepsilon}$$

In the *secant method* the same formula is used as for the Newton – Raphson method, except that the derivative is approximated by using the values from the last two iterates:

$$\frac{\mathrm{d}f}{\mathrm{d}x}\Big|_{x^k} = \frac{f\left(x^k\right) - f\left(x^{k-1}\right)}{x^k - x^{k-1}}$$

This is equivalent to drawing a straight line through the last two iterate values on a plot of $f(x)$ versus x. The Newton – Raphson method is equivalent to drawing a straight line tangent to the curve at the last x. In the *method of false position* (or regula falsi), the secant method is used to obtain x^{k+1}, but the previous value is taken as either x^{k-1} or x^k. The choice is made so that the function evaluated for that choice has the opposite sign to $f(x^{k+1})$. This method is slower than the secant method, but it is more robust and keeps the root between two points at all times. In all these methods, appropriate strategies are required for bounds on the function or when $\mathrm{d}f/\mathrm{d}x = 0$. *Brent's method* combines bracketing, bisection, and an inverse quadratic interpolation to provide a method that is fast and guaranteed to converge, if the root can be bracketed initially [15, p. 251].

In the *method of bisection*, if a root lies between x_1 and x_2 because $f(x_1) < 0$ and $f(x_2) > 0$, then the function is evaluated at the center, $x_c = 0.5\,(x_1 + x_2)$. If $f(x_c) > 0$, the root lies between x_1 and x_c. If $f(x_c) < 0$, the root lies between x_c and x_2. The process is then repeated. If $f(x_c) = 0$, the root is x_c. If $f(x_1) > 0$ and $f(x_2) > 0$, more than one root may exist between x_1 and x_2 (or no roots).

For systems of equations the Newton – Raphson method is widely used, especially for equations arising from the solution of differential equations.

$$f_i(\{x_j\}) = 0,\ 1 \leq i, j \leq n,$$

$$\text{where } \{x_j\} = (x_1, x_2, \ldots, x_n) = \boldsymbol{x}$$

Then, an expansion in several variables occurs:

$$\boldsymbol{f}_i\left(\boldsymbol{x}^{k+1}\right)=f_i\left(\boldsymbol{x}^k\right)+\sum_{j=1}^{n}\frac{\partial f_i}{\partial x_j}|_{\boldsymbol{x}^k}\left(x_j^{k+1}-x_j^k\right)+\cdots$$

The Jacobian matrix is defined as

$$\boldsymbol{J}_{ij}^k=\left.\frac{\partial f_i}{\partial x_j}\right|_{\boldsymbol{x}^k}$$

and the Newton – Raphson method is

$$\sum_{j=1}^{n}\boldsymbol{J}_{ij}^k\left(x^{k+1}-x^k\right)=-f_i\left(\boldsymbol{x}^k\right)$$

For convergence, the norm of the inverse of the Jacobian must be bounded, the norm of the function evaluated at the initial guess must be bounded, and the second derivative must be bounded [14, p. 115], [3, p. 12].

A review of the usefulness of solution methods for nonlinear equations is available [16]. This review concludes that the Newton – Raphson method may not be the most efficient. Broyden's method approximates the inverse to the Jacobian and is a good all-purpose method, but a good initial approximation to the Jacobian matrix is required. Furthermore, the rate of convergence deteriorates for large problems, for which the Newton – Raphson method is better. Brown's method [16] is very attractive, whereas Brent's is not worth the extra storage and computation. For large systems of equations, efficient software is available [11 – 13].

Homotopy methods can be used to ensure finding the solution when the problem is especially complicated. Suppose an attempt is made to solve $f(\boldsymbol{x}) = 0$, and it fails; however, $g(\boldsymbol{x}) = 0$ can be solved easily, where $g(\boldsymbol{x})$ is some function, perhaps a simplification of $f(\boldsymbol{x})$. Then, the two functions can be embedded in a homotopy by taking

$$\boldsymbol{h}(\boldsymbol{x},t)=t\,f(\boldsymbol{x})+(1-t)\,g(x)$$

In this equation, $\boldsymbol{h}$ can be a $n\times n$ matrix for problems involving n variables; then x is a vector of length n. Then $h(x,t) = 0$ can be solved for $t = 0$ and t gradually changes until at $t = 1$, $\boldsymbol{h}(\boldsymbol{x}, 1) = f(\boldsymbol{x})$. If the Jacobian of $\boldsymbol{h}$ with respect to x is nonsingular on the homotopy path (as t varies), the method is guaranteed to work. In classical methods, the interval from $t = 0$ to $t = 1$ is broken up into N subdivisions. Set $\Delta t = 1/N$ and solve for $t = 0$, which is easy by the choice of $g(x)$. Then set $t = \Delta t$ and use the Newton – Raphson method to solve for x. Since the initial guess is presumably pretty good, this has a high chance of being successful. That solution is then used as the initial guess for $t = 2\,\Delta t$ and the process is repeated by moving stepwise to $t = 1$. If the Newton – Raphson method does not converge, then Δt must be reduced and a new solution attempted.

Another way of using homotopy is to create an ordinary differential equation by differentiating the homotopy equation along the path (where $\boldsymbol{h} = 0$).

$$\frac{\mathrm{d}\boldsymbol{h}\left[\boldsymbol{x}(t),t\right]}{\mathrm{d}t}=\frac{\partial\boldsymbol{h}}{\partial\boldsymbol{x}}\frac{\mathrm{d}\boldsymbol{x}}{\mathrm{d}t}+\frac{\partial\boldsymbol{h}}{\partial t}=0$$

This can be expressed as an ordinary differential equation for $x(t)$:

$$\frac{\partial\boldsymbol{h}}{\partial\boldsymbol{x}}\frac{\mathrm{d}\boldsymbol{x}}{\mathrm{d}t}=-\frac{\partial\boldsymbol{h}}{\partial t}$$

If Euler's method is used to solve this equation, a value x^0 is used, and $\mathrm{d}x/\mathrm{d}t$ from the above equation is solved for. Then

$$\boldsymbol{x}^{1,0}=\boldsymbol{x}^0+\Delta t\frac{\mathrm{d}\boldsymbol{x}}{\mathrm{d}t}$$

is used as the initial guess and the homotopy equation is solved for $\boldsymbol{x}^1$.

$$\frac{\partial\boldsymbol{h}}{\partial x}\left(\boldsymbol{x}^{1,k+1}-\boldsymbol{x}^{1,k}\right)=-\boldsymbol{h}\left(\boldsymbol{x}^{1,k},t\right)$$

Then t is increased by Δt and the process is repeated.

In arc-length parameterization, both x and t are considered parameterized by a parameter s, which is thought of as the arc length along a curve. Then the homotopy equation is written along with the arc-length equation.

$$\frac{\partial\boldsymbol{h}}{\partial\boldsymbol{x}}\frac{\mathrm{d}x}{\mathrm{d}s}+\frac{\partial\boldsymbol{h}}{\partial t}\frac{\mathrm{d}t}{\mathrm{d}s}=0$$

$$\frac{\mathrm{d}\boldsymbol{x}^{\mathrm{T}}}{\mathrm{d}s}\frac{\mathrm{d}\boldsymbol{x}}{\mathrm{d}s}+\left(\frac{\mathrm{d}t}{\mathrm{d}s}\right)^2=1$$

The initial conditions are

$$\boldsymbol{x}(0)=\boldsymbol{x}^0$$
$$t(0)=0$$

The advantage of this approach is that it works even when the Jacobian of $\boldsymbol{h}$ becomes singular because the full matrix is rarely singular. Illustrations applied to chemical engineering are available [17]. Software to perform these computations is available (called LOCA) [18].

1.4. Linear Difference Equations

Difference equations arise in chemical engineering from staged operations, such as distillation or extraction, as well as from differential equations modeling adsorption and chemical reactors. The value of a variable in the n-th stage is noted by a subscript n. For example, if $y_{n,i}$ denotes the mole fraction of the i-th species in the vapor phase on the n-th stage of a distillation column, $x_{n,i}$ is the corresponding liquid mole fraction, R the reflux ratio (ratio of liquid returned to the column to product removed from the condenser), and $K_{n,i}$ the equilibrium constant, then the mass balances about the top of the column give

$$y_{n+1,i} = \frac{R}{R+1} x_{n,i} + \frac{1}{R+1} x_{0,i}$$

and the equilibrium equation gives

$$y_{n,i} = K_{n,i} x_{n,i}$$

If these are combined,

$$K_{n+1,i} x_{n+1,i} = \frac{R}{R+1} x_{n,i} + \frac{1}{R+1} x_{0,i}$$

is obtained, which is a linear difference equation. This particular problem is quite complicated, and the interested reader is referred to [19, Chap. 6]. However, the form of the difference equation is clear. Several examples are given here for solving difference equations. More complete information is available in [20].

An equation in the form

$$x_{n+1} - x_n = f_{n+1}$$

can be solved by

$$x_n = \sum_{i=1}^{n} f_i$$

Usually, difference equations are solved analytically only for linear problems. When the coefficients are constant and the equation is linear and homogeneous, a trial solution of the form

$$x_n = \varphi^n$$

is attempted; φ is raised to the power n. For example, the difference equation

$$c x_{n-1} + b x_n + a x_{n+1} = 0$$

coupled with the trial solution would lead to the equation

$$a\varphi^2 + b\varphi + c = 0$$

This gives

$$\varphi_{1,2} = \frac{-b \pm \sqrt{b^2 - 4ac}}{2a}$$

and the solution to the difference equation is

$$x_n = A\varphi_1^n + B\varphi_2^n$$

where A and B are constants that must be specified by boundary conditions of some kind.

When the equation is nonhomogeneous, the solution is represented by the sum of a particular solution and a general solution to the homogeneous equation.

$$x_n = x_{n,P} + x_{n,H}$$

The general solution is the one found for the homogeneous equation, and the particular solution is any solution to the nonhomogeneous difference equation. This can be found by methods analogous to those used to solve differential equations: the method of undetermined coefficients and the method of variation of parameters. The last method applies to equations with variable coefficients, too. For a problem such as

$$x_{n+1} - f_n x_n = 0$$
$$x_0 = c$$

the general solution is

$$x_n = c \prod_{i=1}^{n} f_{i-1}$$

This can then be used in the method of variation of parameters to solve the equation

$$x_{n+1} - f_n x_n = g_n$$

1.5. Eigenvalues

The $n \times n$ matrix $\boldsymbol{A}$ has n eigenvalues λ_i, $i = 1, \ldots, n$, which satisfy

$$\det(\boldsymbol{A} - \lambda_i \boldsymbol{I}) = 0$$

If this equation is expanded, it can be represented as

$$P_n(\lambda) = (-\lambda)^n + a_1(-\lambda)^{n-1} + a_2(-\lambda)^{n-2} + \cdots + a_{n-1}(-\lambda) + a_n = 0$$

If the matrix $\boldsymbol{A}$ has real entries then a_i are real numbers, and the eigenvalues either are real numbers or occur in pairs as complex numbers with their complex conjugates (for definition of complex numbers, see Chap. 3). The Hamilton – Cayley theorem [19, p. 127] states that the matrix $\boldsymbol{A}$ satisfies its own characteristic equation.

$$P_n(\boldsymbol{A})=(-\boldsymbol{A})^n+a_1(-\boldsymbol{A})^{n-1}+a_2(-\boldsymbol{A})^{n-2}+\cdots +a_{n-1}(-\boldsymbol{A})+a_n\boldsymbol{I}=0$$

A laborious way to find the eigenvalues of a matrix is to solve the n-th order polynomial for the λ_i—far too time consuming. Instead the matrix is transformed into another form whose eigenvalues are easier to find. In the *Givens method* and the *Housholder method* the matrix is transformed into the tridiagonal form; then, in a fixed number of calculations the eigenvalues can be found [15]. The Givens method requires 4 n^3/3 operations to transform a real symmetric matrix to tridiagonal form, whereas the Householder method requires half that number [14]. Once the tridiagonal form is found, a Sturm sequence is applied to determine the eigenvalues. These methods are especially useful when only a few eigenvalues of the matrix are desired.

If all the eigenvalues are needed, the $\boldsymbol{Q}\,\boldsymbol{R}$ algorithm is preferred [21].

The eigenvalues of a certain tridiagonal matrix can be found analytically. If $\boldsymbol{A}$ is a tridiagonal matrix with

$$a_{ii}=p,\,a_{i,i+1}=q,a_{i+1,i}=r,\,qr>0$$

then the eigenvalues of $\boldsymbol{A}$ are [22]

$$\lambda_i=p+2(qr)^{1/2}\cos\frac{i\pi}{n+1}\quad i=1,2,\ldots,n$$

This result is useful when finite difference methods are applied to the diffusion equation.

2. Approximation and Integration

2.1. Introduction

Two types of problems arise frequently:

1) A *function* is known exactly at a set of points and an interpolating function is desired. The interpolant may be exact at the set of points, or it may be a "best fit" in some sense. Alternatively it may be desired to represent a function in some other way.
2) *Experimental data* must be fit with a mathematical model. The data have experimental error, so some uncertainty exists. The parameters in the model as well as the uncertainty in the determination of those parameters is desired.

These problems are addressed in this chapter. Section 2.2 gives the properties of polynomials defined over the whole domain and Section 2.3 of polynomials defined on segments of the domain. In Section 2.4, quadrature methods are given for evaluating an integral. Least-squares methods for parameter estimation for both linear and nonlinear models are given in Sections 2.5. Fourier transforms to represent discrete data are described in Section 2.7. The chapter closes with extensions to two-dimensional representations.

2.2. Global Polynomial Approximation

A *global* polynomial $P_m(x)$ is defined over the entire region of space

$$P_m(x)=\sum_{j=0}^{m}c_jx^j$$

This polynomial is of degree m (highest power is x^m) and order $m+1$ ($m+1$ parameters $\{c_j\}$). If a set of $m+1$ points is given,

$$y_1=f(x_1),\,y_2=f(x_2),\ldots,y_{m+1}=f(x_{m+1})$$

then Lagrange's formula yields a polynomial of degree m that goes through the $m+1$ points:

$$P_m(x)=\frac{(x-x_2)(x-x_3)\cdots(x-x_{m+1})}{(x_1-x_2)(x_1-x_3)\cdots(x_1-x_{m+1})}y_1+\frac{(x-x_1)(x-x_3)\cdots(x-x_{m+1})}{(x_2-x_1)(x_2-x_3)\cdots(x_2-x_{m+1})}y_2+\cdots+\frac{(x-x_1)(x-x_2)\cdots(x-x_m)}{(x_{m+1}-x_1)(x_{m+1}-x_2)\cdots(x_{m+1}-x_m)}y_{m+1}$$

Note that each coefficient of y_j is a polynomial of degree m that vanishes at the points $\{x_j\}$ (except for one value of j) and takes the value of 1.0 at that point, i.e.,

$$P_m(x_j)=y_j\quad j=1,2,\ldots,m+1$$

If the function $f(x)$ is known, the error in the approximation is [23]

$$|\mathrm{error}(x)|\leq\frac{|x_{m+1}-x_1|^{m+1}}{(m+1)!}\max_{x_1\leq x\leq x_{m+1}}\left|f^{(m+1)}(x)\right|$$

The evaluation of $P_m(x)$ at a point other than the defining points can be made with Neville's algorithm [15]. Let P_1 be the value at x of the unique function passing through the point (x_1, y_1); i.e., $P_1 = y_1$. Let P_{12} be the value at x of the unique polynomial passing through the points x_1 and x_2. Likewise, $P_{ijk\ldots r}$ is the unique polynomial passing through the points x_i, x_j, x_k, ..., x_r. The following scheme is used:

$$\begin{array}{llllll} x_1 & y_1 = P_1 & & & & \\ & & P_{12} & & & \\ x_2 & y_2 = P_2 & & P_{123} & & \\ & & P_{23} & & P_{1234} & \\ x_3 & y_3 = P_3 & & P_{234} & & \\ & & P_{34} & & & \\ x_4 & y_4 = P_4 & & & & \end{array}$$

These entries are defined by using

$$P_{i(i+1)\cdots(i+m)} = \frac{(x - x_{i+m}) P_{i(i+1)\cdots(i+m-1)} + (x_i - x) P_{(i+1)(i+2)\cdots(i+m)}}{x_i - x_{i+m}}$$

Consider P_{1234}: the terms on the right-hand side of the equation involve P_{123} and P_{234}. The "parents," P_{123} and P_{234}, already agree at points 2 and 3. Here $i = 1$, $m = 3$; thus, the parents agree at $x_{i+1}, \ldots, x_{i+m-1}$ already. The formula makes $P_{i(i+1)\ldots(i+m)}$ agree with the function at the additional points x_{i+m} and x_i. Thus, $P_{i(i+1)\ldots(i+m)}$ agrees with the function at all the points $\{x_i, x_{i+1}, \ldots, x_{i+m}\}$.

Orthogonal Polynomials. Another form of the polynomials is obtained by defining them so that they are orthogonal. It is required that $P_m(x)$ be orthogonal to $P_k(x)$ for all $k = 0, \ldots, m-1$.

$$\int_a^b W(x)\, P_k(x)\, P_m(x)\, \mathrm{d}x = 0$$

$$k = 0,1,2,\ldots, m-1$$

The orthogonality includes a nonnegative weight function, $W(x) \geq 0$ for all $a \leq x \leq b$. This procedure specifies the set of polynomials to within multiplicative constants, which can be set either by requiring the leading coefficient to be one or by requiring the norm to be one.

$$\int_a^b W(x)\, P_m^2(x)\, \mathrm{d}x = 1$$

The polynomial $P_m(x)$ has m roots in the closed interval a to b.

The polynomial

$$p(x) = c_0 P_0(x) + c_1 P_1(x) + \cdots c_m P_m(x)$$

minimizes

$$I = \int_a^b W(x) \left[f(x) - p(x) \right]^2 \mathrm{d}x$$

when

$$c_j = \frac{\int_a^b W(x)\, f(x) P_j(x) \mathrm{d}x}{W_j},$$

$$W_j = \int_a^b W(x)\, P_j^2(x)\, \mathrm{d}x$$

Note that each c_j is independent of m, the number of terms retained in the series. The minimum value of I is

$$I_{\min} = \int_a^b W(x)\, f^2(x)\, \mathrm{d}x - \sum_{j=0}^{m} W_j c_j^2$$

Such functions are useful for continuous data, i.e., when $f(x)$ is known for all x.

Typical orthogonal polynomials are given in Table 1. Chebyshev polynomials are used in spectral methods (see Chap. 8). The last two rows of Table 1 are widely used in the orthogonal collocation method in chemical engineering.

Table 1. Orthogonal polynomials [15, 23]

a	b	$W(x)$	Name	Recursion relation
-1	1	1	Legendre	$(i+1) P_{i+1} = (2i+1) x P_i - i P_{i-1}$
-1	1	$\frac{1}{\sqrt{1-x^2}}$	Chebyshev	$T_{i+1} = 2x T_i - T_{i-1}$
0	1	$x^{q-1}(1-x)^{p-q}$	Jacobi (p, q)	
$-\infty$	∞	e^{-x^2}	Hermite	$H_{i+1} = 2x H_i - 2i H_{i-1}$
0	∞	$x^c \mathrm{e}^{-x}$	Laguerre (c)	$(i+1) L_{i+1}^c = (-x+2i+c+1) L_i^c - (i+c) L_{i-1}^c$
0	1	1	shifted Legendre	
0	1	1	shifted Legendre, function of x^2	

The last entry (the shifted Legendre polynomial as a function of x^2) is defined by

$$\int_0^1 W\left(x^2\right) P_k\left(x^2\right) P_m\left(x^2\right) x^{a-1} \mathrm{d}x = 0$$

$$k = 0,1,\ldots,m-1$$

where a = 1 is for planar, a = 2 for cylindrical, and a = 3 for spherical geometry. These functions are useful if the solution can be proved to be an even function of x.

Rational Polynomials. Rational polynomials are ratios of polynomials. A rational polynomial $R_{i(i+1)\ldots(i+m)}$ passing through m + 1 points

$$y_i = f\left(x_i\right),\; i = 1,\ldots,m+1$$

is

$$R_{i(i+1)\cdots(i+m)} = \frac{P_\mu(x)}{Q_\nu(x)} = \frac{p_0+p_1x+\cdots+p_\mu x^\mu}{q_0+q_1x+\cdots+q_\nu x^\nu},$$

$$m+1 = \mu+\nu+1$$

An alternative condition is to make the rational polynomial agree with the first m + 1 terms in the power series, giving a Padé approximation, i.e.,

$$\frac{\mathrm{d}^k R_{i(i+1)\cdots(i+m)}}{\mathrm{d}x^k} = \frac{\mathrm{d}^k f(x)}{\mathrm{d}x^k} \quad k = 0,\ldots,m$$

The Bulirsch – Stoer recursion algorithm can be used to evaluate the polynomial:

$$R_{i(i+1)\cdots(i+m)} = R_{(i+1)\cdots(i+m)}$$

$$+\frac{R_{(i+1)\cdots(i+m)}-R_{i(i+1)\cdots(i+m-1)}}{\text{Den}}$$

$$\text{Den} = \left(\frac{x-x_i}{x-x_{i+m}}\right)$$

$$\left(1-\frac{R_{(i+1)\cdots(i+m)} R_{i(i+1)\cdots(i+m-1)}}{R_{(i+1)\cdots(i+m)}-R_{(i+1)\cdots(i+m-1)}}\right)-1$$

Rational polynomials are useful for approximating functions with poles and singularities, which occur in Laplace transforms (see Section 4.2).

Fourier series are discussed in Section 4.1. Representation by sums of exponentials is also possible [24].

In summary, for discrete data, Legendre polynomials and rational polynomials are used. For continuous data a variety of orthogonal polynomials and rational polynomials are used. When the number of conditions (discrete data points) exceeds the number of parameters, then see Section 2.5.

2.3. Piecewise Approximation

Piecewise approximations can be developed from difference formulas [3]. Consider a case in which the data points are equally spaced

$$x_{n+1}-x_n = \Delta x$$

$$y_n = y\left(x_n\right)$$

forward differences are defined by

$$\Delta y_n = y_{n+1}-y_n$$

$$\Delta^2 y_n = \Delta y_{n+1}-\Delta y_n = y_{n+2}-2y_{n+1}+y_n$$

Then, a new variable is defined

$$\alpha = \frac{x_\alpha-x_0}{\Delta x}$$

and the finite interpolation formula through the points $y_0, y_1, \ldots, y_n$ is written as follows:

$$y_\alpha = y_0+\alpha\Delta y_0+\frac{\alpha(\alpha-1)}{2!}\Delta^2 y_0+\cdots+ \frac{\alpha(\alpha-1)\cdots(\alpha-n+1)}{n!}\Delta^n y_0 \tag{1}$$

Keeping only the first two terms gives a straight line through $(x_0, y_0) - (x_1, y_1)$; keeping the first three terms gives a quadratic function of position going through those points plus (x_2, y_2). The value $\alpha = 0$ gives $x = x_0$; $\alpha = 1$ gives $x = x_1$, etc.

Backward differences are defined by

$$\nabla y_n = y_n-y_{n-1}$$

$$\nabla^2 y_n = \nabla y_n-\nabla y_{n-1} = y_n-2y_{n-1}+y_{n-2}$$

The interpolation polynomial of order n through the points $y_0, y_{-1}, y_{-2}, \ldots$ is

$$y_\alpha = y_0+\alpha\nabla y_0+\frac{\alpha(\alpha+1)}{2!}\nabla^2 y_0+\cdots+ \frac{\alpha(\alpha+1)\cdots(\alpha+n-1)}{n!}\nabla^n y_0$$

The value $\alpha = 0$ gives $x = x_0$; $\alpha = -1$ gives $x = x_{-1}$. Alternatively, the interpolation polynomial of order n through the points $y_1, y_0, y_{-1}, \ldots$ is

$$y_\alpha = y_1+(\alpha-1)\nabla y_1+\frac{\alpha(\alpha-1)}{2!}\nabla^2 y_1+\cdots+ \frac{(\alpha-1)\alpha(\alpha+1)\cdots(\alpha+n-2)}{n!}\nabla^n y_1$$

Now $\alpha = 1$ gives $x = x_1$; $\alpha = 0$ gives $x = x_0$.

The *finite element method* can be used for piecewise approximations [3]. In the finite element method the domain $a \leq x \leq b$ is divided into elements as shown in Figure 3. Each function N_i (x) is zero at all nodes except x_i; N_i $(x_i) = 1$. Thus, the approximation is

$$y(x)=\sum_{i=1}^{NT}c_iN_i(x)=\sum_{i=1}^{NT}y(x_i)N_i(x)$$

where $c_i = y(x_i)$. For convenience, the trial functions are defined within an element by using new coordinates:

$$u=\frac{x-x_i}{\Delta x_i}$$

The Δx_i need not be the same from element to element. The trial functions are defined as N_i (x) (Fig. 3 A) in the global coordinate system and N_I (u) (Fig. 3 B) in the local coordinate system (which also requires specification of the element). For $x_i < x < x_{i+1}$

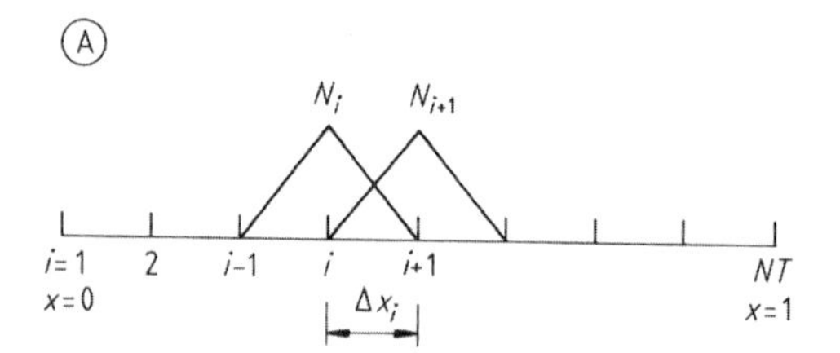

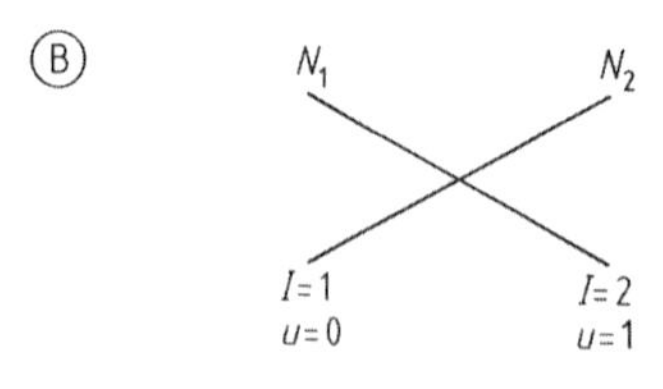

Figure 3. Galerkin finite element method – linear functions
A) Global numbering system; B) Local numbering system

$$y(x)=\sum_{i=1}^{NT}c_iN_i(x)=c_iN_i(x)+c_{i+1}N_{i+1}(x)$$

because all the other trial functions are zero there. Thus

$$y(x)=c_iN_{I=1}(u)+c_{i+1}N_{I=2}(u),$$

$$x_i<x<x_{i+1},\ 0<u<1$$

Then

$$N_{I=1}=1-u,\ N_{I=2}=u$$

and the expansion is rewritten as

$$y(x)=\sum_{I=1}^{2}c_I^eN_I(u) \tag{2}$$

x in e-th element and $c_i = c_I^e$ within the element e. Thus, given a set of points (x_i, y_i), a finite element approximation can be made to go through them.

Quadratic approximations can also be used within the element (see Fig. 4). Now the trial functions are

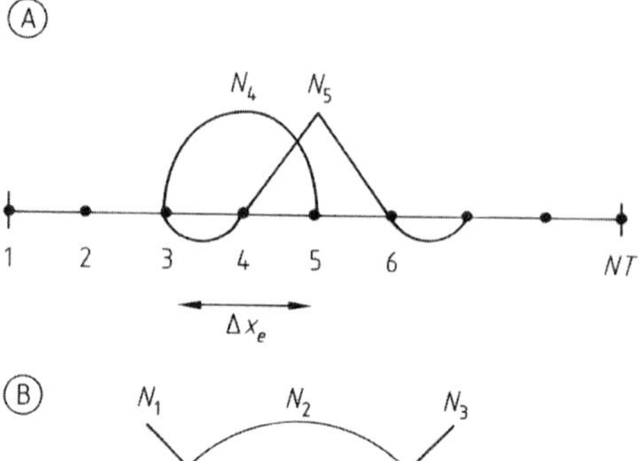

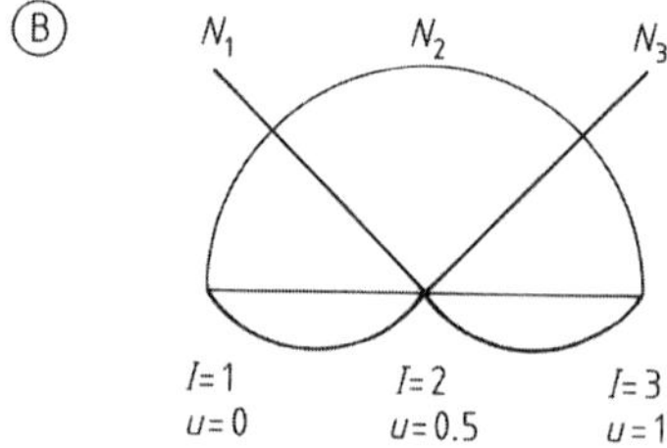

Figure 4. Finite elements approximation – quadratic elements
A) Global numbering system; B) Local numbering system

$$\begin{aligned}N_{I=1}&=2(u-1)\left(u-\tfrac{1}{2}\right)\\N_{I=2}&=4u(1-u)\\N_{I=3}&=2u\left(u-\tfrac{1}{2}\right)\end{aligned} \tag{3}$$

The approximation going through an odd number of points (x_i, y_i) is then

$$y(x)=\sum_{I=1}^{3}c_I^eN_I(u)\quad x\text{ in }e\text{–th element}$$

$$\text{with } c_I^e=y(x_i),i=(e-1)2+I$$
$$\text{in the } e\text{–th element}$$

Hermite cubic polynomials can also be used; these are continuous and have continuous first derivatives at the element boundaries [3].

Splines. Splines are functions that match given values at the points $x_1, \ldots, x_{NT}$, shown in Figure 5, and have continuous derivatives up to some order at the knots, or the points $x_2, \ldots, x_{NT-1}$. *Cubic splines* are most common. In this case the function is represented by a cubic polynomial within each interval and has continuous first and second derivatives at the knots.

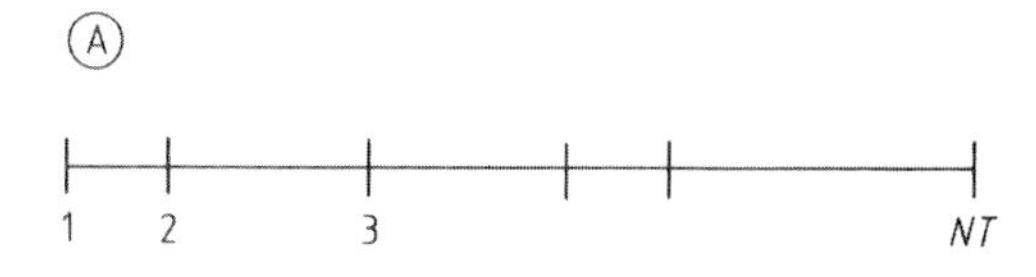

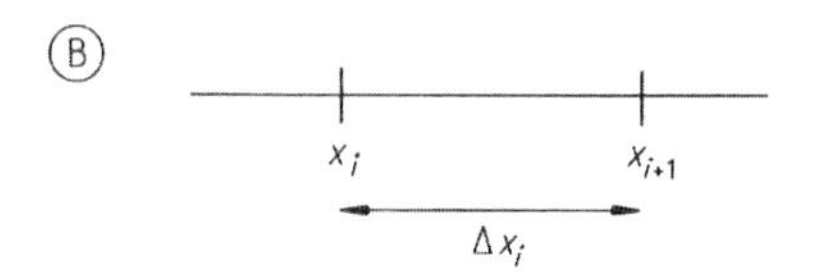

Figure 5. Finite elements for cubic splines
A) Notation for spline knots. B) Notation for one element

Consider the points shown in Figure 5 A. The notation for each interval is shown in Figure 5 B. Within each interval the function is represented as a cubic polynomial.

$$C_i(x) = a_{0i} + a_{1i}x + a_{2i}x^2 + a_{3i}x^3$$

The interpolating function takes on specified values at the knots.

$$C_{i-1}(x_i) = C_i(x_i) = f(x_i)$$

Given the set of values $\{x_i, f(x_i)\}$, the objective is to pass a smooth curve through those points, and the curve should have continuous first and second derivatives at the knots.

$$C'_{i-1}(x_i) = C'_i(x_i)$$

$$C''_{i-1}(x_i) = C''_i(x_i)$$

The formulas for the cubic spline are derived as follows for one region. Since the function is a cubic function the third derivative is constant and the second derivative is linear in x. This is written as

$$C''_i(x) = C''_i(x_i) + \left[C''_i(x_{i+1}) - C''_i(x_i)\right]\frac{x - x_i}{\Delta x_i}$$

and integrated once to give

$$C'_i(x) = C'_i(x_i) + C''_i(x_i)(x - x_i) + \left[C''_i(x_{i+1}) - C''_i(x_i)\right]\frac{(x - x_i)^2}{2\Delta x_i}$$

and once more to give

$$C_i(x) = C_i(x_i) + C'_i(x_i)(x - x_i) + C''_i(x_i)\frac{(x - x_i)^2}{2} + \left[C''_i(x_{i+1}) - C''_i(x_i)\right]\frac{(x - x_i)^3}{6\Delta x_i}$$

Now

$$y_i = C_i(x_i), y'_i = C'_i(x_i), y''_i = C''_i(x_i)$$

is defined so that

$$C_i(x) = y_i + y'_i(x - x_i) + \frac{1}{2}y''_i(x - x_i)^2 + \frac{1}{6\Delta x_i}\left(y''_{i+1} - y''_i\right)(x - x_i)^3$$

A number of algebraic steps make the interpolation easy. These formulas are written for the i-th element as well as the $i-1$-th element. Then the continuity conditions are applied for the first and second derivatives, and the values y'_i and y'_{i-1} are eliminated [15]. The result is

$$y''_{i-1}\Delta x_{i-1} + y''_i\, 2(\Delta x_{i-1} + \Delta x_i) + y''_{i+1}\Delta x_i = -6\left(\frac{y_i - y_{i-1}}{\Delta x_{i-1}} - \frac{y_{i+1} - y_i}{\Delta x_i}\right)$$

This is a tridiagonal system for the set of $\{y''_i\}$ in terms of the set of $\{y_i\}$. Since the continuity conditions apply only for $i = 2, \ldots, NT - 1$, only $NT - 2$ conditions exist for the NT values of y''_i. Two additional conditions are needed, and these are usually taken as the value of the second derivative at each end of the domain, y''_1, y''_{NT}. If these values are zero, the *natural cubic splines* are obtained; they can also be set to achieve some other purpose, such as making the first derivative match some desired condition at the two ends. With these values taken as zero, in the natural cubic spline, an $NT - 2$ system of tridiagonal equations exists, which is easily solved. Once the second derivatives are known at each of the knots, the first derivatives are given by

$$y'_i = \frac{y_{i+1} - y_i}{\Delta x_i} - y''_i\frac{\Delta x_i}{3} - y''_{i+1}\frac{\Delta x_i}{6}$$

The function itself is then known within each element.

Orthogonal Collocation on Finite Elements. In the method of orthogonal collocation on finite elements the solution is expanded in a polynomial of order $NP = NCOL + 2$ within each element [3]. The choice $NCOL = 1$ corresponds to using quadratic polynomials, whereas $NCOL = 2$ gives cubic polynomials. The notation is shown in Figure 6. Set the function to a known value at the two endpoints

$$y_1 = y(x_1)$$

$$y_{NT} = y(x_{NT})$$

and then at the $NCOL$ interior points to each element

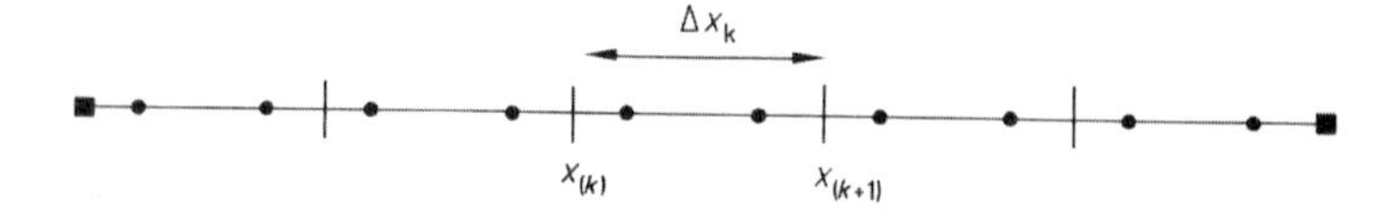

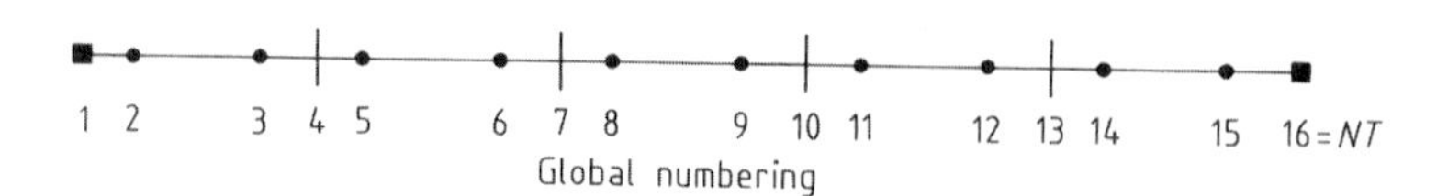

Figure 6. Notation for orthogonal collocation on finite elements
• Residual condition; ■ Boundary conditions; | Element boundary, continuity
NE = total no. of elements.
NT = (*NCOL* + 1) *NE* + 1

$$y_i^e = y_i = y\left(x_i\right), i = \left(NCOL{+}1\right)e{+}I$$

The actual points x_i are taken as the roots of the orthogonal polynomial.

$$P_{NCOL}\left(u\right) = 0 \text{ gives } u_1, u_2, \ldots, u_{NCOL}$$

and then

$$x_i = x_{(e)} + \Delta x_e u_I \equiv x_I^e$$

The first derivatives must be continuous at the element boundaries:

$$\left.\frac{\mathrm{d}y}{\mathrm{d}x}\right|_{x=x(2)-} = \left.\frac{\mathrm{d}y}{\mathrm{d}x}\right|_{x=x(2)+}$$

Within each element the interpolation is a polynomial of degree *NCOL* + 1. Overall the function is continuous with continuous first derivatives. With the choice *NCOL* = 2, the same approximation is achieved as with Hermite cubic polynomials.

2.4. Quadrature

To calculate the value of an integral, the function can be approximated by using each of the methods described in Section 2.3. Using the first three terms in Equation 1 gives

$$\int\limits_{x_0}^{x_0+h} y\left(x\right)\mathrm{d}x = \int\limits_0^1 y_\alpha h \mathrm{d}\alpha$$
$$= \tfrac{h}{2}\left(y_0 + y_1\right) - \tfrac{1}{12} h^3 y_0^{''}\left(\xi\right), x_0 \le \xi \le x_0 + h$$

This corresponds to passing a straight line through the points (x_0, y_0), (x_1, y_1) and integrating under the interpolant. For equally spaced points at $a = x_0$, $a + \Delta x = x_1$, $a + 2\,\Delta x = x_2$, . . . , $a + N\,\Delta x = x_N$, $a + (N + 1)\,\Delta x = b = x_{n+1}$, the trapezoid rule is obtained.

Trapezoid Rule.

$$\int\limits_a^b y\left(x\right)\mathrm{d}x = \tfrac{h}{2}\left(y_0 + 2y_1 + 2y_2 + \cdots + 2y_N\right.$$
$$\left. + y_{N+1}\right) + O\left(h^3\right)$$

The first five terms in Equation 1 are retained and integrated over two intervals.

$$\int\limits_{x_0}^{x_0+2h} y\left(x\right)\mathrm{d}x = \int\limits_0^2 y_\alpha h \mathrm{d}\alpha = \tfrac{h}{3}\left(y_0 + 4y_1 + y_2\right)$$
$$-\tfrac{h^5}{90} y_0^{(\mathrm{IV})}\left(\xi\right), x_0 \le \xi \le x_0 + 2h$$

This corresponds to passing a quadratic function through three points and integrating. For an even number of intervals and an odd number of points, $2\,N + 1$, with $a = x_0$, $a + \Delta x = x_1$, $a + 2\,\Delta x = x_2$, . . . , $a + 2\,N\,\Delta x = b$, Simpson's rule is obtained.

Simpson's Rule.

$$\int\limits_a^b y\left(x\right)\mathrm{d}x = \tfrac{h}{3}\left(y_0 + 4y_1 + 2y_2 + 4y_3 + 2y_4\right.$$
$$\left. + \cdots + 2y_{2N-1} + 4y_{2N} + y_{2N+1}\right) + O\left(h^5\right)$$

Within each pair of intervals the interpolant is continuous with continuous derivatives, but only the function is continuous from one pair to another.

If the *finite element representation* is used (Eq. 2), the integral is

$$\int_{x_i}^{x_{i+1}} y(x)\,\mathrm{d}x = \int_0^1 \sum_{I=1}^{2} c_I^e N_I(u)\,(x_{i+1}-x_i)\,\mathrm{d}u$$
$$= \Delta x_i \sum_{I=1}^{2} c_I^e \int_0^1 N_I(u)\,\mathrm{d}u = \Delta x_i \left(c_1^e \tfrac{1}{2} + c_2^e \tfrac{1}{2}\right)$$
$$= \frac{\Delta x_i}{2}\,(y_i + y_{i+1})$$

Since $c_1^e = y_i$ and $c_2^e = y_{i+1}$, the result is the same as the trapezoid rule. These formulas can be added together to give *linear elements*:

$$\int_a^b y(x)\,\mathrm{d}x = \sum\nolimits_e \frac{\Delta x_e}{2}\,(y_1^e + y_2^e)$$

If the *quadratic* expansion is used (Eq. 3), the endpoints of the element are x_i and x_{i+2}, and x_{i+1} is the midpoint, here assumed to be equally spaced between the ends of the element:

$$\int_{x_i}^{x_{i+2}} y(x)\,\mathrm{d}x = \int_0^1 \sum_{I=1}^{3} c_I^e N_I(u)\,(x_{i+2}-x_i)\,\mathrm{d}u$$
$$= \Delta x_i \sum_{I=1}^{3} c_I^e \int_0^1 N_I(u)\,\mathrm{d}u$$
$$= \Delta x_e \left(c_1^e \tfrac{1}{6} + c_2^e \tfrac{2}{3} + c_3^e \tfrac{1}{6}\right)$$

For many elements, with different Δx^e, *quadratic elements:*

$$\int_a^b y(x) = \sum\nolimits_e \frac{\Delta x_e}{6}\,(y_1^e + 4 y_2^e + y_3^e)$$

If the element sizes are all the same this gives Simpson's rule.

For cubic splines the quadrature rule within one element is

$$\int_{x_i}^{x_{i+1}} C_i(x)\,\mathrm{d}x = \tfrac{1}{2}\Delta x_i\,(y_i + y_{i+1})$$
$$-\tfrac{1}{24}\Delta x_i^3 \left(y_i'' + y_{i+1}''\right)$$

For the entire interval the quadrature formula is

$$\int_{x_1}^{x_{NT}} y(x)\,\mathrm{d}x = \tfrac{1}{2} \sum_{i=1}^{NT-1} \Delta x_i (y_i + y_{i+1})$$
$$-\tfrac{1}{24} \sum_{i=1}^{NT-1} \Delta x_i^3 (y_i'' + y_{i+1}'')$$

with $y_1''= 0$, $y''_{NT} = 0$ for natural cubic splines.

When orthogonal polynomials are used, as in Equation 1, the m roots to $P_m(x) = 0$ are chosen as quadrature points and called points $\{x_j\}$. Then the quadrature is *Gaussian:*

$$\int_0^1 y(x)\,\mathrm{d}x = \sum_{j=1}^{m} W_j y(x_j)$$

The quadrature is exact when y is a polynomial of degree $2m - 1$ in x. The m weights and m Gauss points result in $2m$ parameters, chosen to exactly represent a polynomial of degree $2m - 1$, which has $2m$ parameters. The Gauss points and weights are given in Table 2. The weights can be defined with $W(x)$ in the integrand as well.

Table 2. Gaussian quadrature points and weights *

N	x_i	W_i
1	0.5000000000	0.6666666667
2	0.2113248654	0.5000000000
	0.7886751346	0.5000000000
3	0.1127016654	0.2777777778
	0.5000000000	0.4444444445
	0.8872983346	0.2777777778
4	0.0694318442	0.1739274226
	0.3300094783	0.3260725774
	0.6699905218	0.3260725774
	0.9305681558	0.1739274226
5	0.0469100771	0.1184634425
	0.2307653450	0.2393143353
	0.5000000000	0.2844444444
	0.7692346551	0.2393143353
	0.9530899230	0.1184634425

* For a given N the quadrature points $x_2, x_3, \ldots, x_{NP-1}$ are given above. $x_1 = 0$, $x_{NP} = 1$. For $N = 1$, $W_1 = W_3 = 1/6$ and for $N \geq 2$, $W_1 = W_{NP} = 0$.

For orthogonal collocation on finite elements the quadrature formula is

$$\int_0^1 y(x)\,\mathrm{d}x = \sum\nolimits_e \Delta x_e \sum_{j=1}^{NP} W_j y(x_J^e)$$

Each special polynomial has its own quadrature formula. For example, Gauss – Legendre polynomials give the quadrature formula

$$\int_0^\infty e^{-x} y(x)\,\mathrm{d}x = \sum_{i=1}^{n} W_i y(x_i)$$

(points and weights are available in mathematical tables) [23].

For Gauss – Hermite polynomials the quadrature formula is

$$\int_{-\infty}^{\infty} e^{-x^2} y(x)\,\mathrm{d}x = \sum_{i=1}^{n} W_i y(x_i)$$

(points and weights are available in mathematical tables) [23].

Romberg's method uses extrapolation techniques to improve the answer [15]. If I_1 is the value of the integral obtained by using interval size $h = \Delta x$, I_2 the value of I obtained by using interval size $h/2$, and I_0 the true value of I, then the error in a method is approximately h^m, or

$$I_1 \approx I_0 + ch^m$$

$$I_2 \approx I_0 + c\left(\frac{h}{2}\right)^m$$

Replacing the $\approx$ by an equality (an approximation) and solving for c and I_0 give

$$I_0 = \frac{2^m I_2 - I_1}{2^m - 1}$$

This process can also be used to obtain $I_1, I_2, \ldots$, by halving h each time, calculating new estimates from each pair, and calling them $J_1, J_2, \ldots$ (i.e., in the formula above, I_0 is replaced with J_1). The formulas are reapplied for each pair of J's to obtain $K_1, K_2, \ldots$. The process continues until the required tolerance is obtained.

I_1	I_2	I_3	I_4
	J_1	J_2	J_3
		K_1	K_2
			L_1

Romberg's method is most useful for a low-order method (small m) because significant improvement is then possible.

When the integrand has singularities, a variety of techniques can be tried. The integral may be divided into one part that can be integrated analytically near the singularity and another part that is integrated numerically. Sometimes a change of argument allows analytical integration. Series expansion might be helpful, too. When the domain is infinite, Gauss – Legendre or Gauss – Hermite quadrature can be used. Also a transformation can be made [15]. For example, let $u = 1/x$ and then

$$\int_a^b f(x)\,\mathrm{d}x = \int_{1/b}^{1/a} \frac{1}{u^2} f\left(\frac{1}{u}\right) \mathrm{d}u \quad a, b > 0$$

2.5. Least Squares

When fitting experimental data to a mathematical model, it is necessary to recognize that the experimental measurements contain error; the goal is to find the set of parameters in the model that best represents the experimental data. Reference [23] gives a complete treatment relating the least-squares procedure to maximum likelihood.

In a least-squares parameter estimation, it is desired to find parameters that minimize the sum of squares of the deviation between the experimental data and the theoretical equation

$$\chi^2 = \sum_{i=1}^{N} \left[\frac{y_i - y(x_i; a_1, a_2, \ldots, a_M)}{\sigma_i}\right]^2$$

where y_i is the i-th experimental data point for the value x_i, $y(x_i; a_1, a_2, \ldots, a_M)$ the theoretical equation at x_i, σ_i the standard deviation of the i-th measurement, and the parameters $\{a_1, a_2, \ldots, a_M\}$ are to be determined to minimize χ^2. The simplification is made here that the standard deviations are all the same. Thus, we minimize the variance of the curve fit.

$$\sigma^2 = \sum_{i=1}^{N} \frac{[y_i - y(x_i; a_1, a_2, \ldots, a_M)]^2}{N}$$

Linear Least Squares. When the model is a straight line, one is minimizing

$$\chi^2 = \sum_{i=1}^{N} [y_i - a - bx_i]^2$$

The linear correlation coefficient r is defined by

$$r = \frac{\sum_{i=1}^{N} (x_i - \overline{x})(y_i - \overline{y})}{\sqrt{\sum_{i=1}^{N} (x_i - \overline{x})^2} \sqrt{\sum_{i=1}^{N} (y_i - \overline{y})^2}}$$

and

$$\chi^2 = (1 - r^2) \sum_{i=1}^{N} [y_i - \overline{y}]^2$$

where $\overline{y}$ is the average of y_i values. Values of r near 1 indicate a positive correlation; r near -1 means a negative correlation, and r near zero means no correlation. These parameters are easily found by using standard programs, such as Microsoft Excel.

Polynomial Regression. In polynomial regression, one expands the function in a polynomial in x.

$$y(x)=\sum_{j=1}^{M}a_j x^{j-1}$$

The parameters are easily determined using computer software. In Microsoft Excel, the data is put into columns A and B and the graph is created as for a linear curve fit. Then add a trendline and choose the degree of polynomial desired.

Multiple Regression. In multiple regression, any set of functions can be used, not just polynomials.

$$y(x)=\sum_{j=1}^{M}a_j f_j(x)$$

where the set of functions $\{f_j(x)\}$ is known and specified. Note that the unknown parameters $\{a_j\}$ enter the equation linearly. In this case, the spreadsheet can be expanded to have a column for x, and then successive columns for $f_j(x)$. In Microsoft Excel, choose Regression under Tools/Data Analysis, and complete the form. In addition to the actual correlation, one gets the expected variance of the unknowns, which allows one to assess how accurately they were determined.

Nonlinear Regression. In nonlinear regression, the same procedure is followed except that an optimization routine must be used to find the minimum χ^2. See Chapter 10.

2.6. Fourier Transforms of Discrete Data [15]

Suppose a signal $y(t)$ is sampled at equal intervals

$$y_n=y(n\Delta),n=\ldots,-2,-1,0,1,2,\ldots$$

$$\Delta=\text{sampling rate}$$

$$\text{(e.g., number of samples per second)}$$

The Fourier transform and inverse transform are

$$Y(\omega)=\int_{-\infty}^{\infty}y(t)\,e^{\mathrm{i}\omega t}\mathrm{d}t$$

$$y(t)=\frac{1}{2\pi}\int_{-\infty}^{\infty}Y(\omega)\,e^{-\mathrm{i}\omega t}\mathrm{d}\omega$$

(For definition of i, see Chap. 3.) The Nyquist critical frequency or critical angular frequency is

$$f_c=\frac{1}{2\Delta},\omega_c=\frac{\pi}{\Delta}$$

If a function $y(t)$ is bandwidth limited to frequencies smaller than f_c, i.e.,

$$Y(\omega)=0\ \text{for}\ \omega>\omega_c$$

then the function is completely determined by its samples y_n. Thus, the entire information content of a signal can be recorded by sampling at a rate $\Delta^{-1}=2f_c$. If the function is not bandwidth limited, then aliasing occurs. Once a sample rate Δ is chosen, information corresponding to frequencies greater than f_c is simply aliased into that range. The way to detect this in a Fourier transform is to see if the transform approaches zero at $\pm f_c$; if not, aliasing has occurred and a higher sampling rate is needed.

Next, for N samples, where N is even

$$y_k=y(t_k),\ t_k=k\Delta,\ k=0,1,2,\ldots,N-1$$

and the sampling rate is Δ; with only N values $\{y_k\}$ the complete Fourier transform $Y(\omega)$ cannot be determined. Calculate the value $Y(\omega_n)$ at the discrete points

$$\omega_n=\frac{2\pi n}{N\Delta},n=-\frac{N}{2},\ldots,0,\ldots,\frac{N}{2}$$

$$Y_n=\sum_{k=0}^{N-1}y_k e^{2\pi ikn/N}$$

$$Y(\omega_n)=\Delta Y_n$$

The discrete inverse Fourier transform is

$$y_k=\frac{1}{N}\sum_{k=0}^{N-1}Y_n e^{-2\pi ikn/N}$$

The *fast Fourier transform* (*FFT*) is used to calculate the Fourier transform as well as the inverse Fourier transform. A discrete Fourier transform of length N can be written as the sum of two discrete Fourier transforms, each of length $N/2$, and each of these transforms is separated into two halves, each half as long. This continues until only one component is left. For this reason, N is taken as a power of 2, $N=2^p$.

The vector $\{y_j\}$ is filled with zeroes, if need be, to make $N=2^p$ for some p. For the computer program, see [15, p. 381]. The standard

Fourier transform takes N^2 operations to calculate, whereas the fast Fourier transform takes only $N \log_2 N$. For large N the difference is significant; at $N = 100$ it is a factor of 15, but for $N = 1000$ it is a factor of 100.

The discrete Fourier transform can also be used for *differentiating* a function; this is used in the spectral method for solving differential equations. Consider a grid of equidistant points:

$$x_n = n\Delta x, n = 0,1,2,\ldots, 2N-1, \Delta x = \frac{L}{2N}$$

the solution is known at each of these grid points $\{Y(x_n)\}$. First, the Fourier transform is taken:

$$y_k = \frac{1}{2N} \sum_{n=0}^{2N-1} Y(x_n)\,\mathrm{e}^{-2ik\pi x_n/L}$$

The inverse transformation is

$$Y(x) = \frac{1}{L} \sum_{k=-N}^{N} y_k \mathrm{e}^{2ik\pi x/L}$$

which is differentiated to obtain

$$\frac{\mathrm{d}Y}{\mathrm{d}x} = \frac{1}{L} \sum_{k=-N}^{N} y_k \frac{2\pi i k}{L} \mathrm{e}^{2ik\pi x/L}$$

Thus, at the grid points

$$\left.\frac{\mathrm{d}Y}{\mathrm{d}x}\right|_n = \frac{1}{L} \sum_{k=-N}^{N} y_k \frac{2\pi i k}{L} \mathrm{e}^{2ik\pi x_n/L}$$

The process works as follows. From the solution at all grid points the Fourier transform is obtained by using FFT $\{y_k\}$. This is multiplied by $2\pi i k/L$ to obtain the Fourier transform of the derivative:

$$y'_k = y_k \frac{2\pi i k}{L}$$

The inverse Fourier transform is then taken by using FFT, to give the value of the derivative at each of the grid points:

$$\left.\frac{\mathrm{d}Y}{\mathrm{d}x}\right|_n = \frac{1}{L} \sum_{k=-N}^{N} y'_k \mathrm{e}^{2ik\pi x_n/L}$$

Any nonlinear term can be treated in the same way: evaluate it in real space at N points and take the Fourier transform. After processing using this transform to get the transform of a new function, take the inverse transform to obtain the new function at N points. This is what is done in direct numerical simulation of turbulence (DNS).

2.7. Two-Dimensional Interpolation and Quadrature

Bicubic splines can be used to interpolate a set of values on a regular array, $f(x_i, y_j)$. Suppose NX points occur in the x direction and NY points occur in the y direction. Press et al. [15] suggest computing NY different cubic splines of size NX along lines of constant y, for example, and storing the derivative information. To obtain the value of f at some point x, y, evaluate each of these splines for that x. Then do one spline of size NY in the y direction, doing both the determination and the evaluation.

Multidimensional integrals can also be broken down into one-dimensional integrals. For example,

$$\int_a^b \int_{f_1(x)}^{f_2(x)} z(x,y)\,\mathrm{d}x\mathrm{d}y = \int_a^b G(x)\,\mathrm{d}x;$$

$$G(x) = \int_{f_1(x)}^{f_2(x)} z(x,y)\,\mathrm{d}x$$

3. Complex Variables [25 – 31]

3.1. Introduction to the Complex Plane

A complex number is an ordered pair of real numbers, x and y, that is written as

$$z = x + iy$$

The variable i is the imaginary unit which has the property

$$i^2 = -1$$

The real and imaginary parts of a complex number are often referred to:

$$\mathrm{Re}\,(z) = x, \mathrm{Im}\,(z) = y$$

A complex number can also be represented graphically in the complex plane, where the real part of the complex number is the abscissa and the imaginary part of the complex number is the ordinate (see Fig. 7).

Another representation of a complex number is the *polar form*, where r is the magnitude and θ is the argument.

$$r = |x + i\,y| = \sqrt{x^2 + y^2},\ \theta = \arg(x + i\,y)$$

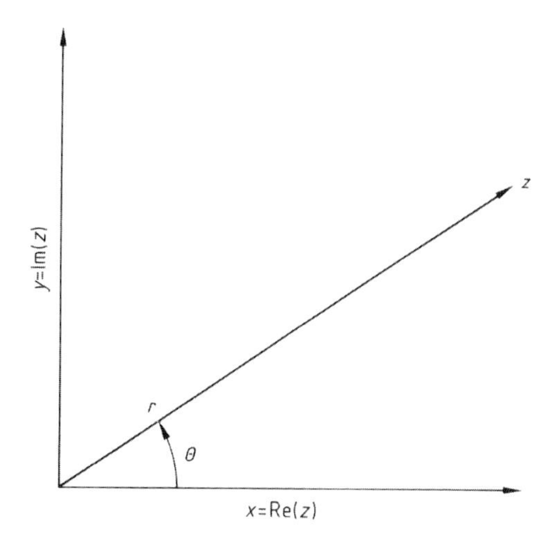

Figure 7. The complex plane

Write

$$z = x + \mathrm{i}\, y = r\,(\cos\theta + \mathrm{i}\,\sin\theta)$$

so that

$$x = r\cos\theta,\ y = r\sin\theta$$

and

$$\theta = \arctan\frac{y}{x}$$

Since the arctangent repeats itself in multiples of π rather than $2\,\pi$, the argument must be defined carefully. For example, the θ given above could also be the argument of $-(x + \mathrm{i}\, y)$. The function $z = \cos\theta + \mathrm{i}\sin\theta$ obeys $|z| = |\cos\theta + \mathrm{i}\sin\theta| = 1$.

The rules of equality, addition, and multiplication are

$$z_1 = x_1 + \mathrm{i}\, y_1,\ z_2 = x_2 + \mathrm{i}\, y_2$$

Equality :
$z_1 = z_2$ if and only if $x_1 = x_2$ and $y_1 = y_2$
Addition :
$z_1 + z_2 = (x_1 + x_2) + \mathrm{i}\,(y_1 + y_2)$
Multiplication :
$z_1 z_2 = (x_1 x_2 - y_1 y_2) + \mathrm{i}\,(x_1 y_2 + x_2 y_1)$

The last rule can be remembered by using the standard rules for multiplication, keeping the imaginary parts separate, and using $\mathrm{i}^2 = -1$. In the complex plane, addition is illustrated in Figure 8. In polar form, multiplication is

$$z_1 z_2 = r_1 r_2\,[\cos(\theta_1 + \theta_2) + \mathrm{i}\sin(\theta_1 + \theta_2)]$$

The magnitude of $z_1 + z_2$ is bounded by

$$|z_1 \pm z_2| \le |z_1| + |z_2| \text{ and } |z_1| - |z_2| \le |z_1 \pm z_2|$$

as can be seen in Figure 8. The magnitude and arguments in multiplication obey

$$|z_1 z_2| = |z_1|\,|z_2|,\arg(z_1 z_2) = \arg z_1 + \arg z_2$$

The *complex conjugate* is $z^* = x - \mathrm{i}\, y$ when $z = x + \mathrm{i}\, y$ and $|z^*| = |z|$, $\arg z^* = -\arg z$

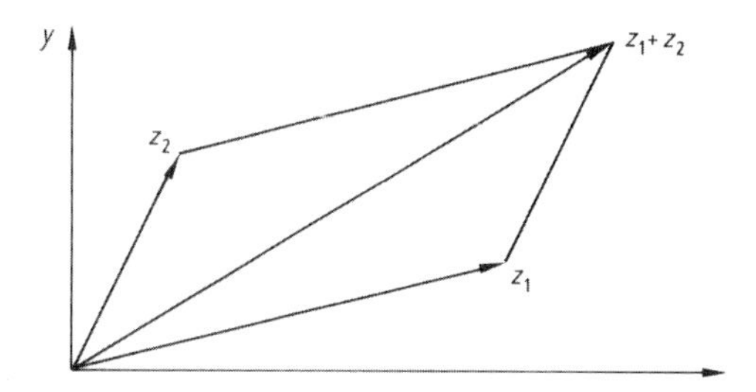

Figure 8. Addition in the complex plane

For complex conjugates then

$$z^*\, z = |z|^2$$

The reciprocal is

$$\frac{1}{z} = \frac{z^*}{|z|^2} = \frac{1}{r}\,(\cos\theta - \mathrm{i}\,\sin\theta),\ \arg\left(\frac{1}{z}\right) = -\arg z$$

Then

$$\frac{z_1}{z_2} = \frac{x_1 + \mathrm{i}\, y_1}{x_2 + \mathrm{i}\, y_2} = (x_1 + \mathrm{i}\, y_1)\,\frac{x_2 - \mathrm{i}\, y_2}{x_2^2 + y_2^2}$$

$$= \frac{x_1 x_2 + y_1 y_2}{x_2^2 + y_2^2} + \mathrm{i}\,\frac{x_2 y_1 - x_1 y_2}{x_2^2 + y_2^2}$$

and

$$\frac{z_1}{z_2} = \frac{r_1}{r_2}\,[\cos(\theta_1 - \theta_2) + \mathrm{i}\sin(\theta_1 - \theta_2)]$$

3.2. Elementary Functions

Properties of elementary functions of complex variables are discussed here [32]. When the polar form is used, the argument must be specified because the same physical angle can be achieved with arguments differing by $2\,\pi$. A complex number taken to a real power obeys

$$u = z^n,\ |z^n| = |z|^n,\ \arg\ (z^n) = n\arg z\ (\mathrm{mod} 2\pi)$$

$$u = z^n = r^n\,(\cos n\theta + \mathrm{i}\sin n\theta)$$

Roots of a complex number are complicated by careful accounting of the argument

$$z = w^{1/n} \text{ with } w = R\left(\cos\Theta + \mathrm{i}\sin\Theta\right),\ 0 \leq \Theta \leq 2\pi$$

then

$$z_k = R^{1/n}\{\cos\left[\frac{\Theta}{n} + (k-1)\frac{2\pi}{n}\right]$$

$$+\mathrm{i}\sin\left[\frac{\Theta}{n} + (k-1)\frac{2\pi}{n}\right]\}$$

such that

$$(z_k)^n = w \text{ for every } k$$

$$z = r\left(\cos\theta + \mathrm{i}\sin\theta\right)$$
$$r^n = R,\ n\theta = \Theta\ (\mathrm{mod}\ 2\pi)$$

The exponential function is

$$\mathrm{e}^z = \mathrm{e}^x\left(\cos y + \mathrm{i}\sin y\right)$$

Thus,

$$z = r\left(\cos\theta + \mathrm{i}\sin\theta\right)$$

can be written

$$z = r\mathrm{e}^{\mathrm{i}\theta}$$

and

$$|\mathrm{e}^z| = \mathrm{e}^x,\ \arg\mathrm{e}^z = y\ (\mathrm{mod}\ 2\pi)$$

The exponential obeys

$$\mathrm{e}^z \neq 0 \text{ for every finite } z$$

and is periodic with period 2 π:

$$e^{z+2\pi\,\mathrm{i}} = \mathrm{e}^z$$

Trigonometric functions can be defined by using

$$e^{iy} = \cos y + \mathrm{i}\sin y, \text{ and } \mathrm{e}^{-\mathrm{i}y} = \cos y - \mathrm{i}\sin y$$

Thus,

$$\cos y = \frac{\mathrm{e}^{\mathrm{i}\,y} + \mathrm{e}^{-\mathrm{i}\,y}}{2} = \cosh \mathrm{i}\,y$$

$$\sin y = \frac{\mathrm{e}^{\mathrm{i}\,y} - \mathrm{e}^{-\mathrm{i}\,y}}{2\mathrm{i}} = -\mathrm{i}\sinh \mathrm{i}\,y$$

The second equation follows from the definitions

$$\cosh z \equiv \frac{\mathrm{e}^z + \mathrm{e}^{-z}}{2},\ \sinh z \equiv \frac{\mathrm{e}^z - \mathrm{e}^{-z}}{2}$$

The remaining hyperbolic functions are

$$\tanh z \equiv \frac{\sinh z}{\cosh z},\ \coth z \equiv \frac{1}{\tanh z}$$

$$\mathrm{sech}\, z \equiv \frac{1}{\cosh z},\ \mathrm{csch}\, z \equiv \frac{1}{\sinh z}$$

The circular functions with complex arguments are defined

$$\cos z = \frac{\mathrm{e}^{\mathrm{i}\,z} + \mathrm{e}^{-\mathrm{i}\,z}}{2},\ \sin z = \frac{\mathrm{e}^{\mathrm{i}\,z} - \mathrm{e}^{-\mathrm{i}\,z}}{2},$$

$$\tan z = \frac{\sin z}{\cos z}$$

and satisfy

$$\sin(-z) = -\sin z,\ \cos(-z) = \cos z$$

$$\sin(\mathrm{i}\,z) = \mathrm{i}\sinh z,\ \cos(\mathrm{i}\,z) = -\cosh z$$

All trigonometric identities for real, circular functions with real arguments can be extended without change to complex functions of complex arguments. For example,

$$\sin^2 z + \cos^2 z = 1,$$

$$\sin(z_1 + z_2) = \sin z_1 \cos z_2 + \cos z_1 \sin z_2$$

The same is true of hyperbolic functions. The absolute boundaries of sin z and cos z are not bounded for all z.

Trigonometric identities can be defined by using

$$e^{i\theta} = \cos\theta + \mathrm{i}\sin\theta$$

For example,

$$\mathrm{e}^{\mathrm{i}(\alpha+\beta)} = \cos(\alpha+\beta) + \mathrm{i}\sin(\alpha+\beta)$$

$$= \mathrm{e}^{\mathrm{i}\alpha}\mathrm{e}^{\mathrm{i}\beta} = (\cos\alpha + \mathrm{i}\sin\alpha)$$
$$(\cos\beta + \mathrm{i}\sin\beta)$$
$$= \cos\alpha\cos\beta - \sin\alpha\sin\beta$$
$$+\mathrm{i}\ (\cos\alpha\sin\beta + \cos\beta\sin\alpha)$$

Equating real and imaginary parts gives

$$\cos(\alpha+\beta) = \cos\alpha\cos\beta - \sin\alpha\sin\beta$$

$$\sin(\alpha+\beta) = \cos\alpha\sin\beta + \cos\beta\sin\alpha$$

The logarithm is defined as

$$\ln z = \ln|z| + \mathrm{i}\arg z$$

and the various determinations differ by multiples of 2 π i. Then,

$$\mathrm{e}^{\ln z} = z$$

$$\ln(\mathrm{e}^z) - z \equiv 0\ (\mathrm{mod}\ 2\,\pi\mathrm{i})$$

Also,

$$\ln(z_1\, z_2) - \ln z_1 - \ln z_2 \equiv 0\ (\mathrm{mod}\ 2\,\pi\mathrm{i})$$

is always true, but

$$\ln(z_1\, z_2) = \ln z_1 + \ln z_2$$

holds only for some determinations of the logarithms. The principal determination of the argument can be defined as $-\pi < \arg \leq \pi$.

3.3. Analytic Functions of a Complex Variable

Let $f(z)$ be a single-valued continuous function of z in a domain D. The function $f(z)$ is differentiable at the point z_0 in D if

$$\lim_{h\to 0}\frac{f(z_0+h)-f(z_0)}{h}$$

exists as a finite (complex) number and is independent of the direction in which h tends to zero. The limit is called the derivative, $f'(z_0)$. The derivative now can be calculated with h approaching zero in the complex plane, i.e., anywhere in a circular region about z_0. The function $f(z)$ is differentiable in D if it is differentiable at all points of D; then $f(z)$ is said to be an analytic function of z in D. Also, $f(z)$ is analytic at z_0 if it is analytic in some ε neighborhood of z_0. The word analytic is sometimes replaced by holomorphic or regular.

The *Cauchy – Riemann equations* can be used to decide if a function is analytic. Set

$$f(z)=f(x+\mathrm{i}\,y)=u(x,y)+\mathrm{i}\,v(x,y)$$

Theorem [30, p. 51]. Suppose that $f(z)$ is defined and continuous in some neighborhood of $z = z_0$. A necessary condition for the existence of $f'(z_0)$ is that $u(x, y)$ and $v(x, y)$ have first-order partials and that the Cauchy – Riemann conditions (see below) hold.

$$\frac{\partial u}{\partial x}=\frac{\partial v}{\partial y}\text{ and }\frac{\partial u}{\partial y}=-\frac{\partial v}{\partial x}\text{ at }z_0$$

Theorem [30, p. 61]. The function $f(z)$ is analytic in a domain D if and only if u and v are continuously differentiable and satisfy the Cauchy – Riemann conditions there.

If $f_1(z)$ and $f_2(z)$ are analytic in domain D, then $\alpha_1 f_1(z) + \alpha_2 f_2(z)$ is analytic in D for any (complex) constants α_1, α_2.

$$f_1(z)+f_2(z)\text{ is analytic in D}$$

$$f_1(z)/f_2(z)\text{ is analytic in D except where }f_2(z)=0$$

An analytic function of an analytic function is analytic. If $f(z)$ is analytic, $f'(z)\neq 0$ in D, $f(z_1)\neq f(z_2)$ for $z_1\neq z_2$, then the inverse function $g(w)$ is also analytic and

$$g'(w)=\frac{1}{f'(z)}\text{ where }w=f(z),$$

$$g(w)=g[f(z)]=z$$

Analyticity implies continuity but the converse is not true: $z^* = x - \mathrm{i}\,y$ is continuous but, because the Cauchy – Riemann conditions are not satisfied, it is not analytic. An entire function is one that is analytic for all finite values of z. Every polynomial is an entire function. Because the polynomials are analytic, a ratio of polynomials is analytic except when the denominator vanishes. The function $f(z) = |z^2|$ is continuous for all z but satisfies the Cauchy – Riemann conditions only at $z = 0$. Hence, $f'(z)$ exists only at the origin, and $|z|^2$ is nowhere analytic. The function $f(z) = 1/z$ is analytic except at $z = 0$. Its derivative is $-1/z^2$, where $z\neq 0$. If $\ln z = \ln|z| + \mathrm{i}\arg z$ in the cut domain $-\pi < \arg z \leq \pi$, then $f(z) = 1/\ln z$ is analytic in the same cut domain, except at $z = 1$, where $\log z = 0$. Because e^z is analytic and $\pm\,\mathrm{i}\,z$ are analytic, $\mathrm{e}^{\pm \mathrm{i}z}$ is analytic and linear combinations are analytic. Thus, the sine and cosine and hyperbolic sine and cosine are analytic. The other functions are analytic except when the denominator vanishes.

The derivatives of the elementary functions are

$$\frac{\mathrm{d}}{\mathrm{d}z}\mathrm{e}^z=\mathrm{e}^z,\ \frac{\mathrm{d}}{\mathrm{d}z}z^n=n\,z^{n-1}$$

$$\frac{\mathrm{d}}{\mathrm{d}z}(\ln z)=\frac{1}{z},\ \frac{\mathrm{d}}{\mathrm{d}z}\sin z=\cos z,$$

$$\frac{\mathrm{d}}{\mathrm{d}z}\cos z=-\sin z$$

In addition,

$$\frac{\mathrm{d}}{\mathrm{d}z}(f\,g)=f\frac{\mathrm{d}g}{\mathrm{d}z}+g\frac{\mathrm{d}f}{\mathrm{d}z}$$

$$\frac{\mathrm{d}}{\mathrm{d}z}f[g(z)]=\frac{\mathrm{d}f}{\mathrm{d}g}\frac{\mathrm{d}g}{\mathrm{d}z}$$

$$\frac{\mathrm{d}}{\mathrm{d}z}\sin w=\cos w\frac{\mathrm{d}w}{\mathrm{d}z},\ \frac{\mathrm{d}}{\mathrm{d}z}\cos w=-\sin w\frac{\mathrm{d}w}{\mathrm{d}z}$$

Define $z^a = \mathrm{e}^{a\ln z}$ for complex constant a. If the determination is $-\pi < \arg z \leq \pi$, then z^a is analytic on the complex plane with a cut on the negative real axis. If a is an integer n, then $\mathrm{e}^{2\pi in} = 1$ and z^n has the same limits approaching the cut from either side. The function can be made continuous across the cut and the function is analytic there, too. If $a = 1/n$ where n is an integer, then

$$z^{1/n}=\mathrm{e}^{(\ln z)/n}=|z|^{1/n}\mathrm{e}^{\mathrm{i}(\arg z)/n}$$

So $w = z^{1/n}$ has n values, depending on the choice of argument.

Laplace Equation. If $f(z)$ is analytic, where

$$f(z) = u(x,y) + \mathrm{i}\, v(x,y)$$

the Cauchy – Riemann equations are satisfied. Differentiating the Cauchy – Riemann equations gives the Laplace equation:

$$\frac{\partial^2 u}{\partial x^2} = \frac{\partial^2 v}{\partial x \partial y} = \frac{\partial^2 v}{\partial y \partial x} = -\frac{\partial^2 u}{\partial y^2} \text{ or}$$

$$\frac{\partial^2 u}{\partial x^2} + \frac{\partial^2 u}{\partial y^2} = 0$$

Similarly,

$$\frac{\partial^2 v}{\partial x^2} + \frac{\partial^2 v}{\partial y^2} = 0$$

Thus, general solutions to the Laplace equation can be obtained from analytic functions [30, p. 60]. For example,

$$\ln \frac{1}{|z - z_0|}$$

is analytic so that a solution to the Laplace equation is

$$\ln \left[(x-a)^2 + (y-b)^2 \right]^{-1/2}$$

A solution to the Laplace equation is called a harmonic function. A function is harmonic if, and only if, it is the real part of an analytic function. The imaginary part is also harmonic. Given any harmonic function u, a conjugate harmonic function v can be constructed such that $f = u + \mathrm{i}\, v$ is locally analytic [30, p. 290].

Maximum Principle. If $f(z)$ is analytic in a domain D and continuous in the set consisting of D and its boundary C, and if $|f(z)| \leq M$ on C, then $|f(z)| < M$ in D unless $f(z)$ is a constant [30, p. 134].

3.4. Integration in the Complex Plane

Let C be a rectifiable curve in the complex plane

$$C: z = z(t),\ 0 \leq t \leq 1$$

where $z(t)$ is a continuous function of bounded variation; C is oriented such that $z_1 = z(t_1)$ precedes the point $z_2 = z(t_2)$ on C if and only if $t_1 < t_2$. Define

$$\int_C f(z)\, \mathrm{d}z = \int_0^1 f[z(t)]\, \mathrm{d}z(t)$$

The integral is linear with respect to the integrand:

$$\int_C [\alpha_1 f_1(z) + \alpha_2 f_2(z)]\, \mathrm{d}z$$

$$= \alpha_1 \int_C f_1(z)\, \mathrm{d}z + \alpha_2 \int_C f_2(z)\, \mathrm{d}z$$

The integral is additive with respect to the path. Let curve C_2 begin where curve C_1 ends and $C_1 + C_2$ be the path of C_1 followed by C_2. Then,

$$\int_{C_1 + C_2} f(z)\, \mathrm{d}z = \int_{C_1} f(z)\, \mathrm{d}z + \int_{C_2} f(z)\, \mathrm{d}z$$

Reversing the orientation of the path replaces the integral by its negative:

$$\int_{-C} f(z)\, \mathrm{d}z = -\int_C f(z)\, \mathrm{d}z$$

If the path of integration consists of a finite number of arcs along which $z(t)$ has a continuous derivative, then

$$\int_C f(z)\, \mathrm{d}z = \int_0^1 f[z(t)]\, z'(t)\, \mathrm{d}t$$

Also if $s(t)$ is the arc length on C and $l(C)$ is the length of C

$$\left| \int_C f(z)\, \mathrm{d}z \right| \leq \max_{z \in C} |f(z)|\, l(C)$$

and

$$\left| \int_C f(z)\, \mathrm{d}z \right| \leq \int_C |f(z)|\, |\mathrm{d}z| = \int_0^1 |f[z(t)]|\, \mathrm{d}s(t)$$

Cauchy's Theorem [25, 30, p. 111]. Suppose $f(z)$ is an analytic function in a domain D and C is a simple, closed, rectifiable curve in D such that $f(z)$ is analytic inside and on C. Then

$$\oint_C f(z)\, \mathrm{d}z = 0 \qquad (4)$$

If D is simply connected, then Equation 4 holds for every simple, closed, rectifiable curve C in D. If D is simply connected and if a and b are any two points in D, then

$$\int_a^b f(z)\, \mathrm{d}z$$

is independent of the rectifiable path joining a and b in D.

Cauchy's Integral. If C is a closed contour such that $f(z)$ is analytic inside and on C, z_0 is a point inside C, and z traverses C in the counterclockwise direction,

$$f(z_0) = \frac{1}{2\pi i}\oint_C \frac{f(z)}{z-z_0}dz$$

$$f'(z_0) = \frac{1}{2\pi i}\oint_C \frac{f(z)}{(z-z_0)^2}dz$$

Under further restriction on the domain [30, p. 127],

$$f^{(m)}(z_0) = \frac{m!}{2\pi i}\oint_C \frac{f(z)}{(z-z_0)^{m+1}}dz$$

Power Series. If $f(z)$ is analytic interior to a circle $|z - z_0| < r_0$, then at each point inside the circle the series

$$f(z) = f(z_0) + \sum_{n=1}^{\infty} \frac{f^{(n)}(z_0)}{n!}(z-z_0)^n$$

converges to $f(z)$. This result follows from Cauchy's integral. As an example, e^z is an entire function (analytic everywhere) so that the MacLaurin series

$$e^z = 1 + \sum_{n=1}^{\infty}\frac{z^n}{n!}$$

represents the function for all z.

Another result of Cauchy's integral formula is that if $f(z)$ is analytic in an annulus R, $r_1 < |z - z_0| < r_2$, it is represented in R by the Laurent series

$$f(z) = \sum_{n=-\infty}^{\infty} A_n(z-z_0)^n,\ r_1 < |z-z_0| \le r_2$$

where

$$A_n = \frac{1}{2\pi i}\int_C \frac{f(z)}{(z-z_0)^{n+1}}dz,$$

$$n = 0, \pm 1, \pm 2, \ldots,$$

and C is a closed curve counterclockwise in R.

Singular Points and Residues [33, p. 159, 30, p. 180]. If a function in analytic in every neighborhood of z_0, but not at z_0 itself, then z_0 is called an isolated singular point of the function. About an isolated singular point, the function can be represented by a Laurent series.

$$f(z) = \cdots + \frac{A_{-2}}{(z-z_0)^2} + \frac{A_{-1}}{z-z_0} + A_0 + A_1(z-z_0) + \cdots 0 < |z-z_0| \le r_0 \quad (5)$$

In particular,

$$A_{-1} = \frac{1}{2\pi i}\oint_C f(z)\,dz$$

where the curve C is a closed, counterclockwise curve containing z_0 and is within the neighborhood where $f(z)$ is analytic. The complex number A_{-1} is the residue of $f(z)$ at the isolated singular point z_0; $2\pi i A_{-1}$ is the value of the integral in the positive direction around a path containing no other singular points.

If $f(z)$ is defined and analytic in the exterior $|z - z_0| > R$ of a circle, and if

$$v(\zeta) = f\left(z_0 + \frac{1}{\zeta}\right) \text{ obtained by } \zeta = \frac{1}{z-z_0}$$

has a removable singularity at $\zeta = 0$, $f(z)$ is analytic at infinity. It can then be represented by a Laurent series with nonpositive powers of $z - z_0$.

If C is a closed curve within which and on which $f(z)$ is analytic except for a finite number of singular points $z_1, z_2, \ldots, z_n$ interior to the region bounded by C, then the residue theorem states

$$\oint_C f(z)\,dz = 2\pi i(\varrho_1 + \varrho_2 + \cdots \varrho_n)$$

where ϱ_n denotes the residue of $f(z)$ at z_n.

The series of negative powers in Equation 5 is called the principal part of $f(z)$. If the principal part has an infinite number of nonvanishing terms, the point z_0 is an essential singularity. If $A_{-m} \neq 0$, $A_{-n} = 0$ for all $m < n$, then z_0 is called a pole of order m. It is a simple pole if $m = 1$. In such a case,

$$f(z) = \frac{A_{-1}}{z-z_0} + \sum_{n=0}^{\infty} A_n(z-z_0)^n$$

If a function is not analytic at z_0 but can be made so by assigning a suitable value, then z_0 is a removable singular point.

When $f(z)$ has a pole of order m at z_0,

$$\varphi(z) = (z-z_0)^m f(z),\ 0 < |z-z_0| < r_0$$

has a removable singularity at z_0. If $\varphi(z_0) = A_{-m}$, then $\varphi(z)$ is analytic at z_0. For a simple pole,

$$A_{-1} = \varphi(z_0) = \lim_{z \to z_0}(z-z_0)\,f(z)$$

Also $|f(z)| \to \infty$ as $z \to z_0$ when z_0 is a pole. Let the function $p(z)$ and $q(z)$ be analytic at z_0, where $p(z_0) \neq 0$. Then

$$f(z) = \frac{p(z)}{q(z)}$$

has a simple pole at z_0 if, and only if, $q(z_0) = 0$ and $q'(z_0) \neq 0$. The residue of $f(z)$ at the simple pole is

$$A_{-1} = \frac{p(z_0)}{q'(z_0)}$$

If $q^{(\mathrm{i}-1)}(z_0) = 0$, i = 1, . . . , m, then z_0 is a pole of $f(z)$ of order m.

Branch [33, p. 163]. A branch of a multiple-valued function $f(z)$ is a single-valued function that is analytic in some region and whose value at each point there coincides with the value of $f(z)$ at the point. A branch cut is a boundary that is needed to define the branch in the greatest possible region. The function $f(z)$ is singular along a branch cut, and the singularity is not isolated. For example,

$$z^{1/2} = f_1(z) = \sqrt{r}\left(\cos\frac{\theta}{2} + \mathrm{i}\sin\frac{\theta}{2}\right)$$

$$-\pi < \theta < \pi, \; r > 0$$

is double valued along the negative real axis. The function tends to $\sqrt{r}\mathrm{i}$ when $\theta \to \pi$ and to $-\sqrt{r}\mathrm{i}$ when $\theta \to -\pi$; the function has no limit as $z \to -r$ $(r > 0)$. The ray $\theta = \pi$ is a branch cut.

Analytic Continuation [33, p. 165]. If $f_1(z)$ is analytic in a domain D_1 and domain D contains D_1, then an analytic function $f(z)$ may exist that equals $f_1(z)$ in D_1. This function is the analytic continuation of $f_1(z)$ onto D, and it is unique. For example,

$$f_1(z) = \sum_{n=0}^{\infty} z^n, \; |z| < 1$$

is analytic in the domain $D_1 : |z| < 1$. The series diverges for other z. Yet the function is the MacLaurin series in the domain

$$f_1(z) = \frac{1}{1-z}, \; |z| < 1$$

Thus,

$$f_1(z) = \frac{1}{1-z}$$

is the analytic continuation onto the entire z plane except for $z = 1$.

An extension of the Cauchy integral formula is useful with Laplace transforms. Let the curve C be a straight line parallel to the imaginary axis and z_0 be any point to the right of that (see Fig. 9). A function $f(z)$ is of order z^k as $|z| \to \infty$ if positive numbers M and r_0 exist such that

$$\left|z^{-k} f(z)\right| < M \text{ when } |z| > r_0, \text{ i.e.,}$$

$$|f(z)| < M|z|^k \text{ for } |z| \text{ sufficiently large}$$

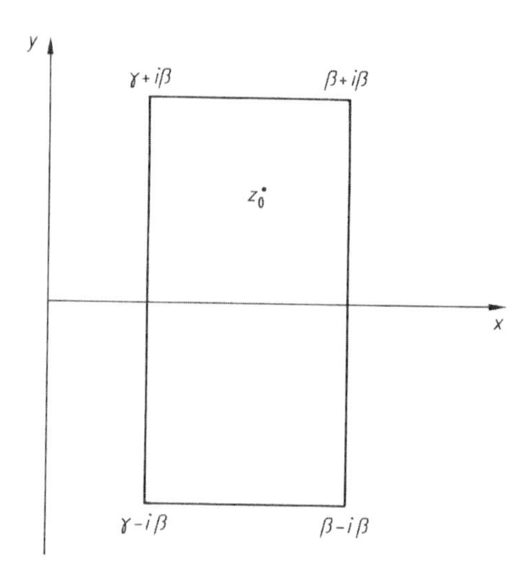

Figure 9. Integration in the complex plane

Theorem [33, p. 167]. Let $f(z)$ be analytic when $R(z) \geq \gamma$ and $O(z^{-k})$ as $|z| \to \infty$ in that half-plane, where γ and k are real constants and $k > 0$. Then for any z_0 such that $R(z_0) > \gamma$

$$f(z_0) = -\frac{1}{2\pi\mathrm{i}} \lim_{\beta\to\infty} \int_{\gamma-\mathrm{i}\beta}^{\gamma+\mathrm{i}\beta} \frac{f(z)}{z-z_0} \mathrm{d}z,$$

i.e., integration takes place along the line $x = \gamma$.

3.5. Other Results

Theorem [32, p. 84]. Let $P(z)$ be a polynomial of degree n having the zeroes $z_1, z_2, \ldots, z_n$ and let π be the least convex polygon containing the zeroes. Then $P'(z)$ cannot vanish anywhere in the exterior of π.

If a polynomial has real coefficients, the roots are either real or form pairs of complex conjugate numbers.

The radius of convergence R of the Taylor series of $f(z)$ about z_0 is equal to the distance from z_0 to the nearest singularity of $f(z)$.

Conformal Mapping. Let $u(x, y)$ be a harmonic function. Introduce the coordinate transformation

$$x = \hat{x}(\xi, \eta), \; y = \hat{y}(\xi, \eta)$$

It is desired that

$$U(\xi, \eta) = u[\hat{x}(\xi, \eta), \hat{y}(\xi, \eta)]$$

be a harmonic function of ξ and η.

Theorem [30, p. 284]. The transformation

$$z = f(\zeta) \tag{6}$$

takes all harmonic functions of x and y into harmonic functions of ξ and η if and only if either $f(\zeta)$ or $f^*(\zeta)$ is an analytic function of $\zeta = \xi + \mathrm{i}\,\eta$.

Equation 6 is a restriction on the transformation which ensures that

$$\text{if } \frac{\partial^2 u}{\partial x^2} + \frac{\partial^2 u}{\partial y^2} = 0 \text{ then } \frac{\partial^2 U}{\partial \zeta^2} + \frac{\partial^2 U}{\partial \eta^2} = 0$$

Such a mapping with $f(\zeta)$ analytic and $f'(\zeta) \neq 0$ is a conformal mapping.

If Laplace's equation is to be solved in the region exterior to a closed curve, then the point at infinity is in the domain D. For flow in a long channel (governed by the Laplace equation) the inlet and outlet are at infinity. In both cases the transformation

$$\zeta = \frac{a\,z + b}{z - z_0}$$

takes z_0 into infinity and hence maps D into a bounded domain D*.

4. Integral Transforms [34 – 39]

4.1. Fourier Transforms

Fourier Series [40]. Let $f(x)$ be a function that is periodic on $-\pi < x < \pi$. It can be expanded in a Fourier series

$$f(x) = \frac{a_0}{2} + \sum_{n=1}^{\infty} (a_n \cos n\,x + b_n \sin n\,x)$$

where

$$a_0 = \frac{1}{\pi} \int_{-\pi}^{\pi} f(x)\,\mathrm{d}x, \; a_n = \frac{1}{\pi} \int_{-\pi}^{\pi} f(x) \cos n\,x\,\mathrm{d}x,$$

$$b_n = \frac{1}{\pi} \int_{-\pi}^{\pi} f(x) \sin n\,x\,\mathrm{d}x$$

The values $\{a_n\}$ and $\{b_n\}$ are called the finite cosine and sine transform of f, respectively. Because

$$\cos n\,x = \frac{1}{2}\left(\mathrm{e}^{\mathrm{i}nx} + \mathrm{e}^{-\mathrm{i}nx}\right)$$

$$\text{and } \sin n\,x = \frac{1}{2\mathrm{i}}\left(\mathrm{e}^{\mathrm{i}nx} - \mathrm{e}^{-\mathrm{i}nx}\right)$$

the Fourier series can be written as

$$f(x) = \sum_{n=-\infty}^{\infty} c_n \mathrm{e}^{-\mathrm{i}nx}$$

where

$$c_n = \begin{cases} \frac{1}{2}(a_n + \mathrm{i}\,b_n) & \text{for} \quad n \geq 0 \\ \frac{1}{2}(a_{-n} - \mathrm{i}\,b_{-n}) & \text{for} \quad n < 0 \end{cases}$$

and

$$c_n = \frac{1}{2\pi} \int_{-\pi}^{\pi} f(x)\,\mathrm{e}^{\mathrm{i}nx}\,\mathrm{d}x$$

If f is real

$$c_{-n} = c_n^*.$$

If f is continuous and piecewise continuously differentiable

$$f'(x) = \sum_{-\infty}^{\infty} (-\mathrm{i}\,n)\,c_n \mathrm{e}^{-\mathrm{i}nx}$$

If f is twice continuously differentiable

$$f''(x) = \sum_{-\infty}^{\infty} \left(-n^2\right) c_n \mathrm{e}^{-\mathrm{i}nx}$$

Inversion. The Fourier series can be used to solve linear partial differential equations with constant coefficients. For example, in the problem

$$\frac{\partial T}{\partial t} = \frac{\partial^2 T}{\partial x^2}$$

$$T(x, 0) = f(x)$$

$$T(-\pi, t) = T(\pi, t)$$

Let

$$T=\sum_{-\infty}^{\infty} c_n(t)\,\mathrm{e}^{-inx}$$

Then,

$$\sum_{-\infty}^{\infty}\frac{\mathrm{d}c_n}{\mathrm{d}t}\,\mathrm{e}^{-inx}=\sum_{-\infty}^{\infty} c_n(t)\left(-n^2\right)\mathrm{e}^{-inx}$$

Thus, $c_n(t)$ satisfies

$$\frac{\mathrm{d}c_n}{\mathrm{d}t}=-n^2\,c_n,\ \text{or}\ c_n=c_n(0)\,\mathrm{e}^{-n^2t}$$

Let $c_n(0)$ be the Fourier coefficients of the initial conditions:

$$f(x)=\sum_{-\infty}^{\infty} c_n(0)\,\mathrm{e}^{-inx}$$

The formal solution to the problem is

$$T=\sum_{-\infty}^{\infty} c_n(0)\,\mathrm{e}^{-n^2t}\mathrm{e}^{-inx}$$

Fourier Transform [40]. When the function $f(x)$ is defined on the entire real line, the Fourier transform is defined as

$$\boldsymbol{F}[f]\equiv\hat{f}(\omega)=\int_{-\infty}^{\infty} f(x)\,\mathrm{e}^{\mathrm{i}\omega x}\,\mathrm{d}x$$

This integral converges if

$$\int_{-\infty}^{\infty}|f(x)|\,\mathrm{d}x$$

does. The inverse transformation is

$$f(x)=\frac{1}{2\pi}\int_{-\infty}^{\infty}\hat{f}(\omega)\,\mathrm{e}^{-\mathrm{i}\omega x}\,\mathrm{d}\omega$$

If $f(x)$ is continuous and piecewise continuously differentiable,

$$\int_{-\infty}^{\infty} f(x)\,\mathrm{e}^{\mathrm{i}\omega x}\,\mathrm{d}x$$

converges for each ω, and

$$\lim_{x\to\pm\infty} f(x)=0$$

then

$$\boldsymbol{F}\left[\frac{\mathrm{d}f}{\mathrm{d}x}\right]=-\mathrm{i}\omega\boldsymbol{F}[f]$$

If f is real $F[f(-\omega)]=F[f(\omega)^*]$. The real part is an even function of ω and the imaginary part is an odd function of ω.

A function $f(x)$ is *absolutely integrable* if the improper integral

$$\int_{-\infty}^{\infty}|f(x)|\,\mathrm{d}x$$

has a finite value. Then the improper integral

$$\int_{-\infty}^{\infty} f(x)\,\mathrm{d}x$$

converges. The function is *square integrable* if

$$\int_{-\infty}^{\infty}|f(x)|^2\mathrm{d}x$$

has a finite value. If $f(x)$ and $g(x)$ are square integrable, the product $f(x)\,g(x)$ is absolutely integrable and satisfies the Schwarz inequality:

$$\left|\int_{-\infty}^{\infty} f(x)\,g(x)\,\mathrm{d}x\right|^2 \le \int_{-\infty}^{\infty}|f(x)|^2\mathrm{d}x\int_{-\infty}^{\infty}|g(x)|^2\mathrm{d}x$$

The triangle inequality is also satisfied:

$$\left\{\int_{-\infty}^{\infty}|f+g|^2\mathrm{d}x\right\}^{1/2} \le\left\{\int_{-\infty}^{\infty}|f|^2\mathrm{d}x\right\}^{1/2}\left\{\int_{-\infty}^{\infty}|g|^2\mathrm{d}x\right\}^{1/2}$$

A sequence of square integrable functions $f_n(x)$ converges in the mean to a square integrable function $f(x)$ if

$$\lim_{n\to\infty}\int_{-\infty}^{\infty}|f(x)-f_n(x)|^2\mathrm{d}x=0$$

The sequence also satisfies the Cauchy criterion

$$\lim_{\substack{n\to\infty\\ m\to\infty}}\int_{-\infty}^{\infty}|f_n-f_m|^2\mathrm{d}x=0$$

Theorem [40, p. 307]. If a sequence of square integrable functions $f_n(x)$ converges to a function $f(x)$ uniformly on every finite interval $a\le x\le b$, and if it satisfies Cauchy's criterion, then $f(x)$ is square integrable and $f_n(x)$ converges to $f(x)$ in the mean.

Theorem (Riesz – Fischer) [40, p. 308]. To every sequence of square integrable functions $f_n(x)$ that satisfy Cauchy's criterion, there corresponds a square integrable function $f(x)$ such that $f_n(x)$ converges to $f(x)$ in the mean. Thus, the limit in the mean of a sequence of functions is defined to within a null function.

Square integrable functions satisfy the Parseval equation.

$$\int_{-\infty}^{\infty} \left|\hat{f}(\omega)\right|^2 \mathrm{d}\omega = 2\pi \int_{-\infty}^{\infty} |f(x)|^2 \mathrm{d}x$$

This is also the total power in a signal, which can be computed in either the time or the frequency domain. Also

$$\int_{-\infty}^{\infty} \hat{f}(\omega)\,\hat{g}*(\omega)\,\mathrm{d}\omega = 2\pi \int_{-\infty}^{\infty} f(x)\,g*(x)\,\mathrm{d}x$$

Fourier transforms can be used to solve differential equations too. Then it is necessary to find the inverse transformation. If $f(x)$ is square integrable, the Fourier transform of its Fourier transform is $2\pi f(-x)$, or

$$f(x) = \boldsymbol{F}\left[\hat{f}(\omega)\right] = \frac{1}{2\pi}\int_{-\infty}^{\infty} \hat{f}(\omega)\,\mathrm{e}^{-\mathrm{i}\omega x}\mathrm{d}\omega$$

$$= \frac{1}{2\pi}\int_{-\infty}^{\infty}\int_{-\infty}^{\infty} f(x)\,\mathrm{e}^{\mathrm{i}\omega x}\mathrm{d}x e^{-\mathrm{i}\omega x}\mathrm{d}\omega$$

$$f(x) = \frac{1}{2\pi}\boldsymbol{F}\left[\boldsymbol{F}f(-x)\right] \text{ or } f(-x) = \frac{1}{2\pi}\boldsymbol{F}\left[\boldsymbol{F}[f]\right]$$

Properties of Fourier Transforms [40, p. 324], [15].

$$\boldsymbol{F}\left[\frac{\mathrm{d}f}{\mathrm{d}x}\right] = -\mathrm{i}\,\omega\,\boldsymbol{F}[f] = \mathrm{i}\,\omega\hat{f}$$

$$\boldsymbol{F}[\mathrm{i}\,x\,f(x)] = \frac{\mathrm{d}}{\mathrm{d}\omega}\boldsymbol{F}[f] = \frac{\mathrm{d}}{\mathrm{d}\omega}\hat{f}$$

$$\boldsymbol{F}[f(a\,x-b)] = \frac{1}{|a|}\mathrm{e}^{\mathrm{i}\omega b/a}\hat{f}\left(\frac{\omega}{a}\right)$$

$$\boldsymbol{F}\left[\mathrm{e}^{\mathrm{i}cx}f(x)\right] = \hat{f}(\omega+c)$$

$$\boldsymbol{F}[\cos\omega_0 x\,f(x)] = \frac{1}{2}\left[\hat{f}(\omega+\omega_0)+\hat{f}(\omega-\omega_0)\right]$$

$$\boldsymbol{F}[\sin\omega_0 x\,f(x)] = \frac{1}{2\mathrm{i}}\left[\hat{f}(\omega+\omega_0)-\hat{f}(\omega-\omega_0)\right]$$

$$\boldsymbol{F}\left[\mathrm{e}^{-\mathrm{i}\omega_0 x}f(x)\right] = \hat{f}(\omega-\omega_0)$$

If $f(x)$ is real, then $f(-\omega) = \hat{f}*(\omega)$. If $f(x)$ is imaginary, then $\hat{f}(-\omega) = -\hat{f}*(\omega)$. If $f(x)$ is even, then $\hat{f}(\omega)$ is even. If $f(x)$ is odd, then $\hat{f}(\omega)$ is odd. If $f(x)$ is real and even, then $\hat{f}(\omega)$ is real and even. If $f(x)$ is real and odd, then $\hat{f}(\omega)$ is imaginary and odd. If $f(x)$ is imaginary and even, then $\hat{f}(\omega)$ is imaginary and even. If $f(x)$ is imaginary and odd, then $\hat{f}(\omega)$ is real and odd.

Convolution [40, p. 326].

$$f*h(x_0) \equiv \int_{-\infty}^{\infty} f(x_0-x)\,h(x)\,\mathrm{d}x$$

$$= \frac{1}{2\pi}\int_{-\infty}^{\infty} \mathrm{e}^{\mathrm{i}\omega x_0}\hat{f}(-\omega)\,\hat{h}(\omega)\,\mathrm{d}\omega$$

Theorem. The product

$$\hat{f}(\omega)\,\hat{h}(\omega)$$

is the Fourier transform of the convolution product $f*h$. The convolution permits finding inverse transformations when solving differential equations. To solve

$$\frac{\partial T}{\partial t} = \frac{\partial^2 T}{\partial x^2}$$

$$T(x,0) = f(x),\, -\infty<x<\infty$$

T bounded

take the Fourier transform

$$\frac{\mathrm{d}\hat{T}}{\mathrm{d}t}+\omega^2\hat{T} = 0$$

$$\hat{T}(\omega,0) = \hat{f}(\omega)$$

The solution is

$$\hat{T}(\omega,t) = \hat{f}(\omega)\,\mathrm{e}^{-\omega^2 t}$$

The inverse transformation is

$$T(x,t) = \frac{1}{2\pi}\int_{-\infty}^{\infty} \mathrm{e}^{-\mathrm{i}\omega x}\,\hat{f}(\omega)\,\mathrm{e}^{-\omega^2 t}\mathrm{d}\omega$$

Because

$$\mathrm{e}^{-\omega^2 t} = \boldsymbol{F}\left[\frac{1}{\sqrt{4\pi t}}\mathrm{e}^{-x^2/4t}\right]$$

the convolution integral can be used to write the solution as

$$T(x,t) = \frac{1}{\sqrt{4\pi t}}\int_{-\infty}^{\infty} f(y)\,\mathrm{e}^{-(x-y)^2/4t}\mathrm{d}y$$

Finite Fourier Sine and Cosine Transform [41]. In analogy with finite Fourier transforms (on $-\pi$ to π) and Fourier transforms (on $-\infty$ to $+\infty$), finite Fourier sine and cosine transforms (0 to π) and Fourier sine and cosine transforms (on 0 to $+\infty$) can be defined.

The finite Fourier sine and cosine transforms are

$$f_s(n) = \boldsymbol{F}_s^n[f] = \frac{2}{\pi}\int_0^\pi f(x)\sin nx\,\mathrm{d}x,$$

$$n = 1, 2, \ldots,$$

$$f_c(n) = \boldsymbol{F}_c^n[f] = \frac{2}{\pi}\int_0^\pi f(x)\cos nx\,\mathrm{d}x$$

$$n = 0, 1, 2, \ldots$$

$$f(x) = \sum_{n=1}^{\infty} f_s(n)\sin nx,$$

$$f(x) = \frac{1}{2}f_c(0) + \sum_{n=1}^{\infty} f_c(n)\cos nx$$

They obey the operational properties

$$\boldsymbol{F}_s^n\left[\frac{\mathrm{d}^2 f}{\mathrm{d}x^2}\right] = -n^2\,\boldsymbol{F}_s^n[f]$$

$$+\frac{2n}{\pi}\left[f(0) - (-1)^n f(\pi)\right]$$

f, f' are continuous, f'' is piecewise continuous on $0 \le x \le \pi$.

$$\left.\begin{aligned} &f_s(n)\cos nk \\ &= \boldsymbol{F}_s^n\left[\tfrac{1}{2}f_1(x-k) + \tfrac{1}{2}f_1(x+k)\right] \\ &f_s(n)(-1)^{n+1} = \boldsymbol{F}_s^n[f(\pi - x)] \end{aligned}\right\} \begin{aligned} &f \text{ is piecewise} \\ &\text{continuous on} \\ &0 \le x \le \pi \end{aligned}$$

and f_1 is the extension of f, k is a constant

$$f_1(x) = f(x) \quad 0 < x < \pi$$

$$\left.\begin{aligned} f_1(-x) &= -f_1(x) \\ f_1(x+2\pi) &= f_1(x) \end{aligned}\right\} -\infty < x < \infty$$

Also,

$$\boldsymbol{F}_c^n\left[\frac{\mathrm{d}^2 f}{\mathrm{d}x^2}\right] = -n^2\boldsymbol{F}_c^n[f] - \frac{2}{\pi}\frac{\mathrm{d}f}{\mathrm{d}x}(0)$$

$$+(-1)^n\frac{2}{\pi}\frac{\mathrm{d}f}{\mathrm{d}x}(\pi)$$

$$f_c(n)\cos nk = \boldsymbol{F}_c^n\left[\tfrac{1}{2}f_2(x-k) + \tfrac{1}{2}f_2(x+k)\right]$$

$$f_c(n)(-1)^n = \boldsymbol{F}_c^n[f(\pi - x)]$$

and f_2 is the extension of f.

$$f_2(x) = f(x) \quad 0 < x < \pi$$

$$\left.\begin{aligned} f_2(-x) &= f_2(x) \\ f_2(x+2\pi) &= f_2(x) \end{aligned}\right\} -\infty < x < \infty$$

Table 3. Finite sine transforms [41]

$f_s(n) = \int_0^\pi F(x)\sin nx\,\mathrm{d}x \;(n = 1, 2, \ldots)$	$F(x) \;(0<x<\pi)$
$(-1)^{n+1} f_s(n)$	$F(\pi - x)$
$\frac{1}{n}$	$\frac{(\pi - x)}{\pi}$
$\frac{(-1)^{n+1}}{n}$	$\frac{x}{\pi}$
$\frac{1-(-1)^n}{n}$	1
$\frac{\pi}{n^2}\sin nc \;(0<c<\pi)$	$\begin{cases} (\pi - c)\,x & (x \le c) \\ c\,(\pi - x) & (x \ge c) \end{cases}$
$\frac{\pi}{n}\cos nc \;(0<c<\pi)$	$\begin{cases} -x & (x<c) \\ \pi - x & (x) \end{cases}$
$\frac{\pi^2(-1)^{n-1}}{n} - \frac{2[1-(-1)^n]}{n^3}$	x^2
$\pi(-1)^n\left(\frac{6}{n^3} - \frac{\pi^2}{n}\right)$	x^3
$\frac{n}{n^2+c^2}\left[1-(-1)^n e^{c\pi}\right]$	e^{cx}
$\frac{n}{n^2+c^2}$	$\frac{\sinh c(\pi - x)}{\sinh c\pi}$
$\frac{n}{n^2-k^2} \;(\lvert k\rvert \ne 0, 1, 2, \ldots)$	$\frac{\sin k(\pi - x)}{\sin k\pi}$
$0 \;(n \ne m);\; f_s(m) = \frac{\pi}{2}$	$\sin mx \;(m = 1, 2, \ldots)$
$\frac{n}{n^2-k^2}\left[1-(-1)^n\cos kx\right]$	$\cos kx \;(\lvert k\rvert \ne 1, 2, \ldots)$
$\frac{n}{n^2-m^2}\left[1-(-1)^{n+m}\right],\; (n \ne m);\; f_s(m) = 0$	$\cos mx \;(m = 1, 2, \ldots)$

Also,

$$\boldsymbol{F}_s^n\left[\frac{\mathrm{d}f}{\mathrm{d}x}\right] = -n\boldsymbol{F}_c^n[f]$$

$$\boldsymbol{F}_c^n\left[\frac{\mathrm{d}f}{\mathrm{d}x}\right] = n\boldsymbol{F}_s^n[f] - \frac{2}{\pi}f(0) + (-1)^n\frac{2}{\pi}f(\pi)$$

When two functions $F(x)$ and $G(x)$ are defined on the interval $-2\pi < x < 2\pi$, the function

$$F(x) * G(x) = \int_{-\pi}^{\pi} f(x-y)\,g(y)\,\mathrm{d}y$$

is the convolution on $-\pi < x < \pi$. If F and G are both even or both odd, the convolution is even; it is odd if one function is even and the other odd. If F and G are piecewise continuous on $0 \le x \le \pi$, then

$$f_s(n)\,g_s(n) = \boldsymbol{F}_c^n\left[-\tfrac{1}{2}F_1(x) * G_1(x)\right]$$

$$f_s(n)\,g_c(n) = \boldsymbol{F}_s^n\left[\tfrac{1}{2}F_1(x) * G_2(x)\right]$$

$$f_c(n)\,g_c(n) = \boldsymbol{F}_c^n\left[\tfrac{1}{2}F_2(x) * G_2(x)\right]$$

where F_1 and G_1 are odd extensions of F and G, respectively, and F_2 and G_2 are even extensions of F and G, respectively. Finite sine and cosine transforms are listed in Tables 3 and 4.

Table 4. Finite cosine transforms [41]

$f_c(n) = \int_0^\pi F(x)\cos n\,x\,\mathrm{d}x \ (n=0,1,\ldots)$	$F(x)\ (0<x<\pi)$
$(-1)^n f_c(n)$	$F(\pi-x)$
0 when n=1,2, . . . ; $f_c(0)=\pi$	1
$\frac{2}{n}\sin n\,c;\ f_c(0)=2\,c-\pi$	$\begin{cases}1 & (0<x<c)\\ -1 & (c<x<\pi)\end{cases}$
$-\frac{1-(-1)^n}{n^2};\ f_c(0)=\frac{\pi^2}{2}$	x
$\frac{(-1)^n}{n^2};\ f_c(0)=\frac{\pi^2}{6}$	$\frac{x^2}{2\,\pi}$
$\frac{1}{n^2};\ f_c(0)=0$	$\frac{(\pi-x)^2}{2\,\pi}-\frac{\pi}{6}$
$\frac{(-1)^n e^{cx}-1}{n^2+c^2}$	$\frac{1}{c}e^{cx}$
$\frac{1}{n^2+c^2}$	$\frac{\cosh c(\pi-x)}{c\sinh c\,\pi}$
$\frac{(-1)^n\cos k\,\pi-1}{n^2-k^2}\ (\lvert k\rvert\neq 0,1,\ldots)$	$\frac{1}{k}\sin k\,x$
$\frac{(-1)^{n+m}-1}{n^2-m^2};\ f_c(m)=0\ (m=1,\ldots)$	$\frac{1}{m}\sin m\,x$
$\frac{1}{n^2-k^2}\ (\lvert k\rvert\neq 0,1,\ldots)$	$-\frac{\cos k(\pi-x)}{k\sin k\,x}$
$0\ (n\neq m);\ f_c(m)=\frac{\pi}{2}\ (m=1,2,\ldots)$	$\cos m\,x$

On the semi-infinite domain, $0<x<\infty$, the Fourier sine and cosine transforms are

$$F_s^\omega[f]\equiv\int_0^\infty f(x)\sin\omega\,x\,\mathrm{d}x,$$
$$F_c^\omega[f]\equiv\int_0^\infty f(x)\cos\omega\,x\,\mathrm{d}x \text{ and}$$
$$f(x)=\frac{2}{\pi}F_s^{\omega\prime}[F_s^\omega[f]],\ f(x)=\frac{2}{\pi}F_c^{\omega\prime}[F_c^\omega[f]]$$

The sine transform is an odd function of ω, whereas the cosine function is an even function of ω. Also,

$$F_s^\omega\left[\frac{\mathrm{d}^2 f}{\mathrm{d}x^2}\right]=f(0)\,\omega-\omega^2F_s^\omega[f]$$

$$F_c^\omega\left[\frac{\mathrm{d}^2 f}{\mathrm{d}x^2}\right]=-\frac{\mathrm{d}f}{\mathrm{d}x}(0)-\omega^2F_c^\omega[f]$$

provided $f(x)$ and $f'(x)\to 0$ as $x\to\infty$. Thus, the sine transform is useful when $f(0)$ is known and the cosine transform is useful when $f'(0)$ is known.

Hsu and Dranoff [42] solved a chemical engineering problem by applying finite Fourier transforms and then using the fast Fourier transform (see Chap. 2).

4.2. Laplace Transforms

Consider a function $F(t)$ defined for $t>0$. The Laplace transform of $F(t)$ is [41]

$$\boldsymbol{L}[F]=f(s)=\int_0^\infty e^{-st}F(t)\,\mathrm{d}t$$

The Laplace transformation is linear, that is,

$$\boldsymbol{L}[F+G]=\boldsymbol{L}[F]+\boldsymbol{L}[G]$$

Thus, the techniques described herein can be applied only to linear problems. Generally, the assumptions made below are that $F(t)$ is at least piecewise continuous, that it is continuous in each finite interval within $0<t<\infty$, and that it may take a jump between intervals. It is also of exponential order, meaning $e^{-\alpha t}\lvert F(t)\rvert$ is bounded for all $t>T$, for some finite T.

The unit step function is

$$S_k(t)=\begin{cases}0 & 0\le t<k\\ 1 & t>k\end{cases}$$

and its Laplace transform is

$$\boldsymbol{L}[S_k(t)]=\frac{e^{-ks}}{s}$$

In particular, if $k=0$ then

$$\boldsymbol{L}[1]=\frac{1}{s}$$

The Laplace transforms of the first and second derivatives of $F(t)$ are

$$\boldsymbol{L}\left[\frac{\mathrm{d}F}{\mathrm{d}t}\right]=s\,f(s)-F(0)$$

$$\boldsymbol{L}\left[\frac{\mathrm{d}^2F}{\mathrm{d}t^2}\right]=s^2 f(s)-s\,F(0)-\frac{\mathrm{d}F}{\mathrm{d}t}(0)$$

More generally,

$$\boldsymbol{L}\left[\frac{\mathrm{d}^nF}{\mathrm{d}t^n}\right]=s^n f(s)-s^{n-1}F(0)$$
$$-s^{n-2}\frac{\mathrm{d}F}{\mathrm{d}t}(0)-\cdots-\frac{\mathrm{d}^{n-1}F}{\mathrm{d}t^{n-1}}(0)$$

The inverse Laplace transformation is

$$F(t)=\boldsymbol{L}^{-1}[f(s)]\ \text{where}\ f(s)=\boldsymbol{L}[F]$$

The inverse Laplace transformation is not unique because functions that are identical except for isolated points have the same Laplace transform. They are unique to within a null function. Thus, if

$$\boldsymbol{L}[F_1] = f(s) \text{ and } \boldsymbol{L}[F_2] = f(s)$$

it must be that

$$F_2 = F_1 + N(t)$$

$$\text{where } \int_0^T (N(t))\,\mathrm{d}t = 0 \text{ for every } T$$

Laplace transforms can be inverted by using Table 5, but knowledge of several rules is helpful.

Table 5. Laplace transforms (see [23] for a more complete list)

$L[F]$	$F(t)$
$\frac{1}{s}$	1
$\frac{1}{s^2}$	t
$\frac{1}{s^n}$	$\frac{t^{n-1}}{(n-1)!}$
$\frac{1}{\sqrt{s}}$	$\frac{1}{\sqrt{\pi t}}$
$s^{-3/2}$	$2\sqrt{t/\pi}$
$\frac{\Gamma(k)}{s^k}\ (k>0)$	t^{k-1}
$\frac{1}{s-a}$	e^{at}
$\frac{1}{(s-a)^n}\ (n=1,2,\ldots)$	$\frac{1}{(n-1)!}t^{n-1}e^{at}$
$\frac{\Gamma(k)}{(s-a)^k}\ (k>0)$	$t^{k-1}\,e^{at}$
$\frac{1}{(s-a)(s-b)}$	$\frac{1}{a-b}\left(e^{at}-e^{bt}\right)$
$\frac{s}{(s-a)(s-b)}$	$\frac{1}{a-b}\left(ae^{at}-be^{bt}\right)$
$\frac{1}{s^2+a^2}$	$\frac{1}{a}\sin a\,t$
$\frac{s}{s^2+a^2}$	$\cos a\,t$
$\frac{1}{s^2-a^2}$	$\frac{1}{a}\sinh a\,t$
$\frac{s}{s^2-a^2}$	$\cosh a\,t$
$\frac{s}{(s^2+a^2)^2}$	$\frac{t}{2a}\sin a\,t$
$\frac{s^2-a^2}{(s^2+a^2)^2}$	$t\cos a\,t$
$\frac{1}{(s-a)^2+b^2}$	$\frac{1}{b}e^{a\,t}\sin b\,t$
$\frac{s-a}{(s-a)^2+b^2}$	$e^{at}\cos b\,t$

Substitution.

$$f(s-a) = \boldsymbol{L}\left[e^{at}\,F(t)\right]$$

This can be used with polynomials. Suppose

$$f(s) = \frac{1}{s} + \frac{1}{s+3} = \frac{2\,s+3}{s\,(s+3)}$$

Because

$$\boldsymbol{L}[1] = \frac{1}{s}$$

Then

$$F(t) = 1 + e^{-3t},\ t \geq 0$$

More generally, translation gives the following.

Translation.

$$f(a\,s-b) = f\left[a\left(s-\frac{b}{a}\right)\right] = \boldsymbol{L}\left[\frac{1}{a}\,e^{bt/a}\,F\left(\frac{t}{a}\right)\right],$$

$$a>0$$

The step function

$$S(t) = \begin{cases} 0 & 0 \leq t < \frac{1}{h} \\ 1 & \frac{1}{h} \leq t < \frac{2}{h} \\ 2 & \frac{2}{h} \leq t < \frac{3}{h} \end{cases}$$

has the Laplace transform

$$\boldsymbol{L}[S(t)] = \frac{1}{s}\frac{1}{1-e^{-hs}}$$

The Dirac delta function $\delta(t-t_0)$ (see Equation 116 in → Mathematical Modeling) has the property

$$\int_0^\infty \delta(t-t_0)\,F(t)\,\mathrm{d}t = F(t_0)$$

Its Laplace transform is

$$\boldsymbol{L}[\delta(t-t_0)] = e^{-st_0},\ t_0 \geq 0,\ s>0$$

The square wave function illustrated in Figure 10 has Laplace transform

$$\boldsymbol{L}[F_c(t)] = \frac{1}{s}\tanh\frac{c\,s}{2}$$

The triangular wave function illustrated in Figure 11 has Laplace transform

$$L[T_c(t)] = \frac{1}{s^2}\tanh\frac{c\,s}{2}$$

Other Laplace transforms are listed in Table 5.

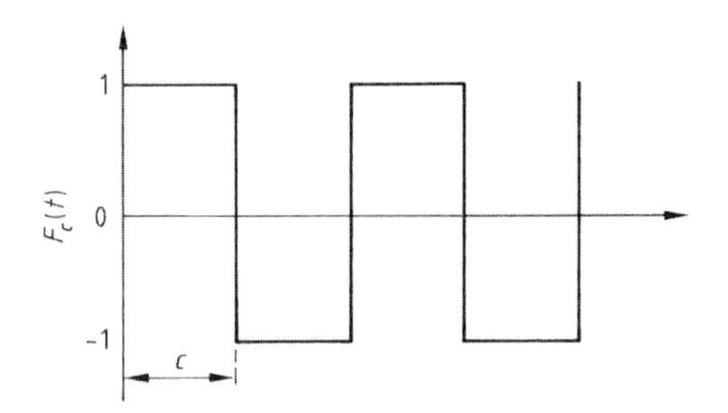

Figure 10. Square wave function

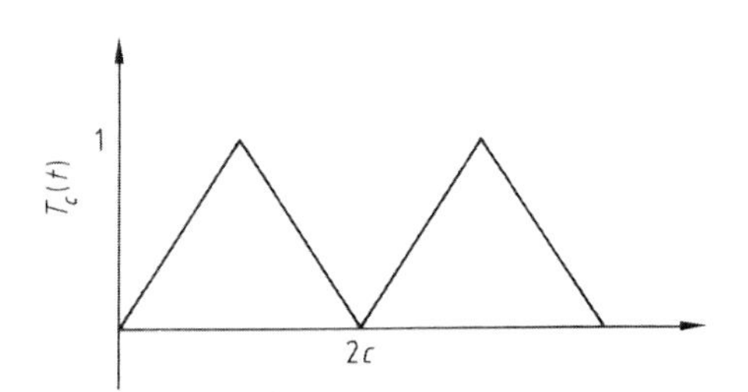

Figure 11. Triangular wave function

Convolution properties are also satisfied:

$$F(t)*G(t)=\int_0^t F(\tau)\,G(t-\tau)\,\mathrm{d}\tau$$

and

$$f(s)\,g(s)=\boldsymbol{L}\,[F(t)*G(t)]$$

Derivatives of Laplace Transforms. The Laplace integrals $\boldsymbol{L}\,[F(t)]$, $\boldsymbol{L}\,[t\,F(t)]$, $\boldsymbol{L}\,[t^2\,F(t)]$, . . . are uniformly convergent for $s_1 \geq \alpha$ and

$$\lim_{s\to\infty} f(s)=0,\ \lim_{s\to\infty} \boldsymbol{L}\,[t^n\,F(t)]=0,\ n=1,2,\ldots$$

and

$$\frac{\mathrm{d}^n f}{\mathrm{d}s^n}=\boldsymbol{L}\,[(-t)^n\,F(t)]$$

Integration of Laplace Transforms.

$$\int_s^\infty f(\xi)\,\mathrm{d}\xi=\boldsymbol{L}\left[\frac{F(t)}{t}\right]$$

If $F(t)$ is a periodic function, $F(t)=F(t+a)$, then

$$f(s)=\frac{1}{1-\mathrm{e}^{-as}}\int_0^a \mathrm{e}^{-st}F(t)\,\mathrm{d}t,$$

where $F(t)=F(t+a)$

Partial Fractions [43]. Suppose $q(s)$ has m factors

$$q(s)=(s-a_1)(s-a_2)\cdots(s-a_m)$$

All the factors are linear, none are repeated, and the a_n are all distinct. If $p(s)$ has a smaller degree than $q(s)$, the Heaviside expansion can be used to evaluate the inverse transformation:

$$\boldsymbol{L}^{-1}\left[\frac{p(s)}{q(s)}\right]=\sum_{i=1}^m \frac{p(a_i)}{q'(a_i)}\,\mathrm{e}^{a_i t}$$

If the factor $(s-a)$ is repeated m times, then

$$f(s)=\frac{p(s)}{q(s)}=\frac{A_m}{(s-a)^m}+\frac{A_{m-1}}{(s-a)^{m-1}}+\cdots+\frac{A_1}{s-a}+h(s)$$

where

$$\varphi(s)\equiv\frac{(s-a)^m\,p(s)}{q(s)}$$

$$A_m=\varphi(a),\ A_k=\frac{1}{(m-k)!}\frac{\mathrm{d}^{m-k}\varphi(s)}{\mathrm{d}s^{m-k}}\Big|_a,$$

$$k=1,\ldots,m-1$$

The term $h(s)$ denotes the sum of partial fractions not under consideration. The inverse transformation is then

$$F(t)=\mathrm{e}^{at}\left(A_m\frac{t^{m-1}}{(m-1)!}+A_{m-1}\frac{t^{m-2}}{(m-2)!}+\cdots+A_2\frac{t}{1!}+A_1\right)+H(t)$$

The term in $F(t)$ corresponding to

$s-a$ in $q(s)$ is $\varphi(a)\,\mathrm{e}^{a\,t}$

$(s-a)^2$ in $q(s)$ is $[\varphi'(a)+\varphi(a)\,t]\,\mathrm{e}^{a\,t}$

$(s-a)^3$ in $q(s)$ is $\frac{1}{2}\left[\varphi''(a)+2\varphi'(a)\,t+\varphi(a)\,t^2\right]\mathrm{e}^{a\,t}$

For example, let

$$f(s)=\frac{1}{(s-2)(s-1)^2}$$

For the factor $s-2$,

$$\varphi(s)=\frac{1}{(s-1)^2},\ \varphi(2)=1$$

For the factor $(s-1)^2$,

$$\varphi(s)=\frac{1}{s-2},\ \varphi'(s)=-\frac{1}{(s-2)^2}$$

$$\varphi(1)=-1,\ \varphi'(1)=-1$$

The inverse Laplace transform is then

$$F(t)=\mathrm{e}^{2t}+[-1-t]\,\mathrm{e}^t$$

Quadratic Factors. Let $p(s)$ and $q(s)$ have real coefficients, and $q(s)$ have the factor

$$(s-a)^2+b^2, b>0$$

where a and b are real numbers. Then define $\varphi(s)$ and $h(s)$ and real constants A and B such that

$$f(s)=\frac{p(s)}{q(s)}=\frac{\varphi(s)}{(s-a)^2+b^2}$$

$$=\frac{A\,s+B}{(s-a)^2+b^2}+h(s)$$

Let φ_1 and φ_2 be the real and imaginary parts of the complex number $\varphi(a+\mathrm{i}\,b)$.

$$\varphi(a+\mathrm{i}\,b)\equiv\varphi_1+\mathrm{i}\,\varphi_2$$

Then

$$f(s)=\frac{1}{b}\frac{(s-a)\varphi_2+b\varphi_1}{(s-a)^2+b^2}+h(s)$$

$$F(t)=\frac{1}{b}\,\mathrm{e}^{at}\,(\varphi_2\cos b\,t+\varphi_1\sin b\,t)+H(t)$$

To solve ordinary differential equations by using these results:

$$Y''(t)-2Y'(t)+Y(t)=\mathrm{e}^{2t}$$

$$Y(0)=0, Y'(0)=0$$

Taking Laplace transforms

$$\boldsymbol{L}\left[Y''(t)\right]-2\boldsymbol{L}\left[Y'(t)\right]+\boldsymbol{L}[Y(t)]=\frac{1}{s-2}$$

using the rules

$$s^2\,y(s)-s\,Y(0)-Y'(0)-2\,[s\,y(s)-Y(0)]+y(s)=\frac{1}{s-2}$$

and combining terms

$$\left(s^2-2\,s+1\right)\,y(s)=\frac{1}{s-2}$$

$$y(s)=\frac{1}{(s-2)(s-1)^2}$$

lead to

$$Y(t)=\mathrm{e}^{2t}-(1+t)\,\mathrm{e}^t$$

To solve an integral equation:

$$Y(t)=a+2\int_0^t Y(\tau)\cos(t-\tau)\,\mathrm{d}\tau$$

it is written as

$$Y(t)=a+Y(t)*\cos t$$

Then the Laplace transform is used to obtain

$$y(s)=\frac{a}{s}+2\,y(s)\,\frac{s}{s^2+1}$$

$$\text{or } y(s)=\frac{a\left(s^2+1\right)}{s\,(s-1)^2}$$

Taking the inverse transformation gives

$$Y(t)=s\left[1+2t\mathrm{e}^t\right]$$

Next, let the variable s in the Laplace transform be complex. $F(t)$ is still a real-valued function of the positive real variable t. The properties given above are still valid for s complex, but additional techniques are available for evaluating the integrals. The real-valued function is O [exp (x_0t)]:

$$|F(t)|<M\mathrm{e}^{x_0t},\ z_0=x_0+\mathrm{i}\,y_0$$

The Laplace transform

$$f(s)=\int_0^\infty \mathrm{e}^{-st}\,F(t)\,\mathrm{d}t$$

is an analytic function of s in the half-plane $x>x_0$ and is absolutely convergent there; it is uniformly convergent on $x\geq x_1>x_0$.

$$\frac{\mathrm{d}^n f}{\mathrm{d}s^n}=\boldsymbol{L}\left[(-t)^n\,F(t)\right]\ \ n=1,2,\ldots,x>x_0$$

$$\text{and } f^*(s)=f(s^*)$$

The functions $|f(s)|$ and $|x\,f(s)|$ are bounded in the half-plane $x\geq x_1>x_0$ and $f(s)\to 0$ as $|y|\to\infty$ for each fixed x. Thus,

$$|f(x+\mathrm{i}\,y)|<M,\ \ |x\,f(x+\mathrm{i}\,y)|<M,$$

$$x\geq x_1>x_0$$

$$\lim_{y\to\pm\infty} f(x+\mathrm{i}\,y)=0,\ x>x_0$$

If $F(t)$ is continuous, $F'(t)$ is piecewise continuous, and both functions are O [exp (x_0t)], then $|f(s)|$ is $O(1/s)$ in each half-plane $x\geq x_1>x_0$.

$$|s\,f(s)|<M$$

If $F(t)$ and $F'(t)$ are continuous, $F''(t)$ is piecewise continuous, and all three functions are O [exp (x_0t)], then

$$\left|s^2 f(s)-s\,F(0)\right|<M, x\geq x_1>x_0$$

The additional constraint $F(0)=0$ is necessary and sufficient for $|f(s)|$ to be $O(1/s^2)$.

Inversion Integral [41]. Cauchy's integral formula for $f(s)$ analytic and $O(s^{-k})$ in a half-plane $x\geq y$, $k>0$, is

$$f(s)=\frac{1}{2\,\pi\,\mathrm{i}}\lim_{\beta\to\infty}\int_{\gamma-\mathrm{i}\beta}^{\gamma+\mathrm{i}\beta}\frac{f(z)}{s-z}\mathrm{d}z,\ \mathrm{Re}\,(s)>\gamma$$

Applying the inverse Laplace transformation on either side of this equation gives

$$F\ (t) = \frac{1}{2\,\pi \mathrm{i}} \lim_{\beta \to \infty} \int\limits_{\gamma - \mathrm{i}\beta}^{\gamma + \mathrm{i}\beta} \mathrm{e}^{zt}\, f\ (z)\, \mathrm{d}z$$

If $F\ (t)$ is of order $O\ [\exp\ (x_0 t)]$ and $F\ (t)$ and $F'\ (t)$ are piecewise continuous, the inversion integral exists. At any point t_0, where $F\ (t)$ is discontinuous, the inversion integral represents the mean value

$$F\ (t_0) = \lim_{\varepsilon \to \infty} \frac{1}{2} \left[F\ (t_0 + \varepsilon) + F\ (t_0 - \varepsilon)\right]$$

When $t = 0$ the inversion integral represents 0.5 $F\ (O\ +)$ and when $t < 0$, it has the value zero.

If $f\ (s)$ is a function of the complex variable s that is analytic and of order $O\ (s^{-k-m})$ on $R\ (s) \geq x_0$, where $k > 1$ and m is a positive integer, then the inversion integral converges to $F\ (t)$ and

$$\frac{\mathrm{d}^n F}{\mathrm{d}t^n} = \frac{1}{2\,\pi \mathrm{i}} \lim_{\beta \to \infty} \int\limits_{\gamma - \mathrm{i}\beta}^{\gamma + \mathrm{i}\beta} \mathrm{e}^{zt} z^n\, f\ (z)\, \mathrm{d}z,$$

$n = 1,2,\ldots, m$

Also $F\ (t)$ and its n derivatives are continuous functions of t of order $O\ [\exp\ (x_0 t)]$ and they vanish at $t = 0$.

$$F\ (0) = F'\ (0) = \cdots F^{(m)}\ (0) = 0$$

Series of Residues [41]. Let $f\ (s)$ be an analytic function except for a set of isolated singular points. An isolated singular point is one for which $f\ (z)$ is analytic for $0 < |z - z_0| < \varrho$ but z_0 is a singularity of $f\ (z)$. An isolated singular point is either a pole, a removable singularity, or an essential singularity. If $f\ (z)$ is not defined in the neighborhood of z_0 but can be made analytic at z_0 simply by defining it at some additional points, then z_0 is a *removable singularity*. The function $f\ (z)$ has a *pole* of order $k \geq 1$ at z_0 if $(z - z_0)^k f\ (z)$ has a removable singularity at z_0 whereas $(z - z_0)^{k-1} f\ (z)$ has an unremovable isolated singularity at z_0. Any isolated singularity that is not a pole or a removable singularity is an *essential* singularity.

Let the function $f\ (z)$ be analytic except for the isolated singular point $s_1, s_2, \ldots, s_n$. Let $\varrho_n\ (t)$ be the residue of $\mathrm{e}^{zt} f\ (z)$ at $z = s_n$ (for definition of residue, see Section 3.4). Then

$$F\ (t) = \sum_{n=1}^{\infty} \varrho_n\ (t)$$

When s_n is a simple pole

$$\varrho_n\ (t) = \lim_{z \to s_n} (z - s_n)\, \mathrm{e}^{zt}\, f\ (z)$$

$$= \mathrm{e}^{s_n t} \lim_{z \to s_n} (z - s_n)\ f\ (z)$$

When

$$f\ (z) = \frac{p\ (z)}{q\ (z)}$$

where $p\ (z)$ and $q\ (z)$ are analytic at $z = s_n$, $p\ (s_n) \neq 0$, then

$$\varrho_n\ (t) = \frac{p\ (s_n)}{q'\ (s_n)} \mathrm{e}^{s_n t}$$

If s_n is a removable pole of $f\ (s)$, of order m, then

$$\varphi_n\ (z) = (z - s_n)^m\ f\ (z)$$

is analytic at s_n and the residue is

$$\varrho_n\ (t) = \frac{\Phi_n\ (s_n)}{(m-1)!} \text{ where } \Phi_n\ (z) = \frac{\partial^{m-1}}{\partial z^{m-1}} \left[\varphi_n\ (z)\, \mathrm{e}^{zt}\right]$$

An important inversion integral is when

$$f\ (s) = \frac{1}{s} \exp\left(-s^{1/2}\right)$$

The inverse transform is

$$F\ (t) = 1 - \operatorname{erf}\left(\frac{1}{2\sqrt{t}}\right) = \operatorname{erfc}\left(\frac{1}{2\sqrt{t}}\right)$$

where erf is the error function and erfc the complementary error function.

4.3. Solution of Partial Differential Equations by Using Transforms

A common problem facing chemical engineers is to solve the heat conduction equation or diffusion equation

$$\varrho\, C_p \frac{\partial T}{\partial t} = k \frac{\partial^2 T}{\partial x^2} \text{ or } \frac{\partial c}{\partial t} = D \frac{\partial^2 c}{\partial x^2}$$

The equations can be solved on an infinite domain $-\infty < x < \infty$, a semi-infinite domain $0 \leq x < \infty$, or a finite domain $0 \leq x \leq L$. At a boundary, the conditions can be a fixed temperature $T\ (0, t) = T_0$ (boundary condition of the first kind, or Dirichlet condition), or a fixed flux $-k \frac{\partial T}{\partial x}\ (0,t) = q_0$ (boundary condition of the second kind, or Neumann condition), or a combination $-k\frac{\partial T}{\partial x}(0,t) = h[T(0,t) - T_0]$ (boundary condition of the third kind, or Robin condition).

The functions T_0 and q_0 can be functions of time. All properties are constant (ϱ, C_p, k, D, h), so that the problem remains linear. Solutions are presented on all domains with various boundary conditions for the heat conduction problem.

$$\frac{\partial T}{\partial t} = \alpha \frac{\partial^2 T}{\partial x^2}, \ \alpha = \frac{k}{\varrho C_p}$$

Problem 1. Infinite domain, on $-\infty < x < \infty$.

$$T(x,0) = f(x), \text{ initial conditions}$$

$$T(x,t) \text{ bounded}$$

Solution is via Fourier transforms

$$\hat{T}(\omega,t) = \int_{-\infty}^{\infty} T(x,t)\,e^{i\omega x}\,dx$$

Applied to the differential equation

$$\boldsymbol{F}\left[\frac{\partial^2 T}{\partial x^2}\right] = -\omega^2\,\alpha \boldsymbol{F}\,[T]$$

$$\frac{\partial \hat{T}}{\partial t} + \omega^2\,\alpha \hat{T} = 0, \hat{T}(\omega,0) = \hat{f}(\omega)$$

By solving

$$\hat{T}(\omega,t) = \hat{f}(\omega)\,e^{-\omega^2 \alpha t}$$

the inverse transformation gives [40, p. 328], [44, p. 58]

$$T(x,t) = \frac{1}{2\pi}\lim_{L\to\infty}\int_{-L}^{L} e^{-i\omega x}\,\hat{f}(\omega)\,e^{-\omega^2 \alpha t} d\omega$$

Another solution is via Laplace transforms; take the Laplace transform of the original differential equation.

$$s\,t(s,x) - f(x) = \alpha \frac{\partial^2 t}{\partial x^2}$$

This equation can be solved with Fourier transforms [40, p. 355]

$$t(s,x) = \frac{1}{2\sqrt{s\alpha}}\int_{-\infty}^{\infty} e\left\{-\sqrt{\frac{s}{\alpha}}\,|x-y|\right\} f(y)\,dy$$

The inverse transformation is [40, p. 357], [44, p. 53]

$$T(x,t) =$$

$$\frac{1}{2\sqrt{\pi\,\alpha\,t}}\int_{-\infty}^{\infty} e^{-(x-y)^2/4\alpha t}\,f(y)\,dy$$

Problem 2. Semi-infinite domain, boundary condition of the first kind, on $0 \le x \le \infty$

$$T(x,0) = T_0 = \text{constant}$$

$$T(0,t) = T_1 = \text{constant}$$

The solution is

$$T(x,t) = T_0 + [T_1 - T_0]\left[1 - \text{erf}\left(x/\sqrt{4\alpha\,t}\right)\right]$$

$$\text{or } T(x,t) = T_0 + (T_1 - T_0)\ \text{erfc}\left(x/\sqrt{4\alpha\,t}\right)$$

Problem 3. Semi-infinite domain, boundary condition of the first kind, on $0 \le x < \infty$

$$T(x,0) = f(x)$$

$$T(0,t) = g(t)$$

The solution is written as

$$T(x,t) = T_1(x,t) + T_2(x,t)$$

where

$$T_1(x,0) = f(x), T_2(x,0) = 0$$

$$T_1(0,t) = 0, T_2(0,t) = g(t)$$

Then T_1 is solved by taking the sine transform

$$U_1 = F_s^\omega[T_1]$$

$$\frac{\partial U_1}{\partial t} = -\omega^2\,\alpha\,U_1$$

$$U_1(\omega,0) = F_s^\omega[f]$$

Thus,

$$U_1(\omega,t) = F_s^\omega[f]\,e^{-\omega^2 \alpha t}$$

and [40, p. 322]

$$T_1(x,t) = \frac{2}{\pi}\int_0^\infty F_s^\omega[f]\,e^{-\omega^2 \alpha t}\sin\omega x\,d\omega$$

Solve for T_2 by taking the sine transform

$$U_2 = F_s^\omega[T_2]$$

$$\frac{\partial U_2}{\partial t} = -\omega^2\,\alpha\,U_2 + \alpha\,g(t)\,\omega$$

$$U_2(\omega,0) = 0$$

Thus,

$$U_2(\omega,t)=\int_0^t e^{-\omega^2\alpha(t-\tau)}\alpha\omega\, g(\tau)\,d\tau$$

and [40, p. 435]

$$T_2(x,t)=\frac{2\alpha}{\pi}\int_0^\infty \omega\sin\omega x\int_0^t e^{-\omega^2\alpha(t-\tau)}g(\tau)\,d\tau d\omega$$

The solution for T_1 can also be obtained by Laplace transforms.

$$t_1=\boldsymbol{L}[T_1]$$

Applying this to the differential equation

$$st_1-f(x)=\alpha\frac{\partial^2 t_1}{\partial x^2},\ t_1(0,s)=0$$

and solving gives

$$t_1=\frac{1}{\sqrt{s\alpha}}\int_0^x e^{-\sqrt{s/\alpha}(x'-x)}f(x')\,dx'$$

and the inverse transformation is [40, p. 437], [44, p. 59]

$$T_1(x,t)= \tag{7}$$

$$\frac{1}{\sqrt{4\pi\alpha t}}\int_0^\infty\left[e^{-(x-\xi)^2/4\alpha t}-e^{-(x+\xi)^2/4\alpha t}\right]f(\xi)\,d\xi$$

Problem 4. Semi-infinite domain, boundary conditions of the second kind, on $0\le x<\infty$.

$$T(x,0)=0$$

$$-k\frac{\partial T}{\partial x}(0,t)=q_0=\text{constant}$$

Take the Laplace transform

$$t(x,s)=L[T(x,t)]$$

$$s\,t=\alpha\frac{\partial^2 t}{\partial x^2}$$

$$-k\frac{\partial t}{\partial x}=\frac{q_0}{s}$$

The solution is

$$t(x,s)=\frac{q_0\sqrt{\alpha}}{k\,s^{3/2}}e^{-x\sqrt{s/\alpha}}$$

The inverse transformation is [41, p. 131], [44, p. 75]

$$T(x,t)=\frac{q_0}{k}\left[2\sqrt{\frac{\alpha t}{\pi}}e^{-x^2/4\alpha t}-x\,\text{erfc}\left(\frac{x}{\sqrt{4\alpha t}}\right)\right]$$

Problem 5. Semi-infinite domain, boundary conditions of the third kind, on $0\le x<\infty$

$$T(x,0)=f(x)$$

$$k\frac{\partial T}{\partial x}(0,t)=h\,T(0,t)$$

Take the Laplace transform

$$s\,t-f(x)=\alpha\frac{\partial^2 t}{\partial x^2}$$

$$k\frac{\partial t}{\partial x}(0,s)=h\,t(0,s)$$

The solution is

$$t(x,s)=\int_0^\infty f(\xi)\,g(x,\xi,s)\,d\xi$$

where [41, p. 227]

$$2\sqrt{s/\alpha}\,g(x,\xi,s)=\exp\left(-|x-\xi|\sqrt{s/\alpha}\right)$$
$$+\frac{\sqrt{s}-h\sqrt{\alpha}/k}{\sqrt{s}+h\sqrt{\alpha}/k}\exp\left[-(x+\xi)\sqrt{s/\alpha}\right]$$

One form of the inverse transformation is [41, p. 228]

$$T(x,t)=\frac{2}{\pi}\int_0^\infty e^{-\beta^2\alpha t}\cos[\beta x-\mu(\beta)]\int_0^\infty f(\xi)$$
$$\cos[\beta\xi-\mu(\beta)]\,d\xi d\beta$$

$$\mu(\beta)=\arg(\beta+i\,h/k)$$

Another form of the inverse transformation when $f=T_0$= constant is [41, p. 231], [44, p. 71]

$$T(x,t)=T_0\left[\text{erf}\left(\frac{x}{\sqrt{4\alpha t}}\right)+e^{hx/k}e^{h^2\alpha t/k^2}\right.$$
$$\left.\text{erfc}\left(\frac{h\sqrt{\alpha t}}{k}+\frac{x}{\sqrt{4\alpha t}}\right)\right]$$

Problem 6. Finite domain, boundary condition of the first kind

$$T(x,0)=T_0=\text{constant}$$

$$T(0,t)=T(L,t)=0$$

Take the Laplace transform

$$s\,t(x,s)-T_0=\alpha\frac{\partial^2 t}{\partial x^2}$$

$$t(0,s)=t(L,s)=0$$

The solution is

$$t(x,s)=-\frac{T_0}{s}\frac{\sinh\sqrt{\frac{s}{\alpha}}x}{\sinh\sqrt{\frac{s}{\alpha}}L}-\frac{T_0}{s}\frac{\sinh\sqrt{\frac{s}{\alpha}}(L-x)}{\sinh\sqrt{\frac{s}{\alpha}}L}+\frac{T_0}{s}$$

The inverse transformation is [41, p. 220], [44, p. 96]

$$T\ (x,t)=\frac{2}{\pi}\ T_0\sum\nolimits_{n=1,3,5,\ldots}\frac{2}{n}\ \mathrm{e}^{-n^2\pi^2\alpha t/L^2}\sin\frac{n\,\pi\,x}{L}$$

or (depending on the inversion technique) [40, pp. 362, 438]

$$T\ (x,t)=\frac{T_0}{\sqrt{4\,\pi\,\alpha\,t}}\int\limits_0^L\sum_{n=-\infty}^{\infty}\left[\mathrm{e}^{-[(x-\xi)+2nL]^2/4\alpha t}\right.$$

$$\left.-\mathrm{e}^{-[(x+\xi)+2nL]^2/4\alpha t}\right]\mathrm{d}\xi$$

Problem 7. Finite domain, boundary condition of the first kind

$$T\,(x,0)=0$$

$$T\,(0,t)=0$$

$$T\,(L,0)=T_0=\text{constant}$$

Take the Laplace transform

$$s\,t\ (x,s)=\alpha\ \frac{\partial^2 t}{\partial x^2}$$

$$t\ (0,s)=0,t\ (L,s)=\frac{T_0}{s}$$

The solution is

$$t\ (x,s)=\frac{T_0}{s}\frac{\sinh\frac{x}{L}\sqrt{s/\alpha}}{\sinh\sqrt{s/\alpha}}$$

and the inverse transformation is [41, p. 201], [44, p. 313]

$$T\ (x,t)=T_0\left[\frac{x}{L}+\frac{2}{\pi}\sum_{n=1}^{\infty}\frac{(-1)^n}{n}\ \mathrm{e}^{-n^2\pi^2\alpha t/L^2}\sin\frac{n\,\pi\,x}{L}\right]$$

An alternate transformation is [41, p. 139], [44, p. 310]

$$T\ (x,t)=T_0\sum_{n=0}^{\infty}\left[\mathrm{erf}\left(\frac{(2\,n{+}1)\ L{+}x}{\sqrt{4\,\alpha\,t}}\right)\right.$$

$$\left.-\mathrm{erf}\left(\frac{(2\,n{+}1)\ L{-}x}{\sqrt{4\,\alpha\,t}}\right)\right]$$

Problem 8. Finite domain, boundary condition of the second kind

$$T\,(x,0)=T_0$$

$$\frac{\partial\,T}{\partial x}\ (0,t)=0,T\ (L,t)=0$$

Take the Laplace transform

$$s\,t\ (x,s)-T_0=\alpha\frac{\partial^2 t}{\partial x^2}$$

$$\frac{\partial t}{\partial x}\ (0,s)=0,t\ (L,s)=0$$

The solution is

$$t\ (x,s)=\frac{T_0}{s}\left[1-\frac{\cosh x\,\sqrt{s/\alpha}}{\cosh L\,\sqrt{s/\alpha}}\right]$$

Its inverse is [41, p. 138]

$$T\ (x,t)=T_0-T_0\sum_{n=0}^{\infty}(-1)^n\left[\mathrm{erfc}\left(\frac{(2n{+}1)\ L{-}x}{\sqrt{4\,\alpha\,t}}\right)\right.$$

$$\left.+\mathrm{erfc}\left(\frac{(2n{+}1)\ L{+}x}{\sqrt{4\,\alpha\,t}}\right)\right]$$

5. Vector Analysis

Notation. A *scalar* is a quantity having magnitude but no direction (e.g., mass, length, time, temperature, and concentration). A *vector* is a quantity having both magnitude and direction (e.g., displacement, velocity, acceleration, force). A second-order *dyadic* has magnitude and two directions associated with it, as defined precisely below. The most common examples are the stress dyadic (or tensor) and the velocity gradient (in fluid flow). Vectors are printed in boldface type and identified in this chapter by lower-case Latin letters. Second-order dyadics are printed in boldface type and are identified in this chapter by capital or Greek letters. Higher order dyadics are not discussed here. Dyadics can be formed from the components of tensors including their directions, and some of the identities for dyadics are more easily proved by using tensor analysis, which is not presented here (see also, → Transport Phenomena, Chap. 1.1.4). Vectors are also first-order dyadics.

Vectors. Two vectors $\boldsymbol{u}$ and $\boldsymbol{v}$ are equal if they have the same magnitude and direction. If they have the same magnitude but the opposite direction, then $\boldsymbol{u}=-\,\boldsymbol{v}$. The sum of two vectors is identified geometrically by placing the vector $\boldsymbol{v}$ at the end of the vector $\boldsymbol{u}$, as shown in Figure 12. The product of a scalar m and a vector $\boldsymbol{u}$ is a vector $m\,\boldsymbol{u}$ in the same direction as $\boldsymbol{u}$ but with a magnitude that equals the magnitude of $\boldsymbol{u}$ times the scalar m. Vectors obey commutative and associative laws.

$\boldsymbol{u}+\boldsymbol{v}=\boldsymbol{v}+\boldsymbol{u}$	Commutative law for addition
$\boldsymbol{u}+(\boldsymbol{v}+\boldsymbol{w})=(\boldsymbol{u}+\boldsymbol{v})+\boldsymbol{w}$	Associative law for addition
$m\,\boldsymbol{u}=\boldsymbol{u}\,m$	Commutative law for scalar multiplication
$m\,(n\,\boldsymbol{u})=(m\,n)\,\boldsymbol{u}$	Associative law for scalar multiplication
$(m+n)\,\boldsymbol{u}=m\,\boldsymbol{u}+n\,\boldsymbol{u}$	Distributive law
$m\,(\boldsymbol{u}+\boldsymbol{v})=m\,\boldsymbol{u}+m\,\boldsymbol{v}$	Distributive law

The same laws are obeyed by dyadics, as well.

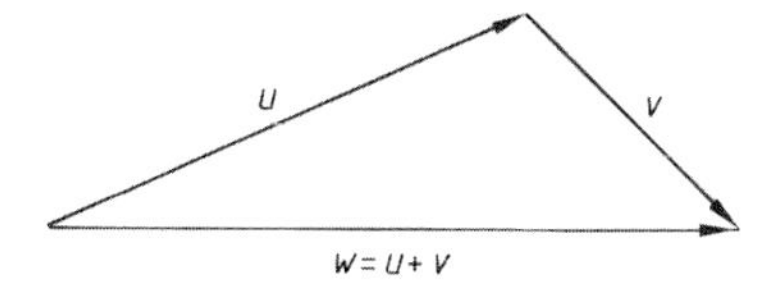

Figure 12. Addition of vectors

A unit vector is a vector with magnitude 1.0 and some direction. If a vector has some magnitude (i.e., not zero magnitude), a unit vector $\boldsymbol{e}_u$ can be formed by

$$\boldsymbol{e}_u=\frac{\boldsymbol{u}}{|\boldsymbol{u}|}$$

The original vector can be represented by the product of the magnitude and the unit vector.

$$\boldsymbol{u}=|\boldsymbol{u}|\,\boldsymbol{e}_u$$

In a cartesian coordinate system the three principle, orthogonal directions are customarily represented by *unit vectors*, such as $\{\boldsymbol{e}_x, \boldsymbol{e}_y, \boldsymbol{e}_z\}$ or $\{\boldsymbol{i}, \boldsymbol{j}, \boldsymbol{k}\}$. Here, the first notation is used (see Fig. 13). The coordinate system is right handed; that is, if a right-threaded screw rotated from the x to the y direction, it would advance in the z direction. A vector can be represented in terms of its components in these directions, as illustrated in Figure 14. The vector is then written as

$$\boldsymbol{u}=u_x\boldsymbol{e}_x+u_y\boldsymbol{e}_y+u_z\boldsymbol{e}_z$$

The magnitude is

$$|\boldsymbol{u}|=\sqrt{u_x^2+u_y^2+u_z^2}$$

The position vector is

$$\boldsymbol{r}=x\boldsymbol{e}_x+y\boldsymbol{e}_y+z\boldsymbol{e}_z$$

with magnitude

$$|\boldsymbol{r}|=\sqrt{x^2+y^2+z^2}$$

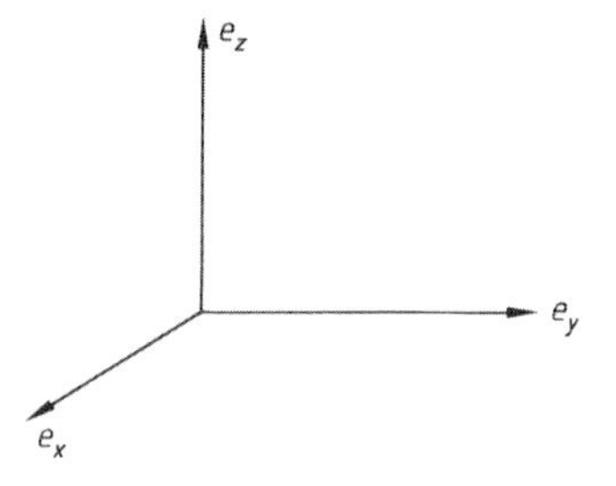

Figure 13. Cartesian coordinate system

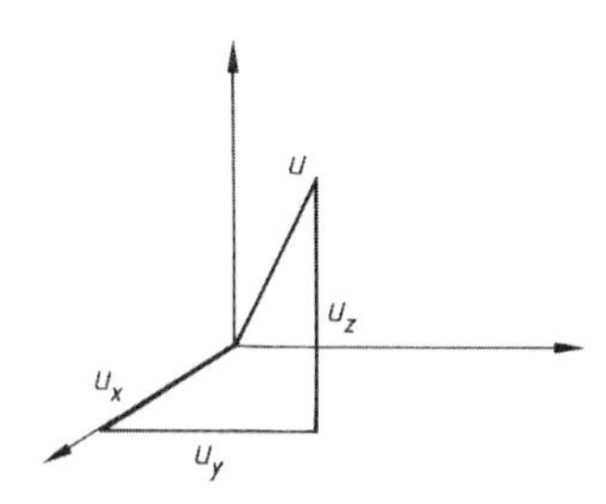

Figure 14. Vector components

Dyadics. The dyadic $\boldsymbol{A}$ is written in component form in cartesian coordinates as

$$\boldsymbol{A}=A_{xx}\boldsymbol{e}_x\boldsymbol{e}_x+A_{xy}\boldsymbol{e}_x\boldsymbol{e}_y+A_{xz}\boldsymbol{e}_x\boldsymbol{e}_z$$

$$+A_{yx}\boldsymbol{e}_y\boldsymbol{e}_x+A_{yy}\boldsymbol{e}_y\boldsymbol{e}_y+A_{yz}\boldsymbol{e}_y\boldsymbol{e}_z$$

$$+A_{zx}\boldsymbol{e}_z\boldsymbol{e}_x+A_{zy}\boldsymbol{e}_z\boldsymbol{e}_y+A_{zz}\boldsymbol{e}_z\boldsymbol{e}_z$$

Quantities such as $\boldsymbol{e}_x\boldsymbol{e}_y$ are called *unit dyadics*. They are second-order dyadics and have two directions associated with them, $\boldsymbol{e}_x$ and $\boldsymbol{e}_y$; the order of the pair is important. The components $A_{xx}, \ldots, A_{zz}$ are the components of the tensor A_{ij} which here is a 3×3 matrix of numbers that are transformed in a certain way when variables undergo a linear transformation. The $y\,x$ momentum flux can be defined as the flux of x momentum across an area with unit normal in the y direction. Since two directions are involved, a second-order dyadic (or tensor) is needed to represent it, and because the y momentum across an area with unit normal in the x direction may not be the same thing, the order of the indices must be kept straight. The dyadic $\boldsymbol{A}$ is said to be symmetric if

$$\boldsymbol{A}_{ij}=\boldsymbol{A}_{ji}$$

Here, the indices i and j can take the values x, y, or z; sometimes $(x, 1)$, $(y, 2)$, $(x, 3)$ are identified and the indices take the values 1, 2, or 3. The dyadic $\boldsymbol{A}$ is said to be antisymmetric if

$$\boldsymbol{A}_{ij} = -\boldsymbol{A}_{ji}$$

The transpose of $\boldsymbol{A}$ is

$$\boldsymbol{A}_{ij}^{\mathrm{T}} = \boldsymbol{A}_{ji}$$

Any dyadic can be represented as the sum of a symmetric portion and an antisymmetric portion.

$$\boldsymbol{A}_{ij} = \boldsymbol{B}_{ij}+\boldsymbol{C}_{ij}, \boldsymbol{B}_{ij}\equiv\frac{1}{2}(\boldsymbol{A}_{ij}+\boldsymbol{A}_{ji}),$$

$$\boldsymbol{C}_{ij}\equiv\frac{1}{2}(\boldsymbol{A}_{ij}-\boldsymbol{A}_{ji})$$

An ordered pair of vectors is a second-order dyadic.

$$\boldsymbol{u}\,\boldsymbol{v} = \sum_i\sum_j \boldsymbol{e}_i\boldsymbol{e}_j u_i\, v_j$$

The transpose of this is

$$(\boldsymbol{u}\,\boldsymbol{v})^{\mathrm{T}} = \boldsymbol{v}\,\boldsymbol{u}$$

but

$$\boldsymbol{u}\,\boldsymbol{v}\neq\boldsymbol{v}\,\boldsymbol{u}$$

The Kronecker delta is defined as

$$\delta_{ij} = \begin{Bmatrix} 1 \text{ if } i=j \\ 0 \text{ if } i\neq j \end{Bmatrix}$$

and the unit dyadic is defined as

$$\delta = \sum_i\sum_j \boldsymbol{e}_i\boldsymbol{e}_j\,\delta_{ij}$$

Operations. The *dot* or *scalar product* of two vectors is defined as

$$\boldsymbol{u}\cdot\boldsymbol{v} = |\boldsymbol{u}|\,|\boldsymbol{v}|\cos\theta,\ 0\leq\theta\leq\pi$$

where θ is the angle between $\boldsymbol{u}$ *and* $\boldsymbol{v}$. The scalar product of two vectors is a scalar, not a vector. It is the magnitude of $\boldsymbol{u}$ multiplied by the projection of $\boldsymbol{v}$ on $\boldsymbol{u}$, or vice versa. The scalar product of $\boldsymbol{u}$ with itself is just the square of the magnitude of $\boldsymbol{u}$.

$$\boldsymbol{u}\cdot\boldsymbol{u} = \left|\boldsymbol{u}^2\right| = u^2$$

The following laws are valid for scalar products

$\boldsymbol{u}\cdot\boldsymbol{v} = \boldsymbol{v}\cdot\boldsymbol{u}$	Commutative law for scalar products
$\boldsymbol{u}\cdot(\boldsymbol{v}+\boldsymbol{w}) = \boldsymbol{u}\cdot\boldsymbol{v}+\boldsymbol{u}\cdot\boldsymbol{w}$	Distributive law for scalar products

$$\boldsymbol{e}_x\cdot\boldsymbol{e}_x = \boldsymbol{e}_y\cdot\boldsymbol{e}_y = \boldsymbol{e}_z\cdot\boldsymbol{e}_z = 1$$
$$\boldsymbol{e}_x\cdot\boldsymbol{e}_y = \boldsymbol{e}_x\cdot\boldsymbol{e}_z = \boldsymbol{e}_y\cdot\boldsymbol{e}_z = 0$$

If the two vectors $\boldsymbol{u}$ and $\boldsymbol{v}$ are written in component notation, the scalar product is

$$\boldsymbol{u}\cdot\boldsymbol{v} = u_x\,v_x+u_y\,v_y+u_z\,v_z$$

If $\boldsymbol{u}\cdot\boldsymbol{v} = 0$ and $\boldsymbol{u}$ and $\boldsymbol{v}$ are not null vectors, then $\boldsymbol{u}$ and $\boldsymbol{v}$ are perpendicular to each other and $\theta = \pi/2$.

The *single dot product* of two dyadics is

$$\boldsymbol{A}\cdot\boldsymbol{B} = \sum_i\sum_j \boldsymbol{e}_i\boldsymbol{e}_j\left(\sum_k A_{ik}\,B_{kj}\right)$$

The *double dot product* of two dyadics is

$$\boldsymbol{A}{:}\boldsymbol{B} = \sum_i\sum_j A_{ij}\,B_{ji}$$

Because the dyadics may not be symmetric, the order of indices and which indices are summed are important. The order is made clearer when the dyadics are made from vectors.

$$(\boldsymbol{u}\,\boldsymbol{v})\cdot(\boldsymbol{w}\,\boldsymbol{x}) = \boldsymbol{u}\,(\boldsymbol{v}\cdot\boldsymbol{w})\,\boldsymbol{x} = \boldsymbol{u}\,\boldsymbol{x}\,(\boldsymbol{v}\cdot\boldsymbol{w})$$

$$(\boldsymbol{u}\,\boldsymbol{v}):(\boldsymbol{w}\,\boldsymbol{x}) = (\boldsymbol{u}\cdot\boldsymbol{x})\,(\boldsymbol{v}\cdot\boldsymbol{w})$$

The dot product of a dyadic and a vector is

$$\boldsymbol{A}\cdot\boldsymbol{u} = \sum_i \boldsymbol{e}_i\left(\sum_j A_{ij}\,u_j\right)$$

The *cross* or *vector product* is defined by

$$\boldsymbol{c} = \boldsymbol{u}\times\boldsymbol{v} = \boldsymbol{a}\,|u|\,|v|\sin\theta,\ 0\leq\theta\leq\pi$$

where $\boldsymbol{a}$ is a unit vector in the direction of $\boldsymbol{u}\times\boldsymbol{v}$. The direction of $\boldsymbol{c}$ is perpendicular to the plane of $\boldsymbol{u}$ and $\boldsymbol{v}$ such that $\boldsymbol{u}$, $\boldsymbol{v}$, and $\boldsymbol{c}$ form a right-handed system. If $\boldsymbol{u} = \boldsymbol{v}$, or $\boldsymbol{u}$ is parallel to $\boldsymbol{v}$, then $\theta = 0$ and $\boldsymbol{u}\times\boldsymbol{v} = 0$. The following laws are valid for cross products.

$\boldsymbol{u}\times\boldsymbol{v} = -\boldsymbol{v}\times\boldsymbol{u}$	Commutative law fails for vector product
$\boldsymbol{u}\times(\boldsymbol{v}\times\boldsymbol{w}) \neq (\boldsymbol{u}\times\boldsymbol{v})\times\boldsymbol{w}$	Associative law fails for vector product
$\boldsymbol{u}\times(\boldsymbol{v}+\boldsymbol{w}) = \boldsymbol{u}\times\boldsymbol{v}+\boldsymbol{u}\times\boldsymbol{w}$	Distributive law for vector product

$$\boldsymbol{e}_x\times\boldsymbol{e}_x = \boldsymbol{e}_y\times\boldsymbol{e}_y = \boldsymbol{e}_z\times\boldsymbol{e}_z = 0$$
$$\boldsymbol{e}_x\times\boldsymbol{e}_y = \boldsymbol{e}_z,\ \boldsymbol{e}_y\times\boldsymbol{e}_z = \boldsymbol{e}_x,\ \boldsymbol{e}_z\times\boldsymbol{e}_x = \boldsymbol{e}_y$$

$$\boldsymbol{u}\times\boldsymbol{v} = \det\begin{bmatrix} \boldsymbol{e}_x & \boldsymbol{e}_y & \boldsymbol{e}_z \\ u_x & u_y & u_z \\ v_x & v_y & v_z \end{bmatrix}$$
$$= \boldsymbol{e}_x\,(u_y\,v_z-v_y\,u_z)+\boldsymbol{e}_y\,(u_z\,v_z-u_x\,v_z)$$
$$+\boldsymbol{e}_z\,(u_x\,v_y-u_y\,v_x)$$

This can also be written as

$$\boldsymbol{u}\times\boldsymbol{v}=\sum_i\sum_j \varepsilon_{kij}\, u_i\, v_j \boldsymbol{e}_k$$

where

$$\varepsilon_{ijk}=\begin{cases}1 & \text{if } i,j,k \text{ is an even permutation of } 123\\ -1 & \text{if } i,j,k \text{ is an odd permutation of } 123\\ 0 & \text{if any two of } i,j,k \text{ are equal}\end{cases}$$

Thus ε_{123} = 1, ε_{132} =- 1, ε_{312} = 1, ε_{112} = 0, for example.

The magnitude of $\boldsymbol{u}\times\boldsymbol{v}$ is the same as the area of a parallelogram with sides $\boldsymbol{u}$ and $\boldsymbol{v}$. If $\boldsymbol{u}\times\boldsymbol{v}$ = **0** and $\boldsymbol{u}$ *and* $\boldsymbol{v}$ are not null vectors, then $\boldsymbol{u}$ *and* $\boldsymbol{v}$ are parallel. Certain triple products are useful.

$$(\boldsymbol{u}\cdot\boldsymbol{v})\,\boldsymbol{w}\neq\boldsymbol{u}\,(\boldsymbol{v}\cdot\boldsymbol{w})$$

$$\boldsymbol{u}\cdot(\boldsymbol{v}\times\boldsymbol{w})=\boldsymbol{v}\cdot(\boldsymbol{w}\times\boldsymbol{u})=\boldsymbol{w}\cdot(\boldsymbol{u}\times\boldsymbol{v})$$

$$\boldsymbol{u}\times(\boldsymbol{v}\times\boldsymbol{w})=(\boldsymbol{u}\cdot\boldsymbol{w})\,v-(\boldsymbol{u}\cdot\boldsymbol{v})\,\boldsymbol{w}$$

$$(\boldsymbol{u}\times\boldsymbol{v})\times\boldsymbol{w}=(\boldsymbol{u}\cdot\boldsymbol{w})\,v-(\boldsymbol{v}\cdot\boldsymbol{w})\,\boldsymbol{u}$$

The cross product of a dyadic and a vector is defined as

$$\boldsymbol{A}\times\boldsymbol{u}=\sum_i\sum_j \boldsymbol{e}_i\boldsymbol{e}_j\left(\sum_k\sum_l \varepsilon_{klj}\, A_{ik}\, u_l\right)$$

The magnitude of a dyadic is

$$|\boldsymbol{A}|=A=\sqrt{\frac{1}{2}\,(\boldsymbol{A}:\boldsymbol{A}^{\mathrm{T}})}=\sqrt{\frac{1}{2}\sum_i\sum_j A_{ij}^2}$$

There are three *invariants* of a dyadic. They are called invariants because they take the same value in any coordinate system and are thus an intrinsic property of the dyadic. They are the trace of $\boldsymbol{A}, \boldsymbol{A}^2, \boldsymbol{A}^3$ [45].

$$\mathrm{I}=\mathrm{tr}\boldsymbol{A}=\sum_i A_{ii}$$

$$\mathrm{II}=\mathrm{tr}\boldsymbol{A}^2=\sum_i\sum_j A_{ij}\, A_{ji}$$

$$\mathrm{III}=\mathrm{tr}\boldsymbol{A}^3=\sum_i\sum_j\sum_k A_{ij}\, A_{jk}\, A_{ki}$$

The invariants can also be expressed as

$$\mathrm{I}_1=\mathrm{I}$$

$$\mathrm{I}_2=\tfrac{1}{2}\left(\mathrm{I}^2-\mathrm{II}\right)$$

$$\mathrm{I}_3=\tfrac{1}{6}\left(\mathrm{I}^3-3\,\mathrm{I}\cdot\mathrm{II}+2\,\mathrm{III}\right)=\det\boldsymbol{A}$$

Invariants of two dyadics are available [46]. Because a second-order dyadic has nine components, the *characteristic equation*

$$\det\,(\lambda\,\boldsymbol{\delta}-\boldsymbol{A})=0$$

can be formed where λ is an eigenvalue. This expression is

$$\lambda^3-I_1\lambda^2+I_2\lambda-I_3=0$$

An important theorem of HAMILTON and CAYLEY [47] is that a second-order dyadic satisfies its own characteristic equation.

$$\boldsymbol{A}^3-I_1\boldsymbol{A}^2+I_2\boldsymbol{A}-I_3\boldsymbol{\delta}=0 \tag{7}$$

Thus $\boldsymbol{A}^3$ can be expressed in terms of $\boldsymbol{\delta}$, $\boldsymbol{A}$, and $\boldsymbol{A}^2$. Similarly, higher powers of $\boldsymbol{A}$ can be expressed in terms of $\boldsymbol{\delta}$, $\boldsymbol{A}$, and $\boldsymbol{A}^2$. Decomposition of a dyadic into a symmetric and an antisymmetric part was shown above. The antisymmetric part has zero trace. The symmetric part can be decomposed into a part with a trace (the isotropic part) and a part with zero trace (the deviatoric part).

$$\boldsymbol{A}=\underbrace{1/3\,\boldsymbol{\delta\delta}:\boldsymbol{A}}_{\text{Isotropic}}+\underbrace{1/2\,[\boldsymbol{A}+\boldsymbol{A}^{\mathrm{T}}-2/3\,\boldsymbol{\delta\delta}:\boldsymbol{A}]}_{\text{Deviatoric}}+\underbrace{1/2\,[\boldsymbol{A}-\boldsymbol{A}^{\mathrm{T}}]}_{\text{Antisymmetric}}$$

Differentiation. The derivative of a vector is defined in the same way as the derivative of a scalar. Suppose the vector $\boldsymbol{u}$ depends on t. Then

$$\frac{\mathrm{d}\boldsymbol{u}}{\mathrm{d}t}=\lim_{\Delta t\to 0}\frac{\boldsymbol{u}\,(t+\Delta t)-\boldsymbol{u}\,(t)}{\Delta t}$$

If the vector is the position vector $\boldsymbol{r}\,(t)$, then the difference expression is a vector in the direction of $\Delta\,\boldsymbol{r}$ (see Fig. 15). The derivative

$$\frac{\mathrm{d}\boldsymbol{r}}{\mathrm{d}t}=\lim_{\Delta t\to 0}\frac{\Delta\boldsymbol{r}}{\Delta t}=\lim_{\Delta t\to 0}\frac{\boldsymbol{r}\,(t+\Delta t)-\boldsymbol{r}\,(t)}{\Delta t}$$

is the velocity. The derivative operation obeys the following laws.

$$\frac{\mathrm{d}}{\mathrm{d}t}\,(\boldsymbol{u}+\boldsymbol{v})=\frac{\mathrm{d}\boldsymbol{u}}{\mathrm{d}t}+\frac{\mathrm{d}\boldsymbol{v}}{\mathrm{d}t}$$

$$\frac{\mathrm{d}}{\mathrm{d}t}\,(\boldsymbol{u}\cdot\boldsymbol{v})=\frac{\mathrm{d}\boldsymbol{u}}{\mathrm{d}t}\cdot\boldsymbol{v}+\boldsymbol{u}\cdot\frac{\mathrm{d}\boldsymbol{v}}{\mathrm{d}t}$$

$$\frac{\mathrm{d}}{\mathrm{d}t}\,(\boldsymbol{u}\times\boldsymbol{v})=\frac{\mathrm{d}\boldsymbol{u}}{\mathrm{d}t}\times\boldsymbol{v}+\boldsymbol{u}\times\frac{\mathrm{d}\boldsymbol{v}}{\mathrm{d}t}$$

$$\frac{\mathrm{d}}{\mathrm{d}t}\,(\varphi\boldsymbol{u})=\frac{\mathrm{d}\varphi}{\mathrm{d}t}\boldsymbol{u}+\varphi\frac{\mathrm{d}\boldsymbol{u}}{\mathrm{d}t}$$

If the vector $\boldsymbol{u}$ depends on more than one variable, such as x, y and z, partial derivatives are defined in the usual way. For example,

if $\boldsymbol{u}\,(x,y,z)$, then

$$\frac{\partial\boldsymbol{u}}{\partial x}=\lim_{\Delta t\to 0}\frac{\boldsymbol{u}(x+\Delta x,\,y,\,z)-\boldsymbol{u}(x,\,y,\,z)}{\Delta x}$$

Rules for differentiation of scalar and vector products are

$$\frac{\partial}{\partial x}(\boldsymbol{u}\cdot\boldsymbol{v})=\frac{\partial \boldsymbol{u}}{\partial x}\cdot\boldsymbol{v}+\boldsymbol{u}\cdot\frac{\partial \boldsymbol{v}}{\partial x}$$

$$\frac{\partial}{\partial x}(\boldsymbol{u}\times\boldsymbol{v})=\frac{\partial \boldsymbol{u}}{\partial x}\times\boldsymbol{v}+\boldsymbol{u}\times\frac{\partial \boldsymbol{v}}{\partial x}$$

Differentials of vectors are

$$\mathrm{d}\boldsymbol{u}=\mathrm{d}u_x\boldsymbol{e}_x+\mathrm{d}u_y\boldsymbol{e}_y+\mathrm{d}u_z\boldsymbol{e}_z$$

$$\mathrm{d}(\boldsymbol{u}\cdot\boldsymbol{v})=\mathrm{d}\boldsymbol{u}\cdot\boldsymbol{v}+\boldsymbol{u}\cdot\mathrm{d}\boldsymbol{v}$$

$$\mathrm{d}(\boldsymbol{u}\times\boldsymbol{v})=\mathrm{d}\boldsymbol{u}\times\boldsymbol{v}+\boldsymbol{u}\times\mathrm{d}\boldsymbol{v}$$

$$\mathrm{d}\boldsymbol{u}=\frac{\partial \boldsymbol{u}}{\partial x}\mathrm{d}x+\frac{\partial \boldsymbol{u}}{\partial y}\mathrm{d}y+\frac{\partial \boldsymbol{u}}{\partial z}\mathrm{d}z$$

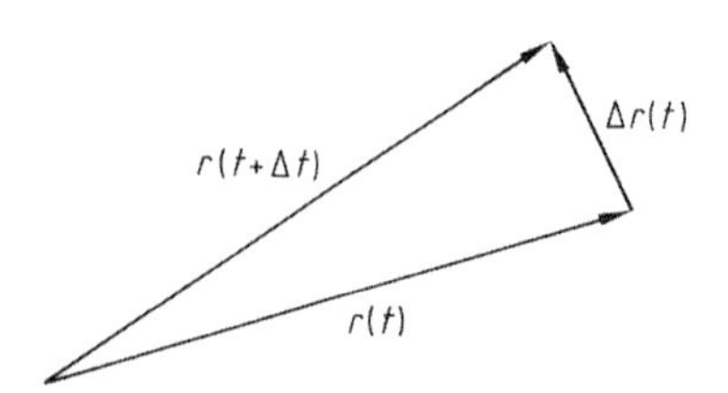

Figure 15. Vector differentiation

If a curve is given by $\boldsymbol{r}(t)$, the length of the curve is [43]

$$L=\int_a^b\sqrt{\frac{\mathrm{d}\boldsymbol{r}}{\mathrm{d}t}\cdot\frac{\mathrm{d}\boldsymbol{r}}{\mathrm{d}t}}\mathrm{d}t$$

The arc-length function can also be defined:

$$s(t)=\int_a^t\sqrt{\frac{\mathrm{d}\boldsymbol{r}}{\mathrm{d}t^*}\cdot\frac{\mathrm{d}\boldsymbol{r}}{\mathrm{d}t^*}}\mathrm{d}t^*$$

This gives

$$\left(\frac{\mathrm{d}s}{\mathrm{d}t}\right)^2=\frac{\mathrm{d}r}{\mathrm{d}t}\cdot\frac{\mathrm{d}r}{\mathrm{d}t}=\left(\frac{\mathrm{d}x}{\mathrm{d}t}\right)^2+\left(\frac{\mathrm{d}y}{\mathrm{d}t}\right)^2+\left(\frac{\mathrm{d}z}{\mathrm{d}t}\right)^2$$

Because

$$d\boldsymbol{r}=\mathrm{d}x\boldsymbol{e}_x+\mathrm{d}y\boldsymbol{e}_y+\mathrm{d}z\boldsymbol{e}_z$$

then

$$ds^2=\mathrm{d}\boldsymbol{r}\cdot\mathrm{d}\boldsymbol{r}=\mathrm{d}x^2+\mathrm{d}y^2+\mathrm{d}z^2$$

The derivative $\mathrm{d}\boldsymbol{r}/\mathrm{d}t$ is tangent to the curve in the direction of motion

$$\boldsymbol{u}=\frac{\frac{\mathrm{d}r}{\mathrm{d}t}}{\left|\frac{\mathrm{d}r}{\mathrm{d}t}\right|}$$

Also,

$$\boldsymbol{u}=\frac{\mathrm{d}\boldsymbol{r}}{\mathrm{d}s}$$

Differential Operators. The vector differential operator (del operator) ∇ is defined in cartesian coordinates by

$$\nabla=\boldsymbol{e}_x\frac{\partial}{\partial x}+\boldsymbol{e}_y\frac{\partial}{\partial y}+\boldsymbol{e}_z\frac{\partial}{\partial z}$$

The *gradient* of a scalar function is defined

$$\nabla\varphi=\boldsymbol{e}_x\frac{\partial\varphi}{\partial x}+\boldsymbol{e}_y\frac{\partial\varphi}{\partial y}+\boldsymbol{e}_z\frac{\partial\varphi}{\partial z}$$

and is a vector. If φ is height or elevation, the gradient is a vector pointing in the uphill direction. The steeper the hill, the larger is the magnitude of the gradient.

The *divergence* of a vector is defined by

$$\nabla\cdot\boldsymbol{u}=\frac{\partial u_x}{\partial x}+\frac{\partial u_y}{\partial y}+\frac{\partial u_z}{\partial z}$$

and is a scalar. For a volume element ΔV, the net outflow of a vector $\boldsymbol{u}$ over the surface of the element is

$$\int_{\Delta S}\boldsymbol{u}\cdot\boldsymbol{n}\mathrm{d}S$$

This is related to the divergence by [48, p. 411]

$$\nabla\cdot\boldsymbol{u}=\lim_{\Delta V\to 0}\frac{1}{\Delta V}\int_{\Delta S}\boldsymbol{u}\cdot\boldsymbol{n}\mathrm{d}S$$

Thus, the divergence is the net outflow per unit volume.

The *curl* of a vector is defined by

$$\nabla\times\boldsymbol{u}=\left(\boldsymbol{e}_x\frac{\partial}{\partial x}+\boldsymbol{e}_y\frac{\partial}{\partial y}+\boldsymbol{e}_z\frac{\partial}{\partial z}\right)$$

$$\times(\boldsymbol{e}_xu_x+\boldsymbol{e}_yu_y+\boldsymbol{e}_zu_z)$$

$$=\boldsymbol{e}_x\left(\frac{\partial u_z}{\partial y}-\frac{\partial u_y}{\partial z}\right)+\boldsymbol{e}_y\left(\frac{\partial u_x}{\partial z}-\frac{\partial u_z}{\partial x}\right)$$

$$+\boldsymbol{e}_z\left(\frac{\partial u_y}{\partial x}-\frac{\partial u_x}{\partial y}\right)$$

and is a vector. It is related to the integral

$$\int_C\boldsymbol{u}\cdot\mathrm{d}\boldsymbol{s}=\int_C u_s\mathrm{d}s$$

which is called the circulation of $\boldsymbol{u}$ around path C. This integral depends on the vector and the contour C, in general. If the circulation does not depend on the contour C, the vector is said to be irrotational; if it does, it is rotational. The relationship with the curl is [48, p. 419]

$$\boldsymbol{n}\cdot(\nabla\times\boldsymbol{u}) = \lim_{\Delta\to 0}\frac{1}{\Delta S}\int_C \boldsymbol{u}\cdot d\boldsymbol{s}$$

Thus, the normal component of the curl equals the net circulation per unit area enclosed by the contour C.

The gradient, divergence, and curl obey a distributive law but not a commutative or associative law.

$$\nabla(\varphi+\psi) = \nabla\varphi+\nabla\psi$$

$$\nabla\cdot(\boldsymbol{u}+\boldsymbol{v}) = \nabla\cdot\boldsymbol{u}+\nabla\cdot\boldsymbol{v}$$

$$\nabla\times(\boldsymbol{u}+\boldsymbol{v}) = \nabla\times\boldsymbol{u}+\nabla\times\boldsymbol{v}$$

$$\nabla\cdot\varphi\neq\varphi\nabla$$

$$\nabla\cdot\boldsymbol{u}\neq\boldsymbol{u}\cdot\nabla$$

Useful formulas are [49]

$$\nabla\cdot(\varphi\boldsymbol{u}) = \nabla\varphi\cdot\boldsymbol{u}+\varphi\nabla\cdot\boldsymbol{u}$$

$$\nabla\times(\varphi\boldsymbol{u}) = \nabla\varphi\times\boldsymbol{u}+\varphi\nabla\times\boldsymbol{u}$$

$$\nabla\cdot(\boldsymbol{u}\times\boldsymbol{v}) = \boldsymbol{v}\cdot(\nabla\times\boldsymbol{u})-\boldsymbol{u}\cdot(\nabla\times\boldsymbol{v})$$

$$\nabla\times(\boldsymbol{u}\times\boldsymbol{v}) = \boldsymbol{v}\cdot\nabla\boldsymbol{u}-\boldsymbol{v}(\nabla\cdot\boldsymbol{u})-\boldsymbol{u}\cdot\nabla\boldsymbol{v}+\boldsymbol{u}(\nabla\cdot\boldsymbol{v})$$

$$\nabla\times(\nabla\times\boldsymbol{u}) = \nabla(\nabla\cdot\boldsymbol{u})-\nabla^2\boldsymbol{u}$$

$$\nabla\cdot(\nabla\varphi) = \nabla^2\varphi = \frac{\partial^2\varphi}{\partial x^2}+\frac{\partial^2\varphi}{\partial y^2}+\frac{\partial^2\varphi}{\partial z^2}, \text{ where } \nabla^2$$

is called the Laplacian operator. $\nabla\times(\nabla\varphi) = \mathbf{0}$. The curl of the gradient of φ is zero.

$$\nabla\cdot(\nabla\times\boldsymbol{u}) = 0$$

The divergence of the curl of $\boldsymbol{u}$ is zero. Formulas useful in fluid mechanics are

$$\nabla\cdot(\nabla\boldsymbol{v})^{\mathrm{T}} = \nabla(\nabla\cdot\boldsymbol{v})$$

$$\nabla\cdot(\tau\cdot\boldsymbol{v}) = \boldsymbol{v}\cdot(\nabla\cdot\tau)+\tau:\nabla\boldsymbol{v}$$

$$\boldsymbol{v}\cdot\nabla\boldsymbol{v} = \tfrac{1}{2}\nabla(\boldsymbol{v}\cdot\boldsymbol{v})-\boldsymbol{v}\times(\nabla\times\boldsymbol{v})$$

If a coordinate system is transformed by a rotation and translation, the coordinates in the new system (denoted by primes) are given by

$$\begin{pmatrix} x' \\ y' \\ z' \end{pmatrix} = \begin{pmatrix} l_{11} & l_{12} & l_{13} \\ l_{21} & l_{22} & l_{23} \\ l_{31} & l_{32} & l_{33} \end{pmatrix}$$

$$\begin{pmatrix} x \\ y \\ z \end{pmatrix} + \begin{pmatrix} a_1 \\ a_2 \\ a_3 \end{pmatrix}$$

Any function that has the same value in all coordinate systems is an invariant. The gradient of an invariant scalar field is invariant; the same is true for the divergence and curl of invariant vectors fields.

The gradient of a vector field is required in fluid mechanics because the velocity gradient is used. It is defined as

$$\nabla\boldsymbol{v} = \sum_i\sum_j \boldsymbol{e}_i\boldsymbol{e}_j\frac{\partial v_j}{\partial x_i} \text{ and}$$

$$(\nabla\boldsymbol{v})^{\mathrm{T}} = \sum_i\sum_j \boldsymbol{e}_i\boldsymbol{e}_j\frac{\partial v_i}{\partial x_j}$$

The divergence of dyadics is defined

$$\nabla\cdot\boldsymbol{\tau} = \sum_i \boldsymbol{e}_i\left(\sum_j\frac{\partial\tau_{ji}}{\partial x_j}\right) \text{ and}$$

$$\nabla\cdot(\varphi\boldsymbol{u}\,\boldsymbol{v}) = \sum_i \boldsymbol{e}_i\left[\sum_j\frac{\partial}{\partial x_j}(\varphi u_j\,v_i)\right]$$

where $\boldsymbol{\tau}$ is any second-order dyadic.

Useful relations involving dyadics are

$$(\varphi\,\boldsymbol{\delta}:\nabla\boldsymbol{v}) = \varphi(\nabla\cdot\boldsymbol{v})$$

$$\nabla\cdot(\varphi\,\boldsymbol{\delta}) = \nabla\varphi$$

$$\nabla\cdot(\varphi\boldsymbol{\tau}) = \nabla\varphi\cdot\boldsymbol{\tau}+\varphi\nabla\cdot\boldsymbol{\tau}$$

$$\boldsymbol{n}\,\boldsymbol{t}:\boldsymbol{\tau} = \boldsymbol{t}\cdot\boldsymbol{\tau}\cdot\boldsymbol{n} = \boldsymbol{\tau}:\boldsymbol{n}\,\boldsymbol{t}$$

A surface can be represented in the form

$$f(x,y,z) = c = \text{constant}$$

The normal to the surface is given by

$$\boldsymbol{n} = \frac{\nabla f}{|\nabla f|}$$

provided the gradient is not zero. Operations can be performed entirely within the surface. Define

$$\boldsymbol{\delta}_{\mathrm{II}}\equiv\boldsymbol{\delta}-\boldsymbol{n}\,\boldsymbol{n},\ \nabla_{\mathrm{II}}\equiv\boldsymbol{\delta}_{\mathrm{II}}\cdot\nabla,\ \frac{\partial}{\partial n}\equiv\boldsymbol{n}\cdot\nabla$$

$$\boldsymbol{v}_{\mathrm{II}}\equiv\boldsymbol{\delta}_{\mathrm{II}}\cdot\boldsymbol{v},\ v_n\equiv\boldsymbol{n}\cdot\boldsymbol{v}$$

Then a vector and del operator can be decomposed into

$$\boldsymbol{v} = \boldsymbol{v}_{\mathrm{II}}+\boldsymbol{n}v_n,\ \nabla = \nabla_{\mathrm{II}}+\boldsymbol{n}\frac{\partial}{\partial n}$$

The velocity gradient can be decomposed into

$$\nabla\boldsymbol{v} = \nabla_{\mathrm{II}}\boldsymbol{v}_{\mathrm{II}}+(\nabla_{\mathrm{II}}\boldsymbol{n})\,v_n+\boldsymbol{n}\nabla_{\mathrm{II}}v_n+\boldsymbol{n}\frac{\partial\boldsymbol{v}_{\mathrm{II}}}{\partial n}+\boldsymbol{n}\,\boldsymbol{n}\frac{\partial v_n}{\partial n}$$

The surface gradient of the normal is the negative of the curvature dyadic of the surface.

$$\nabla_{\mathrm{II}}\boldsymbol{n} = -\boldsymbol{B}$$

The surface divergence is then

$$\nabla_{\mathrm{II}}\cdot\boldsymbol{v} = \boldsymbol{\delta}_{\mathrm{II}}:\nabla\boldsymbol{v} = \nabla_{\mathrm{II}}\cdot\boldsymbol{v}_{\mathrm{II}}-2\,H\,v_n$$

where H is the mean curvature.

$$H = \frac{1}{2} \boldsymbol{\delta}_{\mathrm{II}} : \boldsymbol{B}$$

The surface curl can be a scalar

$$\nabla_{\mathrm{II}} \times \boldsymbol{v} = -\boldsymbol{\varepsilon}_{\mathrm{II}} : \nabla \boldsymbol{v} = -\boldsymbol{\varepsilon}_{\mathrm{II}} : \nabla_{\mathrm{II}} \boldsymbol{v}_{\mathrm{II}} = -\boldsymbol{n} \cdot (\nabla \times \boldsymbol{v}),$$

$$\boldsymbol{\varepsilon}_{\mathrm{II}} = \boldsymbol{n} \cdot \boldsymbol{\varepsilon}$$

or a vector

$$\nabla_{\mathrm{II}} \times \boldsymbol{v} \equiv \nabla_{\mathrm{II}} \times \boldsymbol{v}_{\mathrm{II}} = \boldsymbol{n} \nabla_{\mathrm{II}} \times \boldsymbol{v} = \boldsymbol{n}\, \boldsymbol{n} \cdot \nabla \times v$$

Vector Integration [48, pp. 206 – 212]. If $\boldsymbol{u}$ is a vector, then its integral is also a vector.

$$\int \boldsymbol{u}(t)\,\mathrm{d}t = \boldsymbol{e}_x \int u_x(t)\,\mathrm{d}t + \boldsymbol{e}_y \int u_y(t)\,\mathrm{d}t$$

$$+\boldsymbol{e}_z \int u_z(t)\,\mathrm{d}t$$

If the vector $\boldsymbol{u}$ is the derivative of another vector, then

$$\boldsymbol{u} = \frac{\mathrm{d}\boldsymbol{v}}{\mathrm{d}t}, \int \boldsymbol{u}(t)\,\mathrm{d}t = \int \frac{\mathrm{d}\boldsymbol{v}}{\mathrm{d}t}\,\mathrm{d}t = \boldsymbol{v} + \text{constant}$$

If $\boldsymbol{r}(t)$ is a position vector that defines a curve C, the line integral is defined by

$$\int_C \boldsymbol{u} \cdot \mathrm{d}\boldsymbol{r} = \int_C (u_x \mathrm{d}x + u_y \mathrm{d}y + u_z \mathrm{d}z)$$

Theorems about this line integral can be written in various forms.

Theorem [43]. If the functions appearing in the line integral are continuous in a domain D, then the line integral is independent of the path C if and only if the line integral is zero on every simple closed path in D.

Theorem [43]. If $\boldsymbol{u} = \nabla \varphi$ where φ is single-valued and has continuous derivatives in D, then the line integral is independent of the path C and the line integral is zero for any closed curve in D.

Theorem [43]. If f, g, and h are continuous functions of x, y, and z, and have continuous first derivatives in a simply connected domain D, then the line integral

$$\int_C (f\,\mathrm{d}x + g\mathrm{d}y + h\mathrm{d}z)$$

is independent of the path if and only if

$$\frac{\partial h}{\partial y} = \frac{\partial g}{\partial z}, \frac{\partial f}{\partial z} = \frac{\partial h}{\partial x}, \frac{\partial g}{\partial x} = \frac{\partial f}{\partial y}$$

or if f, g, and h are regarded as the x, y, and z components of a vector $\boldsymbol{v}$:

$$\nabla \times \boldsymbol{v} = 0$$

Consequently, the line integral is independent of the path (and the value is zero for a closed contour) if the three components in it are regarded as the three components of a vector and the vector is derivable from a potential (or zero curl). The conditions for a vector to be derivable from a potential are just those in the third theorem. In two dimensions this reduces to the more usual theorem.

Theorem [48, p. 207]. If M and N are continuous functions of x and y that have continuous first partial derivatives in a simply connected domain D, then the necessary and sufficient condition for the line integral

$$\int_C (M\mathrm{d}x + N\mathrm{d}y)$$

to be zero around every closed curve C in D is

$$\frac{\partial M}{\partial y} = \frac{\partial N}{\partial x}$$

If a vector is integrated over a surface with incremental area dS and normal to the surface $\boldsymbol{n}$, then the surface integral can be written as

$$\int\!\!\int_S \boldsymbol{u} \cdot \mathrm{d}\boldsymbol{S} = \int\!\!\int_S \boldsymbol{u} \cdot \boldsymbol{n} \mathrm{d}S$$

If $\boldsymbol{u}$ is the velocity then this integral represents the flow rate past the surface S.

Divergence Theorem [48, 49]. If V is a volume bounded by a closed surface S and $\boldsymbol{u}$ is a vector function of position with continuous derivatives, then

$$\int_V \nabla \cdot \boldsymbol{u} \mathrm{d}V = \int_S \boldsymbol{n} \cdot \boldsymbol{u} \mathrm{d}S = \int_S \boldsymbol{u} \cdot \boldsymbol{n} \mathrm{d}S = \int_S \boldsymbol{u} \cdot \mathrm{d}\boldsymbol{S}$$

where $\boldsymbol{n}$ is the normal pointing outward to S. The normal can be written as

$$\boldsymbol{n} = \boldsymbol{e}_x \cos(x, n) + \boldsymbol{e}_y \cos(y, n) + \boldsymbol{e}_z \cos(z, n)$$

where, for example, cos (x, n) is the cosine of the angle between the normal n and the x axis. Then the divergence theorem in component form is

$$\int_V \left(\frac{\partial u_x}{\partial x} + \frac{\partial u_y}{\partial y} + \frac{\partial u_z}{\partial z} \right) \mathrm{d}x\mathrm{d}y\mathrm{d}z =$$

$$\int_S \left[u_x \cos(x, n) + u_y \cos(y, n) + u_z \cos(z, n) \right] \mathrm{d}S$$

If the divergence theorem is written for an incremental volume

$$\nabla\cdot\boldsymbol{u}=\lim_{\Delta V\to 0}\frac{1}{\Delta V}\int\limits_{\Delta S} u_n \mathrm{d}S$$

the divergence of a vector can be called the integral of that quantity over the area of a closed volume, divided by the volume. If the vector represents the flow of energy and the divergence is positive at a point P, then either a source of energy is present at P or energy is leaving the region around P so that its temperature is decreasing. If the vector represents the flow of mass and the divergence is positive at a point P, then either a source of mass exists at P or the density is decreasing at the point P. For an incompressible fluid the divergence is zero and the rate at which fluid is introduced into a volume must equal the rate at which it is removed.

Various theorems follow from the divergence theorem.

Theorem. If φ is a solution to Laplace's equation

$$\nabla^2\varphi=0$$

in a domain D, and the second partial derivatives of φ are continuous in D, then the integral of the normal derivative of φ over any piecewise smooth closed orientable surface S in D is zero. Suppose $\boldsymbol{u} = \varphi\, \nabla\psi$ satisfies the conditions of the divergence theorem: then *Green's theorem* results from use of the divergence theorem [49].

$$\int\limits_V \left(\varphi\nabla^2\psi+\nabla\varphi\cdot\nabla\psi\right)\mathrm{d}V=\int\limits_S \varphi\frac{\partial\psi}{\partial n}\mathrm{d}S$$

and

$$\int\limits_V \left(\varphi\nabla^2\psi-\psi\nabla^2\varphi\right)\mathrm{d}V=\int\limits_S \left(\varphi\frac{\partial\psi}{\partial n}-\psi\frac{\partial\varphi}{\partial n}\right)\mathrm{d}S$$

Also if φ satisfies the conditions of the theorem and is zero on S then φ is zero throughout D. If two functions φ and ψ both satisfy the Laplace equation in domain D, and both take the same values on the bounding curve C, then $\varphi = \psi$; i.e., the solution to the Laplace equation is unique.

The divergence theorem for dyadics is

$$\int\limits_V \nabla\cdot\boldsymbol{\tau}\mathrm{d}V=\int\limits_S \boldsymbol{n}\cdot\boldsymbol{\tau}\,\mathrm{d}S$$

Stokes Theorem [48, 49]. Stokes theorem says that if S is a surface bounded by a closed, nonintersecting curve C, and if $\boldsymbol{u}$ has continuous derivatives then

$$\oint\limits_C \boldsymbol{u}\cdot\mathrm{d}\boldsymbol{r}=\int\!\!\int\limits_S (\nabla\times\boldsymbol{u})\cdot\boldsymbol{n}\mathrm{d}S=\int\!\!\int\limits_S (\nabla\times\boldsymbol{u})\cdot\mathrm{d}\boldsymbol{S}$$

The integral around the curve is followed in the counterclockwise direction. In component notation, this is

$$\oint\limits_C \left[u_x\cos(x,s)+u_y\cos(y,s)+u_z\cos(z,s)\right]\mathrm{d}s=$$
$$\int\!\!\int\limits_S \left[\left(\frac{\partial u_z}{\partial y}-\frac{\partial u_y}{\partial z}\right)\cos(x,n)\right.$$
$$+\left(\frac{\partial u_x}{\partial z}-\frac{\partial u_z}{\partial x}\right)\cos(y,n)$$
$$\left.+\left(\frac{\partial u_y}{\partial x}-\frac{\partial u_x}{\partial y}\right)\cos(z,n)\right]\mathrm{d}S$$

Applied in two dimensions, this results in Green's theorem in the plane:

$$\oint\limits_C (M\mathrm{d}x+N\mathrm{d}y)=\int\!\!\int\limits_S \left(\frac{\partial N}{\partial x}-\frac{\partial M}{\partial y}\right)\mathrm{d}x\mathrm{d}y$$

The formula for dyadics is

$$\int\!\!\int\limits_S \boldsymbol{n}\cdot(\nabla\times\boldsymbol{\tau})\,\mathrm{d}S=\oint\limits_C \boldsymbol{\tau}^{\mathrm{T}}\cdot\mathrm{d}\boldsymbol{r}$$

Representation. Two theorems give information about how to represent vectors that obey certain properties.

Theorem [48, p. 422]. The necessary and sufficient condition that the curl of a vector vanish identically is that the vector be the gradient of some function.

Theorem [48, p. 423]. The necessary and sufficient condition that the divergence of a vector vanish identically is that the vector is the curl of some other vector.

Leibniz Formula. In fluid mechanics and transport phenomena, an important result is the derivative of an integral whose limits of integration are moving. Suppose the region $V(t)$ is moving with velocity $\boldsymbol{v}_s$. Then Leibniz's rule holds:

$$\frac{\mathrm{d}}{\mathrm{d}t}\int\!\!\int\!\!\int\limits_{V(t)} \varphi\mathrm{d}V=\int\!\!\int\!\!\int\limits_{V(t)} \frac{\partial\varphi}{\partial t}\mathrm{d}V+\int\!\!\int\limits_S \varphi\boldsymbol{v}_s\cdot\boldsymbol{n}\mathrm{d}S$$

Curvilinear Coordinates. Many of the relations given above are proved most easily by using tensor analysis rather than dyadics. Once proven, however, the relations are perfectly general in any coordinate system. Displayed here are the specific results for cylindrical and spherical geometries. Results are available for a few other geometries: parabolic cylindrical, paraboloidal, elliptic cylindrical, prolate spheroidal, oblate spheroidal, ellipsoidal, and bipolar coordinates [45, 50].

For *cylindrical coordinates*, the geometry is shown in Figure 16. The coordinates are related to cartesian coordinates by

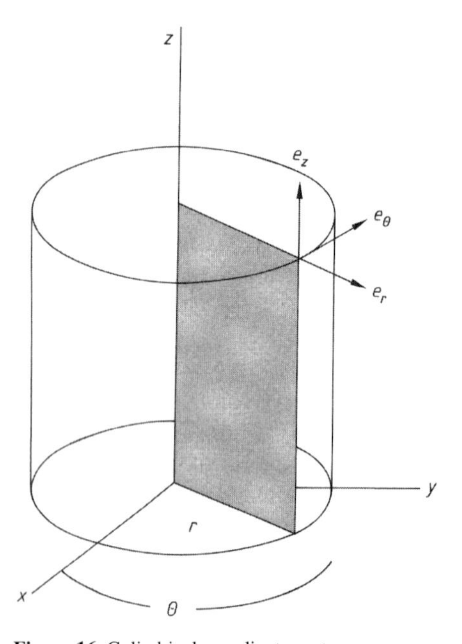

Figure 16. Cylindrical coordinate system

$$x = r\cos\theta \qquad r = \sqrt{x^2+y^2}$$
$$y = r\sin\theta \qquad \theta = \arctan\left(\frac{y}{x}\right)$$
$$z = z \qquad z = z$$

The unit vectors are related by

$$\boldsymbol{e}_r = \cos\theta\boldsymbol{e}_x + \sin\theta\boldsymbol{e}_y \qquad \boldsymbol{e}_x = \cos\theta\boldsymbol{e}_r - \sin\theta\boldsymbol{e}_\theta$$
$$\boldsymbol{e}_\theta = -\sin\theta\boldsymbol{e}_x + \cos\theta\boldsymbol{e}_y \qquad \boldsymbol{e}_y = \sin\theta\boldsymbol{e}_r + \cos\theta\boldsymbol{e}_\theta$$
$$\boldsymbol{e}_z = \boldsymbol{e}_z \qquad \boldsymbol{e}_z = \boldsymbol{e}_z$$

Derivatives of the unit vectors are

$$\mathrm{d}\boldsymbol{e}_\theta = -\boldsymbol{e}_r\mathrm{d}\theta, \mathrm{d}\boldsymbol{e}_r = \boldsymbol{e}_\theta\mathrm{d}\theta, \mathrm{d}\boldsymbol{e}_z = 0$$

Differential operators are given by [45]

$$\nabla = \boldsymbol{e}_r\frac{\partial}{\partial r}+\frac{\boldsymbol{e}_\theta}{r}\frac{\partial}{\partial\theta}+\boldsymbol{e}_z\frac{\partial}{\partial z},$$
$$\nabla\varphi = \boldsymbol{e}_r\frac{\partial\varphi}{\partial r}+\frac{\boldsymbol{e}_\theta}{r}\frac{\partial\varphi}{\partial\theta}+\boldsymbol{e}_z\frac{\partial\varphi}{\partial z}$$
$$\nabla^2\varphi = \frac{1}{r}\frac{\partial}{\partial r}\left(r\frac{\partial\varphi}{\partial r}\right)+\frac{1}{r^2}\frac{\partial^2\varphi}{\partial\theta^2}+\frac{\partial^2\varphi}{\partial z^2}$$
$$= \frac{\partial^2\varphi}{\partial r^2}+\frac{1}{r}\frac{\partial\varphi}{\partial r}+\frac{1}{r^2}\frac{\partial^2\varphi}{\partial\theta^2}+\frac{\partial^2\varphi}{\partial z^2}$$
$$\nabla\cdot\boldsymbol{v} = \frac{1}{r}\frac{\partial}{\partial r}(r\,v_r)+\frac{1}{r}\frac{\partial v_\theta}{\partial\theta}+\frac{\partial v_z}{\partial z}$$
$$\nabla\times\boldsymbol{v} = \boldsymbol{e}_r\left(\frac{1}{r}\frac{\partial v_z}{\partial\theta}-\frac{\partial v_\theta}{\partial z}\right)+\boldsymbol{e}_\theta\left(\frac{\partial v_r}{\partial z}-\frac{\partial v_z}{\partial r}\right)$$
$$+\boldsymbol{e}_z\left[\frac{1}{r}\frac{\partial}{\partial r}(r\,v_\theta)-\frac{1}{r}\frac{\partial v_r}{\partial\theta}\right]$$
$$\nabla\cdot\boldsymbol{\tau} = \boldsymbol{e}_r\left[\frac{1}{r}\frac{\partial}{\partial r}(r\tau_{rr})+\frac{1}{r}\frac{\partial\tau_{\theta r}}{\partial\theta}+\frac{\partial\tau_{zr}}{\partial z}-\frac{\tau_{\theta\theta}}{r}\right]+$$
$$+\boldsymbol{e}_\theta\left[\frac{1}{r^2}\frac{\partial}{\partial r}(r^2\tau_{r\theta})+\frac{1}{r}\frac{\partial\tau_{\theta\theta}}{\partial\theta}+\frac{\partial\tau_{z\theta}}{\partial z}+\frac{\tau_{\theta r}-\tau_{r\theta}}{r}\right]+$$
$$+\boldsymbol{e}_z\left[\frac{1}{r}\frac{\partial}{\partial r}(r\tau_{rz})+\frac{1}{r}\frac{\partial\tau_{\theta z}}{\partial\theta}+\frac{\partial\tau_{zz}}{\partial z}\right]$$
$$\nabla\boldsymbol{v} = \boldsymbol{e}_r\boldsymbol{e}_r\frac{\partial v_r}{\partial r}+\boldsymbol{e}_r\boldsymbol{e}_\theta\frac{\partial v_\theta}{\partial r}+\boldsymbol{e}_r\boldsymbol{e}_z\frac{\partial v_z}{\partial r}+$$
$$+\boldsymbol{e}_\theta\boldsymbol{e}_r\left(\frac{1}{r}\frac{\partial v_r}{\partial\theta}-\frac{v_\theta}{r}\right)+\boldsymbol{e}_\theta\boldsymbol{e}_\theta\left(\frac{1}{r}\frac{\partial v_\theta}{\partial\theta}+\frac{v_r}{r}\right)+$$
$$+\boldsymbol{e}_\theta\boldsymbol{e}_z\frac{1}{r}\frac{\partial v_z}{\partial\theta}+\boldsymbol{e}_z\boldsymbol{e}_r\frac{\partial v_r}{\partial z}+\boldsymbol{e}_z\boldsymbol{e}_\theta\frac{\partial v_\theta}{\partial z}+\boldsymbol{e}_z\boldsymbol{e}_z\frac{\partial v_z}{\partial z}$$
$$\nabla^2\boldsymbol{v} = \boldsymbol{e}_r\left[\frac{\partial}{\partial r}\left(\frac{1}{r}\frac{\partial}{\partial r}(r\,v_r)\right)+\frac{1}{r^2}\frac{\partial^2 v_r}{\partial\theta^2}+\frac{\partial^2 v_r}{\partial z^2}\right.$$
$$\left.-\frac{2}{r^2}\frac{\partial v_\theta}{\partial\theta}\right]+$$
$$+\boldsymbol{e}_\theta\left[\frac{\partial}{\partial r}\left(\frac{1}{r}\frac{\partial}{\partial r}(r\,v_\theta)\right)+\frac{1}{r^2}\frac{\partial^2 v_\theta}{\partial\theta^2}+\frac{\partial^2 v_\theta}{\partial z^2}+\frac{2}{r^2}\frac{\partial v_r}{\partial\theta}\right]+$$
$$+\boldsymbol{e}_z\left[\frac{1}{r}\frac{\partial}{\partial r}\left(r\frac{\partial v_z}{\partial r}\right)+\frac{1}{r^2}\frac{\partial^2 v_z}{\partial\theta^2}+\frac{\partial^2 v_z}{\partial z^2}\right]$$

For *spherical coordinates*, the geometry is shown in Figure 17. The coordinates are related to cartesian coordinates by

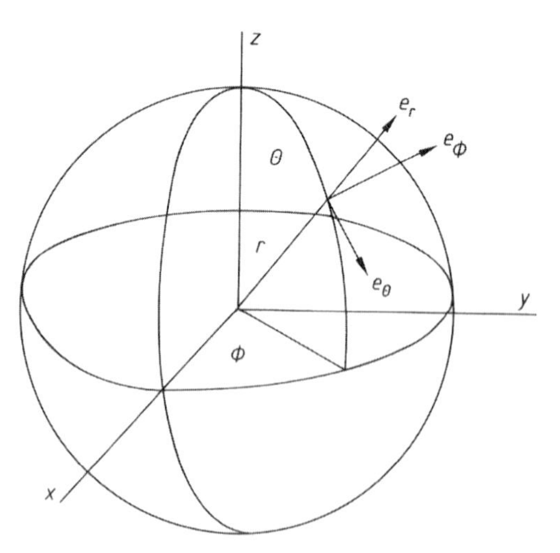

Figure 17. Spherical coordinate system

$$x = r\sin\theta\cos\varphi \qquad r = \sqrt{x^2+y^2+z^2}$$

$$y = r\sin\theta\sin\varphi \qquad \theta = \arctan\left(\sqrt{x^2+y^2}/z\right)$$

$$z = r\cos\theta \qquad \varphi = \arctan\left(\frac{y}{x}\right)$$

The unit vectors are related by

$$\boldsymbol{e}_r = \sin\theta\cos\varphi\boldsymbol{e}_x + \sin\theta\sin\varphi\boldsymbol{e}_y + \cos\theta\boldsymbol{e}_z$$

$$\boldsymbol{e}_\theta = \cos\theta\cos\varphi\boldsymbol{e}_x + \cos\theta\sin\varphi\boldsymbol{e}_y - \sin\theta\boldsymbol{e}_z$$

$$\boldsymbol{e}_\phi = -\sin\varphi\boldsymbol{e}_x + \cos\varphi\boldsymbol{e}_y$$

$$\boldsymbol{e}_x = \sin\theta\cos\varphi\boldsymbol{e}_r + \cos\theta\cos\varphi\boldsymbol{e}_\theta - \sin\varphi\boldsymbol{e}_\phi$$

$$\boldsymbol{e}_y = \sin\theta\sin\varphi\boldsymbol{e}_r + \cos\theta\sin\varphi\boldsymbol{e}_\theta + \cos\varphi\boldsymbol{e}_\phi$$

$$\boldsymbol{e}_z = \cos\theta\boldsymbol{e}_r - \sin\theta\boldsymbol{e}_\theta$$

Derivatives of the unit vectors are

$$\frac{\partial\boldsymbol{e}_r}{\partial\theta} = \boldsymbol{e}_\theta, \ \frac{\partial\boldsymbol{e}_\theta}{\partial\theta} = -\boldsymbol{e}_r$$

$$\frac{\partial\boldsymbol{e}_r}{\partial\varphi} = e_\phi\sin\theta, \ \frac{\partial\boldsymbol{e}_\theta}{\partial\varphi} = \boldsymbol{e}_\phi\cos\theta,$$

$$\frac{\partial\boldsymbol{e}_\phi}{\partial\varphi} = -\boldsymbol{e}_r\sin\theta - \boldsymbol{e}_\theta\cos\theta$$

Others 0

Differential operators are given by [45]

$$\nabla = \boldsymbol{e}_r\frac{\partial}{\partial r} + \boldsymbol{e}_\theta\frac{1}{r}\frac{\partial}{\partial\theta} + \boldsymbol{e}_\phi\frac{1}{r\sin\theta}\frac{\partial}{\partial\varphi},$$

$$\nabla\psi = \boldsymbol{e}_r\frac{\partial\psi}{\partial r} + \boldsymbol{e}_\theta\frac{1}{r}\frac{\partial\psi}{\partial\theta} + \boldsymbol{e}_\phi\frac{1}{r\sin\theta}\frac{\partial\psi}{\partial\varphi}$$

$$\nabla^2\psi = \frac{1}{r^2}\frac{\partial}{\partial r}\left(r^2\frac{\partial\psi}{\partial r}\right) + \frac{1}{r^2\sin\theta}\frac{\partial}{\partial\theta}\left(\sin\theta\frac{\partial\psi}{\partial\theta}\right) +$$

$$+\frac{1}{r^2\sin^2\theta}\frac{\partial^2\psi}{\partial\varphi^2}$$

$$\nabla\cdot\boldsymbol{v} = \frac{1}{r^2}\frac{\partial}{\partial r}\left(r^2 v_r\right) + \frac{1}{r\sin\theta}\frac{\partial}{\partial\theta}\left(v_\theta\sin\theta\right) +$$

$$+\frac{1}{r\sin\theta}\frac{\partial v_\phi}{\partial\varphi}$$

$$\nabla\times\boldsymbol{v} = \boldsymbol{e}_r\left[\frac{1}{r\sin\theta}\frac{\partial}{\partial\theta}\left(v_\phi\sin\theta\right) - \frac{1}{r\sin\theta}\frac{\partial v_\theta}{\partial\varphi}\right] +$$

$$+\boldsymbol{e}_\theta\left[\frac{1}{r\sin\theta}\frac{\partial v_r}{\partial\varphi} - \frac{1}{r}\frac{\partial}{\partial r}\left(r\,v_\phi\right)\right] +$$

$$+\boldsymbol{e}_\phi\left[\frac{1}{r}\frac{\partial}{\partial r}\left(r\,v_\theta\right) - \frac{1}{r}\frac{\partial v_r}{\partial\theta}\right]$$

$$\nabla\cdot\boldsymbol{\tau} = \boldsymbol{e}_r\left[\frac{1}{r^2}\frac{\partial}{\partial r}\left(r^2\tau_{rr}\right)\right.$$

$$\left.+\frac{1}{r\sin\theta}\frac{\partial}{\partial\theta}\left(\tau_{\theta r}\sin\theta\right) + \frac{1}{r\sin\theta}\frac{\partial\tau_{\phi r}}{\partial\varphi} - \frac{\tau_{\theta\theta}+\tau_{\phi\phi}}{r}\right] +$$

$$+\boldsymbol{e}_\theta\left[\frac{1}{r^3}\frac{\partial}{\partial r}\left(r^3\tau_{r\theta}\right) + \frac{1}{r\sin\theta}\frac{\partial}{\partial\theta}\left(\tau_{\theta\theta}\sin\theta\right) +\right.$$

$$\left.\frac{1}{r\sin\theta}\frac{\partial\tau_{\phi\theta}}{\partial\varphi} + \frac{\tau_{\theta r}-\tau_{r\theta}-\tau_{\phi\phi}\cot\theta}{r}\right] +$$

$$+\boldsymbol{e}_\phi\left[\frac{1}{r^3}\frac{\partial}{\partial r}\left(r^3\tau_{r\phi}\right) + \frac{1}{r\sin\theta}\frac{\partial}{\partial\theta}\left(\tau_{\theta\phi}\sin\theta\right) +\right.$$

$$\left.\frac{1}{r\sin\theta}\frac{\partial\tau_{\phi\phi}}{\partial\varphi} + \frac{\tau_{\phi r}-\tau_{r\phi}+\tau_{\phi\theta}\cot\theta}{r}\right]$$

$$\nabla\boldsymbol{v} = \boldsymbol{e}_r\boldsymbol{e}_r\frac{\partial v_r}{\partial r} + \boldsymbol{e}_r\boldsymbol{e}_\theta\frac{\partial v_\theta}{\partial r} + \boldsymbol{e}_r\boldsymbol{e}_\phi\frac{\partial v_\phi}{\partial r} +$$

$$+\boldsymbol{e}_\theta\boldsymbol{e}_r\left(\frac{1}{r}\frac{\partial v_r}{\partial\theta} - \frac{v_\theta}{r}\right) + \boldsymbol{e}_\theta\boldsymbol{e}_\theta\left(\frac{1}{r}\frac{\partial v_\theta}{\partial\theta} + \frac{v_r}{r}\right) +$$

$$+\boldsymbol{e}_\theta\boldsymbol{e}_\phi\frac{1}{r}\frac{\partial v_\phi}{\partial\theta} + \boldsymbol{e}_\phi\boldsymbol{e}_r\left(\frac{1}{r\sin\theta}\frac{\partial v_r}{\partial\varphi} - \frac{v_\phi}{r}\right) +$$

$$+\boldsymbol{e}_\phi\boldsymbol{e}_\theta\left(\frac{1}{r\sin\theta}\frac{\partial v_\theta}{\partial\varphi} - \frac{v_\phi}{r}\cot\theta\right) +$$

$$+\boldsymbol{e}_\phi\boldsymbol{e}_\phi\left(\frac{1}{r\sin\theta}\frac{\partial v_\phi}{\partial\varphi} + \frac{v_r}{r} + \frac{v_\theta}{r}\cot\theta\right)$$

$$\nabla^2\boldsymbol{v} = \boldsymbol{e}_r\left[\frac{\partial}{\partial r}\left(\frac{1}{r^2}\frac{\partial}{\partial r}\left(r^2 v_r\right)\right)\right.$$

$$+\frac{1}{r^2\sin\theta}\frac{\partial}{\partial\theta}\left(\sin\theta\frac{\partial v_r}{\partial\theta}\right)$$

$$\left.+\frac{1}{r^2\sin^2\theta}\frac{\partial^2 v_r}{\partial\varphi^2} - \frac{2}{r^2\sin\theta}\frac{\partial}{\partial\theta}\left(v_\theta\sin\theta\right) - \frac{2}{r^2\sin\theta}\frac{\partial v_\phi}{\partial\varphi}\right]$$

$$+\boldsymbol{e}_\theta\left[\frac{1}{r^2}\frac{\partial}{\partial r}\left(r^2\frac{\partial v_\theta}{\partial r}\right) + \frac{1}{r^2}\frac{\partial}{\partial\theta}\left(\frac{1}{\sin\theta}\frac{\partial}{\partial\theta}\left(v_\theta\sin\theta\right)\right)\right.$$

$$\left.+\frac{1}{r^2\sin^2\theta}\frac{\partial^2 v_\theta}{\partial\varphi^2} + \frac{2}{r^2}\frac{\partial v_r}{\partial\theta} - \frac{2}{r^2}\frac{\cot\theta}{\sin\theta}\frac{\partial v_\phi}{\partial\varphi}\right]$$

$$+\boldsymbol{e}_\phi\left[\frac{1}{r^2}\frac{\partial}{\partial r}\left(r^2\frac{\partial v_\phi}{\partial r}\right) + \frac{1}{r^2}\frac{\partial}{\partial\theta}\left(\frac{1}{\sin\theta}\frac{\partial}{\partial\theta}\left(v_\phi\sin\theta\right)\right)\right.$$

$$\left.+\frac{1}{r^2\sin^2\theta}\frac{\partial^2 v_\phi}{\partial\varphi^2} + \frac{2}{r^2\sin\theta}\frac{\partial v_r}{\partial\varphi} + \frac{2}{r^2}\frac{\cot\theta}{\sin\theta}\frac{\partial v_\theta}{\partial\varphi}\right]$$

6. Ordinary Differential Equations as Initial Value Problems

A differential equation for a function that depends on only one variable (often time) is called an ordinary differential equation. The general solution to the differential equation includes many possibilities; the boundary or initial conditions are required to specify which of those are desired. If all conditions are at one point, the problem is an initial value problem and can be integrated from that point on. If some of the conditions are available at one point and others at another point, the ordinary differential equations become two-point boundary value problems, which are treated in Chapter 7. Initial value problems as ordinary differential equations arise in control of lumped-parameter models, transient models of stirred tank reactors, polymerization reactions and plug-flow reactors, and generally in models where no spatial gradients occur in the unknowns.

6.1. Solution by Quadrature

When only one equation exists, even if it is nonlinear, solving it by quadrature may be possible. For

$$\frac{dy}{dt} = f(y)$$

$$y(0) = y_0$$

the problem can be separated

$$\frac{dy}{f(y)} = dt$$

and integrated:

$$\int_{y_0}^{y} \frac{dy'}{f(y')} = \int_0^t dt = t$$

If the quadrature can be performed analytically then the exact solution has been found.

For example, consider the kinetics problem with a second-order reaction.

$$\frac{dc}{dt} = -kc^2, c(0) = c_0$$

To find the function of the concentration versus time, the variables can be separated and integrated.

$$\frac{dc}{c^2} = -kdt,$$

$$-\frac{1}{c} = -kt+D$$

Application of the initial conditions gives the solution:

$$\frac{1}{c} = kt+\frac{1}{c_0}$$

For other ordinary differential equations an integrating factor is useful. Consider the problem governing a stirred tank with entering fluid having concentration c_{in} and flow rate F, as shown in Figure 18. The flow rate out is also F and the volume of the tank is V. If the tank is completely mixed, the concentration in the tank is c and the concentration of the fluid leaving the tank is also c. The differential equation is then

$$V\frac{dc}{dt} = F(c_{in}-c), \; c(0) = c_0$$

Upon rearrangement,

$$\frac{dc}{dt}+\frac{F}{V}c = \frac{F}{V}c_{in}$$

is obtained. An integrating factor is used to solve this equation. The integrating factor is a function that can be used to turn the left-hand side into an exact differential and can be found by using Fréchet differentials [51]. In this case,

$$\exp\left(\frac{F\,t}{V}\right)\left[\frac{dc}{dt}+\frac{F}{V}c\right] = \frac{d}{dt}\left[\exp\left(\frac{F\,t}{V}\right)c\right]$$

Thus, the differential equation can be written as

$$\frac{d}{dt}\left[\exp\left(\frac{F\,t}{V}\right)c\right] = \exp\left(\frac{F\,t}{V}\right)\left[\frac{F}{V}c_{in}\right]$$

This can be integrated once to give

$$\exp\left(\frac{F\,t}{V}\right)c = c(0)+\frac{F}{V}\int_0^t \exp\left(\frac{F\,t'}{V}\right)c_{in}(t')\,dt'$$

or

$$c(t) = \exp\left(-\frac{F\,t}{V}\right)c_0$$

$$+\frac{F}{V}\int_0^t \exp\left(-\frac{F\,(t-t')}{V}\right)c_{in}(t')\,dt'$$

If the integral on the right-hand side can be calculated, the solution can be obtained analytically. If not, the numerical methods described in the next sections can be used. Laplace transforms can also be attempted. However, an analytic solution is so useful that quadrature and an integrating factor should be tried before resorting to numerical solution.

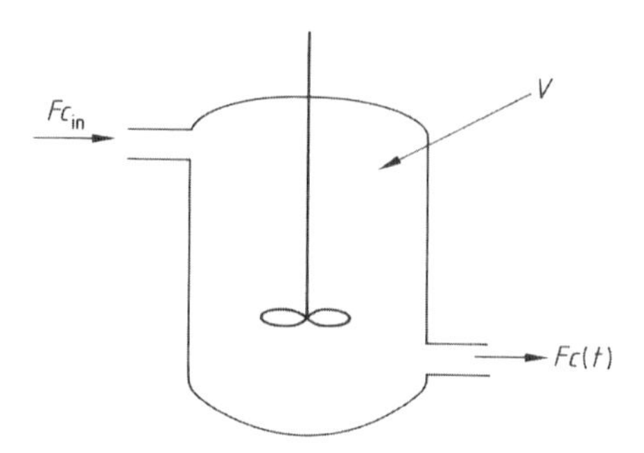

Figure 18. Stirred tank

6.2. Explicit Methods

Consider the ordinary differential equation

$$\frac{dy}{dt} = f(y)$$

Multiple equations that are still initial value problems can be handled by using the same techniques discussed here. A higher order differential equation

$$y^{(n)}+F\left(y^{(n-1)},y^{(n-2)},\ldots,y',y\right)=0$$

with initial conditions

$$G_i\left(y^{(n-1)}(0),y^{(n-2)}(0),\ldots,y'(0),y(0)\right)=0$$

$$i=1,\ldots,n$$

can be converted into a set of first-order equations. By using

$$y_i \equiv y^{(i-1)}=\frac{\mathrm{d}^{(i-1)}y}{\mathrm{d}t^{(i-1)}}=\frac{\mathrm{d}}{\mathrm{d}t}\,y^{(i-2)}=\frac{\mathrm{d}y_{i-1}}{\mathrm{d}t}$$

the higher order equation can be written as a set of first-order equations:

$$\frac{\mathrm{d}y_1}{\mathrm{d}t}=y_2$$

$$\frac{\mathrm{d}y_2}{\mathrm{d}t}=y_3$$

$$\frac{\mathrm{d}y_3}{\mathrm{d}t}=y_4$$

$$\ldots$$

$$\frac{\mathrm{d}y_n}{\mathrm{d}t}=-F\left(y_{n-1},y_{n-2},\ldots,y_2,y_1\right)$$

The initial conditions would have to be specified for variables $y_1(0), \ldots, y_n(0)$, or equivalently $y(0), \ldots, y^{(n-1)}(0)$. The set of equations is then written as

$$\frac{\mathrm{d}\boldsymbol{y}}{\mathrm{d}t}=\boldsymbol{f}\left(\boldsymbol{y},t\right)$$

All the methods in this chapter are described for a single equation; the methods apply to the multiple equations as well. Taking the single equation in the form

$$\frac{\mathrm{d}y}{\mathrm{d}t}=f\left(y\right)$$

multiplying by dt, and integrating once yields

$$\int_{t_n}^{t_{n+1}}\frac{\mathrm{d}y}{\mathrm{d}t'}\,\mathrm{d}t'=\sum_{t_n}^{t_{n+1}}f\left(y\left(t'\right)\right)\mathrm{d}t'$$

This is

$$y^{n+1}=y^n+\int_{t_n}^{t_{n+1}}\frac{\mathrm{d}y}{\mathrm{d}t'}\,\mathrm{d}t'$$

The last substitution gives a basis for the various methods. Different interpolation schemes for $y(t)$ provide different integration schemes; using low-order interpolation gives low-order integration schemes [3].

Euler's method is first order

$$y^{n+1}=y^n+\Delta t f\left(y^n\right)$$

Adams – Bashforth Methods. The *second-order* Adams – Bashforth method is

$$y^{n+1}=y^n+\frac{\Delta t}{2}\left[3f\left(y^n\right)-f\left(y^{n-1}\right)\right]$$

The *fourth-order* Adams – Bashforth method is

$$y^{n+1}=y^n+\frac{\Delta t}{24}\left[55f\left(y^n\right)-59f\left(y^{n-1}\right)\right.$$
$$\left.+37f\left(y^{n-2}\right)-9f\left(y^{n-3}\right)\right]$$

Notice that the higher order explicit methods require knowing the solution (or the right-hand side) evaluated at times in the past. Because these were calculated to get to the current time, this presents no problem except for starting the evaluation. Then, Euler's method may have to be used with a very small step size for several steps to generate starting values at a succession of time points. The error terms, order of the method, function evaluations per step, and stability limitations are listed in Table 6. The advantage of the fourth-order Adams – Bashforth method is that it uses only one function evaluation per step and yet achieves high-order accuracy. The disadvantage is the necessity of using another method to start.

Runge – Kutta Methods. Runge – Kutta methods are explicit methods that use several function evaluations for each time step. The general form of the methods is

$$y^{n+1}=y^n+\sum_{i=1}^{v}w_i\,k_i$$

with

$$k_i=\Delta t\,f\left(t^n+c_i\Delta t,y^n+\sum_{j=1}^{i-1}a_{ij}\,k_j\right)$$

Runge – Kutta methods traditionally have been writen for $f(t, y)$ and that is done here, too. If these equations are expanded and compared with a Taylor series, restrictions can be placed on the parameters of the method to make it first order, second order, etc. Even so, additional parameters can be chosen. A *second-order* Runge – Kutta method is

$$y^{n+1}=y^n+\frac{\Delta t}{2}\left[f^n+f\left(t^n+\Delta t,y^n+\Delta t\,f^n\right)\right]$$

The midpoint scheme is another second-order Runge – Kutta method:

Table 6. Properties of integration methods for ordinary differential equations

Method	Error term	Order	Function evaluations per step	Stability limit, $\lambda\,\Delta t \leq$
Explicit methods				
Euler	$\frac{h^2}{2} y''$	1	1	2.0
Second-order Adams – Bashforth	$\frac{5}{12} h^3 y'''$	2	1	
Fourth-order Adams – Bashforth	$\frac{251}{720} h^5 y^{(5)}$	4	1	0.3
Second-order Runge – Kutta (midpoint)		2	2	2.0
Runge – Kutta – Gill		4	4	2.8
Runge – Kutta – Feldberg	$y^{n+1} - z^{n+1}$	5	6	3.0
Predictor – corrector methods				
Second-order Runge – Kutta		2	2	2.0
Adams, fourth-order		2	2	1.3
Implicit methods, stability limit ∞				
Backward Euler		1	many, iterative	∞ *
Trapezoid rule	$-\frac{1}{12} h^3 y'''$	2	many, iterative	2 *
Fourth-order Adams – Moulton		4	many, iterative	3 *

* Oscillation limit, $\lambda\,\Delta t \leq$.

$$y^{n+1} = y^n + \Delta t\, f\left(t^n + \frac{\Delta t}{2}, y^n + \frac{\Delta t}{2} f^n\right)$$

A popular *fourth-order* method is the Runge – Kutta – Gill method with the formulas

$$k_1 = \Delta t\, f\,(t^n, y^n)$$

$$k_2 = \Delta t\, f\left(t^n + \tfrac{\Delta t}{2}, y^n + \tfrac{k_1}{2}\right)$$

$$k_3 = \Delta t\, f\left(t^n + \tfrac{\Delta t}{2}, y^n + a\,k_1 + b\,k_2\right)$$

$$k_4 = \Delta t\, f\,(t^n + \Delta t, y^n + c\,k_2 + d\,k_3)$$

$$y^{n+1} = y^n + \tfrac{1}{6}(k_1 + k_4) + \tfrac{1}{3}(b\,k_2 + d\,k_3)$$

$$a = \tfrac{\sqrt{2}-1}{2},\; b = \tfrac{2-\sqrt{2}}{2},$$

$$c = -\tfrac{\sqrt{2}}{2},\; d = 1 + \tfrac{\sqrt{2}}{2}$$

Another fourth-order Runge – Kutta method is given by the Runge – Kutta – Feldberg formulas [52]; although the method is fourth-order, it achieves fifth-order accuracy. The popular integration package RKF 45 is based on this method.

$$k_1 = \Delta t\, f\,(t^n, y^n)$$

$$k_2 = \Delta t\, f\left(t^n + \tfrac{\Delta t}{4}, y^n + \tfrac{k_1}{4}\right)$$

$$k_3 = \Delta t\, f\,\left(t^n + \tfrac{3}{8}\Delta t, y^n + \tfrac{3}{32} k_1 + \tfrac{9}{32} k_2\right)$$

$$k_4 = \Delta t f\,\left(t^n + \tfrac{12}{13}\Delta t, y^n + \tfrac{1932}{2197} k_1 - \tfrac{7200}{2197} k_2 + \tfrac{7296}{2197} k_3\right)$$

$$k_5 = \Delta t f\,\left(t^n + \Delta t, y^n + \tfrac{439}{216} k_1 - 8\,k_2 + \tfrac{3680}{513} k_3 - \tfrac{845}{4104} k_4\right)$$

$$k_6 = \Delta t\, f\,\left(t^n + \tfrac{\Delta t}{2}, y^n - \tfrac{8}{27} k_1 + 2k_2 - \tfrac{3544}{2565} k_3 + \tfrac{1859}{4104} k_4 - \tfrac{11}{40} k_5\right)$$

$$y^{n+1} = y^n + \tfrac{25}{216} k_1 + \tfrac{1408}{2565} k_3 + \tfrac{2197}{4104} k_4 - \tfrac{1}{5} k_5$$

$$z^{n+1} = y^n + \tfrac{16}{135} k_1 + \tfrac{6656}{12\,825} k_3 + \tfrac{28\,561}{56\,430} k_4 - \tfrac{9}{50} k_5 + \tfrac{2}{55} k_6$$

The value of $y^{n+1} - z^{n+1}$ is an estimate of the error in y^{n+1} and can be used in step-size control schemes.

Generally, a high-order method should be used to achieve high accuracy. The Runge – Kutta – Gill method is popular because it is high order and does not require a starting method (as does the fourth-order Adams – Bashforth method). However, it requires four function evaluations per time step, or four times as many as the Adams – Bashforth method. For problems in which the function evaluations are a significant portion of the calculation time this might be important. Given the speed of computers and the widespread availability of desktop computers, the efficiency of a method is most important only for very large problems that are going to be solved many times. For other problems the most important criterion for choosing a method is probably the time the user spends setting up the problem.

The stability of an integration method is best estimated by determining the rational polynomial corresponding to the method. Apply this method to the equation

$$\frac{dy}{dt} = -\lambda y, y(0) = 1$$

and determine the formula for r_{mn}:

$$y^{k+1} = r_{mn}(\lambda\,\Delta t)\,y^k$$

The rational polynomial is defined as

$$r_{mn}(z) = \frac{p_n(z)}{q_m(z)} \approx e^{-z}$$

and is an approximation to exp (− z), called a Padé approximation. The stability limits are the largest positive z for which

$$|r_{mn}(z)| \leq 1$$

The method is A acceptable if the inequality holds for $Re\ z > 0$. It is A (0) acceptable if the inequality holds for z real, $z > 0$ [53]. The method will not induce oscillations about the true solution provided

$$r_{mn}(z) > 0$$

A method is L acceptable if it is A acceptable and

$$\lim_{z\to\infty} r_{mn}(z) = 0$$

For example, Euler's method gives

$$y^{n+1} = y^n - \lambda\,\Delta t y^n \text{ or } y^{n+1} = (1-\lambda\,\Delta t)\,y^n$$

$$\text{or } r_{mn} = 1-\lambda\,\Delta t$$

The stability limit is then

$$\lambda\,\Delta t \leq 2$$

The Euler method will not oscillate provided

$$\lambda\,\Delta t \leq 1$$

The stability limits listed in Table 6 are obtained in this fashion. The limit for the Euler method is 2.0; for the Runge – Kutta – Gill method it is 2.785; for the Runge – Kutta – Feldberg method it is 3.020. The rational polynomials for the various explicit methods are illustrated in Figure 19. As can be seen, the methods approximate the exact solution well as $\lambda\,\Delta t$ approaches zero, and the higher order methods give a better approximation at high values of $\lambda\,\Delta t$.

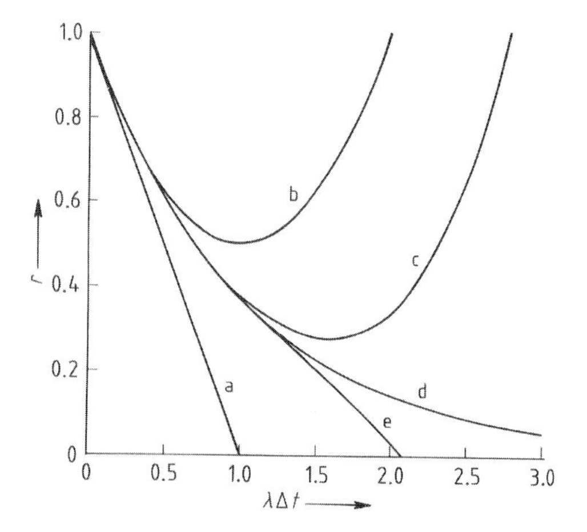

Figure 19. Rational approximations for explicit methods
a) Euler; b) Runge – Kutta – 2; c) Runge – Kutta – Gill; d) Exact curve; e) Runge – Kutta – Feldberg

In solving sets of equations

$$\frac{dy}{dt} = \boldsymbol{A}\,\boldsymbol{y} + \boldsymbol{f}, \boldsymbol{y}(0) = y_0$$

all the eigenvalues of the matrix $\boldsymbol{A}$ must be examined. FINLAYSON [3] and AMUNDSON [54, p. 197 – 199] both show how to transform these equations into an orthogonal form so that each equation becomes one equation in one unknown, for which single equation analysis applies. For linear problems the eigenvalues do not change, so the stability and oscillation limits must be satisfied for every eigenvalue of the matrix $\boldsymbol{A}$. When solving nonlinear problems the equations are linearized about the solution at the local time, and the analysis applies for small changes in time, after which a new analysis about the new solution must be made. Thus, for nonlinear problems the eigenvalues keep changing.

Richardson extrapolation can be used to improve the accuracy of a method. Step forward one step Δt with a p-th order method. Then redo the problem, this time stepping forward from the same initial point but in two steps of length $\Delta t/2$, thus ending at the same point. Call the solution of the one-step calculation y_1 and the solution of the two-step calculation y_2. Then an improved solution at the new time is given by

$$y = \frac{2^p\,y_2 - y_1}{2^p - 1}$$

This gives a good estimate provided Δt is small enough that the method is truly convergent with order p. This process can also be repeated in the same way Romberg's method was used for quadrature (see Section 2.4).

The accuracy of a numerical calculation depends on the step size used, and this is chosen automatically by efficient codes. For example, in the Euler method the local truncation error LTE is

$$\text{LTE} = \frac{\Delta t^2}{2} y_n''$$

Yet the second derivative can be evaluated by using the difference formulas as

$$y_n'' = \nabla (\Delta t \, y_n') = \Delta t \, (y_n' - y_{n-1}') =$$

$$\Delta t \, (f_n - f_{n-1})$$

Thus, by monitoring the difference between the right-hand side from one time step to another, an estimate of the truncation error is obtained. This error can be reduced by reducing Δt. If the user specifies a criterion for the largest local error estimate, then Δt is reduced to meet that criterion. Also, Δt is increased to as large a value as possible, because this shortens computation time. If the local truncation error has been achieved (and estimated) by using a step size Δt_1

$$\text{LTE} = c \, \Delta t_1^p$$

and the desired error is ε, to be achieved using a step size Δt_2

$$\varepsilon = c \, \Delta t_2^p$$

then the next step size Δt_2 is taken from

$$\frac{\text{LTE}}{\varepsilon} = \left(\frac{\Delta t_1}{\Delta t_2} \right)^p$$

Generally, things should not be changed too often or too drastically. Thus one may choose not to increase Δt by more than a factor (such as 2) or to increase Δt more than once every so many steps (such as 5) [55]. In the most sophisticated codes the alternative exists to change the order of the method as well. In this case, the truncation error of the orders one higher and one lower than the current one are estimated, and a choice is made depending on the expected step size and work.

6.3. Implicit Methods

By using different interpolation formulas, involving y^{n+1}, implicit integration methods can be derived. Implicit methods result in a nonlinear equation to be solved for y^{n+1} so that iterative methods must be used. The backward Euler method is a first-order method:

$$y^{n+1} = y^n + \Delta t \, f\left(y^{n+1}\right)$$

The trapezoid rule (see Section 2.4) is a second-order method:

$$y^{n+1} = y^n + \frac{\Delta t}{2} \left[f\left(y^n\right) + f\left(y^{n+1}\right) \right]$$

When the trapezoid rule is used with the finite difference method for solving partial differential equations it is called the Crank – Nicolson method. Adams methods exist as well, and the fourth-order Adams – Moulton method is

$$y^{n+1} = y^n + \frac{\Delta t}{24} \left[9 f\left(y^{n+1}\right) + 19 f\left(y^n\right) - 5 f\left(y^{n-1}\right) + f\left(y^{n-2}\right) \right]$$

The properties of these methods are given in Table 6. The implicit methods are stable for any step size but do require the solution of a set of nonlinear equations, which must be solved iteratively. An application to dynamic distillation problems is given in [56].

All these methods can be written in the form

$$y^{n+1} = \sum_{i=1}^{k} \alpha_i \, y^{n+1-i} + \Delta t \sum_{i=0}^{k} \beta_i \, f\left(y^{n+1-i}\right)$$

or

$$y^{n+1} = \Delta t \, \beta_0 \, f\left(y^{n+1}\right) + w^n$$

where w^n represents known information. This equation (or set of equations for more than one differential equation) can be solved by using successive substitution:

$$y^{n+1,\,k+1} = \Delta t \, \beta_0 f\left(y^{n+1,\,k}\right) + w^n$$

Here, the superscript k refers to an iteration counter. The successive substitution method is guaranteed to converge, provided the first derivative of the function is bounded and a small enough time step is chosen. Thus, if it has not converged within a few iterations, Δt can be reduced and the iterations begun again. The *Newton – Raphson method* (see Section 1.2) can also be used.

In many computer codes, iteration is allowed to proceed only a fixed number of times (e.g., three) before Δt is reduced. Because a good history of the function is available from previous time steps, a good initial guess is usually possible.

The best software packages for stiff equations (see Section 6.4) use Gear's backward difference formulas. The formulas of various orders are [57].

$$1: y^{n+1} = y^n + \Delta t\, f\,\left(y^{n+1}\right)$$

$$2: y^{n+1} = \frac{4}{3}\, y^n - \frac{1}{3}\, y^{n-1} + \frac{2}{3}\, \Delta t\, f\,\left(y^{n+1}\right)$$

$$3: y^{n+1} = \frac{18}{11}\, y^n - \frac{9}{11}\, y^{n-1} + \frac{2}{11}\, y^{n-2}$$

$$+\frac{6}{11}\, \Delta t\, f\,\left(y^{n+1}\right)$$

$$4: y^{n+1} = \frac{48}{25}\, y^n - \frac{36}{25}\, y^{n-1} + \frac{16}{25}\, y^{n-2} - \frac{3}{25}\, y^{n-3}$$

$$+\frac{12}{25}\, \Delta t\, f\,\left(y^{n+1}\right)$$

$$5: y^{n+1} = \frac{300}{137}\, y^n - \frac{300}{137}\, y^{n-1} + \frac{200}{137}\, y^{n-2}$$

$$-\frac{75}{137}\, y^{n-3} + \frac{12}{137}\, y^{n-4}$$

$$+\frac{60}{137}\, \Delta t\, f\,\left(y^{n+1}\right)$$

The stability properties of these methods are determined in the same way as explicit methods. They are always expected to be stable, no matter what the value of Δt is, and this is confirmed in Figure 20.

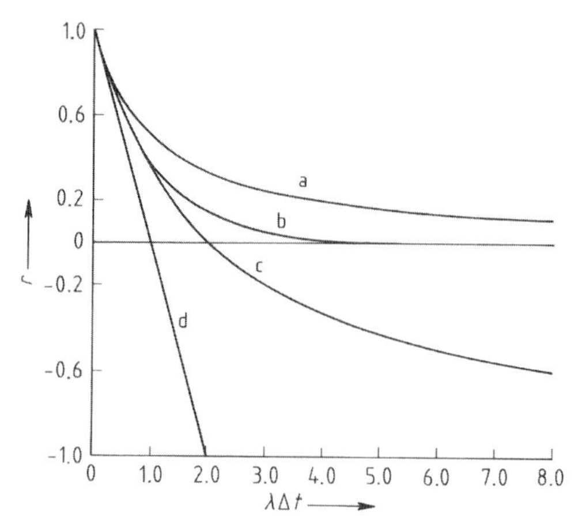

Figure 20. Rational approximations for implicit methods
a) Backward Euler; b) Exact curve; c) Trapezoid; d) Euler

Predictor – corrector methods can be employed in which an explicit method is used to predict the value of y^{n+1}. This value is then used in an implicit method to evaluate $f\,(\,y^{n+1})$.

6.4. Stiffness

Why is it desirable to use implicit methods that lead to sets of algebraic equations that must be solved iteratively whereas explicit methods lead to a direct calculation? The reason lies in the stability limits; to understand their impact, the concept of stiffness is necessary. When modeling a physical situation, the time constants governing different phenomena should be examined. Consider flow through a packed bed, as illustrated in Figure 21.

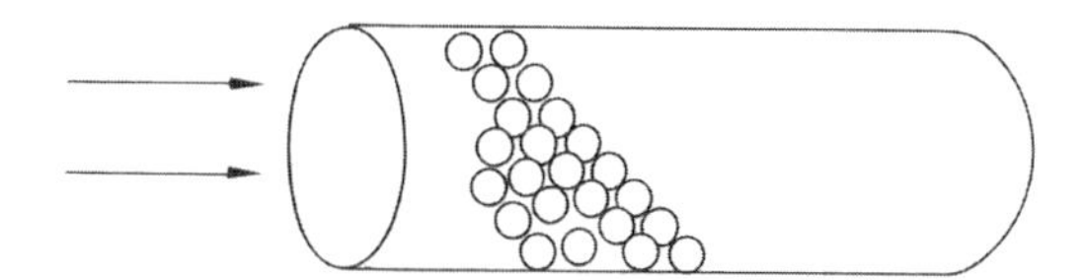

Figure 21. Flow through packed bed

The superficial velocity u is given by

$$u = \frac{Q}{A\,\varphi}$$

where Q is the volumetric flow rate, A is the cross-sectional area, and φ is the void fraction. A time constant for flow through the device is then

$$t_{\text{flow}} = \frac{L}{u} = \frac{\varphi\, A\, L}{Q}$$

where L is the length of the packed bed. If a chemical reaction occurs, with a reaction rate given by

$$\frac{\text{Moles}}{\text{Volume time}} = -k\, c$$

where k is the rate constant (time^{-1}) and c is the concentration (moles/volume), the characteristic time for the reaction is

$$t_{\text{rxn}} = \frac{1}{k}$$

If diffusion occurs inside the catalyst, the time constant is

$$t_{\text{internal diffusion}} = \frac{\varepsilon\, R^2}{D_{\text{e}}}$$

where ε is the porosity of the catalyst, R is the catalyst radius, and D_{e} is the effective diffusion coefficient inside the catalyst. The time constant for heat transfer is

$$t_{\text{internal heat transfer}} = \frac{R^2}{\alpha} = \frac{\varrho_s C_s R^2}{k_e}$$

where ϱ_s is the catalyst density, C_s is the catalyst heat capacity per unit mass, k_e is the effective thermal conductivity of the catalyst, and α is the thermal diffusivity. The time constants for diffusion of mass and heat through a boundary layer surrounding the catalyst are

$$t_{\text{external diffusion}} = \frac{R}{k_g}$$

$$t_{\text{external heat transfer}} = \frac{\varrho_s C_s R}{h_p}$$

where k_g and h_p are the mass-transfer and heat-transfer coefficients, respectively. The importance of examining these time constants comes from realization that their orders of magnitude differ greatly. For example, in the model of an automobile catalytic converter [58] the time constant for internal diffusion was 0.3 s, for internal heat transfer 21 s, and for flow through the device 0.003 s. Flow through the device is so fast that it might as well be instantaneous. Thus, the time derivatives could be dropped from the mass balance equations for the flow, leading to a set of differential-algebraic equations (see below). If the original equations had to be solved, the eigenvalues would be roughly proportional to the inverse of the time constants. The time interval over which to integrate would be a small number (e.g., five) multiplied by the longest time constant. Yet the explicit stability limitation applies to all the eigenvalues, and the largest eigenvalue would determine the largest permissible time step. Here, $\lambda = 1/0.003\ \text{s}^{-1}$. Very small time steps would have to be used, e.g., $\Delta t \leq 2 \times 0.003$ s, but a long integration would be required to reach steady state. Such problems are termed stiff, and implicit methods are very useful for them. In that case the stable time constant is not of any interest, because any time step is stable. What is of interest is the largest step for which a solution can be found. If a time step larger than the smallest time constant is used, then any phenomena represented by that smallest time constant will be overlooked—at least transients in it will be smeared over. However, the method will still be stable. Thus, if the very rapid transients of part of the model are not of interest, they can be ignored and an implicit method used [59].

The idea of stiffness is best explained by considering a system of linear equations:

$$\frac{d\boldsymbol{y}}{dt} = \boldsymbol{A}\,\boldsymbol{y}$$

Let λ_i be the eigenvalues of the matrix $\boldsymbol{A}$. This system can be converted into a system of n equations, each of them having only one unknown; the eigenvalues of the new system are the same as the eigenvalues of the original system [3, pp. 39 – 42], [54, pp. 197 – 199]. Then the stiffness ratio SR is defined as [53, p. 32]

$$\text{SR} = \frac{\max_i |\text{Re}\,(\lambda_i)|}{\min_i |\text{Re}\,(\lambda_i)|}$$

SR = 20 is not stiff, SR = 10^3 is stiff, and SR = 10^6 is very stiff. If the problem is nonlinear, the solution is expanded about the current state:

$$\frac{dy_i}{dt} = f_i\,[\boldsymbol{y}\,(t^n)] + \sum_{j=1}^{n} \frac{\partial f_i}{\partial y_j}\,[y_j - y_j\,(t^n)]$$

The question of stiffness then depends on the eigenvalue of the Jacobian at the current time. Consequently, for nonlinear problems the problem can be stiff during one time period and not stiff during another. Packages have been developed for problems such as these. Although the chemical engineer may not actually calculate the eigenvalues, knowing that they determine the stability and accuracy of the numerical scheme, as well as the step size employed, is useful.

6.5. Differential – Algebraic Systems

Sometimes models involve ordinary differential equations subject to some algebraic constraints. For example, the equations governing one equilibrium stage (as in a distillation column) are

$$M\frac{dx^n}{dt} = V^{n+1}\,y^{n+1} - L^n\,x^n - V^n\,y^n + L^{n-1}\,x^{n-1}$$

$$x^{n-1} - x^n = E^n\,\left(x^{n-1} - x^{*,n}\right)$$

$$\sum_{i=1}^{N} x_i = 1$$

where x and y are the mole fractions in the liquid and vapor, respectively; L and V are liquid and vapor flow rates, respectively; M is the holdup; and the superscript n is the stage number. The efficiency is E, and the concentration in equilibrium with the vapor is x^*. The first equation is an ordinary differential equation for the mass of one component on the stage, whereas the third equation represents a constraint that the mass fractions add to one. As a second example, the following kinetics problem can be considered:

$$\frac{dc_1}{dt} = f\,(c_1, c_2)$$

$$\frac{dc_2}{dt} = k_1\,c_1 - k_2\,c_2^2$$

The first equation could be the equation for a stirred tank reactor, for example. Suppose both k_1 and k_2 are large. The problem is then stiff, but the second equation could be taken at equilibrium. If

$$c_1 \rightleftarrows 2c_2$$

The equilibrium condition is then

$$\frac{c_2^2}{c_1} = \frac{k_1}{k_2} \equiv K$$

Under these conditions the problem becomes

$$\frac{dc_1}{dt} = f\,(c_1, c_2)$$

$$0 = k_l c_1 - k_2 c_2^2$$

Thus, a differential-algebraic system of equations is obtained. In this case, the second equation can be solved and substituted into the first to obtain differential equations, but in the general case that is not possible.

Differential-algebraic equations can be written in the general notation

$$\boldsymbol{F}\left(t, \boldsymbol{y}, \frac{d\boldsymbol{y}}{dt}\right) = 0$$

or the variables and equations may be separated according to whether they come primarily from differential $[y(t)]$ or algebraic equations $[x(t)]$:

$$\frac{d\boldsymbol{y}}{dt} = \boldsymbol{f}\,(t, \boldsymbol{y}, \boldsymbol{x})\,, \boldsymbol{g}\,(t, \boldsymbol{y}, \boldsymbol{x}) = 0$$

Another form is not strictly a differential-algebraic set of equations, but the same principles apply; this form arises frequently when the Galerkin finite element is applied:

$$\boldsymbol{A}\frac{d\boldsymbol{y}}{dt} = \boldsymbol{f}\,(\boldsymbol{y})$$

The computer program DASSL [60, 61] can solve such problems. They can also be solved by writing the differential equation as

$$\frac{d\boldsymbol{y}}{dt} = \boldsymbol{A}^{-1}\boldsymbol{f}\,(\boldsymbol{y})$$

When $\boldsymbol{A}$ is independent of $\boldsymbol{y}$, the inverse (from LU decompositions) need be computed only once.

In actuality, higher order backward-difference Gear methods are used in the computer program DASSL [60, 61].

Differential-algebraic systems are more complicated than differential systems because the solution may not always be defined. PONTELIDES et al. [62] introduced the term "index" to identify possible problems. The index is defined as the minimum number of times the equations must be differentiated with respect to time to convert the system to a set ofordinary differential equations. These higher derivatives may not exist, and the process places limits on which variables can be given initial values. Sometimes the initial values must be constrained by the algebraic equations [62]. For a differential-algebraic system modeling a distillation tower, the index depends on the specification of pressure for the column [62]. Several chemical engineering examples of differential-algebraic systems and a solution for one involving two-phase flow are given in [63].

6.6. Computer Software

Efficient software packages are widely available for solving ordinary differential equations as initial value problems. In each of the packages the user specifies the differential equation to be solved and a desired error criterion. The package then integrates in time and adjusts the step size to achieve the error criterion, within the limitations imposed by stability.

A popular explicit Runge – Kutta package is RKF 45. An estimate of the truncation error at each step is available. Then the step size can be reduced until this estimate is below the user-specified tolerance. The method is thus automatic, and the user is assured of the results. Note, however, that the tolerance is set on the local truncation error, namely, from one step to another, whereas the user is generally interested in the global trunction error, i.e., the error after several steps. The global error is generally made smaller by making the tolerance smaller, but the absolute accuracy is not the same as the tolerance. If the problem is stiff, then very small step sizes are used and the computation becomes very lengthy. The RKF 45 code discovers this and returns control to the user with a message indicating the problem is too hard to solve with RKF 45.

A popular implicit package is LSODE, a version of Gear's method [57] written by ALAN HINDMARSH at Lawrence Livermore Laboratory. In this package, the user specifies the differential equation to be solved and the tolerance desired. Now the method is implicit and, therefore, stable for any step size. The accuracy may not be acceptable, however, and sets of nonlinear equations must be solved. Thus, in practice, the step size is limited but not nearly so much as in the Runge – Kutta methods. In these packages both the step size and the order of the method are adjusted by the package itself. Suppose a k-th order method is being used. The truncation error is determined by the $(k + 1)$-th order derivative. This is estimated by using difference formulas and the values of the right-hand sides at previous times. An estimate is also made for the k-th and $(k + 2)$-th derivative. Then, the errors in a $(k - 1)$-th order method, a k-th order method, and a $(k + 1)$-th order method can be estimated. Furthermore, the step size required to satisfy the tolerance with each of these methods can be determined. Then the method and step size for the next step that achieves the biggest step can be chosen, with appropriate adjustments due to the different work required for each order. The package generally starts with a very small step size and a first-order method—the backward Euler method. Then it integrates along, adjusting the order up (and later down) depending on the error estimates. The user is thus assured that the local truncation error meets the tolerance. A further difficulty arises because the set of nonlinear equations must be solved. Usually a good guess of the solution is available, because the solution is evolving in time and past history can be extrapolated. Thus, the Newton – Raphson method will usually converge. The package protects itself, though, by only doing a few (i.e., three) iterations. If convergence is not reached within these iterations, the step size is reduced and the calculation is redone for that time step. The convergence theorem for the Newton – Raphson method (Chap. 1) indicates that the method will converge if the step size is small enough. Thus, the method is guaranteed to work. Further economies are possible. The Jacobian needed in the Newton – Raphson method can be fixed over several time steps. Then if the iteration does not converge, the Jacobian can be reevaluated at the current time step. If the iteration still does not converge, then the step size is reduced and a new Jacobian is evaluated. The successive substitution method can also be used—wh ich is even faster, except that it may not converge. However, it too will converge if the time step is small enough.

The Runge – Kutta methods give extremely good accuracy, especially when the step size is kept small for stability reasons. If the problem is stiff, though, backward difference implicit methods must be used. Many chemical reactor problems are stiff, necessitating the use of implicit methods. In the MATLAB suite of ODE solvers, ode45 uses a revision of the RKF45 program, while the ode15s program uses an improved backward difference method. Ref. [64] gives details of the programs in MATLAB. Fortunately, many packages are available. On the NIST web page, http://gams.nist.gov/ choose "problem decision tree", and then "differential and integral equations" to find packages which can be downloaded. On the Netlib web site, http://www.netlib.org/, choose "ode" to find packages which can be downloaded. Using Microsoft Excel to solve ordinary differential equations is cumbersome, except for the simplest problems.

6.7. Stability, Bifurcations, Limit Cycles

In this section, bifurcation theory is discussed in a general way. Some aspects of this subject involve the solution of nonlinear equations; other aspects involve the integration of ordinary differential equations; applications include chaos and fractals as well as unusual operation of some chemical engineering equipment. An excellent introduction to the subject and details needed to apply the methods are given in [65]. For more details of the algorithms described below and a concise survey with some chemical engineering examples, see [66] and [67]. Bifurcation results are closely connected with stability of the steady states, which is essentially a transient phenomenon.

Consider the problem

$$\frac{\partial u}{\partial t} = F\,(u,\lambda)$$

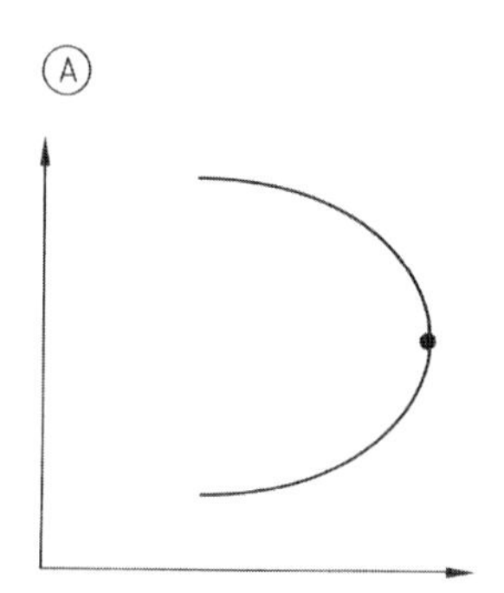

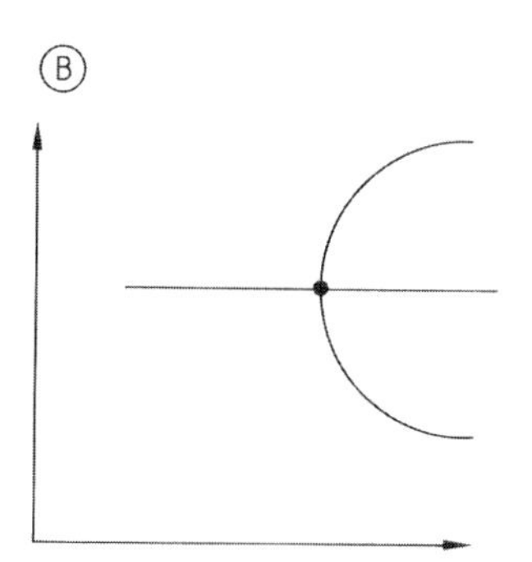

Figure 22. Limit points and bifurcation – limit points
A) Limit point (or turning point); B) Bifurcation-limit point (or singular turning point or bifurcation point)

The variable u can be a vector, which makes F a vector, too. Here, F represents a set of equations that can be solved for the steady state:

$$F\ (u,\lambda)=0$$

If the Newton – Raphson method is applied,

$$F_u^s\,\delta\,u^s=-F\ (u^s,\lambda)$$

$$u^{s+1}=u^s+\delta u^s$$

is obtained, where

$$F_u^s=\frac{\partial F}{\partial u}\,(u^s)$$

is the Jacobian. Look at some property of the solution, perhaps the value at a certain point or the maximum value or an integral of the solution. This property is plotted versus the parameter λ; typical plots are shown in Figure 22. At the point shown in Figure 22 A, the determinant of the Jacobian is zero:

$$\det F_u=0$$

For the limit point,

$$\frac{\partial F}{\partial\lambda}\neq 0$$

whereas for the bifurcation-limit point

$$\frac{\partial F}{\partial\lambda}=0$$

The stability of the steady solutions is also of interest. Suppose a steady solution $u_{\rm ss}$; the function u is written as the sum of the known steady state and a perturbation u':

$$u=u_{\rm ss}+u'$$

This expression is substituted into the original equation and linearized about the steady-state value:

$$\frac{\partial u_{\rm ss}}{\partial t}+\frac{\partial u'}{\partial t}=F\ (u_{\rm ss}+u',\lambda)$$

$$\approx F\ (u_{\rm ss},\lambda)+\frac{\partial F}{\partial u}|_{u_{\rm ss}}u'+\cdots$$

The result is

$$\frac{\partial u'}{\partial t}=F_u^{\rm ss}\,u'$$

A solution of the form

$$u'\ (x,t)={\rm e}^{\sigma t}\ X\ (x)$$

gives

$$\sigma{\rm e}^{\sigma t}\ X\ =F_u^{\rm ss}{\rm e}^{\sigma t}X$$

The exponential term can be factored out and

$$(F_u^{\rm ss}-\sigma\,\delta)\ X=0$$

A solution exists for X if and only if

$$\det\ |F_u^{\rm ss}-\sigma\,\delta|=0$$

The σ are the eigenvalues of the Jacobian. Now clearly if Re $(\sigma)>0$ then u' grows with time, and the steady solution $u_{\rm ss}$ is said to be unstable to small disturbances. If Im $(\sigma)=0$ it is called stationary instability, and the disturbance would grow monotonically, as indicated in Figure 23 A. If Im $(\sigma)\neq 0$ then the disturbance grows in an oscillatory fashion, as shown in Figure 23 B, and is called oscillatory instability. The case in which Re $(\sigma)=0$ is the dividing point between stability and instability. If Re $(\sigma)=0$ and Im $(\sigma)=0$—the point governing the onset of stationary instability—then $\sigma=0$. However, this means that $\sigma=0$ is an eigenvalue of the Jacobian, and the determinant of the Jacobian is zero. Thus, the points at which the determinant of the Jacobian is zero (for limit points

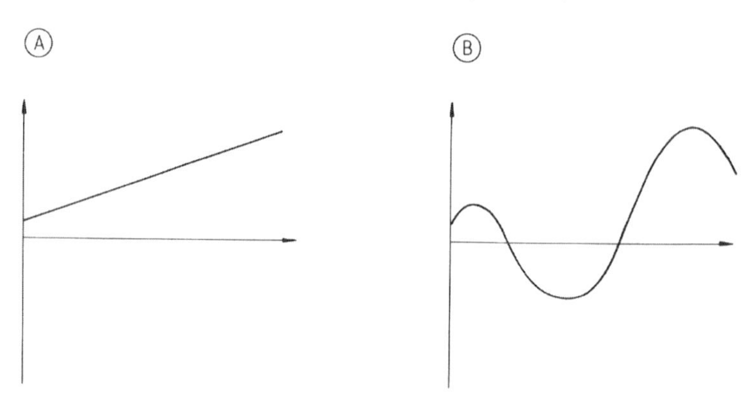

Figure 23. Stationary and oscillatory instability
A) Stationary instability; B) Oscillatory instability

and bifurcation-limit points) are the points governing the onset of stationary instability. When Re $(\sigma) = 0$ but Im $(\sigma) \neq 0$, which is the onset of oscillatory instability, an even number of eigenvalues pass from the left-hand complex plane to the right-hand complex plane. The eigenvalues are complex conjugates of each other (a result of the original equations being real, with no complex numbers), and this is called a Hopf bifurcation. Numerical methods to study Hopf bifurcation are very computationally intensive and are not discussed here [65].

To return to the problem of solving for the steady-state solution: near the limit point or bifurcation-limit point two solutions exist that are very close to each other. In solving sets of equations with thousands of unknowns, the difficulties in convergence are obvious. For some dependent variables the approximation may be converging to one solution, whereas for another set of dependent variables it may be converging to the other solution; or the two solutions may all be mixed up. Thus, solution is difficult near a bifurcation point, and special methods are required. These methods are discussed in [66].

The first approach is to use *natural continuation* (also known as Euler – Newton continuation). Suppose a solution exists for some parameter λ. Call the value of the parameter λ_0 and the corresponding solution u_0. Then

$$F\ (u_0\,,\lambda_0) = 0$$

Also, compute u_λ as the solution to

$$F_u^{\mathrm{ss}}\,u_\lambda = -F_\lambda$$

at this point $[\lambda_0, u_0]$. Then predict the starting guess for another λ using

$$u^0 = u_0 + u_\lambda\,(\lambda - \lambda_0)$$

and apply Newton – Raphson with this initial guess and the new value of λ. This will be a much better guess of the new solution than just u_0 by itself.

Even this method has difficulties, however. Near a limit point the determinant of the Jacobian may be zero and the Newton method may fail. Perhaps no solutions exist at all for the chosen parameter λ near a limit point. Also, the ability to switch from one solution path to another at a bifurcation-limit point is necessary. Thus, other methods are needed as well: arc-length continuation and pseudo-arc-length continuation [66]. These are described in Chapter 1.

6.8. Sensitivity Analysis

Often, when solving differential equations, the solution as well as the sensitivity of the solution to the value of a parameter must be known. Such information is useful in doing parameter estimation (to find the best set of parameters for a model) and in deciding whether a parameter needs to be measured accurately. The differential equation for $y\ (t,\ \alpha)$ where α is a parameter, is

$$\frac{\mathrm{d}y}{\mathrm{d}t} = f\ (y, \alpha)\,,\ y\ (0) = y_0$$

If this equation is differentiated with respect to α, then because y is a function of t and α

$$\frac{\partial}{\partial \alpha}\left(\frac{\mathrm{d}y}{\mathrm{d}t}\right) = \frac{\partial f}{\partial y}\frac{\partial y}{\partial \alpha} + \frac{\partial f}{\partial \alpha}$$

Exchanging the order of differentiation in the first term leads to the ordinary differential equation

$$\frac{\mathrm{d}}{\mathrm{d}t}\left(\frac{\partial y}{\partial \alpha}\right)=\frac{\partial f}{\partial y}\frac{\partial y}{\partial \alpha}+\frac{\partial f}{\partial \alpha}$$

The initial conditions on $\partial y/\partial \alpha$ are obtained by differentiating the initial conditions

$$\frac{\partial}{\partial \alpha}\left[y\left(0,\alpha\right)=y_0\right],\ \text{or}\ \frac{\partial y}{\partial \alpha}\left(0\right)=0$$

Next, let

$$y_1=y,\ y_2=\frac{\partial y}{\partial \alpha}$$

and solve the set of ordinary differential equations

$$\frac{\mathrm{d}y_1}{\mathrm{d}t}=f\left(y_1,\alpha\right) \qquad y_1\left(0\right)=y_0$$

$$\frac{\mathrm{d}y_2}{\mathrm{d}t}=\frac{\partial f}{\partial y}\left(y_1,\alpha\right)y_2+\frac{\partial f}{\partial \alpha} \qquad y_2\left(0\right)=0$$

Thus, the solution $y\ (t,\ \alpha)$ and the derivative with respect to α are obtained. To project the impact of α, the solution for $\alpha = \alpha_1$ can be used:

$$y\left(t,\alpha\right)=y_1\left(t,\alpha_1\right)+\frac{\partial y}{\partial \alpha}\left(t,\alpha_1\right)\left(\alpha-\alpha_1\right)+\cdots$$

$$=y_1\left(t,\alpha_1\right)+y_2\left(t,\alpha_1\right)\left(\alpha-\alpha_1\right)+\cdots$$

This is a convenient way to determine the sensitivity of the solution to parameters in the problem.

6.9. Molecular Dynamics

(→ Molecular Dynamics Simulations)

Special integration methods have been developed for molecular dynamics calculations due to the structure of the equations. A very large number of equations are to be integrated, with the following form based on molecular interactions between molecules

$$m_i\frac{d^2\boldsymbol{r}_i}{dt^2}=\boldsymbol{F}_i\left(\{\boldsymbol{r}\}\right),\boldsymbol{F}_i\left(\{\boldsymbol{r}\}\right)=-\nabla V$$

where m_i is the mass of the i-th particle, $\boldsymbol{r}_i$ is the position of the i-th particle, $\boldsymbol{F}_i$ is the force acting on the i-th particle, and V is the potential energy that depends upon the location of all the particles (but not their velocities). Since the major part of the calculation is in the evaluation of the forces, or potentials, a method must be used that minimizes the number of times the forces are calculated to move from one time to another time. Rewrite this equation in the form of an acceleration.

$$\frac{d^2\boldsymbol{r}_i}{dt^2}=\frac{1}{m_i}\boldsymbol{F}_i\left(\{\boldsymbol{r}\}\right)\equiv\boldsymbol{a}_i$$

In the Verlot method, this equation is written using central finite differences (Eq. 12). Note that the accelerations do not depend upon the velocities.

$$\boldsymbol{r}_i\left(t+\Delta t\right)=2\boldsymbol{r}_i\left(t\right)-\boldsymbol{r}_i\left(t-\Delta t\right)+\boldsymbol{a}_i\left(t\right)\Delta t^2$$

The calculations are straightforward, and no explicit velocity is needed. The storage requirement is modest, and the precision is modest (it is a second-order method). Note that one must start the calculation with values of $\{r\}$ at time t and $t-\Delta t$.

In the Velocity Verlot method, an equation is written for the velocity, too.

$$\frac{d\boldsymbol{v}_i}{dt}=\boldsymbol{a}_i$$

The trapezoid rule (see page 18) is applied to obtain

$$\boldsymbol{v}_i\left(t+\Delta t\right)=\boldsymbol{v}_i\left(t\right)+\frac{1}{2}\left[\boldsymbol{a}_i\left(t\right)+\boldsymbol{a}_i\left(t+\Delta t\right)\right]\Delta t$$

The position of the particles is expanded in a Taylor series.

$$\boldsymbol{r}_i\left(t+\Delta t\right)=\boldsymbol{r}_i\left(t\right)+\boldsymbol{v}_i\Delta t+\frac{1}{2}\boldsymbol{a}_i\left(t\right)\Delta t^2$$

Beginning with values of $\{r\}$ and $\{v\}$ at time zero, one calculates the new positions and then the new velocities. This method is second order in Δt, too. For additional details, see [68 – 72].

7. Ordinary Differential Equations as Boundary Value Problems

Diffusion problems in one dimension lead to boundary value problems. The boundary conditions are applied at two different spatial locations: at one side the concentration may be fixed and at the other side the flux may be fixed. Because the conditions are specified at two different locations the problems are not initial value in character. To begin at one position and integrate directly is impossible because at least one of the conditions is specified somewhere else and not enough conditions are available to begin the calculation. Thus, methods have been developed especially for boundary value problems. Examples include heat and mass transfer in a slab, reaction – diffusion problems in a porous catalyst, reactor with axial dispersion, packed beds, and countercurrent heat transfer.

7.1. Solution by Quadrature

When only one equation exists, even if it is nonlinear, it may possibly be solved by quadrature. For

$$\frac{\mathrm{d}y}{\mathrm{d}t} = f(y)$$

$$y(0) = y_0$$

the problem can be separated

$$\frac{\mathrm{d}y}{f(y)} = \mathrm{d}t$$

and integrated

$$\int_{y_0}^{y} \frac{dy'}{f(y')} = \int_0^t \mathrm{d}t = t$$

If the quadrature can be performed analytically, the exact solution has been found.

As an example, consider the flow of a non-Newtonian fluid in a pipe, as illustrated in Figure 24. The governing differential equation is [73]

$$\frac{1}{r}\frac{\mathrm{d}}{\mathrm{d}r}(r\tau) = -\frac{\Delta p}{L}$$

where r is the radial position from the center of the pipe, τ is the shear stress, Δp is the pressure drop along the pipe, and L is the length over which the pressure drop occurs. The variables are separated once

$$\mathrm{d}(r\tau) = -\frac{\Delta p}{L} r\mathrm{d}r$$

and then integrated to give

$$r\tau = -\frac{\Delta p}{L}\frac{r^2}{2} + c_1$$

Proceeding further requires choosing a constitutive relation relating the shear stress and the velocity gradient as well as a condition specifying the constant. For a Newtonian fluid

$$\tau = -\eta\frac{\mathrm{d}v}{\mathrm{d}r}$$

where v is the velocity and η the viscosity. Then the variables can be separated again and the result integrated to give

$$-\eta v = -\frac{\Delta p}{L}\frac{r^2}{4} + c_1 \ln r + c_2$$

Now the two unknowns must be specified from the boundary conditions. This problem is a two-point boundary value problem because one of the conditions is usually specified at $r = 0$ and the other at $r = R$, the tube radius. However, the technique of separating variables and integrating works quite well.

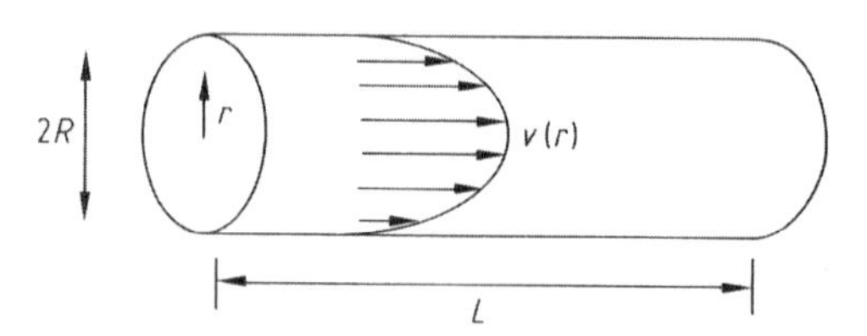

Figure 24. Flow in pipe

When the fluid is non-Newtonian, it may not be possible to do the second step analytically. For example, for the Bird – Carreau fluid [74, p. 171], stress and velocity are related by

$$\tau = \frac{\eta_0}{\left[1 + \lambda\left(\frac{\mathrm{d}v}{\mathrm{d}r}\right)^2\right]^{(1-n)/2}}$$

where η_0 is the viscosity at $v = 0$ and λ the time constant.

Putting this value into the equation for stress as a function of r gives

$$\frac{\eta_0}{\left[1 + \lambda\left(\frac{\mathrm{d}v}{\mathrm{d}r}\right)^2\right]^{(1-n)/2}} = -\frac{\Delta p}{L}\frac{r}{2} + \frac{c_1}{r}$$

This equation cannot be solved analytically for $\mathrm{d}v/\mathrm{d}r$, except for special values of n. For problems such as this, numerical methods must be used.

7.2. Initial Value Methods

An initial value method is one that utilizes the techniques for initial value problems but allows for an iterative calculation to satisfy all the boundary conditions. Suppose the nonlinear boundary value problem

$$\frac{\mathrm{d}^2y}{\mathrm{d}x^2}=f\left(x,y,\frac{\mathrm{d}y}{\mathrm{d}x}\right)$$

with the boundary conditions

$$a_0\,y\,(0)-a_1\frac{\mathrm{d}y}{\mathrm{d}x}\,(0)=\alpha,\ a_{\mathrm{i}}\geq 0$$

$$b_0\,y\,(1)-b_1\frac{\mathrm{d}y}{\mathrm{d}x}\,(1)=\beta,\ b_{\mathrm{i}}\geq 0$$

Convert this second-order equation into two first-order equations along with the boundary conditions written to include a parameter s.

$$\frac{\mathrm{d}u}{\mathrm{d}x}=v$$

$$\frac{\mathrm{d}v}{\mathrm{d}x}=f\,(x,u,v)$$

$$u\,(0)=a_1s-c_1\alpha$$

$$v\,(0)=a_0s-c_0\alpha$$

The parameters c_0 and c_1 are specified by the analyst such that

$$a_1c_0-a_0c_1=1$$

This ensures that the first boundary condition is satisfied for any value of parameter s. If the proper value for s is known, u (0) and u' (0) can be evaluated and the equation integrated as an initial value problem. The parameter s should be chosen iteratively so that the last boundary condition is satisfied.

The model for a *chemical reactor with axial diffusion* is

$$\frac{1}{Pe}\frac{\mathrm{d}c^2}{\mathrm{d}z^2}-\frac{\mathrm{d}c}{\mathrm{d}z}=DaR\,(c)$$

$$-\frac{1}{Pe}\frac{\mathrm{d}c}{\mathrm{d}z}\,(0)+c\,(0)=c_{\mathrm{in}},\ \frac{\mathrm{d}c}{\mathrm{d}z}\,(1)=0$$

where Pe is the Péclet number and Da the Damköhler number.

The boundary conditions are due to Danckwerts [75] and to Wehner and Wilhelm [76]. This problem can be treated by using initial value methods also, but the method is highly sensitive to the choice of the parameter s, as outlined above. Starting at z = 0 and making small changes in s will cause large changes in the solution at the exit, and the boundary condition at the exit may be impossible to satisfy. By starting at z = 1, however, and integrating backwards, the process works and an iterative scheme converges in many cases [77]. However, if the problem is extremely nonlinear the iterations may not converge. In such cases, the methods for boundary value problems described below must be used.

Packages to solve boundary value problems are available on the internet. On the NIST web page, http://gams.nist.gov/ choose "problem decision tree", and then "differential and integral equations", then "ordinary differential equations", "multipoint boundary value problems". On the Netlib web site, http://www.netlib.org/, search on "boundary value problem". Any spreadsheet that has an iteration capability can be used with the finite difference method. Some packages for partial differential equations also have a capability for solving one-dimensional boundary value problems [e.g., Comsol Multiphysics (formerly FEMLAB)].

7.3. Finite Difference Method

To apply the finite difference method, we first spread grid points through the domain. Figure 25 shows a uniform mesh of n points (nonuniform meshes are also possible). The unknown, here c (x), at a grid point x_i is assigned the symbol $c_i = c\,(x_i)$. The finite difference method can be derived easily by using a Taylor expansion of the solution about this point.

$$\begin{aligned}c_{i+1}&=c_i+\left.\frac{\mathrm{d}c}{\mathrm{d}x}\right|_i\Delta x+\left.\frac{\mathrm{d}^2c}{\mathrm{d}x^2}\right|_i\frac{\Delta x^2}{2}+\ldots\\ c_{i-1}&=c_i-\left.\frac{\mathrm{d}c}{\mathrm{d}x}\right|_i\Delta x+\left.\frac{\mathrm{d}^2c}{\mathrm{d}x^2}\right|_i\frac{\Delta x^2}{2}-\ldots\end{aligned}\tag{8}$$

These formulas can be rearranged and divided by Δx to give

$$\left.\frac{\mathrm{d}c}{\mathrm{d}x}\right|_i=\frac{c_{i+1}-c_i}{\Delta x}-\left.\frac{\mathrm{d}^2c}{\mathrm{d}x^2}\right|_i\frac{\Delta x}{2}+\ldots\tag{9}$$

$$\left.\frac{\mathrm{d}c}{\mathrm{d}x}\right|_i=\frac{c_i-c_{i-1}}{\Delta x}-\left.\frac{\mathrm{d}^2c}{\mathrm{d}x^2}\right|_i\frac{\Delta x}{2}+\ldots\tag{10}$$

which are representations of the first derivative. Alternatively the two equations can be subtracted from each other, rearranged and divided by Δx to give

$$\left.\frac{\mathrm{d}c}{\mathrm{d}x}\right|_i=\frac{c_{i+1}-c_{i-1}}{2\Delta x}-\left.\frac{\mathrm{d}^2c}{\mathrm{d}x^3}\right|_i\frac{\Delta x^2}{3!}\tag{11}$$

If the terms multiplied by Δx or Δx^2 are neglected, three representations of the first derivative are possible. In comparison with the Taylor series, the truncation error in the first two expressions is proportional to Δx, and the methods are said to be first order. The truncation error in the last expression is proportional to Δx^2, and the method is said to be second order. Usually, the last equation is chosen to ensure the best accuracy.

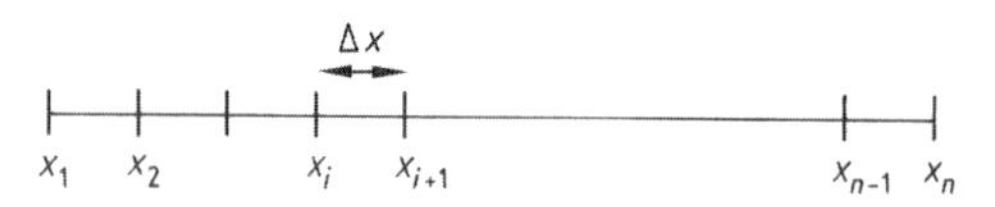

Figure 25. Finite difference mesh; Δx uniform

The finite difference representation of the second derivative can be obtained by adding the two expressions in Equation 8. Rearrangement and division by Δx^2 give

$$\left.\frac{\mathrm{d}^2 c}{\mathrm{d}x^2}\right|_i = \frac{c_{i+1} - 2c_i + c_{i-1}}{\Delta x^2} - \left.\frac{\mathrm{d}^4 c}{\mathrm{d}x^4}\right|_i \frac{\Delta x^2}{4!} + \ldots \quad (12)$$

The truncation error is proportional to Δx^2.

To see how to solve a differential equation, consider the equation for convection, diffusion, and reaction in a tubular reactor:

$$\frac{1}{Pe}\frac{\mathrm{d}^2 c}{\mathrm{d}x^2} - \frac{\mathrm{d}c}{\mathrm{d}x} = Da\, R(c)$$

To evaluate the differential equation at the i-th grid point, the finite difference representations of the first and second derivatives can be used to give

$$\frac{1}{Pe}\frac{c_{i+1} - 2c_i + c_{i-1}}{\Delta x^2} - \frac{c_{i+1} - c_{i-1}}{2\Delta x} = Da\, R \quad (13)$$

This equation is written for $i = 2$ to $n - 1$ (i.e., the internal points). The equations would then be coupled but would involve the values of c_1 and c_n, as well. These are determined from the boundary conditions.

If the boundary condition involves a derivative, the finite difference representation of it must be carefully selected; here, three possibilities can be written. Consider a derivative needed at the point $i = 1$. First, Equation 9 could be used to write

$$\left.\frac{\mathrm{d}c}{\mathrm{d}x}\right|_1 = \frac{c_2 - c_1}{\Delta x} \quad (14)$$

Then a second-order expression is obtained that is one-sided. The Taylor series for the point c_{i+2} is written:

$$c_{i+2} = c_i + \left.\frac{\mathrm{d}c}{\mathrm{d}x}\right|_i 2\Delta x + \left.\frac{\mathrm{d}^2 c}{\mathrm{d}x^2}\right|_i \frac{4\Delta x^2}{2!} + \left.\frac{\mathrm{d}^3 c}{\mathrm{d}x^3}\right|_i \frac{8\Delta x^3}{3!} + \ldots$$

Four times Equation 8 minus this equation, with rearrangement, gives

$$\left.\frac{\mathrm{d}c}{\mathrm{d}x}\right|_i = \frac{-3c_i + 4c_{i+1} - c_{i+2}}{2\Delta x} + O\left(\Delta x^2\right)$$

Thus, for the first derivative at point $i = 1$

$$\left.\frac{\mathrm{d}c}{\mathrm{d}x}\right|_i = \frac{-3c_i + 4c_2 - c_3}{2\Delta x} \quad (15)$$

This one-sided difference expression uses only the points already introduced into the domain. The third alternative is to add a false point, outside the domain, as $c_0 = c\ (x = -\ \Delta x)$. Then the centered first derivative, Equation 11, can be used:

$$\left.\frac{\mathrm{d}c}{\mathrm{d}x}\right|_1 = \frac{c_2 - c_0}{2\Delta x}$$

Because this equation introduces a new variable, another equation is required. This is obtained by also writing the differential equation (Eq. 13), for $i = 1$.

The same approach can be taken at the other end. As a boundary condition, any of three choices can be used:

$$\left.\frac{\mathrm{d}c}{\mathrm{d}x}\right|_n = \frac{c_n - c_{n-1}}{\Delta x}$$

$$\left.\frac{\mathrm{d}c}{\mathrm{d}x}\right|_n = \frac{c_{n-2} - 4c_{n-1} + 3c_n}{2\Delta x}$$

$$\left.\frac{\mathrm{d}c}{\mathrm{d}x}\right|_n = \frac{c_{n+1} - c_{n-1}}{2\Delta x}$$

The last two are of order Δx^2 and the last one would require writing the differential equation (Eq. 13) for $i = n$, too.

Generally, the first-order expression for the boundary condition is not used because the error in the solution would decrease only as Δx, and the higher truncation error of the differential equation (Δx^2) would be lost. For this problem the boundary conditions are

$$-\frac{1}{Pe}\frac{\mathrm{d}c}{\mathrm{d}x}(0) + c(0) = c_{\mathrm{in}}$$

$$\frac{\mathrm{d}c}{\mathrm{d}x}(1) = 0$$

Thus, the three formulations would give first order in Δx

$$-\frac{1}{Pe}\frac{c_2 - c_1}{\Delta x} + c_1 = c_{\mathrm{in}}$$

$$\frac{c_n - c_{n-1}}{\Delta x} = 0$$

plus Equation 13 at points $i = 2$ through $n - 1$; second order in Δx, by using a three-point one-sided derivative

$$-\frac{1}{Pe}\frac{-3c_1+4c_2-c_3}{2\Delta x}+c_1=c_{\text{in}}$$

$$\frac{c_{n-2}-4c_{n-1}+3c_n}{2\Delta x}=0$$

plus Equation 13 at points i = 2 through $n-1$; second order in Δx, by using a false boundary point

$$-\frac{1}{Pe}\frac{c_2-c_0}{2\Delta x}+c_1=c_{\text{in}}$$

$$\frac{c_{n+1}-c_{n-1}}{2\Delta x}=0$$

plus Equation 13 at points i = 1 through n.

The sets of equations can be solved by using the Newton – Raphson method, as outlined in Section 1.2.

Frequently, the transport coefficients (e.g., diffusion coefficient D or thermal conductivity) depend on the dependent variable (concentration or temperature, respectively). Then the differential equation might look like

$$\frac{\mathrm{d}}{\mathrm{d}x}\left(D(c)\frac{\mathrm{d}c}{\mathrm{d}x}\right)=0$$

This could be written as

$$-\frac{\mathrm{d}J}{\mathrm{d}x}=0 \qquad (16)$$

in terms of the mass flux J, where the mass flux is given by

$$J=-D(c)\frac{\mathrm{d}c}{\mathrm{d}x}$$

Because the coefficient depends on c the equations are more complicated. A finite difference method can be written in terms of the fluxes at the midpoints, i + 1/2. Thus,

$$-\frac{J_{i+1/2}-J_{i-1/2}}{\Delta x}=0$$

Then the constitutive equation for the mass flux can be written as

$$J_{i+1/2}=-D\left(c_{i+1/2}\right)\frac{c_{i+1}-c_i}{\Delta x}$$

If these are combined,

$$\frac{D\left(c_{i+1/2}\right)\left(c_{i+1}-c_i\right)-D\left(c_{i-1/2}\right)\left(c_i-c_{i-1}\right)}{\Delta x^2}=0$$

This represents a set of nonlinear algebraic equations that can be solved with the Newton – Raphson method. However, in this case a viable iterative strategy is to evaluate the transport coefficients at the last value and then solve

$$\frac{D\left(c^k_{i+1/2}\right)\left(c^{k+1}_{i+1}-c^{k+1}_i\right)-D\left(c^k_{i-1/2}\right)\left(c^{k+1}_i-c^{k+1}_{i-1}\right)}{\Delta x^2}=0$$

The advantage of this approach is that it is easier to program than a full Newton – Raphson method. If the transport coefficients do not vary radically, the method converges. If the method does not converge, use of the full Newton – Raphson method may be necessary.

Three ways are commonly used to evaluate the transport coefficient at the midpoint. The first one employs the transport coefficient evaluated at the average value of the solutions on either side:

$$D\left(c_{i+1/2}\right)\approx D\left[\frac{1}{2}\left(c_{i+1}+c_i\right)\right]$$

The second approach uses the average of the transport coefficients on either side:

$$D\left(c_{i+1/2}\right)\approx\frac{1}{2}\left[D\left(c_{i+1}\right)+D\left(c_i\right)\right] \qquad (17)$$

The truncation error of these approaches is also Δx^2 [78, Chap. 14], [3, p. 215]. The third approach employs an "upstream" transport coefficient.

$$D\left(C_{i+1/2}\right)\approx D\left(c_{i+1}\right), \text{ when } D\left(c_{i+1}\right)>D\left(c_i\right)$$

$$D\left(c_{i+1/2}\right)\approx\left(c_i\right), \text{ when } D\left(c_{i+1}\right)<D\left(c_i\right)$$

This approach is used when the transport coefficients vary over several orders of magnitude and the "upstream" direction is defined as the one in which the transport coefficient is larger. The truncation error of this approach is only Δx [78, Chap. 14] , [3, p. 253], but this approach is useful if the numerical solutions show unrealistic oscillations [3, 78].

Rigorous error bounds for linear ordinary differential equations solved with the finite difference method are dicussed by Isaacson and Keller [79, p. 431].

7.4. Orthogonal Collocation

The orthogonal collocation method has found widespread application in chemical engineering, particularly for chemical reaction engineering. In the collocation method [3], the dependent variable is expanded in a series.

$$y(x)=\sum_{i=1}^{N+2}a_i y_i(x) \qquad (18)$$

Suppose the differential equation is

$$N[y]=0$$

Then the expansion is put into the differential equation to form the residual:

$$\text{Residual}=N\left[\sum_{i=1}^{N+2}a_i y_i(x)\right]$$

In the collocation method, the residual is set to zero at a set of points called collocation points:

$$N\left[\sum_{i=1}^{N+2}a_i y_i(x_j)\right]=0,\; j=2,\ldots,N+1$$

This provides N equations; two more equations come from the boundary conditions, giving $N + 2$ equations for $N + 2$ unknowns. This procedure is especially useful when the expansion is in a series of orthogonal polynomials, and when the collocation points are the roots to an orthogonal polynomial, as first used by LANCZOS [80, 81]. A major improvement was the proposal by VILLADSEN and STEWART [82] that the entire solution process be done in terms of the solution at the collocation points rather than the coefficients in the expansion. Thus, Equation 18 would be evaluated at the collocation points

$$y(x_j)=\sum_{i=1}^{N+2}a_i y_i(x_j),\; j=1,\ldots,N+2$$

and solved for the coefficients in terms of the solution at the collocation points:

$$a_i=\sum_{j=1}^{N+2}[y_i(x_j)]^{-1}y(x_j),\; i=1,\ldots,N+2$$

Furthermore, if Equation 18 is differentiated once and evaluated at all collocation points, the first derivative can be written in terms of the values at the collocation points:

$$\frac{\mathrm{d}y}{\mathrm{d}x}(x_j)=\sum_{i=1}^{N+2}a_i\frac{\mathrm{d}y_i}{\mathrm{d}x}(x_j),\; j=1,\ldots,N+2$$

This can be expressed as

$$\frac{\mathrm{d}y}{\mathrm{d}x}(x_j)=\sum_{i,k=1}^{N+2}[y_i(x_k)]^{-1}y(x_k)\frac{\mathrm{d}y_i}{\mathrm{d}x}(x_j),\; j=1,\ldots,N+2$$

or shortened to

$$\frac{\mathrm{d}y}{\mathrm{d}x}(x_j)=\sum_{k=1}^{N+2}A_{jk}y(x_k),$$

$$A_{jk}=\sum_{i=1}^{N+2}[y_i(x_k)]^{-1}\frac{\mathrm{d}y_i}{\mathrm{d}x}(x_j)$$

Similar steps can be applied to the second derivative to obtain

$$\frac{\mathrm{d}^2y}{\mathrm{d}x^2}(x_j)=\sum_{k=1}^{N+2}B_{jk}y(x_k),$$

$$B_{jk}=\sum_{i=1}^{N+2}[y_i(x_k)]^{-1}\frac{\mathrm{d}^2y_i}{\mathrm{d}x^2}(x_j)$$

This method is next applied to the differential equation for reaction in a tubular reactor, after the equation has been made nondimensional so that the dimensionless length is 1.0.

$$\frac{1}{Pe}\frac{\mathrm{d}^2c}{\mathrm{d}x^2}-\frac{\mathrm{d}c}{\mathrm{d}x}=Da\,R(c),$$
$$-\frac{\mathrm{d}c}{\mathrm{d}x}(0)=Pe\,[c(0)-c_{\text{in}}],\;\frac{\mathrm{d}c}{\mathrm{d}x}(1)=0 \qquad (19)$$

The differential equation at the collocation points is

$$\frac{1}{Pe}\sum_{k=1}^{N+2}B_{jk}c(x_k)-\sum_{k=1}^{N+2}A_{jk}c(x_k)=Da\,R(c_j) \qquad (20)$$

and the two boundary conditions are

$$-\sum_{k=1}^{N+2}A_{lk}c(x_k)=Pe\,(c_1-c_{\text{in}}),$$
$$\sum_{k=1}^{N+2}A_{N+2,k}c(x_k)=0 \qquad (21)$$

Note that 1 is the first collocation point ($x = 0$) and $N + 2$ is the last one ($x = 1$). To apply the method, the matrices A_{ij} and B_{ij} must be found and the set of algebraic equations solved, perhaps with the Newton – Raphson method. If orthogonal polynomials are used and the collocation points are the roots to one of the orthogonal polynomials, the orthogonal collocation method results.

In the orthogonal collocation method the solution is expanded in a series involving orthogonal polynomials, where the polynomials $P_{i-1}(x)$ are defined in Section 2.2.

$$y=a+bx+x(1-x)\sum_{i=1}^{N}a_iP_{i-1}(x)$$
$$=\sum_{i=1}^{N+2}b_iP_{i-1}(x) \qquad (22)$$

which is also

$$y=\sum_{i=1}^{N+2}d_ix^{i-1}$$

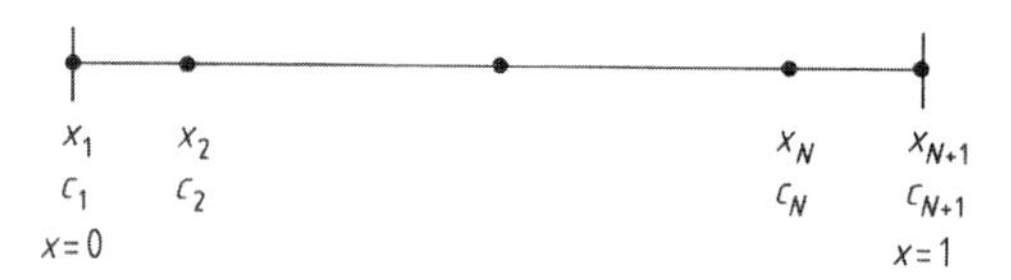

Figure 26. Orthogonal collocation points

The collocation points are shown in Figure 26. There are N interior points plus one at each end, and the domain is always transformed to lie on 0 to 1. To define the matrices A_{ij} and B_{ij} this expression is evaluated at the collocation points; it is also differentiated and the result is evaluated at the collocation points.

$$y(x_j) = \sum_{i=1}^{N+2} d_i x_j^{i-1}$$

$$\frac{\mathrm{d}y}{\mathrm{d}x}(x_j) = \sum_{i=1}^{N+2} d_i (i-1) x_j^{i-2}$$

$$\frac{\mathrm{d}^2 y}{\mathrm{d}x^2}(x_j) = \sum_{i=1}^{N+2} d_i (i-1)(i-2) x_j^{i-3}$$

These formulas are put in matrix notation, where $\boldsymbol{Q}$, $\boldsymbol{C}$, and $\boldsymbol{D}$ are $N+2$ by $N+2$ matrices.

$$\boldsymbol{y} = \boldsymbol{Q}\boldsymbol{d}, \frac{\mathrm{d}\boldsymbol{y}}{\mathrm{d}x} = \boldsymbol{C}\boldsymbol{d}, \frac{\mathrm{d}^2\boldsymbol{y}}{\mathrm{d}x^2} = \boldsymbol{D}\boldsymbol{d}$$

$$Q_{ji} = x_j^{i-1},\ C_{ji} = (i-1) x_j^{i-2},$$

$$D_{ji} = (i-1)(i-2) x_j^{i-3}$$

In solving the first equation for $\boldsymbol{d}$, the first and second derivatives can be written as

$$\boldsymbol{d} = \boldsymbol{Q}^{-1}\boldsymbol{y},\ \frac{\mathrm{d}\boldsymbol{y}}{\mathrm{d}x} = \boldsymbol{C}\boldsymbol{Q}^{-1}\boldsymbol{y} = \boldsymbol{A}\boldsymbol{y},$$

$$\frac{\mathrm{d}^2\boldsymbol{y}}{\mathrm{d}x^2} = \boldsymbol{D}\boldsymbol{Q}^{-1}\boldsymbol{y} = \boldsymbol{B}\boldsymbol{y} \qquad (23)$$

Thus the derivative at any collocation point can be determined in terms of the solution at the collocation points. The same property is enjoyed by the finite difference method (and the finite element method described below), and this property accounts for some of the popularity of the orthogonal collocation method. In applying the method to Equation 19, the same result is obtained; Equations 20 and 21, with the matrices defined in Equation 23. To find the solution at a point that is not a collocation point, Equation 22 is used; once the solution is known at all collocation points, $\boldsymbol{d}$ can be found; and once $\boldsymbol{d}$ is known, the solution for any x can be found.

To use the orthogonal collocation method, the matrices are required. They can be calculated as shown above for small N ($N < 8$) and by using more rigorous techniques, for higher N (see Chap. 2). However, having the matrices listed explicitly for $N = 1$ and 2 is useful; this is shown in Table 7.

For some reaction diffusion problems, the solution can be an even function of x. For example, for the problem

$$\frac{\mathrm{d}^2 c}{\mathrm{d}x^2} = kc,\ \frac{\mathrm{d}c}{\mathrm{d}x}(0) = 0,\ c(1) = 1 \qquad (24)$$

the solution can be proved to involve only even powers of x. In such cases, an orthogonal collocation method, which takes this feature into

Table 7. Matrices for orthogonal collocation

$N = 1$, $a = 0.1666666667$

$$x_j = \begin{pmatrix} 0 \\ 0.5 \\ 1 \end{pmatrix}, W_j = \begin{pmatrix} a \\ 4a \\ a \end{pmatrix}, A_{ji} = \begin{pmatrix} -3 & 4 & -1 \\ -1 & 0 & 1 \\ 1 & -4 & 3 \end{pmatrix}, B_{ji} = \begin{pmatrix} 4 & -8 & 4 \\ 4 & -8 & 4 \\ 4 & -8 & 4 \end{pmatrix}, Q_{ji}^{-1} = \begin{pmatrix} 1 & 0 & 0 \\ -3 & 4 & -1 \\ 2 & -4 & 2 \end{pmatrix}$$

$N = 2$

$$x_j = \begin{pmatrix} 0 \\ e \\ 1-e \\ 1 \end{pmatrix}, W_j = \begin{pmatrix} 0 \\ 0.5 \\ 0.5 \\ 0 \end{pmatrix}, Q_{ji}^{-1} = \begin{pmatrix} 1 & 0 & 0 & 0 \\ -7 & 7+b & -1-b & 1 \\ 12 & -18-f & 12+f & -6 \\ -6 & 10+d & -10-d & 6 \end{pmatrix}$$

$$A_{ji} = \begin{pmatrix} -7 & 7+b & -1-b & 1 \\ -1-a & a & a & 1-a \\ -1+a & -a & -a & 1+a \\ -1 & 1+b & -7-b & 7 \end{pmatrix}, B_{ji} = \begin{pmatrix} 24 & -37-c & 25+c & -12 \\ 16+d & -24 & 12 & -4-d \\ -4-d & 12 & -24 & 16+d \\ -12 & 25+c & -37-c & 24 \end{pmatrix}$$

where $a = 1.732050808$, $b = 1.196152423$, $c = 0.17691454$, $d = 0.392304846$
$e = 0.2113248654$, $f = 0.58845727$

account, is convenient. This can easily be done by using expansions that only involve even powers of x. Thus, the expansion

$$y\left(x^2\right) = y\left(1\right) + \left(1-x^2\right)\sum_{i=1}^{N} a_i P_{i-1}\left(x^2\right)$$

is equivalent to

$$y\left(x^2\right) = \sum_{i=1}^{N+1} b_i P_{i-1}\left(x^2\right) = \sum_{i=1}^{N+1} d_i x^{2i-2}$$

The polynomials are defined to be orthogonal with the weighting function $W\left(x^2\right)$.

$$\int_0^1 W\left(x^2\right) P_k\left(x^2\right) P_m\left(x^2\right) x^{a-1} \mathrm{d}x = 0 \quad (25)$$

$$k \leq m-1$$

where the power on x^{a-1} defines the geometry as planar or Cartesian (a = 1), cylindrical (a = 2), and spherical (a = 3). An analogous development is used to obtain the $(N + 1)\times(N + 1)$ matrices

$$y\left(x_j\right) = \sum_{i=1}^{N+1} d_i x_j^{2i-2}$$

$$\frac{\mathrm{d}y}{\mathrm{d}x}\left(x_j\right) = \sum_{i=1}^{N+1} d_i \left(2i-2\right) x_j^{2i-3}$$

$$\nabla^2 y\left(x_i\right) = \left.\sum_{i=1}^{N+1} d_i \nabla^2\left(x^{2i-2}\right)\right|_{x_j}$$

$$\boldsymbol{y} = \boldsymbol{Qd}, \frac{\mathrm{d}\boldsymbol{y}}{\mathrm{d}x} = \boldsymbol{Cd}, \nabla^2 \boldsymbol{y} = \boldsymbol{Dd}$$

$$Q_{ji} = x_j^{2i-2},\; C_{ji} = \left(2i-2\right) x_j^{2i-3},$$

$$D_{ji} = \nabla^2\left(x^{2i-2}\right)|_{x_j}$$

$$\boldsymbol{d} = \boldsymbol{Q}^{-1}\boldsymbol{y}, \frac{\mathrm{d}\boldsymbol{y}}{\mathrm{d}x} = \boldsymbol{CQ}^{-1}\boldsymbol{y} = \boldsymbol{Ay},$$

$$\nabla^2 \boldsymbol{y} = \boldsymbol{DQ}^{-1}\boldsymbol{y} = \boldsymbol{By}$$

In addition, the quadrature formula is

$$\boldsymbol{WQ} = \boldsymbol{f},\; \boldsymbol{W} = \boldsymbol{fQ}^{-1}$$

where

$$\int_0^1 x^{2i-2} x^{a-1} \mathrm{d}x = \sum_{j=1}^{N+1} W_j x_j^{2i-2}$$

$$= \frac{1}{2i-2+a} \equiv f_i$$

As an example, for the problem

$$\frac{1}{x^{a-1}} \frac{\mathrm{d}}{\mathrm{d}x}\left(x^{a-1}\frac{\mathrm{d}c}{\mathrm{d}x}\right) = \varphi^2 R\left(c\right)$$

$$\frac{\mathrm{d}c}{\mathrm{d}x}\left(0\right) = 0,\; c\left(1\right) = 1$$

orthogonal collocation is applied at the interior points

$$\sum_{i=1}^{N+1} B_{ji} c_i = \varphi^2 R\left(c_j\right),\; j = 1,\ldots,N$$

and the boundary condition solved for is

$$c_{N+1} = 1$$

The boundary condition at x = 0 is satisfied automatically by the trial function. After the solution has been obtained, the effectiveness factor η is obtained by calculating

$$\eta \equiv \frac{\int_0^1 R\left[c\left(x\right)\right] x^{a-1} \mathrm{d}x}{\int_0^1 R\left[c\left(1\right)\right] x^{a-1} \mathrm{d}x} = \frac{\sum_{i=1}^{N+1} W_j R\left(c_j\right)}{\sum_{i=1}^{N+1} W_j R\left(1\right)}$$

Note that the effectiveness factor is the average reaction rate divided by the reaction rate evaluated at the external conditions. Error bounds have been given for linear problems [83, p. 356]. For planar geometry the error is

$$\text{Error in } \eta = \frac{\varphi^{2(2N+1)}}{(2N+1)!\,(2N+2)!}$$

This method is very accurate for small N (and small φ^2); note that for finite difference methods the error goes as $1/N^2$, which does not decrease as rapidly with N. If the solution is desired at the center (a frequent situation because the center concentration can be the most extreme one), it is given by

$$c\left(0\right) = d_1 \sum_{i=1}^{N+1} \left[Q^{-1}\right]_{1i} y_i$$

The collocation points are listed in Table 8. For small N the results are usually more accurate when the weighting function in Equation 25 is $1 - x^2$. The matrices for N = 1 and N = 2 are given in Table 9 for the three geometries. Computer programs to generate matrices and a program to solve reaction diffusion problems, OCRXN, are available [3, p. 325, p. 331].

Orthogonal collocation can be applied to distillation problems. Stewart et al. [84, 85] developed a method using Hahn polynomials that retains the discrete nature of a plate-to-plate distillation column. Other work treats problems with multiple liquid phases [86]. Some of the applications to chemical engineering can be found in [87 – 90].

Table 8. Collocation points for orthogonal collocation with symmetric polynomials and W=1

	Geometry		
N	Planar	Cylindrical	Spherical
1	0.5773502692	0.7071067812	0.7745966692
2	0.3399810436	0.4597008434	0.5384693101
	0.8611363116	0.8880738340	0.9061793459
3	0.2386191861	0.3357106870	0.4058451514
	0.6612093865	0.7071067812	0.7415311856
	0.9324695142	0.9419651451	0.9491079123
4	0.1834346425	0.2634992300	0.3242534234
	0.5255324099	0.5744645143	0.6133714327
	0.7966664774	0.8185294874	0.8360311073
	0.9602898565	0.9646596062	0.9681602395
5	0.1488743390	0.2165873427	0.2695431560
	0.4333953941	0.4803804169	0.5190961292
	0.6794095683	0.7071067812	0.7301520056
	0.8650633667	0.8770602346	0.8870625998
	0.9739065285	0.9762632447	0.9782286581

7.5. Orthogonal Collocation on Finite Elements

In the method of orthogonal collocation on finite elements, the domain is first divided into elements, and then within each element orthogonal collocation is applied. Figure 27 shows the domain being divided into NE elements, with $NCOL$ interior collocation points within each element, and $NP = NCOL + 2$ total points per element, giving $NT = NE * (NCOL + 1) + 1$ total number of points. Within each element a local coordinate is defined

$$u=\frac{x-x_{(k)}}{\Delta x_k},\ \Delta x_k=x_{(k+1)}-x_{(k)}$$

The reaction – diffusion equation is written as

$$\frac{1}{x^{a-1}}\frac{\mathrm{d}}{\mathrm{d}x}\left(x^{a-1}\frac{\mathrm{d}c}{\mathrm{d}x}\right)=\frac{\mathrm{d}^2c}{\mathrm{d}x^2}+\frac{a-1}{x}\frac{\mathrm{d}c}{\mathrm{d}x}=\varphi^2R(c)$$

and transformed to give

$$\frac{1}{\Delta x_k^2}\frac{\mathrm{d}^2}{\mathrm{d}u^2}+\frac{a-1}{x_{(k)}+u\Delta x_k}\frac{1}{\Delta x_k}\frac{\mathrm{d}c}{\mathrm{d}u}=\varphi^2R(c)$$

The boundary conditions are typically

$$\frac{\mathrm{d}c}{\mathrm{d}x}(0)=0,\ -\frac{\mathrm{d}c}{\mathrm{d}x}(1)=Bi_m\left[c(1)-c_B\right]$$

where Bi_m is the Biot number for mass transfer. These become

$$\frac{1}{\Delta x_1}\frac{\mathrm{d}c}{\mathrm{d}u}(u=0)=0,$$

in the first element;

$$-\frac{1}{\Delta x_{NE}}\frac{\mathrm{d}c}{\mathrm{d}u}(u=1)=Bi_m\left[c(u=1)-c_B\right],$$

in the last element. The orthogonal collocation method is applied at each interior collocation point.

$$\frac{1}{\Delta x_k^2}\sum_{J=1}^{NP}B_{IJ}c_J+\frac{a-1}{x_{(k)}+u_I\Delta x_k}\frac{1}{\Delta x_k}\sum_{J=1}^{NP}A_{IJ}c_J=$$
$$=\varphi^2R(c_J),\ I=2,\ldots,NP-1$$

The local points $i = 2, \ldots, NP - 1$ represent the interior collocation points. Continuity of the function and the first derivative between elements is achieved by taking

$$\left.\frac{1}{\Delta x_{k-1}}\sum_{J=1}^{NP}A_{NP,J}c_J\right|_{\text{element }k-1}$$
$$=\left.\frac{1}{\Delta x_k}\sum_{J=1}^{NP}A_{1,J}c_J\right|_{\text{element }k}$$

at the points between elements. Naturally, the computer code has only one symbol for the solution at a point shared between elements, but the derivative condition must be imposed. Finally, the boundary conditions at $x = 0$ and $x = 1$ are applied:

$$\frac{1}{\Delta x_k}\sum_{J=1}^{NP}A_{1,J}c_J=0,$$

in the first element;

$$-\frac{1}{\Delta x_{NE}}\sum_{J=1}^{NP}A_{NP,J}c_J=Bi_m\left[c_{NP}-c_B\right],$$

in the last element.

These equations can be assembled into an overall matrix problem

$$\boldsymbol{A}\,\boldsymbol{A}\boldsymbol{c}=\boldsymbol{f}$$

Table 9. Matrices for orthogonal collocation with symmetric polynomials and $W=1-x^2$

Planar geometry, a = 1

$N = 1$

$$x_j = \begin{pmatrix} 0.447214 \\ 1.000000 \end{pmatrix}, A_{ji} = \begin{pmatrix} -1.118034 & 1.118034 \\ -2.500000 & 2.500000 \end{pmatrix}$$

$$W_j = \begin{pmatrix} 0.833333 \\ 0.166667 \end{pmatrix}, B_{ji} = \begin{pmatrix} -2.5 & 2.5 \\ -2.5 & 2.5 \end{pmatrix}, Q_{ji}^{-1} = \begin{pmatrix} 1.25 & -0.25 \\ -1.25 & 1.25 \end{pmatrix}$$

$N = 2$

$$x_j = \begin{pmatrix} 0.285232 \\ 0.765055 \\ 1.000000 \end{pmatrix}, W_j = \begin{pmatrix} 0.554858 \\ 0.378475 \\ 0.066667 \end{pmatrix}, A_{ji} = \begin{pmatrix} -1.752962 & 2.507614 & -0.754652 \\ -1.370599 & -0.653547 & 2.024146 \\ 1.791503 & -8.791503 & 7.000000 \end{pmatrix}$$

$$B_{ji} = \begin{pmatrix} -4.73987 & 5.67713 & -0.93725 \\ 8.32288 & -23.26013 & 14.93725 \\ 19.07189 & -47.07190 & 28.00000 \end{pmatrix}, Q_{ji}^{-1} = \begin{pmatrix} 1.26430 & -0.38930 & 0.125 \\ -3.42435 & 5.17435 & -1.750 \\ 2.16005 & -4.78505 & 2.625 \end{pmatrix}$$

Cylindrical geometry, a = 2

$N = 1$

$$x_j = \begin{pmatrix} 0.577350 \\ 1.000000 \end{pmatrix}, W_j = \begin{pmatrix} 0.375 \\ 0.125 \end{pmatrix}, A_{ji} = \begin{pmatrix} -1.732051 & 1.732051 \\ -3.000000 & 3.000000 \end{pmatrix}$$

$$B_{ji} = \begin{pmatrix} -6 & 6 \\ -6 & 6 \end{pmatrix}, Q_{ji}^{-1} = \begin{pmatrix} 1.5 & -0.5 \\ -1.5 & 1.5 \end{pmatrix}$$

$N = 2$

$$x_j = \begin{pmatrix} 0.39377 \\ 0.80309 \\ 1.000000 \end{pmatrix}, W_j = \begin{pmatrix} 0.18820 \\ 0.25624 \\ 0.05555 \end{pmatrix}, A_{ji} = \begin{pmatrix} -2.53958 & 3.82562 & -1.28603 \\ -1.37768 & -1.24519 & 2.62287 \\ 1.71548 & -9.71548 & 8.00000 \end{pmatrix}$$

$$B_{ji} = \begin{pmatrix} -9.90238 & 12.29966 & -2.39728 \\ 9.03367 & -32.76429 & 23.73061 \\ 22.7575 & -65.42415 & 42.66667 \end{pmatrix}, Q_{ji}^{-1} = \begin{pmatrix} 1.58808 & -0.89141 & 0.33333 \\ -3.97389 & 6.64056 & -2.66667 \\ 2.41582 & -5.74914 & 3.33333 \end{pmatrix}$$

Spherical geometry, a = 3

$N = 1$

$$x_j = \begin{pmatrix} 0.654654 \\ 1.000000 \end{pmatrix}, W_j = \begin{pmatrix} 0.233333 \\ 0.100000 \end{pmatrix}, A_{ji} = \begin{pmatrix} -2.291288 & 2.291288 \\ -3.500000 & 3.500000 \end{pmatrix}$$

$$B_{ji} = \begin{pmatrix} -10.5 & 10.5 \\ -10.5 & 10.5 \end{pmatrix}, Q_{ji}^{-1} = \begin{pmatrix} 1.75 & -0.75 \\ -1.75 & 1.75 \end{pmatrix}$$

$N = 2$

$$x_j = \begin{pmatrix} 0.46885 \\ 0.83022 \\ 1.00000 \end{pmatrix}, W_j = \begin{pmatrix} 0.09491 \\ 0.19081 \\ 0.04762 \end{pmatrix}, A_{ji} = \begin{pmatrix} -3.19933 & 5.01517 & -1.81584 \\ -1.40870 & -1.80674 & 3.21544 \\ 1.69677 & -10.69677 & 9.00000 \end{pmatrix}$$

$$B_{ji} = \begin{pmatrix} -15.66996 & 20.03488 & -4.36492 \\ 9.96512 & -44.33004 & 34.36492 \\ 26.93285 & -86.93229 & 60.00000 \end{pmatrix}, Q_{ji}^{-1} = \begin{pmatrix} 1.88193 & -1.50693 & 0.625 \\ -4.61225 & 8.36225 & -3.750 \\ 2.73032 & -6.85532 & 4.125 \end{pmatrix}$$

The form of these equations is special and is discussed by FINLAYSON [3, p. 116], who also gives the computer code to solve linear equations arising in such problems. Reaction – diffusion problems are solved by the program OCFERXN [3, p. 337]. See also the program COLSYS described below.

The error bounds of DEBOOR [91] give the following results for second-order problems solved with cubic trial functions on finite elements with

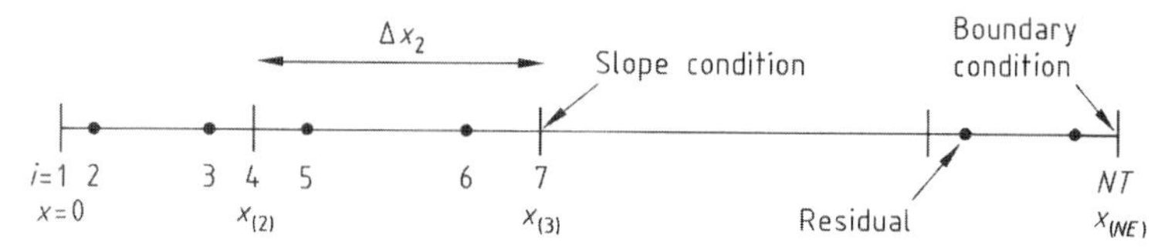

Figure 27. Grid for orthogonal collocation on finite elements

continuous first derivatives. The error at all positions is bounded by

$$\left\| \frac{d^i}{dx^i} (y - y_{\text{exact}}) \right\|_\infty \leq \text{constant} |\Delta x|^2$$

The error at the collocation points is more accurate, giving what is known as superconvergence.

$$\left| \frac{d^i}{dx^i} (y - y_{\text{exact}}) \right|_{\text{collocation points}} \leq \text{constant} |\Delta x|^4$$

7.6. Galerkin Finite Element Method

In the finite element method the domain is divided into elements and an expansion is made for the solution on each finite element. In the Galerkin finite element method an additional idea is introduced: the Galerkin method is used to solve the equation. The Galerkin method is explained before the finite element basis set is introduced.

To solve the problem

$$\frac{1}{x^{a-1}} \frac{d}{dx} \left(x^{a-1} \frac{dc}{dx} \right) = \varphi^2 R(c)$$

$$\frac{dc}{dx}(0) = 0, \ -\frac{dc}{dx}(1) = Bi_m \left[c(1) - c_B \right]$$

the unknown solution is expanded in a series of known functions $\{b_i(x)\}$, with unknown coefficients $\{a_i\}$.

$$c(x) = \sum_{i=1}^{NT} a_i b_i(x)$$

The series (the trial solution) is inserted into the differential equation to obtain the residual:

$$\text{Residual} = \sum_{i=1}^{NT} a_i \frac{1}{x^{a-1}} \frac{d}{dx} \left(x^{a-1} \frac{db_i}{dx} \right) - \varphi^2 R \left[\sum_{i=1}^{NT} a_i b_i(x) \right]$$

The residual is then made orthogonal to the set of basis functions.

$$\int_0^1 b_j(x) \left\{ \sum_{i=1}^{NT} a_i \frac{1}{x^{a-1}} \frac{d}{dx} \left(x^{a-1} \frac{db_i}{dx} \right) - \varphi^2 R \left[\sum_{i=1}^{NT} a_i b_i(x) \right] \right\} x^{a-1} dx = 0$$

$$j = 1, \ldots, NT$$

This process makes the method a Galerkin method. The basis for the orthogonality condition is that a function that is made orthogonal to each member of a complete set is then zero. The residual is being made orthogonal, and if the basis functions are complete, and an infinite number of them are used, then the residual is zero. Once the residual is zero the problem is solved. It is necessary also to allow for the boundary conditions. This is done by integrating the first term of Equation 26 by parts and then inserting the boundary conditions:

$$\begin{aligned}
&\int_0^1 b_j(x) \frac{1}{x^{a-1}} \frac{d}{dx} \left(x^{a-1} \frac{db_i}{dx} \right) x^{a-1} dx = \\
&\int_0^1 \frac{d}{dx} \left[b_j(x) x^{a-1} \frac{db_i}{dx} \right] dx - \int_0^1 \frac{db_j}{dx} \frac{db_i}{dx} x^{a-1} dx \\
&= \left[b_j(x) x^{a-1} \frac{db_i}{dx} \right]_0^1 - \int_0^1 \frac{db_j}{dx} \frac{db_i}{dx} x^{a-1} dx \\
&= -\int_0^1 \frac{db_j}{dx} \frac{db_i}{dx} x^{a-1} dx - Bi_m b_j(1) \left[b_i(1) - c_B \right]
\end{aligned} \tag{27}$$

Combining this with Equation 26 gives

$$\begin{aligned}
&-\sum_{i=1}^{NT} \int_0^1 \frac{db_j}{dx} \frac{db_i}{dx} x^{a-1} dx a_i \\
&- Bi_m b_j(1) \left[\sum_{i=1}^{NT} a_i b_i(1) - c_B \right] \\
&= \varphi^2 \int_0^1 b_j(x) \left[\sum_{i=1}^{NT} a_i b_i(x) \right] x^{a-1} dx \\
&j = 1, \ldots, NT
\end{aligned} \tag{28}$$

This equation defines the Galerkin method, and a solution that satisfies this equation (for all $j = 1, \ldots, \infty$) is called a weak solution. For an approximate solution the equation is written once for each member of the trial function, $j = 1, \ldots, NT$. If the boundary condition is

$$c(1) = c_B$$

then the boundary condition is used (instead of Eq. 28) for $j = NT$,

$$\sum_{i=1}^{NT} a_i b_i (1) = c_B$$

The Galerkin finite element method results when the Galerkin method is combined with a finite element trial function. Both linear and quadratic finite element approximations are described in Chapter 2. The trial functions $b_i(x)$ are then generally written as $N_i(x)$.

$$c(x) = \sum_{i=1}^{NT} c_i N_i(x)$$

Each $N_i(x)$ takes the value 1 at the point x_i and zero at all other grid points (Chap. 2). Thus c_i are the nodal values, $c(x_i) = c_i$. The first derivative must be transformed to the local coordinate system, $u = 0$ to 1 when x goes from x_i to $x_i + \Delta x$.

$$\frac{dN_j}{dx} = \frac{1}{\Delta x_e}\frac{dN_J}{du},\ dx = \Delta x_e du$$

in the e-th element. Then the Galerkin method is

$$\begin{aligned} &-\sum_e \frac{1}{\Delta x_e}\sum_{I=1}^{NP}\int_0^1 \frac{dN_J}{du}\frac{dN_I}{du}(x_e+u\Delta x_e)^{a-1}du c_I^e \\ &-Bi_m \sum_e N_J(1)\left[\sum_{I=1}^{NP} c_I^e N_I(1) - c_1\right] \\ &= \varphi^2 \sum_e \Delta x_e \int_0^1 N_J(u) R\left[\sum_{I=1}^{NP} c_I^e N_I(u)\right] \\ &(x_e+u\Delta x_e)^{a-1}du \end{aligned} \quad (29)$$

The element integrals are defined as

$$B_{JI}^e = -\frac{1}{\Delta x_e}\int_0^1 \frac{dN_J}{du}\frac{dN_I}{du}(x_e+u\Delta x_e)^{a-1}du,$$

$$F_J^e = \varphi^2 \Delta x_e \int_0^1 N_J(u) R\left[\sum_{I=1}^{NP} c_I^e N_I(u)\right] (x_e+u\Delta x_e)^{a-1}du$$

whereas the boundary element integrals are

$$BB_{JI}^e = -Bi_m N_J(1) N_I(1),$$

$$FF_J^e = -Bi_m N_J(1) c_1$$

Then the entire method can be written in the compact notation

$$\sum_e B_{JI}^e c_I^e + \sum_e BB_{JI}^e c_I^e = \sum_e F_J^e + \sum_e FF_J^e$$

The matrices for various terms are given in Table 10. This equation can also be written in the form

$$\boldsymbol{AAc} = \boldsymbol{f}$$

where the matrix $\boldsymbol{AA}$ is sparse. If linear elements are used the matrix is tridiagonal. If quadratic elements are used the matrix is pentadiagonal. Naturally the linear algebra is most efficiently carried out if the sparse structure is taken into account. Once the solution is found the solution at any point can be recovered from

$$c^e(u) = c_{I=1}^e(1-u) + c_{I=2}^e u$$

for linear elements

$$\begin{aligned} c^e(u) &= c_{I=1}^e 2(u-1)\left(u-\tfrac{1}{2}\right) \\ &+ c_{I=2}^e 4u(1-u) + c_{I=3}^e 2u\left(u-\tfrac{1}{2}\right) \end{aligned}$$

for quadratic elements

Table 10. Element matrices for Galerkin method

Linear shape functions	*Quadratic shape functions*
$N_1 = 1-u,\ N_2 = u,\ \frac{dN_1}{du} = -1,\ \frac{dN_2}{du} = 1$	$N_1 = 2(u-1)\left(u-\frac{1}{2}\right),\ N_2 = 4u(1-u),\ N_3 = 2u\left(u-\frac{1}{2}\right),$ $\frac{dN_1}{du} = 4u-3,\ \frac{dN_2}{du} = 4-8u,\ \frac{dN_3}{du} = 4u-1$
$\int_0^1 \frac{dN_J}{du}\frac{dN_I}{du}du = \begin{pmatrix} 1 & -1 \\ -1 & 1 \end{pmatrix},\ \int_0^1 N_J \frac{dN_I}{du}du = \begin{pmatrix} -\frac{1}{2} & \frac{1}{2} \\ -\frac{1}{2} & \frac{1}{2} \end{pmatrix}$	$\int_0^1 \frac{dN_J}{du}\frac{dN_I}{du}du = \frac{1}{3}\begin{pmatrix} 7 & -8 & 1 \\ -8 & 16 & -8 \\ 1 & -8 & 7 \end{pmatrix},\ \int_0^1 N_J \frac{dN_I}{du}du = \frac{1}{6}\begin{pmatrix} -3 & 4 & -1 \\ -4 & 0 & 4 \\ 1 & -4 & 3 \end{pmatrix}$
$\int_0^1 N_J N_I du = \begin{pmatrix} \frac{1}{3} & \frac{1}{6} \\ \frac{1}{6} & \frac{1}{3} \end{pmatrix},\ \int_0^1 N_J du = \begin{pmatrix} \frac{1}{2} \\ \frac{1}{2} \end{pmatrix},\ \int_0^1 N_J u du = \begin{pmatrix} \frac{1}{6} \\ \frac{1}{3} \end{pmatrix}$	$\int_0^1 N_J N_I du = \frac{1}{30}\begin{pmatrix} 4 & 2 & -1 \\ 2 & 16 & 2 \\ -1 & 2 & 4 \end{pmatrix},\ \int_0^1 N_J du = \frac{1}{6}\begin{pmatrix} 1 \\ 4 \\ 1 \end{pmatrix},\ \int_0^1 N_J u du = \frac{1}{6}\begin{pmatrix} 0 \\ 2 \\ 1 \end{pmatrix}$

Because the integrals in Equation 28 may be complicated, they are usually formed by using Gaussian quadrature. If *NG* Gauss points are used, a typical term would be

$$\int_0^1 N_J(u)\, R\left[\sum_{I=1}^{NP} c_I^e N_I(u)\right](x_e+u\Delta x_e)^{a-1}\mathrm{d}u$$

$$=\sum_{k=1}^{NG} W_k N_J(u_k)\, R\left[\sum_{I=1}^{NP} c_I^e N_I(u_k)\right]$$

$$(x_e+u_k\Delta x_e)^{a-1}$$

7.7. Cubic B-Splines

Cubic B-splines have cubic approximations within each element, but first and second derivatives continuous between elements. The functions are the same ones discussed in Chapter 2, and they can be used to solve differential equations, too. See Sincovec [92].

7.8. Adaptive Mesh Strategies

In many two-point boundary value problems, the difficulty in the problem is the formation of a boundary layer region, or a region in which the solution changes very dramatically. In such cases small mesh spacing should be used there, either with the finite difference method or the finite element method. If the region is known a priori, small mesh spacings can be assumed at the boundary layer. If the region is not known though, other techniques must be used. These techniques are known as adaptive mesh techniques. The general strategy is to estimate the error, which depends on the grid size and derivatives of the solution, and refine the mesh where the error is large.

The adaptive mesh strategy was employed by ASCHER et al. [93] and by RUSSELL and CHRISTIANSEN [94]. For a second-order differential equation and cubic trial functions on finite elements, the error in the *i*-th element is given by

$$\|\text{Error}\|_i = c\Delta x_i^4 \|u^{(4)}\|_i$$

Because cubic elements do not have a nonzero fourth derivative, the third derivative in adjacent elements is used [3, p. 166]:

$$a_i=\frac{1}{\Delta x_i^3}\frac{\mathrm{d}^3 c^i}{\mathrm{d}u^3},\; a_{i+1}=\frac{1}{\Delta x_{i+1}^3}\frac{\mathrm{d}^3 c^{i+1}}{\mathrm{d}u^3}$$

$$\|u^{(4)}\|_i\approx\frac{1}{2}\left[\frac{a_i-a_{i-1}}{\frac{1}{2}(x_{i+1}-x_{i-1})}+\frac{a_{i+1}-a_i}{\frac{1}{2}(x_{i+2}-x_i)}\right]$$

Element sizes are then chosen so that the following error bounds are satisfied

$$C\Delta x_i^4\|u^{(4)}\|_i\leq\varepsilon \text{ for all } i$$

These features are built into the code COLSYS (http://www.netlib.org/ode/).

The error expected from a method one order higher and one order lower can also be defined. Then a decision about whether to increase or decrease the order of the method can be made by taking into account the relative work of the different orders. This provides a method of adjusting both the mesh spacing (Δx, sometimes called h) and the degree of polynomial (p). Such methods are called $h-p$ methods.

7.9. Comparison

What method should be used for any given problem? Obviously the error decreases with some power of Δx, and the power is higher for the higher order methods, which suggests that the error is less. For example, with linear elements the error is

$$y(\Delta x)=y_{\text{exact}}+c_2\Delta x^2$$

for small enough (and uniform) Δx. A computer code should be run for varying Δx to confirm this. For quadratic elements, the error is

$$y(\Delta x)=y_{\text{exact}}+c_3\Delta x^3$$

If orthogonal collocation on finite elements is used with cubic polynomials, then

$$y(\Delta x)=y_{\text{exact}}+c_4\Delta x^4$$

However, the global methods, not using finite elements, converge even faster [95], for example,

$$y(N)=y_{\text{exact}}+c_N\left(\frac{1}{NCOL}\right)^{NCOL}$$

Yet the workload of the methods is also different. These considerations are discussed in [3]. Here, only sweeping generalizations are given.

If the problem has a relatively smooth solution, then the orthogonal collocation method is preferred. It gives a very accurate solution, and N can be quite small so the work is small. If the problem has a steep front in it, the finite difference method or finite element method is indicated, and adaptive mesh techniques should probably be employed. Consider the reaction – diffusion problem: as the Thiele modulus φ increases from a small value with no diffusion limitations to a large value with significant diffusion limitations, the solution changes as shown in Figure 28. The orthogonal collocation method is initially the method of choice. For intermediate values of φ, N = 3 – 6 must be used, but orthogonal collocation still works well (for η down to approximately 0.01). For large φ, use of the finite difference method, the finite element method, or an asymptotic expansion for large φ is better. The decision depends entirely on the type of solution that is obtained. For steep fronts the finite difference method and finite element method with adaptive mesh are indicated.

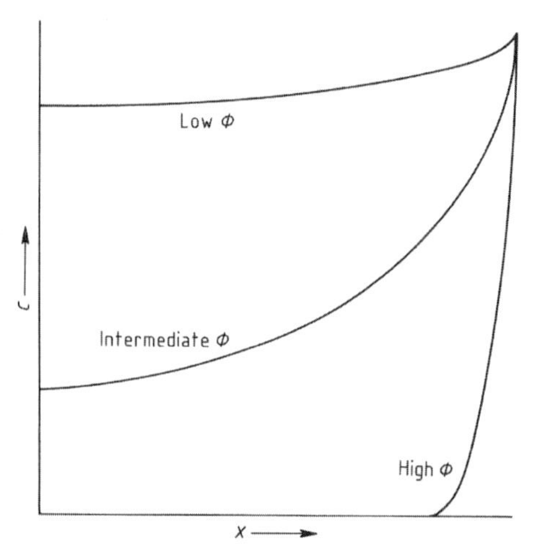

Figure 28. Concentration solution for different values of Thiele modulus

7.10. Singular Problems and Infinite Domains

If the solution being sought has a singularity, a good numerical solution may be hard to find. Sometimes even the location of the singularity may not be known [96, pp. 230 – 238]. One method of solving such problems is to refine the mesh near the singularity, by relying on the better approximation due to a smaller Δx. Another approach is to incorporate the singular trial function into the approximation. Thus, if the solution approaches $f(x)$ as x goes to zero, and $f(x)$ becomes infinite, an approximation may be taken as

$$y(x) = f(x) + \sum_{i=1}^{N} a_i y_i(x)$$

This function is substituted into the differential equation, which is solved for a_i. Essentially, a new differential equation is being solved for a new variable:

$$u(x) \equiv y(x) - f(x)$$

The differential equation is more complicated but has a better solution near the singularity (see [97, pp. 189 – 192], [98, p. 611]).

Sometimes the domain is infinite. Boundary layer flow past a flat plate is governed by the Blasius equation for stream function [99, p. 117].

$$2\frac{\mathrm{d}^3}{\mathrm{d}\eta^3} + f\frac{\mathrm{d}^2}{\mathrm{d}\eta^2} = 0$$

$$f = \frac{\mathrm{d}f}{\mathrm{d}\eta} = 0 \text{ at } \eta = 0$$

$$\frac{\mathrm{d}f}{\mathrm{d}\eta} = 1 \text{ at } \eta \rightarrow \infty$$

Because one boundary is at infinity using a mesh with a constant size is difficult! One approach is to transform the domain. For example, let

$$z = \mathrm{e}^{-\eta}$$

Then η = 0 becomes z = 1 and $\eta = \infty$ becomes z = 0. The derivatives are

$$\frac{\mathrm{d}z}{\mathrm{d}\eta} = -\mathrm{e}^{-\eta} = -z,\ \frac{\mathrm{d}^2 z}{\mathrm{d}\eta^2} = \mathrm{e}^{-\eta} = z$$

$$\frac{\mathrm{d}f}{\mathrm{d}\eta} = \frac{\mathrm{d}f}{\mathrm{d}z}\frac{\mathrm{d}z}{\mathrm{d}\eta} = -z\frac{\mathrm{d}f}{\mathrm{d}z}$$

$$\frac{\mathrm{d}^2 f}{\mathrm{d}\eta^2} = \frac{\mathrm{d}^2 f}{\mathrm{d}z^2}\left(\frac{\mathrm{d}z}{\mathrm{d}\eta}\right)^2 + \frac{\mathrm{d}f}{\mathrm{d}z}\frac{\mathrm{d}^2 z}{\mathrm{d}\eta^2} = z^2\frac{\mathrm{d}^2 f}{\mathrm{d}z^2} + z\frac{\mathrm{d}f}{\mathrm{d}z}$$

The Blasius equation becomes

$$2\left[-z^3\frac{\mathrm{d}^3 f}{\mathrm{d}z^3} - 3z^2\frac{\mathrm{d}^2 f}{\mathrm{d}z^2} - z\frac{\mathrm{d}f}{\mathrm{d}z}\right]$$

$$+f\left[z^2\frac{\mathrm{d}^2}{\mathrm{d}z^2} + z\frac{\mathrm{d}f}{\mathrm{d}z}\right] = 0 \qquad \text{for } 0 \leq z \leq 1.$$

The differential equation now has variable coefficients, but these are no more difficult to handle than the original nonlinearities.

Another approach is to use a variable mesh, perhaps with the same transformation. For example, use $z = e^{-\eta}$ and a constant mesh size in z. Then with 101 points distributed uniformly from $z = 0$ to $z = 1$, the following are the nodal points:

$$z = 0.,0.01,0.02,\ldots,0.99,1.0$$

$$\eta = \infty,4.605,3.912,\ldots,0.010,0$$

$$\Delta\eta = \infty,0.639,\ldots,0.01$$

Still another approach is to solve on a finite mesh in which the last point is far enough away that its location does not influence the solution. A location that is far enough away must be found by trial and error.

8. Partial Differential Equations

Partial differential equations are differential equations in which the dependent variable is a function of two or more independent variables. These can be time and one space dimension, or time and two or more space dimensions, or two or more space dimensions alone. Problems involving time are generally either hyperbolic or parabolic, whereas those involving spatial dimensions only are often elliptic. Because the methods applied to each type of equation are very different, the equation must first be classified as to its type. Then the special methods applicable to each type of equation are described. For a discussion of all methods, see [100 – 103]; for a discussion oriented more toward chemical engineering applications, see [104]. Examples of hyperbolic and parabolic equations include chemical reactors with radial dispersion, pressure-swing adsorption, dispersion of an effluent, and boundary value problems with transient terms added (heat transfer, mass transfer, chemical reaction). Examples of elliptic problems include heat transfer and mass transfer in two and three spatial dimensions and steady fluid flow.

8.1. Classification of Equations

A set of differential equations may be hyperbolic, elliptic, or parabolic, or it may be of mixed type. The type may change for different parameters or in different regions of the flow. This can happen in the case of nonlinear problems; an example is a compressible flow problem with both subsonic and supersonic regions. *Characteristic curves* are curves along which a discontinuity can propagate. For a given set of equations, it is necessary to determine if characteristics exist or not, because that determines whether the equations are hyperbolic, elliptic, or parabolic.

Linear Problems For linear problems, the theory summarized by JOSEPH et al. [105] can be used.

$$\frac{\partial}{\partial t},\frac{\partial}{\partial x_i},\ldots,\frac{\partial}{\partial x_n}$$

is replaced with the Fourier variables

$$i\xi_0,i\xi_1,\ldots,i\xi_n$$

If the m-th order differential equation is

$$P = \sum_{|\alpha|=m} a_\alpha \partial^\alpha + \sum_{|\alpha|<m} b_\alpha \partial^\alpha$$

where

$$\alpha = (\alpha_0,\alpha_1,\ldots,\alpha_n)\,,\ |\alpha| = \sum_{i=0}^{n} \alpha_i$$

$$\partial^\alpha = \frac{\partial^{|\alpha|}}{\partial t^{\alpha_0} \partial x_1^{\alpha_1} \ldots \partial x_n^{\alpha_n}}$$

the characteristic equation for P is defined as

$$\sum_{|\alpha|=m} a_\alpha \sigma^\alpha = 0,\ \sigma = (\sigma_0,\sigma_1,\ldots,\sigma_n)$$
$$\sigma^\alpha = \sigma_0^{\alpha_0} \sigma_1^{\alpha_1} \ldots \sigma_n^{\alpha_n} \tag{30}$$

where σ represents coordinates. Thus only the highest derivatives are used to determine the type. The surface is defined by this equation plus a normalization condition:

$$\sum_{k=0}^{n} \sigma_k^2 = 1$$

The shape of the surface defined by Equation 30 is also related to the type: elliptic equations give rise to ellipses; parabolic equations give rise to parabolas; and hyperbolic equations give rise to hyperbolas.

$$\frac{\sigma_1^2}{a^2} + \frac{\sigma_2^2}{b^2} = 1,\ \text{Ellipse}$$

$$\sigma_0 = a\sigma_1^2,\ \text{Parabola}$$

$$\sigma_0^2 - a\sigma_1^2 = 0,\ \text{Hyperbola}$$

If Equation 30 has no nontrivial real zeroes then the equation is called elliptic. If all the roots are real and distinct (excluding zero) then the operator is hyperbolic.

This formalism is applied to three basic types of equations. First consider the equation arising from steady diffusion in two dimensions:

$$\frac{\partial^2 c}{\partial x^2}+\frac{\partial^2 c}{\partial y^2}=0$$

This gives

$$-\xi_1^2-\xi_2^2=-\left(\xi_2^1+\xi_2^2\right)=0$$

Thus,

$$\sigma_1^2+\sigma_2^2=1 \text{ (normalization)}$$

$$\sigma_1^2+\sigma_2^2=0 \text{ (equation)}$$

These cannot both be satisfied so the problem is elliptic. When the equation is

$$\frac{\partial^2 u}{\partial t^2}-\frac{\partial^2 u}{\partial x^2}=0$$

then

$$-\xi_0^2+\xi_1^2=0$$

Now real ξ_0 can be solved and the equation is hyperbolic

$$\sigma_0^2+\sigma_1^2=1 \text{ (normalization)}$$

$$-\sigma_0^2+\sigma_1^2=0 \text{ (equation)}$$

When the equation is

$$\frac{\partial c}{\partial t}=D\left(\frac{\partial^2 c}{\partial x^2}+\frac{\partial^2 c}{\partial y^2}\right)$$

then

$$\sigma_0^2+\sigma_1^2+\sigma_2^2=1 \text{ (normalization)}$$

$$\sigma_1^2+\sigma_2^2=0 \text{ (equation)}$$

thus we get

$$\sigma_0^2=1 \text{ (for normalization)}$$

and the characteristic surfaces are hyperplanes with t = constant. This is a parabolic case.

Consider next the telegrapher's equation:

$$\frac{\partial T}{\partial t}+\beta\frac{\partial^2 T}{\partial t^2}=\frac{\partial^2 T}{\partial x^2}$$

Replacing the derivatives with the Fourier variables gives

$$i\xi_0-\beta\xi_0^2+\xi_1^2=0$$

The equation is thus second order and the type is determined by

$$-\beta\sigma_0^2+\sigma_1^2=0$$

The normalization condition

$$\sigma_0^2+\sigma_1^2=1$$

is required. Combining these gives

$$1-(1+\beta)\,\sigma_0^2=0$$

The roots are real and the equation is hyperbolic. When $\beta=0$

$$\xi_1^2=0$$

and the equation is parabolic.

First-order quasi-linear problems are written in the form

$$\sum_{l=0}^{n}\boldsymbol{A}_1\frac{\partial \boldsymbol{u}}{\partial x_1}=\boldsymbol{f},\ \boldsymbol{x}=(t,x_1\ldots,x_n) \tag{31}$$

$$\boldsymbol{u}=(u_1,u_2,\ldots,u_k)$$

The matrix entries $\boldsymbol{A}_1$ is a $k\times k$ matrix whose entries depend on $\boldsymbol{u}$ but not on derivatives of $\boldsymbol{u}$. Equation 31 is hyperbolic if

$$\mathrm{A}=\mathrm{A}_\mu$$

is nonsingular and for any choice of real λ_1, $l=0,\ldots,n$, $l\neq\mu$ the roots α_k of

$$\det\left(\alpha\boldsymbol{A}-\sum_{\substack{l=0\\ l\neq\mu}}^{n}\lambda_1\boldsymbol{A}_1\right)=0$$

are real. If the roots are complex the equation is elliptic; if some roots are real and some are complex the equation is of mixed type.

Apply these ideas to the *advection equation*

$$\frac{\partial u}{\partial t}+F(u)\frac{\partial u}{\partial x}=0$$

Thus,

$$\det\left(\alpha\boldsymbol{A}_0-\lambda_1\boldsymbol{A}_1\right)=0 \text{ or } \det\left(\alpha\boldsymbol{A}_1-\lambda_0\boldsymbol{A}_0\right)=0$$

In this case,

$$n=1,\ \boldsymbol{A}_0=1,\ \boldsymbol{A}_1=F(u)$$

Using the first of the above equations gives

$$\det\left(\alpha-\lambda_1 F(u)\right)=0, \text{ or } \alpha=\lambda_1 F(u)$$

Thus, the roots are real and the equation is hyperbolic.

The final example is the heat conduction problem written as

$$\varrho C_p \frac{\partial T}{\partial t} = -\frac{\partial q}{\partial x},\ q = -k\frac{\partial T}{\partial x}$$

In this formulation the constitutive equation for heat flux is separated out; the resulting set of equations is first order and written as

$$\varrho C_p \frac{\partial T}{\partial t} + \frac{\partial q}{\partial x} = 0$$

$$k\frac{\partial T}{\partial x} = -q$$

In matrix notation this is

$$\begin{bmatrix} \varrho C_p & 0 \\ 0 & 0 \end{bmatrix} \begin{bmatrix} \frac{\partial T}{\partial t} \\ \frac{\partial q}{\partial t} \end{bmatrix} + \begin{bmatrix} 0 & 1 \\ k & 0 \end{bmatrix} \begin{bmatrix} \frac{\partial T}{\partial x} \\ \frac{\partial q}{\partial t} \end{bmatrix} = \begin{bmatrix} 0 \\ -q \end{bmatrix}$$

This compares with

$$\boldsymbol{A}_0 \frac{\partial \boldsymbol{u}}{\partial x_0} + \boldsymbol{A}_1 \frac{\partial \boldsymbol{u}}{\partial x_1} = \boldsymbol{f}$$

In this case $\boldsymbol{A}_0$ is singular whereas $\boldsymbol{A}_1$ is nonsingular. Thus,

$$\det\left(\alpha \boldsymbol{A}_1 - \lambda_0 \boldsymbol{A}_0\right) = 0$$

is considered for any real λ_0. This gives

$$\begin{vmatrix} -\varrho C_p \lambda_0 & \alpha \\ k\alpha & 0 \end{vmatrix} = 0$$

or

$$\alpha^2 k = 0$$

Thus the α is real, but zero, and the equation is parabolic.

8.2. Hyperbolic Equations

The most common situation yielding hyperbolic equations involves unsteady phenomena with convection. A prototype equation is

$$\frac{\partial c}{\partial t} + \frac{\partial F(c)}{\partial x} = 0$$

Depending on the interpretation of c and $F(c)$, this can represent accumulation of mass and convection. With $F(c) = u\,c$, where u is the velocity, the equation represents a mass balance on concentration. If diffusive phenomenon are important, the equation is changed to

$$\frac{\partial c}{\partial t} + \frac{\partial F(c)}{\partial x} = D\frac{\partial^2 c}{\partial x^2} \tag{32}$$

where D is a diffusion coefficient. Special cases are the *convective diffusive equation*

$$\frac{\partial c}{\partial t} + u\frac{\partial c}{\partial x} = D\frac{\partial^2 c}{\partial x^2} \tag{33}$$

and *Burgers viscosity equation*

$$\frac{\partial u}{\partial t} + u\frac{\partial u}{\partial x} = \nu\frac{\partial^2 u}{\partial x^2} \tag{34}$$

where u is the velocity and ν is the kinematic viscosity. This is a prototype equation for the Navier – Stokes equations (→ Fluid Mechanics). For *adsorption phenomena* [106, p. 202],

$$\varphi\frac{\partial c}{\partial t} + \varphi u\frac{\partial c}{\partial x} + (1-\varphi)\frac{\mathrm{d}f}{\mathrm{d}c}\frac{\partial c}{\partial t} = 0 \tag{35}$$

where φ is the void fraction and $f(c)$ gives the equilibrium relation between the concentrations in the fluid and in the solid phase. In these examples, if the diffusion coefficient D or the kinematic viscosity ν is zero, the equations are hyperbolic. If D and ν are small, the phenomenon may be essentially hyperbolic even though the equations are parabolic. Thus the numerical methods for hyperbolic equations may be useful even for parabolic equations.

Equations for several methods are given here, as taken from [107]. If the convective term is treated with a centered difference expression the solution exhibits oscillations from node to node, and these vanish only if a very fine grid is used. The simplest way to avoid the oscillations with a hyperbolic equation is to use upstream derivatives. If the flow is from left to right, this would give the following for Equations (40):

$$\frac{\mathrm{d}c_i}{\mathrm{d}t} + \frac{F(c_i) - F(c_{i-1})}{\Delta x} = D\frac{c_{i+1} - 2c_i + c_{i-1}}{\Delta x^2}$$

for Equation 34:

$$\frac{\mathrm{d}u_i}{\mathrm{d}t} + u_i\frac{u_i - u_{i-1}}{\Delta x} = \nu\frac{u_{i+1} - 2u_i + u_{i-1}}{\Delta x^2}$$

and for Equation 35:

$$\varphi\frac{\mathrm{d}c_i}{\mathrm{d}t} + \varphi u_i\frac{c_i - c_{i-1}}{\Delta x} + (1-\varphi)\left.\frac{\mathrm{d}f}{\mathrm{d}c}\right|_i \frac{\mathrm{d}c_i}{\mathrm{d}t} = 0$$

If the flow were from right to left, then the formula would be

$$\frac{\mathrm{d}c_i}{\mathrm{d}t} + \frac{F(c_{i+1}) - F(c_i)}{\Delta x} = D\frac{c_{i+1} - 2c_i + c_{i-1}}{\Delta x^2}$$

If the flow could be in either direction, a local determination must be made at each node i and the appropriate formula used. The effect of using upstream derivatives is to add artificial or numerical diffusion to the model. This can be ascertained by taking the finite difference form of the convective diffusion equation

$$\frac{dc_i}{dt}+u\frac{c_i-c_{i-1}}{\Delta x}=D\frac{c_{i+1}-2c_i+c_{i-1}}{\Delta x^2}$$

and rearranging

$$\frac{dc_i}{dt}+u\frac{c_{i+1}-c_{i-1}}{2\Delta x}$$

$$=\left(D+\frac{u\Delta x}{2}\right)\frac{c_{i+1}-2c_i+c_{i-1}}{\Delta x^2}$$

Thus the diffusion coefficient has been changed from

$$D \text{ to } D+\frac{u\Delta x}{2}$$

Expressed in terms of the cell Peclet number, $Pe_\Delta=u\Delta x/D$, this is D is changed to $D[1+Pe_\Delta/2]$

The cell Peclet number should always be calculated as a guide to the calculations. Using a large cell Peclet number and upstream derivatives leads to excessive (and artificial) smoothing of the solution profiles.

Another method often used for hyperbolic equations is the *MacCormack method.* This method has two steps; it is written here for Equation 33.

$$c_i^{*n+1}=c_i^n-\frac{u\Delta t}{\Delta x}\left(c_{i+1}^n-c_i^n\right)$$

$$+\frac{\Delta tD}{\Delta x^2}\left(c_{i+1}^n-2c_i^n+c_{i-1}^n\right)$$

$$c_i^{n+1}=\frac{1}{2}\left(c_i^n+c_i^{*n+1}\right)-\frac{u\Delta t}{2\Delta x}\left(c_i^{*n+1}-c_{i-1}^{*n+1}\right)$$

$$+\frac{\Delta tD}{2\Delta x^2}\left(c_{i+1}^{*n+1}-2c_i^{*n+1}+c_{i-1}^{*n+1}\right)$$

The concentration profile is steeper for the MacCormack method than for the upstream derivatives, but oscillations can still be present. The flux-corrected transport method can be added to the MacCormack method. A solution is obtained both with the upstream algorithm and the MacCormack method; then they are combined to add just enough diffusion to eliminate the oscillations without smoothing the solution too much. The algorithm is complicated and lengthy but well worth the effort [107 – 109].

If finite element methods are used, an explicit Taylor – Galerkin method is appropriate. For the convective diffusion equation the method is

$$\frac{1}{6}\left(c_{i+1}^{n+1}-c_{i+1}^n\right)+\frac{2}{3}\left(c_i^{n+1}-c_i^n\right)+\frac{1}{6}\left(c_{i-1}^{n+1}-c_{i-1}^n\right)$$

$$=-\frac{u\Delta t}{2\Delta x}\left(c_{i+1}^n-c_{i-1}^n\right)+\left(\frac{\Delta tD}{\Delta x^2}+\frac{u^2\Delta t^2}{2\Delta x^2}\right)\left(c_{i+1}^n\right.$$

$$\left.-2c_i^n+c_{i-1}^n\right)$$

Leaving out the $u^2\ \Delta t^2$ terms gives the Galerkin method. Replacing the left-hand side with

$$c_i^{n+1}-c_i^n$$

gives the Taylor finite difference method, and dropping the $u^2\ \Delta t^2$ terms in that gives the centered finite difference method. This method might require a small time step if reaction phenomena are important. Then the implicit Galerkin method (without the Taylor terms) is appropriate

A stability diagram for the explicit methods applied to the convective diffusion equation is shown in Figure 29. Notice that all the methods require

$$Co=\frac{u\Delta t}{\Delta x}\leq 1$$

where Co is the Courant number. How much Co should be less than one depends on the method and on $r=D\ \Delta t/\Delta x^2$, as given in Figure 29. The MacCormack method with flux correction requires a smaller time step than the MacCormack method alone (curve a), and the implicit Galerkin method (curve e) is stable for all values of Co and r shown in Figure 29 (as well as even larger values).

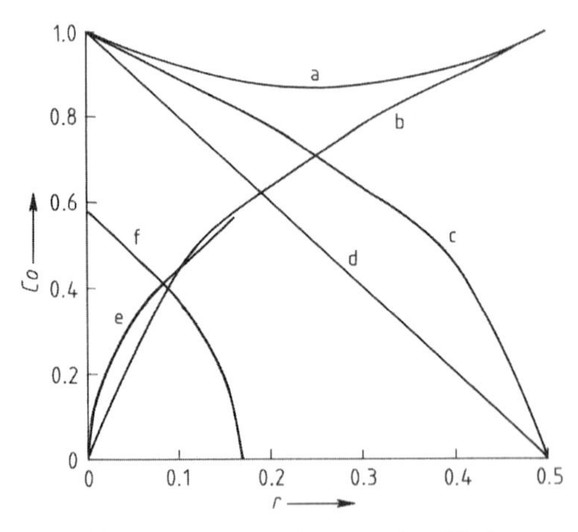

Figure 29. Stability diagram for convective diffusion equation (stable below curve)
a) MacCormack; b) Centered finite difference; c) Taylor finite difference; d) Upstream; e) Galerkin; f) Taylor – Galerkin

Each of these methods tries to avoid oscillations that would disappear if the mesh were fine enough. For the steady convective diffusion equation these oscillations do not occur provided

$$\frac{u\Delta x}{2D}=\frac{Pe_{\Delta}}{2}<=1 \tag{36}$$

For large u, Δx must be small to meet this condition. An alternative is to use a small Δx in regions where the solution changes drastically. Because these regions change in time, the elements or grid points must move. The criteria to move the grid points can be quite complicated, and typical methods are reviewed in [107]. The criteria include moving the mesh in a known way (when the movement is known a priori), moving the mesh to keep some property (e.g., first- or second-derivative measures) uniform over the domain, using a Galerkin or weighted residual criterion to move the mesh, and Euler – Lagrange methods which move part of the solution exactly by convection and then add on some diffusion after that.

The final illustration is for adsorption in a packed bed, or chromatography. Equation 35 can be solved when the adsorption phenomenon is governed by a Langmuir isotherm.

$$f(c)=\frac{\alpha c}{1+Kc}$$

Similar numerical considerations apply and similar methods are available [110 – 112].

8.3. Parabolic Equations in One Dimension

In this section several methods are applied to parabolic equations in one dimension: separation of variables, combination of variables, finite difference method, finite element method, and the orthogonal collocation method. Separation of variables is successful for linear problems, whereas the other methods work for linear or nonlinear problems. The finite difference, the finite element, and the orthogonal collocation methods are numerical, whereas the separation or combination of variables can lead to analytical solutions.

Analytical Solutions. Consider the diffusion equation

$$\frac{\partial c}{\partial t}=D\frac{\partial^2 c}{\partial x^2}$$

with boundary and initial conditions

$$c(x,0)=0$$

$$c(0,t)=1,\ c(L,t)=0$$

A solution of the form

$$c(x,t)=T(t)\,X(x)$$

is attempted and substituted into the equation, with the terms separated to give

$$\frac{1}{DT}\frac{\mathrm{d}T}{\mathrm{d}t}=\frac{1}{X}\frac{\mathrm{d}^2X}{\mathrm{d}x^2}$$

One side of this equation is a function of x alone, whereas the other side is a function of t alone. Thus, both sides must be a constant. Otherwise, if x is changed one side changes, but the other cannot because it depends on t. Call the constant $-\lambda$ and write the separate equations

$$\frac{\mathrm{d}T}{\mathrm{d}t}-\lambda DT,\ \frac{\mathrm{d}^2X}{\mathrm{d}x^2}-\lambda X$$

The first equation is solved easily

$$T(t)=T(0)\,\mathrm{e}^{-\lambda Dt}$$

and the second equation is written in the form

$$\frac{\mathrm{d}^2X}{\mathrm{d}x^2}+\lambda X=0$$

Next consider the boundary conditions. If they are written as

$$c(L,t)=1=T(t)\,X(L)$$

$$c(0,t)=0=T(t)\,X(0)$$

the boundary conditions are difficult to satisfy because they are not homogeneous, i.e. with a zero right-hand side. Thus, the problem must be transformed to make the boundary conditions homogeneous. The solution is written as the sum of two functions, one of which satisfies the nonhomogeneous boundary conditions, whereas the other satisfies the homogeneous boundary conditions.

$$c(x,t)=f(x)+u(x,t)$$

$$u(0,t)=0$$

$$u(L,t)=0$$

Thus, $f(0)=1$ and $f(L)=0$ are necessary. Now the combined function satisfies the boundary conditions. In this case the function $f(x)$ can be taken as

$$f(x) = L - x$$

The equation for u is found by substituting for c in the original equation and noting that the $f(x)$ drops out for this case; it need not disappear in the general case:

$$\frac{\partial u}{\partial t} = D \frac{\partial^2 u}{\partial x^2}$$

The boundary conditions for u are

$$u(0,t) = 0$$

$$u(L,t) = 0$$

The initial conditions for u are found from the initial condition

$$u(x,0) = c(x,0) - f(x) = \frac{x}{L} - 1$$

Separation of variables is now applied to this equation by writing

$$u(x,t) = T(t)\, X(x)$$

The same equation for $T(t)$ and $X(x)$ is obtained, but with $X(0) = X(L) = 0$.

$$\frac{d^2 X}{dx^2} + \lambda X = 0$$

$$X(0) = X(L) = 0$$

Next $X(x)$ is solved for. The equation is an eigenvalue problem. The general solution is obtained by using e^{mx} and finding that $m^2 + \lambda = 0$; thus $m = \pm i\sqrt{\lambda}$. The exponential term

$$e^{\pm i\sqrt{\lambda} x}$$

is written in terms of sines and cosines, so that the general solution is

$$X = B \cos\sqrt{\lambda} x + E \sin\sqrt{\lambda} x$$

The boundary conditions are

$$X(L) = B \cos\sqrt{\lambda} L + E \sin\sqrt{\lambda} L = 0$$

$$X(0) = B = 0$$

If $B = 0$, then $E \neq 0$ is required to have any solution at all. Thus, λ must satisfy

$$\sin\sqrt{\lambda} L = 0$$

This is true for certain values of λ, called eigenvalues or characteristic values. Here, they are

$$\lambda_n = n^2 \pi^2 / L^2$$

Each eigenvalue has a corresponding eigenfunction

$$X_n(x) = E \sin n\, \pi\, x / L$$

The composite solution is then

$$X_n(x)\, T_n(t) = E\, A \sin \frac{n\, \pi\, x}{L}\, e^{-\lambda_n D t}$$

This function satisfies the boundary conditions and differential equation but not the initial condition. To make the function satisfy the initial condition, several of these solutions are added up, each with a different eigenfunction, and $E\,A$ is replaced by A_n.

$$u(x,t) = \sum_{n=1}^{\infty} A_n \sin \frac{n\, \pi\, x}{L}\, e^{-n^2 \pi^2 D t / L^2}$$

The constants A_n are chosen by making $u(x, t)$ satisfy the initial condition.

$$u(x,0) = \sum_{n=1}^{\infty} A_n \sin \frac{n\, \pi\, x}{L} = \frac{x}{L} - 1$$

The residual $R(x)$ is defined as the error in the initial condition:

$$R(x) = \frac{x}{L} - 1 - \sum_{n=1}^{\infty} A_n \sin \frac{n\, \pi\, x}{L}$$

Next, the Galerkin method is applied, and the residual is made orthogonal to a complete set of functions, which are the eigenfunctions.

$$\int_0^L \left(\frac{x}{L} - 1\right) \sin \frac{m\, \pi\, x}{L} dx$$

$$= \sum_{n=1}^{\infty} A_n \int_0^L \sin \frac{m\, \pi\, x}{L} \sin \frac{n\, \pi\, x}{L} dx = \frac{A_m}{2}$$

The Galerkin criterion for finding A_n is the same as the least-squares criterion [3, p. 183]. The solution is then

$$c(x,t) = 1 - \frac{x}{L} + \sum_{n=1}^{\infty} A_n \sin \frac{n\, \pi\, x}{L}\, e^{-n^2 \pi^2 D t / L^2}$$

This is an "exact" solution to the linear problem. It can be evaluated to any desired accuracy by taking more and more terms, but if a finite number of terms are used, some error always occurs. For large times a single term is adequate, whereas for small times many terms are needed. For small times the Laplace transform method is also useful, because it leads to solutions that converge with fewer terms. For small times, the method of combination of variables may be used as well. For nonlinear problems, the method of separation of variables fails and one of the other methods must be used.

The method of combination of variables is useful, particularly when the problem is posed in a semi-infinite domain. Here, only one example is provided; more detail is given in [3, 113, 114]. The method is applied here to the nonlinear problem

$$\frac{\partial c}{\partial t}=\frac{\partial}{\partial x}\left[D\left(c\right)\frac{\partial c}{\partial x}\right]=D\left(c\right)\frac{\partial^2 c}{\partial x^2}+\frac{\mathrm{d}\,D\left(c\right)}{\mathrm{d}c}\left(\frac{\partial c}{\partial x}\right)^2$$

with boundary and initial conditions

$$c\left(x,0\right)=0$$

$$c\left(0,t\right)=1,\;c\left(\infty,t\right)=0$$

The transformation combines two variables into one

$$c\left(x,t\right)=f\left(\eta\right)\;\text{ where }\eta=\frac{x}{\sqrt{4D_0t}}$$

The use of the 4 and D_0 makes the analysis below simpler. The equation for $c\ (x,\ t)$ is transformed into an equation for $f\ (\eta)$

$$\frac{\partial c}{\partial t}=\frac{\mathrm{d}f}{\mathrm{d}\eta}\frac{\partial\eta}{\partial t},\;\frac{\partial c}{\partial x}=\frac{\mathrm{d}f}{\mathrm{d}\eta}\frac{\partial\eta}{\partial x}$$

$$\frac{\partial^2 c}{\partial x^2}=\frac{\mathrm{d}^2 f}{\mathrm{d}\eta^2}\left(\frac{\partial\eta}{\partial x}\right)^2+\frac{\mathrm{d}f}{\mathrm{d}\eta}\frac{\partial^2\eta}{\partial x^2}$$

$$\frac{\partial\eta}{\partial t}=-\frac{x/2}{\sqrt{4D_0t^3}},\;\frac{\partial\eta}{\partial x}=\frac{1}{\sqrt{4D_0t}},\;\frac{\partial^2\eta}{\partial x^2}=0$$

The result is

$$\frac{\mathrm{d}}{\mathrm{d}\eta}\left[K\left(c\right)\frac{\mathrm{d}f}{\mathrm{d}\eta}\right]+2\eta\frac{\mathrm{d}f}{\mathrm{d}\eta}=0$$

$$K\left(c\right)=D\left(c\right)/D_0$$

The boundary conditions must also combine. In this case the variable η is infinite when either x is infinite or t is zero. Note that the boundary conditions on $c\ (x,\ t)$ are both zero at those points. Thus, the boundary conditions can be combined to give

$$f\left(\infty\right)=0$$

The other boundary condition is for $x = 0$ or $\eta = 0$,

$$f\left(0\right)=1$$

Thus, an ordinary differential equation must be solved rather than a partial differential equation. When the diffusivity is constant the solution is the well-known complementary error function:

$$c\left(x,t\right)=1-\operatorname{erf}\eta=\operatorname{erfc}\eta$$

$$\operatorname{erf}\eta=\frac{\int_0^\eta \mathrm{e}^{-\xi^2}\mathrm{d}\xi}{\int_0^\infty \mathrm{e}^{-\xi^2}\mathrm{d}\xi}$$

This is a tabulated function [23].

Numerical Methods. Numerical methods are applicable to both linear and nonlinear problems on finite and semi-infinite domains. The *finite difference method* is applied by using the method of lines [115]. In this method the same equations are used for the spatial variations of the function, but the function at a grid point can vary with time. Thus the linear diffusion problem is written as

$$\frac{\mathrm{d}c_i}{\mathrm{d}t}=D\frac{c_{i+1}-2c_i+c_{i-1}}{\Delta x^2}\tag{37}$$

This can be written in the general form

$$\frac{\mathrm{d}\boldsymbol{c}}{\mathrm{d}t}=\boldsymbol{AA}c$$

This set of ordinary differential equations can be solved by using any of the standard methods. The stability of explicit schemes is deduced from the theory presented in Chapter 6. The equations are written as

$$\frac{\mathrm{d}c_i}{\mathrm{d}t}=D\frac{c_{i+1}-2c_i+c_{i-1}}{\Delta x^2}=\frac{D}{\Delta x^2}\sum_{j=1}^{n+1}B_{ij}c_j$$

where the matrix $\boldsymbol{B}$ is tridiagonal. The stability of the integration of these equations is governed by the largest eigenvalue of $\boldsymbol{B}$. If Euler's method is used for integration,

$$\Delta t\frac{D}{\Delta x^2}\leq\frac{2}{\left|\lambda\right|_{\max}}$$

The largest eigenvalue of $\boldsymbol{B}$ is bounded by the Gerschgorin theorem [14, p. 135].

$$\left|\lambda\right|_{\max}\leq\max\nolimits_{2<j<n}\sum_{i=2}^{n}\left|B_{ji}\right|=4$$

This gives the well-known stability limit

$$\Delta t\frac{D}{\Delta x^2}\leq\frac{1}{2}$$

If other methods are used to integrate in time, then the stability limit changes according to the method. It is interesting to note that the eigenvalues of Equation 37 range from $D\ \pi^2/L^2$ (smallest) to $4\ D/\Delta x^2$ (largest), depending on the boundary conditions. Thus the problem becomes stiff as Δx approaches zero [3, p. 263].

Implicit methods can also be used. Write a finite difference form for the time derivative and average the right-hand sides, evaluated at the old and new times:

$$\frac{c_i^{n+1}-c_i^n}{\Delta t}=D\,(1-\theta)\,\frac{c_{i+1}^n-2c_i^n+c_{i-1}^n}{\Delta x^2}$$

$$+D\theta\frac{c_{i+1}^{n+1}-2c_i^{n+1}+c_{i-1}^{n+1}}{\Delta x^2}$$

Now the equations are of the form

$$-\frac{D\Delta t\theta}{\Delta x^2}c_{i+1}^{n+1}+\left[1+2\frac{D\Delta t\theta}{\Delta x^2}\right]c_i^{n+1}-\frac{D\Delta t\theta}{\Delta x^2}c_{i-1}^{n+1}$$

$$=c_i^n+\frac{D\Delta t(1-\theta)}{\Delta x^2}\left(c_{i+1}^n-2c_i^n+c_{i-1}^n\right)$$

and require solving a set of simultaneous equations, which have a tridiagonal structure. Using $\theta = 0$ gives the Euler method (as above); $\theta = 0.5$ gives the Crank – Nicolson method; $\theta = 1$ gives the backward Euler method. The stability limit is given by

$$\frac{D\Delta t}{\Delta x^2}\leq\frac{0.5}{1-2\theta}$$

whereas the oscillation limit is given by

$$\frac{D\Delta t}{\Delta x^2}\leq\frac{0.25}{1-\theta}$$

If a time step is chosen between the oscillation limit and stability limit, the solution will oscillate around the exact solution, but the oscillations remain bounded. For further discussion, see [3, p. 218].

Finite volume methods are utilized extensively in computational fluid dynamics. In this method, a mass balance is made over a cell accounting for the change in what is in the cell and the flow in and out. Figure 30 illustrates the geometry of the i-th cell. A mass balance made on this cell (with area A perpendicular to the paper) is

$$A\Delta x(c_i^{n+1}-c_i^n)=\Delta tA(J_{j-1/2}-J_{i+1/2})$$

where J is the flux due to convection and diffusion, positive in the $+x$ direction.

$$J=uc-D\frac{\partial c}{\partial x},\;J_{i-1/2}=u_{i-1/2}c_{i-1/2}-D\frac{c_i-c_{i-1/2}}{\Delta x}$$

The concentration at the edge of the cell is taken as

$$c_{i-1/2}=\frac{1}{2}(c_i+c_{i-1})$$

Rearrangement for the case when the velocity u is the same for all nodes gives

$$\frac{c_i^{n+1}-c_i^n}{\Delta t}+\frac{u(c_{i+1}-c_{i-1})}{2\Delta x}=\frac{D}{\Delta x^2}(c_{i+1}-2c_i+c_{i-1})$$

This is the same equation as obtained using the finite difference method. This is not always true, and the finite volume equations are easy to derive. In two- and three-dimensions, the mesh need not be rectangular, as long as it is possible to compute the velocity normal to an edge of the cell. The finite volume method is useful for applications involving filling, such as injection molding, when only part of the cell is filled with fluid. Such applications do involve some approximations, since the interface is not tracked precisely, but they are useful engineering approximations.

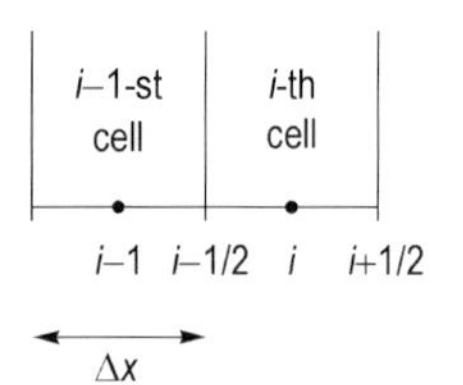

Figure 30.

The *finite element method* is handled in a similar fashion, as an extension of two-point boundary value problems by letting the solution at the nodes depend on time. For the diffusion equation the finite element method gives

$$\sum_e\sum_I C_{JI}^e\frac{\mathrm{d}c_I^e}{\mathrm{d}t}=\sum_e\sum_I B_{JI}^e\,c_I^e$$

with the mass matrix defined by

$$C_{JI}^e=\Delta x_e\int_0^1 N_J\,(u)\,N_I\,(u)\,\mathrm{d}u$$

This set of equations can be written in matrix form

$$\boldsymbol{CC}\frac{\mathrm{d}c}{\mathrm{d}t}=\boldsymbol{AA}c$$

Now the matrix $\boldsymbol{C}\,\boldsymbol{C}$ is not diagonal, so that a set of equations must be solved for each time step, even when the right-hand side is evaluated explicitly. This is not as time-consuming as it seems, however. The explicit scheme is written as

$$CC_{ji}\frac{c_i^{n+1}-c_i^n}{\Delta t}=AA_{ji}c_i^n$$

and rearranged to give

$$CC_{ji}\left(c_i^{n+1}-c_i^n\right)=\Delta tAA_{ji}c_i^n \text{ or}$$

$$\boldsymbol{CC}\left(\boldsymbol{c}^{n+1}-\boldsymbol{c}^n\right)=\Delta t\boldsymbol{AAc}$$

This is solved with an $L\,U$ decomposition (see Section 1.1) that retains the structure of the mass matrix $\boldsymbol{C}\,\boldsymbol{C}$. Thus,

$$\boldsymbol{CC}=\boldsymbol{LU}$$

At each step, calculate

$$\boldsymbol{c}^{n+1}-\boldsymbol{c}^n=\Delta t\boldsymbol{U}^{-1}\boldsymbol{L}^{-1}\boldsymbol{AAc}^n$$

This is quick and easy to do because the inverse of $\boldsymbol{L}$ and $\boldsymbol{U}$ are simple. Thus the problem is reduced to solving one full matrix problem and then evaluating the solution for multiple right-hand sides. For implicit methods the same approach can be used, and the LU decomposition remains fixed until the time step is changed.

The *method of orthogonal collocation* uses a similar extension: the same polynomial of x is used but now the coefficients depend on time.

$$\left.\frac{\partial c}{\partial t}\right|_{x_j}=\frac{\mathrm{d}c\left(x_j,t\right)}{\mathrm{d}t}=\frac{\mathrm{d}c_j}{\mathrm{d}t}$$

Thus, for diffusion problems

$$\frac{\mathrm{d}c_j}{\mathrm{d}t}=\sum_{i=1}^{N+2}B_{ji}c_i,\ j=2,\ldots,N+1$$

This can be integrated by using the standard methods for ordinary differential equations as initial value problems. Stability limits for explicit methods are available [3, p. 204].

The method of orthogonal collocation on finite elements can also be used, and details are provided elsewhere [3, pp. 228 – 230].

The maximum eigenvalue for all the methods is given by

$$|\lambda|_{\max}=\frac{LB}{\Delta x^2} \tag{38}$$

where the values of LB are as follows:

Finite difference	4
Galerkin, linear elements, lumped	4
Galerkin, linear elements	12
Galerkin, quadratic elements	60
Orthogonal collocation on finite elements, cubic	36

Spectral methods employ the discrete Fourier transform (see 2 and Chebyshev polynomials on rectangular domains [116]).

In the Chebyshev collocation method, $N+1$ collocation points are used

$$x_j=\cos\frac{\pi j}{N},\ j=0,1,\ldots,N$$

As an example, consider the equation

$$\frac{\partial u}{\partial t}+f\left(u\right)\frac{\partial u}{\partial x}=0$$

An explicit method in time can be used

$$\frac{u^{n+1}-u^n}{\Delta t}+f\left(u^n\right)\left.\frac{\partial u}{\partial x}\right|^n=0$$

and evaluated at each collocation point

$$\frac{u_j^{n+1}-u_j^n}{\Delta t}+f\left(u_j^n\right)\left.\frac{\partial u}{\partial x}\right|_j^n=0$$

The trial function is taken as

$$u_j\left(t\right)=\sum_{p=0}^{N}a_p\left(t\right)\cos\frac{\pi pj}{N},\ u_j^n=u_j\left(t^n\right) \tag{39}$$

Assume that the values u_j^n exist at some time. Then invert Equation 39 using the fast Fourier transform to obtain $\{a_p\}$ for $p=0, 1, \ldots, N$; then calculate S_p

$$S_p=S_{p+2}+\left(p+1\right)a_{p+1},\ 0\le p\le N-1$$

$$S_N=0,\ S_{N+1}=0$$

and finally

$$a_p^{(1)}=\frac{2S_p}{c_p}$$

Thus, the first derivative is given by

$$\left.\frac{\partial u}{\partial x}\right|_j=\sum_{p=0}^{N}a_p^{(1)}\left(t\right)\cos\frac{\pi pj}{N}$$

This is evaluated at the set of collocation points by using the fast Fourier transform again. Once the function and the derivative are known at each collocation point the solution can be advanced forward to the $n+1$-th time level.

The advantage of the spectral method is that it is very fast and can be adapted quite well to parallel computers. It is, however, restricted in the geometries that can be handled.

8.4. Elliptic Equations

Elliptic equations can be solved with both finite difference and finite element methods. One-dimensional elliptic problems are two-point boundary value problems and are covered in Chapter 7. Two-dimensional elliptic problems are often solved with direct methods, and iterative methods are usually used for three-dimensional problems. Thus, two aspects must be considered: how the equations are discretized to form sets of algebraic equations and how the algebraic equations are then solved.

The prototype elliptic problem is steady-state heat conduction or diffusion,

$$k\left(\frac{\partial^2 T}{\partial x^2}+\frac{\partial^2 T}{\partial y^2}\right)=Q$$

possibly with a heat generation term per unit volume, Q. The boundary conditions can be

Dirichlet or 1st kind:	$T=T_1$ on boundary S_1
Neumann or 2nd kind:	$k\frac{\partial T}{\partial n}=q_2$ on boundary S_2
Robin, mixed, or 3rd kind:	$-k\frac{\partial T}{\partial n}=h\,(T-T_3)$ on boundary S_3

Illustrations are given for constant physical properties k, h, while T_1, q_2, T_3 are known functions on the boundary and Q is a known function of position. For clarity, only a two-dimensional problem is illustrated. The finite difference formulation is given by using the following nomenclature

$$T_{i,j}=T\,(i\Delta x, j\Delta y)$$

The finite difference formulation is then

$$\frac{T_{i+1,j}-2T_{i,j}+T_{i-1,j}}{\Delta x^2}+\frac{T_{i,j+1}-2T_{i,j}+T_{i,j-1}}{\Delta y^2}=Q_{i,j}/k \tag{40}$$

$$T_{i,j}=T_1 \text{ for } i,j \text{ on boundary } S_1$$

$$k\frac{\partial T}{\partial n}\Big|_{i,j}=q_2 \text{ for } i,j \text{ on boundary } S_2$$

$$-k\frac{\partial T}{\partial n}\Big|_{i,j}=h\,(T_{i,j}-T_3) \text{ for } i,j \text{ on boundary } S_3$$

If the boundary is parallel to a coordinate axis the boundary slope is evaluated as in Chapter 7, by using either a one-sided, centered difference or a false boundary. If the boundary is more irregular and not parallel to a coordinate line, more complicated expressions are needed and the finite element method may be the better method.

Equation 40 is rewritten in the form

$$2\left(1+\frac{\Delta x^2}{\Delta y^2}\right)T_{i,j}=T_{i+1,j}+T_{i-1,j}+\frac{\Delta x^2}{\Delta y^2}(T_{i,j+1}+T_{i,j-1})-\Delta x^2\frac{Q_{i,j}}{k}$$

And this is converted to the Gauss–Seidel iterative method.

$$2\left(1+\frac{\Delta x^2}{\Delta y^2}\right)T^{s+1}_{i,j}=T^{s}_{i+1,j}+T^{s+1}_{i-1,j}+\frac{\Delta x^2}{\Delta y^2}\left(T^{s}_{i,j+1}+T^{s+1}_{i,j-1}\right)-\Delta x^2\frac{Q_{i,j}}{k}$$

Calculations proceed by setting a low i, computing from low to high j, then increasing i and repeating the procedure. The relaxation method uses

$$2\left(1+\frac{\Delta x^2}{\Delta y^2}\right)T^{*}_{i,j}=T^{s}_{i+1,j}+T^{s+1}_{i-1,j}+\frac{\Delta x^2}{\Delta y^2}\left(T^{s}_{i,j+1}-T^{s+1}_{i,j-1}\right)-\Delta x^2\frac{Q_{i,j}}{k}$$

$$T^{s+1}_{i,j}=T^{s}_{i,j}+\beta\left(T^{*}_{i,j}-T^{s}_{i,j}\right)$$

If $\beta = 1$, this is the Gauss – Seidel method. If $\beta > 1$, it is overrelaxation; if $\beta < 1$, it is underrelaxation. The value of β may be chosen empirically, $0 < \beta < 2$, but it can be selected theoretically for simple problems like this [117, p. 100], [3, p. 282]. In particular, the optimal value of the iteration parameter is given by

$$-\ln\,(\beta_{\text{opt}}-1)\approx R$$

and the error (in solving the algebraic equation) is decreased by the factor $(1 - R)^N$ for every N iterations. For the heat conduction problem and Dirichlet boundary conditions,

$$R=\frac{\pi^2}{2n^2}$$

(when there are n points in both x and y directions). For Neumann boundary conditions, the value is

$$R=\frac{\pi^2}{2n^2}\,\frac{1}{1+\max\left[\Delta x^2/\Delta y^2,\Delta y^2/\Delta x^2\right]}$$

Iterative methods can also be based on lines (for 2D problems) or planes (for 3D problems).

Preconditioned conjugate gradient methods have been developed (see Chap. 1). In these methods a series of matrix multiplications are done iteration by iteration; and the steps lend themselves to the efficiency available in parallel computers. In the multigrid method the problem is solved on several grids, each more refined than the previous one. In iterating between the solutions on different grids, one converges to the solution of the algebraic equations. A chemical engineering application is given in [118]. Software for a variety of these methods is available, as described below.

The Galerkin finite element method (FEM) is useful for solving elliptic problems and is particularly effective when the domain or geometry is irregular [119 – 125]. As an example, cover the domain with triangles and define a trial function on each triangle. The trial function takes the value 1.0 at one corner and 0.0 at the other corners, and is linear in between (see Fig. 31). These trial functions on each triangle are pieced together to give a trial function on the whole domain. For the heat conduction problem the method gives [3]

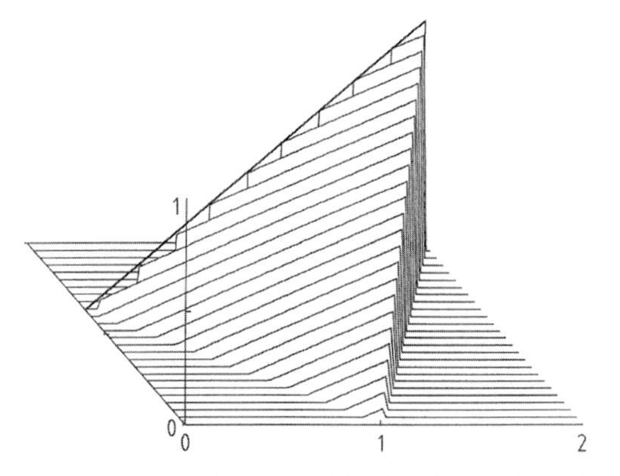

Figure 31. Finite elements trial function: linear polynomials on triangles

$$\sum_e \sum_J A^e_{IJ} T^e_J = \sum_e \sum_J F^e_I \tag{41}$$

where

$$A^e_{IJ} = -\int k \nabla N_I \cdot \nabla N_J \mathrm{d}A - \int_{C_3} h_3 N_I N_J \mathrm{d}C$$

$$F^e_I = \int N_I Q \mathrm{d}A + \int_{C_2} N_I q_2 \mathrm{d}C - \int_{C_3} N_I h_3 T_3 \mathrm{d}C$$

Also, a necessary condition is that

$$T_i = T_1 \text{ on } C_1$$

In these equations I and J refer to the nodes of the triangle forming element e and the summation is made over all elements. These equations represent a large set of linear equations, which are solved using matrix techniques (Chap. 1).

If the problem is nonlinear, e.g., with k or Q a function of temperature, the equations must be solved iteratively. The integrals are given for a triangle with nodes I, J, and K in counterclockwise order. Within an element,

$$T = N_I(x,y)\, T_I + N_J(x,y)\, T_J + N_K(x,y)\, T_K$$

$$N_I = \frac{a_I + b_I x + c_I y}{2\Delta}$$

$$a_I = x_J y_K - x_K y_J$$

$$b_I = y_I - y_K$$

$$c_I = x_K - x_J$$

plus permutation on I,K,J

$$2\Delta = \det \begin{bmatrix} 1 & x_I & y_I \\ 1 & x_J & y_J \\ 1 & x_K & y_K \end{bmatrix} = 2\,(\text{area of triangle})$$

$$a_I + a_J + a_K = 1$$

$$b_I + b_J + b_K = 0$$

$$c_I + c_J + c_K = 0$$

$$A^e_{IJ} = -\frac{k}{4\Delta}(b_I b_J + c_I c_J)$$

$$F^e_{IJ} = \frac{Q}{2}(a_I + b_I \overline{x} + c_I \overline{y}) = \frac{QD}{3}$$

$$\overline{x} = \frac{x_I + x_J + x_K}{3},\ \overline{y} = \frac{y_I + y_J + y_K}{3}$$

$$a_I + b_I \overline{x} + c_I \overline{y} = \frac{2}{3}\Delta$$

The trial functions in the finite element method are not limited to linear ones. Quadratic functions, and even higher order functions, are frequently used. The same considerations hold as for boundary value problems: the higher order trial functions converge faster but require more work. It is possible to refine both the mesh (h) and power of polynomial in the trial function (p) in an hp method. Some problems have constraints on some of the variables. If singularities exist in the solution, it is possible to include them in the basis functions and solve for the difference between the total solution and the singular function [126 – 129].

When applying the Galerkin finite element method, one must choose both the shape of the element and the basis functions. In two dimensions, triangular elements are usually used because it is easy to cover complicated geometries and refine parts of the mesh. However, rectangular elements sometimes have advantages, particularly when some parts of the solution do not change very much and the elements can be long. In three dimensions the same considerations apply: tetrahedral elements are frequently used, but brick elements are also possible. While linear elements (in both two and three dimensions) are usually used, higher accuracy can be obtained by using quadratic or cubic basis functions within the element. The reason is that all methods converge according to the mesh size to some power, and the power is larger when higher order elements are used. If the solution is discontinuous, or has discontinuous first derivatives, then the lowest order basis functions are used because the convergence is limited by the properties of the solution, not the finite element approximation.

One nice feature of the finite element method is the use of natural boundary conditions. In this problem the natural boundary conditions are the Neumann or Robin conditions. When using Equation 41, the problem can be solved on a domain that is shorter than needed to reach some limiting condition (such as at an outflow boundary). The externally applied flux is still applied at the shorter domain, and the solution inside the truncated domain is still valid. Examples are given in [107] and [131]. The effect of this is to allow solutions in domains that are smaller, thus saving computation time and permitting the solution in semi-infinite domains.

8.5. Parabolic Equations in Two or Three Dimensions

Computations become much more lengthy with two or more spatial dimensions, for example, the unsteady heat conduction equation

$$\varrho C_p \frac{\partial T}{\partial t} = k\left(\frac{\partial^2 T}{\partial x^2}+\frac{\partial^2 T}{\partial y^2}\right)-Q$$

or the unsteady diffusion equation

$$\frac{\partial c}{\partial t} = D\left(\frac{\partial^2 T}{\partial x^2}+\frac{\partial^2 c}{\partial y^2}\right)-R(c)$$

In the finite difference method an explicit technique would evaluate the right-hand side at the n-th time level:

$$\rho C_p \frac{T^{n+1}_{i,j}-T^n_{i,j}}{\Delta t}$$

$$= \frac{k}{\Delta x^2}\left(T^n_{i+1,j}-2T^n_{i-1,j}+T^n_{i-1,j}\right)$$

$$+\frac{k}{\Delta y^2}\left(T^n_{i,j+1}-2T^n_{i,j}+T^n_{i,j-1}\right)-Q$$

When $Q = 0$ and $\Delta x = \Delta y$, the time step is limited by

$$\Delta t \leq \frac{\Delta x^2 \varrho C_p}{4k} \text{ or } \frac{\Delta x^2}{4D}$$

These time steps are smaller than for one-dimensional problems. For three dimensions, the limit is

$$\Delta t \leq \frac{\Delta x^2}{6D}$$

To avoid such small time steps, which must be smaller when Δx decreases, an implicit method could be used. This leads to large sparse matrices, rather than convenient tridiagonal matrices.

8.6. Special Methods for Fluid Mechanics

The method of operator splitting is also useful when different terms in the equation are best evaluated by using different methods or as a technique for reducing a larger problem to a series of smaller problems. Here the method is illustrated by using the Navier – Stokes equations. In vector notation the equations are

$$\varrho \frac{\partial \boldsymbol{u}}{\partial t}+\varrho \boldsymbol{u}\cdot\boldsymbol{\nabla}\boldsymbol{u} = \varrho \boldsymbol{f}-\boldsymbol{\nabla} p+\mu\nabla^2\boldsymbol{u}$$

The equation is solved in the following steps

$$\varrho \frac{\boldsymbol{u}^*-\boldsymbol{u}^n}{\Delta t} = -\varrho \boldsymbol{u}^n\cdot\boldsymbol{\nabla}\boldsymbol{u}^n+\varrho \boldsymbol{f}+\mu\nabla^2\boldsymbol{u}^n$$

$$\nabla^2 p^{n+1} = \frac{1}{\Delta t}\boldsymbol{\nabla}\cdot\boldsymbol{u}^*$$

$$\varrho \frac{\boldsymbol{u}^{n+1}-\boldsymbol{u}^*}{\Delta t} = -\boldsymbol{\nabla} p$$

This can be done by using the finite difference [132, p. 162] or the finite element method [133 – 135].

In fluid flow problems solved with the finite element method, the basis functions for pressure and velocity are often different. This is required by the LBB condition (named after LADYSHENSKAYA, BREZZI, and BABUSKA) [134, 135]. Sometimes a discontinuous basis function is used for pressure to meet this condition. Other times a penalty term is added, or the quadrature is done using a small number of quadrature points. Thus, one has to be careful how to apply the finite element method to the Navier – Stokes equations. Fortunately, software exists that has taken this into account.

Level Set Methods. Multiphase problems are complicated because the terms in the equations depend on which phase exists at a particular point, and the phase boundary may move or be unknown. It is desirable to compute on a fixed grid, and the level set formulation allows this. Consider a curved line in a two-dimensional problem or a curved surface in a three-dimensional problem. One defines a level set function ϕ, which is the signed distance function giving the distance from some point to the closest point on the line or surface. It defined to be negative on one side of the interface and positive on the other. Then the curve

$$\phi(x,y,z) = 0$$

represents the location of the interface. For example, in flow problems the level set function is defined as the solution to

$$\frac{\partial \phi}{\partial t} + \boldsymbol{u} \cdot \boldsymbol{\nabla} \phi = 0$$

The physics governing the velocity of the interface must be defined, and this equation is solved along with the other equations representing the problem [130 – 137].

Lattice Boltzmann Methods. Another way to solve fluid flow problems is based on a molecular viewpoint and is called the Lattice Boltzmann method [138 – 141]. The treatment here follows [142]. A lattice is defined, and one solves the following equation for $f_i\ (x,t)$, the probability of finding a molecule at the point x with speed c_i.

$$\frac{\partial f_i}{\partial t} + c_i \cdot \boldsymbol{\nabla} f_i = -\frac{f_i - f_i^{\text{eq}}}{\tau}$$

The right-hand side represents a single time relaxation for molecular collisions, and τ is related to the kinematic viscosity. By means of a simple stepping algorithm, the computational algorithm is

$$f_i(\mathbf{x} + \mathbf{c}_i \Delta t, t + \Delta t) = f_i(\mathbf{x},t) - \frac{\Delta t}{\tau}(f_i - f_i^{\text{eq}})$$

Consider the lattice shown in Figure 32. The various velocities are

$$c_1 = (1,0),\ c_3 = (0,1),\ c_5 = (-1,0),\ c_7 = (0,-1)$$

$$c_2 = (1,1),\ c_4 = (-1,1),\ c_6 = (-1,-1),\ c_8 = (1,-1)$$

$$c_0 = (0,0)$$

The density, velocity, and shear stress (for some k) are given by

$$\rho = \sum_i f_i(\boldsymbol{x},t),\ \rho\mathbf{u} = \sum_i \boldsymbol{c}_i f_i(\boldsymbol{x},t),$$

$$\tau = k \sum_i \boldsymbol{c}_i \boldsymbol{c}_i \left[f_i(\boldsymbol{x},t) - f_i^{\text{eq}}(\boldsymbol{x},t)\right]$$

For this formulation, the kinematic viscosity and speed of sound are given by

$$\nu = \frac{1}{3}\left(\tau - \frac{1}{2}\right),\ c_s = \frac{1}{3}$$

The equilibrium distribution is

$$f_i^{\text{eq}} = \rho w_i \left[1 + \frac{\mathbf{c}_i \cdot \mathbf{u}}{c_s^2} + \frac{(\mathbf{c}_i \cdot \mathbf{u})^2}{2c_s^4} - \frac{\mathbf{u} \cdot \mathbf{u}}{2c_s^2}\right]$$

where the weighting functions are

$$w_0 = \frac{4}{9},\ w_1 = w_3 = w_5 = w_7 = \frac{1}{9},$$
$$w_2 = w_4 = w_6 = w_8 = \frac{1}{36}$$

With these conditions, the solution for velocity is a solution of the Navier—Stokes equation. These equations lead to a large computational problem, but it can be solved by parallel processing on multiple computers.

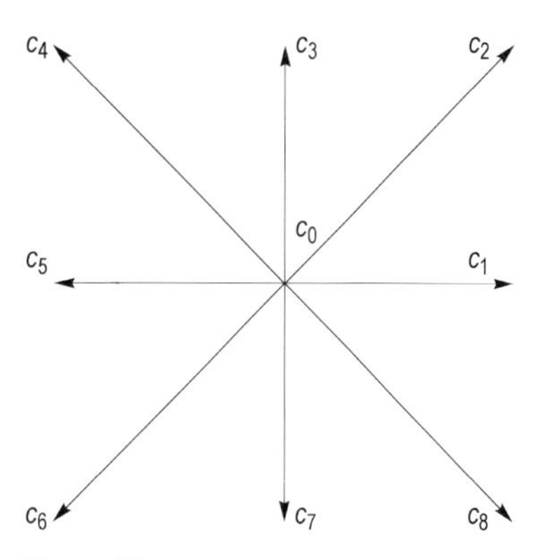

Figure 32.

8.7. Computer Software

A variety of general-purpose computer programs are available commercially. Mathematica (http://www.wolfram.com/), Maple (http://www.maplesoft.com/) and Mathcad (http://www.mathcad.com/) all have the capability of doing symbolic manipulation so that algebraic solutions can be obtained. For example, Mathematica can solve some ordinary and partial differential equations analytically; Maple can make simple graphs and do linear algebra and simple computations, and Mathcad can do simple calculations. In this section, examples are given for the use of Matlab (http://www.mathworks.com/), which is a package of numerical analysis tools, some of which are accessed by simple commands, and some of which are accessed by writing programs in C. Spreadsheets can also be used to solve simple problems. A popular program used in chemical engineering education is Polymath (http://www.polymath-software.com/), which can numerically solve sets of linear or nonlinear equations, ordinary differential equations as initial value problems, and perform data analysis and regression.

The mathematical methods used to solve partial differential equations are described in more detail in [143 – 148]. Since many computer programs are available without cost, consider the following decision points. The first decision is whether to use an approximate, engineering flow model, developed from correlations, or to solve the partial differential equations that govern the problem. Correlations are quick and easy to apply, but they may not be appropriate to your problem, or give the needed detail. When using a computer package to solve partial differential equations, the first task is always to generate a mesh covering the problem domain. This is not a trivial task, and special methods have been developed to permit importation of a geometry from a computer-aided design program. Then, the mesh must be created automatically. If the boundary is irregular, the finite element method is especially well-suited, although special embedding techniques can be used in finite difference methods (which are designed to be solved on rectangular meshes). Another capability to consider is the ability to track free surfaces that move during the computation. This phenomenon introduces the same complexity that occurs in problems with a large Peclet number, with the added difficulty that the free surface moves between mesh points, and improper representation can lead to unphysical oscillations. The method used to solve the equations is important, and both explicit and implicit methods (as described above) can be used. Implicit methods may introduce unacceptable extra diffusion, so the engineer needs to examine the solution carefully. The methods used to smooth unphysical oscillations from node to node are also important, and the engineer needs to verify that the added diffusion or smoothing does not give inaccurate solutions. Since current-day problems are mostly nonlinear, convergence is always an issue since the problems are solved iteratively. Robust programs provide several methods for convergence, each of which is best in some circumstance or other. It is wise to have a program that includes many iterative methods. If the iterative solver is not very robust, the only recourse to solving a steady-state problem may be to integrate the time-dependent problem to steady state. The solution time may be long, and the final result may be further from convergence than would be the case if a robust iterative solver were used.

A variety of computer programs is available on the internet, some of them free. First consider general-purpose programs. On the NIST web page, http://gams.nist.gov/ choose "problem decision tree", and then "differential and integral equations", then "partial differential equations". The programs are organized by type of problem (elliptic, parabolic, and hyperbolic) and by the number of spatial dimensions (one or more than one). On the Netlib web site, http://www.netlib.org/, search on "partial differential equation". The website: http://software.sandia.gov has a variety of programs available. Lau [141, 145] provides many programs in C++ (also see http://www.nr.com/). The multiphysics program Comsol Multiphysics (formerly FEMLAB) also solves many standard equations arising in Mathematical Physics.

Computational fluid dynamics (CFD) (→ Computational Fluid Dynamics) programs are more specialized, and most of them have been designed to solve sets of equations that are appropriate to specific industries. They can then include approximations and correlations for some features that would be difficult to solve for directly. Four widely used major packages are Fluent (http://www.fluent.com/), CFX (now part of ANSYS), Comsol Multiphysics (formerly FEMLAB) (http://www.comsol.com/), and ANSYS (http://www.ansys.com/). Of these, Comsol Multiphysics is particularly useful because it has a convenient graphical user interface, permits easy mesh generation and refinement (including adaptive mesh refinement), allows the user to add in phenomena and additional equations easily, permits solution by continuation methods (thus enhancing convergence), and has extensive graphical output capabilities. Other packages are also available (see http://cfd-online.com/), and these may contain features and correlations specific to the engineer's industry. One important point to note is that for turbulent flow, all the programs contain approximations, using the k-epsilon models of turbulence, or large eddy simulations; the direct numerical simulation of turbulence is too slow to apply it to very big problems, although it does give insight (independent of any approximations) that is useful for interpreting turbulent phenomena. Thus, the method used to include those turbulent correlations is important, and the method also may affect convergence or accuracy.

9. Integral Equations [149 – 155]

If the dependent variable appears under an integral sign an equation is called an integral equation; if derivatives of the dependent variable appear elsewhere in the equation it is called an integrodifferential equation. This chapter describes the various classes of equations, gives information concerning Green's functions, and presents numerical methods for solving integral equations.

9.1. Classification

Volterra integral equations have an integral with a variable limit, whereas Fredholm integral equations have a fixed limit. Volterra equations are usually associated with initial value or evolutionary problems, whereas Fredholm equations are analogous to boundary value problems. The terms in the integral can be unbounded, but still yield bounded integrals, and these equations are said to be weakly singular. A *Volterra equation of the second kind* is

$$y(t) = g(t) + \lambda \int_a^t K(t,s) y(s)\, \mathrm{d}s \qquad (42)$$

whereas a *Volterra equation of the first kind* is

$$y(t) = \lambda \int_a^t K(t,s)\, y(s)\, \mathrm{d}s$$

Equations of the first kind are very sensitive to solution errors so that they present severe numerical problems.

An example of a problem giving rise to a Volterra equation of the second kind is the following heat conduction problem:

$$\varrho C_p \frac{\partial T}{\partial t} = k \frac{\partial^2 T}{\partial x^2},\ 0 \le x < \infty,\ t > 0$$

$$T(x,0) = 0,\ \frac{\partial T}{\partial x}(0,t) = -g(t),$$

$$\lim_{x\to\infty} T(x,t) = 0,\ \lim_{x\to\infty} \frac{\partial T}{\partial x} = 0$$

If this is solved by using Fourier transforms the solution is

$$T(x,t) = \frac{1}{\sqrt{\pi}} \int_0^t g(s) \frac{1}{\sqrt{t-s}} \mathrm{e}^{-x^2/4(t-s)} \mathrm{d}s$$

Suppose the problem is generalized so that the boundary condition is one involving the solution *T*, which might occur with a radiation boundary condition or heat-transfer coefficient. Then the boundary condition is written as

$$\frac{\partial T}{\partial x} = -G(T,t),\ x = 0,\ t > 0$$

The solution to this problem is

$$T(x,t) = \frac{1}{\sqrt{\pi}} \int_0^t G(T(0,s),s) \frac{1}{\sqrt{t-s}}\, \mathrm{e}^{-x^2/4(t-s)} \mathrm{d}s$$

If $T^*(t)$ is used to represent $T(0,t)$, then

$$T^*(t) = \frac{1}{\sqrt{\pi}} \int_0^t G(\boldsymbol{T}(s),s) \frac{1}{\sqrt{t-s}} \mathrm{d}s$$

Thus the behavior of the solution at the boundary is governed by an integral equation. NAGEL and KLUGE [156] use a similar approach to solve for adsorption in a porous catalyst.

The existence and uniqueness of the solution can be proved [151, p. 30, 32].

Sometimes the kernel is of the form

$$K(t,s) = K(t-s)$$

Equations of this form are called convolution equations and can be solved by taking the Laplace transform. For the integral equation

$$Y(t) = G(t) + \lambda \int_0^t K(t-\tau)\, Y(\tau)\, \mathrm{d}\tau$$

$$K(t) * Y(t) \equiv \int_0^t K(t-\tau)\, Y(\tau)\, \mathrm{d}\tau$$

the Laplace transform is

$$y(s) = g(s) + k(s)\, y(s)$$

$$k(s)\, y(s) = \boldsymbol{L}\,[K(t) * Y(t)]$$

Solving this for $y(s)$ gives

$$y(s) = \frac{g(s)}{1-k(s)}$$

If the inverse transform can be found, the integral equation is solved.

A *Fredholm equation of the second kind* is

$$y(x) = g(x) + \lambda \int_a^b K(x,s)\, y(s)\, \mathrm{d}s \tag{43}$$

whereas a *Fredholm equation of the first kind* is

$$\int_a^b K(x,s)\, y(s)\, \mathrm{d}s = g(x)$$

The limits of integration are fixed, and these problems are analogous to boundary value problems. An eigenvalue problem is a homogeneous equation of the second kind.

$$y(x) = \lambda \int_a^b K(x,s)\, y(s)\, \mathrm{d}s \tag{44}$$

Solutions to this problem occur only for specific values of λ, the eigenvalues. Usually the Fredholm equation of the second or first kind is solved for values of λ different from these, which are called regular values.

Nonlinear Volterra equations arise naturally from initial value problems. For the initial value problem

$$\frac{\mathrm{d}y}{\mathrm{d}t} = F(t, y(t))$$

both sides can be integrated from 0 to t to obtain

$$y(t) = y(0) + \int_0^t F(s, y(s))\, \mathrm{d}s$$

which is a nonlinear Volterra equation. The general nonlinear Volterra equation is

$$y(t) = g(t) + \int_0^t K(t, s, y(s))\, \mathrm{d}s \tag{45}$$

Theorem [151, p. 55]. If $g(t)$ is continuous, the kernel $K(t, s, y)$ is continuous in all variables and satisfies a Lipschitz condition

$$|K(t,s,y) - K(t,s,z)| \le L\, |y-z|$$

then the nonlinear Volterra equation has a unique continuous solution.

A successive substitution method for its solution is

$$y_{n+1}(t) = g(t) + \int_0^t K[t, s, y_n(s)]\, \mathrm{d}s$$

Nonlinear Fredholm equations have special names. The equation

$$f(x) = \int_0^1 K[x, y, f(y)]\, \mathrm{d}y$$

is called the Urysohn equation [150 p. 208]. The special equation

$$f(x) = \int_0^1 K[x,y]\, F[y, f(y)]\, \mathrm{d}y$$

is called the Hammerstein equation [150, p. 209]. Iterative methods can be used to solve these equations, and these methods are closely tied to fixed point problems. A fixed point problem is

$$x = F(x)$$

and a successive substitution method is

$$x_{n+1} = F(x_n)$$

Local convergence theorems prove the process convergent if the solution is close enough to the answer, whereas global convergence theorems are valid for any initial guess [150, p. 229 – 231]. The successive substitution method for nonlinear Fredholm equations is

$$y_{n+1}(x) = \int_0^1 K[x, s, y_n(s)]\, \mathrm{d}s$$

Typical conditions for convergence include that the function satisfies a Lipschitz condition.

9.2. Numerical Methods for Volterra Equations of the Second Kind

Volterra equations of the second kind are analogous to initial value problems. An initial value problem can be written as a Volterra equation of the second kind, although not all Volterra equations can be written as initial value problems [151, p. 7]. Here the general nonlinear Volterra equation of the second kind is treated (Eq. 45). The simplest numerical method involves replacing the integral by a quadrature using the trapezoid rule.

$$y_n \equiv y(t_n) = g(t_n) + \Delta t \left\{ \frac{1}{2} K(t_n, t_0, y_0) + \sum_{i=1}^{n-1} K(t_n, t_i, y_i) + \frac{1}{2} K(t_n, t_n, y_n) \right\}$$

This equation is a nonlinear algebraic equation for y_n. Since y_0 is known it can be applied to solve for $y_1, y_2, \ldots$ in succession. For a single integral equation, at each step one must solve a single nonlinear algebraic equation for y_n. Typically, the error in the solution to the integral equation is proportional to Δt^μ, and the power μ is the same as the power in the quadrature error [151, p. 97].

The stability of the method [151, p. 111] can be examined by considering the equation

$$y(t) = 1 - \lambda \int_0^t y(s) \, \mathrm{d}s$$

whose solution is

$$y(t) = \mathrm{e}^{-\lambda t}$$

Since the integral equation can be differentiated to obtain the initial value problem

$$\frac{\mathrm{d}y}{\mathrm{d}t} = -\lambda y, \; y(0) = 1$$

the stability results are identical to those for initial value methods. In particular, using the trapezoid rule for integral equations is identical to using this rule for initial value problems. The method is A-stable.

Higher order integration methods can also be used [151, p. 114, 124]. When the kernel is infinite at certain points, i.e., when the problem has a weak singularity, see [151, p. 71, 151].

9.3. Numerical Methods for Fredholm, Urysohn, and Hammerstein Equations of the Second Kind

Whereas Volterra equations could be solved from one position to the next, like initial value differential equations, Fredholm equations must be solved over the entire domain, like boundary value differential equations. Thus, large sets of equations will be solved and the notation is designed to emphasize that.

The methods are also based on quadrature formulas. For the integral

$$I(\varphi) = \int_a^b \varphi(y) \, \mathrm{d}y$$

a quadrature formula is written:

$$I(\varphi) = \sum_{i=0}^{n} w_i \varphi(y_i)$$

Then the integral Fredholm equation can be rewritten as

$$f(x) - \lambda \sum_{i=0}^{n} w_i K(x, y_i) f(y_i) = g(x), \quad a \le x \le b \tag{46}$$

If this equation is evaluated at the points $x = y_j$,

$$f(y_j) - \lambda \sum_{i=0}^{n} w_i K(y_j, y_i) f(y_i) = g(y_i)$$

is obtained, which is a set of linear equations to be solved for $\{f(y_j)\}$. The solution at any point is then given by Equation 46.

A common type of integral equation has a singular kernel along $x = y$. This can be transformed to a less severe singularity by writing

$$\int_a^b K(x,y) f(y) \, \mathrm{d}y = \int_a^b K(x,y) \left[f(y) - f(x) \right] \mathrm{d}y + \int_a^b K(x,y) f(x) \, \mathrm{d}y = \int_a^b K(x,y) \left[f(y) - f(x) \right] \mathrm{d}y + f(x) H(x)$$

where

$$H(x) = \int_a^b K(x,y) f(x) \, \mathrm{d}y$$

is a known function. The integral equation is then replaced by

$$f(x)=g(x)+\sum_{i=0}^{n} w_i K(x,y_i)\left[f(y_i)-f(x)\right] +f(x)H(x)$$

Collocation methods can be applied as well [149, p. 396]. To solve integral Equation 43 expand f in the function

$$f=\sum_{i=0}^{n} a_i \varphi_i(x)$$

Substitute f into the equation to form the residual

$$\sum_{i=0}^{n} a_i \varphi_i(x)-\lambda \sum_{i=0}^{n} a_i \int_a^b K(x,y)\varphi_i(y)\,\mathrm{d}y=g(x)$$

Evaluate the residual at the collocation points

$$\sum_{i=0}^{n} a_i \varphi_i(x_j)-\lambda \sum_{i=0}^{n} a_i \int_a^b K(x_j,y)\varphi_i(y)\,\mathrm{d}y=g(x_j)$$

The expansion can be in piecewise polynomials, leading to a collocation finite element method, or global polynomials, leading to a global approximation. If orthogonal polynomials are used then the quadratures can make use of the accurate Gaussian quadrature points to calculate the integrals. Galerkin methods are also possible [149, p. 406]. Mills et al. [157] consider reaction – diffusion problems and say the choice of technique cannot be made in general because it is highly dependent on the kernel.

When the integral equation is nonlinear, iterative methods must be used to solve it. Convergence proofs are available, based on Banach's contractive mapping principle. Consider the Urysohn equation, with $g(x)=0$ without loss of generality:

$$f(x)=\int_a^b F\left[x,y,f(y)\right]\mathrm{d}y$$

The kernel satisfies the Lipschitz condition

$$\max_{a\leq x,y\leq b}\left|F\left[x,y,f(y)\right]-F\left[x,z,f(z)\right]\right|\leq K\left|y-z\right|$$

Theorem [150, p. 214]. If the constant K is < 1 and certain other conditions hold, the successive substitution method

$$f_{n+1}(x)=\int_a^b F\left[x,y,f_n(y)\right]\mathrm{d}y, n=0,1\ldots$$

converges to the solution of the integral equations.

9.4. Numerical Methods for Eigenvalue Problems

Eigenvalue problems are treated similarly to Fredholm equations, except that the final equation is a matrix eigenvalue problem instead of a set of simultaneous equations. For example,

$$\sum_{i=1}^{n} w_i K(y_i,y_i) f(y_i)=\lambda f(y_j),$$

$$i=0,1,\ldots,n$$

leads to the matrix eigenvalue problem

$$\boldsymbol{K}\,\boldsymbol{D}\,\boldsymbol{f}=\lambda \boldsymbol{f}$$

Where $\boldsymbol{D}$ is a diagonal matrix with $D_{ii}=w_i$.

9.5. Green's Functions [158 – 160]

Integral equations can arise from the formulation of a problem by using Green's function. For example, the equation governing heat conduction with a variable heat generation rate is represented in differential forms as

$$\frac{\mathrm{d}^2 T}{\mathrm{d}x^2}=\frac{Q(x)}{k},\ T(0)=T(1)=0$$

In integral form the same problem is [149, pp. 57 – 60]

$$T(x)=\frac{1}{k}\int_0^1 G(x,y)\,Q(y)\,\mathrm{d}y$$

$$G(x,y)=\begin{cases}-x(1-y) & x\leq y\\ -y(1-x) & y\leq x\end{cases}$$

Green's functions for typical operators are given below.

For the Poisson equation with solution decaying to zero at infinity

$$\nabla^2\psi=-4\pi\varrho$$

the formulation as an integral equation is

$$\psi(\boldsymbol{r})=\int_V \rho(\boldsymbol{r}_0)\,G(\boldsymbol{r},\boldsymbol{r}_0)\,\mathrm{d}V_0$$

where Green's function is [50, p. 891]

$$G(\boldsymbol{r},\boldsymbol{r}_0)=\frac{1}{r}\text{ in three dimensions}$$

$$=-2\ln r\text{ in two dimensions}$$

where $r=\sqrt{(x-x_0)^2+(y-y_0)^2+(z-z_0)^2}$ in three dimensions

and $r=\sqrt{(x-x_0)^2+(y-x_0)^2}$ in two dimensions

For the problem

$$\frac{\partial u}{\partial t} = D\nabla^2 u,\ u = 0 \text{ on } S,$$

with a point source at x_0, y_0, z_0

Green's function is [44, p. 355]

$$u = \frac{1}{8[\pi D(t-\tau)]^{3/2}}$$
$$\mathrm{e}^{-[(x-x_0)^2+(y-y_0)^2+(z-z_0)^2]/4\,D(t-\tau)}$$

When the problem is

$$\frac{\partial c}{\partial t} = D\nabla^2 c$$

$$c = f\,(x,y,z) \text{ in } S \text{ at } t = 0$$

$$c = \varphi\,(x,y,z) \text{ on } S, t>0$$

the solution can be represented as [44, p. 353]

$$c = \int\int\int (u)_{\tau=0} f\,(x,y,z)\,\mathrm{d}x\mathrm{d}y\mathrm{d}z$$

$$+D\int_0^t\int\int\varphi\,(x,y,z,\tau)\,\frac{\partial u}{\partial n}\mathrm{d}S\mathrm{d}t$$

When the problem is two dimensional,

$$u = \frac{1}{\sqrt{4\pi D(t-\tau)}}\mathrm{e}^{-[(x-x_0)^2+(y-y_0)^2]/4D(t-\tau)}$$

$$c = \int\int (u)_{\tau=0} f\,(x,y)\,\mathrm{d}x\mathrm{d}y$$

$$+D\int_0^t\int\varphi\,(x,y,\tau)\,\frac{\partial u}{\partial n}\mathrm{d}C\mathrm{d}t$$

For the following differential equation and boundary conditions

$$\frac{1}{x^{a-1}}\frac{\mathrm{d}}{\mathrm{d}x}\left(x^{a-1}\frac{\mathrm{d}c}{\mathrm{d}x}\right) = f\,[x,c\,(x)]\,,$$

$$\frac{\mathrm{d}c}{\mathrm{d}x}\,(0) = 0,\ \frac{2}{Sh}\frac{\mathrm{d}c}{\mathrm{d}x}\,(1) + c\,(1) = g$$

where *Sh* is the Sherwood number, the problem can be written as a Hammerstein integral equation:

$$c\,(x) = g - \int_0^1 G\,(x,y,Sh)\,f\,[y,c\,(y)]\,y^{a-1}\mathrm{d}y$$

Green's function for the differential operators are [163]

$a = 1$

$$G\,(x,y,Sh) = \begin{cases} 1+\frac{2}{Sh}-x, & y\leq x \\ 1+\frac{2}{Sh}-y, & x<y \end{cases}$$

$a = 2$

$$G\,(x,y,Sh) = \begin{cases} \frac{2}{Sh}-\ln x, & y\leq x \\ \frac{2}{Sh}-\ln y, & x<y \end{cases}$$

$a = 3$

$$G\,(x,y,Sh) = \begin{cases} \frac{2}{Sh}+\frac{1}{x}-1, & y\leq x \\ \frac{2}{Sh}+\frac{1}{y}-1, & x<y \end{cases}$$

Green's functions for the reaction diffusion problem were used to provide computable error bounds by Ferguson and Finlayson [163].

If Green's function has the form

$$K\,(x,y) = \begin{cases} u\,(x)\,v\,(y) & 0\leq y\leq x \\ u\,(y)\,v\,(x) & x\leq y\leq 1 \end{cases}$$

the problem

$$f\,(x) = \int_0^1 K\,(x,y)\,F\,[y,f\,(y)]\,\mathrm{d}y$$

may be written as

$$f\,(x) - \int_0^x [u\,(x)\,v\,(y) - u\,(y)\,v\,(x)]\cdot$$
$$f\,[y,f\,(y)]\,\mathrm{d}y = \alpha v\,(x)$$

where

$$\alpha = \int_0^1 u\,(y)\,F\,[y,f\,(y)]\,\mathrm{d}y$$

Thus, the problem ends up as one directly formulated as a fixed point problem:

$$f = \Phi\,(f)$$

When the problem is the diffusion – reaction one, the form is

$$c\,(x) = g - \int_0^x [u\,(x)\,v\,(y) - u\,(y)\,v\,(x)]$$
$$f\,[y,c\,(y)]\,y^{a-1}\mathrm{d}y - \alpha v\,(x)$$

$$\alpha = \int_0^1 u\,(y)\,f\,[y,c\,(y)]\,y^{a-1}\mathrm{d}y$$

Dixit and Taularidis [164] solved problems involving Fischer – Tropsch synthesis reactions in a catalyst pellet using a similar method.

9.6. Boundary Integral Equations and Boundary Element Method

The boundary element method utilizes Green's theorem and integral equations. Here, the method is described briefly for the following boundary value problem in two or three dimensions

$$\nabla^2\varphi = 0,\ \varphi = f_1 \text{ on } S_1,\ \frac{\partial\varphi}{\partial n} = f_2 \text{ on } S_2$$

Green's theorem (see page 46) says that for any functions sufficiently smooth

$$\int_V \left(\varphi\nabla^2\psi - \psi\nabla^2\varphi\right) \mathrm{d}V = \int_S \left(\varphi\frac{\partial\psi}{\partial n} - \psi\frac{\partial\varphi}{\partial n}\right) \mathrm{d}S$$

Suppose the function ψ satisfies the equation

$$\nabla^2\psi = 0$$

In two and three dimensions, such a function is

$$\psi = \ln r,\ r = \sqrt{(x-x_0)^2+(y-y_0)^2}$$

in two dimensions

$$\psi = \frac{1}{r},\ r = \sqrt{(x-x_0)^2+(y-y_0)^2+(z-z_0)^2}$$

in three dimensions

where $\{x_0, y_0\}$ or $\{x_0, y_0, z_0\}$ is a point in the domain. The solution φ also satisfies

$$\nabla^2\varphi = 0$$

so that

$$\int_S \left(\varphi\frac{\partial\psi}{\partial n} - \psi\frac{\partial\varphi}{\partial n}\right) \mathrm{d}S = 0$$

Consider the two-dimensional case. Since the function ψ is singular at a point, the integrals must be carefully evaluated. For the region shown in Figure 33, the domain is $S = S_1 + S_2$; a small circle of radius r_0 is placed around the point P at x_0, y_0. Then the full integral is

$$\int_S \left(\varphi\frac{\partial \ln r}{\partial n} - \ln r\frac{\partial\varphi}{\partial n}\right) \mathrm{d}S$$
$$+ \int_{\theta=0}^{\theta=2\pi} \left(\varphi\frac{\partial \ln r_0}{\partial n} - \ln r_0\frac{\partial\varphi}{\partial n}\right) r_0 \mathrm{d}\theta = 0$$

As r_0 approaches 0,

$$\lim_{r_0\to 0} r_0 \ln r_0 = 0$$

and

$$\lim_{r_0\to 0} \int_{\theta=0}^{\theta=2\pi} \varphi\frac{\partial \ln r_0}{\partial n} r_0 \mathrm{d}\theta = -\varphi(P)\, 2\pi$$

Thus for an internal point,

$$\varphi(P) = \frac{1}{2\pi}\int_S \left(\varphi\frac{\partial \ln r}{\partial n} - \ln r\frac{\partial\varphi}{\partial n}\right) \mathrm{d}S \tag{47}$$

If P is on the boundary, the result is [165, p. 464]

$$\varphi(P) = \frac{1}{\pi}\int_S \left(\varphi\frac{\partial \ln r}{\partial n} - \ln r\frac{\partial\varphi}{\partial n}\right) \mathrm{d}S$$

Putting in the boundary conditions gives

$$\pi\varphi(P) = \int_{S_1} \left(f_1\frac{\partial \ln r}{\partial n} - \ln r\frac{\partial\varphi}{\partial n}\right) \mathrm{d}S + \int_{S_2} \left(\varphi\frac{\partial \ln r}{\partial n} - f_2 \ln r\right) \mathrm{d}S \tag{48}$$

This is an integral equation for φ on the boundary. Note that the order is one less than the original differential equation. However, the integral equation leads to matrices that are dense rather than banded or sparse, so some of the advantage of lower dimension is lost. Once this integral equation (Eq. 48) is solved to find φ on the boundary, it can be substituted in Equation 47 to find φ anywhere in the domain.

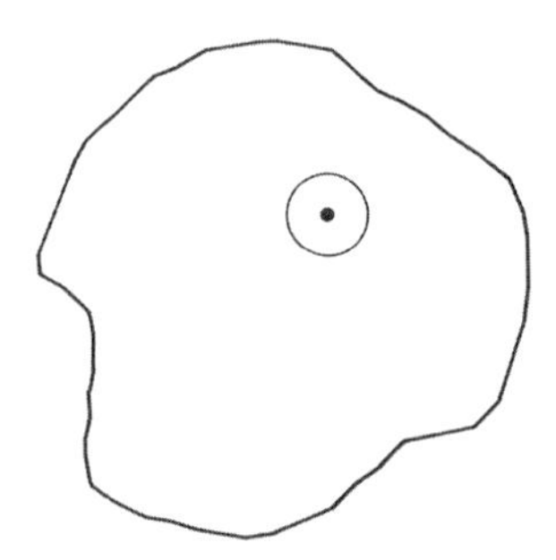

Figure 33. Domain with singularity at P

In the boundary finite element method, both the function and its normal derivative along the boundary are approximated.

$$\varphi = \sum_{j=1}^{N} \varphi_j N_j(\xi),\ \frac{\partial\varphi}{\partial n} = \sum_{j=1}^{N} \left(\frac{\partial\varphi}{\partial n}\right)_j N_j(\xi)$$

One choice of trial functions can be the piecewise constant functions shown in Figure 34. The integral equation then becomes

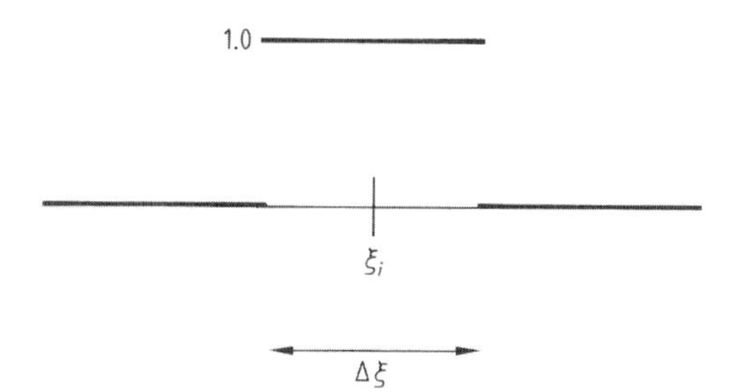

Figure 34. Trial function on boundary for boundary finite element method

$$\pi\varphi_i \sum_{j=1}^{N} \left[\varphi_j \int_{s_j} \frac{\partial \ln r_i}{\partial n} dS - \left(\frac{\partial \varphi_j}{\partial n} \right) \int_{s_j} \ln r_i \right] ds$$

The function φ_j is of course known along s_1, whereas the derivative $\partial\varphi_j/\partial n$ is known along s_2. This set of equations is then solved for φ_i and $\partial\varphi_i/\partial n$ along the boundary. This constitutes the boundary integral method applied to the Laplace equation.

If the problem is Poisson's equation

$$\nabla^2 \varphi = g\,(x,y)$$

Green's theorem gives

$$\int_S \left(\varphi \frac{\partial \ln r}{\partial n} - \ln r \frac{\partial \varphi}{\partial n} \right) dS + \int_A g \ln r \, dA = 0$$

Thus, for an internal point,

$$2\pi\varphi\,(P) = \int_S \left(\varphi \frac{\partial \ln r}{\partial r} - \ln r \frac{\partial \varphi}{\partial n} \right) dS + \int_A g \ln r dA \tag{49}$$

and for a boundary point,

$$\pi\varphi\,(P) = \int_S \left(\varphi \frac{\partial \ln r}{\partial n} - \ln r \frac{\partial \varphi}{\partial n} \right) dS + \int_A g \ln r dA \tag{50}$$

If the region is nonhomogeneous this method can be used [165, p. 475], and it has been applied to heat conduction by HSIEH and SHANG [166]. The finite element method can be applied in one region and the boundary finite element method in another region, with appropriate matching conditions [165, p. 478]. If the problem is nonlinear, then it is more difficult. For example, consider an equation such as Poisson's in which the function depends on the solution as well

$$\nabla^2 \varphi = g\,(x,y,\varphi)$$

Then the integral appearing in Equation 50 must be evaluated over the entire domain, and the solution in the interior is given by Equation 49. For further applications, see [167] and [168].

10. Optimization

We provide a survey of systematic methods for a broad variety of optimization problems. The survey begins with a general classification of mathematical optimization problems involving continuous and discrete (or integer) variables. This is followed by a review of solution methods of the major types of optimization problems for continuous and discrete variable optimization, particularly nonlinear and mixed-integer nonlinear programming. In addition, we discuss direct search methods that do not require derivative information as well as global optimization methods. We also review extensions of these methods for the optimization of systems described by differential and algebraic equations.

10.1. Introduction

Optimization is a key enabling tool for decision making in chemical engineering [306]. It has evolved from a methodology of academic interest into a technology that continues to make significant impact in engineering research and practice. Optimization algorithms form the core tools for a) experimental design, parameter estimation, model development, and statistical analysis; b) process synthesis analysis, design, and retrofit; c) model predictive control and real-time optimization; and d) planning, scheduling, and the integration of process operations into the supply chain [307, 308].

As shown in Figure 35, optimization problems that arise in chemical engineering can be classified in terms of continuous and discrete variables. For the former, nonlinear programming (NLP) problems form the most general case, and widely applied specializations include linear programming (LP) and quadratic programming (QP). An important distinction for NLP is whether the optimization problem is convex or nonconvex. The latter NLP problem may have multiple local optima, and an important question is whether a global solution is required for the NLP. Another important distinction is whether the problem is assumed to be differentiable or not.

Mixed-integer problems also include discrete variables. These can be written as mixed-integer nonlinear programs (MINLP) or as mixed-integer linear programs (MILP) if all variables appear linearly in the constraint and objective functions. For the latter an important case occurs when all the variables are integer; this gives rise to an integer programming (IP) problem. IPs can be further classified into many special problems (e.g., assignment, traveling salesman, etc.), which are not shown in Figure 35. Similarly, the MINLP problem also gives rise to special problem classes, although here the main distinction is whether its relaxation is convex or nonconvex.

The ingredients of formulating optimization problems include a mathematical model of the system, an objective function that quantifies a criterion to be extremized, variables that can serve as decisions, and, optionally, inequality constraints on the system. When represented in algebraic form, the general formulation of discrete/continuous optimization problems can be written as the following mixed-integer optimization problem:

$$\begin{aligned} \text{Min} \quad & f(x,y) \\ \text{s.t.} \quad & h(x,y)=0 \\ & g(x,y)\leq 0 \\ & x\in\Re^n, y\in\{0,1\}^t \end{aligned} \tag{51}$$

where $f(x, y)$ is the objective function (e.g., cost, energy consumption, etc.), $h(x, y)=0$ are the equations that describe the performance of the system (e.g., material balances, production rates), the inequality constraints $g(x, y)\leq 0$ can define process specifications or constraints for feasible plans and schedules, and s.t. denotes subject to. Note that the operator Max $f(x)$ is equivalent to Min $-f(x)$. We define the real n-vector x to represent the continuous variables while the t-vector y represents the discrete variables, which, without loss of generality, are often restricted to take 0–1 values to define logical or discrete decisions, such as assignment of equipment and sequencing of tasks. (These variables can also be formulated to take on other integer values as well.) Problem 51 corresponds to a mixed-integer nonlinear program (MINLP) when any of the functions involved are nonlinear. If all functions are linear it corresponds to a mixed-integer linear program (MILP). If there are no 0–1 variables, then problem 51 reduces to a nonlinear program 52 or linear program 65 depending on whether or not the functions are linear.

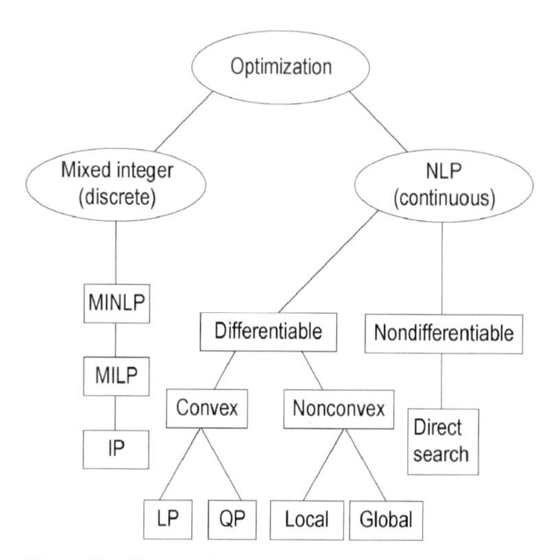

Figure 35. Classes of optimization problems and algorithms

We first start with continuous variable optimization and consider in the next section the solution of NLPs with differentiable objective and constraint functions. If only local solutions are required for the NLP, then very efficient large-scale methods can be considered. This is followed by methods that are not based on local optimality criteria; we consider direct search optimization methods that do not require derivatives, as well as deterministic global optimization methods. Following this, we consider the solution of mixed-integer problems and outline the main characteristics of algorithms for their solution. Finally, we conclude with a discussion of optimization modeling software and its implementation in engineering models.

10.2. Gradient-Based Nonlinear Programming

For continuous variable optimization we consider problem 51 without discrete variables y.

The general NLP problem 52 is presented below:

$$\begin{aligned} \text{Min} \quad & f(x) \\ \text{s.t.} \quad & h(x)=0 \\ & g(x)\leq 0 \end{aligned} \tag{52}$$

and we assume that the functions $f(x)$, $h(x)$, and $g(x)$ have continuous first and second derivatives. A key characteristic of problem 52 is whether the problem is convex or not, i.e., whether it has a convex objective function and a convex feasible region. A function $\phi(x)$ of x in some domain X is convex if and only if for all points $x_1, x_2 \in X$:

$$\phi[\alpha x_1+(1-\alpha)x_2]\leq\alpha\phi(x_1)+(1-\alpha)\phi(x_2) \tag{53}$$

holds for all $\alpha \in (0, 1)$. If $\phi(x)$ is differentiable, then an equivalent definition is:

$$\phi(x_1)+\nabla\phi(x_1)^T(x-x_1)\leq\phi(x) \tag{54}$$

Strict convexity requires that the inequalities 53 and 54 be strict. Convex feasible regions require $g(x)$ to be a convex function and $h(x)$ to be linear. If 52 is a convex problem, than any local solution is guaranteed to be a global solution to 52. Moreover, if the objective function is strictly convex, then this solution x^* is unique. On the other hand, nonconvex problems may have multiple local solutions, i.e., feasible solutions that minimize the objective function within some neighborhood about the solution.

We first consider methods that find only local solutions to nonconvex problems, as more difficult (and expensive) search procedures are required to find a global solution. Local methods are currently very efficient and have been developed to deal with very large NLPs. Moreover, by considering the structure of convex NLPs (including LPs and QPs), even more powerful methods can be applied. To study these methods, we first consider conditions for local optimality.

Local Optimality Conditions – A Kinematic Interpretation. Local optimality conditions are generally derived from gradient information from the objective and constraint functions. The proof follows by identifying a local minimum point that has no feasible descent direction. Invoking a theorem of the alternative (e.g., Farkas – Lemma) leads to the celebrated Karush – Kuhn – Tucker (KKT) conditions [169]. Instead of a formal development of these conditions, we present here a more intuitive, kinematic illustration. Consider the contour plot of the objective function $f(x)$ given in Figure 36 as a smooth valley in space of the variables x_1 and x_2. For the contour plot of this unconstrained problem, Min $f(x)$, consider a ball rolling in this valley to the lowest point of $f(x)$, denoted by x^*. This point is at least a local minimum and is defined by a point with zero gradient and at least nonnegative curvature in all (nonzero) directions p. We use the first derivative (gradient) vector $\nabla f(x)$ and second derivative (Hessian) matrix $\nabla_{xx} f(x)$ to state the necessary first- and second-order conditions for unconstrained optimality:

$$\nabla_x f(x^*)=0 \quad p^T\nabla_{xx}f(x^*)p\geq 0 \quad \text{for all } p\neq 0 \tag{55}$$

These necessary conditions for local optimality can be strengthened to sufficient conditions by making the inequality in the relations 55 strict (i.e., positive curvature in all directions). Equivalently, the sufficient (necessary) curvature conditions can be stated as: $\nabla_{xx} f(x^*)$ has all positive (nonnegative) eigenvalues and is therefore defined as a positive (semi-)definite matrix.

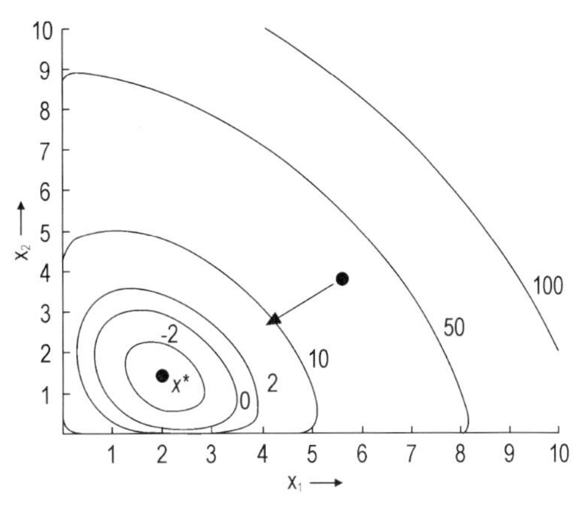

Figure 36. Unconstrained minimum

Now consider the imposition of inequality $[g(x)\leq 0]$ and equality constraints $[h(x)=0]$ in Figure 37. Continuing the kinematic interpretation, the inequality constraints $g(x)\leq 0$ act as "fences" in the valley, and equality constraints

$h(x) = 0$ as "rails". Consider now a ball, constrained on a rail and within fences, to roll to its lowest point. This stationary point occurs when the normal forces exerted by the fences $[-\nabla g(x^*)]$ and rails $[-\nabla h(x^*)]$ on the ball are balanced by the force of "gravity" $[-\nabla f(x^*)]$. This condition can be stated by the following *KKT necessary conditions* for constrained optimality:

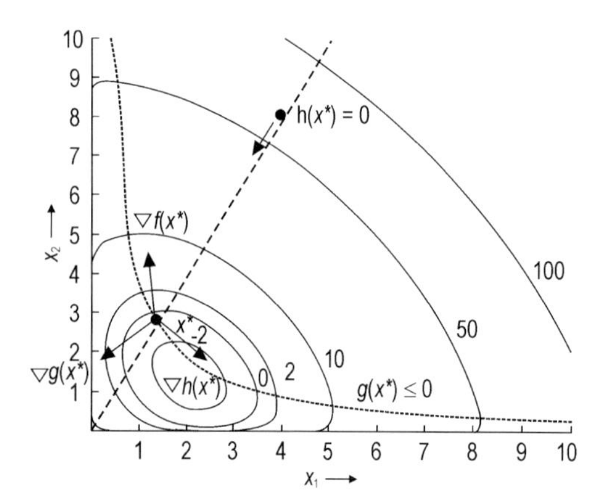

Figure 37. Constrained minimum

Stationarity Condition: It is convenient to define the Lagrangian function $L(x, \lambda, \nu) = f(x) + g(x)^T \lambda + h(x)^T \nu$ along with "weights" or multipliers λ and ν for the constraints. These multipliers are also known as "dual variables" and "shadow prices". The stationarity condition (balance of forces acting on the ball) is then given by:

$$\nabla L(x,\lambda,\nu) = \nabla f(x) + \nabla h(x)\lambda + \nabla g(x)\nu = 0 \quad (56)$$

Feasibility: Both inequality and equality constraints must be satisfied (ball must lie on the rail and within the fences):

$$h(x) = 0, \quad g(x) \le 0 \quad (57)$$

Complementarity: Inequality constraints are either strictly satisfied (active) or inactive, in which case they are irrelevant to the solution. In the latter case the corresponding KKT multiplier must be zero. This is written as:

$$\nu^T g(x) = 0, \quad \nu \ge 0 \quad (58)$$

Constraint Qualification: For a local optimum to satisfy the KKT conditions, an additional regularity condition or constraint qualification (CQ) is required. The KKT conditions are derived from gradient information, and the CQ can be viewed as a condition on the relative influence of constraint curvature. In fact, linearly constrained problems with nonempty feasible regions require no constraint qualification. On the other hand, as seen in Figure 38, the problem Min x_1, s.t. $x_2 \ge 0$, $(x_1)^3 \ge x_2$ has a minimum point at the origin but does not satisfy the KKT conditions because it does not satisfy a CQ.

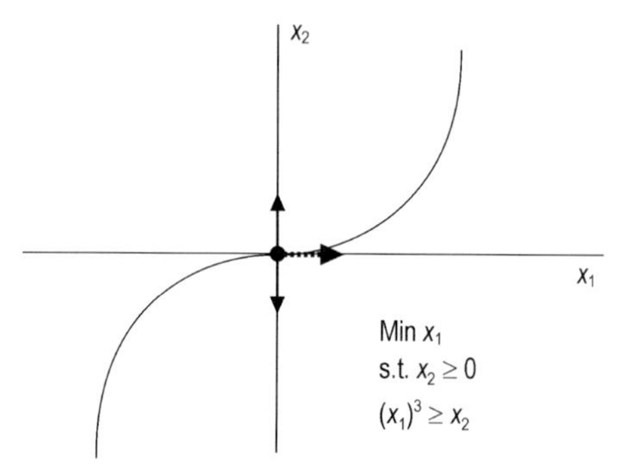

Figure 38. Failure of KKT conditions at constrained minimum (note linear dependence of constraint gradients)

CQs be defined in several ways. For instance, the linear independence constraint qualification (LICQ) requires that the active constraints at x^* be linearly independent, i.e., the matrix $[\nabla h(x^*) \,|\, \nabla g_A(x^*)]$ is full column rank, where g_A is the vector of inequality constraints with elements that satisfy $g_{A,i}(x^*) = 0$. With LICQ, the KKT multipliers (λ, ν) are guaranteed to be unique at the optimal solution. The weaker Mangasarian – Fromovitz constraint qualification (MFCQ) requires only that $\nabla h(x^*)$ have full column rank and that a direction p exist that satisfies $\nabla h(x^*)^T p = 0$ and $\nabla g_A(x^*)^T p > 0$. With MFCQ, the KKT multipliers (λ, ν) are guaranteed to be bounded (but not necessarily unique) at the optimal solution. Additional discussion can be found in [169].

Second Order Conditions: As with unconstrained optimization, nonnegative (positive) curvature is necessary (sufficient) in all of the allowable (i.e., constrained) nonzero directions p. This condition can be stated in several ways. A typical necessary second-order condition requires a point x^* that satisfies LICQ and first-order conditions 56–58 with multipliers (λ, ν) to satisfy the additional conditions given by:

$$p^T \nabla_{xx} L(x^*,\lambda,\nu)p \geq 0 \quad \text{for all}$$

$$p \neq 0, \nabla h(x^*)^T p = 0,$$

$$\nabla g_A(x^*)^T p = 0 \tag{59}$$

The corresponding sufficient conditions require that the inequality in 59 be strict. Note that for the example in Figure 36, the allowable directions *p* span the entire space for *x* while in Figure 37, there are *no* allowable directions *p*.

Example: To illustrate the KKT conditions, consider the following unconstrained NLP:

$$\text{Min } (x_1)^2 - 4x_1 + 3/2(x_2)^2$$
$$-7x_2 + x_1 x_2 + 9 - \ln x_1 - \ln x_2 \tag{60}$$

corresponding to the contour plot in Figure 36. The optimal solution can be found by solving for the first order conditions 54:

$$\nabla f(x) = \begin{bmatrix} 2x_1 - 4 + x_2 - 1/x_1 \\ 3x_2 - 7 + x_1 - 1/x_2 \end{bmatrix} = 0$$

$$\Rightarrow x^* = \begin{bmatrix} 1.3475 \\ 2.0470 \end{bmatrix} \tag{61}$$

and $f(x^*) = -2.8742$. Checking the second-order conditions leads to:

$$\nabla_{xx} f(x*) = \begin{bmatrix} 2 + 1/(x_1^*)^2 & 1 \\ 1 & 3 + 1/(x_2^*)^2 \end{bmatrix} \Rightarrow$$

$$\nabla_{xx} f(x*) = \begin{bmatrix} 2.5507 & 1 \\ 1 & 3.2387 \end{bmatrix} \text{(positive definite)} \tag{62}$$

Now consider the constrained NLP:

$$\begin{aligned} \text{Min} \quad & (x_1)^2 - 4x_1 + 3/2(x_1)^2 - 7x_2 + x_1 x_2 \\ & +9 - \ln x_1 - \ln x_2 \\ \text{s.t.} \quad & 4 - x_1 x_2 \leq 0 \\ & 2x_1 - x_2 = 0 \end{aligned} \tag{63}$$

that corresponds to the plot in Figure 37. The optimal solution can be found by applying the first-order KKT conditions 56–58:

$$\nabla L(x,\lambda,\nu) = \nabla f(x) + \nabla h(x)\lambda + \nabla g(x)\nu =$$

$$\begin{bmatrix} 2x_1 - 4 + x_2 - 1/x_1 \\ 3x_2 - 7 + x_1 - 1/x_2 \end{bmatrix} + \begin{bmatrix} 2 \\ -1 \end{bmatrix} \lambda + \begin{bmatrix} -x_2 \\ -x_1 \end{bmatrix} \nu = 0$$

$$g(x) = 4 - x_1 x_2 \leq 0, \quad h(x) = 2x_1 - x_2 = 0$$

$$g(x)\nu = (4 - x_1 x_2)\nu, \quad \nu \geq 0$$

$$\Downarrow$$

$$x* = \begin{bmatrix} 1.4142 \\ 2.8284 \end{bmatrix}, \lambda* = 1.036, \nu* = 1.068$$

and $f(x^*) = -1.8421$. Checking the second-order conditions 59 leads to:

$$\nabla_{xx} L(x*,\lambda*,\nu*) = \nabla_{xx}\left[f(x*) + h(x*)\lambda* + g(x*)\nu*\right] =$$

$$\begin{bmatrix} 2 + 1/(x_1)^2 & 1 - \nu \\ 1 - \nu & 3 + 1/(x_2)^2 \end{bmatrix} = \begin{bmatrix} 2.5 & 0.068 \\ 0.068 & 3.125 \end{bmatrix}$$

$$[\nabla h(x*) | \nabla g_A(x*)]$$

$$p = \begin{bmatrix} 2 & -2.8284 \\ -1 & -1.4142 \end{bmatrix} p = 0, p \neq 0$$

Note that LICQ is satisfied. Moreover, because $[\nabla h(x^*) | \nabla g_A(x^*)]$ is square and nonsingular, there are no nonzero vectors *p* that satisfy the allowable directions. Hence, the sufficient second-order conditions ($p^T \nabla_{xx} L(x^*,\lambda^*,\nu^*) > 0$, for all allowable *p*) are *vacuously satisfied* for this problem.

Convex Cases of NLP. Linear programs and quadratic programs are special cases of problem 52 that allow for more efficient solution, based on application of KKT conditions 56–59. Because these are convex problems, any locally optimal solution is a global solution. In particular, if the objective and constraint functions in problem 52 are linear then the following linear program (LP):

$$\begin{aligned} \text{Min} \quad & c^T x \\ \text{s.t.} \quad & Ax = b \\ & Cx \leq d \end{aligned} \tag{65}$$

can be solved in a finite number of steps, and the optimal solution lies at a vertex of the polyhedron described by the linear constraints. This is shown in Figure 39, and in so-called primal degenerate cases, multiple vertices can be alternate optimal solutions with the same values of the objective function. The standard method to solve problem 65 is the simplex method, developed in the late 1940s [170], although, starting from KARMARKAR's discovery in 1984, interior point methods have become quite advanced and competitive for large scale problems [171]. The simplex method proceeds by moving successively from vertex to vertex with improved objective function values. Methods to solve problem 65 are well implemented and widely used, especially in planning and logistical applications. They also form the basis for MILP methods (see below). Currently, state-of-the-art LP solvers can handle millions of variables and constraints and the application of further decomposition methods leads to the solution of problems that are two or three orders of magnitude larger than this [172, 173]. Also, the interior point method is described below from the perspective of more general NLPs.

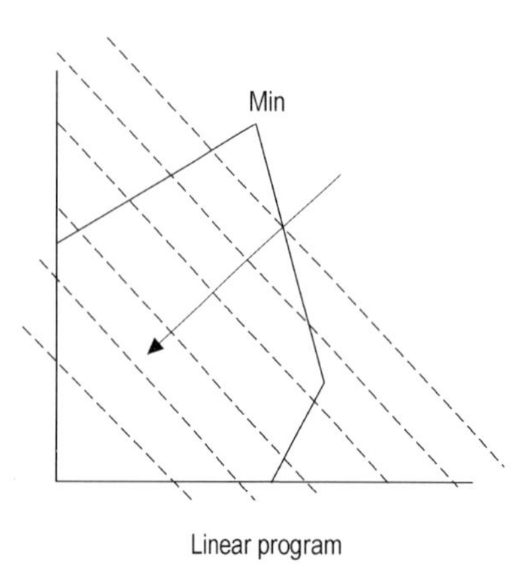

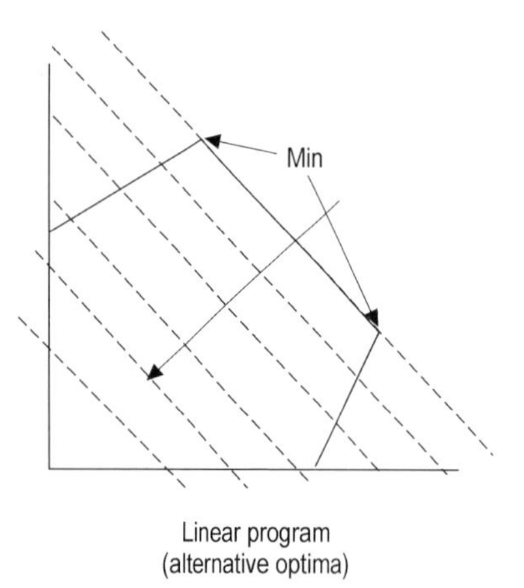

Figure 39. Contour plots of linear programs

Quadratic programs (QP) represent a slight modification of problem 65 and can be stated as:

$$\begin{aligned} \text{Min} \quad & c^T x + \tfrac{1}{2} x^T Q x \\ \text{s.t.} \quad & Ax = b \\ & x \le d \end{aligned} \tag{66}$$

If the matrix Q is positive semidefinite (positive definite), when projected into the null space of the active constraints, then problem 66 is (strictly) convex and the QP is a global (and unique) minimum. Otherwise, local solutions exist for problem 66, and more extensive global optimization methods are needed to obtain the global solution. Like LPs, convex QPs can be solved in a finite number of steps. However, as seen in Figure 40, these optimal solutions can lie on a vertex, on a constraint boundary, or in the interior. A number of active set strategies have been created that solve the KKT conditions of the QP and incorporate efficient updates of active constraints. Popular QP methods include null space algorithms, range space methods, and Schur complement methods. As with LPs, QP problems can also be solved with interior point methods [171].

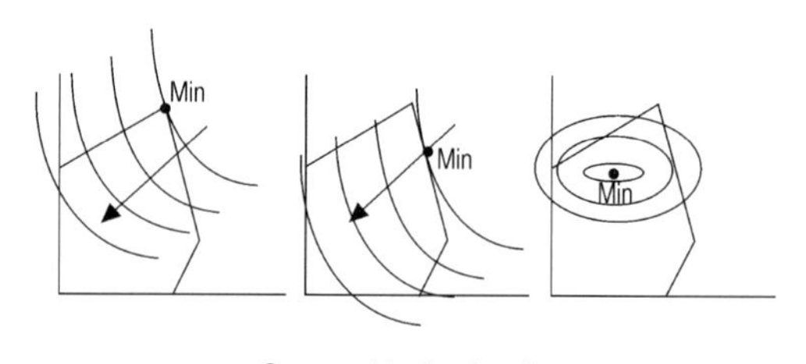

Figure 40. Contour plots of convex quadratic programs

Solving the General NLP Problem. Solution techniques for problem 52 deal with satisfaction of the KKT conditions 56–59. Many NLP solvers are based on successive quadratic programming (SQP) as it allows the construction of a number of NLP algorithms based on the Newton – Raphson method for equation solving. SQP solvers have been shown to require the fewest function evaluations to solve NLPs [174] and they can be tailored to a broad range of process engineering problems with different structure.

The SQP strategy applies the equivalent of a Newton step to the KKT conditions of the nonlinear programming problem, and this leads to a fast rate of convergence. By adding slack variables s the first-order KKT conditions can be rewritten as:

$$\nabla f(x) + \nabla h(x)\lambda + \nabla g(x)\nu = 0 \quad (67a)$$

$$h(x) = 0 \quad (67b)$$

$$g(x) + s = 0 \quad (67c)$$

$$SVe = 0 \quad (67d)$$

$$(s,\nu) \geq 0 \quad (67e)$$

where $e = [1, 1, \ldots, 1]^T$, $S = \mathrm{diag}(s)$ and $V = \mathrm{diag}(\nu)$. SQP methods find solutions that satisfy Equations 67a, 67b, 67c, 67d and 67e by generating Newton-like search directions at iteration k. However, equations 67d and active bounds 67e are dependent and serve to make the KKT system ill-conditioned near the solution. SQP algorithms treat these conditions in two ways. In the *active set strategy*, discrete decisions are made regarding the active constraint set, $i \in I = \{i | g_i(x^*) = 0\}$, and Equation 67d is replaced by $s_i = 0$, $i \in I$, and $v_i = 0$, $i \notin I$. Determining the active set is a combinatorial problem, and a straightforward way to determine an estimate of the active set [and also satisfy 67e] is to formulate, at a point x^k, and solve the following QP at iteration k:

$$\begin{array}{ll} \mathrm{Min} & \nabla f(x^k)^T p + \frac{1}{2} p^T \nabla_{xx} L(x^k, \lambda^k, \nu^k) p \\ \mathrm{s.t.} & h(x^k) + \nabla h(x^k)^T p = 0 \\ & g(x^k) + \nabla g(x^k)^T p + s = 0,\ s \geq 0 \end{array} \quad (68)$$

The KKT conditions of 68 are given by:

$$\nabla f\left(x^k\right) + \nabla^2 L\left(x^k, \lambda^k, \nu^k\right) p + \nabla h\left(x^k\right)\lambda + \nabla g\left(x^k\right)\nu = 0 \quad (69a)$$

$$h\left(x^k\right) + \nabla h\left(x^k\right)^T p = 0 \quad (69b)$$

$$g\left(x^k\right) + \nabla g\left(x^k\right)^T p + s = 0 \quad (69c)$$

$$SVe = 0 \quad (69d)$$

$$(s,\nu) \geq 0 \quad (69e)$$

where the Hessian of the Lagrange function $\nabla_{xx} L(x,\lambda,\nu) = \nabla_{xx}\left[f(x) + h(x)^T \lambda + g(x)^T \nu\right]$ is calculated directly or through a quasi-Newtonian approximation (created by differences of gradient vectors). It is easy to show that 69a–69c correspond to a Newton–Raphson step for 67a–67c applied at iteration k. Also, selection of the active set is now handled at the QP level by satisfying the conditions 69d, 69e. To evaluate and change candidate active sets, QP algorithms apply inexpensive matrix-updating strategies to the KKT matrix associated with the QP 68. Details of this approach can be found in [175, 176].

As alternatives that avoid the combinatorial problem of selecting the active set, interior point (or barrier) methods modify the NLP problem 52 to form problem 70

$$\begin{array}{ll} \mathrm{Min}\ f\left(x^k\right) - \mu \sum_i \ln s_i & \\ \mathrm{s.t.} \quad h\left(x^k\right) = 0 & \\ \qquad g\left(x^k\right) + s = 0 & \end{array} \quad (70)$$

where the solution to this problem has $s > 0$ for the penalty parameter $\mu > 0$, and decreasing μ to zero leads to solution of problem 52. The KKT conditions for this problem can be written as Equation 71

$$\begin{array}{l} \nabla f(x*) + \nabla h(x*)\lambda + \nabla g(x*)\nu = 0 \\ h(x*) = 0 \\ g(x*) + s = 0 \\ SVe = \mu e \end{array} \quad (71)$$

and at iteration k the Newton steps to solve 71 are given by:

$$\begin{bmatrix} \nabla_{xx} L(x_k, \lambda_k, \nu_k) & & \nabla h(x_k) & \nabla g(x_k) \\ & S_k^{-1} V_k & & I \\ \nabla h(x_k)^T & & & \\ \nabla g(x_k)^T & I & & \end{bmatrix} \begin{bmatrix} \Delta x \\ \Delta s \\ \Delta \lambda \\ \Delta \nu \end{bmatrix} = - \begin{bmatrix} \nabla_x L(x_k, \lambda_k, \nu_k) \\ \nu_k - S_k^{-1} \mu e \\ h(x_k) \\ g(x_k) + s_k \end{bmatrix} \quad (72)$$

A detailed description of this algorithm, called IPOPT, can be found in [177].

Both active set and interior point methods have clear trade-offs. Interior point methods may require more iterations to solve problem 70 for various values of μ, while active set methods require the solution of the more expensive QP subproblem 68. Thus, if there are few inequality constraints or an active set is known (say from a good starting guess, or a known QP solution from a previous iteration) then solving problem 68 is not expensive and the active set method is favored. On the other hand, for problems with many inequality constraints, interior point methods are often faster as they avoid the combinatorial problem of selecting the active set. This is especially the case for large-scale problems and when a large number of bounds are active. Examples that demonstrate the performance of these approaches include the solution of model predictive control (MPC) problems [178 – 180] and the solution of large optimal control problems using barrier NLP solvers. For instance, IPOPT allows the solution of problems with more than 10^6 variables and up to 50 000 degrees of freedom [181, 182].

Other Gradient-Based NLP Solvers. In addition to SQP methods, a number of NLP solvers have been developed and adapted for large-scale problems. Generally, these methods require more function evaluations than SQP methods, but they perform very well when interfaced to optimization modeling platforms, where function evaluations are cheap. All of these can be derived from the perspective of applying Newton steps to portions of the KKT conditions.

LANCELOT [183] is based on the solution of bound constrained subproblems. Here an augmented Lagrangian is formed from problem 52 and subproblem 73 is solved.

$$\begin{aligned}&\text{Min } f(x)+\lambda^T h(x)\\&\qquad+\nu\left(g(x)+s\right)+\tfrac{1}{2}\rho\|g(x),g(x)+s\|^2\\&\text{s.t.}\quad s\geq 0\end{aligned}\tag{73}$$

The above subproblem can be solved very efficiently for fixed values of the multipliers λ and v and penalty parameter ρ. Here a gradient-projection, trust-region method is applied. Once subproblem 73 is solved, the multipliers and penalty parameter are updated in an outer loop and the cycle repeats until the KKT conditions for problem 52 are satisfied. LANCELOT works best when exact second derivatives are available. This promotes a fast convergence rate in solving each subproblem and allows a bound constrained trust-region method to exploit directions of negative curvature in the Hessian matrix.

Reduced gradient methods are active set strategies that rely on partitioning the variables and solving Equations 67a, 67b, 67c, 67d and 67e in a nested manner. Without loss of generality, problem 52 can be rewritten as problem 74.

$$\begin{aligned}&\text{Min } f(z)\\&\text{s.t. } c(z)=0\\&a\leq z\leq b\end{aligned}\tag{74}$$

Variables are partitioned as nonbasic variables (those fixed to their bounds), basic variables (those that can be solved from the equality constraints), and superbasic variables (those remaining variables between bounds that serve to drive the optimization); this leads to $z^T=\left[z_N^T,z_S^T,z_B^T\right]$. This partition is derived from local information and may change over the course of the optimization iterations. The corresponding KKT conditions can be written as Equations 75a, 75b, 75c, 75d and 75e

$$\nabla_{\mathrm{N}}f(z)+\nabla_{\mathrm{N}}c(z)\gamma=\beta_{\mathrm{a}}-\beta_{\mathrm{b}}\tag{75a}$$

$$\nabla_{\mathrm{S}}f(z)+\nabla_{\mathrm{S}}c(z)\gamma=0\tag{75b}$$

$$\nabla_{\mathrm{B}}f(z)+\nabla_{\mathrm{B}}c(z)\gamma=0\tag{75c}$$

$$c(z)=0\tag{75d}$$

$$z_{\mathrm{N},j}=a_j\text{ or }b_j,\ \beta_{\mathrm{a},j}\geq 0,\beta_{\mathrm{b},j}=0\text{ or }\beta_{\mathrm{b},j}\geq 0,\beta_{\mathrm{a},j}=0\tag{75e}$$

where λ and β are the KKT multipliers for the equality and bound constraints, respectively, and 75e replaces the complementarity conditions 58. Reduced gradient methods work by nesting equations 75b, 75d within 75a, 75c. At iteration *k*, for fixed values of z_{N}^k and z_{S}^k, we can solve for z_{B} using 75d and for λ using 75b. Moreover, linearization of these equations leads to constrained derivatives or *reduced gradients* (Eq. 76)

$$\frac{df}{dz_{\mathrm{s}}}=\nabla f_{\mathrm{s}}-\nabla c_{\mathrm{s}}(\nabla c_{\mathrm{B}})^{-1}\nabla f_{\mathrm{B}}\tag{76}$$

which indicate how $f(z)$ (and z_{B}) change with respect to z_{S} and z_{N}. The algorithm then proceeds by updating z_{S} using reduced gradients in a Newton-type iteration to solve equation 75c. Following this, bound multipliers β are calculated from 75a. Over the course of the iterations, if the variables z_{B} or z_{S} exceed their bounds or if some bound multipliers β become negative, then the variable partition needs to be changed and the equations 75a, 75b, 75c, 75d and 75e are reconstructed. These reduced gradient methods are embodied in the popular GRG2, CONOPT, and SOLVER codes [173].

The SOLVER code has been incorporated into Microsoft Excel. CONOPT [184] is an efficient and widely used code in several optimization modeling environments.

MINOS [185] is a well-implemented package that offers a variation on reduced gradient strategies. At iteration k, equation 75d is replaced by its linearization (Eq. 77)

$$c(z_{\mathrm{N}}^k, z_{\mathrm{S}}^k, z_{\mathrm{B}}^k) + \nabla_{\mathrm{B}} c\left(z^k\right)^T (z_{\mathrm{B}} - z_{\mathrm{B}}^k) + \nabla_{\mathrm{S}} c\left(z^k\right)^T (z_{\mathrm{S}} - z_{\mathrm{S}}^k) = 0 \quad (77)$$

and (75a–75c, 75e) are solved with Equation 77 as a subproblem using concepts from the reduced gradient method. At the solution of this subproblem, the constraints 75d are relinearized and the cycle repeats until the KKT conditions of 75a, 75b, 75c, 75d and 75e are satisfied. The augmented Lagrangian function 73 is used to penalize movement away from the feasible region. For problems with few degrees of freedom, the resulting approach leads to an extremely efficient method even for very large problems. MINOS has been interfaced to a number of modeling systems and enjoys widespread use. It performs especially well on large problems with few nonlinear constraints. However, on highly nonlinear problems it is usually less reliable than other reduced gradient methods.

Algorithmic Details for NLP Methods. All of the above NLP methods incorporate concepts from the Newton – Raphson Method for equation solving. Essential features of these methods are a) providing accurate derivative information to solve for the KKT conditions, b) stabilization strategies to promote convergence of the Newton-like method from poor starting points, and c) regularization of the Jacobian matrix in Newton's method (the so-called KKT matrix) if it becomes singular or ill-conditioned.

a) *Providing first and second derivatives:* The KKT conditions require first derivatives to define stationary points, so accurate first derivatives are essential to determine locally optimal solutions for differentiable NLPs. Moreover, Newton–Raphson methods that are applied to the KKT conditions, as well as the task of checking second-order KKT conditions, necessarily require information on second derivatives. (Note that second-order conditions are not checked by methods that do not use second derivatives). With the recent development of automatic differentiation tools, many modeling and simulation platforms can provide exact first and second derivatives for optimization. When second derivatives are available for the objective or constraint functions, they can be used directly in LANCELOT, SQP and, less efficiently, in reduced gradient methods. Otherwise, for problems with few superbasic variables, reduced gradient methods and reduced space variants of SQP can be applied. Referring to problem 74 with n variables and m equalities, we can write the QP step from problem 68 as Equation 78.

$$\begin{aligned} &\text{Min} \quad \nabla f\left(z^k\right)^T p + {}^1\!/_2\, p^T \nabla_{xx} L\left(z^k, \gamma^k\right) p \\ &\text{s.t.} \quad c\left(z^k\right) + \nabla c\left(z^k\right)^T p = 0 \\ &\qquad a \le z^k + p \le b \end{aligned} \quad (78)$$

Defining the search direction as $p = Z_k p_Z + Y_k p_Y$, where $\nabla c(x^k)^T Z_k = 0$ and $[Y_k | Z_k]$ is a nonsingular $n \times n$ matrix, allows us to form the following reduced QP subproblem (with $n - m$ variables)

$$\begin{aligned} &\text{Min} \quad \left[Z_k^T \nabla f\left(z^k\right) + w_k\right]^T p_Z + \tfrac{1}{2} p_Z{}^T B_k p_Z \\ &\text{s.t.} \quad a \le z^k + Z_k p_Z + Y_k p_Y \le b \end{aligned} \quad (79)$$

where $p_Y = -[\nabla c(z^k)^T Y_k]^{-1} c(z^k)$. Good choices of Y_k and Z_k, which allow sparsity to be exploited in $\nabla c(z^k)$, are: $Y_k^T = [0|I]$ and $Z_k^T = \left[I | -\nabla_{\mathrm{N,S}} c\left(z^k\right) \nabla_B c\left(z^k\right)^{-1}\right]$. Here we define the reduced Hessian $B_k = Z_z^T \nabla_{zz} L\left(z^k, \gamma^k\right) Z_k$ and $w_k = Z_k^T \nabla_{zz} L\left(z^k, \gamma^k\right) Y_k p_Y$. In the absence of second-derivative information, B_k can be approximated using positive definite quasi-Newton approximations [175]. Also, for the interior point method, a similar reduced space decomposition can be applied to the Newton step given in 72.

Finally, for problems with least-squares functions, as in data reconciliation, parameter estimation, and model predictive control, one can often assume that the values of the objective function and its gradient at the solution are vanishingly small. Under these conditions, one can show that the multipliers (λ, ν) also vanish and $\nabla_{xx} L(x^*, \lambda, \nu)$ can be substituted by $\nabla_{xx}f(x^*)$. This *Gauss – Newton approximation* has been shown to be very efficient for the solution of least-squares problems [175].

b) *Line-search and trust-region methods* are used to promote convergence from poor starting points. These are commonly used with the search directions calculated from NLP subproblems such as problem 68. In a *trust-region approach*, the constraint, $\|p\| \leq \Delta$ is added and the iteration step is taken if there is sufficient reduction of some merit function (e.g., the objective function weighted with some measure of the constraint violations). The size of the trust region Δ is adjusted based on the agreement of the reduction of the actual merit function compared to its predicted reduction from the subproblem [183]. Such methods have strong global convergence properties and are especially appropriate for ill-conditioned NLPs. This approach has been applied in the KNITRO code [186]. *Line-search methods* can be more efficient on problems with reasonably good starting points and well-conditioned subproblems, as in real-time optimization. Typically, once a search direction is calculated from problem 68, or other related subproblem, a step size $\alpha \in (0, 1]$ is chosen so that $x^k + \alpha p$ leads to a sufficient decrease of a merit function. As a recent alternative, a novel *filter-stabilization* strategy (for both line-search and trust-region approaches) has been developed based on a bicriterion minimization, with the objective function and constraint infeasibility as competing objectives [187]. This method often leads to better performance than those based on merit functions.

c) *Regularization of the KKT matrix for the NLP subproblem* (e.g., in Equation 72) is essential for good performance of general purpose algorithms. For instance, to obtain a unique solution to Eqn. 72, active constraint gradients must be full rank, and the Hessian matrix, when projected into the null space of the active constraint gradients, must be positive definite. These properties may not hold far from the solution, and corrections to the Hessian in SQP may be necessary [176]. Regularization methods ensure that subproblems like Eqns. 68 and 72 remain well-conditioned; they include addition of positive constants to the diagonal of the Hessian matrix to ensure its positive definiteness, judicious selection of active constraint gradients to ensure that they are linearly independent, and scaling the subproblem to reduce the propagation of numerical errors. Often these strategies are heuristics built into particular NLP codes. While quite effective, most of these heuristics do not provide convergence guarantees for general NLPs.

Table 11 summarizes the characteristics of a collection of widely used NLP codes. Much more information on widely available codes can also be found on the NEOS server (www-neos.mcs.anl.gov) and the NEOS Software Guide.

Table 11. Representative NLP solvers

Method	Algorithm type	Stabilization	Second-order information
CONOPT [184]	reduced gradient	line search	exact and quasi-Newton
GRG2 [173]	reduced gradient	line search	quasi-Newton
IPOPT [177]	SQP, barrier	line search	exact
KNITRO [186]	SQP, barrier	trust region	exact and quasi-Newton
LANCELOT [183]	aug mented Lagrangian, bound constrained	trust region	exact and quasi-Newton
LOQO [188]	SQP, barrier	line search	exact
MINOS [185]	reduced gradient, augmented Lagrangian	line search	quasi-Newton
NPSOL [189]	SQP, Active set	line search	quasi-Newton
SNOPT [190]	reduced space SQP, active set	line search	quasi-Newton
SOCS [191]	SQP, active set	line search	exact
SOLVER [173]	reduced gradient	line search	quasi-Newton
SRQP [192]	reduced space SQP, active set	line search	quasi-Newton

10.3. Optimization Methods without Derivatives

A broad class of optimization strategies does not require derivative information. These methods have the advantage of easy implementation and little prior knowledge of the optimization problem. In particular, such methods are well suited for "quick and dirty" optimization studies that explore the scope of optimization for new problems, prior to investing effort for more sophisticated modeling and solution strategies. Most of these methods are derived from heuristics that naturally spawn numerous variations. As a result, a very broad literature describes these methods. Here we discuss only a few important trends in this area.

Classical Direct Search Methods. Developed in the 1960s and 1970s, these methods include one-at-a-time search and methods based on experimental designs [193]. At that time, these direct search methods were the most popular optimization methods in chemical engineering. Methods that fall into this class include the pattern search [194], the conjugate direction method [195], simplex and complex searches [196], and the adaptive random search methods [197 – 199]. All of these methods require only objective function values for unconstrained minimization. Associated with these methods are numerous studies on a wide range of process problems. Moreover, many of these methods include heuristics that prevent premature termination (e.g., directional flexibility in the complex search as well as random restarts and direction generation). To illustrate these methods, Figure 41 illustrates the performance of a pattern search method as well as a random search method on an unconstrained problem.

Simulated Annealing. This strategy is related to random search methods and derives from a class of heuristics with analogies to the motion of molecules in the cooling and solidification of metals [200]. Here a "temperature" parameter θ can be raised or lowered to influence the probability of accepting points that do not improve the objective function. The method starts with a base point x and objective value $f(x)$. The next point x' is chosen at random from a distribution. If $f(x') < f(x)$, the move is

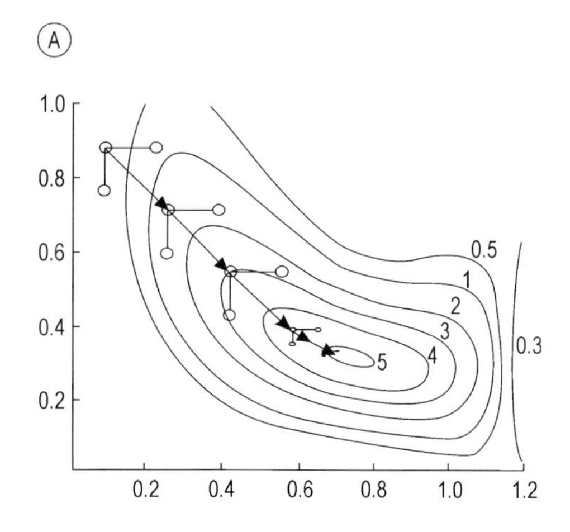

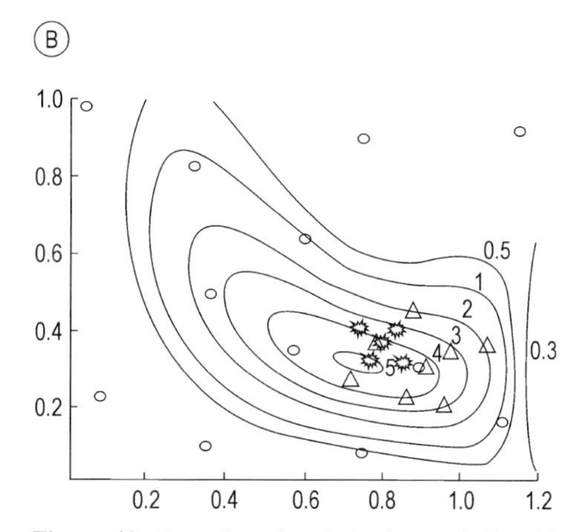

Figure 41. Examples of optimization methods without derivatives
A) Pattern search method; B) Random search method
Circles: 1st phase; Triangles: 2nd phase; Stars: 3rd phase

accepted with x' as the new point. Otherwise, x' is accepted with probability $p(\theta, x', x)$. Options include the Metropolis distribution, $p(\theta, x, x') = \exp[-\{f(x') - f(x)\}/\theta]$, and the Glauber distribution, $p(\theta, x, x') = \exp[-\{f(x') - f(x)\}/(1 + \exp[-\{f(x') - f(x)\}]/\theta)]$. The θ parameter is then reduced and the method continues until no further progress is made.

Genetic Algorithms. This approach, first proposed in [201], is based on the analogy of improving a population of solutions through modifying their gene pool. It also has similar performance characteristics as random search methods and simulated annealing. Two forms of genetic modification, crossover or mutation, are used and the elements of the optimization vector $\boldsymbol{x}$ are represented as binary strings. Crossover deals with random swapping of vector elements (among parents with highest objective function values or other rankings of population) or any linear combinations of two parents. Mutation deals with the addition of a random variable to elements of the vector. Genetic algorithms (GAs) have seen widespread use in process engineering and a number of codes are available. A related GA algorithm is described in [173].

Derivative-Free Optimization (DFO). In the past decade, the availability of parallel computers and faster computing hardware and the need to incorporate complex simulation models within optimization studies have led a number of optimization researchers to reconsider classical direct search approaches. In particular, Dennis and Torczon [202] developed a multidimensional search algorithm that extends the simplex approach [196]. They note that the Nelder–Mead algorithm fails as the number of variables increases, even for very simple problems. To overcome this, their multidimensional pattern-search approach combines reflection, expansion, and contraction steps that act as line search algorithms for a number of linear independent search directions. This approach is easily adapted to parallel computation and the method can be tailored to the number of processors available. Moreover, this approach converges to locally optimal solutions for unconstrained problems and exhibits an unexpected performance synergy when multiple processors are used. The work of Dennis and Torczon [202] has spawned considerable research on analysis and code development for DFO methods. Moreover, Conn et al. [203] construct a multivariable DFO algorithm that uses a surrogate model for the objective function within a trust-region method. Here points are sampled to obtain a well-defined quadratic interpolation model, and descent conditions from trust-region methods enforce convergence properties. A number of trust-region methods that rely on this approach are reviewed in [203]. Moreover, a number of DFO codes have been developed that lead to black-box optimization implementations for large, complex simulation models. These include the DAKOTA package at Sandia National Lab [204], http://endo.sandia.gov/DAKOTA/software.html and FOCUS developed at Boeing Corporation [205].

Direct search methods are easy to apply to a wide variety of problem types and optimization models. Moreover, because their termination criteria are not based on gradient information and stationary points, they are more likely to favor the search for global rather than locally optimal solutions. These methods can also be adapted easily to include integer variables. However, rigorous convergence properties to globally optimal solutions have not yet been discovered. Also, these methods are best suited for unconstrained problems or for problems with simple bounds. Otherwise, they may have difficulties with constraints, as the only options open for handling constraints are equality constraint elimination and addition of penalty functions for inequality constraints. Both approaches can be unreliable and may lead to failure of the optimization algorithm. Finally, the performance of direct search methods scales poorly (and often exponentially) with the number of decision variables. While performance can be improved with the use of parallel computing, these methods are rarely applied to problems with more than a few dozen decision variables.

10.4. Global Optimization

Deterministic optimization methods are available for nonconvex nonlinear programming problems of the form problem 52 that guarantee convergence to the global optimum. More specifically, one can show under mild conditions that they converge to an ε distance to the global optimum on a finite number of steps. These methods are generally more expensive than local NLP methods, and they require the exploitation of the structure of the nonlinear program.

Global optimization of nonconvex programs has received increased attention due to their practical importance. Most of the deterministic global optimization algorithms are based on spatial branch-and-bound algorithm [206], which divides the feasible region of continuous variables and compares lower bound and upper bound for fathoming each subregion. The one that contains the optimal solution is found by eliminating subregions that are proved not to contain the optimal solution.

For nonconvex NLP problems, Quesada and Grossmann [207] proposed a spatial branch-and-bound algorithm for concave separable, linear fractional, and bilinear programs using linear and nonlinear underestimating functions [208]. For nonconvex MINLP, Ryoo and Sahinidis [209] and later Tawarmalani and Sahinidis [210] developed BARON, which branches on the continuous and discrete variables with bounds reduction method. Adjiman et al. [211, 212] proposed the SMIN-αBB and GMIN-αBB algorithms for twice-differentiable nonconvex MINLPs. Using a valid convex underestimation of general functions as well as for special functions, Adjiman et al. [213] developed the αBB method, which branches on both the continuous and discrete variables according to specific options. The branch-and-contract method [214] has bilinear, linear fractional, and concave separable functions in the continuous variables and binary variables, uses bound contraction, and applies the outer-approximation (OA) algorithm at each node of the tree. Smith and Pantelides [215] proposed a reformulation method combined with a spatial branch-and-bound algorithm for nonconvex MINLP and NLP.

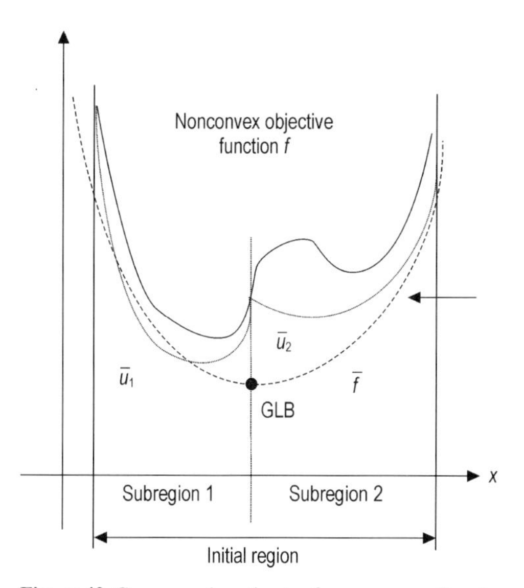

Figure 42. Convex underestimator for nonconvex function

Because in global optimization one cannot exploit optimality conditions like the KKT conditions for a local optimum, these methods work by first partitioning the problem domain (i.e., containing the feasible region) into subregions (see Fig. 42). Upper bounds on the objective function are computed over all subregions of the problem. In addition, lower bounds can be derived from convex underestimators of the objective function and constraints for each subregion. The algorithm then proceeds to eliminate all subregions that have lower bounds that are greater than the least upper bound. After this, the remaining regions are further partitioned to create new subregions and the cycle continues until the upper and lower bounds converge. Below we illustrate the specific steps of the algorithm for nonlinear programs that involve bilinear, linear fractional, and concave separable terms [207, 214].

Nonconvex NLP with Bilinear, Linear Fractional, and Concave Separable Terms. Consider the following specific nonconvex NLP problem,

$$\begin{aligned}&\underset{x}{\text{Min}}\ \ f(x)=\sum_{(i,j)\in BL_0} a_{ij}\, x_i x_j+\sum_{(i,j)\in LF_0} b_{ij}\frac{x_i}{x_j}\\&+\sum_{i\in C_0} g_i(x_i)+h(x)\\&\text{subject to}\\&f_k(x)=\sum_{(i,j)\in BL_k} a_{ijk}x_i x_j+\sum_{(i,j)\in LF_k} b_{ijk}\frac{x_i}{x_j}\\&+\sum_{i\in C_k} g_{i,k}(x_i)+h_k(x)\leq 0\quad k\in K\\&x\in S\cap\Omega_0\subset R^n\end{aligned} \tag{80}$$

where a_{ij}, a_{ijk}, b_{ij}, b_{ijk} are scalars with $i\in I=\{1,2,\cdots,n\}$, $j\in J=\{1,2,\cdots,n\}$, and $k\in K=\{1,2,\cdots,m\}$. BL_0, BL_k, LF_0, LF_k are (i, j)-index sets, with $i\neq j$, that define the bilinear and linear fractional terms present in the problem. The functions $h(x), h_k(x)$ are convex, and twice continuously differentiable. C_0 and C_k are index sets for the univariate twice continuously differentiable concave functions $g_i(x_i), g_{i,k}(x_i)$. The set $S\subset R^n$ is convex, and $\omega_0\subset R^n$ is an n-dimensional hyperrectangle defined in terms of the initial variable bounds $x^{\mathrm{L,in}}$ and $x^{\mathrm{U,in}}$:

$$\begin{aligned}\Omega_0=\{&x\in R^n: 0\leq x^{\mathrm{L,in}}\leq x\leq x^{\mathrm{U,in}},\ x_j^{\mathrm{L,in}}>0\\&\text{if}\ (i,j)\in LF_0\cup LF_k,\quad i\in I, j\in J, k\in K\}\end{aligned}$$

The feasible region of problem 80 is denoted by D. Note that a nonlinear equality constraint of the form $f_k(x)=0$ can be accommodated in problem 51 through the representation by the inequalities $f_k(x) \le 0$ and $-f_k(x) \le 0$, provided $h_k(x)$ is separable.

To obtain a lower bound LB (Ω) for the global minimum of problem 80 over $D \cap \Omega$, where $\Omega = \{x \in R^n : x^{\mathrm{L}} \le x \le x^{\mathrm{U}}\} \subseteq \Omega_0$, the following problem is proposed:

$$\min_{(x,y,z)} \hat{f}(x,y,z) = \sum_{(i,j) \in BL_0} a_{ij} y_{ij} + \sum_{(i,j) \in LF_0} b_{ij} z_{ij} + \sum_{i \in C_0} \hat{g}_i(x_i) + h(x)$$

subject to

$$\hat{f}_k(x,y,z) = \sum_{(i,j) \in BL_k} a_{ijk} y_{ij} + \sum_{(i,j) \in LF_k} b_{ijk} z_{ij} + \sum_{i \in C_k} \hat{g}_{i,k}(x_i) + h_k(x) \le 0 \quad k \in K$$

$$(x,y,z) \in T(\Omega) \subset R^n \times R^{n_1} \times R^{n_2}$$

$$x \in S \cap \Omega \subset R^n, \; y \in R_+^{n_1}, \; z \in R_+^{n_2}, \qquad (81)$$

where the functions and sets are defined as follows:

a) $\hat{g}_i(x_i)$ and $\hat{g}_{i,k}(x_i)$ are the convex envelopes for the univariate functions over the domain $x_i \in [x_i^L, x_i^U]$ [216]:

$$\hat{g}_i(x_i) = g_i\left(x_i^{\mathrm{L}}\right) + \left(\frac{g_i\left(x_i^{\mathrm{U}}\right) - g_i\left(x_i^L\right)}{x_i^{\mathrm{U}} - x_i^{\mathrm{L}}}\right)\left(x_i - x_i^{\mathrm{L}}\right) \le g_i(x_i) \qquad (82)$$

$$\hat{g}_{i,k}(x_i) = g_{i,k}\left(x_i^{\mathrm{L}}\right) + \left(\frac{g_{i,k}\left(x_i^{\mathrm{U}}\right) - g_{i,k}\left(x_i^L\right)}{x_i^{\mathrm{U}} - x_i^{\mathrm{L}}}\right)\left(x_i - x_i^{\mathrm{L}}\right) \le g_{i,k}(x_i) \qquad (83)$$

where $\hat{g}_i(x_i) = g_i(x_i)$ at $x_i = x_i^{\mathrm{L}}$, and $x_i = x_i^{\mathrm{U}}$; likewise, $\hat{g}_{i,k}(x_i) = g_{i,k}(x_i)$ at $x_i = x_i^{\mathrm{L}}$, and $x_i = x_i^{\mathrm{U}}$.

b) $y = \{y_{ij}\}$ is a vector of additional variables for relaxing the bilinear terms in 80, and is used in the following inequalities which determine the convex and concave envelopes of bilinear terms:

$$\begin{array}{ll} y_{ij} \ge x_j^{\mathrm{L}} x_i + x_i^{\mathrm{L}} x_j - x_i^{\mathrm{L}} x_j^{\mathrm{L}} & (i,j) \in BL^+ \\ y_{ij} \ge x_j^{\mathrm{U}} x_i + x_i^{\mathrm{U}} x_j - x_i^{\mathrm{U}} x_j^{\mathrm{U}} & (i,j) \in BL^+ \end{array} \qquad (84)$$

$$\begin{array}{ll} y_{ij} \le x_j^{\mathrm{L}} x_i + x_i^{\mathrm{U}} x_j - x_i^{\mathrm{U}} x_j^{\mathrm{L}} & (i,j) \in BL^- \\ y_{ij} \le x_j^{\mathrm{U}} x_i + x_i^{\mathrm{L}} x_j - x_i^{\mathrm{L}} x_j^{\mathrm{U}} & (i,j) \in BL^- \end{array} \qquad (85)$$

where

$$BL^+ = \{(i,j) : (i,j) \in BL_0 \cup BL_k, \, a_{ij} > 0 \text{ or } a_{ijk} > 0, \, k \in K\}$$

$$BL^- = \{(i,j) : (i,j) \in BL_0 \cup BL_k, \, a_{ij} < 0 \text{ or } a_{ijk} < 0, \, k \in K\}$$

The inequalities 84 were first derived by McCormick [208], and along with the inequalities 85 theoretically characterized by Al-Khayyal and Falk [217, 218].

c) $z = \{z_{ij}\}$ is a vector of additional variables for relaxing the linear fractional terms in problem 80; these variables are used in the following inequalities:

$$\begin{array}{ll} z_{ij} \ge \frac{x_i}{x_j^{\mathrm{L}}} + x_i^{\mathrm{U}}\left(\frac{1}{x_j} - \frac{1}{x_j^{\mathrm{L}}}\right) & (i,j) \in LF^+ \\ z_{ij} \ge \frac{x_i}{x_j^{\mathrm{U}}} + x_i^L\left(\frac{1}{x_j} - \frac{1}{x_j^{\mathrm{U}}}\right) & (i,j) \in LF^+ \end{array} \qquad (86)$$

$$z_{ij} \ge \frac{1}{x_j}\left(\frac{x_i + \sqrt{x_i^{\mathrm{L}} x_i^{\mathrm{U}}}}{\sqrt{x_i^{\mathrm{L}}} + \sqrt{x_i^{\mathrm{U}}}}\right)^2 \quad (i,j) \in LF^+ \qquad (87)$$

$$\begin{array}{ll} z_{ij} \le \frac{1}{x_j^{\mathrm{L}} x_j^{\mathrm{U}}}\left(x_j^{\mathrm{U}} x_i - x_i^{\mathrm{L}} x_j + x_i^{\mathrm{L}} x_j^{\mathrm{L}}\right) & (i,j) \in LF^- \\ z_{ij} \le \frac{1}{x_j^{\mathrm{L}} x_j^{\mathrm{U}}}\left(x_j^{\mathrm{L}} x_i - x_i^{\mathrm{U}} x_j + x_i^{\mathrm{U}} x_j^{\mathrm{U}}\right) & (i,j) \in LF^- \end{array} \qquad (88)$$

where

$$LF^+ = \{(i,j) : (i,j) \in LF_0 \cup LF_k, \, b_{ij} > 0 \text{ or } b_{ijk} > 0, \, k \in K\}$$

$$LF^- = \{(i,j) : (i,j) \in LF_0 \cup LF_k, \, b_{ij} < 0 \text{ or } b_{ijk} < 0, \, k \in K\}$$

The inequalities 86 and 87, 88 are convex underestimators due to Quesada and Grossmann [207, 219] and Zamora and Grossmann [214], respectively.

d) $T(\Omega) = \{(x,y,z) \in R^n \times R^{n_1} \times R^{n_2}$: 82– 87 are satisfied with $x^{\mathrm{L}}, x^{\mathrm{U}}$ as in $\Omega\}$. The feasible region, and the solution of problem 52 are denoted by $M(\Omega)$, and $(\hat{x}, \hat{y}, \hat{z})_\Omega$, respectively. We define the *approximation gap* $\varepsilon(\Omega)$ at a branch-and-bound node as

$$\varepsilon(\Omega) = \begin{cases} \infty & \text{if OUB} = \infty \\ -\mathrm{LB}(\Omega) & \text{if OUB} = 0 \\ \frac{(\mathrm{OUB} - \mathrm{LB}(\Omega))}{|\mathrm{OUB}|} & \text{otherwise} \end{cases} \qquad (89)$$

where the *overall upper bound* (OUB) is the value of $f(x)$ at the best available feasible point $x \in D$; if no feasible point is available, then OUB$=\infty$.

Note that the underestimating problem 81 is a linear program if $LF^{+}=\varnothing$. During the execution of the spatial branch-and-bound algorithm, problem (NBNC) is solved initially over $M(\Omega_0)$ (root node of the branch-and-bound tree). If a better approximation is required, $M(\Omega_0)$ is refined by partitioning Ω_0 into two smaller hyperrectangles Ω_{01} and Ω_{02}, and two children nodes are created with relaxed feasible regions given by $M(\Omega_{01})$ and $M(\Omega_{02})$. Problem 81 might be regarded as a basic underestimating program for the general problem 80. In some cases, however, it is possible to develop additional convex estimators that might strengthen the underestimating problem. See, for instance, the projections proposed by QUESADA and GROSSMANN [207], the reformulation–linearization technique by SHERALI and ALAMEDDINE [220], and the reformulation–convexification approach by SHERALI and TUNCBILEK [221].

The Set of Branching Variables. A set of branching variables, characterized by the index set BV (Ω) defined below, is determined by considering the optimal solution $(\hat{x},\hat{y},\hat{z})_{\Omega}$ of the underestimating problem:

$$\begin{aligned}\mathrm{BV}(\Omega)=\{i,j: &|\hat{y}_{ij}-\hat{x}_i\hat{x}_j|=\zeta_l \text{ or } |\hat{z}_{ij}-\hat{x}_i/\hat{x}_j|=\zeta_l\\ &\text{or } g_i(\hat{x}_i)-\hat{g}_i(\hat{x}_i)=\zeta_l \text{ or } g_{i,k}(\hat{x}_i)-\hat{g}_{i,k}(\hat{x}_i)=\zeta_l,\\ &\text{for } i\in I, j\in J, k\in K, l\in L\}\end{aligned} \tag{90}$$

where, for a prespecified number l_n, $L=\{1,2,\ldots,l_n\}$ and ζ_1 is the magnitude of the largest approximation error for a nonconvex term in problem 80 evaluated at $(\hat{x},\hat{y},\hat{z})_{\Omega}$:

$$\xi_1=\underset{i\in I,j\in J,k\in K}{\mathrm{Max}}\left[|\hat{y}_{ij}-\hat{x}_i\hat{x}_j|,\ |\hat{z}_{ij}-\hat{x}_i/\hat{x}_j|,\ g_i(\hat{x}_i)-\hat{g}_i(\hat{x}_i),\ g_{i,k}(\hat{x}_i)-\hat{g}_{i,k}(\hat{x}_i)\right]$$

Similarly, we define $\xi_l<\xi_{l-1}$ with $l\in L\setminus\{1\}$ as the l-th largest magnitude for an approximation error; for instance, $\xi_2<\xi_1$ is the second largest magnitude for an approximation error. Note that in some cases it might be convenient to introduce weights in the determination of BV (Ω) in order to scale differences in the approximation errors or to induce preferential branching schemes. This might be particularly useful in applications where specific information can be exploited by imposing an order of precedence on the set of complicating variables.

This basic concept in spatial branch-and-bound for global optimization is as follows. *Bounds* are related to the calculation of upper and lower bounds. For the former, any feasible point or, preferably, a locally optimal point in the subregion, can be used. For the lower bound, convex relaxations of the objective and constraint function are derived such as in problem 81. The *refining* step deals with the construction of partitions in the domain and further partitioning them during the search process. Finally, the *selection* step decides on the order of exploring the open subregions. Thus, the feasible region and the objective function are replaced by convex envelopes to form relaxed problems. Solving these convex relaxed problems leads to global solutions that are lower bounds to the NLP in the particular subregion. Finally, we see that gradient-based NLP solvers play an important role in global optimization algorithms, as they often yield the lower and upper bounds for the subregions. The following *spatial branch-and-bound* global optimization algorithm can therefore be given by the following steps:

0. Initialize algorithm: calculate upper bound by obtaining a local solution to problem 80. Calculate a lower bound solving problem 81 over the entire (relaxed) feasible region Ω_0.

For iteration k with a set of partitions $\Omega_{k,j}$ and bounds in each subregion OLB_j and OUB_j:

1) *Bound:* Define best upper bound: OUB = Min_j OUB_j and delete (fathom) all subregions j with lower bounds $\mathrm{OLB}_j \geq$ OUB. If $\mathrm{OLB}_j \geq \mathrm{OUB} - \varepsilon$, stop.
2) *Refine:* Divide the remaining active subregions into partitiions $\Omega_{k,j1}$ and $\Omega_{k,j2}$. (Several branching rules are available for this step.)
3) *Select:* Solve the convex NLP 81 in the new partitions to obtain OLB_{j1} and OLB_{j2}. Delete partition if no feasible solution.
4) *Update:* Obtain upper bounds, OUB_{j1} and OUB_{j2} to new partitions if present. Set $k = k+1$, update partition sets and go to step 1.

Example: To illustrate the spatial branch-and-bound algorithm, consider the global solution of:

$$\begin{aligned}&\mathrm{Min}\, f(x)=5/2\,x^4-20\,x^3+55\,x_2-57\,x\\ &\text{s.t. } 0.5\leq x\leq 2.5\end{aligned} \tag{91}$$

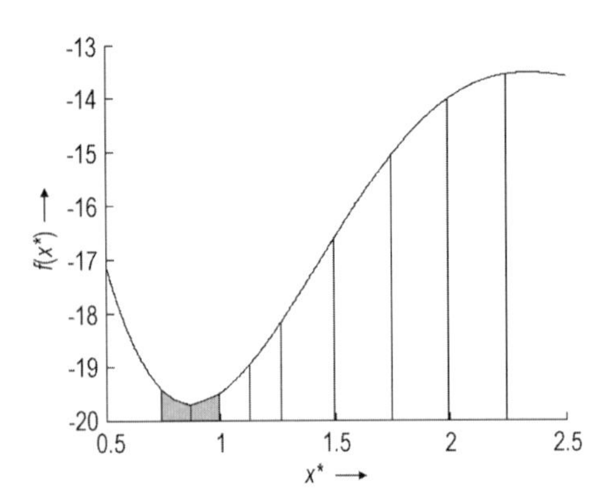

Figure 43. Global optimization example with partitions

As seen in Figure 43, this problem has local solutions at $x^* = 2.5$ and at $x^* = 0.8749$. The latter is also the global solution with $f(x^*) = -19.7$. To find the global solution we note that all but the $-20\ x^3$ term in problem 91 are convex, so we replace this term by a new variable and a linear underestimator within a particular subregion, i.e.:

$$\begin{aligned} &\text{Min} && f_L(x) = 5/2\, x^4 - 20\, w + 55\, x_2 - 57\, x \\ &\text{s.t.} && x_1 \leq x \leq x_u \\ & && w = (x_l)^3 \frac{(x_u - x)}{(x_u - x_l)} + (x_u)^3 \frac{(x - x_l)}{(x_u - x_l)} \end{aligned} \tag{92}$$

In Figure 43 we also propose subregions that are created by simple bisection partitioning rules, and we use a "loose" bounding tolerance of $\varepsilon = 0.2$. In each partition the lower bound, f_{L} is determined by problem 92 and the upper bound f_{U} is determined by the local solution of the original problem in the subregion. Figure 44 shows the progress of the spatial branch-and-bound algorithm as the partitions are refined and the bounds are updated. In Figure 43, note the definitions of the partitions for the nodes, and the sequence numbers in each node that show the order in which the partitions are processed. The grayed partitions correspond to the deleted subregions and at termination of the algorithm we see that $f_{\mathrm{L}j} \geq f_{\mathrm{U}} - \varepsilon$ (i.e., $-19.85 \geq -19.7 - 0.2$), with the gray subregions in Figure 43 still active. Further partitioning in these subregions will allow the lower and upper bounds to converge to a tighter tolerance.

A number of improvements can be made to the bounding, refinement, and selection strategies in the algorithm that accelerate the convergence of this method. A comprehensive discussion of all of these options can be found in [222 – 224]. Also, a number of efficient global optimization codes have recently been developed, including αBB, BARON, LGO, and OQNLP. An interesting numerical comparison of these and other codes can be found in [225].

10.5. Mixed Integer Programming

Mixed integer programming deals with both discrete and continuous decision variables. For simplicity in the presentation we consider the most common case where the discrete decisions are binary variables, i.e., $y_i = 0$ or 1, and we consider the mixed integer problem 51. Unlike local optimization methods, there are no optimality conditions, like the KKT conditions, that can be applied directly.

Mixed Integer Linear Programming. If the objective and constraint functions are all linear in problem 51, and we assume 0–1 binary variables for the discrete variables, then this gives rise to a mixed integer linear programming (MILP) problem given by Equation 93.

$$\begin{aligned} &\text{Min} \quad Z = a^T x + c^T y \\ &\text{s.t.}\ Ax + By \leq b \\ &x \geq 0,\ y \in \{0,1\}^t \end{aligned} \tag{93}$$

As is well known, the (MILP) problem is NP-hard. Nevertheless, an interesting theoretical result is that it is possible to transform it into an LP with the convexification procedures proposed by LOVACZ and SCHRIJVER [226], SHERALI and ADAMS [227], and BALAS et al. [228]. These procedures consist of sequentially lifting the original relaxed $x - y$ space into higher dimension and projecting it back to the original space so as to yield, after a finite number of steps, the integer convex hull. Since the transformations have exponential complexity, they are only of theoretical interest, although they can be used as a basis for deriving cutting planes (e.g. lift and project method by [228]).

As for the solution of problem (MILP), it should be noted that this problem becomes an LP problem when the binary variables are relaxed as continuous variables $0 \leq y \leq 1$. The most common solution algorithms for problem

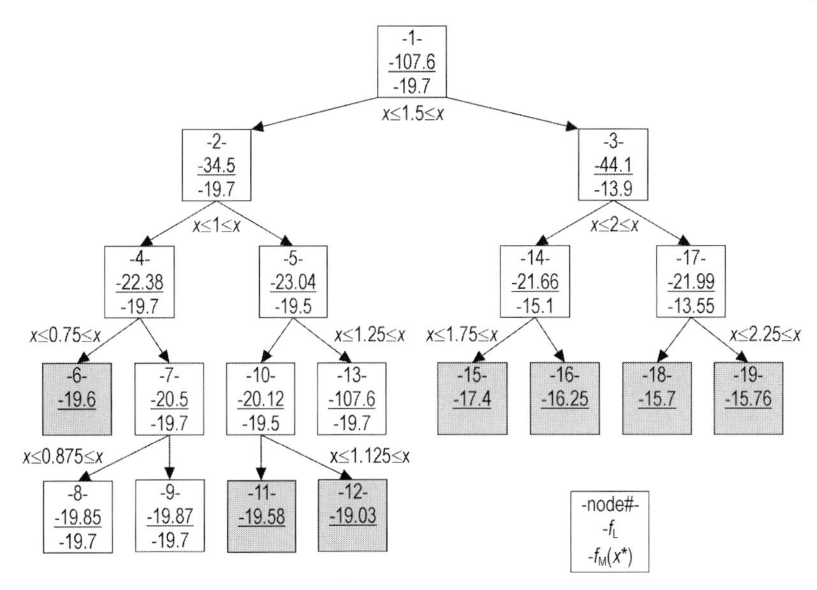

Figure 44. Spatial branch-and-bound sequence for global optimization example

(MILP) are LP-based branch-and-bound methods, which are enumeration methods that solve LP subproblems at each node of the search tree. This technique was initially conceived by LAND and DOIG [229], BALAS [230], and later formalized by DAKIN, [231]. Cutting-plane techniques, which were initially proposed by GOMORY [232], and consist of successively generating valid inequalities that are added to the relaxed LP, have received renewed interest through the works of CROWDER et al. [233], VAN ROY and WOLSEY [234], and especially the lift-and-project method of BALAS et al. [228]. A recent review of branch-and-cut methods can be found in [235]. Finally, Benders decomposition [236] is another technique for solving MILPs in which the problem is successively decomposed into LP subproblems for fixed 0–1 and a master problem for updating the binary variables.

LP-Based Branch and Bound Method. We briefly outline in this section the basic ideas behind the branch-and-bound method for solving MILP problems. Note that if we relax the t binary variables by the inequalities $0 \leq y \leq 1$ then 93 becomes a linear program with a (global) solution that is a lower bound to the MILP 93. There are specific MILP classes in which the LP relaxation of 93 has the same solution as the MILP. Among these problems is the well-known *assignment problem*. Other MILPs that can be solved with efficient special-purpose methods are the *knapsack* problem, the *set-covering* and *set-partitioning* problems, and the *traveling salesman* problem. See [237] for a detailed treatment of these problems.

The branch-and-bound algorithm for solving MILP problems [231] is similar to the spatial branch-and-bound method of the previous section that explores the search space. As seen in Figure 45, binary variables are successively fixed to define the search tree and a number of bounding properties are exploited in order to fathom nodes in to avoid exhaustive enumeration of all the nodes in the tree.

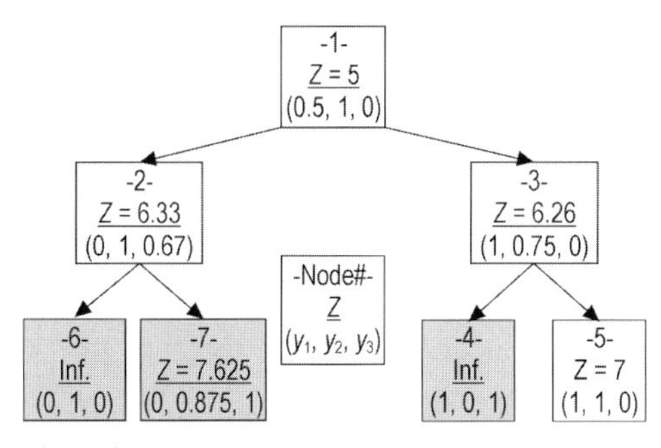

Figure 45. Branch and bound sequence for MILP example

The basic idea in the search is as follows. The top, or root node, in the tree is the solution to the linear programming relaxation of 93. If all the

y variables take on 0–1 values, the MILP problem is solved, and no further search is required. If at least one of the binary variables yields a fractional value, the solution of the LP relaxation yields a lower bound to problem 93. The search then consists of branching on that node by fixing a particular binary variable to 0 or 1, and the corresponding restricted LP relaxations are solved that in turn yield a lower bound for any of their descendant nodes. In particular, the following properties are exploited in the branch-and-bound search:

- Any node (initial, intermediate, leaf node) that leads to feasible LP solution corresponds to a valid upper bound to the solution of the MILP problem 93.
- Any intermediate node with an infeasible LP solution has infeasible leaf nodes and can be fathomed (i.e., all remaining children of this node can be eliminated).
- If the LP solution at an intermediate node is not less than an existing integer solution, then the node can be fathomed.

These properties lead to pruning of the nodes in the search tree. Branching then continues in the tree until the upper and lower bounds converge.

The basic concepts outlined above lead to a branch-and-bound algorithm with the following features. LP solutions at intermediate nodes are relatively easy to calculate since they can be effectively updated with the dual simplex method. The selection of binary variables for branching, known as the *branching rule,* is based on a number of different possible criteria; for instance, choosing the fractional variable closest to 0.5, or the one involving the largest of the smallest pseudocosts for each fractional variable. *Branching strategies* to navigate the tree take a number of forms. More common *depth-first strategies* expand the most recent node to a leaf node or infeasible node and then backtrack to other branches in the tree. These strategies are simple to program and require little storage of past nodes. On the other hand, *breadth-first strategies* expand all the nodes at each level of the tree, select the node with the lowest objective function, and then proceed until the leaf nodes are reached. Here, more storage is required, but generally fewer nodes are evaluated than in depth-first search. In practice, a combination of both strategies is commonly used: branch on the dichotomy 0–1 at each node (i.e., like breadth-first), but expand as in depth-first. Additional description of these strategies can be found in [237].

Example: To illustrate the branch-and-bound approach, we consider the MILP:

$$\begin{aligned}
\text{Min}\quad & Z=x+y_1+2y_2+3y_3\\
\text{s.t.}\; & -x+3y_1+y_2+2y_3\leq 0\\
& -4y_1-8y_2-3y_3\leq -10\\
& x\geq 0, y_1, y_2, y_3=\{0,1\}
\end{aligned}\tag{94}$$

The solution to problem 94 is given by $x=4$, $y_1=1$, $y_2=1$, $y_3=0$, and $Z=7$. Here we use a depth-first strategy and branch on the variables closest to zero or one. Figure 45 shows the progress of the branch-and-bound algorithm as the binary variables are selected and the bounds are updated. The sequence numbers for each node in Figure 45 show the order in which they are processed. The grayed partitions correspond to the deleted nodes and at termination of the algorithm we see that $Z=7$ and an integer solution is obtained at an intermediate node where coincidentally $y_3=0$.

Mixed integer linear programming (MILP) methods and codes have been available and applied to many practical problems for more than twenty years (e.g., [237]. The LP-based branch-and-bound method [231] has been implemented in powerful codes such as OSL, CPLEX, and XPRESS. Recent trends in MILP include the development of branch-and-price [238] and branch-and-cut methods such as the lift-and-project method [228] in which cutting planes are generated as part of the branch-and-bound enumeration. See also [235] for a recent review on MILP. A description of several MILP solvers can also be found in the NEOS Software Guide.

Mixed Integer Nonlinear Programming The most basic form of an MINLP problem when represented in algebraic form is as follows:

$$\begin{aligned}
& \min z=f(x,y)\\
& s.t.\; g_j(x,y)\leq 0\; j\in J \qquad \text{(P1)}\\
& x\in X, y\in Y
\end{aligned}\tag{95}$$

where $f(\cdot)$, $g(\cdot)$ are *convex, differentiable* functions, J is the index set of inequalities, and x and y are the continuous and discrete variables, respectively. The set X is commonly assumed to be a convex compact set, e.g., $X=\{x|x\in\mathbf{R}^n, Dx\leq d, x^L\leq x\leq x^U\}$; the discrete set Y corresponds to a polyhedral set of integer points, $Y=\{y|x\in Z^m, Ay\leq a\}$, which in most applications is restricted to 0–1 values, $y\in\{0,1\}^m$. In most applications of interest the objective and constraint functions $f(\cdot)$, $g(\cdot)$ are linear in y (e.g., fixed cost charges and mixed-logic constraints): $f(x,y)=c^T y+r(x)$, $g(x,y)=By+h(x)$.

Methods that have addressed the solution of problem 95 include the branch-and-bound method (BB) [239 – 243], generalized Benders decomposition (GBD) [244], outer-approximation (OA) [245 – 247], LP/NLP based branch-and-bound [248], and extended cutting plane method (ECP) [249].

There are three basic *NLP subproblems* that can be considered for problem 95:

a) NLP relaxation

$$\begin{aligned}&\text{Min } Z_{LB}^k=f(x,y)\\&s.t.\ g_j(x.y)\leq 0\ j\in J\\&x\in X,\ y\in Y_R \qquad\qquad \text{(NLP1)}\\&y_i\leq\alpha_i^k\ i\in I_{FL}^k\\&y_i\geq\beta_i^k\ i\in I_{FU}^k\end{aligned}\tag{96}$$

where $Y_{\rm R}$ is the continuous relaxation of the set Y, and $I_{\rm FL}^k, I_{\rm FU}^k$ are index subsets of the integer variables y_i, $i\in I$ which are restricted to lower and upper bounds α_i^k, β_i^k at the k-th step of a branch-and-bound enumeration procedure. Note that $\alpha_i^k=\lfloor y_i^l\rfloor, \beta_i^k=\lceil y_i^m\rceil, l<k, m<k$, where y_i^l, y_i^m are noninteger values at a previous step, and $\lfloor\cdot\rfloor$, $\lceil\cdot\rceil$ are the floor and ceiling functions, respectively.

Also note that if $I_{\rm FU}^k=I_{\rm FL}^k=\varnothing\ (k=0)$, problem 96 corresponds to the continuous NLP relaxation of problem 95. Except for few and special cases, the solution to this problem yields in general a noninteger vector for the discrete variables. Problem 96 also corresponds to the k-th step in a branch-and-bound search. The optimal objective function $Z_{\rm LB}^{\rm o}$ provides an absolute lower bound to problem 95; for $m\geq k$, the bound is only valid for $I_{\rm FL}^k\subset I_{\rm FL}^m, I_{\rm FU}^k\subset I_{\rm FL}^m$.

b) NLP subproblem for fixed y^k:

$$\begin{aligned}&\text{Min}\quad Z_{\rm U}^k=f(x,y^k)\\&\text{s.t.}\quad g_j(x,y^k)\leq 0\quad j\in J\\&\qquad x\in X\end{aligned}\tag{97}$$

which yields an upper bound $Z_{\rm U}^k$ to problem 95 provided problem 97 has a feasible solution. When this is not the case, we consider the next subproblem:

c) Feasibility subproblem for fixed y^k:

$$\begin{aligned}&\text{Min } u\\&\text{s.t. } g_j(x,y^k)\leq u\ j\in J\\&\qquad x\in X,\ u\in R^1\end{aligned}\tag{98}$$

which can be interpreted as the minimization of the infinity-norm measure of infeasibility of the corresponding NLP subproblem. Note that for an infeasible subproblem the solution of problem 98 yields a strictly positive value of the scalar variable u.

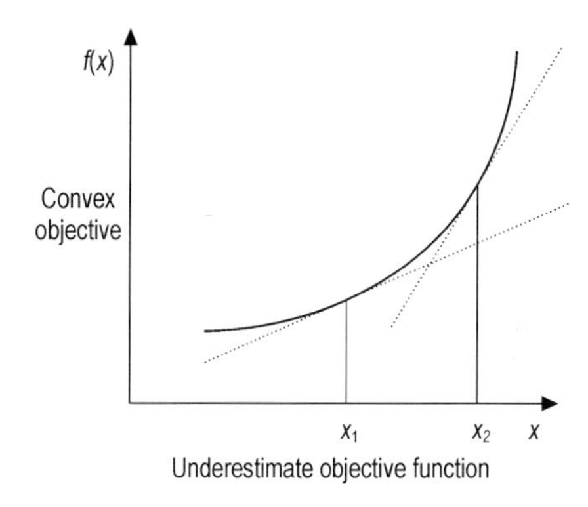

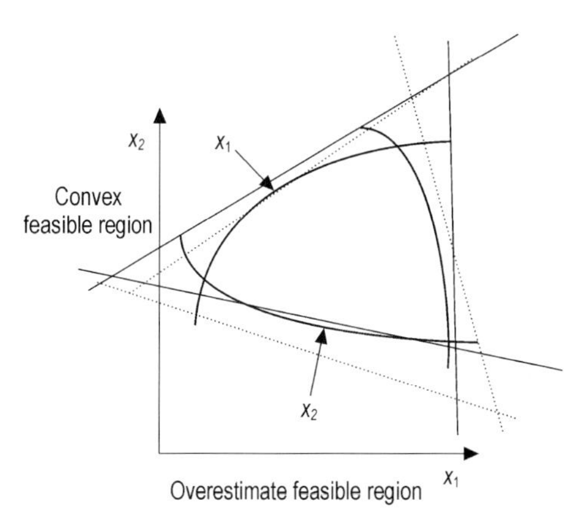

Figure 46. Geometrical interpretation of linearizations in master problem 99

The convexity of the nonlinear functions is exploited by replacing them with supporting hyperplanes, which are generally, but not necessarily, derived at the solution of the NLP subproblems. In particular, the new values y^K (or (x^K, y^K) are obtained from a *cutting-plane MILP problem* that is based on the *K* points (x^k, y^k), *k* = 1. . .*K*, generated at the *K* previous steps:

$$\begin{aligned} &\text{Min } Z_{\mathrm{L}}^{K}=\alpha \\ &\text{st } \left.\begin{aligned} &\alpha \geq f(x^k,y^k) + \nabla f(x^k,y^k)^T \begin{bmatrix} x-x^k \\ y-y^k \end{bmatrix} \\ &g_j(x^k,y^k) + \nabla g_j(x^k,y^k)^T \begin{bmatrix} x-x^k \\ y-y^k \end{bmatrix} \leq 0 \; j \in J^k \end{aligned}\right\} k=1,\ldots K \\ &\quad x \in X,\, y \in Y \end{aligned} \tag{99}$$

where $J^k \subseteq J$. When only a subset of linearizations is included, these commonly correspond to violated constraints in problem 95. Alternatively, it is possible to include all linearizations in problem 99. The solution of 99 yields a valid lower bound Z_{L}^k to problem 95. This bound is nondecreasing with the number of linearization points *K*. Note that since the functions $f(x,y)$ and $g(x,y)$ are convex, the linearizations in problem 99 correspond to outer approximations of the nonlinear feasible region in problem 95. A geometrical interpretation is shown in Figure 46, where it can be seen that the convex objective function is being underestimated, and the convex feasible region overestimated with these linearizations.

Algorithms. The different methods can be classified according to their use of the subproblems 96–98, and the specific specialization of the MILP problem 99, as seen in Figure 47. In the GBD and OA methods (case b), as well in the LP/NLP-based branch-and-bound mehod (case d), problem 98 is solved if infeasible subproblems are found. Each of the methods is explained next in terms of the basic subproblems.

Branch and Bound. While the earlier work in branch and bound (BB) was aimed at linear problems [231] this method can also be applied to nonlinear problems [239 – 243]. The BB method starts by solving first the continuous NLP relaxation. If all discrete variables take integer values the search is stopped. Otherwise, a tree search is performed in the space of the integer variables y_i, $i \in I$. These are successively fixed at the corresponding nodes of the tree, giving rise to relaxed NLP subproblems of the form (NLP1) which yield lower bounds for the subproblems in the descendant nodes. Fathoming of nodes occurs when the lower bound exceeds the current upper bound, when the subproblem is infeasible, or when all integer variables y_i take on discrete values. The last-named condition yields an upper bound to the original problem.

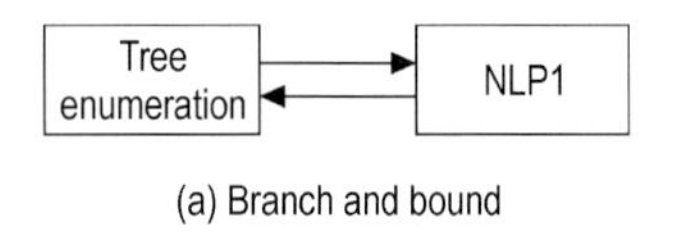

(a) Branch and bound

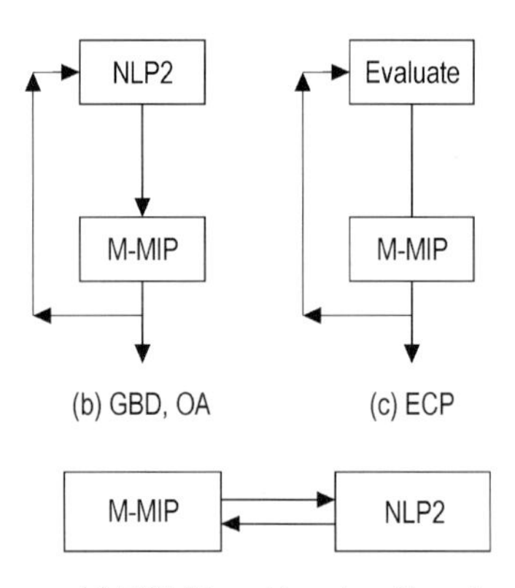

(b) GBD, OA (c) ECP

M-MIP NLP2

(d) LP/NLP based branch and bound

Figure 47. Major steps in the different algorithms

The BB method is generally only attractive if the NLP subproblems are relatively inexpensive to solve, or when only few of them need to be solved. This could be either because of the low dimensionality of the discrete variables, or because the integrality gap of the continuous NLP relaxation of 95 is small.

Outer Approximation [245 – 247]**.** The OA method arises when NLP subproblems 97 and MILP master problems 99 with $J^k = J$ are solved successively in a cycle of iterations to generate the points (x^k, y^k). Since the master problem 99 requires the solution of all feasible discrete variables y^k, the following MILP relaxation is considered, assuming that the solution of K different NLP subproblems ($K = |\mathrm{KFS} \cup \mathrm{KIS}|$), where KFS is a set of solutions from problem 97, and KIS set of solutions from problem 98 is available:

$$\begin{aligned}
&\mathrm{Min}\; Z_{\mathrm{L}}^{K}=\alpha \\
&\mathrm{st}\quad \left.\begin{aligned}
&\alpha \geq f(x^k,y^k) +\nabla f(x^k,y^k)^T \begin{bmatrix} x-x^k \\ y-y^k \end{bmatrix} \\
&g_j(x^k,y^k) +\nabla g_j(x^k,y^k)^T \begin{bmatrix} x-x^k \\ y-y^k \end{bmatrix} \leq 0 \quad j\in J
\end{aligned}\right\} k=1,\ldots K \\
&x\in X,\, y\in Y
\end{aligned} \tag{100}$$

Given the assumption on convexity of the functions $f(x,y)$ and $g(x,y)$, it can be proved that the solution of problem 100 Z_{L}^{k} corresponds to a lower bound of the solution of problem 95. Note that this property can be verified in Figure 46. Also, since function linearizations are accumulated as iterations proceed, the master problems 100 yield a nondecreasing sequence of lower bounds $Z_{\mathrm{L}}^{1}\ldots\leq Z_{\mathrm{L}}^{k}\ldots Z_{\mathrm{L}}^{k}$ since linearizations are accumulated as iterations k proceed.

The OA algorithm as proposed by Duran and Grossmann consists of performing a cycle of major iterations, $k = 1,..K$, in which problem 97 is solved for the corresponding y^K, and the relaxed MILP master problem 100 is updated and solved with the corresponding function linearizations at the point (x^k,y^k) for which the corresponding subproblem NLP2 is solved. If feasible, the solution to that problem is used to construct the first MILP master problem; otherwise a feasibility problem 98 is solved to generate the corresponding continuous point [247]. The initial MILP master problem 100 then generates a new vector of discrete variables. The subproblems 97 yield an upper bound that is used to define the best current solution, $\mathrm{UB}^k=\min\limits_k \left\{Z_{\mathrm{U}}^{k}\right\}$. The cycle of iterations is continued until this upper bound and the lower bound of the relaxed master problem Z_{L}^{k} are within a specified tolerance. One way to avoid solving the feasibility problem 98 in the OA algorithm when the discrete variables in problem 95 are 0–1, is to introduce the following integer cut whose objective is to make infeasible the choice of the previous 0–1 values generated at the K previous iterations [245]:

$$\sum\nolimits_{i\in B^k} y_i-\sum\nolimits_{i\in N^k} y_i\leq \left|B^k\right|-1 \quad k=1,\ldots K \tag{101}$$

where $B^k=\left\{i|y_i^k=1\right\}$, $N^k=\left\{i|y_i^k=0\right\}$, $k=1, \ldots K$. This cut becomes very weak as the dimensionality of the 0–1 variables increases. However, it has the useful feature of ensuring that new 0–1 values are generated at each major iteration. In this way the algorithm will not return to a previous integer point when convergence is achieved. Using the above integer cut the termination takes place as soon as $Z_{\mathrm{L}}^{K}\geq \mathrm{UB}^K$.

The OA algorithm trivially converges in one iteration if $f(x,y)$ and $g(x,y)$ are linear. This property simply follows from the fact that if $f(x,y)$ and $g(x,y)$ are linear in x and y the MILP master problem 100 is identical to the original problem 95. It is also important to note that the MILP master problem need not be solved to optimality.

Generalized Benders Decomposition (GBD) [244]. The GBD method [250] is similar to the outer-approximation method. The difference arises in the definition of the MILP master problem 99. In the GBD method only active inequalities are considered $J^k=\{j|\, g_j(x^k,y^k)=0\}$ and the set $x\in X$ is disregarded. In particular, consider an outer-approximation given at a given point (x^k, y^k)

$$\begin{aligned}
&\alpha\geq f\left(x^k,y^k\right)+\nabla f\left(x^k,y^k\right)^T \begin{bmatrix} x-x^k \\ y-y^k \end{bmatrix} \\
&g\left(x^k,y^k\right)+\nabla g\left(x^k,y^k\right)^T \begin{bmatrix} x-x^k \\ y-y^k \end{bmatrix} \leq 0
\end{aligned} \tag{102}$$

where for a fixed y^k the point x^k corresponds to the optimal solution to problem 97. Making use of the Karush – Kuhn – Tucker conditions and eliminating the continuous variables x, the inequalities in 102 can be reduced as follows [248]:

$$\begin{aligned}
&\alpha\geq f\left(x^k,y^k\right)+\nabla_y f\left(x^k,y^k\right)^T\left(y-y^k\right)+ \\
&\left(\mu^k\right)^T\left[g\left(x^k,y^k\right)+\nabla_y g\left(x^k,y^k\right)^T\left(y-y^k\right)\right]
\end{aligned} \tag{103}$$

which is the Lagrangian cut projected in the y-space. This can be interpreted as a surrogate constraint of the equations in 102, because it is obtained as a linear combination of these.

For the case when there is no feasible solution to problem 97, then if the point x^k is obtained from the feasibility subproblem (NLPF), the following feasibility cut projected in y can be obtained using a similar procedure.

$$\left(\lambda^k\right)^T \left[g\left(x^k,y^k\right)+\nabla_y g\left(x^k,y^k\right)^T\left(y-y^k\right)\right]\leq 0 \tag{104}$$

In this way, problem 99 reduces to a problem projected in the y-space:

$$\begin{aligned}
&\text{Min } Z_{\mathrm{L}}^K=\alpha\\
&\text{st } \alpha\geq f\left(x^k,y^k\right)+\nabla_y f\left(x^k,y^k\right)^T\left(y-y^k\right)+\\
&\left(\mu^k\right)^T\left[g\left(x^k,y^k\right)+\nabla_y g\left(x^k,y^k\right)^T\left(y-y^k\right)\right]\\
&\quad k\in KFS\\
&\left(\lambda^k\right)^T\left[g\left(x^k,y^k\right)+\nabla_y g\left(x^k,y^k\right)^T\left(y-y^k\right)\right]\leq 0\\
&\quad k\in KIS\\
&x\in X,\ \alpha\in R^1
\end{aligned} \tag{105}$$

where KFS is the set of feasible subproblems 97, and KIS the set of infeasible subproblems whose solution is given by problem 98. Also $|\mathrm{KFS}\cup\mathrm{KIS}|=K$. Since master problem 105 can be derived from master problem 100, in the context of problem 95, GBD can be regarded as a particular case of the outer-approximation algorithm. In fact one can prove that given the same set of K subproblems, the lower bound predicted by the relaxed master problem 100 is greater than or equal to that predicted by the relaxed master problem 105 [245]. This proof follows from the fact that the Lagrangian and feasibility cuts 103 and 104 are surrogates of the outer-approximations 102. Given the fact that the lower bounds of GBD are generally weaker, this method commonly requires a larger number of cycles or major iterations. As the number of 0–1 variables increases this difference becomes more pronounced. This is to be expected since only one new cut is generated per iteration. Therefore, user-supplied constraints must often be added to the master problem to strengthen the bounds. Also, it is sometimes possible to generate multiple cuts from the solution of an NLP subproblem in order to strengthen the lower bound [251]. As for the OA algorithm, the trade-off is that while it generally predicts stronger lower bounds than GBD, the computational cost for solving the master problem (M-OA) is greater, since the number of constraints added per iteration is equal to the number of nonlinear constraints plus the nonlinear objective.

If problem 95 has zero integrality gap, the GBD algorithm converges in one iteration once the optimal (x*, y*) is found [252]. This property implies that the only case in which one can expect the GBD method to terminate in one iteration is that in which the initial discrete vector is the optimum, and when the objective value of the NLP relaxation of problem 95 is the same as the objective of the optimal mixed-integer solution.

Extended Cutting Plane (ECP) [249]. The ECP method, which is an extension of Kelley's cutting-plane algorithm for convex NLP [253], does not rely on the use of NLP subproblems and algorithms. It relies only on the iterative solution of problem 99 by successively adding a linearization of the most violated constraint at the predicted point (x^k, y^k):

$$J^k=\left\{\hat{j}\in\arg\left\{\max_{j\in J} g_j\left(x^k,y^k\right)\right\}\right\}$$

Convergence is achieved when the maximum constraint violation lies within the specified tolerance. The optimal objective value of problem 99 yields a nondecreasing sequence of lower bounds. It is of course also possible to either add to problem 99 linearizatons of all the violated constraints in the set J^k, or linearizations of all the nonlinear constraints $j\in J$. In the ECP method the objective must be defined as a linear function, which can easily be accomplished by introducing a new variable to transfer nonlinearities in the objective as an inequality.

Note that since the discrete and continuous variables are converged simultaneously, the ECP method may require a large number of iterations. However, this method shares with the OA method Property 2 for the limiting case when all the functions are linear.

LP/NLP-Based Branch and Bound [248]. This method is similar in spirit to a branch-and-cut method, and avoids the complete solution of the MILP master problem (M-OA) at each major iteration. The method starts by solving an initial NLP subproblem, which is linearized as in (M-OA). The basic idea consists then of performing an LP-based branch-and-bound method for (M-OA) in which NLP subproblems 97 are solved at those nodes in which feasible integer solutions are found. By updating the representation of the master problem in the current open nodes of the tree with the addition of the corresponding linearizations, the need to restart the tree search is avoided.

This method can also be applied to the GBD and ECP methods. The LP/NLP method commonly reduces quite significantly the number of nodes to be enumerated. The trade-off, however, is that the number of NLP subproblems may increase. Computational experience has indicated that often the number of NLP subproblems remains unchanged. Therefore, this method is better suited for problems in which the bottleneck corresponds to the solution of the MILP master problem. LEYFFER [254] has reported substantial savings with this method.

Example: Consider the following MINLP problem, whose objective function and constraints contain nonlinear convex terms

$$
\begin{aligned}
&\text{Min } Z=y_1+1.5y_2+0.5y_3+x_1^2+x_2^2\\
&s.t. \quad (x_1-2)^2-x_2\leq 0\\
&\qquad x_1-2y_1\geq 0\\
&\qquad x_1-x_2-4(1-y_2)\geq 0\\
&\qquad x_2-y_2\geq 0\\
&\qquad x_1+x_2\geq 3y_3\\
&\qquad y_1+y_2+y_3\geq 1\\
&\qquad 0\leq x_1\leq 4, 0\leq x_2\leq 4\\
&\qquad y_1,y_2,y_3=0,1
\end{aligned}
\tag{106}
$$

The optimal solution of this problem is given by $y_1=0$, $y_2=1$, $y_3=0$, $x_1=1$, $x_2=1$, $Z=3.5$. Figure 48 shows the progress of the iterations with the OA and GBD methods, while the table lists the number of NLP and MILP subproblems that are solved with each of the methods. For the case of the MILP problems the total number of LPs solved at the branch-and-bound nodes are also reported.

Extensions of MINLP Methods. Extensions of the methods described above include a quadratic approximation to (RM-OAF) [247] using an approximation of the Hessian matrix. The quadratic approximations can help to reduce the number of major iterations, since an improved representation of the continuous space is obtained. This, however, comes at the price of having to solve an MIQP instead of an MILP at each iteration.

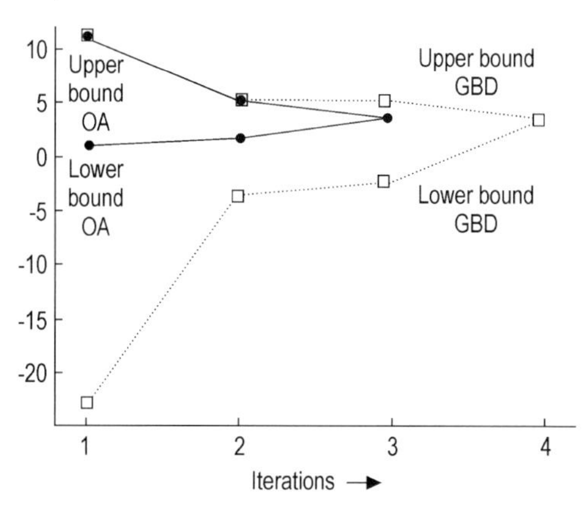

Figure 48. Progress of iterations with OA and GBD methods, and number of subproblems for the BB, OA, GBD, and ECP methods.

The master problem 100 can involve a rather large number of constraints, due to the accumulation of linearizations. One option is to keep only the last linearization point, but this can lead to nonconvergence even in convex problems, since then the monotonic increase of the lower bound is not guaranteed. As shown [248], linear approximations to the nonlinear objective and constraints can be aggregated with an MILP master problem that is a hybrid of the GBD and OA methods.

For the case when linear equalities of the form $h(x, y)=0$ are added to problem 95 there is no major difficulty, since these are invariant to the linearization points. If the equations are nonlinear, however, there are two difficulties. First, it is not possible to enforce the linearized equalities at K points. Second, the nonlinear equations may generally introduce nonconvexities, unless they relax as convex inequalities [255]. KOCIS and GROSSMANN [256] proposed an equality relaxation strategy in which the nonlinear equalities are replaced by the inequalities

$$T^k \nabla h\left(x^k, y^k\right)^T \begin{bmatrix} x-x^k \\ y-y^k \end{bmatrix} \leq 0 \tag{107}$$

where $T^k = \{t_{ii}^k\}$, and $t_{ii}^k = \text{sign}\left(\lambda_i^k\right)$ sign in which λ_i^k is the multiplier associated to the equation $h_i(x, y) = 0$. Note that if these equations relax as the inequalities $h(x, y) \leq 0$ for all y and $h(x, y)$ is convex, this is a rigorous procedure. Otherwise, nonvalid supports may be generated. Also, in the master problem 105 of GBD, no special provision is required to handle equations, since these are simply included in the Lagrangian cuts. However, similar difficulties as in OA arise if the equations do not relax as convex inequalities.

When $f(x,y)$ and $g(x,y)$ are nonconvex in problem 95, or when nonlinear equalities $h(x, y) = 0$ are present, two difficulties arise. First, the NLP subproblems 96 – 98 may not have a unique local optimum solution. Second, the master problem (M-MIP) and its variants (e.g., M-MIPF, M-GBD, M-MIQP) do not guarantee a valid lower bound Z_{L}^K or a valid bounding representation with which the global optimum may be cut off.

Rigorous global optimization approaches for addressing nonconvexities in MINLP problems can be developed when special structures are assumed in the continuous terms (e.g. bilinear, linear fractional, concave separable). Specifically, the idea is to use convex envelopes or underestimators to formulate lower-bounding convex MINLP problems. These are then combined with global optimization techniques for continuous variables [206, 207, 209, 214, 216, 222, 257], which usually take the form of spatial branch-and-bound methods. The lower bounding MINLP problem has the general form,

$$
\begin{aligned}
&\text{Min } Z = \overline{f}\,(x,y) \\
&\text{s.t. } \overline{g}_j\,(x,y) \leq 0 \; j \in J \\
&\qquad x \in X, y \in Y
\end{aligned} \tag{108}
$$

where $\overline{f}, \overline{g}$ are valid convex underestimators such that $\overline{f}\,(x,y) \leq f\,(x,y)$ and the inequalities $\overline{g}\,(x,y) \leq 0$ are satisfied if $g\,(x,y) \leq 0$. A typical example of convex underestimators are the convex envelopes for bilinear terms [208].

Examples of global optimization methods for MINLP problems include the branch-and-reduce method [209, 210], the α-BB method [212], the reformulation/spatial branch-and-bound search method [258], the branch-and-cut method [259], and the disjunctive branch-and-bound method [260]. All these methods rely on a branch-and-bound procedure. The difference lies in how to perform the branching on the discrete and continuous variables. Some methods perform the spatial tree enumeration on both the discrete and continuous variables of problem 108. Other methods perform a spatial branch and bound on the continuous variables and solve the corresponding MINLP problem 108 at each node using any of the methods reviewed above. Finally, other methods branch on the discrete variables of problem 108, and switch to a spatial branch and bound on nodes where a feasible value for the discrete variables is found. The methods also rely on procedures for tightening the lower and upper bounds of the variables, since these have a great effect on the quality of the underestimators. Since the tree searches are not finite (except for ε convergence), these methods can be computationally expensive. However, their major advantage is that they can rigorously find the global optimum. Specific cases of nonconvex MINLP problems have been handled. An example is the work of Pörn and Westerlund [261], who addressed the solution of MINLP problems with pseudoconvex objective function and convex inequalities through an extension of the ECP method.

The other option for handling nonconvexities is to apply a heuristic strategy to try to reduce as much as possible the effect of nonconvexities. While not being rigorous, this requires much less computational effort. We describe here an approach for reducing the effect of nonconvexities at the level of the MILP master problem.

Viswanathan and Grossmann [262] proposed to introduce slacks in the MILP master problem to reduce the likelihood of cutting off feasible solutions. This master problem (augmented penalty/equality relaxation) has the form:

$$
\begin{aligned}
&\min Z^K = \alpha + \sum_{k=1}^{K} w_p^k\, p^k + w_q^k q^k \\
&\text{s.t.} \left.\begin{aligned}
&\alpha \geq f\left(x^k, y^k\right) + \nabla f\left(x^k, y^k\right)^T \begin{bmatrix} x - x^k \\ y - y^k \end{bmatrix} \\
&T^k \nabla h(x^k, y^k)^T \begin{bmatrix} x - x^k \\ y - y^k \end{bmatrix} \leq p^k \\
&g\left(x^k, y^k\right) + \nabla g(x^k, y^k)^T \begin{bmatrix} x - x^k \\ y - y^k \end{bmatrix} \leq q^k
\end{aligned}\right\} k = 1, \ldots K
\end{aligned} \tag{109}
$$

$$\sum_{i\in B^k} y_i - \sum_{i\in N^k} y_i \leq \left|B^k\right| - 1 \; k=1,\ldots K$$
$$x\in X, y\in Y, \alpha\in R^1, p^k, q^k \geq 0$$

where w_p^k, w_q^k are weights that are chosen sufficiently large (e.g., 1000 times the magnitude of the Lagrange multiplier). Note that if the functions are convex then the MILP master problem 109 predicts rigorous lower bounds to problem 95 since all the slacks are set to zero.

Computer Codes for MINLP. Computer codes for solving MINLP problems include the following. The program DICOPT [262] is an MINLP solver that is available in the modeling system GAMS [263]. The code is based on the master problem 109 and the NLP subproblems 97. This code also uses relaxed 96 to generate the first linearization for the above master problem, with which the user need not specify an initial integer value. Also, since bounding properties of problem 109 cannot be guaranteed, the search for nonconvex problems is terminated when there is no further improvement in the feasible NLP subproblems. This is a heuristic that works reasonably well in many problems. Codes that implement the branch-and-bound method using subproblems 96 include the code MINLP_BB, which is based on an SQP algorithm [243] and is available in AMPL, the code BARON [264], which also implements global optimization capabilities, and the code SBB, which is available in GAMS [263]. The code a-ECP implements the extended cutting-plane method [249], including the extension by Pörn and Westerlund [261]. Finally, the code MINOPT [265] also implements the OA and GBD methods, and applies them to mixed-integer dynamic optimization problems. It is difficult to derive general conclusions on the efficiency and reliability of all these codes and their corresponding methods, since no systematic comparison has been made. However, one might anticipate that branch-and-bound codes are likely to perform better if the relaxation of the MINLP is tight. Decomposition methods based on OA are likely to perform better if the NLP subproblems are relatively expensive to solve, while GBD can perform with some efficiency if the MINLP is tight and there are many discrete variables. ECP methods tend to perform well on mostly linear problems.

Logic-Based Optimization. Given difficulties in the modeling and solution of mixed integer problems, the following major approaches based on logic-based techniques have emerged: generalized disjunctive programming 110 [266], mixed-logic linear programming (MLLP) [267], and constraint programming (CP) [268]. The motivations for this logic-based modeling has been to facilitate the modeling, reduce the combinatorial search effort, and improve the handling of nonlinearities. In this section we mostly concentrate on generalized disjunctive programming and provide a brief reference to constraint programming. A general review of logic-based optimization can be found in [269, 309].

Generalized disjunctive programming in 110 [266] is an extension of disjunctive programming [270] that provides an alternative way of modeling MILP and MINLP problems. The general formulation 110 is as follows:

$$\begin{aligned}
&\text{Min } Z = \sum_{k\in K} c_k + f(x) \\
&s.t. \quad g(x) \leq 0 \\
&\bigvee_{j\in J_k} \begin{bmatrix} Y_{jk} \\ h_{jk}(x) \leq 0 \\ c_k = \gamma_{jk} \end{bmatrix}, k\in K \\
&\Omega(Y) = True \\
&x\in R^n, c\in R^m, Y\in \{true, \; false\}^m
\end{aligned} \tag{110}$$

where Y_{jk} are the Boolean variables that decide whether a term j in a disjunction $k \in K$ is true or false, and x are continuous variables. The objective function involves the term $f(x)$ for the continuous variables and the charges c_k that depend on the discrete choices in each disjunction $k \in K$. The constraints $g(x) \leq 0$ hold regardless of the discrete choice, and $h_{jk}(x) \leq 0$ are conditional constraints that hold when Y_{jk} is true in the j-th term of the k-th disjunction. The cost variables c_k correspond to the fixed charges, and are equal to γ_{jk} if the Boolean variable Y_{jk} is true. $\Omega(Y)$ are logical relations for the Boolean variables expressed as propositional logic.

Problem 110 can be reformulated as an MINLP problem by replacing the Boolean variables by binary variables y_{jk},

$$
\begin{aligned}
&\text{Min } Z=\sum_{k\in K}\sum_{j\in J_k}\gamma_{jk}y_{jk}+f(x)\\
&\text{s.t.}\quad g(x)\le 0\\
&h_{jk}(x)\le M_{jk}(1-y_{jk}),\, j\in J_k,\, k\in K \quad \text{(BM)}\\
&\sum_{j\in J_k} y_{jk}=1,\, k\in K\\
&Ay\le a\\
&0\le x\le x^U,\, y_{jk}\in\{0,1\},\, j\in J_k,\, k\in K
\end{aligned}
\qquad (111)
$$

where the disjunctions are replaced by "Big-M" constraints which involve a parameter M_{jk} and binary variables y_{jk}. The propositional logic statements $\Omega(Y)$ = True are replaced by the linear constraints $Ay \le a$ [271] and [272]. Here we assume that x is a nonnegative variable with finite upper bound x^U. An important issue in model 111 is how to specify a valid value for the Big-M parameter M_{jk}. If the value is too small, then feasible points may be cut off. If M_{jk} is too large, then the continuous relaxation might be too loose and yield poor lower bounds. Therefore, finding the smallest valid value for M_{jk} is desired. For linear constraints, one can use the upper and lower bound of the variable x to calculate the maximum value of each constraint, which then can be used to calculate a valid value of M_{jk}. For nonlinear constraints one can in principle maximize each constraint over the feasible region, which is a nontrivial calculation.

Lee and Grossmann [273] have derived the *convex hull relaxation* of problem 110. The basic idea is as follows. Consider a disjunction $k \in K$ that has convex constraints

$$
\underset{j\in J_k}{\vee}\begin{bmatrix} Y_{jk}\\ h_{jk}(x)\le 0\\ c=\gamma_{jk}\end{bmatrix}
\qquad (112)
$$
$$
0\le x\le x^U,\, c\ge 0
$$

where $h_{jk}(x)$ are assumed to be convex and bounded over x. The convex hull relaxation of disjunction 112 [242] is given as follows:

$$
\begin{aligned}
&x=\sum_{j\in J_k} v^{jk},\quad c=\sum_{j\in J}\lambda_{jk}\gamma_{jk}\\
&0\le v^{jk}\le \lambda_{jk}x^U_{jk},\, j\in J_k\\
&\sum_{j\in J_k}\lambda_{jk}=1,\, 0\le\lambda_{jk}\le 1,\, j\in J_k \quad \text{(CH)}\\
&\lambda_{jk}h_{jk}\left(v^{jk}/\lambda_{jk}\right)\le 0,\, j\in J_k\\
&x,c,v^{jk}\ge 0,\, j\in J_k
\end{aligned}
\qquad (113)
$$

where v^{jk} are disaggregated variables that are assigned to each term of the disjunction $k \in K$, and λ_{jk} are the weight factors that determine the feasibility of the disjunctive term. Note that when λ_{jk} is 1, then the j-th term in the k-th disjunction is enforced and the other terms are ignored. The constraints $\lambda_{jk}h_{jk}\left(v^{jk}/\lambda_{jk}\right)$ are convex if $h_{jk}(x)$ is convex [274, p. 160]. A formal proof can be found in [242]. Note that the convex hull 113 reduces to the result by Balas [275] if the constraints are linear. Based on the convex hull relaxation 113, Lee and Grossmann [273] proposed the following convex relaxation program of problem 110.

$$
\begin{aligned}
&\text{Min } Z^L=\sum_{k\in K}\sum_{j\in J_k}\gamma_{jk}\lambda_{jk}+f(x)\\
&\text{s.t.}\quad g(x)\le 0\\
&x=\sum_{j\in J_k} v^{jk},\, \sum_{j\in J_k}\lambda_{jk}=1,\, k\in K \quad \text{(CRP)}\\
&0\le x,\, v^{jk}\le x^U,\, 0\le\lambda_{jk}\le 1,\, j\in J_k,\, k\in K\\
&\lambda_{jk}h_{jk}\left(v^{jk}/\lambda_{jk}\right)\le 0,\, j\in J_k,\, k\in K\\
&A\lambda\le a\\
&0\le x,\, v^{jk}\le x^U,\, 0\le\lambda_{jk}\le 1,\, j\in J_k,\, k\in K
\end{aligned}
\qquad (114)
$$

where x^U is a valid upper bound for x and v. Note that the number of constraints and variables increases in problem 114 compared with problem 110. Problem 114 has a unique optimal solution and it yields a valid lower bound to the optimal solution of problem 110 [273]. Grossmann and Lee [276] proved that problem 114 has the useful property that the lower bound is greater than or equal to the lower bound predicted from the relaxation of problem 111.

Further description of algorithms for disjunctive programming can be found in [277].

Constraint Programming. Constraint programming (CP) [268, 269] is a relatively new modeling and solution paradigm that was originally developed to solve feasibility problems, but it has been extended to solve optimization problems as well. Constraint programming is very expressive, as continuous, integer, and Boolean variables are permitted and, moreover, variables can be indexed by other variables. Constraints can be expressed in algebraic form (e.g., $h(x) \le 0$), as disjunctions (e.g., $[A_1x \le b_1] \vee A_2x \le b_2]$), or as conditional logic statements (e.g., If $g(x) \le 0$ then $r(x) \le 0$). In addition, the language can support special implicit functions such as the all-different $(x_1, x_2, \ldots x_n)$ constraint for assigning different values to the integer variables $x_1, x_2, \ldots x_n$. The language consists of C ++ procedures, although the recent trend has been to provide higher level languages such as OPL. Other commercial CP software packages include ILOG Solver [278], CHIP [279], and ECLiPSe [280].

10.6. Dynamic Optimization

Interest in dynamic simulation and optimization of chemical processes has increased significantly during the last two decades. Chemical processes are modeled dynamically using differential-algebraic equations (DAEs), consisting of differential equations that describe the dynamic behavior of the system, such as mass and energy balances, and algebraic equations that ensure physical and thermodynamic relations. Typical applications include control and scheduling of batch processes; startup, upset, shut-down, and transient analysis; safety studies; and the evaluation of control schemes. We state a general differential-algebraic optimization problem 115 as follows:

$$\begin{aligned}
&\text{Min} && \Phi\left(z\left(t_{\mathrm{f}}\right);y\left(t_{\mathrm{f}}\right);u\left(t_{\mathrm{f}}\right);t_{\mathrm{f}};p\right)\\
&\text{s.t.} && F\left(dz/dt;z\left(t\right);u\left(t\right);t;p\right)=0,\ z\left(0\right)=z_0\\
& && G_{\mathrm{s}}\left[z\left(t_{\mathrm{s}}\right);y\left(t_{\mathrm{s}}\right);u\left(t_{\mathrm{s}}\right);t_{\mathrm{s}};p)\right]=0\\
& && z^{\mathrm{L}}\leq z\left(t\right)\leq x^{\mathrm{U}}\\
& && y^{\mathrm{L}}\leq y\left(t\right)\leq y^{\mathrm{U}}\\
& && u^{\mathrm{L}}\leq u\left(t\right)\leq y^{\mathrm{U}}\\
& && p^{\mathrm{L}}\leq p\leq p^{\mathrm{U}}\\
& && t_{\mathrm{f}}^{t}\leq t_{\mathrm{f}}\leq t_{\mathrm{f}}^{U}
\end{aligned} \tag{115}$$

where Φ is a scalar objective function at final time t_{f}, and F are DAE constraints, G_{s} additional point conditions at times t_{s}, $z(t)$ differential state profile vectors, $y(t)$ algebraic state profile vectors, $u(t)$ control state profile vectors, and p is a time-independent parameter vector.

We assume, without loss of generality, that the index of the DAE system is one, consistent initial conditions are available, and the objective function is in the above Mayer form. Otherwise, it is easy to rcformulate problems to this form. Problem 115 can be solved either by the variational approach or by applying some level of discretization that converts the original continuous time problem into a discrete problem. Early solution strategies, known as indirect methods, were focused on solving the classical variational conditions for optimality. On the other hand, methods that discretize the original continuous time formulation can be divided into two categories, according to the level of discretization. Here we distinguish between the methods that discretize only the control profiles (partial discretization) and those that discretize the state and control profiles (full discretization). Basically, the partially discretized problem can be solved either by dynamic programming or by applying a nonlinear programming (NLP) strategy (direct-sequential). A basic characteristic of these methods is that a feasible solution of the DAE system, for given control values, is obtained by integration at every iteration of the NLP solver. The main advantage of these approaches is that, for the NLP solver, they generate smaller discrete problems than full discretization methods.

Methods that fully discretize the continuous time problem also apply NLP strategies to solve the discrete system and are known as direct-simultaneous methods. These methods can use different NLP and discretization techniques, but the basic characteristic is that they solve the DAE system only once, at the optimum. In addition, they have better stability properties than partial discretization methods, especially in the presence of unstable dynamic modes. On the other hand, the discretized optimization problem is larger and requires large-scale NLP solvers, such as SOCS, CONOPT, or IPOPT.

With this classification we take into account the degree of discretization used by the different methods. Below we briefly present the description of the variational methods, followed by methods that partially discretize the dynamic optimization problem, and finally we consider full discretization methods for problem 115.

Variational Methods. These methods are based on the solution of the first-order necessary conditions for optimality that are obtained from Pontryagin's maximum principle [281, 282]. If we consider a version of problem 115 without bounds, the optimality conditions are formulated as a set of DAEs:

$$\frac{\partial F\left(z,y,u,p,t\right)}{\partial z'}\lambda'=\frac{\partial H}{\partial z}=\frac{\partial F\left(z,y,u,p,t\right)}{\partial z}\lambda \tag{116a}$$

$$F\left(z,y,u,p,t\right)=0 \tag{116b}$$

$$G_f\left(z,y,u,p,t_f\right)=0 \tag{116c}$$

$$G_s\left(z,y,u,p,t_s\right)=0 \tag{116d}$$

$$\frac{\partial H}{\partial y}=\frac{\partial F\left(z,y,u,p,t\right)}{\partial y}\lambda=0 \tag{116e}$$

$$\frac{\partial H}{\partial u}=\frac{\partial F\left(z,y,u,p,t\right)}{\partial u}\lambda=0 \tag{116f}$$

$$\int\limits_0^{t_f}\frac{\partial F\left(z,y,u,p,t\right)}{\partial p}\lambda\, dt=0 \tag{116g}$$

where the Hamiltonian H is a scalar function of the form $H(t)=F(z, y, u, p, y)^T\lambda(t)$ and $\lambda(t)$ is a vector of adjoint variables. Boundary and jump conditions for the adjoint variables are given by:

$$\begin{aligned}&\tfrac{\partial F}{\partial z'}\lambda\left(t_{\mathrm{f}}\right)+\tfrac{\partial \Phi}{\partial z}+\tfrac{\partial G_{\mathrm{f}}}{\partial z}v_{\mathrm{f}}=0\\&\tfrac{\partial F}{\partial z'}\lambda\left(t_{\mathrm{s}}^{-}\right)+\tfrac{\partial G_{\mathrm{s}}}{\partial z}v_{\mathrm{s}}=\tfrac{\partial F}{\partial z'}\lambda\left(t_{\mathrm{s}}^{+}\right)\end{aligned} \tag{117}$$

where v_{f}, v_{s} are the multipliers associated with the final time and point constraints, respectively. The most expensive step lies in obtaining a solution to this boundary value problem. Normally, the state variables are given as initial conditions, and the adjoint variables as final conditions. This formulation leads to boundary value problems (BVPs) that can be solved by a number of standard methods including single shooting, invariant embedding, multiple shooting, or some discretization method such as collocation on finite elements or finite differences. Also the point conditions lead to an additional calculation loop to determine the multipliers v_{f} and v_{s}. On the other hand, when bound constraints are considered, the above conditions are augmented with additional multipliers and associated complementarity conditions. Solving the resulting system leads to a combinatorial problem that is prohibitively expensive except for small problems.

Partial Discretization. With partial discretization methods (also called sequential methods or control vector parametrization), only the control variables are discretized. Given the initial conditions and a given set of control parameters, the DAE system is solved with a differential algebraic equation solver at each iteration. This produces the value of the objective function, which is used by a nonlinear programming solver to find the optimal parameters in the control parametrization ν. The sequential method is reliable when the system contains only stable modes. If this is not the case, finding a feasible solution for a given set of control parameters can be very difficult. The time horizon is divided into time stages and at each stage the control variables are represented with a piecewise constant, a piecewise linear, or a polynomial approximation [283, 284]. A common practice is to represent the controls as a set of Lagrange interpolation polynomials.

For the NLP solver, gradients of the objective and constraint functions with respect to the control parameters can be calculated with the sensitivity equations of the DAE system, given by:

$$\begin{aligned}&\frac{\partial F}{\partial z'}^T s_k{'}+\frac{\partial F}{\partial z}^T s_k+\frac{\partial F}{\partial y}^T w_k+\frac{\partial F}{\partial q_k}^T=0,\\&s_k\left(0\right)=\frac{\partial z\left(0\right)}{\partial q_k}\quad k=1,\ldots N_q\end{aligned} \tag{118}$$

where $s_k\left(t\right)=\frac{\partial z(t)}{\partial q_k}$, $w_k\left(t\right)=\frac{\partial y(t)}{\partial q_k}$, and $q^T=[p^T, \nu^{\mathrm{T}}]$. As can be inferred from Equation 118, the cost of obtaining these sensitivities is directly proportional to N_{q}, the number of decision variables in the NLP. Alternately, gradients can be obtained by integration of the adjoint Equations 116a, 116e, 116g [282, 285, 286] at a cost independent of the number of input variables and proportional to the number of constraints in the NLP.

Methods that are based on this approach cannot treat directly the bounds on state variables, because the state variables are not included in the nonlinear programming problem. Instead, most of the techniques for dealing with inequality path constraints rely on defining a measure of the constraint violation over the entire horizon, and then penalizing it in the objective function, or forcing it directly to zero through an end-point constraint [287]. Other techniques approximate the constraint satisfaction (constraint aggregation methods) by introducing an exact penalty function [286, 288] or a Kreisselmeier–Steinhauser function [288] into the problem.

Finally, initial value solvers that handle path constraints directly have been developed [284].

The main idea is to use an algorithm for constrained dynamic simulation, so that any admissible combination of the control parameters produces an initial value problem that is feasible with respect to the path constraints. The algorithm proceeds by detecting activation and deactivation of the constraints during the solution, and solving the resulting high-index DAE system and their related sensitivities.

Full Discretization. Full discretization methods explicitly discretize all the variables of the DAE system and generate a large-scale nonlinear programming problem that is usually solved with a successive quadratic programming (SQP) algorithm. These methods follow a simultaneous approach (or infeasible path approach); that is, the DAE system is not solved at each iteration; it is only solved at the optimum point. Because of the size of the problem, special decomposition strategies are used to solve the NLP efficiently. Despite this characteristic, the simultaneous approach has advantages for problems with state variable (or path) constraints and for systems where instabilities occur for a range of inputs. In addition, the simultaneous approach can avoid intermediate solutions that may not exist, are difficult to obtain, or require excessive computational effort. There are mainly two different approaches to discretize the state variables explicitly, multiple shooting [289, 290] and collocation on finite elements [181, 191, 291].

With *multiple shooting*, time is discretized into P stages and control variables are parametrized using a finite set of control parameters in each stage, as with partial discretization. The DAE system is solved on each stage, $i = 1, \ldots P$ and the values of the state variables $z(t_i)$ are chosen as additional unknowns. In this way a set of relaxed, decoupled initial value problems (IVP) is obtained:

$$F\left(dz/dt; z\left(t\right); y\left(t\right); \nu_i; p\right) = 0,$$

$$t \in \left[t_{i-1}, t_i\right], z\left(t_{i-1}\right) = z_i \tag{119}$$

$$z_{i+1} - z\left(t_i; z_i; \nu_i; p\right) = 0, i = 1, \ldots P-1$$

Note that continuity among stages is treated through equality constraints, so that the final solution satisfies the DAE system. With this approach, inequality constraints for states and controls can be imposed directly at the grid points, but path constraints for the states may not be satisfied between grid points. This problem can be avoided by applying penalty techniques to enforce feasibility, like the ones used in the sequential methods.

The resulting NLP is solved using SQP-type methods, as described above. At each SQP iteration, the DAEs are integrated in each stage and objective and constraint gradients with respect to p, z_i, and ν_i are obtained using sensitivity equations, as in problem 118. Compared to sequential methods, the NLP contains many more variables, but efficient decompositions have been proposed [290] and many of these calculations can be performed in parallel.

In *collocation methods*, the continuous time problem is transformed into an NLP by approximating the profiles as a family of polynomials on finite elements. Various polynomial representations are used in the literature, including Lagrange interpolation polynomials for the differential and algebraic profiles [291]. In [191] a Hermite–Simpson collocation form is used, while CUTHRELL and BIEGLER [292] and TANARTKIT and BIEGLER [293] use a monomial basis for the differential profiles. All of these representations stem from implicit Runge–Kutta formulae, and the monomial representation is recommended because of smaller condition numbers and smaller rounding errors. Control and algebraic profiles, on the other hand, are approximated using Lagrange polynomials.

Discretizations of problem 115 using collocation formulations lead to the largest NLP problems, but these can be solved efficiently using large-scale NLP solvers such as IPOPT and by exploiting the structure of the collocation equations. BIEGLER et al. [181] provide a review of dynamic optimization methods using simultaneous methods. These methods offer a number of advantages for challenging dynamic optimization problems, which include:

- Control variables can be discretized at the same level of accuracy as the differential and algebraic state variables. The KKT conditions of the discretized problem can be shown to be consistent with the variational conditions of problem 115. Finite elements allow for discontinuities in control profiles.
- Collocation formulations allow problems with unstable modes to be handled in an

efficient and well-conditioned manner. The NLP formulation inherits stability properties of boundary value solvers. Moreover, an elementwise decomposition has been developed that pins down unstable modes in problem 115.
- Collocation formulations have been proposed with moving finite elements. This allows the placement of elements both for accurate breakpoint locations of control profiles as well as accurate DAE solutions.

Dynamic optimization by collocation methods has been used for a wide variety of process applications including batch process optimization, batch distillation, crystallization, dynamic data reconciliation and parameter estimation, nonlinear model predictive control, polymer grade transitions and process changeovers, and reactor design and synthesis. A review of this approach can be found in [294].

10.7. Development of Optimization Models

The most important aspect of a successful optimization study is the formulation of the optimization model. These models must reflect the real-world problem so that meaningful optimization results are obtained, and they also must satisfy the properties of the problem class. For instance, NLPs addressed by gradient-based methods require functions that are defined in the variable domain and have bounded and continuous first and second derivatives. In mixed integer problems, proper formulations are also needed to yield good lower bounds for efficient search. With increased understanding of optimization methods and the development of efficient and reliable optimization codes, optimization practitioners now focus on the *formulation* of optimization models that are realistic, well-posed, and inexpensive to solve. Finally, convergence properties of NLP, MILP, and MINLP solvers require accurate first (and often second) derivatives from the optimization model. If these contain numerical errors (say, through finite difference approximations) then performance of these solvers can deteriorate considerably. As a result of these characteristics, modeling platforms are essential for the formulation task. These are classified into two broad areas: optimization modeling platforms and simulation platforms with optimization.

Optimization modeling platforms provide general purpose interfaces for optimization algorithms and remove the need for the user to interface to the solver directly. These platforms allow the general formulation for all problem classes discussed above with direct interfaces to state of the art optimization codes. Three representative platforms are GAMS (General Algebraic Modeling Systems), AMPL (A Mathematical Programming Language), and AIMMS (Advanced Integrated Multidimensional Modeling Software). All three require problem-model input via a declarative modeling language and provide exact gradient and Hessian information through automatic differentiation strategies. Although possible, these platforms were not designed to handle externally added procedural models. As a result, these platforms are best applied on optimization models that can be developed entirely within their modeling framework. Nevertheless, these platforms are widely used for large-scale research and industrial applications. In addition, the MATLAB platform allows the flexible formulation of optimization models as well, although it currently has only limited capabilities for automatic differentiation and limited optimization solvers. More information on these and other modeling platforms can be found on the NEOS server www-neos.mcs.anl.gov

Simulation platforms with optimization are often dedicated, application-specific modeling tools to which optimization solvers have been interfaced. These lead to very useful optimization studies, but because they were not originally designed for optimization models, they need to be used with some caution. In particular, most of these platforms do not provide exact derivatives to the optimization solver; often they are approximated through finite difference. In addition, the models themselves are constructed and calculated through numerical procedures, instead of through an open declarative language. Examples of these include widely used process simulators such as Aspen/Plus, PRO/II, and Hysys. More recent platforms such as Aspen Custom Modeler and gPROMS include declarative models and exact derivatives.

For optimization tools linked to procedural models, reliable and efficient automatic differentiation tools are available that link to models written, say, in FORTRAN and C, and calculate exact first (and often second) derivatives. Examples of these include ADIFOR, ADOL-C, GRESS, Odyssee, and PADRE. When used with care, these can be applied to existing procedural models and, when linked to modern NLP and MINLP algorithms, can lead to powerful optimization capabilities. More information on these and other automatic differentiation tools can be found on: http://www-unix.mcs.anl.gov/autodiff/AD_Tools/.

Finally, the availability of automatic differentiation and related sensitivity tools for differential equation models allows for considerable flexibility in the formulation of optimization models. In [295] a seven-level modeling hierarchy is proposed that matches optimization algorithms with models that range from completely open (fully declarative model) to fully closed (entirely procedural without sensitivities). At the lowest, fully procedural level, only derivative-free optimization methods are applied, while the highest, declarative level allows the application of an efficient large-scale solver that uses first and second derivatives. Depending on the modeling level, optimization solver performance can vary by several orders of magnitude.

11. Probability and Statistics [15, 296 – 303]

The models treated thus far have been deterministic, that is, if the parameters are known the outcome is determined. In many situations, all the factors cannot be controlled and the outcome may vary randomly about some average value. Then a range of outcomes has a certain probability of occurring, and statistical methods must be used. This is especially true in quality control of production processes and experimental measurements. This chapter presents standard statistical concepts, sampling theory and statistical decisions, and factorial design of experiments or analysis of variances. Multivariant linear and nonlinear regression is treated in Chapter 2.

11.1. Concepts

Suppose N values of a variable y, called $y_1, y_2, \ldots, y_N$, might represent N measurements of the same quantity. The *arithmetic mean* $E\,(y)$ is

$$E\,(y)=\frac{\sum\limits_{i=1}^{N}y_i}{N}$$

The *median* is the middle value (or average of the two middle values) when the set of numbers is arranged in increasing (or decreasing) order. The *geometric mean* $\overline{y}_G$ is

$$\overline{y}_G=(y_1y_2\ldots y_N)^{1/N}$$

The *root-mean-square* or quadratic mean is

$$\text{Root-mean-square}=\sqrt{E\,(y^2)}=\sqrt{\sum\limits_{i=1}^{N}y_i^2/N}$$

The *range* of a set of numbers is the difference between the largest and the smallest members in the set. The *mean deviation* is the mean of the deviation from the mean.

$$\text{Mean-deviation}=\frac{\sum\limits_{i=1}^{N}|y_i-E\,(y)|}{N}$$

The *variance* is

$$var\,(y)=\sigma^2=\frac{\sum\limits_{i=1}^{N}(y_i-E\,(y))^2}{N}$$

and the *standard deviation* σ is the square root of the variance.

$$\sigma=\sqrt{\frac{\sum\limits_{i=1}^{N}(y_i-E\,(y))^2}{N}}$$

If the set of numbers $\{y_i\}$ is a small sample from a larger set, then the *sample average*

$$\overline{y}=\frac{\sum\limits_{i=1}^{n}y_i}{n}$$

is used in calculating the *sample variance*

$$s^2=\frac{\sum\limits_{i=1}^{n}(y_i-\overline{y})^2}{n-1}$$

and the sample standard deviation

$$s=\sqrt{\frac{\sum\limits_{i=1}^{n}(y_i-\overline{y})^2}{n-1}}$$

The value $n-1$ is used in the denominator because the deviations from the sample average must total zero:

$$\sum_{i=1}^{n}(y_i-\overline{y})=0$$

Thus, knowing $n-1$ values of $y_i-\bar{y}$ and the fact that there are n values automatically gives the n-th value. Thus, only $n-1$ degrees of freedom ν exist. This occurs because the unknown mean $E(y)$ is replaced by the sample mean y derived from the data.

If data are taken consecutively, running totals can be kept to permit calculation of the mean and variance without retaining all the data:

$$\sum_{i=1}^{n}(y_i-\overline{y})^2=\sum_{i=1}^{n}y_1^2-2\overline{y}\sum_{i=1}^{n}y_i+(\overline{y})^2$$

$$\overline{y}=\sum_{i=1}^{n}y_i/n$$

Thus,

$$n,\sum_{i=1}^{n}y_1^2,\ \text{and}\sum_{i=1}^{n}y_i$$

are retained, and the mean and variance are computed when needed.

Repeated observations that differ because of experimental error often vary about some central value in a roughly symmetrical distribution in which small deviations occur more frequently than large deviations. In plotting the number of times a discrete event occurs, a typical curve is obtained, which is shown in Figure 49. Then the probability p of an event (score) occurring can be thought of as the ratio of the number of times it was observed divided by the total number of events. A continuous representation of this probability density function is given by the *normal distribution*

$$p(y)=\frac{1}{\sigma\sqrt{2\pi}}\mathrm{e}^{-[y-E(y)]^2/2\sigma^2}. \tag{120}$$

This is called a normal probability distribution function. It is important because many results are insensitive to deviations from a normal distribution. Also, the central limit theorem says that if an overall error is a linear combination of component errors, then the distribution of errors tends to be normal as the number of components increases, almost regardless of the distribution of the component errors (i.e., they need not be normally distributed). Naturally, several sources of error must be present and one error cannot predominate (unless it is normally distributed). The normal distribution function is calculated easily; of more value are integrals of the function, which are given in Table 12; the region of interest is illustrated in Figure 50.

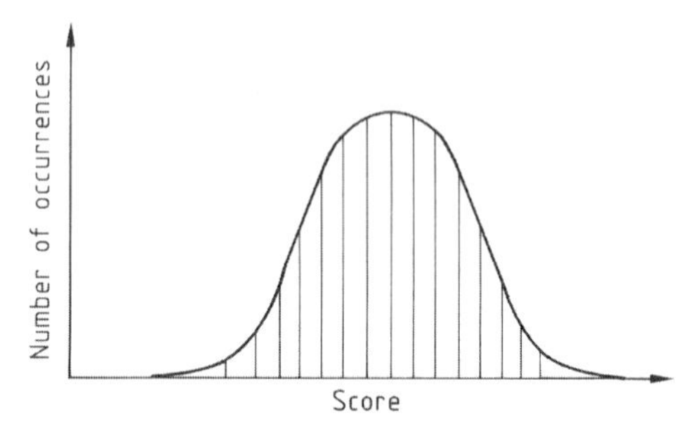

Figure 49. Frequency of occurrence of different scores

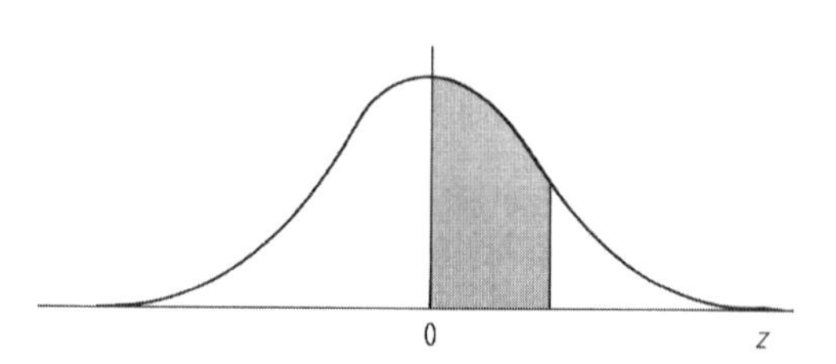

Figure 50. Area under normal curve

For a small sample, the variance can only be estimated with the sample variance s^2. Thus, the normal distribution cannot be used because σ is not known. In such cases *Student's t-distribution*, shown in Figure 51 [303, p. 70], is used:

$$p(y)=\frac{y_0}{\left(1+\frac{t^2}{n-1}\right)^{n/2}},\ t=\frac{\overline{y}-E(y)}{s/\sqrt{n}}$$

and y_0 is chosen such that the area under the curve is one. The number $\nu=n-1$ is the degrees of freedom, and as ν increases, Student's t-distribution approaches the normal distribution. The normal distribution is adequate (rather than the t-distribution) when $\nu>15$, except for the tails of the curve which require larger ν. Integrals of the t-distribution are given in Table 13, the region of interest is shown in Figure 52.

Table 12. Area under normal curve *

$$F(z) = \frac{1}{\sqrt{2\pi}} \int_0^z e^{-z^2/2} dz$$

z	$F(z)''$	z	$F(z)''$
0.0	0.0000	1.5	0.4332
0.1	0.0398	1.6	0.4452
0.2	0.0793	1.7	0.4554
0.3	0.1179	1.8	0.4641
0.4	0.1554	1.9	0.4713
0.5	0.1915	2.0	0.4772
0.6	0.2257	2.1	0.4821
0.7	0.2580	2.2	0.4861
0.8	0.2881	2.3	0.4893
0.9	0.3159	2.4	0.4918
1.0	0.3413	2.5	0.4938
1.1	0.3643	2.7	0.4965
1.2	0.3849	3.0	0.4987
1.3	0.4032	4.0	0.499968
1.4	0.4192	5.0	0.4999997

* Table gives the probability F that a random variable will fall in the shaded region of Figure 50. For a more complete table (in slightly different form), see [23, Table 26.1]. This table is obtained in Microsoft Excel with the function NORMDIST(z,0,1,1)-0.5.

Table 13. Percentage points of area under Students t-distribution *

ν	α=0.10	α=0.05	α=0.01	α=0.001
1	6.314	12.706	63.657	636.619
2	2.920	4.303	9.925	31.598
3	2.353	3.182	5.841	12.941
4	2.132	2.776	4.604	8.610
5	2.015	2.571	4.032	6.859
6	1.943	2.447	3.707	5.959
7	1.895	2.365	3.499	5.408
8	1.860	2.306	3.355	5.041
9	1.833	2.262	3.250	4.781
10	1.812	2.228	3.169	4.587
15	1.753	2.131	2.947	4.073
20	1.725	2.086	2.845	3.850
25	1.708	2.060	2.787	3.725
30	1.697	2.042	2.750	3.646
∞	1.645	1.960	2.576	3.291

* Table gives t values such that a random variable will fall in the shaded region of Figure 52 with probability α. For a one-sided test the confidence limits are obtained for α/2. For a more complet table (in slightly different form), see [23, Table 26.10]. This table is obtained in Microsoft Excel with the function TINV(α,ν).

Other probability distribution functions are useful. Any distribution function must satisfy the following conditions:

$$0 \leq F(x) \leq 1$$

$$F(-\infty) = 0, F(+\infty) = 1$$

$$F(x) \leq F(y) \quad \text{when } x \leq y$$

The *probability density function* is

$$p(x) = \frac{dF(x)}{dx}$$

where

$$dF = p dx$$

is the probability of x being between x and x + dx. The probability density function satisfies

$$p(x) \geq 0$$

$$\int_{-\infty}^{\infty} p(x) dx = 1$$

The *Bernoulli distribution* applies when the outcome can take only two values, such as heads or tails, or 0 or 1. The probability distribution function is

$$p(x = k) = p^k (1-p)^{1-k}, \; k = 0 \text{ or } 1$$

and the mean of a function $g(x)$ depending on x is

$$E[g(x)] = g(1)p + g(0)(1-p)$$

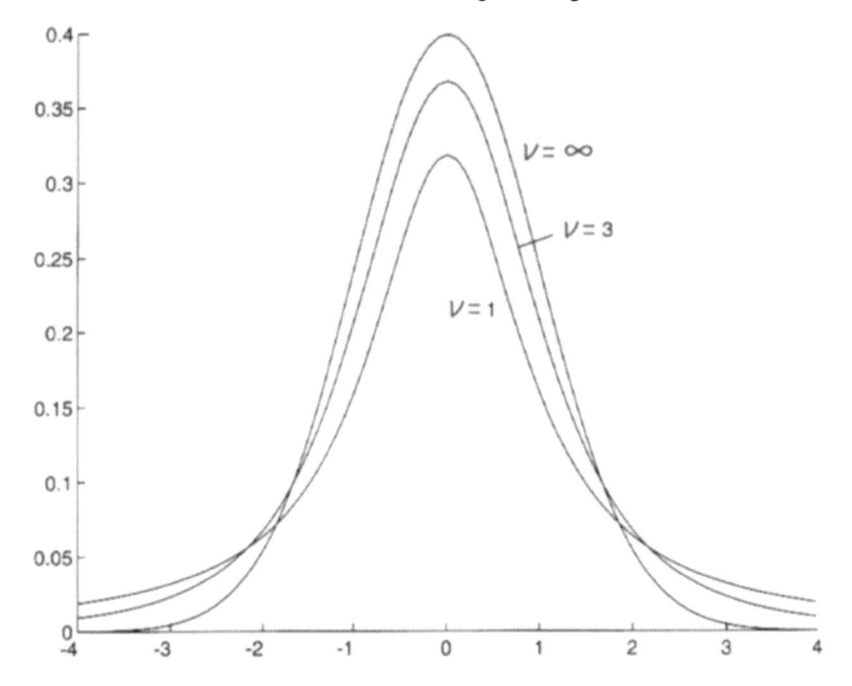

Figure 51. Student's t-distribution.
For explanation of ν see text

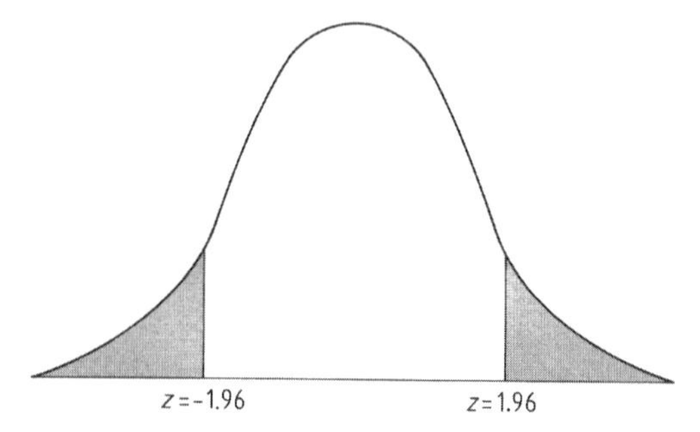

Figure 52. Percentage points of area under Student's t-distribution

The *binomial distribution function* applies when there are n trials of a Bernoulli event; it gives the probability of k occurrences of the event, which occurs with probability p on each trial

$$p(x=k)=\frac{n!}{k(n-k)!}p^k(1-p)^{n-k}$$

The mean and variance are

$$E(x)=np$$

$$var(x)=np(1-p)$$

The *hypergeometric distribution function* applies when there are N objects, of which M are of one kind and $N-M$ are of another kind. Then the objects are drawn one by one, without replacing the last draw. If the last draw had been replaced the distribution would be the binomial distribution. If x is the number of objects of type M drawn in a sample of size n, then the probability of $x = k$ is

$$p(x=k)=$$

$$\frac{M!(N-M)!n!(N-n)!}{k!(M-k)!(n-k)!(N-M-n+k)!N!}$$

The mean and variance are

$$E(x)=\frac{nM}{N}$$

$$var(x)=np(1-p)\frac{N-n}{N-1}$$

The *Poisson distribution* is

$$p(x=k)=\mathrm{e}^{-\lambda}\frac{\lambda^k}{k!}$$

with a parameter λ. The mean and variance are

$$E(x)=\lambda$$

$$var(x)=\lambda$$

The simplest continuous distribution is the uniform distribution. The probability density function is

$$p=\begin{cases}\frac{1}{b-a} & a<x<b\\ 0 & x<a,\ x>b\end{cases}$$

and the probability distribution function is

$$F(x)=\begin{cases}0 & x<a\\ \frac{x-a}{b-a} & a<x<b\\ 1 & b<x\end{cases}$$

The mean and variance are

$$E(x)=\frac{a+b}{2}$$

$$var(x)=\frac{(b-a)^2}{12}$$

The *normal distribution* is given by Equation 120 with variance σ^2.

The *log normal probability density* function is

$$p(x)=\frac{1}{x\sigma\sqrt{2\pi}}\exp\left[-\frac{(\log x-\mu)^2}{2\sigma^2}\right]$$

and the mean and variance are [305, p. 89]

$$E(x)=\exp\left(\mu+\frac{\sigma^2}{2}\right)$$

$$var(x)=\exp\left(\sigma^2-1\right)\exp\left(2\mu+\sigma^2\right)$$

11.2. Sampling and Statistical Decisions

Two variables can be statistically dependent or independent. For example, the height and diameter of all distillation towers are statistically independent, because the distribution of diameters of all columns 10 m high is different from that of columns 30 m high. If y_B is the diameter and y_A the height, the distribution is written as

$$p(y_B|y_A=\text{constant}),\ \text{or here}$$

$$p(y_B|y_A=10)\neq p(y_B|y_A=30)$$

A third variable y_C, could be the age of the operator on the third shift. This variable is probably unrelated to the diameter of the column, and for the distribution of ages is

$$p(y_C|y_A)=p(y_C)$$

Thus, variables y_A and y_C are distributed independently. The joint distribution for two variables is

$$p(y_A,y_B)=p(y_A)\,p(y_B|y_A)$$

if they are statistically dependent, and

$$p(y_A,y_B)=p(y_A)\,p(y_B)$$

if they are statistically independent. If a set of variables y_A, y_B, . . . is independent and identically distributed,

$$p(y_A,y_B,\ldots)=p(y_A)\,p(y_B)\ldots$$

Conditional probabilities are used in hazard analysis of chemical plants.

A measure of the linear dependence between variables is given by the *covariance*

$$Cov(y_A,y_B)=E\{[y_A-E(y_A)]\,[(y_B-E(y_B))]\}$$

$$=\frac{\sum_{i=1}^{N}[y_{Ai}-E(y_A)][y_{Bi}-E(y_B)]}{N}$$

The correlation coefficient ϱ is

$$\varrho(y_A,y_B)=\frac{Cov(y_A,y_B)}{\sigma_A\sigma_B}$$

If y_A and y_B are independent, then $Cov(y_A, y_B)=0$. If y_A tends to increase when y_B decreases then $Cov(y_A, y_B)<0$. The sample correlation coefficient is [15, p. 484]

$$r(y_A,y_B)=\frac{\sum_{i=1}^{n}(y_{Ai}-\overline{y}_A)(y_{Bi}-\overline{y}_B)}{(n-1)\,s_A s_B}$$

If measurements are for independent, identically distributed observations, the errors are independent and uncorrelated. Then $\overline{y}$ varies about $E(y)$ with variance σ^2/n, where n is the number of observations in $\overline{y}$. Thus if something is measured several times today and every day, and the measurements have the same distribution, the variance of the means decreases with the number of samples in each day's measurement n. Of course, other factors (weather, weekends) may cause the observations on different days to be distributed nonidentically.

Suppose Y, which is the sum or difference of two variables, is of interest:

$$Y=y_A\pm y_B$$

Then the mean value of Y is

$$E(Y)=E(y_A)\pm E(y_B)$$

and the variance of Y is

$$\sigma^2(Y)=\sigma^2(y_A)+\sigma^2(y_B)$$

More generally, consider the random variables $y_1, y_2, \ldots$ with means $E(y_1), E(y_2), \ldots$ and variances $\sigma^2(y_1), \sigma^2(y_2), \ldots$ and correlation coefficients ϱ_{ij}. The variable

$$Y = \alpha_1 y_1 + \alpha_2 y_2 + \ldots$$

has a mean

$$E(Y) = \alpha_1 E(y_1) + \alpha_2 E(y_2) + \ldots$$

and variance [303, p. 87]

$$\sigma^2(Y) = \sum_{i=1}^{n} \alpha_i^2 \sigma^2(y_i) + 2\sum_{i=1}^{n} \sum_{j=i+1}^{n} \alpha_i \alpha_j \sigma(y_i)\,\sigma(y_j)\,\varrho_{ij}$$

or

$$\sigma^2(Y) = \sum_{i=1}^{n} \alpha_i^2 \sigma^2(y_i) + 2\sum_{i=1}^{n} \sum_{j=i+1}^{n} \alpha_i \alpha_j Cov(y_i, y_j) \qquad (120)$$

If the variables are uncorrelated and have the same variance, then

$$\sigma^2(Y) = \left(\sum_{i=1}^{n} \alpha_i^2\right) \sigma^2$$

This fact can be used to obtain more accurate cost estimates for the purchased cost of a chemical plant than is true for any one piece of equipment. Suppose the plant is composed of a number of heat exchangers, pumps, towers, etc., and that the cost estimate of each device is $\pm$ 40 % of its cost (the sample standard deviation is 20 % of its cost). In this case the α_i are the numbers of each type of unit. Under special conditions, such as equal numbers of all types of units and comparable cost, the standard deviation of the plant costs is

$$\sigma(Y) = \frac{\sigma}{\sqrt{n}}$$

and is then $\pm\,(40/\sqrt{n})$ %. Thus the standard deviation of the cost for the entire plant is the standard deviation of each piece of equipment divided by the square root of the number of units. Under less restrictive conditions the actual numbers change according to the above equations, but the principle is the same.

Suppose modifications are introduced into the manufacturing process. To determine if the modification causes a significant change, the mean of some property could be measured before and after the change; if these differ, does it mean the process modification caused it, or could the change have happened by chance? This is a statistical decision. A hypothesis H_0 is defined; if it is true, action A must be taken. The reverse hypothesis is H_1; if this is true, action B must be taken. A correct decision is made if action A is taken when H_0 is true or action B is taken when H_1 is true. Taking action B when H_0 is true is called a type I error, whereas taking action A when H_1 is true is called a type II error.

The following test of hypothesis or test of significance must be defined to determine if the hypothesis is true. The level of significance is the maximum probability that an error would be accepted in the decision (i.e., rejecting the hypothesis when it is actually true). Common levels of significance are 0.05 and 0.01, and the test of significance can be either one or two sided. If a sampled distribution is normal, then the probability that the z score

$$z = \frac{y - \overline{y}}{s_y}$$

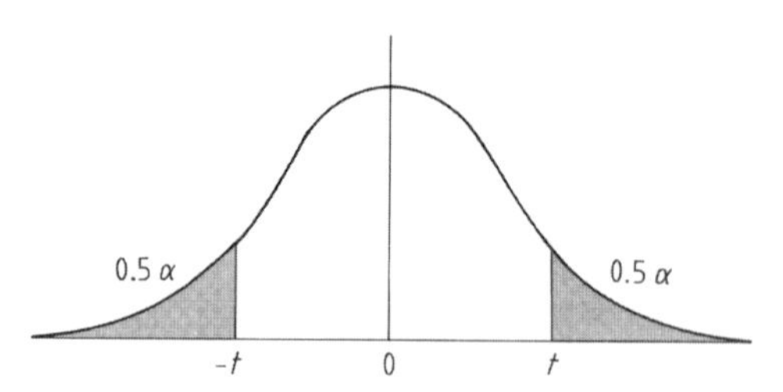

Figure 53. Two-sided statistical decision

is in the unshaded region is 0.95. Because a two-sided test is desired, $F = 0.95/2 = 0.475$. The value given in Table 12 for $F = 0.475$ is $z = 1.96$. If the test was one-sided, at the 5 % level of significance, 0.95 = 0.5 (for negative z) + F (for positive z). Thus, $F = 0.45$ or $z = 1.645$. In the two-sided test (see Fig. 53), if a single sample is chosen and $z < -1.96$ or $z > 1.96$, then this could happen with probability 0.05 if the hypothesis were true. This z would be significantly different from the expected value (based on the chosen level of significance) and the tendency would be to reject the hypothesis. If the value of z was between $-$ 1.96 and 1.96, the hypothesis would be accepted.

The same type of decisions can be made for other distributions. Consider Student's t-distribution. At a 95 % level of confidence, with ν = 10 degrees of freedom, the t values are $\pm$ 2.228. Thus, the sample mean would be expected to be between

$$\overline{y}\pm t_c\frac{s}{\sqrt{n}}$$

with 95 % confidence. If the mean were outside this interval, the hypothesis would be rejected.

The *chi-square distribution* is useful for examining the variance or standard deviation. The statistic is defined as

$$\chi^2=\frac{ns^2}{\sigma^2}$$

$$=\frac{(y_1-\overline{y})^2+(y_2-\overline{y})^2+\ldots+(y_n-\overline{y})^2}{\sigma^2}$$

and the chi-square distribution is

$$p\,(y)=y_0\chi^{\nu-2}\mathrm{e}^{-\chi^2/2}$$

$\nu = n - 1$ is the number of degrees of freedom and y_0 is chosen so that the integral of p (y) over all y is 1. The probability of a deviation larger than χ^2 is given in Table 14; the area in question, in Figure 54. For example, for 10 degrees of freedom and a 95 % confidence level, the critical values of χ^2 are 0.025 and 0.975. Then

$$\frac{s\sqrt{n}}{\chi_{0.975}}<\sigma<\frac{s\sqrt{n}}{\chi_{0.025}}$$

or

$$\frac{s\sqrt{n}}{20.5}<\sigma<\frac{s\sqrt{n}}{3.25}$$

with 95 % confidence.

Tests are available to decide if two distributions that have the same variance have different means [15, p. 465]. Let one distribution be called x, with N_1 samples, and the other be called y, with N_2 samples. First, compute the standard error of the difference of the means:

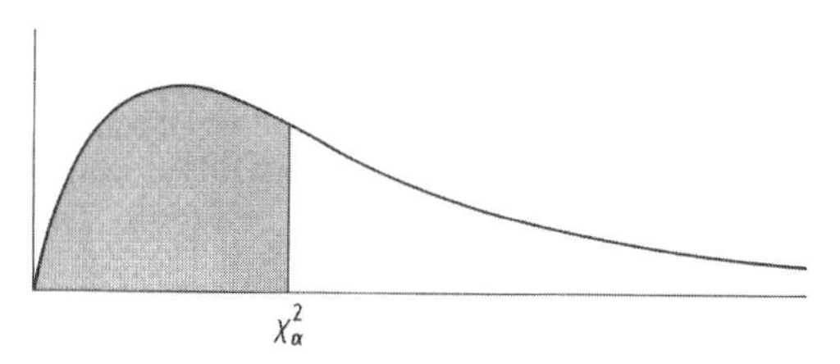

Figure 54. Percentage points of area under chi-squared distribution with ν degrees of freedom

$$s_D=\sqrt{\frac{\sum\limits_{i=1}^{N_1}(x_i-\overline{x})^2+\sum\limits_{i=1}^{N_2}(y_i-\overline{y})^2}{N_1+N_2-2}\left(\frac{1}{N_1}+\frac{1}{N_2}\right)}$$

Next, compute the value of t

$$t=\frac{\overline{x}-\overline{y}}{s_D}$$

and evaluate the significance of t using Student's t-distribution for N_1+N_2-2 degrees of freedom.

If the samples have different variances, the relevant statistic for the t-test is

$$t=\frac{\overline{x}-\overline{y}}{\sqrt{var\,(x)\,/N_1+var\,(y)\,/N_2}}$$

The number of degrees of freedom is now taken approximately as

$$\nu=\frac{\left(\frac{var(x)}{N_1}+\frac{var(y)}{N_2}\right)^2}{\frac{[var(x)/N_1]^2}{N_1-1}+\frac{[var(y)/N_2]^2}{N_2-1}}$$

There is also an F-test to decide if two distributions have significantly different variances. In this case, the ratio of variances is calculated:

$$F=\frac{var\,(x)}{var\,(y)}$$

where the variance of x is assumed to be larger. Then, a table of values is used to determine the significance of the ratio. The table [23, Table 26.9] is derived from the formula [15, p. 169]

$$Q\,(F|\nu_1,\nu_2)=I_{\nu_2/(\nu_2+\nu_1F)}\left(\frac{\nu_2}{2},\frac{\nu_1}{2}\right)$$

where the right-hand side is an incomplete beta function. The F table is given by the Microsoft Excel function FINV(fraction, ν_x, ν_y), where fraction is the fractional value ($\leq$ 1) representing the upper percentage and ν_x and ν_y are the degrees of freedom of the numerator and denominator, respectively.

Example. For two sample variances with 8 degrees of freedom each, what limits will bracket their ratio with a midarea probability of 90 %? FINV(0.95,8,8) = 3.44. The 0.95 is used to get both sides to toal 10 %. Then

$$P\,[1/3.44\leq\mathrm{var}\,(x)\,/\mathrm{var}\,(y)\leq 3.44]=0.90.$$

Table 14. Percentage points of area under chi-square distribution with ν degrees of freedom *

ν	α=0.995	α=0.99	α=0.975	α=0.95	α=0.5	α=0.05	α=0.025	α=0.01	α=0.005
1	7.88	6.63	5.02	3.84	0.455	0.0039	0.0010	0.0002	0.00004
2	10.6	9.21	7.38	5.99	1.39	0.103	0.0506	0.0201	0.0100
3	12.8	11.3	9.35	7.81	2.37	0.352	0.216	0.115	0.072
4	14.9	13.3	11.1	9.49	3.36	0.711	0.484	0.297	0.207
5	16.7	15.1	12.8	11.1	4.35	1.15	0.831	0.554	0.412
6	18.5	16.8	14.4	12.6	5.35	1.64	1.24	0.872	0.676
7	20.3	18.5	16.0	14.1	6.35	2.17	1.69	1.24	0.989
8	22.0	20.1	17.5	15.5	7.34	2.73	2.18	1.65	1.34
9	23.6	21.7	19.0	16.9	8.34	3.33	2.70	2.09	1.73
10	25.2	23.2	20.5	18.3	9.34	3.94	3.25	2.56	2.16
12	28.3	26.2	23.3	21.0	11.3	5.23	4.40	3.57	3.07
15	32.8	30.6	27.5	25.0	14.3	7.26	6.26	5.23	4.60
17	35.7	33.4	30.2	27.6	16.3	8.67	7.56	6.41	5.70
20	40.0	37.6	34.2	31.4	19.3	10.9	9.59	8.26	7.43
25	46.9	44.3	40.6	37.7	24.3	14.6	13.1	11.5	10.5
30	53.7	50.9	47.0	43.8	29.3	18.5	16.8	15.0	13.8
40	66.8	63.7	59.3	55.8	39.3	26.5	24.4	22.2	20.7
50	79.5	76.2	71.4	67.5	49.3	34.8	32.4	29.7	28.0
60	92.0	88.4	83.3	79.1	59.3	43.2	40.5	37.5	35.5
70	104.2	100.4	95.0	90.5	69.3	51.7	48.8	45.4	43.3
80	116.3	112.3	106.6	101.9	79.3	60.4	57.2	53.5	51.2
90	128.3	124.1	118.1	113.1	89.3	69.1	65.6	61.8	59.2
100	140.2	135.8	129.6	124.3	99.3	77.9	74.2	70.1	67.3

* Table value is χ^2_α; $\chi^2 < \chi^2_\alpha$ with probability α. For a more complete table (in slightly different form), see [23, Table 26.8]. The Microsoft Excel function CHIINV(1-α,ν) gives the table value.

11.3. Error Analysis in Experiments

Suppose a measurement of several quantities is made and a formula or mathematical model is used to deduce some property of interest. For example, to measure the thermal conductivity of a solid k, the heat flux q, the thickness of the sample d, and the temperature difference across the sample ΔT must be measured. Each measurement has some error. The heat flux q may be the rate of electrical heat input $\dot{Q}$ divided by the area A, and both quantities are measured to some tolerance. The thickness of the sample is measured with some accuracy, and the temperatures are probably measured with a thermocouple, to some accuracy. These measurements are combined, however, to obtain the thermal conductivity, and the error in the thermal conductivity must be determined. The formula is

$$k = \frac{d}{A \Delta T} \dot{Q}$$

If each measured quantity has some variance, what is the variance in the thermal conductivity?

Suppose a model for Y depends on various measurable quantities, $y_1, y_2, \ldots$ Suppose several measurements are made of $y_1, y_2, \ldots$ under seemingly identical conditions and several different values are obtained, with means $E\,(y_1)$, $E\,(y_2)$, . . . and variances $\sigma_1^2, \sigma_2^2, \ldots$ Next suppose the errors are small and independent of one another. Then a change in Y is related to changes in y_i by

$$\mathrm{d}Y = \frac{\partial Y}{\partial y_1} \mathrm{d}y_1 + \frac{\partial Y}{\partial y_2} \mathrm{d}y_2 + \ldots$$

If the changes are indeed small, the partial derivatives are constant among all the samples. Then the expected value of the change is

$$E\,(\mathrm{d}Y) = \sum_{i=1}^{N} \left(\frac{\partial Y}{\partial y_i} \right) E\,(\mathrm{d}y_i)$$

Naturally $E\,(\mathrm{d}y_i) = 0$ by definition so that $E\,(\mathrm{d}Y) = 0$, too. However, since the errors are independent of each other and the partial derivatives are assumed constant because the errors are small, the variances are given by Equation 121 [296, p. 550]

$$\sigma^2\,(\mathrm{d}Y) = \sum_{i=1}^{N} \left(\frac{\partial Y}{\partial y_i} \right)^2 \sigma_i^2 \tag{121}$$

Thus, the variance of the desired quantity Y can be found. This gives an independent estimate of the errors in measuring the quantity Y from the errors in measuring each variable it depends upon.

11.4. Factorial Design of Experiments and Analysis of Variance

Statistically designed experiments consider, of course, the effect of primary variables, but they also consider the effect of extraneous variables, the interactions among variables, and a measure of the random error. Primary variables are those whose effect must be determined. These variables can be quantitative or qualitative. Quantitative variables are ones that may be fit to a model to determine the model parameters. Curve fitting of this type is discused in Chapter 2. Qualitative variables are ones whose effect needs to be known; no attempt is made to quantify that effect other than to assign possible errors or magnitudes. Qualitative variables can be further subdivided into type I variables, whose effect is determined directly, and type II variables, which contribute to performance variability, and whose effect is averaged out. For example, in studying the effect of several catalysts on yield in a chemical reactor, each different type of catalyst would be a type I variable, because its effect should be known. However, each time the catalyst is prepared, the results are slightly different, because of random variations; thus, several batches may exist of what purports to be the same catalyst. The variability between batches is a type II variable. Because the ultimate use will require using different batches, the overall effect including that variation should be known, because knowing the results from one batch of one catalyst precisely might not be representative of the results obtained from all batches of the same catalyst. A randomized block design, incomplete block design, or Latin square design, for example, all keep the effect of experimental error in the blocked variables from influencing the effect of the primary variables. Other uncontrolled variables are accounted for by introducing randomization in parts of the experimental design. To study all variables and their interaction requires a factorial design, involving all possible combinations of each variable, or a fractional factorial design, involving only a selected set. Statistical techniques are then used to determine the important variables, the important interactions and the error in estimating these effects. The discussion here is a brief overview of [303].

If only two methods exist for preparing some product, to see which treatment is best, the sampling analysis discussed in Section 11.2 can be used to deduce if the means of the two treatments differ significantly. With more treatments, the analysis is more detailed. Suppose the experimental results are arranged as shown in Table 15, i.e., several measurements for each treatment. The objective is to see if the treatments differ significantly from each other, that is, whether their means are different. The samples are assumed to have the same variance. The hypothesis is that the treatments are all the same, and the null hypothesis is that they are different. Deducing the statistical validity of the hypothesis is done by an analysis of variance.

Table 15. Estimating the effect of four treatments

Treatment	1	2	3	4
	–	–	–	–
	–	–	–	–
	–	–	–	–
		–	–	–
			–	–
				–
Treatment average, $\overline{y}_t$	–	–	–	–
Grand average, $\overline{y}$		–		

The data for $k = 4$ treatments are arranged in Table 15. Each treatment has n_t experiments, and the outcome of the i-th experiment with treatment t is called y_{ti}. The treatment average is

$$\overline{y}_t = \frac{\sum\limits_{i=1}^{n_t} y_{ti}}{n_t}$$

and the grand average is

$$\overline{y} = \frac{\sum\limits_{t=1}^{k} n_t \overline{y}_t}{N}, \quad N = \sum_{t=1}^{k} n_t$$

Next, the sum of squares of deviations is computed from the average within the t-th treatment

$$S_t = \sum_{i=1}^{n_t} (y_{ti} - \overline{y}_t)^2$$

Since each treatment has n_t experiments, the number of degrees of freedom is $n_t - 1$. Then the sample variances are

$$s_t^2 = \frac{S_t}{n_t - 1}$$

The within-treatment sum of squares is

$$S_R = \sum_{t=1}^{k} S_t$$

and the within-treatment sample variance is

$$s_R^2 = \frac{S_R}{N-k}$$

Now, if no difference exists between treatments, a second estimate of σ^2 could be obtained by calculating the variation of the treatment averages about the grand average. Thus, the between-treatment mean square is computed:

$$s_T^2 = \frac{S_T}{k-1},\ S_T = \sum_{t=1}^{k} n_t (\overline{y}_t - \overline{y})^2$$

Basically the test for whether the hypothesis is true or not hinges on a comparison between the within-treatment estimate s_R^2 (with $\nu_R = N - k$ degrees of freedom) and the between-treatment estimate s_T^2 (with $\nu_T = k - 1$ degrees of freedom). The test is made based on the F distribution for ν_R and ν_T degrees of freedom [23, Table 26.9], [303, p. 636].

Table 16. Block design with four treatments and five blocks

Treatment	1	2	3	4	Block average
Block 1	–	–	–	–	–
Block 2	–	–	–	–	–
Block 3	–	–	–	–	–
Block 4	–	–	–	–	–
Block 5	–	–	–	–	–
Treatment average	–	–	–	–	grand average

Next consider the case in which *randomized blocking* is used to eliminate the effect of some variable whose effect is of no interest, such as the batch-to-batch variation of the catalysts in the chemical reactor example. With k treatments and n experiments in each treatment, the results from $n\,k$ experiments can be arranged as shown in Table 16; within each block, various treatments are applied in a random order. The block average, the treatment average, and the grand average are computed as before. The following quantities are also computed for the analysis of variance table:

Name	Formula	Degrees of freedom
Average	$S_A = nk\overline{y}^2$	1
Blocks	$S_B = k\sum_{i=1}^{n}(\overline{y}_t - \overline{y})^2$	n-1
Treatments	$S_T = n\sum_{t=1}^{k}(\overline{y}_t - \overline{y})^2$	k-1
Residuals	$S_R = \sum_{t=1}^{k}\sum_{i=1}^{n}(y_{ti} - \overline{y}_i - \overline{y}_t + \overline{y})^2$	(n-1)(k-1)
Total	$S = \sum_{t=1}^{k}\sum_{i=1}^{n} y_{ti}^2$	$N = n\,k$

The key test is again a statistical one, based on the value of

s_T^2/s_R^2, where $s_T^2 = \frac{S_T}{k-1}$

and $s_R^2 = \frac{S_R}{(n-1)(k-1)}$

and the F distribution for ν_R and ν_T degrees of freedom [303, p. 636]. The assumption behind the analysis is that the variations are linear [303, p. 218]. Ways to test this assumption as well as transformations to make if it is not true are provided in [303], where an example is given of how the observations are broken down into a grand average, a block deviation, a treatment deviation, and a residual. For two-way factorial design, in which the second variable is a real one rather than one you would like to block out, see [303, p. 228].

Table 17. Two-level factorial design with three variables

Run	Variable 1	Variable 2	Variable 3
1	–	–	–
2	+	–	–
3	–	+	–
4	+	+	–
5	–	–	+
6	+	–	+
7	–	+	+
8	+	+	+

To measure the effects of variables on a single outcome, a *factorial design* is appropriate. In a two-level factorial design, each variable is considered at two levels only, a high and low value, often designated as a + and a –. The two-level factorial design is useful for indicating trends and showing interactions; it is also the basis for a fractional factorial design. As an example, consider a 2^3 factorial design, with 3 variables and 2 levels for each. The experiments are indicated in Table 17. The main effects are calculated by determining the difference between results from all high values of a variable and all low values of a variable; the result is divided by the number of experiments at each level. For example, for the first variable, calculate

$$\text{Effect of variable } 1 = [(y_2+y_4+y_6+y_8) - [(y_1+y_3+y_5+y_7)]]/4$$

Note that all observations are being used to supply information on each of the main effects and each effect is determined with the precision of a fourfold replicated difference. The advantage of a one-at-a-time experiment is the gain in precision if the variables are additive and the measure of nonadditivity if it occurs [303, p. 313].

Interaction effects between variables 1 and 2 are obtained by comparing the difference between the results obtained with the high and low value of 1 at the low value of 2 with the difference between the results obtained with the high and low value 1 at the high value of 2. The 12-interaction is

$$12 \text{ interaction} = [(y_4-y_3+y_8-y_7) - [(y_2-y_1+y_6-y_5)]]/2$$

The key step is to determine the errors associated with the effect of each variable and each interaction so that the significance can be determined. Thus, standard errors need to be assigned. This can be done by repeating the experiments, but it can also be done by using higher order interactions (such as 123 interactions in a 2^4 factorial design). These are assumed negligible in their effect on the mean but can be used to estimate the standard error [303, pp. 319 – 328]. Then calculated effects that are large compared to the standard error are considered important, whereas those that are small compared to the standard error are considered due to random variations and are unimportant.

In a fractional factorial design, only part of the possible experiments is performed. With k variables, a factorial design requires 2^k experiments. When k is large, the number of experiments can be large; for $k = 5$, $2^5 = 32$. For k this large, Box et al. [296, p. 235] do a fractional factorial design. In the fractional factorial design with $k = 5$, only 8 experiments are chosen. Cropley [298] gives an example of how to combine heuristics and statistical arguments in application to kinetics mechanisms in chemical engineering.

12. Multivariable Calculus Applied to Thermodynamics

Many of the functional relationships required in thermodynamics are direct applications of the rules of multivariable calculus. In this short chapter, those rules are reviewed in the context of the needs of thermodynamics. These ideas were expounded in one of the classic books on chemical engineering thermodynamics [299].

12.1. State Functions

State functions depend only on the state of the system, not on its past history or how one got there. If z is a function of two variables x and y, then $z(x, y)$ is a state function, because z is known once x and y are specified. The differential of z is

$$\mathrm{d}z = M\mathrm{d}x + N\mathrm{d}y$$

The line integral

$$\int_C (M\mathrm{d}x + N\mathrm{d}y)$$

is independent of the path in $x-y$ space if and only if

$$\frac{\partial M}{\partial y} = \frac{\partial N}{\partial x} \tag{122}$$

Because the total differential can be written as

$$\mathrm{d}z = \left(\frac{\partial z}{\partial x}\right)_y \mathrm{d}x + \left(\frac{\partial z}{\partial y}\right)_x \mathrm{d}y \tag{123}$$

for path independence

$$\frac{\partial}{\partial y}\left(\frac{\partial z}{\partial x}\right)_y = \frac{\partial}{\partial x}\left(\frac{\partial z}{\partial y}\right)_x$$

or

$$\frac{\partial^2 z}{\partial y \partial x} = \frac{\partial^2 z}{\partial x \partial y} \tag{124}$$

is needed.

Various relationships can be derived from Equation 123. If z is constant,

$$\left[0 = \left(\frac{\partial z}{\partial x}\right)_y \mathrm{d}x + \left(\frac{\partial z}{\partial y}\right)_x \mathrm{d}y\right]_z$$

Rearrangement gives

$$\left(\frac{\partial z}{\partial x}\right)_y=-\left(\frac{\partial y}{\partial x}\right)_z\left(\frac{\partial z}{\partial y}\right)_x=-\frac{(\partial y/\partial x)_z}{(\partial y/\partial z)_x} \quad (125)$$

Alternatively, if Equation 123 is divided by d*y* while some other variable *w* is held constant,

$$\left(\frac{\partial z}{\partial y}\right)_w=\left(\frac{\partial z}{\partial x}\right)_y\left(\frac{\partial x}{\partial y}\right)_w+\left(\frac{\partial z}{\partial y}\right)_x \quad (126)$$

Dividing both the numerator and the denominator of a partial derivative by d*w* while holding a variable *y* constant yields

$$\left(\frac{\partial z}{\partial x}\right)_y=\frac{(\partial z/\partial w)_y}{(\partial x/\partial w)_y}=\left(\frac{\partial z}{\partial w}\right)_y\left(\frac{\partial w}{\partial x}\right)_y \quad (127)$$

In thermodynamics the state functions include the internal energy *U*, the enthalpy *H*, and the Helmholtz and Gibbs free energies *A* and *G*, respectively, which are defined as follows:

$$H=U+pV$$

$$A=U-TS$$

$$G=H-TS=U+pV-TS=A+pV$$

where *S* is the entropy, *T* the absolute temperature, *p* the pressure, and *V* the volume. These are also state functions, in that the entropy is specified once two variables (e.g., *T* and *p*) are specified. Likewise *V* is specified once *T* and *p* are specified, and so forth.

12.2. Applications to Thermodynamics

All of the following applications are for closed systems with constant mass. If a process is reversible and only *p* – *V* work is done, one form of the first law states that changes in the internal energy are given by the following expression

$$dU=T\mathrm{d}S-p\mathrm{d}V \quad (128)$$

If the internal energy is considered a function of *S* and *V*, then

$$\mathrm{d}U=\left(\frac{\partial U}{\partial S}\right)_V\mathrm{d}S+\left(\frac{\partial U}{\partial V}\right)_S\mathrm{d}V$$

This is the equivalent of Equation 123 and

$$T=\left(\frac{\partial U}{\partial S}\right)_V,\ p=-\left(\frac{\partial U}{\partial V}\right)_S$$

Because the internal energy is a state function, Equation 124 is required:

$$\frac{\partial^2 U}{\partial V\partial S}=\frac{\partial^2 U}{\partial S\partial V}$$

which here is

$$\left(\frac{\partial T}{\partial V}\right)_S=-\left(\frac{\partial p}{\partial S}\right)_V \quad (129)$$

This is one of the Maxwell relations and is merely an expression of Equation 124.

The differentials of the other energies are

$$dH=T\,\mathrm{d}S+V\,\mathrm{d}p \quad (130)$$

$$dA=-S\mathrm{d}T-p\mathrm{d}V \quad (131)$$

$$dG=-S\,\mathrm{d}T+V\,\mathrm{d}p \quad (132)$$

From these differentials, other Maxwell relations can be derived in a similar fashion by applying Equation 124.

$$\left(\frac{\partial T}{\partial p}\right)_S=\left(\frac{\partial V}{\partial S}\right)_p \quad (133)$$

$$\left(\frac{\partial S}{\partial V}\right)_T=\left(\frac{\partial p}{\partial T}\right)_V \quad (134)$$

$$\left(\frac{\partial S}{\partial p}\right)_T=-\left(\frac{\partial V}{\partial T}\right)_p \quad (135)$$

The heat capacity at constant pressure is defined as

$$C_p=\left(\frac{\partial H}{\partial T}\right)_p$$

If entropy and enthalpy are taken as functions of *T* and *p*, the total differentials are

$$\mathrm{d}S=\left(\frac{\partial S}{\partial T}\right)_p\mathrm{d}T+\left(\frac{\partial S}{\partial p}\right)_T\mathrm{d}p$$

$$\mathrm{d}H=\left(\frac{\partial H}{\partial T}\right)_p\mathrm{d}T+\left(\frac{\partial H}{\partial p}\right)_T\mathrm{d}p$$

$$=C_p\mathrm{d}T+\left(\frac{\partial H}{\partial p}\right)_T\mathrm{d}p$$

If the pressure is constant,

$$\mathrm{d}S=\left(\frac{\partial S}{\partial T}\right)_p\mathrm{d}T \text{ and } \mathrm{d}H=C_p\mathrm{d}T$$

When enthalpy is considered a function of *S* and *p*, the total differential is

$$\mathrm{d}H=T\,\mathrm{d}S+V\,\mathrm{d}p$$

When the pressure is constant, this is

$$\mathrm{d}H=T\mathrm{d}S$$

Thus, at constant pressure

$$\mathrm{d}H=C_p\mathrm{d}T=T\mathrm{d}S=T\left(\frac{\partial S}{\partial T}\right)_p\mathrm{d}T$$

which gives

$$\left(\frac{\partial S}{\partial T}\right)_p=\frac{C_p}{T}$$

When p is not constant, using the last Maxwell relation gives

$$\mathrm{d}S=\frac{C_p}{T}\mathrm{d}T-\left(\frac{\partial V}{\partial T}\right)_p\mathrm{d}p \tag{136}$$

Then the total differential for H is

$$\mathrm{d}H=T\mathrm{d}S+V\mathrm{d}p=C_p\mathrm{d}T-T\left(\frac{\partial V}{\partial T}\right)_p\mathrm{d}p+V\mathrm{d}p$$

Rearranging this, when H (T, p), yields

$$\mathrm{d}H=C_p\mathrm{d}T+\left[V-T\left(\frac{\partial V}{\partial T}\right)_p\right]\mathrm{d}p \tag{137}$$

This equation can be used to evaluate enthalpy differences by using information on the equation of state and the heat capacity:

$$\begin{aligned}H\left(T_2,p_2\right)-H\left(T_1,p_1\right)&=\int\limits_{T_1}^{T_2}C_p\left(T,p_1\right)\mathrm{d}T\\&+\int\limits_{p_1}^{p_2}\left[V-T\left(\tfrac{\partial V}{\partial T}\right)_p\right]|_{T_2,p}\mathrm{d}p\end{aligned} \tag{138}$$

The same manipulations can be done for internal energy:

$$\left(\frac{\partial S}{\partial T}\right)_V=\frac{C_v}{T} \tag{139}$$

$$\mathrm{d}S=-\left[\frac{(\partial V/\partial T)_p}{(\partial V/\partial p)_T}\right]\mathrm{d}V+\frac{C_v}{T}\mathrm{d}T \tag{140}$$

$$\mathrm{d}U=C_v\mathrm{d}T-\left[p+T\frac{(\partial V/\partial T)_p}{(\partial V/\partial p)_T}\right]\mathrm{d}V$$

12.3. Partial Derivatives of All Thermodynamic Functions

The various partial derivatives of the thermodynamic functions can be classified into six groups. In the general formulas below, the variables U, H, A, G, or S are denoted by Greek letters, whereas the variables V, T, or p are denoted by Latin letters.

Type 1 (3 possibilities plus reciprocals).

General: $\left(\frac{\partial a}{\partial b}\right)_c$, Specific: $\left(\frac{\partial p}{\partial T}\right)_V$

Equation 125 yields

$$\begin{aligned}\left(\frac{\partial p}{\partial T}\right)_V&=-\left(\frac{\partial V}{\partial T}\right)_p\left(\frac{\partial p}{\partial V}\right)_T\\&=-\frac{(\partial V/\partial T)_p}{(\partial V/\partial p)_T}\end{aligned} \tag{141}$$

This relates all three partial derivatives of this type.

Type 2 (30 possibilities plus reciprocals).

General: $\left(\frac{\partial \alpha}{\partial b}\right)_c$, Specific: $\left(\frac{\partial G}{\partial T}\right)_V$

Using Equation 132 gives

$$\left(\frac{\partial G}{\partial T}\right)_V=-S+V\left(\frac{\partial p}{\partial T}\right)_V$$

Using the other equations for U, H, A, or S gives the other possibilities.

Type 3 (15 possibilities plus reciprocals).

General: $\left(\frac{\partial a}{\partial b}\right)_\alpha$, Specific: $\left(\frac{\partial V}{\partial T}\right)_S$

First the derivative is expanded by using Equation 125, which is called expansion without introducing a new variable:

$$\left(\frac{\partial V}{\partial T}\right)_S=-\left(\frac{\partial S}{\partial T}\right)_V\left(\frac{\partial V}{\partial S}\right)_T=-\frac{(\partial S/\partial T)_V}{(\partial S/\partial V)_T}$$

Then the numerator and denominator are evaluated as type 2 derivatives, or by using Equations (99) and (100):

$$\begin{aligned}\left(\frac{\partial V}{\partial T}\right)_S&=-\frac{C_v/T}{-(\partial V/\partial T)_p(\partial p/\partial V)_T}\\&=\frac{C_v}{T}\frac{\left(\frac{\partial V}{\partial p}\right)_T}{\left(\frac{\partial V}{\partial T}\right)_p}\end{aligned} \tag{142}$$

These derivatives are important for reversible, adiabatic processes (e.g., in an ideal turbine or compressor) because the entropy is constant. Similar derivatives can be obtained for isenthalpic processes, such as a pressure reduction at a valve. In that case, the Joule – Thomson coefficient is obtained for constant H:

$$\left(\frac{\partial T}{\partial p}\right)_H=\frac{1}{C_p}\left[-V+T\left(\frac{\partial V}{\partial T}\right)_p\right]$$

Type 4 (30 possibilities plus reciprocals).

General: $\left(\frac{\partial\alpha}{\partial\beta}\right)_c$, Specific: $\left(\frac{\partial G}{\partial A}\right)_p$

Now, expand through the introduction of a new variable using Equation 127:

$$\left(\frac{\partial G}{\partial A}\right)_p=\left(\frac{\partial G}{\partial T}\right)_p\left(\frac{\partial T}{\partial A}\right)_p=\frac{(\partial G/\partial T)_p}{(\partial A/\partial T)_p}$$

This operation has created two type 2 derivatives. Substitution yields

$$\left(\frac{\partial G}{\partial A}\right)_p=\frac{S}{S+p(\partial V/\partial T)_p}$$

Type 5 (60 possibilities plus reciprocals).

General: $\left(\frac{\partial\alpha}{\partial b}\right)_\beta$, Specific: $\left(\frac{\partial G}{\partial p}\right)_A$

Starting from Equation 132 for dG gives

$$\left(\frac{\partial G}{\partial p}\right)_A=-S\left(\frac{\partial T}{\partial p}\right)_A+V$$

The derivative is a type 3 derivative and can be evaluated by using Equation 125.

$$\left(\frac{\partial G}{\partial p}\right)_A=S\frac{(\partial A/\partial p)_T}{(\partial A/\partial T)_p}+V$$

The two type 2 derivatives are then evaluated:

$$\left(\frac{\partial G}{\partial p}\right)_A=\frac{S_p(\partial V/\partial p)_T}{S+p(\partial V/\partial T)_p}+V$$

These derivatives are also of interest for free expansions or isentropic changes.

Type 6 (30 possibilities plus reciprocals).

General: $\left(\frac{\partial\alpha}{\partial\beta}\right)_\gamma$, Specific: $\left(\frac{\partial G}{\partial A}\right)_H$

Equation 127 is used to obtain two type 5 derivatives.

$$\left(\frac{\partial G}{\partial A}\right)_H=\frac{(\partial G/\partial T)_H}{(\partial A/\partial T)_H}$$

These can then be evaluated by using the procedures for type 5 derivatives.

The difference in molar heat capacities (C_p-C_v) can be derived in similar fashion. Using Equation 139 for C_v yields

$$C_v=T\left(\frac{\partial S}{\partial T}\right)_V$$

To evaluate the derivative, Equation 126 is used to express dS in terms of p and T:

$$\left(\frac{\partial S}{\partial T}\right)_V=-\left(\frac{\partial V}{\partial T}\right)_p\left(\frac{\partial p}{\partial T}\right)_V+\frac{C_p}{T}$$

Substitution for $(\partial p/\partial T)_V$ and rearrangement give

$$C_p-C_v=T\left(\frac{\partial V}{\partial T}\right)_p\left(\frac{\partial p}{\partial T}\right)_V$$
$$=-T\left(\frac{\partial V}{\partial T}\right)_p^2\left(\frac{\partial p}{\partial V}\right)_T$$

Use of this equation permits the rearrangement of Equation 142 into

$$\left(\frac{\partial V}{\partial T}\right)_S=\frac{(\partial V/\partial T)_p^2+\frac{C_p}{T}(\partial V/\partial p)_T}{(\partial V/\partial T)_p}$$

The ratio of heat capacities is

$$\frac{C_p}{C_v}=\frac{T(\partial S/\partial T)_p}{T(\partial S/\partial T)_V}$$

Expansion by using Equation 125 gives

$$\frac{C_p}{C_v}=\frac{-(\partial p/\partial T)_S(\partial S/\partial p)_T}{-(\partial V/\partial T)_S(\partial S/\partial V)_T}$$

and the ratios are then

$$\frac{C_p}{C_v}=\left(\frac{\partial p}{\partial V}\right)_S\left(\frac{\partial V}{\partial p}\right)_T$$

Using Equation 125 gives

$$\frac{C_p}{C_v}=-\left(\frac{\partial p}{\partial V}\right)_S\left(\frac{\partial T}{\partial p}\right)_V\left(\frac{\partial V}{\partial T}\right)_p$$

Entropy is a variable in at least one of the partial derivatives.

13. References

Specific References

1. D. E. Seborg, T. F. Edgar, D. A. Mellichamp: *Process Dynamics and Control*, 2nd ed., John Wiley & Sons, New York 2004.
2. G. Forsyth, C. B. Moler: *Computer Solution of Lineart Algebraic Systems*, Prentice-Hall, Englewood Cliffs 1967.
3. B. A. Finlayson: *Nonlinear Analysis in Chemical Engineering*, McGraw-Hill, New York 1980; reprinted, Ravenna Park, Seattle 2003.

4. S. C. Eisenstat, M. H. Schultz, A. H. Sherman: "Algorithms and Data Structures for Sparse Symmetric Gaussian Elimination," *SIAM J. Sci. Stat. Comput.* **2** (1981) 225 – 237.
5. I. S. Duff: *Direct Methods for Sparse Matrices*, Charendon Press, Oxford 1986.
6. H. S. Price, K. H. Coats, "Direct Methods in Reservoir Simulation," *Soc. Pet. Eng. J.* **14** (1974) 295 – 308.
7. A. Bykat: "A Note on an Element Re-Ordering Scheme," *Int. J. Num. Methods Egn.* **11** (1977) 194 –198.
8. M. R. Hesteness, E. Stiefel: "Methods of conjugate gradients for solving linear systems," *J. Res. Nat. Bur. Stand* **29** (1952) 409 – 439.
9. Y. Saad: *Iterative Methods for Sparse Linear Systems*, 2nd ed., Soc. Ind. Appl. Math., Philadelphia 2003.
10. Y. Saad, M. Schultz: "GMRES: A Generalized Minimal Residual Algorithm for Solving Nonsymmetric Linear Systems." *SIAM J. Sci. Statist. Comput.* **7** (1986) 856–869.
11. http://mathworld.wolfram.com/GeneralizedMinimalResidualMethod.html.
12. http://www.netlib.org/linalg/html_templates/Templates.html.
13. http://software.sandia.gov/.
14. E. Isaacson, H. B. Keller, *Analysis of Numerical Methods*, J. Wiley and Sons, New York 1966.
15. W. H. Press, B. P. Flannery, S. A. Teukolsky, W. T. Vetterling: *Numerical Recipes*, Cambridge University Press, Cambridge 1986.
16. R. W. H. Sargent: "A Review of Methods for Solving Non-linear Algebraic Equations," in R. S. H. Mah, W. D. Seider (eds.): *Foundations of Computer-Aided Chemical Process Design*, American Institute of Chemical Engineers, New York 1981.
17. J. D. Seader: "Computer Modeling of Chemical Processes," *AIChE Monogr. Ser.* **81** (1985) no. 15.
18. http://software.sandia.gov/trilinos/packages/nox/loca_user.html
19. N. R. Amundson: *Mathematical Methods in Chemical Engineering*, Prentice-Hall, Englewood Cliffs, N.J. 1966.
20. R. H. Perry, D. W Green: *Perry's Chemical Engineers' Handbook*, 7th ed., McGraw-Hill, New York 1997.
21. D. S. Watkins: "Understanding the QR Algorithm," *SIAM Rev.* **24** (1982) 427 – 440.
22. G. F. Carey, K. Sepehrnoori: "Gershgorin Theory for Stiffness and Stability of Evolution Systems and Convection-Diffusion," *Comp. Meth. Appl. Mech.* **22** (1980) 23 – 48.
23. M. Abranowitz, I. A. Stegun: *Handbook of Mathematical Functions*, National Bureau of Standards, Washington, D.C. 1972.
24. J. C. Daubisse: "Some Results about Approximation of Functions of One or Two Variables by Sums of Exponentials, " *Int. J. Num. Meth. Eng.* **23** (1986) 1959 – 1967.
25. O. C. McGehee: *An Introduction to Complex Analysis*, John Wiley & Sons, New York 2000.
26. H. A. Priestley: *Introduction to complex analysis*, Oxford University Press, New York 2003.
27. Y. K. Kwok: *Applied complex variables for scientists and engineers,* Cambridge University Press, New York 2002.
28. N. Asmar, G. C. Jones: *Applied complex analysis with partial differential equations*, Prentice Hall, Upper Saddle River, NJ, 2002.
29. M. J. Ablowitz, A. S. Fokas: *Complex variables: Introduction and applications*, Cambridge University Press, New York 2003.
30. J. W. Brown, R. V. Churchill: *Complex variables and applications*, 6th ed., McGraw-Hill, New York 1996; 7th ed. 2003.
31. W. Kaplan: *Advanced calculus*, 5th ed., Addison-Wesley, Redwood City, Calif., 2003.
32. E. Hille: *Analytic Function Theory*, Ginn and Co., Boston 1959.
33. R. V. Churchill: *Operational Mathematics*, McGraw-Hill, New York 1958.
34. J. W. Brown, R. V. Churchill: *Fourier Series and Boundary Value Problems*, 6th ed., McGraw-Hill, New York 2000.
35. R. V. Churchill: *Operational Mathematics*, 3rd ed., McGraw-Hill, New York 1972.
36. B. Davies: *Integral Transforms and Their Applications*, 3rd ed., Springer, Heidelberg 2002.
37. D. G. Duffy: *Transform Methods for Solving Partial Differential Equations*, Chapman & Hall/CRC, New York 2004.
38. A. Varma, M. Morbidelli: *Mathematical Methods in Chemical Engineering*, Oxford, New York 1997.
39. Bateman, H., *Tables of Integral Transforms, vol. I*, McGraw-Hill, New York 1954.
40. H. F. Weinberger: *A First Course in Partial Differential Equations*, Blaisdell, Waltham, Mass. 1965.
41. R. V. Churchill: *Operational Mathematics*, McGraw-Hill, New York 1958.

42. J. T. Hsu, J. S. Dranoff: "Numerical Inversion of Certain Laplace Transforms by the Direct Application of Fast Fourier Transform (FFT) Algorithm," *Comput. Chem. Eng.* **11** (1987) 101 – 110.
43. E. Kreyzig: *Advanced Engineering Mathematics*, 9th ed., John Wiley & Sons, New York 2006.
44. H. S. Carslaw, J. C. Jaeger: *Conduction of Heat in Solids*, 2nd ed., Clarendon Press, Oxford – London 1959.
45. R. B. Bird, R. C. Armstrong, O. Hassager: *Dynamics of Polymeric Liquids*, 2nd ed., Appendix A, Wiley-Interscience, New York 1987.
46. R. S. Rivlin, *J. Rat. Mech. Anal.* **4** (1955) 681 – 702.
47. N. R. Amundson: *Mathematical Methods in Chemical Engineering; Matrices and Their Application*, Prentice-Hall, Englewood Cliffs, N.J. 1966.
48. I. S. Sokolnikoff, E. S. Sokolnikoff: *Higher Mathematics for Engineers and Physicists*, McGraw-Hill, New York 1941.
49. R. C. Wrede, M. R. Spiegel: *Schaum's outline of theory and problems of advanced calculus*, 2nd ed, McGraw Hill, New York 2006.
50. P. M. Morse, H. Feshbach: *Methods of Theoretical Physics*, McGraw-Hill, New York 1953.
51. B. A. Finlayson: *The Method of Weighted Residuals and Variational Principles*, Academic Press, New York 1972.
52. G. Forsythe, M. Malcolm, C. Moler: *Computer Methods for Mathematical Computation*, Prentice-Hall, Englewood Cliffs, N.J. 1977.
53. J. D. Lambert: *Computational Methods in Ordinary Differential Equations*, J. Wiley and Sons, New York 1973.
54. N. R. Amundson: *Mathematical Methods in Chemical Engineering*, Prentice-Hall, Englewood Cliffs, N.J. 1966.
55. J. R. Rice: *Numerical Methods, Software, and Analysis*, McGraw-Hill, New York 1983.
56. M. B. Bogacki, K. Alejski, J. Szymanowski: "The Fast Method of the Solution of Reacting Distillation Problem," *Comput. Chem. Eng.* **13** (1989) 1081 – 1085.
57. C. W. Gear: *Numerical Initial-Value Problems in Ordinary Differential Equations*, Prentice-Hall, Englewood Cliffs, N.J. 1971.
58. N. B. Ferguson, B. A. Finlayson: "Transient Modeling of a Catalytic Converter to Reduce Nitric Oxide in Automobile Exhaust," *AIChE J* **20** (1974) 539 – 550.
59. W. F. Ramirez: *Computational Methods for Process Simulations*, 2nd ed., Butterworth-Heinemann, Boston 1997.
60. U. M. Ascher, L. R. Petzold: *Computer methods for ordinary differential equations and differential-algebraic equations*, SIAM, Philadelphia 1998.
61. K. E. Brenan, S. L. Campbell, L. R. Petzold: *Numerical Solution of Initial-Value Problems in Differential-Algebraic Equations*, Elsevier, Amsterdam 1989.
62. C. C. Pontelides, D. Gritsis, K. R. Morison, R. W. H. Sargent: "The Mathematical Modelling of Transient Systems Using Differential-Algebraic Equations," *Comput. Chem. Eng.* **12** (1988) 449 – 454.
63. G. A. Byrne, P. R. Ponzi: "Differential-Algebraic Systems, Their Applications and Solutions," *Comput. Chem. Eng.* **12** (1988) 377 – 382.
64. L. F. Shampine, M. W. Reichelt: The MATLAB ODE Suite, *SIAM J. Sci. Comp.* **18** (1997) 122.
65. M. Kubicek, M. Marek: *Computational Methods in Bifurcation Theory and Dissipative Structures*, Springer Verlag, Berlin – Heidelberg – New York – Tokyo 1983.
66. T. F. C. Chan, H. B. Keller: "Arc-Length Continuation and Multi-Grid Techniques for Nonlinear Elliptic Eigenvalue Problems," *SIAM J. Sci. Stat. Comput.* **3** (1982) 173 – 194.
67. M. F. Doherty, J. M. Ottino: "Chaos in Deterministic Systems: Strange Attractors, Turbulence and Applications in Chemical Engineering," *Chem. Eng. Sci.* **43** (1988) 139 – 183.
68. M. P. Allen, D. J. Tildesley: *Computer Simulation of Liquids*, Clarendon Press, Oxford 1989.
69. D. Frenkel, B. Smit: *Understanding Molecular Simulation*, Academic Press, San Diego 2002.
70. J. M. Haile: *Molecular Dynamics Simulation*, John Wiley & Sons, New York 1992.
71. A. R. Leach: *Molecular Modelling: Principles and Applications*, Prentice Hall, Englewood Cliffs, NJ, 2001.
72. T. Schlick: *Molecular Modeling and Simulations*, Springer, New York 2002.
73. R. B. Bird, W. E. Stewart, E. N. Lightfoot: *Transport Phenomena*, 2nd ed., John Wiley & Sons, New York 2002.
74. R. B. Bird, R. C. Armstrong, O. Hassager: *Dynamics of Polymeric Liquids*, 2nd ed., Wiley-Interscience, New York 1987.

75. P. V. Danckwerts: "Continuous Flow Systems," *Chem. Eng. Sci.* **2** (1953) 1 – 13.
76. J. F. Wehner, R. Wilhelm: "Boundary Conditions of Flow Reactor," *Chem. Eng. Sci.* **6** (1956) 89 – 93.
77. V. Hlavácek, H. Hofmann: "Modeling of Chemical Reactors-XVI-Steady-State Axial Heat and Mass Transfer in Tubular Reactors. An Analysis of the Uniqueness of Solutions," *Chem. Eng. Sci.* **25** (1970) 173 – 185.
78. B. A. Finlayson: *Numerical Methods for Problems with Moving Fronts*, Ravenna Park Publishing Inc., Seattle 1990.
79. E. Isaacson, H. B. Keller: *Analysis of Numerical Methods*, J. Wiley and Sons, New York 1966.
80. C. Lanczos: "Trigonometric Interpolation of Empirical and Analytical Functions," *J. Math. Phys. (Cambridge Mass.) 17* (1938) 123 – 199.
81. C. Lanczos: *Applied Analysis*, Prentice-Hall, Englewood Cliffs, N.J. 1956.
82. J. Villadsen, W. E. Stewart: "Solution of Boundary-Value Problems by Orthogonal Collocation," *Chem. Eng. Sci.* **22** (1967) 1483 – 1501.
83. M. L. Michelsen, J. Villadsen: "Polynomial Solution of Differential Equations" pp. 341 – 368 in R. S. H. Mah, W. D. Seider (eds.): *Foundations of Computer-Aided Chemical Process Design*, Engineering Foundation, New York 1981.
84. W. E. Stewart, K. L. Levien, M. Morari: "Collocation Methods in Distillation," in A. W. Westerberg, H. H. Chien (eds.): *Proceedings of the Second Int. Conf. on Foundations of Computer-Aided Process Design*, Computer Aids for Chemical Engineering Education (CACHE), Austin Texas, 1984, pp. 535 – 569.
85. W. E. Stewart, K. L. Levien, M. Morari: "Simulation of Fractionation by Orthogonal Collocation," *Chem. Eng. Sci.* **40** (1985) 409 – 421.
86. C. L. E. Swartz, W. E. Stewart: "Finite-Element Steady State Simulation of Multiphase Distillation," *AIChE J.* **33** (1987) 1977 – 1985.
87. K. Alhumaizi, R. Henda, M. Soliman: "Numerical Analysis of a reaction-diffusion-convection system," *Comp. Chem. Eng.* **27** (2003) 579–594.
88. E. F. Costa, P. L. C. Lage, E. C. Biscaia, Jr.: "On the numerical solution and optimization of styrene polymerization in tubular reactors," *Comp. Chem. Eng.* **27** (2003) 1591–1604.
89. J. Wang, R. G. Anthony, A. Akgerman: "Mathematical simulations of the performance of trickle bed and slurry reactors for methanol synthesis," *Comp. Chem. Eng.* **29** (2005) 2474–2484.
90. V. K. C. Lee, J. F. Porter, G. McKay, A. P. Mathews: "Application of solid-phase concentration-dependent HSDM to the acid dye adsorption system," *AIChE J.* **51** (2005) 323–332.
91. C. deBoor, B. Swartz: "Collocation at Gaussian Points," *SIAM J. Num. Anal.* **10** (1973) 582 – 606.
92. R. F. Sincovec: "On the Solution of the Equations Arising From Collocation With Cubic B-Splines," *Math. Comp.* **26** (1972) 893 – 895.
93. U. Ascher, J. Christiansen, R. D. Russell: "A Collocation Solver for Mixed-Order Systems of Boundary-Value Problems," *Math. Comp.* **33** (1979) 659 – 679.
94. R. D. Russell, J. Christiansen: "Adaptive Mesh Selection Strategies for Solving Boundary-Value Problems," *SIAM J. Num. Anal.* **15** (1978) 59 – 80.
95. P. G. Ciarlet, M. H. Schultz, R. S. Varga: "Nonlinear Boundary-Value Problems I. One Dimensional Problem," *Num. Math.* **9** (1967) 394 – 430.
96. W. F. Ames: *Numerical Methods for Partial Differential Equations*, 2nd ed., Academic Press, New York 1977.
97. J. F. Botha, G. F. Pinder: *Fundamental Concepts in The Numerical Solution of Differential Equations*, Wiley-Interscience, New York 1983.
98. W. H. Press, B. P. Flanner, S. A. Teukolsky, W. T. Vetterling: *Numerical Recipes*, Cambridge Univ. Press, Cambridge 1986.
99. H. Schlichting, *Boundary Layer Theory*, 4th ed. McGraw-Hill, New York 1960.
100. R. Aris, N. R. Amundson: *Mathematical Methods in Chemical Engineering, vol. 2, First-Order Partial Differential Equations with Applications*, Prentice-Hall, Englewood Cliffs, NJ, 1973.
101. R. Courant, D. Hilbert: *Methods of Mathematical Physics, vol. I and II*, Intersicence, New York 1953, 1962.
102. P. M. Morse, H. Feshbach: *Methods of Theoretical Physics, vol. I and II*, McGraw-Hill, New York 1953.
103. A. D. Polyanin: *Handbook of Linear Partial Differential Equations for Engineers and*

Scientists, Chapman and Hall/CRC, Boca Raton, FL 2002.
104. D. Ramkrishna, N. R. Amundson: *Linear Operator Methods in Chemical Engineering with Applications to Transport and Chemical Reaction Systems*, Prentice Hall, Englewood Cliffs, NJ, 1985.
105. D. D. Joseph, M. Renardy, J. C. Saut: "Hyperbolicity and Change of Type in the Flow of Viscoelastic Fluids," *Arch. Rational Mech. Anal.* **87** (1985) 213 – 251.
106. H. K. Rhee, R. Aris, N. R. Amundson: *First-Order Partial Differential Equations*, Prentice-Hall, Englewood Cliffs, N.J. 1986.
107. B. A. Finlayson: *Numerical Methods for Problems with Moving Fronts*, Ravenna Park Publishing Inc., Seattle 1990.
108. D. L. Book: *Finite-Difference Techniques for Vectorized Fluid Dynamics Calculations*, Springer Verlag, Berlin – Heidelberg – New York – Tokyo 1981.
109. G. A. Sod: *Numerical Methods in Fluid Dynamics*, Cambridge University Press, Cambridge 1985.
110. G. H. Xiu, J. L. Soares, P. Li, A. E. Rodriques: "Simulation of five-step one-bed sorption-endanced reaction process," *AIChE J.* **48** (2002) 2817–2832.
111. A. Malek, S. Farooq: "Study of a six-bed pressure swing adsorption process," *AIChE J.* **43** (1997) 2509–2523.
112. R. J. LeVeque: *Numerical Methods for Conservation Laws*, Birkhäuser, Basel 1992.
113. W. F. Ames: "Recent Developments in the Nonlinear Equations of Transport Processes," *Ind. Eng. Chem. Fundam.* **8** (1969) 522 – 536.
114. W. F. Ames: *Nonlinear Partial Differential Equations in Engineering*, Academic Press, New York 1965.
115. W. E. Schiesser: *The Numerical Method of Lines*, Academic Press, San Diego 1991.
116. D. Gottlieb, S. A. Orszag: *Numerical Analysis of Spectral Methods: Theory and Applications*, SIAM, Philadelphia, PA 1977.
117. D. W. Peaceman: *Fundamentals of Numerical Reservoir Simulation*, Elsevier, Amsterdam 1977.
118. G. Juncu, R. Mihail: "Multigrid Solution of the Diffusion-Convection-Reaction Equations which Describe the Mass and/or Heat Transfer from or to a Spherical Particle," *Comput. Chem. Eng.* **13** (1989) 259 – 270.
119. G. R. Buchanan: *Schaum's Outline of Finite Element Analysis*," McGraw-Hill, New York 1995.
120. M. D. Gunzburger: *Finite Element Methods for Viscous Incompressible Flows*, Academic Press, San Diego 1989.
121. H. Kardestuncer, D. H. Norrie: *Finite Element Handbook*, McGraw-Hill, New York 1987.
122. J. N. Reddy, D. K. Gartling: *The Finite Element Method in Heat Transfer and Fluid Dynamics*, 2nd ed., CRC Press, Boca Raton, FL 2000.
123. O. C. Zienkiewicz, R. L. Taylor, J. Z. Zhu: *The Finite Element Method: Its Basis & Fundamentals*, vol. **1**, 6th ed., Elsevier Butterworth-Heinemann, Burlington, MA 2005.
124. O. C. Zienkiewicz, R. L. Taylor: *The Finite Element Method*, Solid and Structural Mechanics, vol. **2**, 5th ed., Butterworth-Heinemann, Burlington, MA 2000.
125. O. C. Zienkiewicz, R. L. Taylor: *The Finite Element Method*, Fluid Dynamics, vol. **3**, 5th ed., Butterworth-Heinemann, Burlington, MA 2000.
126. Z. C. Li: *Combined Methods for Elliptic Equations with Singularities, Interfaces and Infinities*, Kluwer Academic Publishers, Boston, MA, 1998.
127. G. J. Fix, S. Gulati, G. I. Wakoff: "On the use of singular functions with finite element approximations," *J. Comp. Phys.* **13** (1973) 209–238.
128. N. M. Wigley: "On a method to subtract off a singularity at a corner for the Dirichlet or Neumann problem," *Math. Comp.* **23** (1968) 395–401.
129. H. Y. Hu, Z. C. Li, A. H. D. Cheng: "Radial Basis Collocation Methods for Elliptic Boundary Value Problems," *Comp. Math. Applic.* **50** (2005) 289–320.
130. S. J. Osher, R. P. Fedkiw:, *Level Set Methods and Dynamic Implicit Surfaces*, Springer, New York 2002.
131. M. W. Chang, B. A. Finlayson: "On the Proper Boundary Condition for the Thermal Entry Problem," *Int. J. Num. Methods Eng.* **15** (1980) 935 – 942.
132. R. Peyret, T. D. Taylor: *Computational Methods for Fluid Flow*, Springer Verlag, Berlin – Heidelberg – New York – Tokyo 1983.
133. P. M. Gresho, S. T. Chan, C. Upson, R. L. Lee: "A Modified Finite Element Method for Solving the Time-Dependent, Incompressible Navier – Stokes Equations," *Int. J. Num.*

Method. Fluids (1984) "Part 1. Theory": 557 – 589; "Part 2: Applications": 619 – 640.
134. P. M. Gresho, R. L. Sani: *Incompressible Flow and the Finite Element Method*, Advection-Diffusion, vol. **1**, John Wiley & Sons, New York 1998.
135. P. M. Gresho, R. L. Sani: *Incompressible Flow and the Finite Element Method* , Isothermal Laminar Flow, vol. **2**, John Wiley & Sons, New York 1998.
136. J. A. Sethian, *Level Set Methods*, Cambridge University Press, Cambridge 1996.
137. C. C. Lin, H. Lee, T. Lee, L. J. Weber: "A level set characteristic Galerkin finite element method for free surface flows," *Int. J. Num. Methods Fluids* **49** (2005) 521–547.
138. S. Succi: *The Lattice Boltzmann Equation for Fluid Dynamics and Beyond*, Oxford University Press, Oxford 2001.
139. M. C. Sukop, D. T. Thorne, Jr.: *Lattice Boltzmann Modeling: An Introduction for Geoscientists and Engineers*, Springer, New York 2006.
140. S. Chen, S. G. D. Doolen: "Lattice Boltzmann method for fluid flows," *Annu. Rev. Fluid Mech.* **30** (2001) 329–364.
141. H. T. Lau: *Numerical Library in C for Scientists and Engineers*, CRC Press, Boca Raton, FL 1994.
142. Y. Y. Al-Jaymany, G. Brenner, P. O. Brum: "Comparative study of lattice-Boltzmann and finite volume methods for the simulation of laminar flow through a 4:1 contraction," *Int. J. Num. Methods Fluids* **46** (2004) 903–920.
143. R. L. Burden, R. L J. D. Faires, A. C. Reynolds: *Numerical Analysis*, 8th ed., Brooks Cole, Pacific Grove, CA 2005.
144. S. C. Chapra, R. P. Canal: *Numerical Methods for Engineers*," 5th ed., McGraw-Hill, New York 2006.
145. H. T. Lau: *Numerical Library in Java for Scientists and Engineers*, CRC Press, Boca Raton, FL 2004.
146. K. W. Morton, D. F. Mayers: *Numerical Solution of Partial Differential Equations*, Cambridge University Press, New York 1994.
147. A. Quarteroni, A. Valli: *Numerical Approximation of Partial Differential Equations*, Springer, Heidelberg 1997.
148. F. Scheid: "Schaum's Outline of Numerical Analysis," 2nd ed., McGraw-Hill, New York 1989.
149. C. T. H. Baker: *The Numerical Treatment of Integral Equations*, Clarendon Press, Oxford 1977.
150. L. M. Delves, J. Walsh (eds.): *Numerical Solution of Integral Equations*, Clarendon Press, Oxford 1974.
151. P. Linz: *Analytical and Numerical Methods for Volterra Equations*, SIAM Publications, Philadelphia 1985.
152. M. A. Golberg (ed.): *Numerical Solution of Integral Equations*, Plenum Press, New York 1990.
153. C. Corduneanu: *Integral Equations and Applications*, Cambridge Univ. Press, Cambridge 1991.
154. R. Kress: *Linear Integral Equations*, 2nd ed. Springer, Heidelberg 1999.
155. P. K. Kythe P. Purl: *Computational Methods for Linear Integral Equations*, Birkhäuser, Basel 2002.
156. G. Nagel, G. Kluge: "Non-Isothermal Multicomponent Adsorption Processes and Their Numerical Treatment by Means of Integro-Differential Equations," *Comput. Chem. Eng.* **13** (1989) 1025 – 1030.
157. P. L. Mills, S. Lai, M. P. Duduković, P. A. Ramachandran: "A Numerical Study of Approximation Methods for Solution of Linear and Nonlinear Diffusion-Reaction Equations with Discontinuous Boundary Conditions," *Comput. Chem. Eng.* **12** (1988) 37 – 53.
158. D. Duffy: *Green's Functions with Applications*, Chapman and Hall/CRC, New York 2001.
159. I. Statgold: *Green's Functions and Boundary Value Problems*, 2nd ed., Interscience, New York 1997.
160. B. Davies: *Integral Transforms and Their Applications*, 3rd ed., Springer, New York 2002.
161. J. M. Bownds, "Theory and performance of a subroutine for solving Volterra integral equations," *J. Comput.* **28** 317–332 (1982).
162. J. G. Blom, H. Brunner, "Algorithm 689; discretized collocation and iterated collocation for nonlinear Volterra integral equations of the second kind," *ACM Trans. Math. Software* **17** (1991) 167–177.
163. N. B. Ferguson, B. A. Finlayson: "Error Bounds for Approximate Solutions to Nonlinear Ordinary Differential Equations," *AIChE J.* **18** (1972) 1053 – 1059.
164. R. S. Dixit, L. L. Tauralidis: "Integral Method of Analysis of Fischer – Tropsch Synthesis Reactions in a Catalyst Pellet," *Chem. Eng. Sci.* **37** (1982) 539 – 544.

165. L. Lapidus, G. F. Pinder: *Numerical Solution of Partial Differential Equations in Science and Engineering*, Wiley-Interscience, New York 1982.
166. C. K. Hsieh, H. Shang: "A Boundary Condition Dissection Method for the Solution of Boundary-Value Heat Conduction Problems with Position-Dependent Convective Coefficients," *Num. Heat Trans. Part B: Fund.* **16** (1989) 245 – 255.
167. C. A. Brebbia, J. Dominguez: *Boundary Elements – An Introductory Course*, 2nd ed., Computational Mechanics Publications, Southhamtpon 1992.
168. J. Mackerle, C. A. Brebbia (eds.): *Boundary Element Reference Book*, Springer Verlag, Berlin – Heidelberg – New York – Tokyo 1988.
169. O. L. Mangasarian: *Nonlinear Programming*, McGraw-Hill, New York 1969.
170. G. B. Dantzig: *Linear Programming and Extensions*, Princeton University Press, Princeton, N.J, 1963.
171. S. J. Wright: *Primal-Dual Interior Point Methods*, SIAM, Philadelphia 1996.
172. F. Hillier, G. J. Lieberman: *Introduction to Operations Research*, Holden-Day, San Francisco, 1974.
173. T. F. Edgar, D. M. Himmelblau, L. S. Lasdon: *Optimization of Chemical Processes*, McGraw-Hill Inc., New York 2002.
174. K. Schittkowski: *More Test Examples for Nonlinear Programming Codes, Lecture notes in economics and mathematical systems no. 282*, Springer-Verlag, Berlin 1987.
175. J. Nocedal, S. J. Wright: *Numerical Optimization*, Springer, New York 1999.
176. R. Fletcher: *Practical Optimization*, John Wiley & Sons, Ltd., Chichester, UK 1987
177. A. Wächter, L. T. Biegler: "On the Implementation of an Interior Point Filter Line Search Algorithm for Large-Scale Nonlinear Programming," **106** *Mathematical Programming* (2006) no. 1, 25 – 57.
178. C. V. Rao, J. B. Rawlings S. Wright: On the Application of Interior Point Methods to Model Predictive Control. *J. Optim. Theory Appl.* **99** (1998) 723.
179. J. Albuquerque, V. Gopal, G. Staus, L. T. Biegler, B. E. Ydstie: "Interior point SQP Strategies for Large-scale Structured Process Optimization Problems," *Comp. Chem. Eng.* **23** (1997) 283.
180. T. Jockenhoevel, L. T. Biegle, A. Wächter: "Dynamic Optimization of the Tennessee Eastman Process Using the OptControlCentre," *Comp. Chem. Eng.* **27** (2003) no. 11, 1513–1531.
181. L. T. Biegler, A. M. Cervantes, A. Wächter: "Advances in Simultaneous Strategies for Dynamic Process Optimization," *Chemical Engineering Science* **57** (2002) no. 4, 575–593.
182. C. D. Laird, L. T. Biegler, B. van Bloemen Waanders, R. A. Bartlett: "Time Dependent Contaminant Source Determination for Municipal Water Networks Using Large Scale Optimization," *ASCE Journal of Water Resource Management and Planning* **131** (2005) no. 2, 125.
183. A. R. Conn, N. Gould, P. Toint: *Trust Region Methods*, SIAM, Philadelphia 2000.
184. A. Drud: "CONOPT – A Large Scale GRG Code," *ORSA Journal on Computing* **6** (1994) 207–216.
185. B. A. Murtagh, M. A. Saunders: MINOS 5.1 User's Guide, Technical Report SOL 83-20R, Stanford University, Palo Alto 1987.
186. R. H. Byrd, M. E. Hribar, J. Nocedal: An Interior Point Algorithm for Large Scale Nonlinear Programming, *SIAM J. Opt.* **9** (1999) no. 4, 877.
187. R. Fletcher, N. I. M. Gould, S. Leyffer, Ph. L. Toint, A. Waechter: "Global Convergence of a Trust-region (SQP)-filter Algorithms for General Nonlinear Programming," *SIAM J. Opt.* **13** (2002) no. 3, 635–659.
188. R. J. Vanderbei, D. F. Shanno: An Interior Point Algorithm for Non-convex Nonlinear Programming. Technical Report SOR-97-21, CEOR, Princeton University, Princeton, NJ, 1997.
189. P. E. Gill, W. Murray, M. Wright: *Practical Optimization*, Academic Press, New York 1981.
190. P. E. Gill, W. Murray, M. A. Saunders: *User's guide for SNOPT: A FORTRAN Package for Large-scale Nonlinear Programming, Technical report, Department of Mathematics*, University of California, San Diego 1998.
191. J. T. Betts: *Practical Methods for Optimal Control Using Nonlinear Programming, Advances in Design and Control 3*, SIAM, Philadelphia 2001.
192. gPROMS User's Guide, PSE Ltd., London 2002.
193. G. E. P. Box: "Evolutionary Operation: A Method for Increasing Industrial Productivity," *Applied Statistics* **6** (1957) 81–101.
194. R. Hooke, T. A. Jeeves: "Direct Search Solution of Numerical and Statistical Problems," *J. ACM* **8** (1961) 212.

195. M. J. D. Powell: "An Efficient Method for Finding the Minimum of a Function of Several Variables without Calculating Derivatives, *Comput. J.* **7** (1964) 155.
196. J. A. Nelder, R. Mead: "A Simplex Method for Function Minimization," *Computer Journal* **7** (1965) 308.
197. R. Luus, T. H. I. Jaakola: "Direct Search for Complex Systems," *AIChE J* **19** (1973) 645–646.
198. R. Goulcher, J. C. Long: "The Solution of Steady State Chemical Engineering Optimization Problems using a Random Search Algorithm," *Comp. Chem. Eng.* **2** (1978) 23.
199. J. R. Banga, W. D. Seider: "Global Optimization of Chemical Processes using Stochastic Algorithms," C. Floudas and P. Pardalos (eds.): *State of the Art in Global Optimization*, Kluwer, Dordrecht 1996, p. 563.
200. P. J. M. van Laarhoven, E. H. L. Aarts: *Simulated Annealing: Theory and Applications*, Reidel Publishing, Dordrecht 1987.
201. J. H. Holland: *Adaptations in Natural and Artificial Systems*, University of Michigan Press, Ann Arbor 1975.
202. J. E. Dennis, V. Torczon:, "Direct Search Methods on Parallel Machines," *SIAM J. Opt.* **1** (1991) 448.
203. A. R. Conn, K. Scheinberg, P. Toint: "Recent Progress in Unconstrained Nonlinear Optimization without Derivatives," *Mathemtatical Programming, Series B*, **79** (1997) no. 3, 397.
204. M. Eldred: "DAKOTA: A Multilevel Parallel Object-Oriented Framework for Design Optimization, Parameter Estimation, Uncertainty Quantification, and Sensitivity Analysis," 2002, http://endo.sandia.gov/DAKOTA/software.html
205. A. J. Booker et al.: "A Rigorous Framework for Optimization of Expensive Functions by Surrogates," CRPC Technical Report 98739, Rice University, Huston TX, February 1998.
206. R. Horst, P. M. Tuy: "Global Optimization: Deterministic Approaches," 3rd ed., Springer Verlag, Berlin 1996.
207. I. Quesada, I. E. Grossmann: "A Global Optimization Algorithm for Linear Fractional and Bilinear Programs," *Journal of Global Optimization* **6** (1995) no. 1, 39–76.
208. G. P. McCormick: "Computability of Global Solutions to Factorable Nonconvex Programs: Part I – Convex Underestimating Problems," *Mathematical Programming* **10** (1976) 147–175.
209. H. S. Ryoo, N. V. Sahinidis: "Global Optimization of Nonconvex NLPs and MINLPs with Applications in Process Design," *Comp. Chem. Eng.* **19** (1995) no. 5, 551–566.
210. M. Tawarmalani, N. V. Sahinidis: "Global Optimization of Mixed Integer Nonlinear Programs: A Theoretical and Computational Study," *Mathematical Programming* **99** (2004) no. 3, 563–591.
211. C. S. Adjiman., I.P. Androulakis, C.A. Floudas: "Global Optimization of MINLP Problems in Process Synthesis and Design. *Comp. Chem. Eng.*, **21** (1997) Suppl., S445–S450.
212. C.S. Adjiman, I.P. Androulakis and C.A. Floudas, "Global Optimization of Mixed-Integer Nonlinear Problems," *AIChE Journal*, **46** (2000) no. 9, 1769–1797.
213. C. S. Adjiman, I.P. Androulakis, C.D. Maranas, C. A. Floudas: "A Global Optimization Method, αBB, for Process Design," *Comp. Chem. Eng.*, **20** (1996) Suppl., S419–S424.
214. J. M. Zamora, I. E. Grossmann: "A Branch and Contract Algorithm for Problems with Concave Univariate, Bilinear and Linear Fractional Terms," *Journal of Gobal Optimization* **14** (1999) no. 3, 217–249.
215. E. M. B. Smith, C. C. Pantelides: "Global Optimization of Nonconvex NLPs and MINLPs with Applications in Process Design," *Comp. Chem. Eng.*, **21** (1997) no. 1001, S791–S796.
216. J. E. Falk, R. M. Soland: "An Algorithm for Separable Nonconvex Programming Problems," *Management Science* **15** (1969) 550–569.
217. F. A. Al-Khayyal, J. E. Falk: "Jointly Constrained Biconvex Programming," *Mathematics of Operations Research* **8** (1983) 273–286.
218. F. A. Al-Khayyal: "Jointly Constrained Bilinear Programs and Related Problems: An Overview," *Computers and Mathematics with Applications*, **19** (1990) 53–62.
219. I. Quesada, I. E. Grossmann: "A Global Optimization Algorithm for Heat Exchanger Networks," *Ind. Eng. Chem. Res.* **32** (1993) 487–499.
220. H. D. Sherali, A. Alameddine: "A New Reformulation-Linearization Technique for

Bilinear Programming Problems," *Journal of Global Optimization* **2** (1992) 379–410.
221. H. D. Sherali, C. H. Tuncbilek: "A Reformulation-Convexification Approach for Solving Nonconvex Quadratic Programming Problems," *Journal of Global Optimization* **7** (1995) 1–31.
222. C. A. Floudas: *Deterministic Global Optimization: Theory, Methods and Applications*, Kluwer Academic Publishers, Dordrecht, The Netherlands, 2000.
223. M. Tawarmalani, N. Sahinidis: *Convexification and Global Optimization in Continuous and Mixed-Integer Nonlinear Programming: Theory, Algorithms, Software, and Applications*, Kluwer Academic Publishers, Dordrecht 2002.
224. R. Horst, H. Tuy: *Global optimization: Deterministic approaches*. Springer-Verlag, Berlin 1993.
225. A. Neumaier, O. Shcherbina, W. Huyer, T. Vinko: "A Comparison of Complete Global Optimization Solvers," *Math. Programming B* **103** (2005) 335–356.
226. L. Lovász A. Schrijver: "Cones of Matrices and Set-functions and 0-1 Optimization," *SIAM J. Opt.* **1** (1991) 166–190.
227. H. D. Sherali, W.P. Adams: "A Hierarchy of Relaxations Between the Continuous and Convex Hull Representations for Zero-One Programming Problems," *SIAM J. Discrete Math.* **3** (1990) no. 3, 411–430.
228. E. Balas, S. Ceria, G. Cornuejols: "A Lift-and-Project Cutting Plane Algorithm for Mixed 0-1 Programs," *Mathematical Programming* **58** (1993) 295–324.
229. A. H. Land, A.G. Doig: "An Automatic Method for Solving Discrete Programming Problems," *Econometrica* **28** (1960) 497–520.
230. E. Balas: "An Additive Algorithm for Solving Linear Programs with Zero-One Variables," *Operations Research* **13** (1965) 517–546.
231. R. J. Dakin: "A Tree Search Algorithm for Mixed-Integer Programming Problems", *Computer Journal* **8** (1965) 250–255.
232. R. E. Gomory: "Outline of an Algorithm for Integer Solutions to Linear Programs," *Bulletin of the American Mathematics Society* **64** (1958) 275–278.
233. H. P. Crowder, E. L. Johnson, M. W. Padberg: "Solving Large-Scale Zero-One Linear Programming Problems," *Operations Research* **31** (1983) 803–834.
234. T. J. Van Roy, L. A. Wolsey: "Valid Inequalities for Mixed 0-1 Programs," *Discrete Applied Mathematics* **14** (1986) 199–213.
235. E. L. Johnson, G. L. Nemhauser, M. W. P. Savelsbergh: "Progress in Linear Programming Based Branch-and-Bound Algorithms: Exposition," *INFORMS Journal of Computing* **12** (2000) 2–23.
236. J. F. Benders: "Partitioning Procedures for Solving Mixed-variables Programming Problems," *Numeri. Math.* **4** (1962) 238–252.
237. G. L. Nemhauser, L. A. Wolsey: *Integer and Combinatorial Optimization*, Wiley-Interscience, New York 1988.
238. C. Barnhart, E. L. Johnson, G. L. Nemhauser, M. W. P. Savelsbergh, P. H. Vance: "Branch-and-price: Column Generation for Solving Huge Integer Programs," *Operations Research* **46** (1998) 316–329.
239. O. K. Gupta, V. Ravindran: "Branch and Bound Experiments in Convex Nonlinear Integer Programming," *Management Science* **31** (1985) no. 12, 1533–1546.
240. S. Nabar, L. Schrage: "Modeling and Solving Nonlinear Integer Programming Problems," Presented at Annual AIChE Meeting, Chicago 1991.
241. B. Borchers, J.E. Mitchell: "An Improved Branch and Bound Algorithm for Mixed Integer Nonlinear Programming," *Computers and Operations Research* **21** (1994) 359–367.
242. R. Stubbs, S. Mehrotra: "A Branch-and-Cut Method for 0-1 Mixed Convex Programming," *Mathematical Programming* **86** (1999) no. 3, 515–532.
243. S. Leyffer: "Integrating SQP and Branch and Bound for Mixed Integer Noninear Programming," *Computational Optimization and Applications* **18** (2001) 295–309.
244. A. M. Geoffrion: "Generalized Benders Decomposition," *Journal of Optimization Theory and Applications* **10** (1972) no. 4, 237–260.
245. M. A. Duran, I.E. Grossmann: "An Outer-Approximation Algorithm for a Class of Mixed-integer Nonlinear Programs," *Math Programming* **36** (1986) 307.
246. X. Yuan, S. Zhang, L. Piboleau, S. Domenech : "Une Methode d'optimisation Nonlineare en Variables Mixtes pour la Conception de Procedes," *RAIRO* **22** (1988) 331.
247. R. Fletcher, S. Leyffer: "Solving Mixed Integer Nonlinear Programs by Outer Approximation," *Math Programming* **66** (1974) 327.

248. I. Quesada, I.E. Grossmann: "An LP/NLP Based Branch and Bound Algorithm for Convex MINLP Optimization Problems," *Comp. Chem. Eng.* **16** (1992) 937–947.
249. T. Westerlund, F. Pettersson: "A Cutting Plane Method for Solving Convex MINLP Problems," *Comp. Chem. Eng.* **19** (1995) S131–S136.
250. O. E. Flippo, A. H. G. R. Kan: "Decomposition in General Mathematical Programming," *Mathematical Programming* **60** (1993) 361–382.
251. T. L. Magnanti, R. T. Wong: "Acclerated Benders Decomposition: Algorithm Enhancement and Model Selection Criteria," *Operations Research* **29** (1981) 464–484.
252. N. V. Sahinidis, I. E. Grossmann: "Convergence Properties of Generalized Benders Decomposition," *Comp. Chem. Eng.* **15** (1991) 481.
253. J. E. Kelley Jr.: "The Cutting-Plane Method for Solving Convex Programs," *J. SIAM* **8** (1960) 703–712.
254. S. Leyffer: "Deterministic Methods for Mixed-Integer Nonlinear Programming," Ph.D. thesis, Department of Mathematics and Computer Science, University of Dundee, Dundee 1993.
255. M. S. Bazaraa, H. D. Sherali, C. M. Shetty: *Nonlinear Programming*, John Wiley & Sons, Inc., New York 1994.
256. G. R. Kocis, I. E. Grossmann: "Relaxation Strategy for the Structural Optimization of Process Flowsheets," *Ind. Eng. Chem. Res.* **26** (1987) 1869.
257. I. E. Grossmann: "Mixed-Integer Optimization Techniques for Algorithmic Process Synthesis", *Advances in Chemical Engineering*, vol. **23**, Process Synthesis, Academic Press, London 1996, pp. 171–246.
258. E. M. B. Smith, C. C. Pantelides: "A Symbolic Reformulation/Spatial Branch and Bound Algorithm for the Global Optimization of Nonconvex MINLPs," *Comp. Chem. Eng.* **23** (1999) 457–478.
259. P. Kesavan P. P. I. Barton: "Generalized Branch-and-cut Framework for Mixed-integer Nonlinear Optimization Problems," *Comp. Chem. Eng.* **24** (2000) 1361–1366.
260. S. Lee I. E. Grossmann: "A Global Optimization Algorithm for Nonconvex Generalized Disjunctive Programming and Applications to Process Systems," *Comp. Chem. Eng.* **25** (2001) 1675–1697.
261. R. Pörn, T. Westerlund: "A Cutting Plane Method for Minimizing Pseudo-convex Functions in the Mixed-integer Case," *Comp. Chem. Eng.* **24** (2000) 2655–2665.
262. J. Viswanathan, I. E. Grossmann: "A Combined Penalty Function and Outer-Approximation Method for MINLP Optimization," *Comp. Chem. Eng.* **14** (1990) 769.
263. A. Brooke, D. Kendrick, A. Meeraus, R. Raman: GAMS – A User's Guide, www.gams.com, 1998.
264. N. V. A. Sahinidis, A. Baron: "A General Purpose Global Optimization Software Package," *Journal of Global Optimization* **8** (1996) no.2, 201–205.
265. C. A. Schweiger, C. A. Floudas: "Process Synthesis, Design and Control: A Mixed Integer Optimal Control Framework," Proceedings of DYCOPS-5 on Dynamics and Control of Process Systems, Corfu, Greece 1998, pp. 189–194.
266. R. Raman, I. E. Grossmann: "Modelling and Computational Techniques for Logic Based Integer Programming," *Comp. Chem. Eng.* **18** (1994) no.7, 563.
267. J. N. Hooker, M. A. Osorio: "Mixed Logical.Linear Programming," *Discrete Applied Mathematics* **96-97** (1994) 395–442.
268. P. V. Hentenryck: *Constraint Satisfaction in Logic Programming*, MIT Press, Cambridge, MA, 1989.
269. J. N. Hooker: *Logic-Based Methods for Optimization: Combining Optimization and Constraint Satisfaction*, John Wiley & Sons, New York 2000.
270. E. Balas: "Disjunctive Programming," *Annals of Discrete Mathematics* **5** (1979) 3–51.
271. H. P. Williams: *Mathematical Building in Mathematical Programming*, John Wiley, Chichester 1985.
272. R. Raman, I. E. Grossmann: "Relation between MILP Modelling and Logical Inference for Chemical Process Synthesis," *Comp. Chem. Eng.* **15** (1991) no. 2, 73.
273. S. Lee, I. E. Grossmann: "New Algorithms for Nonlinear Generalized Disjunctive Programming," *Comp. Chem. Eng.* **24** (2000) no. 9-10, 2125–2141.
274. J. Hiriart-Urruty, C. Lemaréchal: *Convex Analysis and Minimization Algorithms*, Springer-Verlag, Berlin, New York 1993.
275. E. Balas: "Disjunctive Programming and a Hierarchy of Relaxations for Discrete

Optimization Problems," *SIAM J. Alg. Disc. Meth.* **6** (1985) 466–486.
276. I. E. Grossmann, S. Lee: "Generalized Disjunctive Programming: Nonlinear Convex Hull Relaxation and Algorithms," *Computational Optimization and Applications* **26** (2003) 83–100.
277. S. Lee, I. E. Grossmann: "Logic-based Modeling and Solution of Nonlinear Discrete/Continuous Optimization Problems," *Annals of Operations Research* **139** 2005 267–288.
278. ILOG, Gentilly Cedex, France 1999, www.ilog.com,
279. M. Dincbas, P. Van Hentenryck, H. Simonis, A. Aggoun, T. Graf, F. Berthier: The Constraint Logic Programming Language CHIP, FGCS-88: Proceedings of International Conference on Fifth Generation Computer Systems, Tokyo, 693–702.
280. M. Wallace, S. Novello, J. Schimpf: "ECLiPSe: a Platform for Constraint Logic Programming," *ICL Systems Journal* **12** (1997) no.1, 159–200.
281. V. Pontryagin, V. Boltyanskii, R. Gamkrelidge, E. Mishchenko: *The Mathematical Theory of Optimal Processes*, Interscience Publishers Inc., New York, NY, 1962.
282. A. E. Bryson, Y. C. Ho: "Applied Optimal Control: Optimization, Estimation, and Control," Ginn and Company, Waltham, MA, 1969.
283. V. Vassiliadis, PhD Thesis, Imperial College, University of London 1993.
284. W. F. Feehery, P. I. Barton: "Dynamic Optimization with State Variable Path Constraints," *Comp. Chem. Eng.*, **22** (1998) 1241–1256.
285. L. Hasdorff:. *Gradient Optimization and Nonlinear Control*, Wiley-Interscience, New York, NY, 1976.
286. R. W. H. Sargent, G. R. Sullivan: "Development of Feed Changeover Policies for Refinery Distillation Units," *Ind. Eng. Chem. Process Des. Dev.* **18** (1979) 113–124.
287. V. Vassiliadis, R. W. H. Sargent, C. Pantelides: "Solution of a Class of Multistage Dynamic Optimization Problems," *I & EC Research* **33** (1994) 2123.
288. K. F. Bloss, L. T. Biegler, W. E. Schiesser: "Dynamic Process Optimization through Adjoint Formulations and Constraint Aggregation,". *Ind. Eng. Chem. Res.* **38** (1999) 421–432.
289. H. G. Bock, K. J. Plitt: A Multiple Shooting Algorithm for Direct Solution of Optimal Control Problems, 9th IFAC World Congress, Budapest 1984.
290. D. B. Leineweber, H. G. Bock, J. P. Schlöder, J. V. Gallitzendörfer, A. Schäfer, P. Jansohn: "A Boundary Value Problem Approach to the Optimization of Chemical Processes Described by DAE Models," *Computers & Chemical Engineering*, April 1997. (IWR-Preprint 97-14, Universität Heidelberg, March 1997.
291. J. E. Cuthrell, L. T. Biegler: "On the Optimization of Differential- algebraic Process Systems," *AIChE Journal* **33** (1987) 1257–1270.
292. A. Cervantes, L. T. Biegler: "Large-scale DAE Optimization Using Simultaneous Nonlinear Programming Formulations. *AIChE Journal* **44** (1998) 1038.
293. P. Tanartkit, L. T. Biegler: "Stable Decomposition for Dynamic Optimization," *Ind. Eng. Chem. Res.* **34** (1995) 1253–1266.
294. S. Kameswaran, L. T. Biegler: "Simultaneous Dynamic Optimization Strategies: Recent Advances and Challenges," *Chemical Process Control – 7*, to appear 2006.
295. B. van Bloemen Waanders, R. Bartlett, K. Long, P. Boggs, A. Salinger: "Large Scale Non-Linear Programming for PDE Constrained Optimization," Sandia Technical Report SAND2002-3198, October 2002.
296. G. P. Box, J. S. Hunter, W. G. Hunter: *Statistics for Experimenters: Design, Innovation, and Discovery*, 2nd ed., John Wiley & Sons, New York 2005.
297. D. C. Baird: *Experimentation: An Introduction to Measurement Theory and Experiment Design*, 3rd ed., Prentice Hall, Engelwood Cliffs, NJ, 1995.
298. J. B. Cropley: "Heuristic Approach to Comples Kinetics," *ACS Symp. Ser.* **65** (1978) 292 – 302.
299. S. Lipschutz, J. J. Schiller, Jr: *Schaum's Outline of Theory and Problems of Introduction to Probability and Statistics*, McGraw-Hill, New York 1988.
300. D. S. Moore, G. P. McCabe: *Introduction to the Practice of Statistics*," 4th ed., Freeman, New York 2003.
301. D. C. Montgomery, G. C. Runger: *Applied Statistics and Probability for Engineers*, 3 rd ed., John Wiley & Sons, New York 2002.
302. D. C. Montgomery, G. C. Runger, N. F. Hubele: *Engineering Statistics*, 3rd ed., John Wiley & Sons, New York 2004.

303. G. E. P. Box, W. G. Hunter, J. S. Hunter: *Statistics for Experimenters*, John Wiley & Sons, New York 1978.
304. B. W. Lindgren: *Statistical Theory*, Macmillan, New York 1962.
305. O. A. Hougen, K. M. Watson, R. A. Ragatz: *Chemical Process Principles*, 2nd ed., part II, "Thermodynamics," J. Wiley & Sons, New York 1959.
306. L. T. Biegler, I. E. Grossmann, A. W. Westerberg: *Systematic Methods for Chemical Process Design*, Prentice-Hall Englewood Cliffs, NJ 1997.
307. I. E. Grossmann (ed.), "Global Optimization in Engineering Design", Kluwer, Dordrecht 1996.
308. J. Kallrath, "Mixed Integer Optimization in the Chemical Process Industry: Experience, Potential and Future," *Trans. I .Chem E.*, **78** (2000) Part A, 809–822.
309. J. N. Hooker: *Logic-Based Methods for Optimization*, John Wiley & Sons, New York 1999.

Model Reactors and Their Design Equations

VLADIMIR HLAVACEK, Laboratory for Ceramic and Reaction Engineering, Department of Chemical Engineering, University of Buffalo, Buffalo, NY, USA

JAN A. PUSZYNSKI, Chemical and Biological Engineering Department, South Dakota School of Mines and Technology, Rapid City, SD, USA

HENDRIK J. VILJOEN, Department of Chemical and Biomolecular Engineering, University of Nebraska, Lincoln, NE, USA

JORGE E. GATICA, Department of Chemical and Biomedical Engineering, Cleveland State University, Cleveland, OH, USA

1. Introduction 153
2. Batch Reactors 154
2.1. Homogeneous Systems 154
2.1.1. Isothermal Reactors 154
2.1.2. Nonisothermal Reactors 155
2.2. Nonhomogeneous Systems 157
2.2.1. Gas – Liquid Reactions 158
2.2.2. Solid – Solid Reactions 159
2.2.3. Solid – Gas Reactions 161
3. Continuous Stirred-Tank Reactors (CSTR) 162
3.1. Isothermal Homogeneous System 162
3.2. Isothermal Heterogeneous Reactors 166
3.3. Nonisothermal Continuous Stirred-Tank Reactors 167
3.4. Cascade of Tank Reactors 170
4. Packed-Bed Reactors 171
4.1. Mass and Energy Balances 172
4.1.1. The Single-Particle Model 172
4.1.2. The Two-Phase Model 173
4.1.3. The One-Phase Model 174
4.1.4. Boundary Conditions 175
4.2. Values of the Parameters 176
4.2.1. Average Bed Porosity 176
4.2.2. Effective Transport Coefficients 176
4.2.3. Wall Heat-Transfer Coefficient 178
4.3. Further Simplifications 178
4.4. Parametric Sensitivity 180
4.5. Flow Field Description 183
4.5.1. Boundary Conditions 183
4.5.2. Permeability and Inertia Coefficient 183
4.5.3. Thermal-Expansion Coefficient 184
4.5.4. Mass-Expansion Coefficient 184
4.5.5. Pressure-Drop Effects in Packed Beds 184
4.6. Thermomechanical Effects in the Reaction System 186
4.7. Numerical Simulation 186
4.7.1. Discretization of the Physical Domain 187
4.7.2. Discretization of the Governing Equations 187
5. Optimization of Chemical Reactors 189
5.1. The Objective Function for Chemical Reactors 189
5.2. Elementary Optimization Problems 190
5.2.1. Optimization of a Batch System 190
5.2.2. Optimization of a Continuous System (CSTR) 190
5.2.3. Optimization of Complex Systems: Illustrative Example 190
5.3. Optimization of Reactors by the Search Method 191
5.4. Optimum Temperature Profile and Optimization of Multistage Reactors 194
5.5. Optimization of a Multibed Adiabatic Reactor with Heat Exchange Between Catalytic Stages 195
6. References 196

Symbols

a	activity
a	thermal diffusivity, m^2/s
a	interfacial area per unit volume, m^2/m^3
A	heat transfer area, m^2
B, B^*	dimensionless adiabatic temperature rise
c	concentration, $kmol/m^3$
c_J	molar concentration of species J, $kmol/m^3$

Ullmann's Modeling and Simulation

ISBN: 978-3-527-31605-2

c_p, c_v heat capacity per unit mass at constant pressure and volume, respectively, J kg^{-1} °C^{-1}
c_{pm} specific heat capacity of the cooling/heating medium
C reduced concentration, c/c_0
C_p, C_v molar heat capacity at constant pressure and volume, respectively, J $kmol^{-1}$ °C^{-1}
d characteristic dimension, diameter, m
d_I diameter of impeller
D diameter of reactor, m
D coefficient of molecular diffusion, m^2/s
e particle emissivity
E_a activation energy
$E(t)$ residence time distribution function or frequency function, s^{-1}
F molar flow rate
$\mathbf{g}$ gravitational acceleration, m/s^2
h overall heat-transfer coefficient, W m^{-2} K^{-1}
H enthalpy, J
H_J molar enthalpy of species J, J/kmol
ΔH_r heat of reaction at constant pressure as associated with the stoichiometric equation
$(\Delta H_r)_J$ heat of reaction at constant pressure for the conversion of or to one molar unit of J, J/kmol
J component
J molar flux, kmol m^{-2} s^{-1}
J_J molar flux of species J, kmol m^{-2} s^{-1}
k homogeneous reaction velocity constant, dimension depends on the kinetics; for nth order: $m^{3(n-1)}/(kmol^{(n-1)}\ s)$
K chemical equilibrium constant
K_D ratio between two concentrations
K_L mass-transfer coefficient, m/s
L length, m
m total mass of a system, kg
m_J mass of species J in a system, kg
M_J molar mass of species J, kg/kmol
n sequence number of a tank reactor in a cascade or of a bed in a multibed reactor
$\mathbf{n}$ outward unit vector on S
N number of revolutions (impeller) per unit time
N_{QI} dimensionless constant
Nu $\alpha d/\lambda$; Nusselt number
p total pressure, Pa
P permeability
ΔP_r pressure drop across the reactor
P_e total pressure in the bulk stream
Pe $\langle v \rangle L/D_l$; Péclet number for longitudinal dispersion
Pe_h $\langle v \rangle L \varrho c_p/\lambda$; Péclet number for heat dispersion
Pe_t $\langle v \rangle d_t/D_t$; Péclet number for transverse dispersion
Pr $\mu c_p/\lambda$; Prandtl number
Q amount of heat, J
$\dot{Q}$ heat flow, W
r cylindrical or spherical coordinate, m
R rate of reaction
R gas constant, 8.314 J mol^{-1} K^{-1}
R_J molar rate of production of J per unit volume of the reaction phase, kmol m^{-2} s^{-1}
$|R|_J$ molar rate of conversion per unit volume of the reaction phase, kmol m^{-3} s^{-1}
Re v d/ν; Reynolds number
Re_p $\langle v \rangle d_p/\nu$; Reynolds number in packed bed related to particle diameter
s ratio of reaction rate constant
S surface, m^2
Sc ν/D; Schmidt number
St $\alpha/\varrho c_p \langle v \rangle$; Stanton number for heat transfer
St $K_G/\langle v \rangle$ or $K_L/\langle v \rangle$; Stanton number for mass transfer
St' $UA/\varrho c_p \Phi_v$; modified Stanton number
t time, s
t' off-line time, s
t^* reaction time, s
T temperature, K or °C
T^* optimum reaction temperature, K or °C
$\langle T \rangle$ average radial temperature for a two dimensional model
ΔT_{ad} adiabatic temperature rise of a reaction mixture after complete conversion, °C
T^e equilibrium temperature
u internal energy per unit mass, J/kg
$\langle u \rangle$ average internal energy (over the reactor volume), J/kg
$\mathbf{u}$ linear velocity, m/s
$\mathbf{u}_s$ control volume velocity, m/s
U_J molar internal energy of species J, J/kmol

ΔU_r heat of reaction at constant volume as associated with a stoichiometric equation, J
$(\Delta U_r)_J$ heat of reaction at constant reaction volume for the conversion of or to one molar unit of J, J/kmol
v velocity, m/s
v_F front velocity, m/s
v_r relative velocity; velocity of one phase relative to another phase, m/s
V volume, m^3
V_r volume of reaction mixture, m^3
V_R reactor volume, m^3
w_J mass fraction of species J
W mass of the catalyst
W amount of work, J
$\dot{W}$ rate of work done on the surroundings, W
x direction of propagation, m
x_J mole fraction of species J in any phase
$\mathbf{x}$ coordinate, m
y_J mole fraction of species J in any gas phase only used in connection with x_J for the liquid phase
z flow direction, m
Z dimensionless coordinate in the direction of flow, z/L
α heat-transfer coefficient, W m^{-2} K^{-1}
$\overline{\beta}$ dimensionless heat-transfer coefficient
γ ratio between kinetic and mass-transfer rates
γ temperature factor E/RT_a; index a varies, depending on situation
$\overline{\gamma}$ dimensionless activation energy
$\overline{\delta}$ dimensionless parameter
$\boldsymbol{\delta}$ displacement vector
ε porosity
$\overline{\varepsilon}$ dimensionless parameter
ϑ dimensionless temperature
Θ dimensionless time
Θ_c dimensionless cooling temperature
ζ relative degree of conversion, fraction of reactant converted
$\overline{\zeta}$ average conversion
κ ratio between reaction rate constant
$\boldsymbol{\lambda}_e$ effective thermal conductivity, W m^{-1} K^{-1}
λ thermal conductivity, W m^{-1} K^{-1}
λ Lamé constant
μ Lamé constant
μ dynamic viscosity, Ns/m^2
ν kinematic viscosity, m^2/s
ν stoichiometric coefficient
ξ_J degree of conversion of species J
η effectiveness factor
ϱ density, specific mass, kg/m^3
σ Stefan – Boltzmann constant, 5.67×10^{-8} J m^{-2} K^{-4} s^{-1}
τ average residence time in reactor system, s
Φ flow rate
Φ_P production rate, kg/s; kmol/s
Φ_{mJ} mass flow rate for species J, kg/s
Φ_c circulation rate
Φ_v volumetric flow rate, m^3/s
ψ shape factor

Subcripts

0 initial
av average
A, B reactants
c coolant, continuous
e equilibrium
ext extern
f fluid
fus fusion
F fluid phase
g gas
G gas phase
i interface
ign ignition
J arbitrary species
l liquid, longitudinal
L liquid phase
L at the outlet of a tubular reactor of length L
m medium, mass
n output of the n-th tank reactor in a cascade, or of a segment n
r reaction, radial direction
s solid
S solid phase
t transverse
v volumetric
w wall
z axial direction

1. Introduction

Reactor modeling represents a simultaneous solution and analysis of mass, energy, and momentum transfer equations along with reaction rate and equilibrium data. Sometimes additional information (e.g., of economic nature in reactor optimization calculations or mechanical interactions) must be also included in the reactor model.

Important goals of reactor modeling are [1]:

1) Scale-up of experimental laboratory units to pilot-plant size
2) Scale-up of pilot-plant units to full scale size
3) Prediction of behavior of different reactor feedstocks and new catalysts
4) Prediction of transients in reactors for better control
5) Optimization of steady-state operating conditions
6) Better understanding of the system that may lead to process and design improvements

Reactor modeling in the hands of a practitioner is a powerful tool and can result in faster, more economical process design. It should always be coupled with pilot-plant experiments and process development.

The most important parts of reactor modeling are described in this article; namely, simulation of batch and continuous stirred-tank reactors, tubular systems, and packed beds. Some elements of optimization of chemical reactors are presented as well.

2. Batch Reactors

2.1. Homogeneous Systems

2.1.1. Isothermal Reactors

Batch reactors are usually operated as closed systems with no net inflow or outflow of mass. The systems are also well mixed and no spatial gradients exist. The *material balance* for a component J in a batch reactor which meets these conditions, is:

$$\frac{\mathrm{d}m_{\mathrm{J}}}{\mathrm{d}t} = m\frac{\mathrm{d}\xi_{\mathrm{J}}}{\mathrm{d}t} = M_{\mathrm{J}}R_{\mathrm{J}}V_{\mathrm{r}} \tag{2.1}$$

where V_{r} is the volume of the reaction mixture, m the total mass of the system (kg), m_{J} the mass of component J in the system (kg), ξ_{J} the degree of conversion of J, M_{J} the molar mass of component J (kg/kmol), and R_{J} the molar rate of production of J per unit volume of the reaction phase (kmol m^{-3} s^{-1}). In the isothermal case, R_{J} is only a function of the reactants for an irreversible reaction or a function of reactants and products for a reversible reaction. In terms of the density of the reaction mixture, $\varrho = m/V_{\mathrm{r}}$, the material balance can be written in integral form (Eq. 2.2) and the reaction time t that is required for a certain degree of conversion can be obtained by integrating Equation (2.1).

$$t = \int_0^{\xi_{\mathrm{J}}} \frac{\varrho \mathrm{d}\xi_{\mathrm{J}}}{M_{\mathrm{J}}R_{\mathrm{J}}} \tag{2.2}$$

If the density of the reaction mixture remains constant during the course of the reaction, the integration is simplified. In the case of *irreversible nth-order reactions* integration gives

$$t = \frac{(1-\xi_{\mathrm{J}})^{1-n}-1}{k\,(n-1)\,c_{\mathrm{J}_0}^{n-1}}, n\neq 1. \tag{2.3}$$

$$t = \frac{\ln(1-\xi_{\mathrm{J}})}{-k}, n = 1 \tag{2.4}$$

The rate equations of *reversible first-order reactions* with initial concentrations c_{J_0} and c_{P_0} and rate constants k_1 and k_2 for the forward and backward reactions respectively, can also be integrated:

$$t = \frac{\ln\left[(1-\xi_{\mathrm{J}})\,(k_1+k_2) - k_2\frac{c_{\mathrm{J}_0}+c_{\mathrm{P}0}}{c_{\mathrm{J}0}}\right]}{-\,(k_1+k_2)} \tag{2.5}$$

The general problem in reactor design is to calculate the volume of a reactor for a certain average rate of production. The volume can be obtained from the required degree of conversion and the corresponding reaction times. In batch operation, a certain time is used for filling, heating, cooling, and cleaning and during this time no production occurs. If t' denotes this off-line time and Φ_{P} denotes the desired production of product P, the reactor volume is

$$V_{\mathrm{R}} = \frac{\nu_{\mathrm{A}}\Phi_{\mathrm{P}}\,(t+t')}{\nu_{\mathrm{P}}M_{\mathrm{P}}c_{\mathrm{A}_0}\xi_{\mathrm{A}}} \tag{2.6}$$

where ν_{A} and ν_{P} are the stoichiometric coefficients of reactant A and product P, respectively.

Biochemical reactions play an increasingly important role, particularly in biomedical engineering. In vivo and in vitro reactions can be modeled on the basis of chemical reaction engineering. Enzymes can perform complex functions and their mathematical modeling requires an alternative approach. One of the best examples of such a complex enzyme-catalyzed reaction is template-directed nucleic acid synthesis

[2]. An alternative approach [3] is used to derive rate expressions for enzymatic reactions. For conventional reaction schemes, there is no benefit in this alternative approach. However, it is much better suited to handling complexities such as nucleotide selection and editing. The principle idea is to track a single enzyme – template complex over time and then determine its average behavior. The approach is illustrated for the classical Michaelis – Menten reaction scheme for enzymatic catalysis:

$$\mathrm{E{+}S} \underset{k_{-1}}{\overset{k_1}{\Leftrightarrow}} \mathrm{E{\cdot}S} \overset{k_2}{\rightarrow} \mathrm{E{+}P} \tag{2.7}$$

but a more succinct notation makes it easier to describe the enzyme's state:

$$\mathrm{A1} \underset{k_{-1}}{\overset{k_1}{\Leftrightarrow}} \mathrm{A0} \overset{k_2}{\rightarrow} \mathrm{A2} \tag{2.8}$$

Following the reaction pathway from left to right, the enzyme changes from free to bound and back to free. The average time to go from state A1 to state A0 is

$$t_1 = \frac{1}{k_1\,[S]} \tag{2.9}$$

and the average time to go from state A0 to either state A1 or state A2 is

$$t_2 = \frac{1}{k_{-1}+k_2} \tag{2.10}$$

The passage time $\langle t\rangle = t_1 + t_2$ is the average time that elapses from the moment an enzyme leaves the free state until it enters the free state again. The rate of reaction is the probability of going from A0 to A2 (not to A1), divided by the passage time. Since the probability for the successful forward reaction $(\mathrm{A0}\rightarrow\mathrm{A2})$ is

$$P_S = \frac{k_2}{k_{-1}+k_2} \tag{2.11}$$

the production rate of one enzyme molecule is

$$\nu_1 = \frac{P_S}{\langle t\rangle} = \frac{k_1 k_2\,[S]}{k_1\,[S]+k_2+k_{-1}} \tag{2.12}$$

If the total enzyme concentration is $[\mathrm{E_0}]$, then the macroscopic rate expression is

$$\nu = \frac{k_1 k_2\,[\mathrm{S}]\,[\mathrm{E_0}]}{k_1\,[\mathrm{S}]+k_2+k_{-1}} \tag{2.13}$$

The alternative approach can be used to derive a rate expression for nucleic acid synthesis on a template that captures the enzyme functions of nucleotide selection and editing [4, 5].

2.1.2. Nonisothermal Reactors

All chemical reactions are accompanied by evolution or absorption of heat. The most important considerations which determine the choice of temperature in the reactor are:

1) Chemical equilibrium and the conversion rate of the desired reaction; the maximum conversion of a reversible exothermic reaction is reduced when the temperature is increased, but the kinetic rate increases, therefore an optimum temperature exists.
2) Undesirable side reactions (which also depend on temperature and their role in the overall equilibrium) can influence the maximum yield of product.
3) Phase transformations of reactants or products; there are examples where it is better to carry out the reaction in the liquid instead of the gaseous phase to achieve maximum conversion and mass transfer. An example is the formation of methyl *tert*-butyl ether from isobutene and methanol.
4) Cost implications of heating reactants to the reacting temperature; energy cost must be compared with the gain in yield at increased temperature as maximum yield and maximum profitability will not always be at the same temperature [6].

Constant Pressure. At constant pressure, the energy balance in a well-mixed batch reactor is:

$$\varrho c_p \frac{\mathrm{d}T}{\mathrm{d}t} + (\Delta H_\mathrm{r})_\mathrm{J} R_\mathrm{J} = \frac{\dot{Q}}{V_\mathrm{r}} \tag{2.14}$$

where c_p denotes the specific heat capacity, $(\Delta H_\mathrm{r})_\mathrm{J}$ the reaction enthalpy for the conversion of or to one molar unit of J, and $\dot{Q}$ denotes the rate of heat flow to or from the reactor. In the case of adiabatic operation, $\dot{Q} = 0$. Otherwise, $\dot{Q}$ can be determined as follows:

$$\dot{Q} = hA\left(T - T_{\mathrm{c,H}}\right) \tag{2.15}$$

where h is the overall heat-transfer coefficient and A the heat-transfer area. The external temperature can be cooler (T_c) or hotter (T_H) than T, depending on whether cooling or heating must be accomplished. The overall heat-transfer coefficient h is calculated from correlations [7] and it depends on the temperature of the cooling or heating medium. Thus, a heat-transfer coefficient h_{av} is calculated at the average medium temperature. If Φ_m is the mass flow rate and c_{pm} is the specific heat capacity of the cooling or heating medium, then h is corrected according to:

$$h = \frac{h_{av}}{1+\frac{h_{av}A}{2\Phi_m c_{pm}}} \qquad (2.16)$$

Example. In a batch reactor with a volume of 5 m^3, an exothermic reaction A → P is carried out in the liquid phase. The density of the mixture does not change during the reaction. The rate of reaction is given by

$$R_A = -kc_A$$

with $k = 4\times10^6 \exp(-7500/T)\ s^{-1}$. The heat of reaction is $(-\Delta H_r)_A = 1.67\times10^6$ J/kg, the initial concentration of A is c_{A_0} = 1 kmol/m^3, M_A = 100 kg/kmol and $\varrho C_p = 4.2\times10^6\ J\ m^{-3}\ K^{-1}$. The initial temperature of the reaction mixture is 298 K. The reactor is fitted with a heat-exchange coil with a surface area of 3.3 m^2. The heat exchanger is operated with steam (T_H = 393 K, h = 1000 W $m^{-2}\ K^{-1}$) and with cooling water (T_c = 290 K, h = 1400 W $m^{-2}\ K^{-1}$). Filling and emptying of the reactor take 600 s and 900 s, respectively.

For a required conversion $\xi_A \geq 0.95$, the temperature profile over one reaction cycle and the duration of a reaction cycle shall be determined when the following policy of operation is followed: heat up to 350 K, the reaction then proceeds under isothermal conditions until ξ_A = 0.95, followed by cooling down to 310 K.

Solution. The following equations must be solved simultaneously for the first part of the reaction cycle:

$$\frac{d\xi_A}{dt} = k(1-\xi_A)$$

$$\varrho c_p \frac{dT}{dt} = (-\Delta H_r)_A c_{A_0} M_A k(1-\xi_A) + \frac{hA_s}{V_r}[T_s - T]$$

Starting from the initial concentration and temperature, a small temperature interval ΔT is selected. Average values for T and k over this interval are then calculated and used together with the initial value for ξ in the second equation to solve for Δt. This value is then used in the first equation to solve for ξ. An average value for ξ over this interval can then be calculated. With the average values of T, k, and ξ, an improved Δt value can be calculated from the second equation and the process is repeated until the required degree of accuracy is reached. In Figure 1 the temperature profile is shown. Once the conversion is 0.95, cooling commences. The cooling period can be calculated by integrating the following energy equation from T = 350 K to T = 310 K:

$$\varrho c_p \frac{dT}{dt} = (-\Delta H_r)_A c_{A_0} M_A k(1-\xi_A) + \frac{hA_c}{V_r}[T_c - T]$$

Integrating the first two equations from the initial conditions until T is 350 K gives a preheating period of 2384 s. The temperature is then kept constant and only the first equation is integrated until the conversion reaches 0.95; this period takes 1000 s. For the cooling-off period the third equation is integrated and this gives 5426 s. Adding 1500 s for filling and emptying gives a total period of one cycle of 10 310 s.

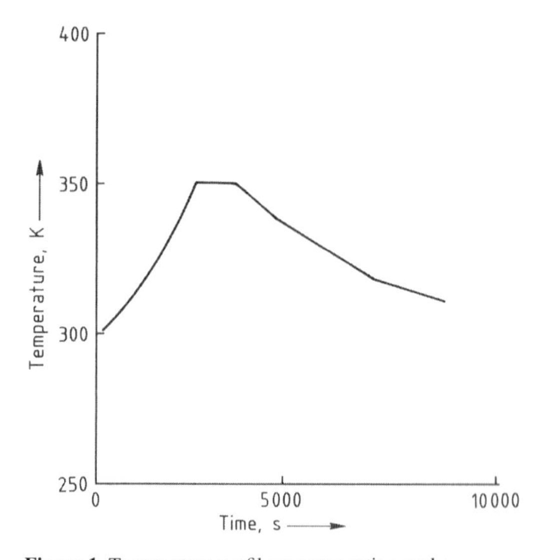

Figure 1. Temperature profile over a reaction cycle

Constant Volume. The energy balance for a constant volume process is

$$\varrho c_v \frac{dT}{dt} + (\Delta U_r)_J R_J = \frac{\dot{Q}}{V_r} \qquad (2.17)$$

Since ΔU_r is the change of internal energy at constant volume and temperature, for the case of an ideal gas ΔH_r and ΔU_r are related as follows:

$$\Delta H_r = \Delta U_r + RT\Delta n \tag{2.18}$$

This relation will be different for reaction mixtures with another equation of state.

Example. A closed tank with an internal volume of 250 m^3 contains a mixture of air with 4 % butane at 18 °C and a pressure of 100 kPa. The mixture is ignited and the reaction proceeds according to:

$$C_4H_{10} + 6.5\,O_2 \rightarrow 4\,CO_2 + 5\,H_2O$$

The heat of combustion at 18 °C and 100 kPa is $(-\Delta H_r)$ = 2880 kJ/mol. The pressure in the vessel shall be determined after the explosion.

Solution. Since the reaction proceeds adiabatically and at constant volume, the total internal energy of the reaction mixture does not change. The temperature rise is calculated from:

$$\int_{T_0}^{T} c_v \mathrm{d}T + \int_{\xi_{B0}}^{\xi_B} \frac{(\Delta U_r)_B}{M_B} \mathrm{d}\xi_B = 0$$

where the index B denotes butane.

To calculate $(\Delta U_r)_B$, the relationship between U and H and the ideal gas law will be used:

$$\Delta H_r = \Delta U_r + RT\Delta n$$

The total change in number of moles is (4 + 5 – 7.5 = 1.5), hence:

$$RT\Delta n = 1.5 \times 8.314\,\mathrm{J\,mol^{-1}K^{-1}} \times 291\mathrm{K} = 3.63 \times 10^3\ \mathrm{J/mol}$$

Since O_2 is the limiting reactant (i.e., O_2 is present in substoichiometric amounts), not all butane will react. The composition of the final mixture can then be calculated. In Table 1 the initial and final compositions are given. The specific heat capacity at constant volume (c_v) of the product mixture is a function of temperature and can be found in thermodynamic tables. In this example the average value of c_v is 1.06 kJ/kg. The mole fractions of butane before and after the reaction were 0.04 and 0.0089, respectively. These values can be converted to mass fractions by multiplying the mole fraction of each component in Table 1 by its molecular mass and normalizing. The mass fractions of butane before and after reaction are 0.077 and 0.017, respectively.

Table 1. Initial and final gas compositions

Compound	Mole fraction before reaction	Mole fraction after reaction
Butane	0.04	0.0089
Nitrogen	0.758	0.708
Oxygen	0.2028	
Carbon dioxide		0.182
Water		0.093

The temperature rise can now be determined:

$$c_v \times (T - T_0) + \left[\frac{(-\Delta H_r)_B - RT\Delta n}{M_B}\right] \times (\Delta\xi_B) = 0$$

so:

$$1.06\,\mathrm{kJ/kg} \times (T - 291\mathrm{K}) + \left(\frac{-2880\,\mathrm{kJ/mol} - 3.63\,\mathrm{kJ/mol}}{58\,\mathrm{g/mol}}\right) \times (0.077 - 0.017) = 0$$

$$T = 3103\mathrm{K} = 2830°\mathrm{C}$$

Applying the ideal gas law, the pressure can be determined:

$$P \times 250\mathrm{m}^3 = 1.046 \times n_0 \times 8.314\,\mathrm{J\,mol^{-1}K^{-1}} \times (3103\mathrm{K}) = 1115.4\,\mathrm{kPa}$$

where $1.046 \times n_0$ is the total number of moles present after the explosion. The last column in Table 1 is the mole fraction based on kmol/kmol original mixture and adding this column will give the relative change in the number of moles in the mixture.

2.2. Nonhomogeneous Systems

Batch systems are spatially homogeneous when the reactant mixture is well mixed during the course of the reaction. Nonideal mixing can cause settling or separation of phases in liquid –gas mixtures, mixtures of nonmiscible liquids or liquid – solid mixtures. However, even reactants which are well-mixed, do not necessarily react in a homogeneous way. Even if a single phase is present, the system can still be spatially nonhomogeneous.

2.2.1. Gas – Liquid Reactions

In the following example, the design of a batch reactor, where both a liquid and a gas phase are present, will be demonstrated.

Example. A batch reactor must be designed for the production of 1×10^5 kg/d of fats containing monounsaturated fatty acids by hydrogenation of cottonseed oil. The reaction is carried out on a nickel catalyst that is dispersed in hard stearin. Hydrogen is bubbled through the oil, thus two phases are present. The following equations describe the conversion and formation of acid moieties with diunsaturated (B), *cis*-monounsaturated (R_1), *trans*-monounsaturated (R_2), and saturated (S) aliphatic chains:

$$\frac{dc_B}{dt} = -(k_1+k_2)\,c_{H_s}^{0.5}c_B$$

$$\frac{dc_{R_1}}{dt} = k_1c_{H_s}^{0.5}c_B - k_3c_{H_s}^{0.5}c_{R_1} + k_4c_{H_s}^{0.5}c_{R_2} - k_5c_{H_s}c_{R_1}$$

$$\frac{dc_{R_2}}{dt} = k_2c_{H_s}^{0.5}c_B + k_3c_{H_s}^{0.5}c_{R_1} - k_4c_{H_s}^{0.5}c_{R_2} - k_6c_{H_s}c_{R_2}$$

$$\frac{dc_S}{dt} = k_5c_{H_s}c_{R_1} + k_6c_{H_s}c_{R_2}$$

The hydrogen concentration in the gas phase (c_{H_i}) is in equilibrium with the hydrogen on the catalyst surface, c_{H_s}. Since the gas is essentially pure hydrogen, c_{H_i} is constant. Furthermore it is assumed that the rate of consumption of adsorbed hydrogen on the catalyst surface is balanced by the rate of mass transfer from the gas phase. Let $K_D = c_{H_s}/c_{H_i}$ denote the ratio between the two concentrations, then the following relation exists:

$$K_D^{0.5} = \frac{-\left[1+\frac{1}{s_2}\right][1-\xi_B]}{2\left(\frac{1}{\gamma}+\frac{Y_{R_1}}{s_5}+\frac{Y_{R_2}}{s_6}\right)} + \frac{\left[\left(1+\frac{1}{s_2}\right)^2(1-\xi_B)^2+\frac{4}{\gamma}\left(\frac{1}{\gamma}+\frac{Y_{R_1}}{s_5}+\frac{Y_{R_2}}{s_6}\right)\right]^{0.5}}{2\left(\frac{1}{\gamma}+\frac{Y_{R_1}}{s_5}+\frac{Y_{R_2}}{s_6}\right)}$$

where $Y_{R_1} = c_{R_1}/Y_{B_0}$, $Y_{R_2} = C_{R_2}/c_{B_0}$, $\gamma = k_1c_{B_0}/K_L\,a_v\,c_{H_i}^{0.5}$, $s_2 = k_1/k_2$, $s_3 = k_1/k_3$, $s_4 = k_1/k_4$, $s_5 = k_1/k_5c_{H_i}^{0.5}$, $s_6 = k_1/k_6c_{H_i}^{0.5}$. The product of the mass-transfer coefficient and the bubble area a_v was determined experimentally as $K_L a_v = 0.022\ \mathrm{s}^{-1}$. Whereas the s_i values denote the ratios of reaction rate constants, γ is a ratio between kinetic and mass transfer rates.

Experiments on the hydrogenation of cottonseed oil using a Rufert nickel catalyst gave the following values for the reaction rate constants [8]: $k_1 = 0.254\sqrt{\mathrm{L/mol}}/\mathrm{min}$, s_2= 2.89, s_3= 5.05, s_4= 15.1, s_5= 3.38, s_6= 3.38. The hydrogen concentration in the gas phase is c_{H_i} = 0.0129 mol/L. The feedstock composition (wt %) consists of 25.6 % S, 27 % R_1, 0.4 % R_2 and 47 % B. The molar concentration of B, c_B is 1.45 mol/L. The value for γ can now be calculated:

$$\gamma = \frac{0.254\sqrt{\mathrm{L/mol}}/\mathrm{min}\times1.45\ \mathrm{mol/L}}{(60\times0.022)\ \mathrm{min}^{-1}\times0.1136\sqrt{\mathrm{mol/L}}/\mathrm{min}} = 2.45$$

The reactions are irreversible, hence c_S does not appear in the rate expressions for components B, R_1, and R_2 and they can be integrated independently from the balancing equation for S. Time can be substituted in terms of the conversion of B:

$$\frac{dY_{R_1}}{d\xi_B} = \frac{(1-\xi_B)-\frac{Y_{R_1}}{s_3}+\frac{Y_{R_2}}{s_4}-\frac{Y_{R_1}}{s_5}K_D^{0.5}}{\left(1+\frac{1}{s_2}\right)(1-\xi_B)}$$

$$\frac{dY_{R_2}}{d\xi_B} = \frac{\left(\frac{1-\xi_B}{s_2}\right)-\frac{Y_{R_1}}{s_3}+\frac{Y_{R_2}}{s_4}-\frac{Y_{R_1}}{s_5}K_D^{0.5}}{\left(1+\frac{1}{s_2}\right)(1-\xi_B)}$$

These two equations, together with the algebraic equation for K_D can now be integrated, using the initial conditions Y_{R_1} = 27/47, Y_{R_2} = 0.4/47, and $\xi_B = 0$. In Figure 2 A, the total yield of monounsaturates is shown as a function of ξ_B. A maximum is obtained at $\xi_B = 0.84$ and the time it takes to obtain 84 % conversion of B is 115 min (see Fig. 2 B). The processing time, excluding the reaction time, is best ascertained from plant studies; for this example 2.5 h are taken. Hence the total time (conversion plus processing) is 4.42 h. Thus, five batches can be processed per day, each batch producing 20×10^3 kg. At 84 % conversion the product mixture consists of 67 kg monounsaturates per 100 kg of mixture, hence the reactor volume is:

$$V_r = 1.05\times\frac{100\ \mathrm{kg}}{67\ \mathrm{kg}}\times20\,000\ \mathrm{kg}\times\frac{1}{800\ \mathrm{kg/m^3}} = 39.2\mathrm{m}^3$$

where the factor 1.05 is introduced because 5 % was added for head space and 800 kg/m^3 is the density of cottonseed oil. Using a height to diameter ratio of two, the specification of the vessel is 2.92×5.84 m.

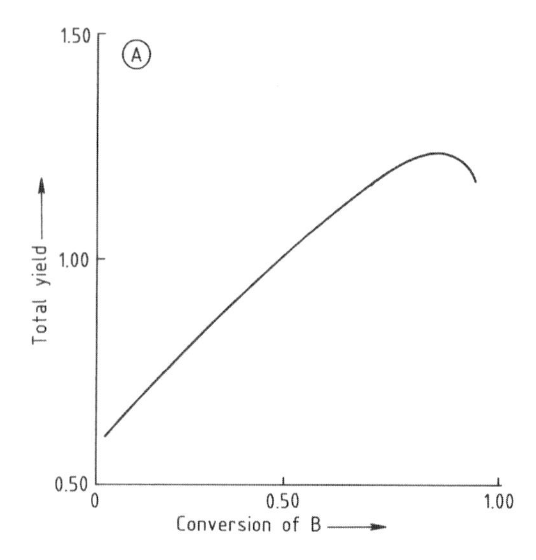

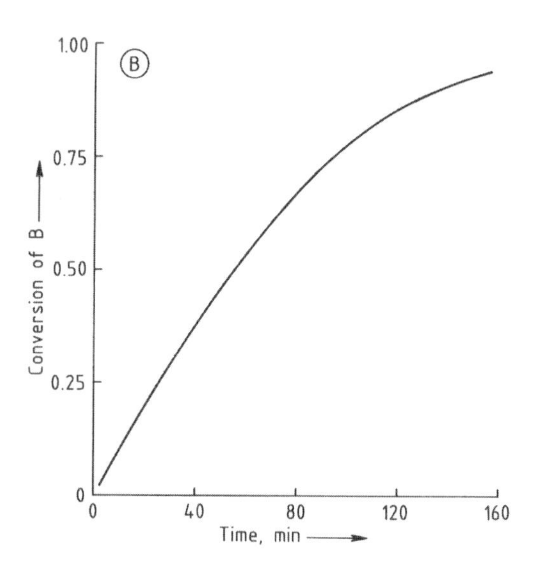

Figure 2. Total yield of monounsaturates in the hydrogenation of cottonseed oil (A) and conversion time of diunsaturates (B)

2.2.2. Solid – Solid Reactions

Even when only one phase is present, the reaction can progress in a nonhomogeneous way. Reactions with large activation energies require a certain amount of preheating before they begin. If the temperature profile in a batch system is not uniform, the reaction rate will also be spatially distributed and if the heat of reaction is large, it will further contribute to the concentration of high temperatures in local areas inside the reactor.

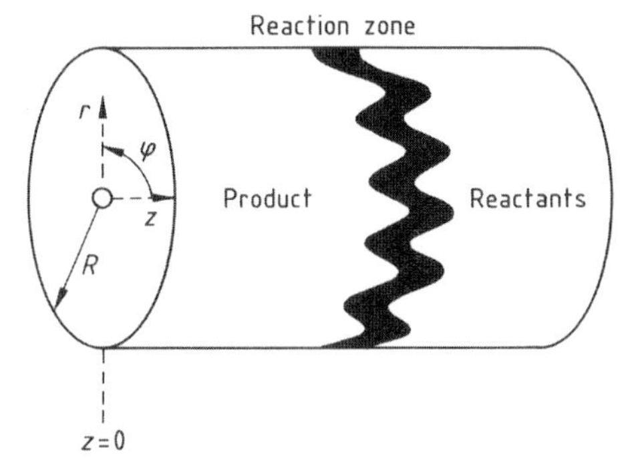

Figure 3. Typical configuration of solid – solid batch reactions

Solid – solid reactions that are associated with large activation energies and large heats of reaction often take place in the combustion mode. The combustion mode is associated with strong spatial gradients and modeling these type of reactions requires special techniques. Examples of solid – solid reactions which occur in the combustion mode are those between Ti and C, Mo and Si, and Ta and C. The well-mixed reactants are usually prepared as a preform that is ignited at one end of the sample (see Fig. 3). Once the combustion wave has traversed the length of the sample, the reaction is complete. Conversion is high in these combustion reactions and under ideal conditions complete conversion can be accomplished. Two factors determine the degree of conversion. If the system loses a lot of heat during the reaction, the degree of conversion drops. Thus, the closer the system can be operated to adiabatic conditions, the higher will be the degree of conversion. The second factor is incomplete mixing of the reactants. In these types of reactions it is important to have intimate contact between the different species on the micro-level. Therefore, mixing of the reactants is one of the most time-consuming and expensive steps in the process.

The mass and energy balances are

$$\frac{\partial c_J}{\partial t} = -R_J \tag{2.19}$$

$$\varrho c_p \frac{\partial T}{\partial t} = \nabla \cdot \lambda \nabla T + (-\Delta H_r)_J R_J \tag{2.20}$$

where λ denotes the thermal conductivity. Note that diffusion of the solid reactant is negligible. The geometry of the system is determined by the shape of the reactant mixture. A popular practice is to press the reactant mixture into a cylindrical preform. For this case, Equation (2.20) must be consistent with a cylindrical geometry, the dimensions of which are determined by the preform. The temperature at the reaction front can be very high and Equation (2.20) must be amended to take phase transformations into account. In the following section some aspects of the maximum temperature rise are discussed.

Maximum Temperature Rise. For $T_{ad} < T_{fus}$ the adiabatic temperature T_{ad} rise of a system where reactant A is transformed to product P, is given by:

$$T_{ad} = T_0 + \frac{(-\Delta H_r)\, c_A}{\varrho c_p} \tag{2.21}$$

When $T_{ad} > T_{fus}$ and $(-\Delta H_r) < \int_{T_0}^{T_{fus}} C_p dT + (-\Delta H_{fus})$ the following equation holds:

$$T_{ad} = T_0 + \frac{(-\Delta H_r)\, c_A}{\varrho c_p} - \frac{\nu\,(-\Delta H_{fus})}{c_p} \tag{2.22}$$

Otherwise:

$$T_{ad} = T_{fus} + \frac{(-\Delta H_r) c_A}{\varrho c_{pl}} - \frac{(-\Delta H_{fus})}{c_{pl}} - \frac{c_p}{c_{pl}} [T_{fus} - T_0] \tag{2.23}$$

where c_{p1} denotes the specific heat capacity of the liquid. Average properties over the temperature range must be used and the densities of the solid and liquid phases are assumed to be the same. The final temperature can be decreased by adding product to the reactant mixture to decrease c_A. There is a bound on the degree of dilution; if the system is too dilute, it will not ignite and self-propagation of the combustion front is not possible [9].

Equations (2.21) – (2.23) are derived on the basis of thermodynamic arguments, but the batch process is transient and the maximum temperatures in the transient process can be different from those obtained from Equations (2.21) – (2.23). Unfortunately no rigorous bounds on the maximum temperature of a transient process are known and they must be determined by numerical methods. Furthermore, these transient maxima can exceed the adiabatic temperature and Equations (2.21) – (2.23) cannot be relied upon to give upper estimates of the reaction temperature.

Example. Determine the maximum temperature of the following reaction:

$$\mathrm{Ti} + \mathrm{C} \rightarrow \mathrm{TiC}$$

A stoichiometric mixture of the reactants is pressed into a cylinder and simultaneously ignited at both ends. The cylinder is 2 cm long and perfectly insulated. The necessary properties of the system are listed below:

Density	3240 kg/m^3
Heat capacity	1113.48 J kg^{-1} K^{-1}
Thermal conductivity	8.37 W m^{-1} K^{-1}
Heat of formation at 25 °C	1.84×10^5 J/mol
Activation energy of formation reaction	8.29×10^5 J/mol
Frequency factor in expression for rate constant of formation reaction	1.05×10^5 s^{-1}

Solution. Equations (2.19), (2.20) are integrated numerically and the maximum temperature at the center of the Ti – C cylinder is shown in Figure 4. The adiabatic temperature for this reaction is obtained from Equation (2.21):

$$T_{ad} = 298\ \mathrm{K} + \frac{(1.8 \times 10^8)\ \mathrm{J/kmol}}{60\ \mathrm{kg/kmol} \times 1113.48\ \mathrm{J\ kg^{-1} K^{-1}}} = 3052\ \mathrm{K}$$

The maximum transient temperature is 5100 K. The temperature overshoot is 2000 K. Since the melting point of TiC is 3420 K, the transient maximum temperature will exceed the melting point.

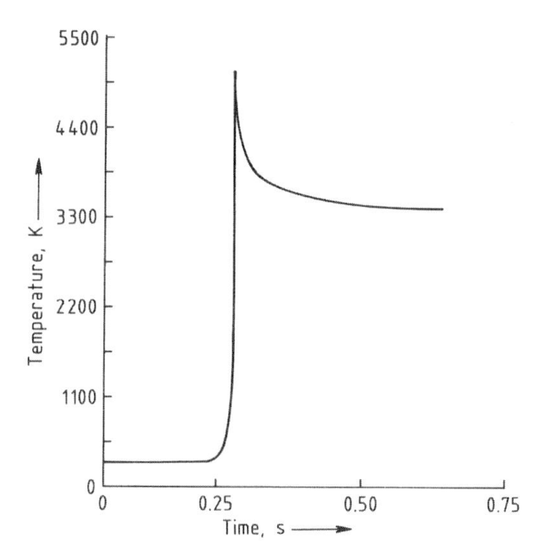

Figure 4. Temperature at the center of the cylindrical TiC sample

Reaction Wave Propagation. When the sample is ignited at one point by an external heat source, the reaction front can traverse the sample in a number of ways. The ideal propagation mode, from a processing point of view, is a stable front that propagates at a constant speed. If this front velocity is denoted by v_F, a similarity transformation can be made between time t and the direction of propagation x to render Equations (2.19), (2.20) time independent:

$$z = x + v_F t \tag{2.24}$$

The integrity of this planar front is not preserved when external heat losses are present. Heat losses also reduce ν_F. If heat losses are large enough, the system is quenched.

Other modes of front propagation also exist [10]. The planar front can, for example, become unstable and collapse into hot spots which are seated at different positions from the centerline. These hot spots propagate in a helical way around the centerline, leaving behind regions of incomplete conversion.

2.2.3. Solid – Gas Reactions

In solid – gas systems, combustion takes place between a solid and a gaseous component. A very important factor is the flow of gas through the porous solid by means of filtration or permeation. Hence this type of reaction is often termed *filtration combustion*. In Figure 5 various configurations for filtration combustion are shown.

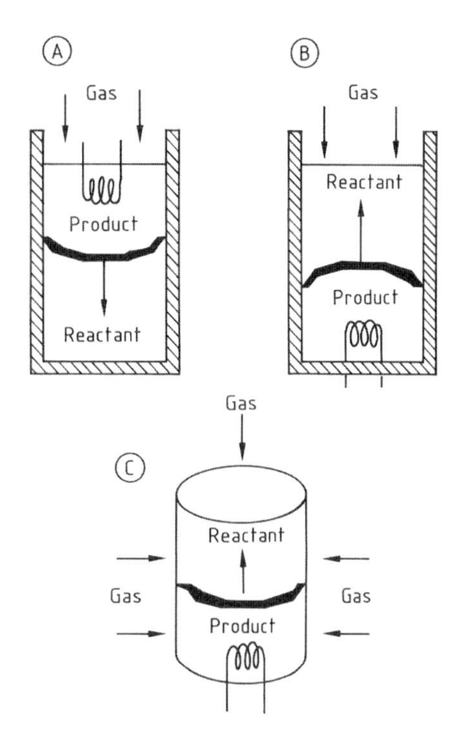

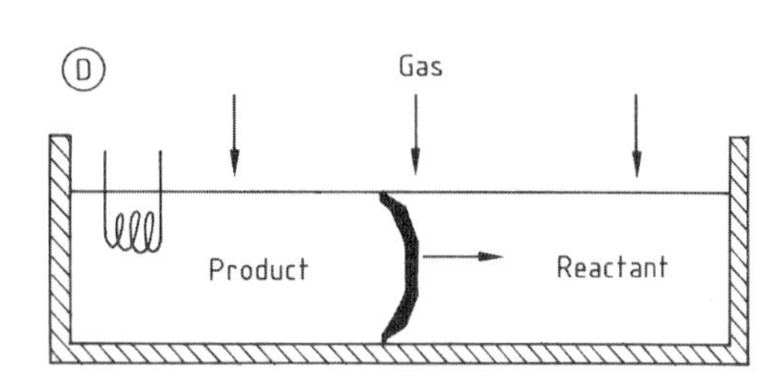

Figure 5. Different configurations of batch solid – gas reactions
A) Reaction initiated at top of container; B) Reaction initiated from bottom of container; C) Free-standing sample with gas penetrating from all sides; D) Design with both counter- and cross-flow of reactant gases towards reaction front

The first two configurations are similar. In Figure 5 A, the reaction was initiated at the top of the container and the front propagates away from the open end. In Figure 5 B the reaction was initiated at the bottom of the container and the front propagates countercurrently to the flow of gas. An advantage of the second option is that the diffusion length for the gas decreases with increasing degree of conversion. Product materials also tend to sinter partially during the reaction, which makes them less permeable than the fresh material, which is another advantage of the second configuration. The sample shown in Figure 5 C is free standing and gas can penetrate from all sides into the sample. However, the reactant powder must be compressed to give the sample the necessary strength, therefore the powder's permeability will decrease. The configuration depicted in Figure 5 D is the most practical, because both counter- and cross-flow of reactant gases towards the reaction front can be maintained without limiting the flow of the gaseous reactant through the solid medium. The filtration combustion model is described by the following set of equations:

$$\frac{\partial \varepsilon \varrho_{\mathrm{g}}}{\partial t} = -\nabla \left(\varepsilon \varrho_{\mathrm{g}} v\right) - \nu_{\mathrm{g}} R_{\mathrm{s}} \tag{2.25}$$

$$\frac{\partial \left[\varepsilon \varrho_{\mathrm{g}} c_{p\mathrm{g}} + (1-\varepsilon)\, \varrho_{\mathrm{s}} c_{p\mathrm{s}}\right]}{\partial t} = \nabla \cdot \nabla \left(\lambda T\right) - \nabla \left(\varepsilon \varrho_{\mathrm{g}} v T\right) + (-\Delta H)_{\mathrm{s}} R_{\mathrm{s}} \tag{2.26}$$

$$p = \frac{\varrho_{\mathrm{g}} R T}{M_{\mathrm{g}}} \tag{2.27}$$

$$v = -P \nabla_p \tag{2.28}$$

$$R_{\mathrm{g}} = -\varrho_{\mathrm{B}}^{0} p^{n} f\left(c_{\mathrm{s}}\right) k \tag{2.29}$$

where ε is the porosity of the bed, P the permeability in the bed, ϱ_{B}^{0} the initial bulk density of the solid, and $f\left(c_{\mathrm{s}}\right)$ the dependence of the reaction on the concentration of the solid phase. The porosity and permeability of the bed change as the reaction progresses, mainly because of sintering and melting; this affects the flow of gas to the front. The effects of sintering and melting can be incorporated into the model.

Example. A rectangular container is filled with titanium powder and placed in a chamber filled with nitrogen. The reaction is ignited at the top right of the container and the pressure in the chamber is kept constant. Low and high chamber pressures affect the conversion and in Figure 6 the results for both cases at different times (time increases from top to bottom) are shown.

The front propagates faster when the overhead pressure is higher and penetrates to the bottom of the container. When the overhead pressure is lower, conversion is incomplete and the reaction time is also much longer. When the depth of the container is increased, the chamber pressure must be further increased, but the exact values required can only be established through a numerical simulation.

3. Continuous Stirred-Tank Reactors (CSTR)

3.1. Isothermal Homogeneous System

The continuous, ideally mixed, stirred-tank reactor (CISTR) is a flow reactor in which all elements of the fluid have the same composition as the outlet stream of the reactor.

The continuous stirred-tank reactor is normally run at steady state, and is usually operated such that it is well mixed. As a result, the CISTR is generally modeled as having no spatial variations in concentration, temperature, or reaction rate throughout the vessel. In systems where mixing is highly nonideal, the well-mixed model is inadequate and other techniques, such as residence-time distributions, must be used to obtain meaningful results.

The CISTR has the following unique characteristics:

1) Temperature and heat transfer can be relatively easily controlled
2) Fine catalyst particles can be effectively suspended in a liquid reacting system
3) In a system with parallel reactions, a reaction with lower order is favored
4) In polymerization reactions, wider molecular mass distributions are expected
5) Very economical when long residence time is required
6) Recommended for liquid rather than gaseous reacting systems

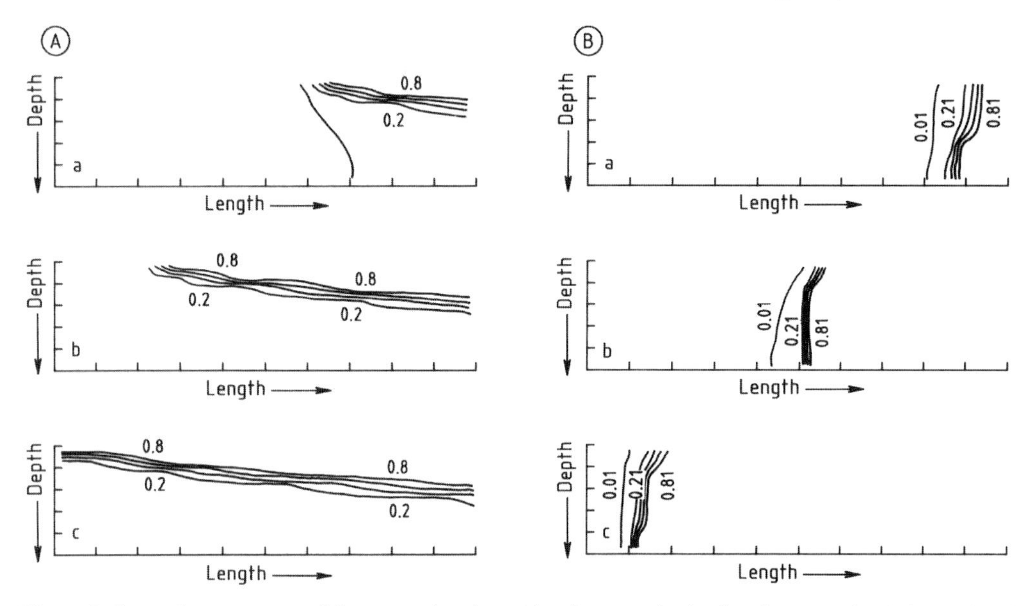

Figure 6. Conversion contours at different reaction times (time increases in the direction a – c) for A) low and B) high chamber pressures

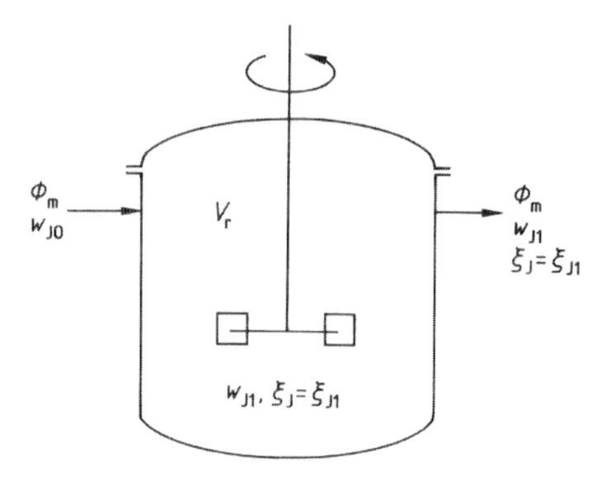

Figure 7. CISTR reactor
For explanation of symbols see text

Material Balance. The overall material balance for a CISTR which is schematically shown in Figure 7 reads:

$$\frac{d\,(m\xi_\mathrm{J})}{\mathrm{d}t} = -\Delta\Phi_\mathrm{m}\xi_\mathrm{J} + M_\mathrm{J}|R|_\mathrm{J}V_\mathrm{r} \tag{3.1}$$

where m is the total mass in the system; $-\Delta\Phi_\mathrm{m}$ the net inflow of mass by convection and diffusion; ξ_J the degree of conversion of component J; $M_\mathrm{J}\,|R|_\mathrm{J}$ the mass rate of production of component J per unit volume; and V_r the volume of the reaction mixture.

When a CISTR is operated in a steady state, the required reactor volume is

$$V_\mathrm{r} = \Phi_\mathrm{m}\frac{\xi_\mathrm{J}}{M_\mathrm{J}|R|_\mathrm{J}} \tag{3.2}$$

or in the case of constant density

$$V_\mathrm{r} = \Phi_v\frac{c_{\mathrm{J}_0}\zeta_\mathrm{J}}{R_\mathrm{J}} \tag{3.3}$$

where Φ_v is the volumetric flow rate; ζ_J the relative degree of conversion; c_{J_0} the molar concentration of species J; and R_J the molar rate of production of J per unit volume of the reaction phase. The mass balance equations for various reaction orders are listed in Table 2.

Table 2. Ratio of outlet to inlet concentrations in a CISTR for reactions of different order

Reaction order, n	Reaction rate equation, R_A	Concentration ratio, $c_\mathrm{A}/c_{\mathrm{A}_0}$
-1	kc_A^{-1}	$\frac{1}{2}\left[1+\sqrt{\left(1-\frac{4k\tau}{c_{\mathrm{A}_0}^2}\right)}\right]$
0	k	$1-\frac{k\tau}{c_{\mathrm{A}_0}}$
$\frac{1}{2}$	$k\sqrt{c_\mathrm{A}}$	$\sqrt{\left(1+\frac{k2\tau2}{4c_{\mathrm{A}_0}}\right)-\frac{k\tau}{2\sqrt{c_{\mathrm{A}_0}}}}$
1	$k\,c_\mathrm{A}$	$\frac{1}{1+k\tau}$
2	kc_A^2	$\frac{1}{2kc_{\mathrm{A}_0}\tau}\lvert\sqrt{(1+4kc_{\mathrm{A}_0})}-1\rvert$

Most important in designing a homogeneous CISTR system is to select proper equipment and stirring procedures that will assure essentially perfect mixing. Choice of a suitable type of impeller and reactor arrangement, the required power, the mode of adding and removing reactants and products, and the cooling or heating rates are all essential concerns. For more details, see [11, 12].

Mixing. For a processing vessel to be effective, the volume of fluid or slurry that is circulated by the impeller must be sufficient to sweep out the entire vessel in a reasonable time. Also, the velocity of the stream leaving the impeller must be high enough to carry the currents to the remotest parts of the tank. In addition, an adequate level of turbulence is required in order to maintain the high effectiveness of the process. The proper selection of an impeller, its speed, and the geometrical dimensions of the reactor are very important design parameters.

In practice, for moderately fast reactions, a tank reactor may be approximately considered as ideally mixed if the discharge rate of the impeller Φ_c (the circulation rate) is at least five to ten times the feed rate Φ_m i.e.,

$$\frac{\Phi_c}{\Phi_m} > 5-10 \qquad (3.4)$$

The impeller discharge rate depends on the type of impeller, the tank geometry, and operating conditions. This discharge rate can be estimated from:

$$\frac{\Phi_c}{N d_I^3} = N_{QI} \qquad (3.5)$$

where d_I is the impeller diameter, N the number of impeller revolutions per unit time, and N_{QI} is a dimensionless constant. The value of this constant depends on the axial blade width of the impeller, the number of blades, curvature or pitch, and the ratio of impeller velocity to fluid rotational velocity [12]. The following values are recommended for agitators in baffled tanks which are commonly used in industry:

Marine-Type Impeller:

$$N_{QI} = 0.5$$

Turbine (with six blades and width-to-diameter ratio of 1 : 5):

$$N_{QI} = 0.93 D/d_I \text{ for } Re_{imp} > 10^4$$

where $Re_{imp} = \varrho N d_I^2/\mu$ is the impeller Reynolds number, and D the diameter of the reactor.

Many reactions require rapid dispersal of reactants in order to avoid even momentary buildup of high concentrations in the system. In homogeneous reacting systems, the most rapid dispersal can be accomplished by discharging the feed as close as possible to the center of the impeller. Discharge ports for product removal should be located at the opposite end of the flow pattern from the inlet nozzle. In general, the flow of a reacting fluid through a reactor is a very complex process. The interaction of flow pattern and chemical reaction cannot be analyzed rigorously even in the simple case of a single isothermal reaction. Therefore simplified and/or idealized models are necessary. Presently, computer-aided three-dimensional modeling of fluid and dispersed phase velocities, temperature, and concentration profiles in CSTRs are conducted with commercial software packages. The most widely used software was developed by the Fluent Company. Commonly, dynamic flow patterns are calculated from conservation equations for mass, energy, and momentum in combination with the algebraic stress model (ASM) for turbulent Reynolds stresses. Numerical solutions of model equations correctly predict blend time and reactor product distribution. These models are very useful for scaleup and finding optimal impeller speed, reactor geometry, and feed location [23].

Residence-Time Distribution (RTD) The frequency function often used in statistics (called the E diagram in RTD description) can be derived for a continuous stirred-tank reactor from a mass balance of the reactor with an ideal pulse perturbation at the inlet stream. The pulse is immediately distributed over the reactor, moreover the product concentration remains the same everywhere and equals the outlet concentration. The $E\,(t)$ diagram reads:

$$E\,(t) = \frac{1}{\tau} e^{-t/\tau} \qquad (3.6)$$

For a continuously operated mixing vessel which shows a RTD of an ideal mixer (Eq. 3.6), the different states of micromixing are

1) CISTR: the ideal mixer (mixing on molecular level)
2) CSSTR: the completely segregated mixer where the fluid consists of lumps which move through the reactor without exchange of matter with their surroundings
3) CSTR: the partially segregated mixer

A detailed description of the degree of micromixing with one parameter is generally not possible. For non-first-order reactions, apart from the reaction rate not only the intensity of mass exchange between fluid elements is important, but also the residence time of a molecule in the reactor.

In the case of maximum micromixing and constant density of the reacting system, the conversion is identical to that calculated for the CISTR. When segregated flow takes place, the situation is different. Each aggregate in the outlet stream can be considered as a batch reactor with a reaction time equal to its residence time. The fraction of aggregates having a residence time between t and $t + \Delta t$ is $E\,\Delta t$. If the remaining reactant concentration after conversion in a batch reactor with residence time t is $c_A^{Batch}(t)$, the average concentration at the outlet of the segregated continuous flow reactor CSSTR can be calculated from:

$$\frac{\bar{c}_A}{c_{A0}} = \frac{1}{c_{A0}} \int_0^\infty c_A^{Batch}(t)\, E\,dt \qquad (3.7)$$

or

$$\frac{\bar{c}_A}{c_{A0}} = \frac{1}{c_{A0}} \int_0^\infty c_A^{Batch}(t)\, \frac{1}{\tau} e^{-t/\tau}\,dt \qquad (3.8)$$

The average outlet concentrations for different reaction kinetic expressions for both ideal micromixing and segregated flow are listed in Table 3 [14]. Outlet concentrations are identical in both cases if a first-order reaction takes place.

Example [14]. Reactant A (M_A = 104 kg/kmol) is to be polymerized in a dispersed phase in a continuous stirred-tank reactor. Based on batch-scale experiments, the following reaction rate expression was found:

$$R_A = -k c_A^{1.5}$$

with $k = 2.5 \times 10^{-4}\,s^{-1}\,kmol^{-0.5}\,m^{1.5}$

Table 3. Design equations for ideally mixed CISTR and for completely segregated fluids CSSTR

Kinetics	CISTR
$R_A = -k c_A$	$\frac{c_A}{c_{A0}} = \frac{1}{1+N_r}$
$N_r = k\tau$	
$R_A = -k$	$\frac{c_A}{c_{A0}} = 1 - N_r,\ N_r \le 1;\ \frac{c_A}{c_{A0}} = 0,\ N_r \ge 1$
$N_r = \frac{k\tau}{c_{A0}}$	
$R_A = -k c_A^{0.5}$	$\frac{c_A}{c_{A0}} = 1 - \frac{N_r^2}{2}\left[-1 + \sqrt{\left(1 + \frac{4}{N_r^2}\right)}\right]$
$N_r = \frac{k\tau}{\sqrt{c_{A0}}}$	
$R_A = -k c_A^2$	$\frac{c_A}{c_{A0}} = \frac{-1 + \sqrt{(1 + 4N_r)}}{2N_r}$
$N_r = k\tau c_{A0}$	
$R_A = -k c_A^n$	$\left(\frac{c_A}{c_{A0}}\right)^n N_r + \frac{c_A}{c_{A0}} - 1 = 0$
$N_r = c_{A0}^{n-1} k\tau$	

Kinetics	CSSTR
$R_A = -k c_A$	$\frac{\bar{c}_A}{c_{A0}} = \frac{1}{1+N_r}$
$N_r = k\tau$	
$R_A = -k$	$\frac{\bar{c}_A}{c_{A0}} = 1 - N_r + N_r e^{-\frac{1}{N_r}}$
$N_r = \frac{k\tau}{c_{A0}}$	
$R_A = -k c_A^{0.5}$	$\frac{\bar{c}_A}{c_{A0}} = 1 - N_r\left[1 - \frac{N_r}{2}\left(1 - e^{-\frac{2}{N_r}}\right)\right]$
$N_r = \frac{k\tau}{\sqrt{c_{A0}}}$	
$R_A = -k c_A^2$	$\frac{\bar{c}_A}{c_{A0}} = -\frac{1}{N_r} e^{\frac{1}{N_r}} Ei\left(-\frac{1}{N_r}\right)^*$
$N_r = k\tau c_{A0}$	
$R_A = -k c_A^n$	$\frac{\bar{c}_A}{c_{A0}} = \frac{1}{\tau}\int_0^\infty [1 + (n-1) c_{A0}^{n-1} k t]^{\frac{1}{1-n}} e^{-t/\tau} dt$
$N_r = c_{A0}^{n-1} k\tau$	

$* \; Ei(-x) = \int_x^\infty \frac{e^{-y}}{y} dy$

The required size of the tank reactor shall be calculated for a production of 10 t/d of polymer with a relative degree of conversion $\zeta_A = 0.9$, under the assumption that coalescence between the liquid drops does not occur. The feed of the dispersed phase consists of pure A, and the density of this phase remains constant at a value of 832 kg/m^3 during the reaction. The volume fraction of the dispersed phase in the reactor is 0.165.

Solution. The average conversion $\bar{\zeta}_A$ at the outlet of the CSSTR is obtained by rearrangement of Equation (3.8):

$$\bar{\zeta}_A = \int_0^\infty \zeta_A^{\text{Batch}}(t)\,\frac{1}{\tau}\,e^{-t/\tau}\,dt \qquad (3.8a)$$

The material balance for a batch system together with the rate expression yields:

$$\bar{\zeta}_A^{\text{Batch}} = 1 - \left(\frac{2}{2 + kt\sqrt{c_{A0}}}\right)^2$$

By integrating Equation (3.8 a) a value of $k\tau\sqrt{c_{A_0}} = 15.4$ is found for $\bar{\zeta}_A = 0.9$. The calculated residence time τ is 6 h and the required volume of the dispersed liquid $\Phi_m \tau / 0.9\, \varrho_A$ is equal to 3.34 m^3. The total required volume in the reactor is 3.34 m^3/0.165 = 20.2 m^3.

Start-Up of a CISTR. For non-steady-state operations, the accumulation term in Equation (3.1) cannot be neglected. Two different periods of non-steady-state operations in a CISTR can be distinguished:

1) Filling a reactor with reactants (variable reactor volume)
2) Non-steady-state operation at a constant reactor volume

In the case of an irreversible liquid reaction (A $\rightarrow$ B) which takes place in a CISTR under isothermal and constant density conditions, the instant concentration in the reactor can be calculated as follows: For $t \leq \tau$, no liquid leaves the reactor and therefore the material balance for a limiting reactant can be expressed as follows:

$$\frac{-d(c_A V)}{dt} = -\Phi_v c_{A_0} + k c_A V \qquad (3.9)$$

where the reaction volume is $V = \Phi_v\, t$. Integration of Equation (3.9) gives:

$$c_A = \frac{c_{A0}}{kt}\left(1 - e^{-kt}\right) \quad \text{for } t \leq \frac{V}{\Phi_v} \qquad (3.10)$$

For $t > \tau$ the reaction volume V is constant and equals V_r and the product flow out of the reactor equals the feed rate Φ_v. Now the material balance for reactant A becomes:

$$\frac{-d(c_A V_r)}{dt} = -V_r \frac{dc_A}{dt} = -\Phi_v\left(c_{A_0} - c_A\right) + k c_A V_r \qquad (3.11)$$

Integration yields:

$$\frac{c_{A_0} - (1 + k\tau)\, c_A}{c_{A_0} - (1 + k\tau)\, c_{A_\tau}} = e^{-(1 + k\tau)(t - \tau)/\tau} \quad \text{for } t > \tau \qquad (3.12)$$

where $c_{A\tau} = \frac{c_{A0}}{k\tau}\left(1 - e^{-k\tau}\right)$

After a long time ($t \rightarrow \infty$) the right-hand side of Equation (3.12) approaches zero, so that ultimately the concentration c_A in the reactor reaches the value $c_{A\infty}$. A conservative estimate for the time required to reach steady state can be made by substituting $c_{A_\tau} = c_{A_0}$ in Equation (3.12) for the required degree of conversion.

3.2. Isothermal Heterogeneous Reactors

Contrary to reactions in a single phase where mixing takes place on a molecular scale, in a multiphase system not all reactive molecules present may be available for chemical reaction due to mass transfer limitations. For a reacting system consisting of two phases L and G, with a reaction taking place in phase L only, the mass balance equations are:

$$\frac{dm_{JL}}{dt} = -\Delta\left(\Phi_{mL} w_{JL}\right) + J_{JL} M_J a V_r + R_J M_J (1 - \varepsilon) V_r \qquad (3.13)$$

and:

$$\frac{dm_{JG}}{dt} = -\Delta\left(\Phi_{mG} w_{JG}\right) + J_{JG} M_J a V_r \qquad (3.14)$$

The term $R_J\, M_J\, (1 - \varepsilon)\, V_r$ is the production rate of J by a homogeneous chemical reaction in phase L where $(1 - \varepsilon)$ is the volume fraction of the reaction phase per unit volume of a reactor. The terms $(-\Delta\Phi_m\, w_J)$ and $(J_J\, M_J\, a)$ represent the net supply of component J by convection and the mass flow of J from the interface to the bulk of the phase, respectively.

When a heterogeneous reaction takes place in a system consisting of two immiscible or partially miscible liquid phases, the overall process can be controlled by a chemical reaction and/or mass transfer between both phases. In the case of a *chemically controlled process*, the overall rate expression in Equation (3.1) is the sum of reaction rates in the individual phases.

When *mass transfer* is the rate-determining step, phase equilibrium cannot be achieved and the reaction rate is fast compared to the rate of mass transfer. Thus, the reaction proceeds as the reactants diffuse through the interface into the bulk of the other phase. If reactant A is a major component of phase I while phase II consists of reactant B and if equilibrium is attained at the interface, the rate of mass transfer of A from phase I to phase II would be equal to the rate of disappearance of A by chemical reaction in phase II. For an irreversible, second-order reaction between A and B, mass transfer of A from phase I to phase II can be described as:

$$J_{\mathrm{AII}} = k_{\mathrm{II}} a \left(\frac{c_{\mathrm{AI}}}{K_{\mathrm{A}}} - c^0_{\mathrm{AII}} \right) (V_{\mathrm{I}} + V_{\mathrm{II}}) \tag{3.15}$$

where k_{II} is the mass-transfer coefficient in phase II per unit volume of both phases (m/s), a is the interfacial area per unit volume of both phases (m^2/m^3), c_{AI} is the concentration of reactant A in phase I, K_{A} is the distribution coefficient of A between both phases, and c^0_{AII} is the steady-state concentration of A in phase II.

The value of the product of mass-transfer coefficient and interfacial area strongly depends on the reactor arrangement and empirical correlations must be used to scale-up such reacting systems [15, 16].

3.3. Nonisothermal Continuous Stirred-Tank Reactors

All chemical reactions are in principle accompanied by evolution or absorption of heat and these effects often have to be taken into account. For a flow process, the energy balance equation reads:

$$\frac{\mathrm{d}\left(\langle u \rangle m\right)}{\mathrm{d}t} = -\Delta\left(h \Phi_{\mathrm{m}}\right) + \dot{Q} - \dot{W} \tag{3.16}$$

where h is the enthalpy per unit mass of the reaction mixture and $\langle u \rangle$ is the average internal energy over the reactor volume. For the steady-state operation of a continuous tank reactor provided with a heat exchange area A, the energy balance becomes:

$$\Phi_{\mathrm{m}}\left(h_1 - h_0\right) = UA\left(T_{\mathrm{c}} - T_1\right) \tag{3.17}$$

where the subscripts 1 and 0 refer to the conditions at the outlet and inlet, respectively; T_{c} is the temperature of the cooling or heating medium. The difference in enthalpy ($h_1 - h_0$) can be calculated from:

$$h_1 - h_0 \int_{T_{0,\mathrm{o}}}^{T_1 \xi_{\mathrm{J},1}} \left[c_p \mathrm{d}T + \left(\Delta h_{\mathrm{r}}\right)_{\mathrm{J}} \mathrm{d}\xi_{\mathrm{J}}\right] \tag{3.18}$$

where c_p is the specific heat capacity at a constant pressure and (Δh_{r}) the enthalpy of reaction at the constant pressure per unit mass of J converted. If the density and heat capacity of the system are constant, the energy balance for reactant A gives:

$$\begin{aligned} &-\left(\Delta H_{\mathrm{r}}\right)_{\mathrm{A}}\left(c_{\mathrm{A}_0} - c_{\mathrm{A}_1}\right) \\ &\quad = c_p \varrho \left(T_1 - T_0\right) + \frac{UA}{\Phi_v}\left(T_1 - T_{\mathrm{c}}\right) \end{aligned} \tag{3.19}$$

where $(\Delta H_{\mathrm{r}})_{\mathrm{A}}$ is the heat of reaction per mole of A converted. In the case of a first-order irreversible reaction Equations (3.1) and (3.19) can be rewritten to:

$$\begin{aligned} &-\left(\Delta H_{\mathrm{r}}\right)_{\mathrm{A}} c_{\mathrm{A}_0} \frac{k\tau}{1 + k\tau} \\ &\quad = \varrho c_p \left(T_1 - T_0\right) + \frac{UA}{\Phi_v}\left(T_1 - T_{\mathrm{c}}\right) \end{aligned} \tag{3.20}$$

In order to find the maximum temperature T_1, Equation (3.20) must be solved by a trial and error procedure, either numerically or graphically, because k is an exponential function of temperature. The left-hand side of Equation (3.20) represents the heat produced per unit volume of a reaction mixture. The right-hand side represents the heat removed per unit volume of the reaction mixture as a result of the heat absorbed by the cold feed and the heat transferred to the cooling medium. The maximum temperature rise can be reached when the system is adiabatic (no heat transfer to the surroundings) and complete conversion is achieved:

$$\Delta T_{\mathrm{ad}} = \frac{-\left(\Delta H_{\mathrm{r}}\right)_{\mathrm{A}} c_{\mathrm{A}_0}}{\varrho c_p} \tag{3.21}$$

Autothermal Reactor Operation. The heat released during an exothermic reaction is often used for preheating inlet reactant streams. A reactor system, in which such a feedback of reaction heat to the incoming reactant stream takes place is said to operate under autothermal conditions. The theory of steady-state behavior of autothermal reactors is described in [17, 18].

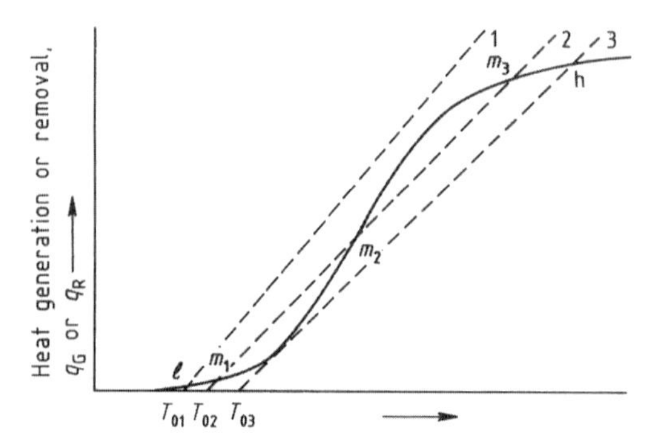

Figure 8. Heat production (——) and heat removal (– – –) in a nonisothermal CISTR for an irreversible exothermic reaction A → R

Multiplicity In Figure 8 both right- (heat removal) and left-hand sides (heat production) are plotted for different inlet and operating conditions. At the intersection of the heat production (solid line) and heat removal curves (dashed lines), Equation (3.20) is obeyed. For a certain range of inlet parameters several solutions may exist (multiplicity of steady states). If the heat removal is represented by line 1 in Figure 8, the reactor is cooled so effectively that steady-state operation is possible only at a low temperature and a very low degree of conversion. If the cooling rate is lowered or the inlet temperature of reactants increased (line 2), three points of intersection occur with the heat generation curve. The two intersection points, the lowest and highest reaction temperature, represent stable conditions due to the fact that the rate of heat removal is higher than that generated in the system. The intermediate intersection point corresponds to unstable steady state. Since the slope of the heat production curve at the intermediate point is greater than that of the heat removal line, any positive temperature perturbation will be amplified until the reactor reaches the upper stable operating point, and negative temperature deviation leads to extinction (lower stable operating point). If the feed temperature T_0 and the cooling capacity are such that the heat removal line 3 prevails (see Fig. 8), the reactor operates in an upper stable state and no special measures need be taken to start up a process.

In Figure 9 a multiplicity region is presented as a function of the maximum temperature increase in the reactor, $\left(\frac{\Delta T_{ad}}{1+St'}\right)$. In the area enveloped by the curves multiplicity will occur. Outside, only one (stable) operating point is possible. Operating points located above the multiplicity region are distinguished by a high degree of conversion, those below the multiplicity region by a low degree of conversion. The point with the lowest value of $\frac{\Delta T_{ad}}{1+St'}$ where multiplicity occurs is often called the *trifurcation point*, and the curves departing from it, the *bifurcation lines*.

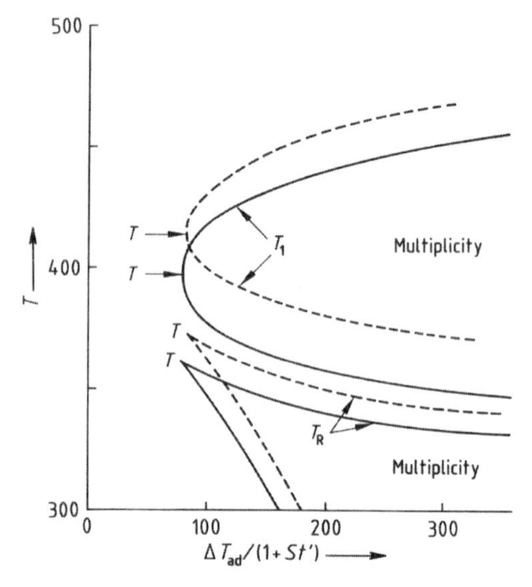

Figure 9. Regions of multiplicity in a CISTR [9]
T = Trifurcation point; ——— $\tau = 1$ s; – – – $\tau = 0.5$ s

HLAVACEK et al. derived the criteria for existence of multiplicity of steady states for first-order reactions [19]. Multiplicity occurs when the dimensionless adiabatic temperature rise $B = \frac{(-\Delta H_r) c_{A_0}}{\varrho c_p T_0} \frac{E}{R T_0}$ exceeds a critical value of B^*:

1) Adiabatic case ($\overline{\beta} = 0$)

$$B^* = \frac{4\overline{\gamma}}{\overline{\gamma}-4}$$

2) Adiabatic case ($\overline{\beta} = 0$); $\overline{\gamma} \to \infty$

$$B^* = 4$$

3) Nonadiabatic case; $\overline{\gamma} \to \infty$

$$B^* = 4\overline{\delta}$$

4) Nonadiabatic case

$$B^* = \frac{4\left(\overline{\varepsilon}+\overline{\delta}\overline{\gamma}\right)^2}{\overline{\gamma}\overline{\delta}\left(\overline{\gamma}-4\right)-4\overline{\varepsilon}}$$

where $\overline{\gamma} = E/RT_0$ is a dimensionless activation energy, $\overline{\beta} = UA/\Phi_v\ \varrho\ c_p$ is a dimensionless heat transfer coefficient, $\overline{\delta} = 1 + \overline{\beta}$, $\overline{\varepsilon} = \overline{\beta}\ \overline{\Theta}_c$, and $\Theta_c = \frac{E}{RT_0^2}(T_c - T_0)$ is a dimensionless cooling temperature.

Hysteresis in Autothermal Systems. If the feed temperature T_0 is increased at a constant feed rate, the reaction will be ignited at a certain feed temperature $(T_0)_{ign}$ and the reactor will operate at the upper stable steady state. If T_0 is then lowered again, the reactor continues operating at a high conversion level until extinction takes place at a feed temperature $(T_0)_{ext}$.

Stability of Autothermal CISTR. The transient behavior of strongly exothermic reacting systems may lead to unexpected, undesired changes in a reactor. This is a result of coupling between the material and energy balances (Eqs. 3.1 and 3.16). Detailed information on dynamic behavior of autothermal systems can be found in [20 – 23].

The mass and energy balance Equations (3.1) and (3.16) can be rewritten to:

$$\frac{d\zeta}{d\Theta} = -\zeta + \frac{\tau R_A}{c_{A_0}} \tag{3.22}$$

and:

$$\frac{dT}{d\Theta} = T_0 - T - \Delta T_{ad}\frac{R_A \tau}{c_{A_0}} + St'\left(T_c - T\right) \tag{3.23}$$

where $\tau = \varrho\ V_r/\Phi_m$ is the average residence time, $\Theta = t/\tau$ is a dimensionless time, and St' a modified Stanton number ($St' = UA/c_p\ \Phi_v$).

In the case of a first-order reaction Equations (3.22) and (3.23) can be rewritten to:

$$\frac{d\zeta}{d\Theta} = -\zeta + k_s\tau\left(1-\zeta\right)e^{\Delta\vartheta} \tag{3.24}$$

$$\frac{d\vartheta}{d\Theta} = -\left(1+St'\right)\Delta\vartheta + \Delta\vartheta_{ad}k_s\tau\left(1-\zeta\right)e^{\Delta\vartheta} + \Delta\vartheta_0 + St'\Delta\vartheta_c \tag{3.25}$$

where:

$$\Delta\vartheta_0 = \left(T_0 - T_s\right)\frac{E}{RT_s^2}$$

$$\Delta\vartheta_c = \left(T_c - T_s\right)\frac{E}{RT_s^2}$$

$$\Delta\vartheta_{ad} = \Delta T_{ad}\frac{E}{RT_s^2}$$

and k_s is the reaction rate constant at temperature T_s.

When a small perturbation is imposed on a steady state, Equations (3.24) and (3.25) can be linearized as follows:

$$\frac{d\Delta\zeta}{d\Theta} = -\frac{1}{1-\zeta_s}\Delta\zeta + \zeta_s\Delta\vartheta \equiv a_{11}\Delta\zeta + a_{12}\Delta\vartheta \tag{3.26}$$

and:

$$\frac{d\Delta\vartheta}{d\Theta} = \frac{\Delta\vartheta_{ad}\zeta_s}{1-\zeta_s}\Delta\zeta + \left[\Delta\vartheta_{ad}\zeta_s - \left(1+St'\right)\right]\Delta\vartheta \equiv a_{21}\Delta\zeta + a_{22}\Delta\vartheta \tag{3.27}$$

where $\Delta\zeta = \zeta - \zeta_s$ and $\Delta\vartheta = \vartheta - \vartheta_s$; ζ_s and ϑ_s are the conversion and the dimensionless temperature at steady state.

The solution of Equations (3.26) and (3.27) yields:

$$\Delta\zeta = b_{11}e^{p_1\Theta} + b_{12}e^{p_2\Theta} \tag{3.28}$$

$$\Delta\vartheta = b_{21}e^{p_1\Theta} + b_{22}e^{p_2\Theta} \tag{3.29}$$

where:

$$p_{1,2} = \frac{\left(a_{11}+a_{22}\right)}{2}\left[1 \pm \sqrt{\left(1 - \frac{4\left(a_{11}a_{22} - a_{12}a_{21}\right)}{\left(a_{11}+a_{22}\right)^2}\right)}\right]$$

The chemical process is stable when:

$$\frac{1+St'}{\Delta\vartheta_{ad}} > \zeta_s\left(1-\zeta_s\right) \tag{3.30}$$

and:

$$\frac{1}{1-\zeta_s} - \Delta\vartheta_{ad}\zeta_s + 1 + St' > 0 \tag{3.31}$$

where $\zeta_s = k\ \tau/(1 + k\ \tau)$ in the case of first-order reaction.

If the second condition (Eq. 3.31) is met, the reactor is dynamically stable and operates at its steady state (ζ_s, T_s). However, if this condition is not satisfied the reactor is dynamically unstable and operates in a fixed *limit cycle* around the steady-state point (ζ_s, T_s) (Fig. 10).

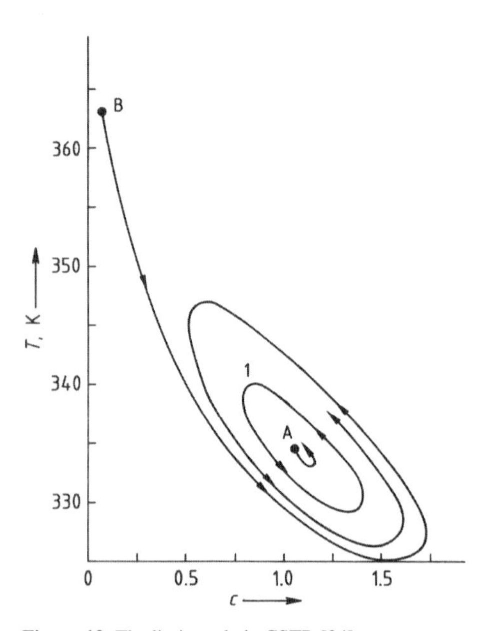

Figure 10. The limit cycle in CSTR [24]

3.4. Cascade of Tank Reactors [27]

Continuous stirred-tank reactors in series (Fig. 11) are used when the required residence time is long and more than one vessel is necessary for efficient, economical mixing. Conversions, yields, and residence times of the individual stirred-tank reactors are calculated analogously to those for a single CISTR. The steady-state *material balance* over the *n*th reactor for a component J reads:

$$\xi_{\mathrm{J}n}-\xi_{\mathrm{J}(n-1)}=\frac{V_{\mathrm{r}n}}{\Phi_{\mathrm{m}}}M_{\mathrm{J}}R_{\mathrm{J}n} \tag{3.32}$$

In most cases, the set of equations describing the degree of conversion in a cascade cannot be solved analytically. For a system consisting of *n* equal-sized CISTRs in series an analytical solution can be obtained for first-order irreversible reactions [25]:

$$c_{\mathrm{A}n}=c_{\mathrm{A}_0}(1+k\tau)^{-n} \tag{3.33}$$

and for second-order reaction: $c_{\mathrm{A}n}$ = 14 k τ_i

$$c_{\mathrm{A}n}=\frac{1}{4k\tau_i}(-2+2\cdot\sqrt{-1\ldots+\sqrt{-1+2\sqrt{1+4kc_{\mathrm{A}_0}\tau_i\cdot s^n}}}) \tag{3.34}$$

In other cases a solution has to be found by an algebraic stepwise method or by a graphical technique [14, 26].

Example. The second-order, reversible reaction $2\,\mathrm{A} \rightleftharpoons \mathrm{C}+\mathrm{D}$ is carried out in a single CISTR with total volume of 2.665 m^3. The volumetric flow rate of reactant A is 7.866×10^{-4} m^3/s, and its inlet and outlet concentrations are 24.03 and 6.94 kmol/m^3, respectively. The forward rate constant at the reaction temperature is 1.156×10^{-4} $\mathrm{m}^3\,\mathrm{kmol}^{-1}\,\mathrm{s}^{-1}$ and the equilibrium constant *K* is 16.0.

It now becomes necessary to replace the working reactor with a series of identical CISTRs. The volume of each reactor in a new reactor arrangement should be one-tenth of the current one. How many CISTRs are required?

Solution. The reaction rate can be expressed as follows:

$$R_{\mathrm{A}}=k_1\left[c_{\mathrm{A}}^2-\frac{(c_{\mathrm{A}_0}-c_{\mathrm{A}})^2}{4K}\right]$$

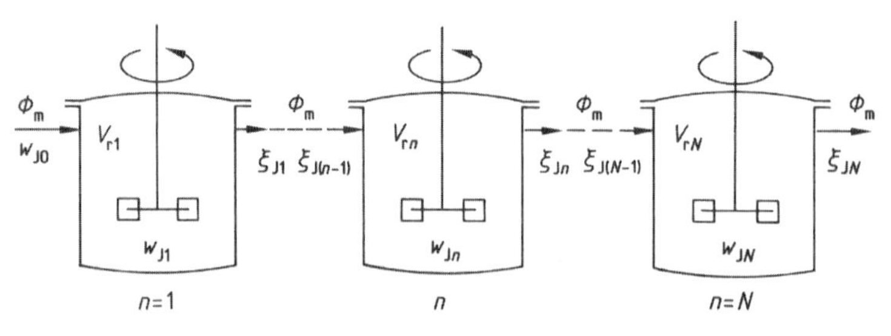

Figure 11. Continuous stirred-tank reactors in series (cascade)

The relation between R_A and c_A described by the above equation is shown in Figure 12. In addition, the material balance equation is also valid for each reactor in a series:

$$R_A = \frac{\Phi_v}{V_r}\left[c_{An} - c_{A(n-1)}\right]$$

The slope Φ_v/V_r in the case of a new arrangement is $2.95 \times 10^{-3}\ s^{-1}$. A straight line through c_{A_0} = 24.03 kmol/m^3 with this slope can be drawn to the intersection with the reaction rate curve. This process has to be repeated until c_{An} = 6.94 kmol/mol is reached. From Figure 12, the number of stages required is 3.8. Hence four CSTRs should be used in this case.

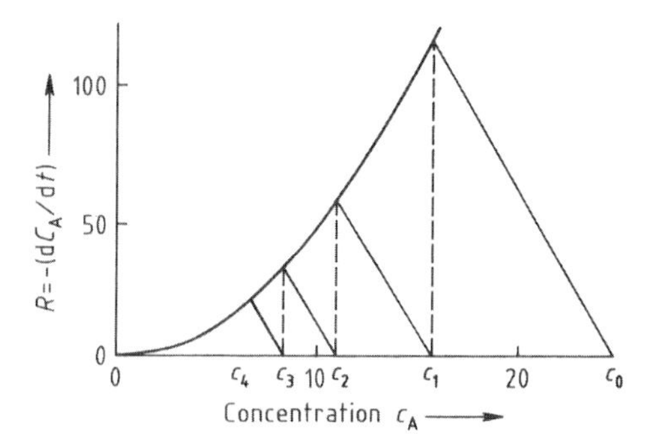

Figure 12. Graphical estimation of required number of CSTRs in series

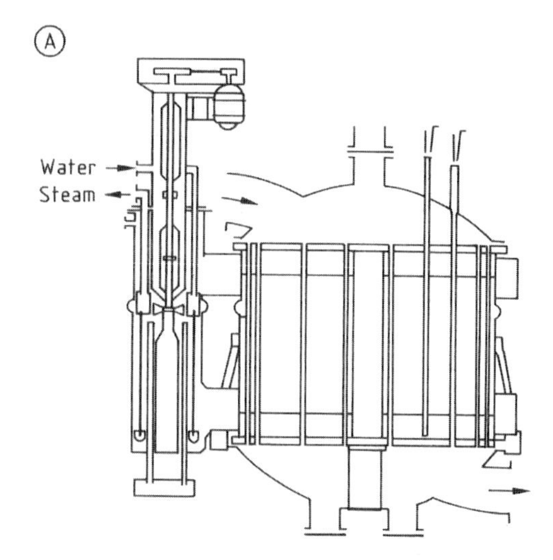

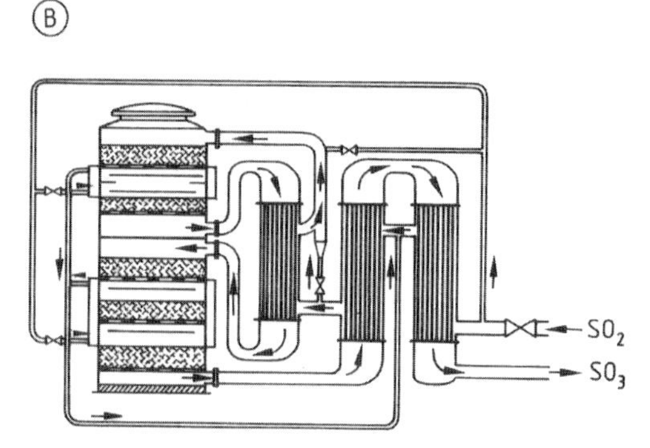

Figure 13. Different configurations for packed-bed reactors A) Multitube heat exchanger; B) Multistage adiabatic reactor

According to "Chemical Reactor Analysis and Design" G. F. Froment, K. B. Bischoff, John Wiley & Sons, New York 1990.

4. Packed-Bed Reactors

The packed-bed reactor constitutes the keystone in the production of several important chemicals. Several reactor configurations are encountered in practice (e.g., Fig. 13). The *heat-exchanger type* of reactor (Fig. 13 A) is used to carry out highly exothermic reactions. The *adiabatic multistage reactor* (Fig. 13 B) with cooling between stages is used to carry out exothermic equilibrium reactions. The *autothermal configuration*, an interesting alternative, is a heat-exchanger type of reactor in which the reactants circulate alternatively through shell and tube arrangements with the double purpose of being preheated and keeping the reactor operating within a prespecified thermal regime by cooling it down.

The analysis of transport phenomena for systems with chemical reaction is usually based on the transport equations resulting from the differential balance laws. To predict global effects, detailed information about the velocity profiles, and temperature and concentration fields is required. This information is extracted from the solution of the associated transport equations, subject to pertinent boundary conditions. When flow through a complex structure such as a porous medium is involved, these governing equations are valid even inside the pores. However, the geometric complexity of a randomly ordered particulate system prevents any general solution of detailed concentration, temperature, and flow fields. Instead, some form of macroscopic balances based on the average over a small volumetric element must be employed. A common practice is, with the help of some empirical relations, to replace the microscopic momentum, mass, and energy balances with corresponding macroscopic equations.

Due to the distributed nature of the system, a dynamic model contains two or more independent variables (time and spatial coordinates). Simultaneously, it is important in principle to include all relevant transport resistance (see Fig. 14) to avoid unrealistic results. Since a very complex model may be too difficult to solve, a number of simplifying assumptions are often adopted in practice. However, care must be taken when choosing the adequate degree of complexity for the reactor model, so as to guarantee sufficient accuracy for a given practical application.

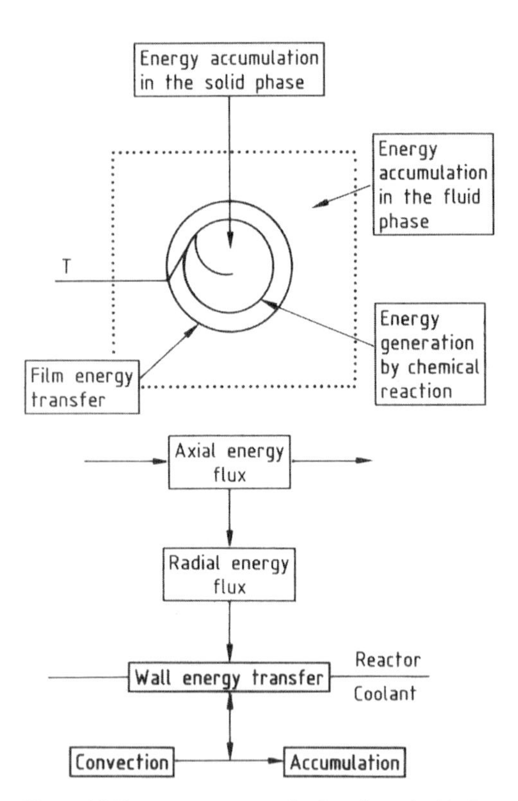

Figure 14. Energy transport mechanisms in packed beds

4.1. Mass and Energy Balances

Mathematical models for adiabatic and nonadiabatic packed-bed reactors have been systematically listed in reviews by FROMENT [28], HLAVACEK [29], and HOFMANN [30]. In these reviews an attempt is made to classify the models according to their complexity. Basically, the mathematical modeling of fixed-bed reactors can be performed either through the finite-stage or the continuum approach [31]. The continuum approach leads to a system of partial differential equations in which mixing effects are accounted for in the form of mass diffusion and heat conduction terms. This approach is being increasingly favored, mainly due to the continuous progress in computer software and numerical modeling.

In general, for the mathematical treatment of the heat and mass transfer processes in packed beds, with or without chemical reaction, three types of continuous models have been developed:

1) The single-particle model, where intra particle phenomena are accounted for separately
2) The two-phase (heterogeneous) model in which both phases, solid and fluid, exchange heat and/or mass, and such transport processes are accounted for
3) The one-phase (pseudohomogeneous) model in which the reactor is approximated as a quasi-continuum model

4.1.1. The Single-Particle Model

Basically the single-particle model is also a heterogeneous model. However, it differs from the two-phase model in that the solid phase is not regarded as a continuum since the concentration and temperature distribution inside the single pellet are accounted for.

Consider a catalyst pellet positioned in a flowing stream (Fig. 15). Three different regions can be distinguished: the bulk fluid phase, the boundary layer surrounding the particle, and the particle itself.

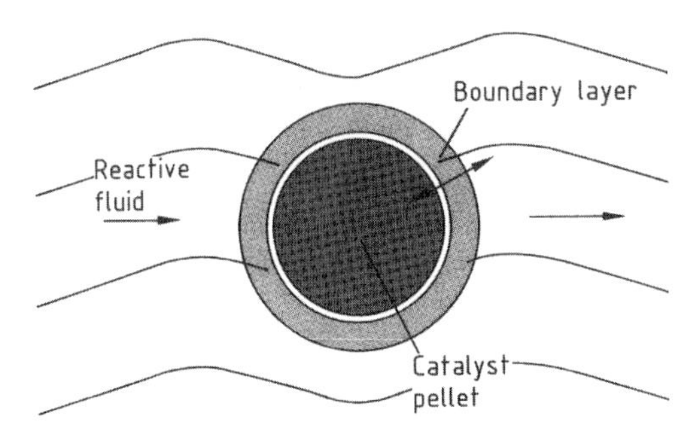

Figure 15. Schematic of a single particle in a flowing stream

Bulk Fluid Phase. The bulk fluid phase can be modeled as a constant temperature and constant concentration field. Perturbations at the inlet of the reactor propagate through the fluid and the conditions at the outer edge of the boundary layer change accordingly; this dynamic evolution of the boundary condition can be solved by using a heterogeneous model (see Section 4.1.2).

Boundary Layer. Temperature and concentration conditions in the boundary layer are frequently assumed to be identical to those of the bulk fluid phase. This is not always the case and, particularly if it is desired to trace the dynamic behavior of the concentration and temperature fields inside the catalyst pellets, the temperature and concentration in the boundary layer must be obtained by solving the corresponding mass and energy balances. The first step in solving these fields is to determine the extent of the boundary layer. This can be accomplished by performing overall energy and mass balances across the boundary layer. Once the thickness of the boundary layer has been found (e.g., Δr), the governing equations and their boundary conditions will be defined by:

Mass balance:

$$\frac{\partial C_\mathrm{f}}{\partial t} = D_\mathrm{f} \nabla^2 C_\mathrm{f} \tag{4.1}$$

Energy balance:

$$\varrho_\mathrm{g} C_{p,\mathrm{f}} \frac{\partial T_\mathrm{f}}{\partial t} = \lambda_\mathrm{f} \nabla^2 T_\mathrm{f} \tag{4.2}$$

subject to:

$$\left.\begin{aligned} C_\mathrm{f} &= C_\mathrm{f}(t) \\ T_\mathrm{f} &= T_\mathrm{f}(t) \end{aligned}\right\} \text{ at } r = R + \Delta r \tag{4.3}$$

$$\left.\begin{aligned} &\frac{\partial C_\mathrm{f}}{\partial r} = -\eta R(C_\mathrm{s}, T_\mathrm{s}) \\ &\lambda_\mathrm{s} \left.\frac{\partial T_\mathrm{s}}{\partial r}\right|_{r \to R^-} - \lambda_f \left.\frac{\partial T_f}{\partial r}\right|_{r \to R^+} \\ &= \eta(-\Delta H_r) R(C_\mathrm{s}, T_\mathrm{s}) \end{aligned}\right\} \text{ at } r = R \tag{4.4}$$

where η is the effectiveness factor, the subscript f denotes fluid, and $R\,(C_\mathrm{s}, T_\mathrm{s}) = k_0^* C_\mathrm{s} \exp(-E_\mathrm{a}/RT)$ stands for the equation for the reaction rate [32].

Catalyst Pellet. Similar to the analysis of the boundary layer, the pellet itself is modeled in terms of mass and energy transport. The pellet can be either porous or nonporous; in the latter case there is no mass diffusion of species into the pellet and only an energy balance is necessary. For certain classes of reactions, the assumption of nonporous catalyst is justifiable, e.g., for reactions whose high exothermicity prevents the reaction front from penetrating deep into the pellet even if it were porous. Under these assumptions the energy balance for the catalyst pellet gives:

$$\varrho_\mathrm{s} C_{p,\mathrm{s}} \frac{\partial T_\mathrm{s}}{\partial t} = \lambda_\mathrm{s} \nabla^2 T_\mathrm{s} \tag{4.5}$$

subject to:

$$\begin{aligned} &\frac{\partial T_\mathrm{s}}{\partial \tau} = 0 \text{ at } r = 0 \\ &k_\mathrm{s} \left.\frac{\partial T_\mathrm{s}}{\partial r}\right|_{\mathrm{r} \to R^-} - \lambda_f \left.\frac{\partial T_f}{\partial r}\right|_{r \to R^+} \\ &\quad = \eta\,(-\Delta H_r)\, R\,(C_\mathrm{s}, T_\mathrm{s}) \text{ at } r = R \end{aligned} \tag{4.6}$$

4.1.2. The Two-Phase Model

Of the two remaining models, the two-phase model is not only more realistic but it also constitutes the basis on which the derivation of the pseudohomogeneous approximation relies.

The most important feature of this model is that temperature and concentration differences are considered between solid and fluid phases. The mass and energy transport processes are governed by the following equations:

Fluid phase:

$$\varepsilon\frac{\partial\left(C_{\mathrm{i,f}}\right)}{\partial t}=-\varepsilon\nabla\cdot\left(\boldsymbol{u}C_{\mathrm{i,f}}\right)+\varepsilon\nabla\cdot D_{\mathrm{f}}\cdot\nabla C_{\mathrm{i,f}}+a_{v}\lambda_{\mathrm{f,s}}\left(C_{\mathrm{i,s}}-C_{\mathrm{i,f}}\right) \quad (4.7)$$

$$\varepsilon\varrho_{\mathrm{f}}C_{\mathrm{p,f}}\frac{\partial T_{\mathrm{f}}}{\partial t}=-\varepsilon\varrho_{\mathrm{f}}C_{p,\mathrm{f}}\boldsymbol{u}\cdot\nabla T_{\mathrm{f}}+\varepsilon\nabla\cdot\lambda_{\mathrm{f}}\cdot\nabla T_{\mathrm{f}}+a_{v}h_{\mathrm{f,s}}\left(T_{\mathrm{s}}-T_{\mathrm{f}}\right) \quad (4.8)$$

Solid phase:

$$\left(1-\varepsilon\right)\varepsilon_{\mathrm{p}}\frac{\partial\left(C_{\mathrm{i,s}}\right)}{\partial t}=-\eta\left(1-\varepsilon\right)\varrho_{\mathrm{s}}R\left(C_{1,\mathrm{s}},\ldots,C_{\mathrm{i,s}},\ldots,C_{\mathrm{S,s}},T_{\mathrm{s}}\right)+a_{v}\lambda_{\mathrm{f,s}}\left(C_{\mathrm{i,f}}-C_{\mathrm{i,s}}\right) \quad (4.9)$$

$$\left(1-\varepsilon\right)\varrho_{\mathrm{s}}C_{p,\mathrm{s}}\frac{\partial T_{\mathrm{s}}}{\partial t}=\eta\left(1-\varepsilon\right)\varrho_{\mathrm{s}}\left(-\Delta H_{r}\right)R\left(C_{1,\mathrm{s}},\ldots,C_{\mathrm{i,s}},\ldots,C_{\mathrm{S,s}},T_{\mathrm{s}}\right)+\left(1-\varepsilon\right)\boldsymbol{\nabla}\cdot\lambda_{\mathrm{s}}\cdot\boldsymbol{\nabla}T_{\mathrm{s}}+a_{v}h_{\mathrm{f,s}}\left(T_{\mathrm{f}}-T_{\mathrm{s}}\right) \quad (4.10)$$

4.1.3. The One-Phase Model

By introduction of the effective transport concept heterogeneous fluid – solid systems can be treated as quasi-continuum media [33]. In the pseudohomogeneous model, additional transfer mechanisms are superimposed on the contributions that originate from the global movement of the fluid. These additional mechanisms are based upon the observation that the fluid traveling between two points of the packed bed undergoes a tortuous journey that is composed of a large number of random steps. It is assumed that the transfer mechanisms resulting from this process can be treated as diffusion- and conduction-like phenomena. The fact that the modeled system is actually heterogeneous is accounted for by defining transport coefficients where both contributions due to the solid and fluid phases are considered.

Although the two-phase models fulfill the obvious physical requirements of considering a system where two phases exchange mass and/or energy, they also contain a number of approximations. For instance, separation of thermal dispersion effects, which in fact are interconnected, is very difficult. An even more difficult problem arises when heat losses must be accounted for; in this case the wall heat transfer coefficient has to be resolved into its solid- and fluid-phase components. RAMKRISHNA and ARCE discussed the validity of the pseudohomogeneous model to represent the heterogeneous nature of packed-beds [34, 35].

The simplest, most obvious pseudohomogeneous model was derived by YAGI et al. [36]. These authors made use of the fact that, at low Reynolds number, temperature differences between the solid and fluid phases are negligible and both phases can be assumed to have the same temperature. This approach has been generalized, and the one-phase model is adopted, even for high Reynolds number, based on the assumption of equal temperatures in both phases (e.g., [37]).

The idea of equivalence was extended by VORTMEYER et al. [38, 39] for arbitrary Reynolds numbers. These authors combined the energy balances for the fluid and solid phases (Eqs. 4.8 and 4.10) by assuming that the driving forces for heat conduction were similar in both phases. The main difference to the approach followed by YAGI and coworkers is that VORTMEYER and coworkers derived a dispersion term which reflects the exchange of energy between both phases; this term is added to the thermal conductivity of the solid phase to form a quantity which is known as the effective thermal conductivity.

An additional assumption concerns the mass transport between the solid and fluid phases. The resistance to mass transfer of the film adjacent to the solid particles can often be considered negligible; thus, the surface reactant concentration can be considered similar to that in the bulk of the fluid phase.

Under these assumptions and by combining both energy balances (Eqs. 4.8 and 4.10), the transport processes in the packed-bed reactor can be considered as governed by the following set of equations:

Mass balance:

$$\varepsilon\frac{\partial\left(C_{\mathrm{i}}\right)}{\partial t}=-\varepsilon\boldsymbol{\nabla}\cdot\left(\boldsymbol{u}C_{\mathrm{i}}\right)+\varepsilon\boldsymbol{\nabla}\cdot\boldsymbol{D}_{\mathrm{e}}\cdot\boldsymbol{\nabla}C_{\mathrm{i}}-\eta\left(1-\varepsilon\right)\varrho_{\mathrm{s}}R\left(C_{1},\ldots,C_{\mathrm{i}},\ldots,C_{\mathrm{S}},T\right) \quad (4.11)$$

Energy balance:

$$\overline{\varrho C_{p}}\frac{\partial T}{\partial t}=-\varepsilon\varrho_{\mathrm{f}}C_{p,\mathrm{f}}\boldsymbol{u}\cdot\boldsymbol{\nabla}T+\boldsymbol{\nabla}\cdot\lambda_{\mathrm{e}}\cdot\boldsymbol{\nabla}T+\eta\left(1-\varepsilon\right)\varrho_{\mathrm{s}}\left(-\Delta H_{r}\right)R\left(C_{1},\ldots,C_{\mathrm{i}},\ldots,C_{\mathrm{S}},T\right) \quad (4.12)$$

where:

$$\overline{\varrho C_\mathrm{p}} = \varepsilon \varrho_\mathrm{f} C_{p,\mathrm{f}} + (1-\varepsilon)\, \varrho_\mathrm{s} C_{p,\mathrm{s}}$$

An alternative two-phase model was proposed by CARBERRY and WHITE [40]. This model combines the pseudohomogeneous and heterogeneous models; it is based on the catalyst particle but uses effective transport properties. As a hybrid between the two previous models it has been termed the "pseudoheterogeneous" model. By considering the solid phase to be discontinuous, this model appears to perform better when matching models. A thorough analysis of this model is given in [41].

4.1.4. Boundary Conditions

Since the classical works by HULBURT [42], DANCKWERTS [43], and WEHNER and WILHELM [44] appeared, boundary conditions in packed-bed reactors have received much attention. The formulation of the boundary conditions proposed by DANCKWERTS is the most often followed. This formulation meets the limiting mixing conditions of the continuous stirred-tank reactor (CSTR) and the plug-flow reactor (PFR). This was proven for isothermal [45] and nonisothermal [46] systems. These boundary conditions have also been shown to be adequate for systems with a multiplicity of steady states [47]. They also predict the behavior of nonadiabatic systems satisfactorily [48] even though the number of possible steady states in these systems could change if the WEHNER and WILHELM approach is followed [49].

Even though the formulation of the boundary conditions as suggested by WEHNER and WILHELM [44] appears more logical from a physical point of view, their correspondence with DANCKWERTS' suggestion for steady-state operation of adiabatic systems has been proven [50]. Modifications have been proposed when radial dispersion is also taken into account [51], and the boundary conditions have also been modified to treat the transient situation [52, 53].

For selection of the proper boundary and initial conditions for flow systems see [54 – 56]. A simple pseudohomogeneous model without radial dispersion is now considered. The reactor is visualized as a tube with three distinguishable sections (see Fig. 16): the reaction zone packed with the catalyst ($0 \le z \le L$), and a fore ($z < 0$) and aft ($z > L$) sections of inert packing. A general formulation of the inlet and exit boundary conditions can be written as:

Inlet ($z = 0$):

$$\left[D_{\mathrm{e},z}\frac{\partial C_\mathrm{i}}{\partial z} - u_z C_\mathrm{i}\right]_{0-} = \left[D_{\mathrm{e},z}\frac{\partial C_\mathrm{i}}{\partial z} - u_z C_\mathrm{i}\right]_{0+} \tag{4.13}$$

$$\left[\lambda_{\mathrm{e},z}\frac{\partial T}{\partial z} - u_z \varrho_\mathrm{f} C_{p,\mathrm{f}} T\right]_{0-} = \left[\lambda_{\mathrm{e},z}\frac{\partial T}{\partial z} - u_z \varrho_\mathrm{f} C_{p,\mathrm{f}} T\right]_{0+} \tag{4.14}$$

Exit ($z = L$):

$$\left[D_{\mathrm{e},z}\frac{\partial C_\mathrm{i}}{\partial z}\right]_{L-} = \left[D_{\mathrm{e},z}\frac{\partial C_\mathrm{i}}{\partial z}\right]_{L+} \tag{4.15}$$

$$\left[\lambda_{\mathrm{e},z}\frac{\partial T}{\partial z}\right]_{L-} = \left[\lambda_{\mathrm{e},z}\frac{\partial T}{\partial z}\right]_{L+} \tag{4.16}$$

where a given approach is followed by selecting the diffusivities/conductivities as indicated in Table 4 ("yes" means parameter is included in the model, "no" means not included).

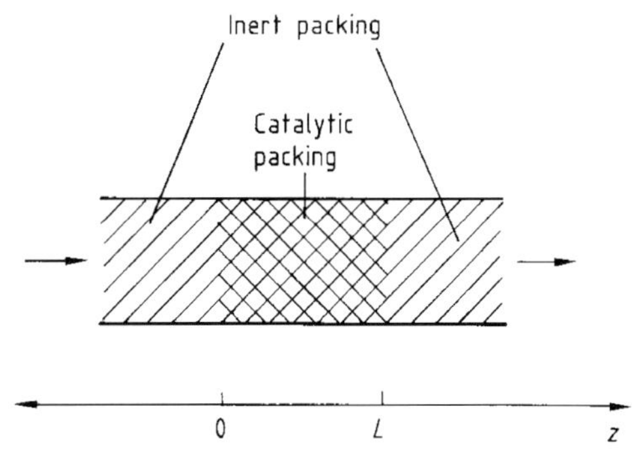

Figure 16. Schematic of the flow system in a packed-bed reactor

Table 4. Dispersion coefficients for different boundary conditions

Parameter	Hulburt	Danckwerts	Wehner and Wilhelm
$D/\lambda_{\mathrm{e},z}\,\vert_{0-}$	no	no	yes
$D/\lambda_{\mathrm{e},z}\,\vert_{0+}$	no	yes	yes
$D/\lambda_{\mathrm{e},z}\,\vert_{L-}$	yes	yes	yes
$D/\lambda_{\mathrm{e},z}\,\vert_{L+}$	no	no	yes

The conditions for the mass balance at the lateral walls are such that it is assumed that no mass flux occurs because the lateral walls are often impermeable (radial reactors are an exception). The boundary conditions for the energy balance, however, vary depending on whether the reactor is operated adiabatically or nonadiabatically. For a pseudohomogeneous model with radial dispersion these boundary conditions are:

Reactor wall (r=R):

$$\frac{\partial C_{\mathrm{i}}}{\partial r}=0 \tag{4.17}$$

$$\lambda_{\mathrm{e},r}\frac{\partial T}{\partial r}=h_w\left(T_w-T\right) \tag{4.18}$$

Reactor centerline (r=0):

$$\frac{\partial C_{\mathrm{i}}}{\partial r}=0 \tag{4.19}$$

$$\frac{\partial T}{\partial r}=0 \tag{4.20}$$

4.2. Values of the Parameters

Various steps have been taken towards the derivation of an integral model for the packed-bed reactor by considering the individual steps involved in the transport of a given property in a packed bed. In this section only few correlations for typical applications will be mentioned, for more details, see [31, 57 – 60].

4.2.1. Average Bed Porosity

The existence of oscillatory radial variations of the void fraction in packed beds has been recognized since the early work of Roblee et al. [61] and Benenati and Brosilow [62, 63]. This nonuniform distribution of the porosity causes the flow to be nonuniformly distributed, particularly near the reactor walls [64, 65]. The inclusion of radial porosity and velocity profiles improves the agreement between experimental and theoretical results [66, 67]. It has been found [68], however, that the predictions obtained by assuming constant average porosity and main-flow velocity for the entire packed bed will be qualitatively correct in most cases.

In order to use a constant average porosity, the following integral relation must be fulfilled:

$$\varepsilon=\frac{1}{\pi R^2}\int_0^R\varepsilon\left(r\right)r\mathrm{d}r$$

There is no unique functional relation between the radius and the porosity, but many functionalities have been proposed in the literature. To find a realistic value for the average porosity, the approach suggested by Martin can be followed [69]. The bed is considered to be formed by two concentric rings with their own properties; a central core with an average porosity ε_0 and a one-particle-diameter-thick concentric shell with porosity ε_{w}. The average porosity, as defined above, must satisfy the relation:

$$\varepsilon_0 A_0+\varepsilon_{\mathrm{w}} A_{\mathrm{w}}=\varepsilon A$$

Then, from geometric considerations, the following equation results:

$$\varepsilon=\varepsilon_{\mathrm{w}}-\left(\varepsilon_{\mathrm{w}}-\varepsilon_0\right)\left(1-d/D\right)^2$$

where the porosity for the outer shell (ε_{w}) is estimated as [69]:

$$\varepsilon_{\mathrm{w}}=1-\left(1-\varepsilon_{\min}\right)\frac{2}{3}\left(\frac{D/d-7/8}{D/d-1/2}\right)$$

where $\varepsilon_{\min}$ is the minimum value observed for the void fraction ($\varepsilon_{\min}$ = 0.23 and ε_0 = 0.39 are taken from [62]).

4.2.2. Effective Transport Coefficients

Since the effective transport coefficients are mainly determined by the flow characteristics, in general, packed beds are anisotropic for effective transport, so that the radial transport component is different from its analogue in the axial direction. For the sake of simplicity, effective transport coefficients are often considered to be unaffected by secondary flow and dependent on the main flow (forced) only.

Mass Transport. A correlation for the mass transport in both spatial directions is [70]:

$$D_{\mathrm{e}(r,z)}=\langle u\rangle_z d\times\left[\frac{0.7}{ReSc}+\frac{C_1}{1+\left(\frac{C_2}{ReSc}\right)}\right]$$

where the constants C_1 and C_2 depend on the geometry of the packing and the fluid flowing through the packed bed as listed in Table 5.

Table 5. Constants for the effective mass-transport coefficients [70]

Constant	Liquids		Gases	
	Spherical packing	Irregular packing	Spherical packing	Irregular packing
C_1 (longitudinal)	2.5	2.5	0.7	4
C_1 (radial)	0.08	0.08	0.12	0.12
C_2 (longitudinal)	8.8	7.7	5.8	5.1
C_2 (radial)			78 ± 20	

A more heuristic approach is to assume a constant value for the axial and radial Bodenstein (or mass Péclet) numbers, i.e.,

$$Pe_{\mathrm{m},z/r} = \frac{\langle u \rangle_z d}{D_{\mathrm{e},z/r}} = \text{constant}$$

where z/r denotes that the expression is valid for both axial and radial transport.

Experimental evidence has shown that *axial dispersion* in packed beds is typically characterized by $Pe_{\mathrm{m},z} \approx 2$. In other words, CSTR conditions (small $Pe_{\mathrm{m},\ z}$) are reached in the void within each void cell, and the packed-bed reactor can be viewed as $(L\ /d)$ CSTRs in series [71]. Therefore, axial mass diffusion can be neglected for ratios $L\ /d > 50$. This criterion applies to systems with particle Reynolds number larger than unity (i.e., $Re_\mathrm{p} = \varrho_\mathrm{f} \langle u \rangle_z\ d/\mu_\mathrm{f} > 1$).

Radial mass dispersion is characterized by a constant radial mass Péclet number:

$$Pe_{\mathrm{m},r} = \frac{\langle u \rangle_z d}{D_{\mathrm{e},r}} \approx 8{-}10$$

Energy Transport. Energy transport is supplemented by radiation and conduction in the solid phase. Therefore, it is more difficult to define constant values for the heat Péclet numbers. The problem of equivalence between the heterogeneous and pseudohomogeneous models has been addressed in [57, 72].

In general the effective thermal conductivity $\lambda_{\mathrm{e}(r,\ z)}$ is formed by two contributions: the quiescent bed conductivity λ_s^0 and a contribution due to the flow, i.e.,

$$\lambda_{\mathrm{e}(r,z)} = \lambda_\mathrm{s}^0 + C\lambda_\mathrm{f} Re Pr$$

where the conductivity for the quiescent bed λ_s^0 is obtained from the correlation proposed by Zehner and Schlünder [73], while the contribution due to the flow can be found from the analysis by Vortmeyer and Berninger [72]. The constant in the above equation is chosen as $C = 0.1$ for the radial (r) direction and $C = 0.1 - 0.2$ for the axial (z) direction.

The equivalence between the heterogeneous and pseudohomogeneous models has been proved for nonreactive conditions [57]. For further information on energy transport in packed beds see [57, 58, 74 – 77].

Effect of Radiation. Transport by radiation is frequently ignored when estimating heat transfer coefficients. This transport mechanism depends on the temperature range for the reactor operation; for a temperature of 600 K at the reactor wall, radiation has been found to contribute up to 20 % to the overall radial heat flux [78].

The contribution of radiation to effective energy transport in packed beds has been analyzed [78, 79]. In the model presented the heat transfer is considered to be controlled by two parallel mechanisms and the overall effective thermal conductivity is assumed as

$$\lambda_{\mathrm{e},r} = \lambda_{\mathrm{e},r}^0 + \lambda^r ad_{\mathrm{e},r}$$

where the contribution due to radiation $\lambda^r\ ad_{\mathrm{e},\ r}$ is

$$\begin{aligned} \lambda^r ad_{\mathrm{e},r} &= 4\psi\sigma \mathrm{d}T^3 \\ &= \left(\frac{8}{2/e - 0.264}\right)\sigma \mathrm{d}T^3 = C_r T^3 \end{aligned}$$

where σ is the Stefan – Boltzmann constant (5.67×10^{-8} J m^{-2} K^{-4} s^{-1}), ψ is a shape factor, and e is the particle emissivity. The treatment of this contribution is carried out by adding an additional term to the conduction term in the energy balance, i.e.,

$$\begin{aligned} \nabla\cdot\lambda_\mathrm{e}\cdot\nabla T &= \frac{1}{r}\frac{\partial}{\partial r}[\left(r\lambda_{\mathrm{e},r}^0 \frac{\partial T}{\partial r}\right) \\ &+ \left(rC_r T^3 \frac{\partial T}{\partial r}\right)] + \ldots \\ &= \lambda_{\mathrm{e},r}\frac{1}{r}\frac{\partial}{\partial r}\left(r\frac{\partial T}{\partial r}\right) + 3C_r T^2 \left(\frac{\partial T}{\partial r}\right)^2 + \ldots \end{aligned}$$

It is apparent that this contribution can become important with increasing particle emissivity and its omission would lead to conservative temperature predictions.

4.2.3. Wall Heat-Transfer Coefficient

The wall heat-transfer coefficient can be estimated from many correlations. The correlations available, however, predict heat transfer coefficients which are largely scattered. For instance, for a particle Reynolds number $Re = 100$, the heat transfer coefficient can differ by almost one order of magnitude [59]. ODENDAAL et al. analyzed the available correlations and compared the radial temperature profiles predicted for the solid and fluid phases. Based on the physical restriction that both profiles can never cross, most of them were discarded and only a few could be considered reliable [59].

The most often accepted correlation has been presented by DIXON et al. [80]. A thorough analysis of the estimation of this coefficient has been presented [81]. The recommended correlation has the form:

$$Nu_{\mathrm{f,w}} = \frac{\alpha_{\mathrm{f,w}} d}{k_{\mathrm{f}}} = \left[1-1.5(D/d)^{-1.5}\right] Pr^{1/3} Re^{0.6} \tag{4.21}$$

where $\alpha_{\mathrm{f,\,w}}$ is the fluid-to-wall heat transfer coefficient.

4.3. Further Simplifications

Despite its simplicity, the pseudohomogeneous model has proven to be a very reliable representation for transport processes in packed beds [72]. If care is taken when evaluating the transport coefficients, predictions obtained from a one-phase model will not differ significantly from those yielded by a two-phase representation. Every model has its deficiencies, thus, the heterogeneous model cannot capture the radial temperature and concentration gradients present in the pellets under reactive conditions, and the pseudohomogeneous model lacks an adequate representation of the solid phase. WINDES et al. [60] have suggested an approach to achieve equivalence between the pseudohomogeneous and heterogeneous models.

The questions concerning equivalence stem from the fact that packed beds behave differently under reactive and nonreactive conditions. This was initially interpreted as a change in the heat transport coefficients under reactive conditions [82]. WIJNGAARDEN and WESTERTERP [83] have addressed the subject in detail and provided a comprehensive explanation for the difference between reactive and nonreactive situations. For the mixing coefficients independence of chemical reaction was demonstrated by GUNN and VORTMEYER [84].

Although radial temperature profiles can be minimized in the design of the reactor, it is not necessary to suppress radial dependence of temperature to obtain reliable predictions with one-dimensional models. In general, the sophistication of the model should never exceed that of the experimental data. For this reason, one-dimensional models have become increasingly useful.

BEEK and SINGER [85] addressed this point by assuming a weak parabolic radial dependence of the temperature profiles on the radial dimension. Then it can be shown that predictions from a one-dimensional model equal the average radial temperature $\langle T \rangle$ for a two-dimensional model [31]. The radial temperature distribution is then obtained as [31]:

$$T(r/R) \approx T_w + \left[\frac{1-(r/R)^2+(2/B_{\mathrm{i}})}{1+(4/B_{\mathrm{i}})}\right] 2\left(\langle T \rangle - T_w\right)$$

The equivalence between the one-dimensional model and the radial average temperature for a two-dimensional model is obtained by a proper computation of the overall heat-transfer coefficient. For a parabolic temperature profile:

$$\frac{1}{U} = \frac{1}{h_w} + \frac{D}{8k_{\mathrm{e},r}}$$

A more refined approach was followed by CRIDER and FOSS [86], who solved the heat transfer problem analytically under nonreactive conditions. The equivalence of one/two-dimensional models was achieved by combining the radial heat-transfer parameters as follows:

$$\frac{1}{U} = \frac{1}{h_w} + \frac{D}{6.133\lambda_{\mathrm{e},r}}$$

This was later confirmed by HLAVACEK et al. [87] who solved the linearized problem and found identical results. An important theoretical conclusion can be drawn from this last observation; as long as only small radial temperature gradients exist, the chemical reaction only seems to have a significant effect on the axial temperature dependence. Under these circumstances, the radial temperature dependence can be described via a radially lumped model.

Example. The above analysis is illustrated by solving for the steady-state temperature and concentration profiles in a typical industrial reactor with the partial oxidation of *o*-xylene to phthalic anhydride as an example. This highly exothermic process is carried out in heat-exchanger-type reactors. The reactor consists of a bundle of tubes with a cooling fluid circulating around the tube arrangement. The process (of German origin, but used worldwide) employs V_2O_5 on silica gel pellets as catalyst. The geometric description and kinetic information have been taken from [88].

If the pseudohomogeneous, two-dimensional model is lumped radially, i.e., if the equations are integrated across the cross-sectional area, the following steady-state, one-dimensional model results:

$$0=-\frac{\partial}{\partial z}\left(u_z C_{\mathrm{X}}\right)+D_{\mathrm{e},z}\frac{\partial^2 C_{\mathrm{X}}}{\partial z^2}-(1-\varepsilon)\,\varrho_{\mathrm{s}}R\left(C_{\mathrm{X}},\langle T\rangle\right) \qquad (4.22)$$

$$0=-\frac{\partial}{\partial z}\left(\varrho_{\mathrm{f}}C_{p,\mathrm{f}}u_z\langle T\rangle\right)+\lambda_{\mathrm{e},z}\frac{\partial^2\langle T\rangle}{\partial z^2}-\frac{4U}{D}\left(\langle T\rangle-T_w\right)+\left(-\Delta H_r\right)(1-\varepsilon)\,\varrho_{\mathrm{s}}R\left(C_{\mathrm{X}},\langle T\rangle\right) \qquad (4.23)$$

where $R\left(C_{\mathrm{X}},\langle T\rangle\right)=k_0\,p_{\mathrm{O}}\,p_{\mathrm{X}}\exp\left(-E_{\mathrm{a}}/R_{\mathrm{u}}\langle T\rangle\right)$ represents the rate of reaction, and p_{X} and p_{O} represent the partial pressures of *o*-xylene and oxygen, respectively. The reaction is performed in excess oxygen (i.e., $p_{\mathrm{O}}/p_{\mathrm{tot}}=0.21\approx$ constant). Although the reaction mechanism for this process is completed by parallel and consecutive reactions, only the main reaction is addressed here.

The results obtained with the one-dimensional (Eqs. 4.22 and 4.23) and two-dimensional models are presented in Figure 17 for two different operating conditions. The one-dimensional model predicts reactor behavior satisfactorily for the conditions of Figure 17 A (T_w = 630 K). For the conditions of Figure 17 B (T_w = 645 K), however, the differences between this model and the two-dimensional models are significant. This can be better understood by analyzing the results presented in Figure 18 where the radial temperature profile is shown as a function of radial position *r*/*R* at different axial locations. The high-temperature situation (Fig. 18 B) shows the nonlinear effect of the reaction rate and the temperature profile deviates markedly from a parabolic dependence(as shown Fig. 18 A). These results show that, for mild radial dependence of the temperature distribution, one-dimensional models can be used as a reliable design tool.

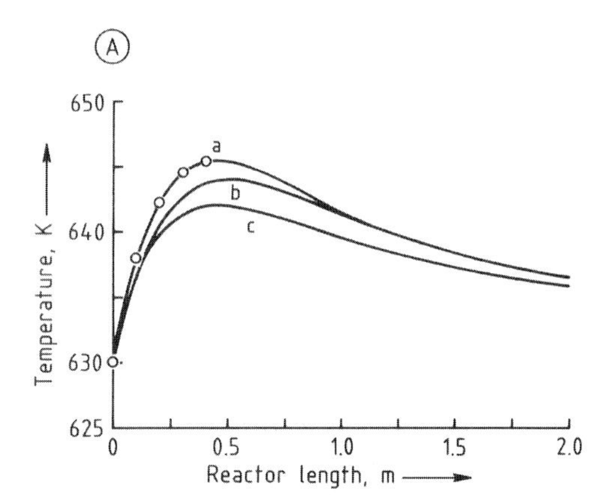

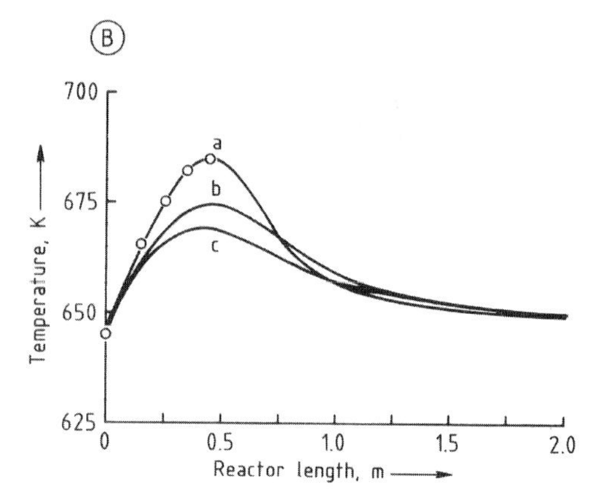

Figure 17. Comparison between one- and two-dimensional models
A) T_w = 630 K, $x_{\mathrm{X},\,0}$ = 0.00924; B) T_w = 645 K, $x_{\mathrm{X},\,0}$ = 0.00924
a) Two-dimensional model; b) Crider and Foss model (one-dimensional; c) Beek and Singer model (one-dimensional)
Open circles in Figures 17 A and 17 B denote axial locations for curves shown in Figures 18 A and 18 B, respectively.

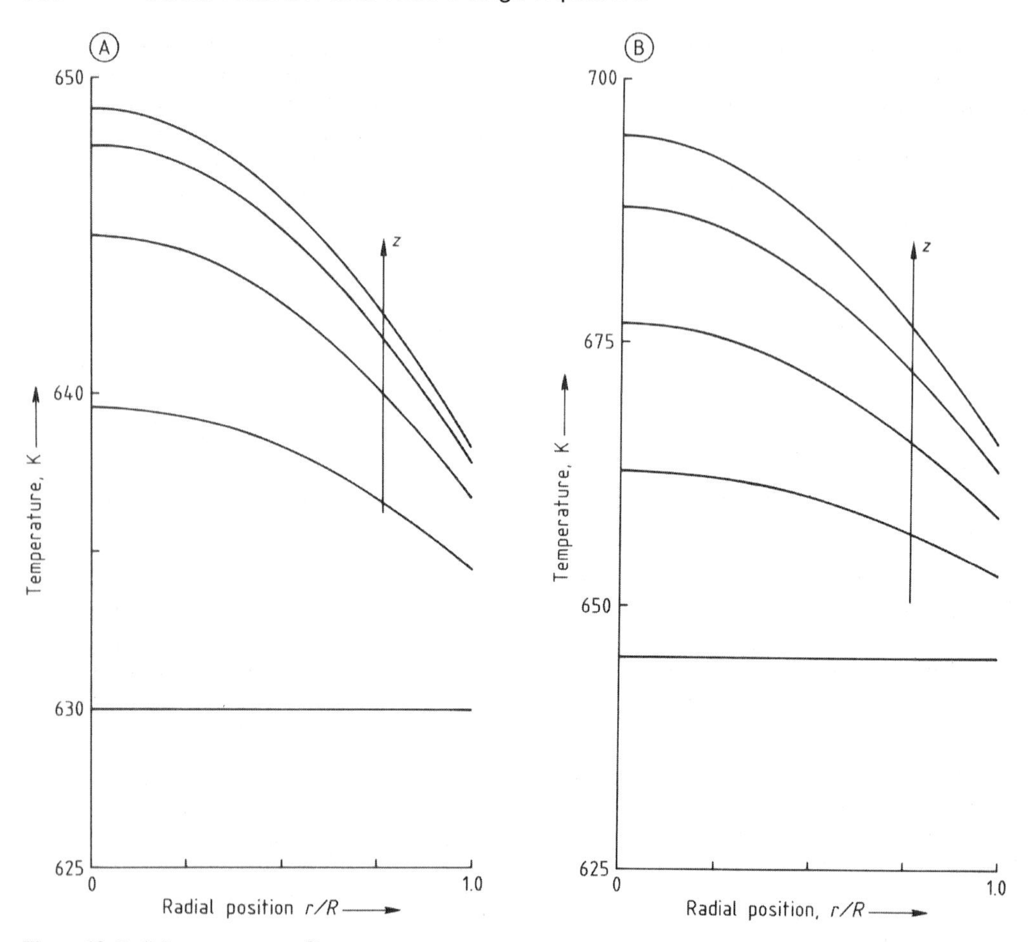

Figure 18. Radial temperature profiles
A) $T_w = 630$ K, $x_{X,0} = 0.00924$; B) $T_w = 645$ K, $x_{X,0} = 0.00924$

4.4. Parametric Sensitivity

Since the 1960s, great improvements have been made in the design and operation of catalytic reactors. In industrial conditions, however, due to the high exothermicity of the chemical reaction, the temperature rises sharply in the catalytic bed towards a maximum or hot spot (usually located near the reactor inlet). These sharp axial temperature gradients can cause poor reaction selectivity and extreme temperatures can be responsible for rapid deactivation or even deterioration of the catalyst. Therefore, the hot spot must be kept within permissible bounds. Furthermore, the attainable conversion frequently has an upper safety limit at levels where the temperature profile becomes extremely sensitive to changes in operational and/or physicochemical parameters. BILOUS and AMUNDSON [89] termed such a reactor condition as "parametric sensitivity". Many studies have been reported on the prediction of this phenomenon, which in turn can lead the reactor to runaway operation [29, 90]. The danger inherent in running such systems is widely recognized, and the final reactor design must guarantee a safe mode of operation.

The simplest design criteria derived using this concept were formulated by simplifying the temperature-dependence in reaction rate term as

$$\frac{E}{RT} = \frac{E}{RT_S}\left[1 - \left(\frac{T-T_s}{T}\right)\right] \approx \frac{E}{RT_S}\left[1 - \left(\frac{T-T_s}{T_S}\right)\right]$$

Where T_S can be viewed as a reference value.

BARKELEW [92], who first suggested this approximation, selected the coolant temperature as the reference value and simplified the concentration term as a correction to first-order kinetics.

These assumptions enabled BARKELEW to find an explicit expression for the maxima locus:

$$(1-\zeta_{\mathrm{m}})\, g\,(\zeta_{\mathrm{m}})=\frac{St'}{B}\Delta\vartheta_{\mathrm{m}}\mathrm{e}^{-\Delta\vartheta_{\mathrm{m}}}$$

which is shown below (Fig. 19) for different values of the parameters; here $g(\zeta_{\mathrm{m}})$ represents the correction for kinetics other than first-order.

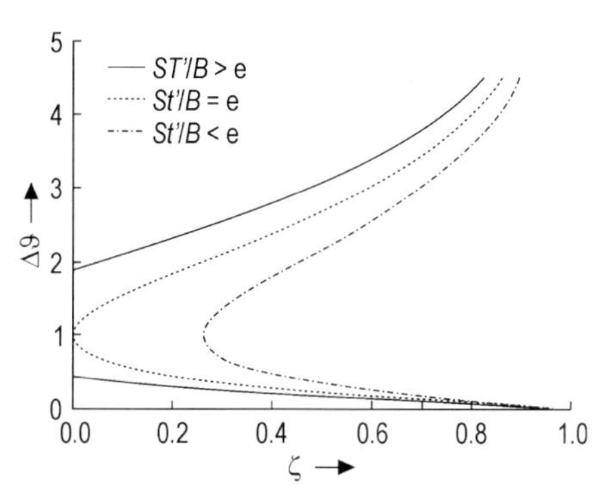

Figure 19. Effect of packing configuration on reactor performance

Analysis of the figure suggests two design criteria, which have been termed as first and second BARKELEW's criteria, namely

$$\Delta\vartheta_{\mathrm{o}}=\frac{E}{RT_{\mathrm{c}}^{2}}\,(T_{\mathrm{o}}-T_{\mathrm{c}})\leq 1$$

and

$$\frac{St'}{B}>\mathrm{e}$$

These criteria, although useful, have been found sometimes to be overly conservative. Estimates of the error in BARKELEW's approximation have been derived and analyzed for different situations; details can be found in [2].

A less conservative criterion was derived by VAN WELSENAERE and FROMENT [93], who assumed the hot-spot temperature T_m as the reference temperature T_S, which can be proven to introduce a smaller error in the simplification of the reaction rate term (particularly if employed for predictions in the vicinity of the hot spot). The resulting equation, however, is no longer explicit and a numerical solution is required. The first criterion remained identical to BARKELEW's, i.e.

$$\Delta\vartheta_{\mathrm{c}}=\frac{E}{RT_{\mathrm{m}}^{2}}\,(T_{\mathrm{m}}-T_{\mathrm{c}})\leq 1$$

The second criterion was derived from numerical analysis, resulting in

$$\frac{B}{\Delta\vartheta_{\mathrm{m}}}\leq 1+\sqrt{\frac{St'}{\mathrm{e}}}+\frac{St'}{\mathrm{e}}$$

where the similarities with BARKELEW's results are apparent. Application of these criteria to several industrial scenarios can be found in [2].

Most studies on parametric sensitivity have assumed constant temperature for the cooling medium [91 – 97]. This is a suitable approach for perfectly mixed coolant, or reactors employing a boiling liquid or a fluid of abnormally large heat capacity as cooling medium. However, in the more common case of molten salt circulation, the thermal gradients in the shell side cannot be neglected in the reactor model. Few papers have studied the nonisothermal situation. For more information; see [98 – 112].

Example. Consider again, the production of phthalic anhydride via partial oxidation of *o*-xylene in a multitube heat-exchanger-type reactor. The steady-state operation will be described by the one-dimensional pseudohomogeneous model (Eqs. 4.22 and 4.23). Parametric sensitivity can be estimated as originally proposed by BILOUS and AMUNDSON [89]. It is computed as the derivative of the variable to be analyzed with respect to one of the physicochemical parameters of the process. Thus, the parametric sensitivity (S_{ij}) of the variable χ_i with respect to the parameter π_j is defined as:

$$S_{ij}=\frac{\partial\chi_i}{\partial\pi_j}$$

In what follows the effect of the reactor wall temperature on the reactor thermal condition is analyzed, i.e., $\pi = T_w$ and $\chi = \langle T \rangle$. This is a typical problem found in industrial practice where the thermal state of the reactor is monitored (i.e., the observed variable would be the readouts of the reactor thermocouples) and the operator acts on the cooling fluid (i.e., the manipulated variable would be the temperature or coolant flow rate) to keep the reactor operating near the desired state.

In order to obtain numerical values for the parametric sensitivity of the reactant concentration (S_C) and the bed temperature (S_T), the governing equations are differentiated with respect to the wall temperature, i.e.,

$$0 = -\frac{\partial}{\partial z}(u_z S_C) + D_{e,z}\frac{\partial^2 S_C}{\partial z^2} - (1-\varepsilon)\,\varrho_s \frac{\partial R}{\partial T_w} \tag{4.24}$$

$$0 = -\frac{\partial}{\partial z}(\varrho_f C_{p,f} u_z S_T) + k_{e,z}\frac{\partial^2 S_T}{\partial z^2} - \frac{4U}{D}(S_T - 1) + (-\Delta H_r)(1-\varepsilon)\,\varrho_s \frac{\partial R}{\partial T_w} \tag{4.25}$$

where:

$$\frac{\partial R}{\partial T_\omega} = k_0 p_O \left(S_C + p_X \frac{E_a}{R_u \langle T \rangle^2} S_T \right) \exp\left(-E_a / R_u \langle T \rangle\right)$$

If Equations (4.24) and (4.25) are integrated simultaneously with the reactor model (i.e., Eqs. 4.22 and 4.23), the sensitivity of the reactant conversion and bed temperature to changes in the coolant temperature is obtained at any point of the reactor length. Of these two sensitivity profiles, only $S_T(z)$ guarantees a safe mode of operation.

The four additional boundary conditions needed are derived from the original boundary conditions, i.e.,

Inlet, $z = 0$:

$$0 = u_z S_C - D_{e,z}\frac{\partial S_C}{\partial z} \tag{4.26}$$

$$\varrho_f C_{p,f} u_z = \varrho_f C_{p,f} u_z S_T - k_{e,z}\frac{\partial S_T}{\partial z} \tag{4.27}$$

Exit, $z = L$:

$$0 = \frac{\partial S_C}{\partial z} = \frac{\partial S_T}{\partial z} \tag{4.28}$$

In Figure 20 the axial temperature profiles for the previous example are shown for two different inlet temperatures and the normalized temperature sensitivity associated with the lower inlet temperature. Figure 20 B illustrates the concept of parametric sensitivity; a 1 K rise in the inlet temperature is sufficient to drive the reactor away from a safe operation regime. Figure 20 A, on the other hand, only shows a small difference in the reactor thermal state after a change of 5 K in the inlet temperature. Indeed the 5 K temperature increase is transmitted almost linearly along the reactor. These two situations can be better understood by inspecting the normalized temperature sensitivity associated with the lower inlet temperature operating condition. In Figure 20 A the temperature sensitivity remains almost constant along the reactor, i.e., the temperature profile should experience an almost constant increase for a temperature rise at the inlet of the reactor. In Figure 20 B, on the other hand, the temperature is predicted to grow dramatically in the vicinity of the hot spot. Furthermore, the sensitivity crosses to negative values after the hot spot axial location; this suggests a temperature overshoot promoting a high degree of conversion and mostly cooling afterwards. Indeed, the temperature profiles corresponding to the higher inlet temperature confirm the predictions extracted from analyzing the temperature sensitivity.

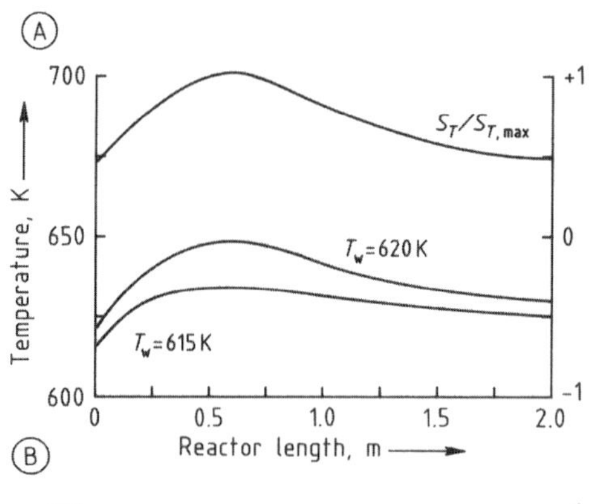

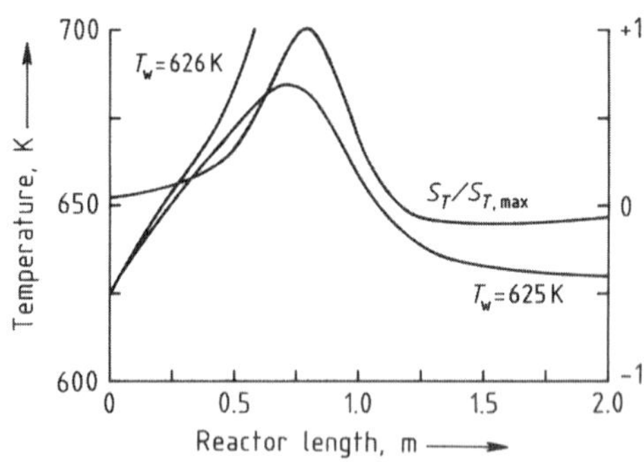

Figure 20. Parametric sensitivity in packed-bed reactors for different inlet temperatures and $x_{X,0} = 0.019$
A) T_w = 620 K and 615 K; B) T_w = 626 K and 625 K

Nowadays the analysis of packed-bed reactors may involve complex mechanisms instead of an overall reaction. When the complexity of these problems is compounded by the sensitivity equations, a computationally demanding stiff system of differential equations results. Specialized software has developed to take advantage of the sparse nature of parametric sensitivity problems and to efficiently deal with the stiff systems of equations [113, 114].

4.5. Flow Field Description

Usually the flow distribution in unbounded porous media is assumed to be well represented by a linear relation between the pressure drop and the fluid velocity (Darcy's law). When using such an approximation, however, two main points are disregarded:

1) Effect of the boundaries on the flow field
2) Increasing importance of the inertial effects as the flow speed increases

If the fluid obeys the Boussinesq approximation, the flow field will be governed by [118]:

$$\varrho_\mathrm{f}\frac{\partial \mathbf{u}}{\partial t} = -\boldsymbol{\nabla} p - \frac{\mu_\mathrm{f}}{\varkappa}\mathbf{u} - \varrho_\mathrm{f}\left[\beta_1\left(T-T_0\right)+\beta_2\frac{1}{C_0}\left(C_0-C\right)\right]\boldsymbol{g} \quad (4.29)$$

$$\boldsymbol{\nabla}\cdot\mathbf{u} = 0 \quad (4.30)$$

which is known as the Darcy – Oberbeck – Boussinesq model for flow through a porous medium [119].

As the flow speed increases, the *inertial terms* become important and the results of Darcy's model are inefficient in describing the momentum transfer in packed beds. The flow through porous media is now said to be described by a "*modified*" *Darcy's law*:

$$\varrho_\mathrm{f}\left(\frac{\partial \mathbf{u}}{\partial t}+\frac{1}{e}\mathbf{u}\cdot\boldsymbol{\nabla}\boldsymbol{u}\right) = -\boldsymbol{\nabla} p - \frac{\mu_\mathrm{f}}{\varkappa}\boldsymbol{u} - \varrho_\mathrm{f}\left[\beta_1\left(T-T_0\right)+\beta_2\frac{1}{C_0}\left(C_0-C\right)\right]\boldsymbol{g} \quad (4.31)$$

where the inertial forces are represented by the term $\mathbf{u}\cdot\nabla\mathbf{u}$. Even though this term arises from a formal volume averaging in the point field equations [120], its inclusion may lead to inconsistencies between boundary conditions and governing equations [121]. FORCHHEIMER [122] first proposed addition of higher order inertial terms to the relation between pressure drop and fluid velocity and modified Darcy's equation to:

$$\frac{\mathrm{d}p}{\mathrm{d}z}+a_1 u_z+a_2(u_z)^2 = 0$$

Later he added a third-order term to fit experimental data. FORCHHEIMER's modification for the one-dimensional flow can be formalized for two and three dimensions as [123, 124]:

$$\boldsymbol{\nabla} p-\varrho_\mathrm{f}\boldsymbol{g}+\frac{\mu_\mathrm{f}}{\varkappa}\mathbf{u}+\frac{\varrho_\mathrm{f}}{\varkappa}b\mathbf{u}\left|\mathbf{u}\right| = 0$$

where b can be considered a structure property associated with inertia effects in the porous matrix. Then the momentum equation becomes:

$$\varrho_\mathrm{f}\frac{\partial \mathbf{u}}{\partial t} = -\boldsymbol{\nabla} p - \frac{\mu_\mathrm{f}}{\varkappa}\mathbf{u} - \frac{\varrho_\mathrm{f}}{\varkappa}b\mathbf{u}\left|\mathbf{u}\right| - \varrho_\mathrm{f}\left[\beta_1\left(T-T_0\right)+\beta_2\frac{1}{C_0}\left(C_0-C\right)\right]\boldsymbol{g} \quad (4.32)$$

4.5.1. Boundary Conditions

Only two boundary conditions are necessary to solve the momentum equations. The classical nonslip boundary conditions are imposed on the solid walls and a given velocity profile is imposed at the entrance and exit of the reactor, i.e.,

Reactor wall ($r = R$):

$$\mathbf{u} = [0{,}0] \quad (4.33)$$

Reactor centerline ($r = 0$):

$$\mathbf{u} = [0{,}u_z] \quad (4.34)$$

Inlet ($z = 0$):

$$\mathbf{u} = [0{,}u_z\left(r\right)] \quad (4.35)$$

Exit ($z = L$):

$$\mathbf{u} = [0{,}u_z\left(r\right)] \quad (4.36)$$

4.5.2. Permeability and Inertia Coefficient

The pressure drop in a porous medium can be approximated by the Ergun correlation, a linear combination of the Carman – Kozeny and Burke – Plummer equations [125]:

$$\frac{\Delta p}{H} = \frac{150(1-\varepsilon)^2}{\varepsilon^3 d^2}\mu u_z+\frac{1.75\left(1-\varepsilon\right)}{\varepsilon^3 d}\varrho u_z^2$$

However, for a certain range of fluid velocities, a linear dependence of the pressure drop on the fluid velocity (namely, the Carman – Kozeny term) provides a reliable representation description. Thus, as Darcy's law proposes a linear dependence between pressure drop and fluid velocity, the above equation can be used to estimate the packing permeability $\varkappa$ and the Forchheimer correction term b as:

$$\varkappa = \frac{d^2\varepsilon^3}{150(1-\varepsilon)^2}$$

$$b = \frac{1.75d}{150\,(1-\varepsilon)}$$

4.5.3. Thermal-Expansion Coefficient

The thermal-expansion coefficients are tabulated for most of the fluids of interest. When not available, a satisfactory estimation can be obtained for gaseous fluids by resorting to the ideal gas law. For instance, the expansion coefficient for temperature variations can be approximated as:

$$\varrho = \frac{pM}{RT}$$

$$\frac{\partial\varrho}{\partial T} = \frac{\varrho}{T}$$

then:

$$\beta_1 \approx \frac{1}{T_0} \approx 10^{-3}$$

4.5.4. Mass-Expansion Coefficient

The expansion coefficient for concentration changes is not as readily obtained as its temperature analogue. A satisfactory estimation can still be obtained by assuming ideal gas behavior. For instance, for a binary mixture A + B where the reaction A $\rightarrow$ B takes place, the fluid density can be written as:

$$\varrho = \varrho_0\left(\frac{1}{1+cC_{\mathrm{A}}y}\right)$$

where c stands for a combination of the initial and final molar masses:

$$c = \frac{M_{\mathrm{i}}-M_{\mathrm{f}}}{M_{\mathrm{f}}}$$

If $c < 0$, then a lower molar mass product is formed in the chemical reaction (for instance, a decomposition reaction). In contrast $c > 0$ would indicate the formation of a product with a higher molar mass (for instance, a polymerization reaction).

If a perturbation is considered in a situation where no conversion has been achieved (i.e., $\xi \rightarrow 0$), the gas density can be represented in a linear approximation as:

$$\varrho = \varrho_0\,(1-\beta_2 C_{\mathrm{A}}\xi)$$

Taking into account that $c < 1$ and $\xi < 1$, the dependence of the density on the conversion can be written as:

$$\varrho \approx \varrho_0\,(1-cC_{\mathrm{A}}\xi)$$

then:

$$\beta_2 \approx c$$

In the case of a reaction in the liquid phase, the density must be determined from an appropriate equation of state, and the expansion coefficients should be found experimentally.

4.5.5. Pressure-Drop Effects in Packed Beds

The effect of reactor pressure on gas-phase reactions or reactions with phase change can be significant. In homogeneous gas-phase reactions pressure-drop effects are typically negligible and can be safely ignored in design calculations.

For gas-phase reactions in packed-bed reactors, on the other hand, pressure drop can be considerable, even to the extent of making a particular design condition unrealizable.

Pressure drop in packed beds is typically modeled through the differential form of the Ergun equation (cf. 4.5.2):

$$\frac{\mathrm{d}p}{\mathrm{d}z} = \frac{150(1-\varepsilon)^2}{\varepsilon^3 d^2}\mu u_z + \frac{1.75\,(1-\varepsilon)}{\varepsilon^3 d}\rho u_z^2$$

Which can be solved subject to

$$p = P^0 \text{ at } z = 0$$

together with the mass and energy balances (Equations 4.11 and 4.12).

Example. To appreciate the importance of this effect on reactor design, consider ethylene oxide production by vapor-phase catalytic oxidation of ethylene

$$2C_2H_4+O_2\rightarrow 2CH_2OCH_2$$

The reaction is to be carried out under isothermal conditions in a packed-bed reactor at 533 K. Ethylene and air are fed in stoichiometric ratio at 10 bar. Under these conditions, the reaction can be assumed to follow the power-law kinetics

$$r_{C_2H_4}=-kp_{C_2H_4}^{1/3}p_{CH_2OCH_2}$$

with $k_{533K} = 14.1$ moles atm^{-1} kg$_{cat}^{-1}$ h^{-1}.

Let us examine the conversion [and reactor pressure] as a function of catalyst weight for a traditional packed-bed reactor arrangement with and without pressure drop being accounted for.

Figure 21 shows the conversion and reduced pressure as a function of the catalyst weight for a single tube (multitube reactor, cf. Fig. 1). A pressure drop of almost 50 % has occurred at 60 % conversion. The conversions predicted without accounting for the pressure drop are significantly higher. Details of the reactor configuration and physical properties for this example can be found in [126].

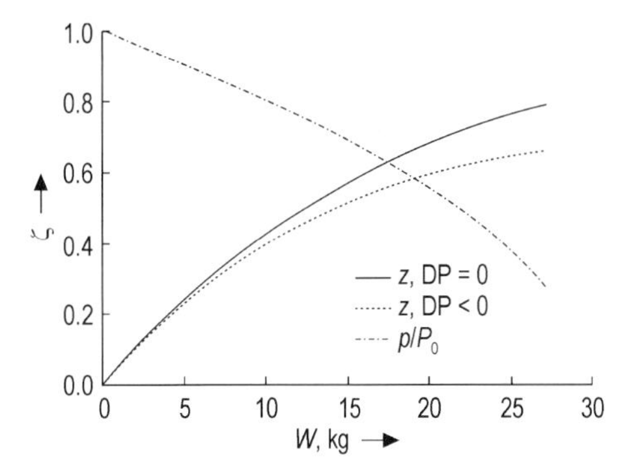

Figure 21. Effect of pressure drop on reactor performance. Conversion z and reduced pressure p/P_o as a function of catalyst weight W for a single tube (multitube reactor).

The consequences of ignoring the pressure drop are further reaching than just overpredicting the conversions realizable in a given reactor. Indeed, pressure-drop effects must be accounted for in the specification of the pumps. Sustaining high inlet pressures in industrial units can be one of the largest expenses in reactor operation.

Since reducing the catalyst particle size might be beneficial from the effectiveness point of view, but detrimental with regard to the costs involved in pressurization, operation of packed-bed reactors is frequently posing a dilemma for the plant engineer. An alternative solution has been to redesign the configuration of the packing to minimize or diminish the effects of pressure losses. Two alternatives have found widespread use in today's industrial environment: (1) spherical and (2) radial-flow reactors (see Fig. 22).

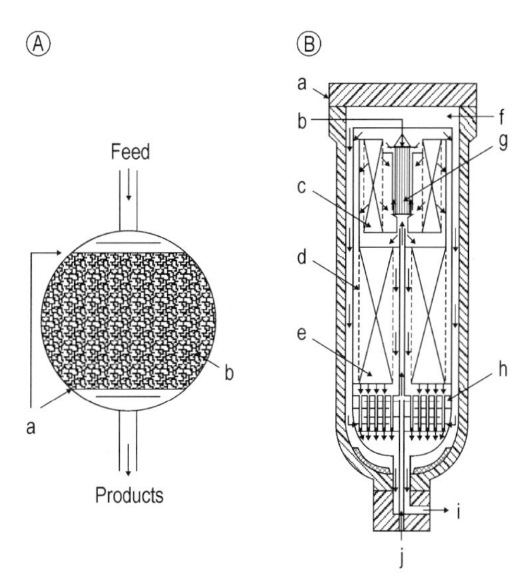

Figure 22. A) Schematic of a spherical reactor. Adapted from [126].
a) Screens; b) Catalyst
B) Radial-flow reactor (Haldør Topsoe S-200 ammonia converter)
a) Pressure shell; b) Indirect quench gas inlet; c) First catalyst bed; d)Annulus around catalyst bed; e) Second catalyst bed; f) Main gas inlet; g) Interbed heat exchanger; h) Lower heat exchanger; i) Gas outlet; j) Cold bypass

The benefits of changing the flow configuration are apparent if a comparison is made for the example discussed above. Figure 23 shows conversion and reduced pressure as a function of catalyst weight for a bank of 40 tubes (i.e., cylindrical configuration) and a single spherical reactor.

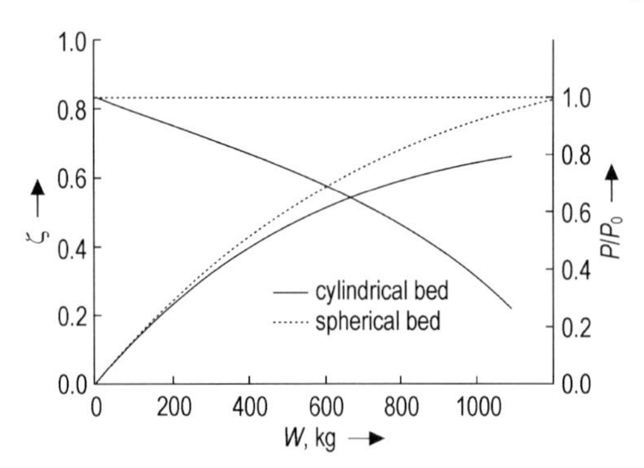

Figure 23. Conversion ζ and reduced pressure p/P_o as a function of catalyst weight W for a bank of 40 tubes (i.e., cylindrical configuration) and a single spherical reactor

4.6. Thermomechanical Effects in the Reaction System

Several phenomena can be observed when analyzing the dynamic behavior of catalytic and noncatalytic packed-bed reactors, e.g., "parametric sensitivity", "wrong-way" behavior, traveling reaction fronts, and ignition and extinction phenomena. All these temperature excursions may disappear or be too small to drive the system out of control. However, the development of thermal stresses during these temperature excursions adds a new factor to the reactor design and operation. The mechanical behavior of the catalyst pellet can be analyzed following the basic theory of elastic behavior of solids (see [127]). For a general situation the equations can be written as:

Energy balance:

$$\varrho_s C_{p,s} \frac{\partial T}{\partial t} = k_s \nabla^2 T - (3\lambda+2\mu)\, \alpha T_0 \frac{\partial}{\partial t} (\nabla \cdot \mathbf{u}) \tag{4.37}$$

Thermomechanical equations:

$$\varrho_s \frac{\partial^2 \delta}{\partial t^2} = \mu \nabla^2 \delta + (\lambda+\mu)\, \nabla\, (\nabla \cdot \delta) - (3\lambda+2\mu)\, \alpha \nabla T \tag{4.38}$$

subject to:

$$\begin{aligned} \delta_r &= 0 \text{ at } r = 0 \\ \sigma_{rr} &= 0 \text{ at } r = \mathrm{R} \end{aligned} \tag{4.39}$$

where λ and μ are the Lamé constants, α is the coefficient of linear expansion, T_0 the temperature of stressless condition, $\boldsymbol{\delta}$ is the displacement vector, and σ_{rr} are the radial stresses.

Thus, the energy balance has been augmented from its previous form (i.e., Eq. 4.5 in Section 4.1.1), by an additional term. This term represents what is known as thermoelastic dissipation. Nevertheless, the temperature fluctuations due to deformations are usually small and, for most situations, they can be neglected. A more systematic criterion is to neglect the thermoelastic dissipation term whenever the inequality below holds:

$$\frac{(3\lambda+2\mu)^2 \alpha^2 T_0}{(\lambda+2\mu)\; \varrho_s C_{p,s}} \ll 1$$

Another noticeable feature of the above equations is the hyperbolic nature of the thermoelastic equations. There is a close connection between the coupling discussed previously and neglecting the inertia term in the equation of motion. Indeed, the possibility of omitting the coupling terms does not depend only on satisfying the previous indicated inequality, but also on the fact that the strain rates must be of the same order as the cooling or heating rates. This implies that the temporal behavior of the displacement vector closely follows the dynamic behavior of the temperature field. In other words, no pronounced lag or vibrations in the motion of the body must arise. Under these circumstances, the inertia term can be neglected and the equations become decoupled [128].

This model is completed by the compability and equilibrium equations (see [127, Chap. 8]).

4.7. Numerical Simulation (→ Mathematics in Chemical Engineering, Chap. 7)

Due to the strongly nonlinear nature of the equations describing reactive flows, up to now analytical techniques have failed to produce solutions for the full form of the governing equations. Nevertheless, considerable progress in understanding reaction systems has been made possible through the application of analytical techniques to some limiting situations. For instance, since 1980 a wealth of information has been accumulated that shows results based on the use of large activation energy asymptotics. Although credit should be given to similar approaches for providing physical insight into many reaction problems, the models described are simplified and limitations on the techniques clearly exist. To overcome these limitations the systems of governing equations, together with the appropriate boundary conditions, must normally be solved by numerical techniques.

For the application of numerical methods, a number of grid points are distributed over the physical domain or the domain is cut up into small pieces (cells). The discretization of the governing equations yields a system of algebraic equations expressing relationships between values of the different variables at neighboring grid points.

Because of finite computer resources, it is not possible to monitor the solution at every point of the region in the space on which the problem has been defined. A finite number of grid points is therefore distributed over the field. Each of these points acts as an observer, monitoring the solution locally and relaying information about it to its neighbors. In most situations a regular, logical connectivity exists between the grid points; thus, in a two-dimensional situation each grid point is surrounded by eight neighbors. Because of this structure, a duality exists. Instead of looking at grid points, every four-point group could be seen as a quadrilateral cell. Discretization of the physical domain will thus be seen as constructing a collection of cells covering the entire domain.

However, efficiency in distributing these observers should be the guiding principle; a limited, fixed number of grid points or grid cells should be arranged in such a way that the largest possible amount of information can be extracted from their use. First, the geometry of the physical domain should be approximated in an accurate way and, second, the grid points should be able to monitor all essential features of the evolving solution without leaving gaps in any region of importance.

4.7.1. Discretization of the Physical Domain

In recent years the method of Orthogonal Collocation (OC) [129] (→ Mathematics in Chemical Engineering, Chap. 7.5, Chap. 7.6) has been used successfully in cases of moderate values of the Damköhler number [37] even when the system is highly sensitive to changes in the operating conditions [68]. However, this approach gives inappropriate results at large values of the Péclet and/or Damköhler numbers when boundary layers associated with the flow and/or chemical reaction exist.

Certainly, in combustion-type problems most of the important physics is often restricted to small parts (or "boundary layers") inside the domain. For problems of this type Finite-Difference (FD) methods are superior to OC methods because a large number of grid points can be used and computations can still be effectively performed through tri-, penta-, and/or nano-diagonal matrix calculations.

By combining the low truncation error properties of OC methods and the ability of FD to locate grid points where needed, Carey and Finlayson [130] proposed a piecewise representation of the profiles. This method known as Orthogonal Collocation on Finite Elements (OCFE) divides the domain into elements where the properties are expanded locally in terms of orthogonal polynomials. The OCFE method is very efficient when dealing with single boundary layers [130, 131] and one-dimensional propagation phenomena [132]. Although the OCFE methods possess several advantages, they have some undesirable features such as the drastic increase of the bandwidth with the consequence of high memory and computer time requirements for some two-dimensional problems. In particular, this method is clearly outclassed by FD methods when multiple two-dimensional boundary layers displace and deform in time. In such situations, a generalized mesh is usually the proper domain discretization. To discretize the governing equations on generalized meshes, FD schemes have proven simpler and exhibit several advantages for efficient programming.

4.7.2. Discretization of the Governing Equations

Once the physical domain has been discretized appropriately and a way has been chosen to change the grid points locations in response to the evolving solution, the governing equations can be approximated on the so constructed generalized mesh. The use of a generalized mesh makes approximation of the governing partial differential equations more complicated than it would be on an orthogonal Cartesian or polar grid. Two different approaches have been followed to discretize the equations in a generalized coordinate system, each having its own merits [133].

The first approach is based on the transformation of the governing equations to the generalized curvilinear coordinates. Since the transformation involves additional terms ("cross derivatives") and variable coefficients ("metrics"), the resulting equations become more complicated. The additional terms appear because of the nonorthogonality of the curvilinear coordinates in the physical space. Depending on the rearrangement of the different terms after substitution of the transformation relations, the resulting transformed equations can be obtained in a conservative or nonconservative form. For more information, see [134].

The second approach is known as the Control Volume Formation (CVF). It takes into account the fact that the governing equations frequently represent the conservation of a transported quantity for an infinitesimally small control volume; thus, the conservation equations can easily be written for any arbitrary control volume of finite size. The approach consists of dividing the physical domain into smaller "cells" and rewriteing the governing equations for these finite-size control volumes. In what follows, compressibility effects are neglected and the fluid and solid heat capacities are assumed to be constant.

The conservation law (the energy balance, for instance) applied to a control volume V is

$$\frac{\partial}{\partial t}\int_V \overline{\varrho C_p} T \mathrm{d}V = -\int_V \boldsymbol{\nabla}\cdot[\varrho_\mathrm{f} C_{p,\mathrm{f}} \mathbf{u} T - \boldsymbol{\lambda}_\mathrm{e}\cdot\boldsymbol{\nabla} T]\mathrm{d}V + \int_V (-\Delta H_r)\, R\,(C_1,\ldots,C_\mathrm{i},\ldots C_\mathrm{s},T)\,\mathrm{d}V \quad (4.40)$$

The convective and conductive transport terms can be rewritten by using the Gauss divergence theorem as:

$$-\int_V \boldsymbol{\nabla}\cdot\left[\varrho_\mathrm{f} C_p^{;\mathrm{f}} \mathbf{u} T - \lambda_\mathrm{e}\cdot\boldsymbol{\nabla} T\right]\mathrm{d}V = \oint_s \mathbf{n}\cdot\left[\varrho_\mathrm{f} C_{p,\mathrm{f}} \mathbf{u} T - \lambda_\mathrm{e}\cdot\boldsymbol{\nabla} T\right]\mathrm{d}S \quad (4.41)$$

where S is the surface enclosing the control volume V and $\mathbf{n}$ is the outward unit normal vector on S.

If the control volume moves and deforms in time, and a Lagrangian approach is followed, the integral of the time-dependent term becomes:

$$\frac{\partial}{\partial t}\int_V \overline{\varrho C_p} T \mathrm{d}V = \frac{\mathrm{d}}{\mathrm{d}t}\int_V \overline{\varrho C_p} T \mathrm{d}V - \oint_s \mathbf{n}\cdot\mathbf{u}_\mathrm{s}\left(\overline{\varrho C_p} T\right)\mathrm{d}S \quad (4.42)$$

where $\mathbf{u}_\mathrm{s}$ is the velocity at which the enclosing surface S moves and/or deforms.

Then, the integral form of the governing equations for the control volume V can be rewritten as:

Mass balance:

$$\varepsilon\frac{\mathrm{d}}{\mathrm{d}t}\int_V C_\mathrm{i}\mathrm{d}V - \oint_\mathrm{S} \mathbf{n}\cdot[(\boldsymbol{u}-\mathbf{u}_\mathrm{s})\, C_\mathrm{i}]\,\mathrm{d}S + \varepsilon\oint_\mathrm{S} \boldsymbol{n}\cdot\mathbf{D}_\mathrm{e}\cdot\boldsymbol{\nabla} C_\mathrm{i}\mathrm{d}S - \int_V R\,(C_1,\ldots,C_\mathrm{i},\ldots,C_\mathrm{S},T)\,\mathrm{d}V \quad (4.43)$$

Energy balance:

$$\frac{\mathrm{d}}{\mathrm{d}t}\int_V \overline{\varrho C_p} T \mathrm{d}V = -\oint_\mathrm{S} \mathbf{n}\cdot\left[(\mathbf{u}-\mathbf{u}_\mathrm{s})\,\varepsilon\varrho_\mathrm{f} C_{p,\mathrm{f}} T\right]\mathrm{d}S + \oint_\mathrm{S} \mathbf{n}\cdot\left[(\mathbf{u}_\mathrm{s}\,(1-\varepsilon)\,\varrho_\mathrm{s} C_{p,s} T)\,\mathrm{d}S - \oint_S \mathbf{n}\cdot\lambda_\mathrm{e}\cdot\boldsymbol{\nabla} T\mathrm{d}S\right] + \int_V (-\Delta H_r)\, R\,(C_1,\ldots,C_\mathrm{i},\ldots,C_\mathrm{S},T)\,\mathrm{d}V \quad (4.44)$$

The discretization of the governing equations now consists of approximating the integral form on the generalized mesh formed by quadrilateral cells. These last forms of the equations are extremely useful since continuity of fluxes through the boundaries of neighboring control volumes can be easily ensured. Conservation of the transported quantity, even for the discrete numerical approximation, will thus be maintained.

Further details of numerical approximation methods are discussed in detail in → Mathematics in Chemical Engineering, Chap. 2.

5. Optimization of Chemical Reactors

Optimization means calculation of the "best way" of operating a particular unit where a chemical reaction takes place. A number of parameters must be chosen during the design process of a chemical reactor; for instance, inlet temperature and concentration of reacting components, residence time, temperature of the cooling liquid, size of the catalyst particles, and dimension of the reactor tube. All these parameters may have a critical effect on a particular design. The optimization process usually includes economical factors in a direct or indirect way. In practical calculations there are many physical or technical constraints. For example, the composition of the inlet mixture of reacting gases must be outside of explosion limits; the maximum temperature in the reactor is given by catalyst deactivation (heterogeneous systems) or by mechanical properties of the reactor tube (homogeneous systems); maximum feed in the reactor may be limited by the capacity of downstream distillation columns, etc. The optimum design of the reactor must maximize the performance and consider the process constraints. The parameters entering into the optimization calculation can have different effects, some may have only a marginal effect while others can be critical for the optimization.

The selection of the criterion of optimality (also known as *objective function, goal function*, or *performance index*) is crucial for design but somewhat arbitrary so the decision to select a certain criterion will always have a "fuzzy" character.

5.1. The Objective Function for Chemical Reactors

The objective function of a chemical reactor can have different forms. A list of functions as suggested by EDGAR and HIMMELBLAU [135] is given below:

1) Maximum conversion in a given equipment
2) Maximum production per batch
3) Minimum time of operation for a prescribed conversion
4) Maximum profit with respect to volume
5) Minimum consumption of energy
6) Minimum reactor volume with respect to certain concentrations
7) Maximum concentration by adjusting optimal temperature profile
8) Maximum selectivity for given feed
9) Minimum operating cost for a running chemical reactor

Typical independent variables are: pressure, inlet temperature and concentration, flow rate of the feed, temperature of the cooling system, and diameter of the catalyst pellets.

Frequently, the objective function has a complex character. It may be desirable, for example, to maximize profit and minimize pollution of the environment. Here two objective functions are introduced and the problem is called optimization of a vector objective function. To solve this problem an efficient point is defined [136]. The system is said to be working at an efficient point when a particular scalar objective function cannot be optimized further without degrading the optima reached for the remaining functions.

Mathematically, the optimization problem can be formulated in the following way. A given objective function must be maximized:

$$F\,(x_1,x_2,\ldots,x_n) = 0$$

subject to the following relations:

$$g_i\,(x_1,x_2,\ldots,x_n) = 0 \text{ with } i = 1,2,\ldots,r<n$$

and constraints:

$$h_j\,(x_1,x_2,\ldots,x_n) \geq 0 \text{ with } j = 1,2,\ldots,m$$

Mathematically, the description of the problem is represented by algebraic equations (e.g., for the nonisothermal CSTR), ordinary differential equations of initial value type (for the nonisothermal packed-bed reactor), nonlinear differential equations of boundary value type (adiabatic packed-bed reactor with external heat exchanger), or parabolic partial differential equations (packed-bed tubular reactor). Frequently, in reaction engineering, these descriptions are highly nonlinear because of the complex kinetic expressions and exponential dependence on temperature of kinetic, adsorption, and equilibrium constants.

The mathematical techniques used to solve the optimization problems in reactor design most often reported in the literature are [137]:

1) Differential calculus
2) Nonlinear programming
3) Variational methods and maximum principle
4) Dynamic programming
5) Optimization via repeated simulations

An important aspect in optimum design of chemical reactors is the parametric sensitivity of the solution. The multiparametric problems are often so complex that direct optimization can be an unaffordable, time-consuming task. However, qualitative analysis of the problem could indicate that only few independent variables significantly affect the solution and that the optimum solution is rather insensitive towards some of the other variables. Very frequently, in the industrial environment, a quasi-optimization of the reactor problem is sought which only considers the important parameters. A quasi-optimal solution can be very close to the optimum one, and has the advantage that the operating parameters can be easily adjusted and safely controlled.

5.2. Elementary Optimization Problems

5.2.1. Optimization of a Batch System

A mass balance for the consecutive reaction:

$$\mathrm{A} \xrightarrow{k_1} \mathrm{B} \xrightarrow{k_2} \mathrm{C}$$

can be easily integrated and the solution for the concentration of B, C_{B} is:

$$C_{\mathrm{B}} = C_{\mathrm{A}}^0 \frac{k_1}{k_2 - k_1} \left[\mathrm{e}^{-k_1 t} - \mathrm{e}^{-k_2 t} \right] \tag{5.1}$$

where C_{A}^0 is the initial concentration of A. The maximum concentration of B, C_{B}^*, can be calculated by differentiation and solving for:

$$\frac{\partial C_{\mathrm{B}}}{\partial t} = C_{\mathrm{A}}^0 \frac{k_1}{k_2 - k_1} \left[k_2 \mathrm{e}^{-k_2 t} - k_1 \mathrm{e}^{-k_1 t} \right] = 0 \tag{5.2}$$

i.e.,

$$t^* = \frac{1}{k_2 - k_1} \ln \left(\frac{k_2}{k_1} \right) \tag{5.3}$$

where t^* is the time for the batch operation resulting in maximum production of B. The maximum concentration of B achievable under t^* conditions can be calculated by substituting Equation (5.3) into Equation (5.1), i.e.,

$$C_{\mathrm{B}}^* = C_{\mathrm{A}}^0 \left(\frac{k_2}{k_1} \right)^{k_2/(k_1 - k_2)} \tag{5.4}$$

If the activation energies of both reactions are different, the maximum selectivity is also a function of the temperature T. The maximum concentration of B can then be calculated from the condition:

$$\frac{\partial C_{\mathrm{B}}}{\partial t} = \frac{\partial C_{\mathrm{B}}}{\partial T} = 0 \tag{5.5}$$

This problem cannot be solved analytically and Equation (5.5) must be solved numerically by a suitable method (e.g., Newton – Raphson) to obtain t^* and T^* (where T^* is the reaction temperature which results in a maximum production of B). The maximum concentration C_{B}^* can then be calculated from Equation (5.1) after inserting t^* and T^*. The procedure described above can be followed for any arbitrary sequence of first-order equations that can be integrated analytically. For a complex nonlinear reaction rate scheme, a simple search approach can be followed; for a given temperature the problem is integrated numerically as an initial value problem and t^* is determined for $\partial C_B/\partial t = 0$. Repeating the calculation for different values of T, a one-dimensional search for the optimum condition in T will result in the values of t^*, T^*, and C_{B}^*.

5.2.2. Optimization of a Continuous System (CSTR)

If the consecutive reactions described in Section 5.2.1 are carried out in a CSTR, in a similar way it can be found that:

$$t^* = \frac{1}{\sqrt{k_1 k_2}} \tag{5.6}$$

$$C_{\mathrm{B}}^* = C_{\mathrm{A}}^0 \frac{1}{\left(\sqrt{k_2/k_1} + 1 \right)^2} \tag{5.7}$$

5.2.3. Optimization of Complex Systems: Illustrative Example

Consider the following reaction scheme:

$$\mathrm{A} \xrightarrow{k_1} \mathrm{B} \xrightarrow{k_4} \mathrm{C} \xrightarrow{k_5} \mathrm{D}$$

$$\mathrm{A} \xrightarrow{k_2} \mathrm{D}, \qquad \mathrm{A} \xrightarrow{k_3} \mathrm{D}$$

All reactions are first order with the rate constants k_i given in the form:

$$k_i = k_{0,i}\exp\left[-\frac{E_i}{R_u}\left(\frac{1}{T}-\frac{1}{658\text{K}}\right)\right]$$

The preexponential factors and activation energies are given in Table 6.

Table 6. Values of preexponential factors and activation energies

Kinetic values	Reaction				
	1	2	3	4	5
$k_{0,\,i}$, s^{-1}	1.02	0.93	0.39	3.28	0.08
E_i, kJ/mol	67.00	58.70	62.80	41.90	62.80

The optimization problem consists of finding the optimal residence time and operation temperature to maximize the production of C, C_C^*. An analytical expression for C_C can be found as:

$$C_C = \frac{k_1 k_4}{k_A - k_4}\cdot \left[\frac{1-e^{-(k_4-k_5)t}}{k_4-k_5} - \frac{1-e^{-(k_A-k_5)t}}{k_A-k_5}\right] e^{-k_5 t} \qquad (5.8)$$

where $k_A = k_1 + k_2 + k_3$. The problem can be solved by a search method via, for instance, the gradient method (→ Mathematics in Chemical Engineering, Chap. 10). The results have been summarized in Table 7.

As can be seen from Table 7, the gradient method converges slowly and the maximum is rather flat in t and T. Evidently, t = 0.40 s, and T = 773 K can be chosen as reaction conditions. In this case only 0.014 is lost in selectivity; the reaction, however, can be carried out at a temperature that is 200 K lower than the optimum temperature. This represents a major energy saving. On the other hand, the residence time is five times higher than at optimum temperature which represents higher investment cost.

5.3. Optimization of Reactors by the Search Method

The optimization of a multiparametric reaction system can be accomplished by the search method. For more information, see [135].

The simplest gradient method is sufficient to illustrate the principle of the search technique. A new approximation is always selected in the gradient direction ∇F (**x**), i.e.,

$$\mathbf{x}^{k+1} = \mathbf{x}^k + h^k \nabla F\left(\mathbf{x}^k\right) \qquad (5.9)$$

where the value of h^k is selected in such a way that:

$$F\left(\mathbf{x}^{k+1}\right) < F\left(\mathbf{x}^k\right) \qquad (5.10)$$

This method can be easily used for problems where the gradient of the function can be calculated analytically. Otherwise, the gradient must be calculated numerically by finite differences.

The rate of convergence of the gradient method is good if the initial point is far from the optimum; close to the optimum, however, the rate of convergence becomes very slow. The search by gradient method can be used easily for parameters with continuous values, but complications arise for integer-value parameters (e.g., number of adiabatic reactor stages). The simplest way to overcome this problem is to perform a gradient search for a sequence of fixed integer value parameters and compare the results.

Illustrative Example [138]. Optimum operating conditions are to be found for a reactor in which nitrogen monoxide is oxidized to nitrogen dioxide. A dilute mixture of nitrogen monoxide (1.5 vol % NO) in air is fed into the reactor at a rate of 50 t NO/d. The following parameters should be optimized: catalyst type, feed composition, dew point of the inlet mixture, feed temperature, temperature control in the reactor, pressure, recirculation ratio, pressure drop, degree of conversion, mass velocity, diameter of the catalyst pellets, number of catalytic stages, and length and diameter of the catalyst bed.

A number of parameters listed here can be chosen a priori based on the information about the process.

Initial Concentration. The initial concentration is 1.5 vol % NO. This is an equilibrium concentration which results from synthesis of NO from N_2 and O_2 at ca. 2200 °C. The concentration of NO in the inlet gas can be increased by liquefaction and subsequent distillation. This modification would increase the processing costs.

Catalyst Type. Experimental results indicate that activated coke is a very good catalyst because it significantly increases the reaction rate. Silica gel is also a suitable catalyst, however, the reactor volume must then be 21 times larger. The

Table 7. Course of iterations using the gradient method

Iteration no.	T, K	t, s	C_C, kmol/m^3	T, K	t, s	C_C, kmol/m^3
0	1073	0.50	0.019	773	0.5	0.406
1	909	0.10	0.414	773	0.45	0.409
5	915	0.12	0.422	773	0.4	0.409
16	983	0.076	0.423	858	0.19	0.419
29	983	0.076	0.423	983	0.076	0.423
33	983	0.076	0.423	983	0.076	0.423

homogeneous reactor (without catalyst) would require a volume 500 times larger. A disadvantage of activated coke is that there is the possibility of combustion in the presence of NO. This can be prevented by careful reactor temperature control, i.e., the adiabatic temperature increase in the reactor must be kept low.

Feed Temperature. The reaction rate is described by a Langmuir – Hinshelwood kinetic expression, thus the reaction rate is not very dependent on temperature. The rate of the surface reaction increases with the temperature; however, the surface concentration is then lowered because of desorption. The rate of the reverse reaction:

$$2\,NO \rightarrow N_2 + O_2$$

can be neglected at temperatures below 200 °C. The exit temperature from the high-temperature reactor is 250 °C, the outlet gas is cooled by water to 20 – 30 °C. Cooling the inlet gas to the reactor below 20 °C would require new cooling equipment which would increase the investment costs.

Dew Point of Feed. Water is a catalyst poison for the oxidation of nitrogen monoxide both on silica gel and activated coke. For a reaction gas having a dew point 0 °C the amount of catalyst must be doubled compared with dry gas feed. For a dew point of − 30 °C the amount of catalyst required increases only by 10 %. Water can be removed from the inlet gas by drying it in an adsorption drying column packed with silica gel.

Temperature Control in the Reactor. The adiabatic temperature increase for 98 % conversion of 1.5 vol % NO in the feed is only 28 °C. Part of the reaction already occurs in the drying column (50 % conversion). The temperature increase involved (14 °C) does not promote undesirable catalyst oxidation. An adiabatic multistage reactor can thus be used for the reaction.

Recirculation. A simple calculation indicates that the reaction is completed after one pass so recirculation is not necessary.

Reactor Pressure. Doubling the pressure approximately doubles the reaction rate and halves the amount of catalyst required. However, since the mixture is strongly diluted, compression will be more expensive than the saving due to a smaller reaction volume. Therefore, the inlet pressure is chosen such that it only compensates for the pressure drop in the reactor.

Data for Calculation. The rate of oxidation of NO is described by Langmuir – Hinshelwood kinetics as:

$$R = \frac{p_{NO}^2 p_{O_2}}{a + b p_{NO}^2 + c p_{NO_2} + \omega p_{H_2O}}$$

The governing equations are described by the one-phase model (see Section 4.1.3) and the pressure drop by the Ergun correlation (Section 4.5.2).

The *reactor costs* C_R are given by:

$$C_R = 0.99\left[128.5 t_{sw} D H_t + 77.2 t_{dh} D^2\right]^{0.648}$$

where t_{sw} is the thickness of the cylindrical side wall, D the internal diameter of the reactor, H_t the total length of the reactor, and t_{dh} the thickness of the dished reactor head. The reactor is designed to withstand 140 kPa overpressure.

The *ventilator costs* C_V are:

$$C_V = 22900 \frac{\Delta P_r}{(\Delta P_r + P_e - 0.579)^{0.915}}$$

where ΔP_r is the pressure drop across the reactor and P_e is the total pressure in the bulk stream at the reactor exit.

The *turbine costs* C_T are:

$$C_T = 122 p_H^{0.6}\left(\frac{\Delta P_r}{\Delta P_r + P_e - 0.579}\right)$$

where:

$$p_H = 22900\left[(\Delta P_r + P_e + 0.421)^{0.283} - 1\right]$$

To estimate the *steam costs* (C_S) it is assumed that the turbine uses steam with 2.76 MPa pressure. The effectiveness of the turbine is 0.68; the turbine works 8000 h/a and steam is generated for \$ 1.10/1000 kg (1966). Then:

$$C_S = 54p_H \left(\frac{\Delta P_r}{\Delta P_r + P_e - 0.579}\right)$$

To estimate the *catalyst cost* (C_C), the life of the catalyst is assumed to be five years. Therefore, for \$ 0.2 per kilogram of catalyst the cost is

$$C_C = 0.018W$$

where W is the mass of the catalyst.

The loss C_N that is due to unconverted NO (\$ 2.40/100 kg NO) is

$$C_N = 3.59 \times 10^5 (1 - \xi)$$

where ξ is the degree of NO conversion in the feed.

The *overall cost* is

$$Y = C_R + C_V + C_T + C_S + C_C + C_N$$

The optimization calculation performed here was based on repetitive calculations of the reactor using the Box – Wilson method [139] in which the base point of the first set of trials should be arbitrarily chosen. The range of the first set of trial points in the factorial design should be quite narrow. Preliminary calculations made by varying one factor at a time indicate certain ranges within which the optimum point probably lies. These ranges are used directly as the upper and lower levels in a two-level factorial experiment (see Table 8). The half-replicate experiments, the radial line trials, and the factor levels are given in Tables 9, 10, 11, respectively.

Table 8. Two-level factorial experimental design

Standarized variable	Natural variable	Factor level	
		− 1	+ 1
x_1	y	0.97	0.99
x_2	G [a]	200	400
x_3	D_p [b]	0.025	0.050
x_4	Z [c]	6	8

[a] G = mass velocity.
[b] D_p = diameter of catalyst particle.
[c] Z = number of catalyst layers.

Table 9. Half-replicate experiments

Trial	x_1	x_2	x_3	x_4	Y
101	− 1	− 1	− 1	− 1	20 692
102	+ 1	− 1	− 1	+ 1	17 081
103	− 1	+ 1	− 1	+ 1	21 019
104	+ 1	+ 1	− 1	− 1	22 862
105	− 1	− 1	+ 1	+ 1	21 147
106	+ 1	− 1	+ 1	− 1	18 453
107	− 1	+ 1	+ 1	− 1	20 636
108	+ 1	+ 1	+ 1	+ 1	22 018
Improved trial					
109	0	0	0	0	17 733

Table 10. Radial line trials

Trial	x_1	x_2	x_3	x_4	Y
110	+ 1/2	− 1/2	− 1/2	+ 1/2	16 509
111	+ 1/4	− 1/4	− 1/4	+ 1/4	17 077
112	+ 3/4	− 3/4	− 3/4	+ 3/4	17 173

Table 11. Factor levels

	− 2	− 1	0	+ 1	+ 2
x_1	0.98	0.9825	0.985	0.9875	0.99
x_2	200.0	225.0	250.0	275.0	300.0
x_3	0.025	0.028125	0.03125	0.034375	0.375

The average value of trials 101 – 108 is 20 488. Since the value at the center of the response surface is much less than the average of the circumferential points, the existence of curvature in the explored region is obvious. The results of trials 109 – 112 along the radial line indicate the occurrence of a minimum in trial 110. The results of the optimum design are as follows:

Catalyst	activated coke
Size of catalyst	≈ 0.01 m
Composition of gas entering bed	1.50 vol % NO
Number of layers	8
Thickness of each layer	≈ 0.80 m
Mass velocity	≈ 0.34 kg m^{-2} s^{-1}
Diameter of bed	≈ 4.0 m
Temperature of entering gas	293 K
Temperature of leaving gas	≈ 331.50 K
Pressure of entering gas	≈ 11.69 kPa
Pressure of leaving gas	≈ 11.0 kPa
Fraction conversion	0.986
Dew point, entering gas	243 K
Weight of catalyst	≈ 37 850 kg

Annual costs (1966)	
Reactor	\$ 8230
Blower	\$ 180
Turbine	\$ 120
Steam	\$ 1330
Catalyst	\$ 1500
Nitrogen monoxide loss	\$ 5210
Total	\$ 16 570

5.4. Optimum Temperature Profile and Optimization of Multistage Reactors

For reversible exothermic catalytic reactions (e.g., water – gas shift synthesis, oxidation of SO_2 to SO_3, synthesis of methanol and ammonia), the optimum temperature profile plays an important role in optimum design.

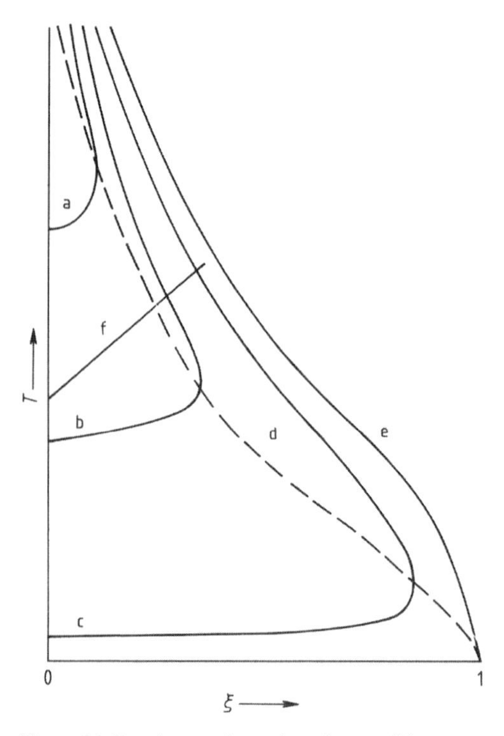

Figure 24. Reaction rate in exothermic reversible systems a) – c) Loci of constant reaction rate; d) Optimum temperature profile; e) Equilibrium curve; f) Adiabatic line

Reaction rate isolines in the $\xi - T$ phase plane are presented in Figure 24. Three generic lines (a – c) representing three different reaction rate values (for an exothermic reversible reaction) in decreasing order are shown. The locus of optimum conditions (d) connects the points of maximum conversion for a given value of the reaction rate, the equilibrium line (e) represents the locus of reaction rate $R(C, T) = 0$, and the adiabatic line (f) represents the history of the reaction rate for a reaction under adiabatic conditions.

For an equilibrium reaction:

$$\nu_1 \mathrm{A} + \nu_2 \mathrm{B} \leftrightharpoons \nu_3 \mathrm{P} + \nu_4 \mathrm{Q}$$

the reaction rate is given by:

$$R = k_1 C_\mathrm{A}^{\nu_1} C_\mathrm{B}^{\nu_2} - k_2 C_\mathrm{P}^{\nu_3} C_\mathrm{Q}^{\nu_4} \tag{5.11}$$

and for the optimum temperature where the maximum rate can be achieved, the following relation holds:

$$\frac{\partial r}{\partial T} = C_\mathrm{A}^{\nu_1} C_\mathrm{B}^{\nu_2} \frac{\partial k_1}{\partial T} - C_\mathrm{P}^{\nu_3} C_\mathrm{Q}^{\nu_4} \frac{\partial k_2}{\partial T} = 0 \tag{5.12}$$

If k_1 and k_2 show an Arrhenius dependence on temperature then:

$$\frac{C_\mathrm{P}^{\nu_3} C_\mathrm{Q}^{\nu_4}}{C_\mathrm{A}^{\nu_1} C_\mathrm{B}^{\nu_2}} = \left(\frac{k_1 E_1}{k_2 E_2}\right)_{T^*} \tag{5.13}$$

where T^* is the optimum reaction temperature.

For equilibrium conditions the following expression holds:

$$\frac{C_\mathrm{P}^{\nu_3} C_\mathrm{Q}^{\nu_4}}{C_\mathrm{A}^{\nu_1} C_\mathrm{B}^{\nu_2}} = \left(\frac{k_1}{k_2}\right)_{T^\mathrm{e}} \tag{5.14}$$

where T^e is the equilibrium temperature.

Combining Equations (5.13) and (5.14), gives:

$$\frac{T^*}{T^\mathrm{e}} = \frac{1}{1 + \frac{R_n T^\mathrm{e}}{E_2 - E_1} \ln\left(\frac{E_2}{E_1}\right)} \tag{5.15}$$

The mass of catalyst W required for processing a given molar flow rate F of reactants in an adiabatic packed bed can be calculated from:

$$\frac{W}{F} = Z = \int_0^y \frac{\mathrm{d}y}{R} \tag{5.16}$$

For the optimum temperature profile in the reactor the integral in Equation (5.16) is minimum, i.e.,

$$\frac{\partial Z}{\partial T} = \int_0^y \left(\frac{\partial r}{\partial T}\right) \frac{\mathrm{d}y}{r^2} = 0 \tag{5.17}$$

The optimum temperature profile is presented in Figure 24 (curve d).

For adiabatic tubular reactors the optimum temperature profile cannot be easily adjusted. Figure 24 reveals that optimum reaction rate is a decreasing function in the ξ vs. T diagram while the adiabatic line is an increasing function. To reconcile the different trends of optimum temperature profile and adiabatic line, heat must be withdrawn in the axial direction. The simplest configuration is cooling between adiabatic layers. This strategy of approximation the optimum temperature profile is shown in Figure 25.

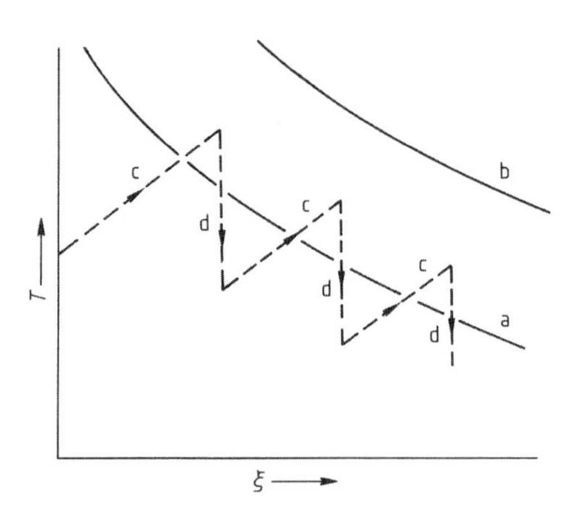

Figure 25. Approximation to the optimum temperature profile
a) Optimum profile; b) Equilibrium curve; c) Adiabatic operation; d) Intermediate cooling

5.5. Optimization of a Multibed Adiabatic Reactor with Heat Exchange Between Catalytic Stages

A schematic for optimization of a n stage adiabatic packed-bed reactor is shown in Figure 26. The mass of the catalyst W is a function of the inlet temperatures of the individual stages and the degrees of conversion obtained in the different stages:

$$W = W\left(T_1',T_2',\ldots,T_n',\xi_1',\xi_2',\ldots,\xi_n'\right)$$

Assume that the degree of conversion in the nth stage is given. It is necessary to calculate $2n-1$ variables in such a way so that the mass of the catalyst is minimal. These parameters can be calculated by differential calculus, i.e.,

$$\left(\frac{\partial Z}{\partial T_j'}\right)_{\xi_i'',T_i'(i\neq j)} = 0 \text{ for } i,j = 1,2,\ldots,n \quad (5.18)$$

$$\left(\frac{\partial Z}{\partial \zeta_j}\right)_{T_i',\xi_i''(i\neq j)} = 0 \text{ for } i,j = 1,2,\ldots,n \quad (5.19)$$

where ξ' and T' denote inlet conversion and temperature to a tray and ξ'' and T'' denote outlet conversion and temperature. After differentiation:

$$\frac{\partial Z}{\partial T_j'} = \int_{\zeta_j'}^{\zeta_j''} \frac{\partial}{\partial T}\left(\frac{1}{R}\right)\mathrm{d}\zeta = 0 \text{ for } i,j = 1,2,\ldots,n \quad (5.20)$$

$$\frac{\partial Z}{\partial \zeta_j'} = \frac{\partial}{\partial \zeta_j''} = \left[\int_{\zeta_j'}^{\zeta_j''} \frac{\mathrm{d}\zeta}{R\left(T_j'\right)} + \int_{\zeta_{j+1}'}^{\zeta_{j+1}''} \frac{\mathrm{d}\zeta}{R\left(T_{j+1}'\right)}\right] \quad (5.21)$$

$$= \frac{1}{R\left(T_j',\zeta_j''\right)} - \frac{1}{R\left(T_{j+1}',\zeta_{j+1}''\right)} = 0 \quad (5.22)$$

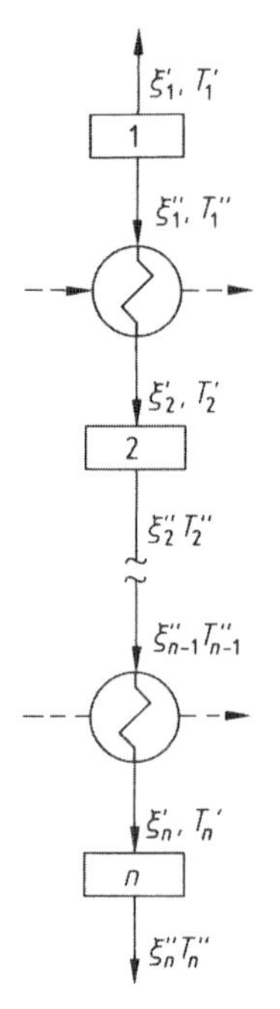

Figure 26. Schematic of a multibed adiabatic reactor with intermediate cooling
ξ' and T' denote inlet, ξ'' and T'' denote outlet conditions

i.e.,

$$R\left(T_j',\zeta_j''\right) = R\left(T_{j+1}',\zeta_{j+1}''\right) \text{ for } i,j = 1,2,\ldots,n \quad (5.23)$$

There are $2n-1$ equations (Eqs. 5.20 and 5.23) to calculate the values of the optimum parameters. Numerically, Equation (5.20) can be solved easily by transforming it to a differential equation:

$$\frac{\partial Z}{\partial \zeta} = \frac{\partial}{\partial T}\left(\frac{1}{R}\right) = f(\zeta) \quad (5.24)$$

subject to the initial condition:

$$Z = 0 \text{ for } \zeta = 0 \quad (5.25)$$

Equation (5.24) is integrated with Equation (5.25) until $Z = 0$. The calculated exit degree of conversion and exit temperature provide a value of reaction rate R. The inlet temperature in the second stage is calculated from Equation (5.23). The process is repeated for the remaining $n-1$ stages. In a similar way adiabatic reactors can be optimized with "cold shot", i.e., cooling between stages with fresh gas [140].

6. References

1. R. Shinnar, "Chemical Reaction Modeling – The Desirable and the Achievable," in: *Chemical Reaction Engineering – Houston,* ACS, Washington, D.C., 1978, p. 1.
2. R. K. Saiki et al.: "Enzymatic Amplification of Beta-Globin Genomic Sequences and Restriction Site Analysis for Diagnosis of Sickle Cell Anemia," *Science* **230** (1985).
3. J. Ninio, "Alternative to the Steady-State Method: Derivation of Reaction Rates from First-Passage Times and Pathway Probabilities," *Proc Natl Acad Sci USA* **84** (1987) 663–667.
4. S. V. Viljoen, M. Griep,M. Nelson, H. J. Viljoen, "A Macroscopic Kinetic Model for DNA Polymerase Elongation and High Fidelity Nucleotide Selection," *Comput. Biol. Chem.* **29** (2005) 101–110.
5. M. Griep, S. Whitney, M. Nelson, H. J. Viljoen, "DNA Polymerase Chain Reaction: A Model of Error Frequencies and Extension Rates", *AIChE J.* **52** (2006) 384–392.
6. K. R. Westerterp, W. P. M. van Swaaij, A. A. C. M. Beenackers: *Chemical Reactor Desgin and Operation*, J. Wiley & Sons, New York 1982.
7. R. B. Bird, W. E. Stewart, E. N. Lightfoot: *Transport Phenomena*, J. Wiley & Sons, New York 1960.
8. H. F. Rase: *Chemical Reactor Design for Process Plants*, J. Wiley & Sons, New York 1977.
9. V. V. Aleksandrov, A. A. Davydenko, Y. A. Kovalenko, N. P. Poddudnyi: "Influence of Two-Dimensionality of the Front with Heat Losses on the Limits of Stationary Gasless Combustion," *Fiz. Goreniya Vzryva* **23** (1987) 2 – 14.
10. J. Puszynski, J. Degreve, V. Hlavacek, "Modeling of Exothermic Non-Catalytic Reactions," *Ind. Eng. Chem. Prod. Res. Dev.* **26** (1987) 1424 – 1434.
11. F. A. Holland, F. S. Chapman: *Liquid Mixing and Processing in Stirred Tanks*, Reinhold, New York 1966.
12. V. W. Uhl, J. B. Gray (eds.): *Mixing Theory and Practice,* vols. I and II, Academic Press, New York 1966.
13. A. Bakker, J. B. Fosano: "Turbulent Mixing and Chemical Reaction in Stirred Tanks", The Online CFM Book, 2000 (http://www.baker.org/cfmbook.reaction.pdf).
14. K. R. Westerterp, W. P. M. Van Swaaij, A. A. C. M. Beenackers: *Chemical Reactor Design and Operation*, J. Wiley & Sons, New York 1984.
15. G. Astarita: *Mass Transfer with Chemical Reaction*, Elsevier, Amsterdam 1967.
16. R. H. : *Perry's Engineering Handbook*, 6th ed. McGraw Hill, New York 1984.
17. C. van Heerden: "Autothermic Processes; Properties and Reactor Design", *Ind. Eng. Chem.* **45** (1953) 1242.
18. C. van Heerden: "The Character of the Stationary State of Exothermic Processes," *Chem. Eng. Sci.* **8** (1958) 133.
19. V. Hlavacek, M. Kubicek, J. Jelinek: "Modeling of Chemical Reactors – XVIII, Stability and Oscillatory Behavior of the CSTR," *Chem. Eng. Sci.* **25** (1970) 1441.
20. R. Aris, N. R. Amundson: "An Analysis of Chemical Reactor Stability and Control," *Chem. Eng. Sci.* **7** (1958) 121.
21. D. D. Perlmutter: *Stability of Chemical Reactors*, Prentice Hall, Englewood Cliffs, NJ, 1972.
22. R. A. Schmitz: "Multiplicity, Stability and Sensitivity of States in Chemically Reacting Systems," *Chem. React. Eng. Rev.* **148** (1975) 156.

23. A. Varma, M. Morbidelli, H. Wu: *Parametric Sensitivity in Chemical Systems*, Cambridge University Press, New York 1999.
24. A. H. Heemskerk, W. R. Dammers, J. M. H. Fortuin: "Limit Cycles Measured in a Liquid-Phase Reaction System," *Chem. Eng. Sci.* **35** (1980) 439.
25. O. Levenspiel: *Chemical Reaction Engineering*, 2nd ed., J. Wiley & Sons, New York 1972.
26. J. M. Eldridge, E. L. Piret: "Continuous Flow Stirred-Tank Reactor Systems," *Chem. Eng. Prog.* **46** (1950) 290.
27. N. H. Chen: *Process Reactor Design,* Allyn and Bacon, Newton, Mass. 1983.
28. G. F. Froment: "Packed Bed Reactors. Steady State Conditions," Review A 5 in *Chemical Reactor Engineering,* Proceedings of the Fifth European/Second International Symposium on Chemical Reaction Engineering, Amsterdam, Elsevier, Amsterdam 1972.
29. V. Hlavacek: "Aspects in Design of Packed Catalytic Reactors," *Ind. Eng. Chem.* **62** (1970) no. 7, 8 – 26.
30. H. Hofmann: "Probleme bei der mathematischen Modellierung von Schüttschicht-Reaktoren," *Chem.-Ing-Tech.* **46** (1974) no. 6, 236 – 242.
31. V. Hlavacek, J. Votruba: "Steady-State Operation of Fixed-Bed Reactors and Monolithic Structures," in L. Lapidus, N. R. Amundson (eds.): *Chemical Reactor Theory. A Review,* Chap. 6, Prentice Hall, Englewood Cliffs, N.J., 1977.
32. S. S. E. Elnashaie, D. L. Cresswell: "Dynamic Behaviour and Stability of Non-Porous Catalyst Particles," *Chem. Eng. Sci.* **28** (1973) 1387 – 1399.
33. G. F. Froment, K. B. Bischoff: *Chemical Reactor Analysis and Design*, J. Wiley & Sons, New York 1979.
34. D. Ramkrishna, P. Arce: "Can Pseudohomogeneous Reactor Models be Valid?," *Chem. Eng. Sci.* **44** (1989) 1949 – 1966.
35. P. Arce, D. Ramkrishna: "Pattern Formation in Catalytic Systems: The Role of Fluid Mixing," *AIChE J.* **37** (1991) no. 1, 98 – 110.
36. S. Yagi, D. Kunii, N. Wakao: "Studies on Axial Effective Thermal Conductivities in Packed Beds," *AIChE J.* **6** (1960) no. 4, 543 – 546.
37. A. Jutan et.al.: "Multivariable Computer Control of a Butane Hydrogenolysis Reactor: Part I. State Space Reactor Modeling," *AIChE J.* **23** (1977) no. 5, 732 – 742.
38. D. Vortmeyer, K. J. Dietrich, K. O. Ring: "Comparison of One- and Two-Phase Model Predictions for Adiabatic Packed-Bed Chemical Reactors," *Adv. Chem. Ser.* **133** (1974) 588 – 599.
39. D. Vortmeyer, R. J. Schaefer: "Equivalence of One- and Two-Phase Models for Heat Transfer Processes in Packed Beds: One Dimensional Theory," *Chem. Eng. Sci.* **29** (1974) 485 – 491.
40. J. J. Carberry, D. White: "On the Role of Transport Phenomena in Catalytic Reactor Behavior," *Ind. Eng. Chem.* **61** (1969) 27 – 31.
41. S. Feyo de Azevedo, M. A. Romero-Ogawa, A. P. Wardle: "Modelling of Tubular Fixed-Bed Catalytic Reactors: A Brief Review," *Chem. Eng. Res. Des.* **68 (A)** (1990) 483 – 502.
42. H. M. Hulburt: "Chemical Processes in Continuous-Flow Systems. Reaction Kinetics," *Ind. Eng. Chem.* **36** (1944) no. 11, 1012 – 1017.
43. P. V. Danckwerts: "Continuous Flow Systems. Distribution of Residence Times," *Chem. Eng. Sci.* **2** (1953) no. 1, 1 – 18.
44. J. F. Wehner, R. H. Wilhelm: "Boundary Conditions of Flow Reactor," *Chem. Eng. Sci.* **6** (1956) 89 – 93.
45. L. T. Fan, Y. K. Ahn: "Critical Evaluation of Boundary Conditions for Tubular Flow Reactors," *Ind. Eng. Chem. Process Des. Dev.* **3** (1962) no. 1, 190 – 195.
46. V. Hlavacek, M. Marek: "Axialer Stoff- und Wärmetransport im adiabatischen Rohrreaktor – II. Numerische Untersuchung – Ablauf einer einfachen Reaktion bzw. einer Folgereaktion," *Chem. Eng. Sci.* **21** (1966) 501 – 514.
47. V. Hlavacek, H. Hofmann: "Modeling of Chemical Reactors – XVII. Steady State Axial Heat and Mass Transfer in Tubular Reactors. Numerical Investigations of Multiplicity," *Chem. Eng. Sci.* **25** (1970) 187 – 199.
48. V. Hlavacek et.al.: "Boundary Conditions for a Tubular Nonisothermal Nonadiabatic Packed Bed Reactor with Wall Heat Transfer in the Fore and Aft Sections," *Chem. Eng. Commun.* **3** (1979) 451–460.
49. J. Puszynski: "Experimental and Numerical Studies of Catalytic Nonadiabatic Reactors" Ph. D. Thesis, Institute for Chemical Technology, Prague, Czechoslovakia 1980.
50. K. B. Bischoff: "A Note on Boundary Conditions for Flow Reactors," *Chem. Eng. Sci.* **16** (1961) no. 1, 131 – 133.
51. L. C. Young, B. A. Finlayson: "Axial Dispersion in Nonisothermal Packed Bed

Chemical Reactors," *Ind. Eng. Chem. Fundam.* **12** (1973) no. 4, 412 – 422.
52. A. Rasmuson: "Exact Solution of a Model for Diffusion in Particles and Longitudinal Dispersion in Packed Beds: Numerical Evaluation," *AIChE J.* **31** (1985) 518, 519.
53. A. Rasmuson: " Exact Solution of Some Models for the Dynamics of Fixed Beds using Danckwerts' Inlet Condition," *Chem. Eng. Sci.* **41** (1986) no. 3, 599, 600.
54. R. A. Novy, H. T. Davis, L. E. Scriven: "Upstream and Downstream Boundary Conditions for Continuous-Flow Systems," *Chem. Eng. Sci.* **45** (1990) no. 6, 1515 – 1524.
55. R. A. Novy, H. T. Davis, L. E. Scriven: "A Comparison of Synthetic Boundary Conditions for Continuous-Flow Systems," *Chem. Eng. Sci.* **46** (1991) 57 – 68.
56. R. A. Novy, H. T. Davis, L. E. Scriven: "A Comparison of Initial Conditions for Continuous-Flow Systems," *Chem. Eng. Sci.* **46** (1991) no. 7, 1725–1737.
57. A. G. Dixon, D. L. Cresswell: "Theoretical Prediction of Effective Transfer Parameters in Packed Beds," *AIChE J.* **25** (1979) no. 4, 663 – 676.
58. D. Vortmeyer, R. Berninger in A. G. Dixon, D. L. Cresswell (eds.): "Comments on the Paper, Theoretical Prediction of Effective Heat Transfer Parameters in Packed Beds," *AIChE J.* **28** (1982) no. 3, 508 – 510.
59. W. Odendaal, W. Gobie, J. J. Carberry: "Thermal Parameter Sensitivity in the Simulation of the Nonisothermal, Nonadiabatic Fixed Bed Reactor – The Two-Dimensional Heterogeneous Model," *Chem. Eng. Commun.* **58** (1987) 37 – 62.
60. L. C. Windes, M. J. Schwedock, W. H. Ray: "Steady State and Dynamic Modeling of a Packed Bed Reactor for the Partial Oxidation of Methanol to Formaldehyde. I. Model Development," *Chem. Eng. Commun.* **78** (1989) 1 – 43.
61. L. H. S. Roblee, R. M. Baird, J. W. Tierney: "Radial Porosity Variations in Packed Beds," *AIChE J.* **4** (1958) 460 – 464.
62. R. F. Benenati, C. B. Brosilow: "Void Fraction Distribution in Beds of Spheres," *AIChE J.* **8** (1962) no. 3, 359 – 361.
63. V. M. H. Govindarao, G. F. Froment: "Voidage Profiles in Packed Beds of Spheres," *Chem. Eng. Sci.* **41** (1986) no. 3, 533 – 539.
64. B. C. Chandrasekhara, D. Vortmeyer: "Flow Model for Velocity Distribution in Fixed Beds Under Isothermal Conditions, " *Wärme Stoffübertrag.* **12** (1979) 105 – 111.
65. D. Vortmeyer, J. Schuster: "Evaluation of Steady Flow Profiles in Rectangular and Circular Packed Beds by a Variational Method," *Chem. Eng. Sci.* **38** (1983) no. 10, 1691 – 1699.
66. O. Kalthoff, D. Vortmeyer: "Ignition/Extinction Phenomena in a Wall Cooled Fixed Bed Reactor. Experiments and Model Calculations Including Radial Porosity and Velocity Distributions," *Chem. Eng. Sci.* **35** (1980) 1637 – 1643.
67. D. Vortmeyer, R. P. Winter: "Improvements in Reactor Analysis Incorporating Porosity and Velocity Profiles," *Ger. Chem. Eng. (Engl. Transl.)* **7** (1984) 19 – 25.
68. J. E. Gatica et al.: "Steady and Non-Steady State Modeling of Tubular Fixed Bed Reactors," *Chem. Eng. Commun.* **78** (1989) 73 – 96.
69. H. Martin: "Low Péclet Number Particle-to-Fluid Heat and Mass Transfer in Packed Beds," *Chem. Eng. Sci.* **33** (1978) 913 – 919.
70. C. L. de Ligny: "Coupling between Diffusion and Convection in Radial Dispersion of Matter by Fluid Flow through Packed Beds," *Chem. Eng. Sci.* **25** (1970) 1177 – 1181.
71. J. J. Carberry: *Chemical and Catalytic Reaction Engineering*, McGraw-Hill, New York 1976.
72. D. Vortmeyer, R. Berninger: "Comparison of Dispersion Models for Calculation of Temperature Profiles in Packed Beds," *Ger. Chem. Eng. (Engl. Transl.)* **6** (1983) 9 – 14.
73. P. Zehner, E. P. Schlunder: "Wärmeleitfähigkeit von Schüttungen bei mäßigen Temperaturen," *Chem. Ing. Tech.* **42** (1970) 933 – 940.
74. A. G. Dixon: "Thermal Resistance Models of Packed-Bed Effective Heat Transfer Parameters," *AIChE J.* **31** (1985) no. 5, 826 – 834.
75. A. G. Dixon: "Wall and Particle-Shape Effects on Heat Transfer in Packed Beds," *Chem. Eng. Commun.* **71** (1988) 217 – 237.
76. D. L. Cresswell, A. G. Dixon: in [58] 511 – 513.
77. A. G. Dixon, D. L. Cresswell: "Effective Heat Transfer Parameters for Transient PackedBed Models, " *AIChE J.* **32** (1986) no. 5, 809–819.
78. D. Vortmeyer: "Radiation in Packed Solids," *Ger. Chem. Eng. (Engl. Transl.)* **3** (1980) 124 – 138.

79. C. L. Tien: "Thermal Radiation in Packed and Fluidized Beds," *J. Heat Transfer* **110** (1988) 1230–1242.
80. A. G. Dixon, M. A. DiConstanzo, B. A. Soucy: "Fluid-Phase Radial Transport in Packed Beds of Low Tube-to-Particle Diameter Ratio," *Int. J. Heat Mass Transfer* **27** (1984) no. 10, 1701 – 1713.
81. E. Tsotsas, E.-U. Schlünder: "Heat Transfer in Packed Beds with Fluid Flow: Remarks on the Meaning and the Calculation of a Heat Transfer Coefficient at the Wall," *Chem. Eng. Sci.* **45** (1989) no. 4, 819 – 837.
82. R. E. Chao, R. A. Caban, M. M. Irizarry: "Wall Heat Transfer to Chemical Reactors," *Can. J. Chem. Eng.* **51** (1973) 67 – 70.
83. R. J. Wijngaarden, K. R. Westerterp: "Do the Effective Heat Conductivity and the Heat Transfer Coefficient at the Wall Inside a Packed Bed Depend on a Chemical Reaction? Weakness and Applicability of Current Models," *Chem. Eng. Sci.* **44** (1989) no. 8, 1653 – 1663.
84. D. J. Gunn, D. Vortmeyer: "The Reaction-Independence of Mixing Coefficients in Fixed-Bed Reactors," *AIChE J.* **36** (1990) no. 9, 1449 – 1451.
85. J. Beek, E. Singer: "A Procedure for Scaling-up a Catalytic Reactor," *Chem. Eng. Prog.* **47** (1951) 534 – 540.
86. J. E. Crider, A. S. Foss: "Effective Wall Heat Transfer Coefficients and Thermal Resistances in Mathematical Models of Packed Beds," *AIChE J.* **11** (1965) no. 6, 1012 – 1018.
87. V. Hlavacek, M. Marek, T. M. John: "Tubular Non-Isothermal Nonadiabatic Packed Bed Reactor. An Analysis of the One-Dimensional Case," *Collect. Czech. Chem. Commun.* **34** (1969) 3664–3675.
88. G. F. Froment: "Fixed Bed Catalytic Reactors. Current Design Status," *Ind. Eng. Chem.* **59** (1967) no. 2, 18 – 27.
89. O. Bilous, N. R. Amundson: "Chemical Reactor Stability and Sensitivity: II. Effects of Parameters on Sensitivity of Empty Tubular Reactors," *AIChE J.* **2** (1956) no. 1, 117 – 126.
90. G. F. Froment: "Progress in the Fundamental Design of Fixed Bed Reactors," in L. K. Doraiswamy, R. A. Mashelkar (eds.): *Frontiers in Chemical Reaction Engineering 1,* Wiley Eastern, New Delhi 1984, p. 12.
91. P. L. Chambré: "On the Characteristics of a Nonisothermal Chemical Reactor," *Chem. Eng. Sci.* **5** (1956) 209 – 216.
92. C. H. Barkelew: "Stability of Chemical Reactors," *Chem. Eng. Prog. Symp. Ser.* **55** (1959) no. 25, 37–46.
93. R. J. van Welsenaere, G. F. Froment: "Parametric Sensitivity and Runaway in Fixed Bed Catalytic Reactors," *Chem. Eng. Sci.* **25** (1970) 1503 – 1516.
94. C. McGreavy, C. I. Adderley: "Parametric Sensitivity and Temperature Runaway in Heterogeneous Fixed Bed Reactors," *Adv. Chem. Ser.* **133** (1974) 519 – 531.
95. R. A. Rajadhyaksha, K. Vasudeva, L. K. Doraiswamy: "Parametric Sensitivity in Fixed Bed Reactors," *Chem. Eng. Sci.* **30** (1975) 1399 – 1408.
96. A. Oroskar, S. A. Stern: " Stability in Chemical Reactors," *AIChE J.* **25** (1979) no. 5, 903 – 905.
97. M. Morbidelli, A. Varma: "Parametric Sensitivity and Runaway in Tubular Reactors," *AIChE J.* **28** (1982) no. 5, 705 – 713.
98. T. F. Degnan, J. Wei: "The Cocurrent Reactor Heat Exchanger: I. Theory," *AIChE J.* **25** (1979) no. 2, 338 – 344.
99. L. M. Akella, H. H. Lee: "A Design Approach Based on Phase Plane Analysis: Countercurrent Reactor/Heat Exchanger with Parametric Sensitivity," *AIChE J.* **29** (1983) no. 1, 87 – 94.
100. A. Soria Lopez, H. I. DeLasa, J. A. Porras: "Parametric Sensitivity of a Fixed Bed Catalytic Reactor: Cooling Fluid Flow Influence," *Chem. Eng. Sci.* **36** (1981) 285 – 291.
101. H. I. DeLasa: "Application of the Pseudoadiabatic Operation to Catalytic Fixed Bed Reactors," *Can. J. Chem. Eng.* **61** (1983) 710 – 718.
102. G. P. Henning, G. A. Pérez: "Parametric Sensitivity in Fixed-Bed Catalytic Reactors," *Chem. Eng. Sci.* **41** (1986) no. 1, 83 – 88.
103. G. P. Henning, G. A. Pérez: "A Generalization of Two Intrinsic Criteria for Runaway in Fixed Bed Catalytic Reactors: Part I. Development of the Criteria," *Chem. Eng. Commun.* **59** (1987) 107–125.
104. E. Bauman et al.: "Parametric Sensitivity in Tubular Reactors with Co-Current External Cooling," *Chem. Eng. Sci.* **45** (1990) no. 5, 1301 – 1307.
105. G. P. Henning, G. A. Pérez: "A Generalization of Two Intrinsic Criteria for Runaway in Fixed Bed Catalytic Reactors: Part II. Criteria Performance Analysis of Co-Current vs.

Counter-Current Operation," *Chem. Eng. Commun.* **59** (1987) 127 – 136.
106. D. O. Borio, J. E. Gatica, J. A. Porras: "Wall-Cooled Fixed-Bed Reactors: Parametric Sensitivity as a Design Criterion," *AIChE J.* **35** (1989) no. 2, 287 – 292.
107. D. O. Borio et. al. "Cocurrently-Cooled Fixed-Bed Reactors: A Simple Approach to Optimal Cooling Design," *AIChE J.* **35** (1989) no. 11, 1899 – 1902.
108. M. Morbidelli, A. Varma: "Parametric Sensitivity and Runaway in Fixed-Bed Catalytic Reactors," *Chem. Eng. Sci.* **41** (1986) 1063 – 1071.
109. M. Morbidelli, A. Varma: "On Parametric Sensitivity and Runaway Criteria of Pseudohomogeneous Tubular Reactors," *Chem. Eng. Sci.* **40** (1985) no. 11, 2165 – 2168.
110. M. Morbidelli, A. Varma: "A Generalized Criterion for Parametric Sensitvity: Application to Thermal Explosion Theory," *Chem. Eng. Sci.* **43** (1988) 91 – 102.
111. M. Morbidelli, A. Varma: "A Generalized Criterion for Parametric Sensitivity: Application to a Pseudohomogeneous Tubular Reactor with Consecutive or Parallel Reactions," *Chem. Eng. Sci.* **44** (1989) no. 8, 1675 – 1696.
112. V. Balakotahia: "Simple Runaway Criteria for Cooled Reactors," *AIChE J.* **35** (1989) no. 6, 1039–1043.
113. A. C. Hindmarsh: "LSODE and LSODI, Two New Initial Value Ordinary Differential Equation Solvers", *ACM-SIGNUM Newsl.* **15** (1980) no. 4, 10–11.
114. J. R. Leis, M. A. Kramer: "The Simultaneous Solution and Sensitivity Analysis of Systems Described by Ordinary Differential Equations", *ACM Trans. Math. Software (TOMS)* **14** (1988) March, 45–60.
115. H. C. Brinkman: "A Calculation of the Viscous Force Exerted by a Flowing Fluid on a Dense Swarm of Particles," *Appl. Sci. Res. Lect. A*: **1** (1947) 27 – 34.
116. D. A. Nield: "The Boundary Correction for the Rayleigh-Darcy Problem: Limitations of the Brinkman Equation," *J. Fluid Mech.* **128** (1983) 37 – 46.
117. K. Vafai, C. L. Tien: "Boundary and Inertia Effects on Flow and Heat Transfer in Porous Media," *Int. J. Heat Mass Transfer* **24** (1981) 195 – 203.
118. D. D. Gray, A. Giorgini: "The Validity of the Boussinesq Approximation for Liquid and Gases," *Int. J. Heat Mass Transfer* **19** (1976) 545 – 551.
119. D. D. Joseph: *Stability of Fluid Motions,* vol. 2, chap. X and XII, Springer Verlag, Berlin 1976.
120. D. Drew, L. Segel: "Averaged Equations for Two-Phase Flow," *Stud. Appl. Math.* **50** (1971) no. 3, 205 – 231.
121. J. L. Beck: "Convection in a Box of Porous Material; Saturated with Fluid," *Phys. Fluids* **15** (1972) no. 8, 1377 – 1383.
122. Ph. Forchheimer: "Wasserbewegung durch Boden," *VDJZ* **45** (1901) no. 50, 1781 – 1788.
123. M. Choudhary, M. Propster, J. Szekely: "On the Importance of the Inertial Terms in the Modeling of Flow Maldistribution in Packed Beds," *AIChE J.* **22** (1976) no. 3, 600 – 603.
124. M. Choudhary, J. Szekely, S. W. Weller: "The Effect of Flow Maldistribution on Conversion in a Catalytic Packed-Bed Reactor: Part I. Analysis," *AIChE J.* **22** (1976) no. 6, 1021 – 1027.
125. S. Ergun: "Fluid Flow through Packed Columns," *Chem. Eng. Prog.* **48** (1952) 89 – 94.
126. H. S. Fogler: *Elements of Chemical Reaction Engineering*, 4th ed., Prentice Hall PTR, Upper Saddle River, NJ, 2004.
127. B. A. Boley, J. H. Weiner: *Theory of Thermal Stresses*, J. Wiley & Sons, New York 1960.
128. B. A. Boley, I. S. Tollins: "Transient Coupled Thermoelastic Boundary Value Problems in the Half Space," *J. Appl. Mech.* **29** (1962) no. 4, 637 – 646.
129. J. V. Villadsen, W. E. Stewart: "Solution of Boundary-Value Problems by Orthogonal Collocation," *Chem. Eng. Sci.* **22** (1967) 1483 – 1495.
130. C. G. Carey, B. A. Finlayson: "Orthogonal Collocation on Finite Elements," *Chem. Eng. Sci.* **30** (1975) 587 – 596.
131. K. F. Jensen, W. H. Ray: "The Bifurcation Behavior of Tubular Reactors," *Chem. Eng. Sci.* **37** (1982) 199 – 222.
132. J. E. Gatica, J. Puszynski, V. Hlavacek: "Reaction Front Propagation in Nonadiabatic Exothermic Reaction Flow Systems," *AIChE J.* **33** (1987) no. 5, 819 – 833.
133. D. A. Anderson, J. C. Tannehill, R. H. Pletcher: *Computational Fluid Mechanics and Heat Transfer*, McGraw-Hill, New York 1984.
134. R. G. Hindman: "Generalized Coordinate Forms of Governing Fluid Equations and Associated Geometrically Induced Errors," *AIAA J.* **16** (1982) 1359 – 1367.

135. T. F. Edgar, D. M. Himmelblau: *Optimization of Chemical Processes*, McGraw-Hill, New York 1988.
136. J. H. Seinfeld, W. L. McBride: "Optimization with Multiple Performance Criteria," *Ind. Eng. Chem. Process Des. Dev.* **9** (1970) 53.
137. D. J. Wilde: *Optimum Seeking Methods*, Prentice Hall, Englewood Cliffs, NJ, 1964.
138. C. Chu, O. A. Hougen: "Optimum Design of a Catalytic Nitric Oxide Reactor," *Chem. Eng. Prog.* **57** (1961) no. 6, 51.
139. H. H. Rosenbrock, C. Storey: *Computational Techniques for Chemical Engineers*, Pergamon Press, New York 1966.
140. J. Caha, V. Hlavacek, M. Kubicek: "Study of the Optimization of Chemical Engineering Equipment. Numerical Solution of the Optimization of an Adiabatic Reactor," *Int. Chem. Eng.* **13** (1973) 466.

Mathematical Modeling

Henning Bockhorn, Technische Hochschule, Darmstadt, Federal Republic of Germany

See also: Mathematics in Chemical Engineering

1. Introduction . . . 207
1.1. Terminology . . . 207
1.2. Application Areas of Mathematical Modeling in Industrial Chemistry and Chemical Engineering . . . 207
1.3. Limitations of Mathematical Models . . . 208
2. Construction and Classification of Mathematical Models . . . 209
2.1. Construction of Mathematical Models . . . 209
2.2. Classification of Mathematical Models . . . 210
3. Empirical Models . . . 213
3.1. Linear Empirical Models . . . 213
3.1.1. Linear Empirical Models with One Variable . . . 214
3.1.1.1. Parameter Estimation (Linear Regression) . . . 214
3.1.1.2. Assessment of Estimated Parameter Values . . . 215
3.1.1.3. Sensitivity Analysis . . . 216
3.1.1.4. Example of a Linear Model . . . 217
3.1.1.5. Concluding Remarks . . . 219
3.1.2. Linear Empirical Models with Several Variables . . . 219
3.1.2.1. Parameter Estimation . . . 220
3.1.2.2. Assessment of Estimated Parameter Values . . . 221
3.1.2.3. Concluding Remarks . . . 222
3.2. Nonlinear Empirical Models . . . 222
3.2.1. Parameter Estimation (Nonlinear Regression) . . . 222
3.2.1.1. Transformation of the Model into Linear Form . . . 223
3.2.1.2. Direct Search Methods . . . 223
3.2.1.3. Gradient Methods . . . 225
3.2.2. Assessment of Estimated Parameter Values . . . 227
3.2.3. Example of a Nonlinear Model . . . 227
3.2.4. Concluding Remarks . . . 228
3.3. Further Calculation Methods and Model Types . . . 228
3.3.1. Further Methods of Parameter Estimation . . . 229
3.3.2. Other Types of Models . . . 229
3.3.2.1. Models with Constraints on the Parameters . . . 229
3.3.2.2. Models Based on Differential Equations . . . 231
4. Models Based on Transport Equations for Probability Density Functions . . . 232
4.1. Terminology . . . 232
4.1.1. One-Dimensional Distribution and Probability Density Functions . . . 232
4.1.2. Multidimensional Distribution and Probability Density Functions . . . 236
4.1.3. Conditional Probability Density Functions . . . 236
4.2. Transport Equations for Single-Point Probability Density Functions . . . 237
4.2.1. General form of the Transport Equations for Probability Density Functions . . . 238
4.2.2. Limitations of Single-Point Probability Density Functions . . . 241
4.3. Examples of Calculating Probability Density Functions . . . 242
4.3.1. Solutions for Deterministic Systems . . . 242
4.3.1.1. Age (Residence Time) Distributions in Chemical Reactors . . . 242
4.3.1.2. Size Distribution in Continuously Operating Crystallizers . . . 246
4.3.2. Solutions for Statistical Systems . . . 247
4.3.2.1. Closure of the Transport Equation for Probability Density Functions . . . 247
4.3.2.2. Solution Methods of the Transport Equation for Probability Density Functions . . . 250
4.3.2.3. Example: Combustion of Propane in a Turbulent Diffusion Flame . . . 252
5. Models Based on Physicochemical Principles (Transport Phenomena) . . . 254
5.1. Application of the Principle of Conservation of Momentum . . . 254
5.1.1. Laminar Tube Flow . . . 255
5.1.2. Turbulent, Nonreactive Free Jets . . . 255
5.1.2.1. Models for Reynolds Stresses . . . 256

Ullmann's Modeling and Simulation

ISBN: 978-3-527-31605-2

5.1.2.2. Solution Method of the Resulting System of Partial Differential Equations 258
5.1.2.3. Example: Turbulent Flow of Nitrogen in Air 261
5.2. Applications of the Principle of Conservation of Enthalpy 262
5.2.1. Heat Conduction 262
5.2.2. Heat Transfer 263
5.2.2.1. Exact Solution for a Boundary Layer Problem 263
5.2.2.2. General Principles of Modeling Heat Transfer 266
5.3. Applications of the Law of Conservation of Mass 267
5.3.1. Mass Transfer without Chemical Reaction 267
5.3.1.1. Exact Solution for a Boundary Layer Problem 267
5.3.1.2. General Principles for Modeling Mass Transfer 269
5.3.1.3. Examples of Simultaneous Mass and Heat Transfer: Dynamic Models 269
5.3.1.4. Examples of Simultaneous Mass and Heat Transfer: Static Models 270
5.3.2. Mass Transfer with Chemical Reactions 274
5.3.3. Chemical Reactions in the Homogeneous Phase 276
5.3.3.1. Isothermal Reactors with Frictionless Flow, Constant Density, and Reactions Without Volume Changes 277
5.3.3.1.1. Stability of Isothermal Reactors . 280
5.3.3.1.2. Sensitivity Analysis 286
5.3.3.2. Nonisothermal Reactors 287
5.3.3.2.1. Heterogeneous Catalytic Reactions 287
5.3.3.2.2. Stability Analysis of Nonisothermal Reactors 294
5.3.3.2.3. Use on Statistical Processes ... 298
6. References 303

Symbols

a general coefficient, temperature conductivity
$\boldsymbol{a}$ general matrix
A chemical component
A general coefficient, source term in momentum conservation equation
b general coefficient, general model parameter (estimate)
$\boldsymbol{b}$ general model parameter in vector notation (estimate)
B chemical component
B general coefficient
$\boldsymbol{B}$ general matrix
Bi Biot number
Bo Bodenstein number
c concentration, general coefficient
$\boldsymbol{c}$ general matrix
c_p specific heat capacity
C chemical component
C model parameter in turbulence models
C_1, C_2, C_3, C_4 integration constants
D chemical component
D diameter, diffusion coefficient
D_{R} tube diameter
$Da_{\mathrm{I}}, Da_{\mathrm{II}}$ Damköhler number (of 1st and 2nd kind)
E statistical event
$\boldsymbol{E}$ statistical error in vector notation
E_{a} activation energy
f dimensionless flow function, mixture fraction (normalized mass fraction of an element in a mixture)
F cross-sectional area, feed flow of a liquid, value of the F-distribution
F distribution function of a statistical variable
g acceleration due to gravity
G mass flow of gas
h enthalpy
$\boldsymbol{h}$ step size in vector notation
Δh_{v} specific latent heat of vaporization
H Heaviside step function, height
Ha Hatta number
HTU height of one transfer unit
$\boldsymbol{I}$ unit matrix
I_u, I_Φ, I_p class of integrals
j total flux of a scalar quantity
J flux of scalar quantities per unit area
$\boldsymbol{J}$ Jacobi matrix
k general parameter, rate coefficient, total heat-transfer coefficient, turbulence energy
K "adsorption" coefficient
K_p class of integrals

L length, likelihood function, mass flow of liquid, number of grid nodes in the x-direction, size, transport operator
Le Lewis number
m deterministic moment, mass fraction, number of occurrences of an event
M number of grid nodes in the y- or r-direction
M_i molecular mass
n number of data sets, number of units
Nu Nusselt number
NTU number of transfer units
O order of magnitude
p general parameter, pressure, statistical weight
$\boldsymbol{p}$ statistical weight in vector notation
P chemical component
P probability density function of a statistical variable
Pe Péclet number
Pr Prandtl number
q general parameter, number of independent variables in multivariable models, volumetric flow rate
Q general function
r general parameter, radial distance, rate of reaction
R gas constant, radius, ratio of enthalpies of vaporization
Re Reynolds number
s specific surface area, general parameter, estimate for standard deviation
S sink flow of a liquid, source term in conservation equation, surface area
s_r^2 mean square deviation about regression
$s^2{}_{\varepsilon}$ mean square deviation of data set
$s^2{}_{\bar{Y}_i}$ mean square deviation about $\bar{Y}_i$
Sc Schmidt number
Sh Sherwood number
t time, value of student t-variable
T temperature
u velocity
$\boldsymbol{u}$ velocity in vector notation, i.e., $\boldsymbol{u}=(u_1, u_2, u_3)$; $\boldsymbol{u}=(u, v, w)$
U velocity in a turbulent (statistical) flow
$\boldsymbol{U}$ velocity in a turbulent (statistical) flow in vector notation, i.e., $\boldsymbol{U}=(U_1, U_2, U_3)$; $\boldsymbol{U}=(\boldsymbol{U}, \boldsymbol{V}, \boldsymbol{W})$
V volume
w residence time distribution, velocity
W residence time distribution function, probability, velocity in a turbulent (statistical) flow
x distance, general independent variable
$\boldsymbol{x}$ distance in vector notation, i.e., $\boldsymbol{x}=(x_1, x_2, x_3)$; $\boldsymbol{x}=(x, y, z)$, general independent variable in vector notation
X partial derivatives of expected values with respect to model parameters in matrix notation
y general dependent variable
$\boldsymbol{y}$ general dependent variable in vector notation
Y general statistical variable
$\boldsymbol{Y}$ general statistical variable in vector notation

Greek Symbols

α age, general coefficient, heat-transfer coefficient, normalized mass fraction, significance level
β general coefficient, general model parameter, mass-transfer coefficient, normalized mass fraction
$\boldsymbol{\beta}$ general model parameter in vector notation
β_r acceleration factor
β_H Prater number
γ Arrhenius number, correlation coefficient, scaling factor
γ_r reciprocal Arrhenius number
δ Dirac delta function
δ film thickness, Kronecker symbol
ε dissipation rate, intergranular volume, normalized rate coefficient, statistical error
ζ phase shift
η effectiveness factor, normalized distance, similarity coordinate
$\boldsymbol{\eta}$ expected values in vector notation
θ general variable, normalized age, normalized residence time, normalized temperature
Θ general statistical variable, source term in scalar quantities equation
κ permeability, normalized rate coefficient
λ eigenvalues, Lagrangian multipliers, mean free path, scaling factor, thermal conductivity, wavelength
μ dynamic viscosity, normalized mass fraction

μ_t apparent turbulent viscosity
μ_n nth moment of a probability density function
ν frequency, kinematic viscosity, stoichiometric coefficient
ξ general variable
Ξ general statistical variable
π normalized mass fraction
ϱ density, scaling factor
σ standard deviation
σ_φ turbulent Schmidt/Prandtl numbers
τ normalized time, residence time, turbulent time scale, viscous stress
τ_N cooling time
φ general variable, normalized mass fraction
Φ general variable, general statistical variable
$\boldsymbol{\Phi}$ general variable and general statistical variable in vector notation, sum of squares of deviations
Φ_K Thiele modulus
ψ general variable
Ψ general variable, general statistical variable, stream function
$\boldsymbol{\Psi}$ general variable and general statistical variable in vector notation
Ω general function, general operator

Subscripts

a, b reference states a and b
A ambient, chemical component
ad adiabatic
B chemical component
Bf bifurcation
c central
C cooling
Cat catalyst
diff diffusion
e east, end, expansion
eff effective
ext external
G gas phase
h enthalpy
i component of a vector, chemical component
int internal
j component of a vector, chemical component
k component of a vector, chemical component, contraction, turbulence energy, summation according to the Einstein summation convention
l number of elements in a difference equation
L liquid phase
L total length
m mean
m mass
n north
o upper limit
p pressure, pressure-related
Ph phase boundary
qs quasi-steady-state
r reaction, reflected
ref reference
s south
S surface
t turbulent
T temperature
tot total
u lower limit
u velocity
w west
W wall
Wi inner wall
Wo outer wall
x local
α a component
ε dissipation rate
φ variable
ω chemical component, summation according to the Einstein summation convention
0 initial, reference state

Superscripts

$*$ transformed or normalized variable or parameter
$'$ scaled variable
$\sim$ weighted or normalized variable
$-$ mean
$\hat{}$ calculated model response
A, B, C, D chemical component
eq equilibrium
f fluctuating
G gas phase
h enthalpy
L liquid phase
m mass
(n) number of iterations
T transpose of a vector or matrix

1. Introduction

1.1. Terminology

Models are used in almost every branch of science and technology. It is difficult to encompass the many meanings of the term "model" with one single definition. The classical definition is "An object M (an object, physical or ideal system, or process) is a model if analogies exist between M and another object O that permit conclusions to be made about O" [1]. This definition describes the representation of the object O and presupposes its isolation from the whole of reality. Thus a model represents reality or a part of it. Since only certain conclusions are required, the model is reduced to parts or individual aspects of reality. The restriction to analogies between model M and object O results in limited conformity of their function, structure, and behavior. The representation of the object O by the model M may have limited accuracy and the model M can be based on a different scale than the object O. The model is satisfactory as long as important variables and phenomena are correctly represented for the specific context or investigation.

An adequate model according to the above definition can rarely explain reality in the form of a theory. Information that has been lost in the representation cannot be retrieved from the model. Attempts at theoretically interpreting the object O with the help of model M must therefore be discussed according to the aspect, scale, and accuracy of the representation.

Analogies between model M and object O can be established in the form of mathematical equations. A mathematical model therefore represents a set of algebraic and/or differential and/or integral equations which are used to describe the behavior of the object O. The terms variables and parameters are used in this article in accordance with their definitions given in mathematical relationships, and not as defined in process control, see for example [2].

1.2. Application Areas of Mathematical Modeling in Industrial Chemistry and Chemical Engineering

The two major tasks facing engineers and scientists in industrial chemistry are (1) the operation and optimization of existing processes and (2) the design of new or improved ones. In this article the term process means a sequence of operations or treatments and transformations performed on specific materials.

For the first task engineers and scientists must obtain exact information about the particular process under consideration. Qualitative aspects and criteria must be quantified. The basic variables and parameters of the relationships which describe the individual parts of the process must be worked out. The individual parts of the process must be combined. Mathematical models are very important in this respect.

For the second task mathematical models help in applying existing processes to new or modified plants and in the definition of safer, more economically viable operating conditions. Data for the construction of new plants cannot be obtained from an operating process by running it to its technical limits; this entails a high degree of risk. In contrast, a mathematical model of a process is easy to manipulate. Unusual operating conditions can easily be simulated. The plant can even be modeled under hazardous conditions to define the limits of operating parameters or risk areas.

The main applications of mathematical models in industrial chemistry are summarized as follows:

1) *Simple Experimentation.* By means of mathematical models, existing processes can be investigated more quickly, economically, and thoroughly than in operating industrial plants. Mathematical models can simulate the plant in time lapse or slow motion. The results from the model are therefore available more quickly or have a higher time resolution.
2) *Repetition of Experiments.* Simulation of processes with mathematical models allows the influence of variations in variables or parameters to be investigated. Errors can easily be implanted into or removed from the model. In a running industrial process, this type of empirical sensitivity analysis has the disadvantage that it requires a high degree of technical effort and entails a high degree of risk.
3) *Sensitivity Analysis.* With a mathematical model, sensitivity analysis does not have to

be carried out on an empirical basis by repetition of tests with systematic changes of the variables and parameters. Sensitivity analysis as a mathematical tool allows calculation of the gradients of dependent variables with respect to the model parameters.

4) *Control and Operation.* Mathematical models can be used as an economical way of estimating the stability of industrial processes or subsystems as a prerequisite for effective control or operation. Again, use of the industrial process for direct experimentation must be ruled out because of the high risk involved.
5) *Optimization.* Description of a complex system in terms of a mathematical model allows optimization objectives for the process to be developed. The optimization objectives can easily be adjusted in accordance with changing requirements. Optimum values of operating variables or parameters can also be determined easily.
6) *Extrapolation.* A suitable mathematical model can be used to test extreme operating conditions that are not possible or practical in the running process. Surfaces of operating variables and parameters can then be constructed and the optimum conditions for running the plant can be extrapolated. Extrapolability must, however, always be discussed in the light of the limits of the model and the part of the object O being modeled.

1.3. Limitations of Mathematical Models

Mathematical models have some significant limitations. The first is the type, quantity, and accuracy of the available data. The success and results of mathematical models depend largely on the information about the object O being modeled. A mathematical model cannot be better than the physical or chemical data on which it is based. In many cases available data are not sufficient and engineers and scientists have to obtain sufficient data for a model by experimental analysis. The alternative is to simplify the model adequately. The two following examples illustrate the associated problems.

The number of theoretical plates necessary for solving a separation problem (e.g., the number of theoretical plates of a rectification column) can usually be calculated more accurately than the effectiveness factor of a separation plate in a real piece of equipment. The latter value depends on the fluid dynamics of the system and on the operating conditions of the column. However, the inaccuracy in the estimate of the effectiveness factor strongly affects the commercial viability of a separation process with a large number of theoretical plates.

Another example concerns the kinetic data for industrial processes that are mainly derived from laboratory experiments. On a laboratory scale, side reactions may be much less noticeable than in an industrial plant. Lack of information about side reactions can lead to rate data that are unsuitable for industrial process design. A further problem is the impurities of the reactants used in an industrial plant; impurities can lead to changes in the reaction rate that cannot be predicted with the pure chemicals used in laboratory experiments.

The accuracy required for the individual parameters of the model depends on the sensitivity with which the results of the model respond to changes in these parameters (see Sections 3.1.1.3 and 5.3). Those parameters which exert the greatest influence on the results of the model must be determined with the greatest accuracy.

A second restriction in the use of mathematical models lies in the mathematical tools that are available for solving the resulting system of equations. Complex structures can often be formulated mathematically. However, the mathematical and numerical methods required to solve the resulting system of equations are either not yet developed or are beyond the capacity of available computers. The rapid development of supercomputers in recent years has enabled increasingly complex problems to be solved.

A further restriction concerns the interpretation of the results of the model and the extrapolability of the representation. This limitation is illustrated in the simple example in Figure 1. The solid curve shows the dynamics of a physicochemical quantity. The object could be, for example, an intermediate in consecutive chemical reactions, φ represents its concentration:

$$\varphi = \frac{k_1}{k_2 - k_1} \psi_0 \left[\exp\left(-k_1 t\right) - \exp\left(-k_2 t\right)\right] \tag{1}$$

where ψ_0 is the initial concentration of starting compound from which φ is formed, t is the time, and k_1 and k_2 are the rate coefficients of two consecutive first-order reactions, $k_1 \neq k_2$. The available data cover the range $0 < t < t_1$ and are shown in Figure 1 as squares. An obvious empirical model based on these data is $\varphi = kt$ and is given by the dashed line. In fact the object is represented by this model in the range $k_1 t \ll 1$ and $k_2 t \ll 1$. If the exponential functions in Equation (1) are developed for the above condition, it follows that:

$$\varphi \approx \frac{k_1}{k_2 - k_1} \psi_0 \left[(k_2 - k_1)\, t \right] = k_1 \psi_0 t \qquad (2)$$

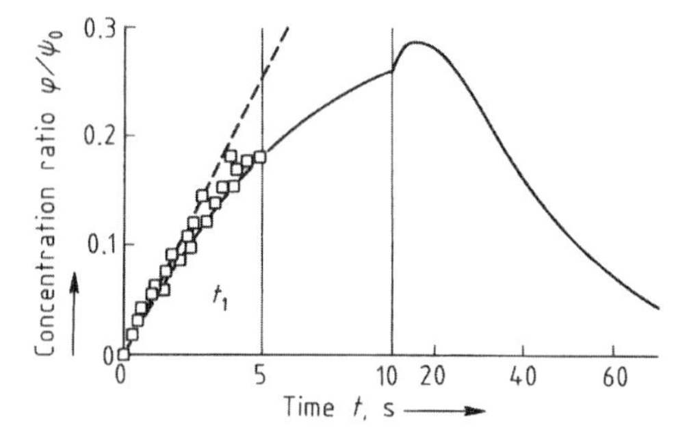

Figure 1. Relationship between a model and object with limited data

It can be seen that for the range covered by experimental data ($0 < t < t_1$), an analogy exists between the model and the object and the phenomenon of interest is also adequately reproduced within this region. The representation has a coarser scale, i.e., $\varphi = kt$ instead of $\varphi = k_1 \psi_0 t$. However, the model does not explain the physicochemical characteristics of the object – an intermediate in a system of consecutive chemical reactions. Finally, extrapolation of this model beyond the range covered by the available data leads to a false representation.

2. Construction and Classification of Mathematical Models

2.1. Construction of Mathematical Models

Based on the terminology used in Section 1.1, the construction of a mathematical model consists of setting up a consistent set of mathematical relationships. In the ideal case these would be identical to the relationships between the process variables. Often, only certain phenomena of particular interest can be incorporated in the model due to the complexity of the real process.

The strategy of analysis used for developing mathematical models generally contains the following steps (see also [3–10]):

1) Formulation of the problem and compilation of objectives and decision criteria
2) Investigation of the process and its classification with the aim of splitting it up into subsystems (process elements)
3) Determination of the relationships between the subsystems
4) Analysis of the variables and relationships between the variables for the individual process elements
5) Setting up of mathematical relationships with variables and parameters; data acquisition
6) Investigation of the representation of the process by the model; comparison of the simulation with real process data
7) Installation of the model; interpretation and collation of the results

For further evaluation of the model the above steps can be complemented by:

1) Sensitivity analysis of the model: identification of the parameters with strong and weak influences on the response of the model
2) Model simplification

If the above steps are compiled in a flow diagram, then steps 4 – 9 must be repeated, possibly several times, until the interpretation of the results of the model still makes sense within the framework of the formulated objectives and expected solutions to the problem.

Industrial processes are very complex. The success of mathematical models in describing them often depends on whether they can be subdivided into process elements. The entire process can then be built up from the individual elements if the relationships between them have been worked out. Subdivision into process elements is not necessarily based on the physical principles of the entire process. For example, a nonideal tubular reactor can be simulated by a series of ideal stirred tank reactors although such units may not exist in the industrial process. A packed absorption column can be represented by a number of theoretical plates even though mass transfer occurs continuously and does not result in stepwise attainment of phase equilibrium.

By correct treatment and adjustment of the process elements, modeling strives to represent the entire process as accurately as possible on the basis of simple, established principles of the individual elements.

Industrial processes are not only very complex, they can also consist of a number of qualitatively different process elements. A chemical process may, for example, encompass process elements for the preparation of the reactants, the chemical reactor, heat exchangers, and finally process elements for separation and possibly processing of the product. The different qualities of the process elements require different mathematical models depending on the physical or physicochemical principles on which they are based. An industrial heat exchanger is represented by a different type of model from that required for a rectification column, the latter has a different mathematical description from a chemical reactor, and so on. Due to the underlying physicochemical principles, mathematical modeling and the construction of mathematical models employ a wide variety of mathematical methods and methods of solution. For example, the design of distillation columns with models based on theoretical plates leads to a system of linear equations which is often solved graphically. The representation of a nonideal reactor by a series of ideal stirred tank reactors leads to a system of algebraic equations which can be highly nonlinear depending on the chemical reaction under consideration. Finally, a reactor model for a homogeneous, turbulent-flow reactor based on the modeled Navier–Stokes equations leads to a system of coupled partial differential equations that can be solved by various methods.

Mathematical modeling is used to formulate widely differing physical and physicochemical phenomena: transfer of heat, mass, and momentum, as well as chemical reactions in homogeneous and heterogeneous systems. Mathematical modeling is thus used in the design of mass-transfer operations, calculation of heat exchangers, chemical reaction engineering, and finally process control. A wide range of methods are used to formulate models and to solve the resulting systems of equations.

For this reason further formalization of the steps for building mathematical models and a further division into individual steps will be avoided. The following section examines the classification of models based on the physical background of the model, the type of system of equations, and the necessary methods for their solution.

2.2. Classification of Mathematical Models

Mathematical models can be classified according to their physical background, the type of systems of equations, and the corresponding methods of solution. Alternatives are discussed in [3], [8–12]. Figure 2 shows three classes of models with different physical backgrounds: models based on physicochemical principles, probability density function (pdf) models, and empirical models.

The first category comprises models based mainly on the mathematical formulation of transport phenomena. Use of this principle requires that the process can be subdivided into process elements that can be described by the laws governing the transport of mass, momentum, and energy, i.e., their conservation principles.

Such models are subdivided into deterministic and statistical models. Deterministic models or model elements have a determined value or set of values for each variable or parameter for any given set of conditions. In contrast, the variables and parameters used in statistical models (or elements of statistical models) are statistical quantities. They can only be given with a certain probability or in terms of moments of probability density functions. If, for example, the probability density function $P(Y)$ holds for the statistical variable Y, then $P(Y)\,\mathrm{d}Y$ is the probability for the variable lying in the range $\mathrm{d}Y$ around Y. Normally the full statistical information is not necessary for technical purposes; therefore statistical variables are often described by moments of the probability density functions. These moments are defined as

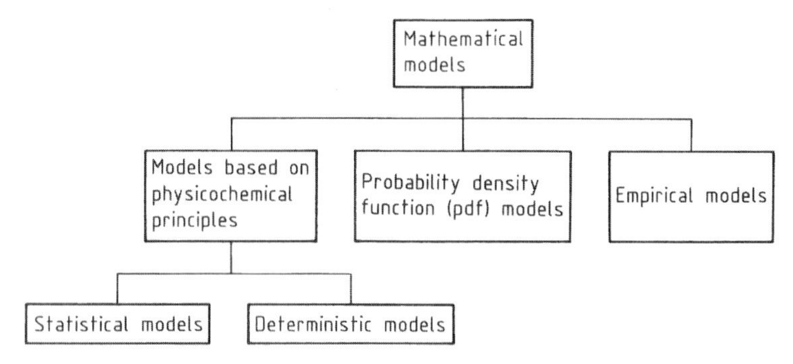

Figure 2. Classification of mathematical models according to their physical principles

$$\mu_n \equiv \int_{-\infty}^{+\infty} Y^n P(Y)\,\mathrm{d}Y$$

For example the first moment

$$\mu_1 \equiv \int_{-\infty}^{+\infty} Y P(Y)\,\mathrm{d}Y$$

is the mean value of the statistical variable Y, the second moment about the mean value is its variance Var (Y):

$$\mu_2' \equiv \int_{-\infty}^{+\infty} (Y - \mu_1)^2 P(Y)\,\mathrm{d}Y, \text{ etc.}$$

A statistical mathematical model is the mathematical description of a process, e.g., in terms of the moments defined above. Figure 3 shows some possibilities for statistical processes. Normally a deterministic process leads to deterministic results (Fig. 3 A). If a statistical error is added to the process results (Fig. 3 B), it leads to a statistical result. The same result is obtained if the deterministic process becomes unstable (Fig. 3 C); a simple example for this is buoyancy-driven turbulence. If the input process variables are statistical variables then the results of the process are also statistical variables (Fig. 3 D). Figure 3 shows that the macroscopic phenomena in statistical processes change but that the physicochemical background of the process and hence of the process model remains the same. The principles of the conservation of mass, momentum, and energy are also valid for statistical variables; the mathematical overhead for handling these cases is, however, more complicated. The transport equations used to describe a statistical process by means of statistical moments contain, for example, a number of statistical hypotheses and model assumptions for their closure but the character of the model equations does not change. For these reasons statistical models are classified as shown in Figure 3. Methods for representing statistical processes will be discussed in Chapter 5.

A further subdivision of models based on physicochemical principles according to the type of model equations is shown in Figure 4. The complexity of the method of solution decreases from right to left. The division is based on the most commonly used types of models but is by no means definitive. Thus, multidimensional models can be formulated in the form of algebraic relationships; models which can be described by algebraic relationships do not have to be steady-state models. Steady-state models describe processes in which the accumulation terms (the changes in the variables over time) disappear. Nonsteady-state models include changes in the variables over time. In models with constant variables, the properties and state are not functions of space so that the system is homogeneous. Description with locally distributed variables considers local changes in the dependent variables. (The nomenclature of process analysis does not make the distinction between variables and parameters introduced in Section 1.1; the latter two types of models are then classified as models with constant or concentrated parameters or as models with distributed parameters [10], [13–15].) Figure 4 also shows that the description of processes by differential equations and difference equations is frequently equivalent. Differential equations must normally be solved by numerical methods. In order to achieve this they are transformed into difference equations. These methods, which are really mathematical tools, have their physical analogy in the description of processes with models based on finite changes (see Sections 4.3.1 and 5.3.3.1).

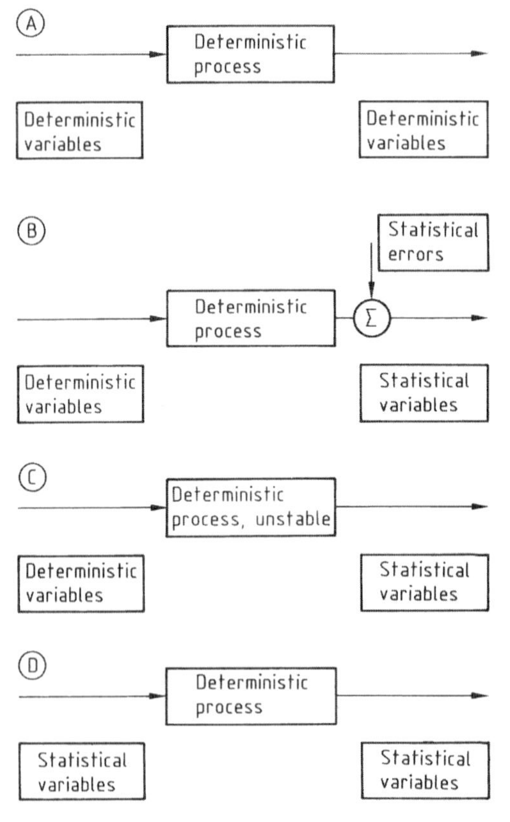

Figure 3. Statistical and deterministic processes

Models based on transport equations for probability density functions generally give their results in the form of functionals $P(\Phi_1, \ldots, \Phi_n)$. These are defined in such a way that the probability of finding the dependent variables $(\Phi_1, \ldots, \Phi_n)$ in the small range $d\Phi_1, \ldots, d\Phi_n$ around the functions $\Phi_1(\boldsymbol{x}, t), \ldots, \Phi_n(\boldsymbol{x}, t)$ is $P(\Phi_1, \ldots, \Phi_n)\, d\Phi_1, \ldots, d\Phi_n$. Probability density functions $P(\Phi_1, \ldots, \Phi_n; \boldsymbol{x}, t)$ that are defined analogously for given values of the independent variables $\boldsymbol{x}$, t are more common. They give the probability that the dependent variables at a fixed distance $\boldsymbol{x} = (x_1, x_2, x_3)$ and time t lie in the range $d\Phi_1, \ldots, d\Phi_n$ around the values $\Phi_1, \ldots, \Phi_n$ as $P(\Phi_1, \ldots, \Phi_n; \boldsymbol{x}, t)\, d\Phi_1, \ldots, d\Phi_n$. These models provide the complete statistical information for statistical processes or give distribution functions of particular process variables. They are closely related to the description of statistical processes by the statistical moments discussed above, the difference is that the transport equations for the probability density functions are formulated.

Classical examples for models based on transport equations for probability density functions are to be found in statistical mechanics and kinetic gas theory. This concept was introduced in industrial systems to describe flow and mixing in nonideally mixed reactors [16]. One-dimensional models which represent the "macromixing" in a chemical reactor in terms of age (residence time) distributions are often sufficient to allow estimation of its behavior. Other simple pdf models are used in other areas, for example to model the crystal size distribution in crystallization, the activity distribution of catalyst pellets, or the age and size distribution of microbiological cultures [17]. This concept has been used to describe reacting turbulent flow [18], [19].

Many industrial processes cannot be described with pdf models or models based on physicochemical principles because of their complexity. In these cases empirical models are used. In such cases the response of the process to variation of one or more process variables is known and the model correlates the process results. The simplest example of this is the fitting of a polynomial to experimental data. Another example in process control is the description of a process response in the form of transfer functions in the time or frequency domain. Empirical models are statistical (Fig. 3 B) because the necessary data have to be obtained experimentally and contain statistical errors. Empirical models have only limited value for describing processes or process elements, e.g., if predictions are to be made outside the range covered by experimental data or if a theoretical approach is to be verified. In spite of this empirical mathematical models are still preferred in many cases.

Discussion on mathematical modeling in the following sections is based on the classification of mathematical models shown in Figures 3 and 4. Empirical models will be discussed first, fol-

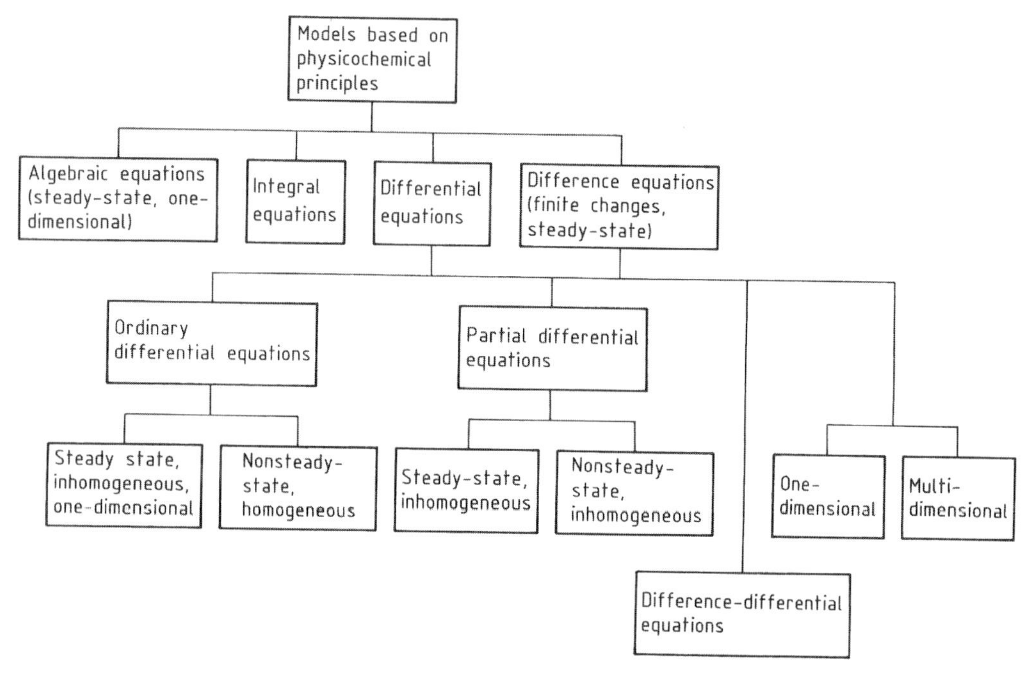

Figure 4. Classification of deterministic models based on physicochemical principles according to the type of resulting equations

lowed by models based on the transport equations for probability density functions, and finally models based on physicochemical principles (transport phenomena). Presentation of the important principles of mathematical modeling will be given priority over a complete description.

Nonstandard numerical and mathematical methods required for solving the resulting set of equations for the developed mathematical models are discussed in more depth. More detailed classifications of industrial processes or process variables can be found in the literature [2], [10], [13–15]. Relevant aspects of such classifications will be discussed where necessary.

3. Empirical Models

The fundamental variables of a process are known, for example, from a series of experiments. If, however, the complexity of the process prevents the formulation of a model on the basis of physicochemical principles, the process can be described by empirical models. The fitting of a polynomial or a similar function to a set of measured values is the simplest empirical model. An empirical model represents a general relationship:

$$y=f(\boldsymbol{x},\boldsymbol{\beta}) \quad (3)$$

where $\boldsymbol{y}$ is the vector of the dependent variables (model responses, process responses), $\boldsymbol{x}$ is the vector of the independent variables (model variables, process variables), and $\boldsymbol{\beta}$ is the vector of the parameters from which the functional relationship is constructed. This relationship is empirical and must be found on the basis of existing experimental data. In the following sections empirical models will be introduced through several examples.

3.1. Linear Empirical Models

If Equation (3) can be represented by the relation

$$y=\beta x \quad (4)$$

the model is linear. Equation (4) is linear with respect to both its parameters and its independent variables. This implies that the first derivatives with respect to the components of the parameter vector and the vector of the independent variables disappear. The simplest relationship of type (4) is a linear model with one dependent variable:

$$y = \beta_0 + \beta_1 x \quad (5)$$

where the estimates b_0 and b_1 of the parameters β_0 and β_1 have to be determined from experimental data. This will be outlined briefly in the following. Only the essential steps of the calculation will be presented. Detailed derivations and proofs are given in [2], [4], [5], [7], [20].

3.1.1. Linear Empirical Models with One Variable

The determination of the estimates of the parameters β_0 and β_1 will be discussed in detail for a linear model with one dependent variable. This model is based on the following assumptions:

1) The dependent variable is a random variable so that the model in Equation (5) can be given in the following form:

$$\langle \bar{Y}_i \rangle = \bar{Y}_i - \varepsilon_i = \beta'_0 + \beta_1 (x_i - \bar{x}) \quad (6)$$

where $\bar{Y}_i$ is the mean value of the dependent variables of n data sets with repeated measurements at x_i; ε_i is the random error in $\bar{Y}_i$. Hence $(\bar{Y}_i - \langle \bar{Y}_i \rangle) = \varepsilon_i$ has an expected value of zero and a constant variance. The angular brackets denote the expected values or mathematical expectation.

2) The variance of Y_i is also constant, has a normal distribution, and is equal to the variance of ε_i.

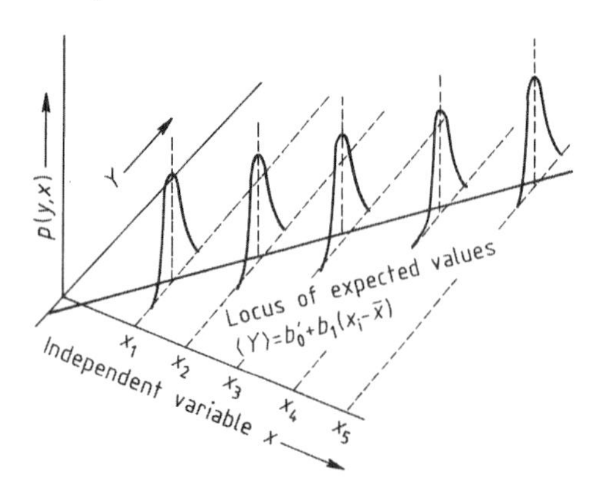

Figure 5. Representation of the statistical process described in Section 3.1.1

3) The observations of Y are statistically independent so that the random errors ε_i are also statistically independent.
4) The variable x is not a random variable.

A process based on these assumptions is shown in Figure 5 and is discussed as an example in Section 3.1.1.4.

3.1.1.1. Parameter Estimation (Linear Regression)

Estimates b'_0 and b_1 of the parameters β'_0 and β_1 are found by minimizing the sum of the squares of the deviations between the observed values Y_i and the expected values of $\bar{Y}_i$, $\langle \bar{Y}_i \rangle$, i.e., the variance

$$\sum_{i=1}^{n} (\bar{Y}_i - \langle \bar{Y}_i \rangle)^2 \stackrel{!}{=} \text{Min} \quad (7)$$

This is known as least squares estimation.

The problem can also be posed in other ways. For the overdetermined system of equations for β'_0 and β_1

$$\bar{Y}_1 - \beta'_0 - \beta_1 (x_1 - \bar{x}) = 0 \quad (8a)$$

$$\bar{Y}_2 - \beta'_0 - \beta_1 (x_2 - \bar{x}) = 0 \quad (8b)$$

$$\vdots$$

$$\bar{Y}_n - \beta'_0 - \beta_1 (x_n - \bar{x}) = 0 \quad (8n)$$

the estimated values b_0 and b_1 have to be found that reduce the residuals on the left-hand sides to as close to zero as possible.

The problem is the minimization of the sum:

$$\Phi = \sum_{i=1}^{n} p_i (\bar{Y}_i - \langle \bar{Y}_i \rangle)^2$$

$$= \sum_{i=1}^{n} p_i \left[\bar{Y}_i - \beta'_0 - \beta_1 (x_i - \bar{x})\right]^2 \quad (9)$$

where p_i are statistical weights determined by the number of measurements made for each data set, i.e., the number of measured values of the dependent variables at each value of the independent variable x. If Equation (9) is partially differentiated with respect to the parameters β'_0

and β_1 and the derivatives set to zero, rearrangement gives the normal equations

$$\sum_{i=1}^{n} p_i \bar{Y}_i = b'_0 \sum_{i=1}^{n} p_i + b_1 \sum_{i=1}^{n} p_i (x_i - \bar{x}) \qquad (10a)$$

and

$$\sum_{i=1}^{n} p_i \bar{Y}_i (x_i - \bar{x}) = b'_0 \sum_{i=1}^{n} p_i (x_i - \bar{x}) + b_1 \sum_{i=1}^{n} p_i (x_i - \bar{x})^2 \qquad (10b)$$

in which the parameters β_0 and β_1 are replaced by their estimated values b'_0 and b_1. Since

$$\sum_{i=1}^{n} p_i (x_i - \bar{x}) \equiv 0 \qquad (11)$$

the coupling between Equations (10 a) and (10 b) disappears and they can be solved separately for b'_0 and b_1:

$$b'_0 = \frac{\sum_{i=1}^{n} p_i \bar{Y}_i}{\sum_{i=1}^{n} p_i} = \bar{Y} \qquad (12a)$$

$$b_1 = \frac{\sum_{i=1}^{n} p_i \bar{Y}_i (x_i - \bar{x})}{\sum_{i=1}^{n} p_i (x_i - \bar{x})^2} \qquad (12b)$$

By forming the second partial derivatives, it can easily be shown that Equations (10 a) and (10 b) actually represent a minimum of the sum $\boldsymbol{\Phi}$. The simple uncoupled form only arises using the scaled form given in Equation (6).

3.1.1.2. Assessment of Estimated Parameter Values

Although the estimation of the parameter values is important, the quality of the empirical model is of particular interest (see points 6 and 7, Section 2.1). The first assessment of the empirical model is the test of the hypothesis $\langle \bar{Y}_i \rangle = \beta'_0 + \beta_1 (x_i - \bar{x})$. For this the mean square deviations

$$s_r^2 = \frac{1}{n-2} \sum_{i=1}^{n} p_i \left(\bar{Y}_i - \hat{Y}_i \right)^2 \qquad (13)$$

and

$$s_\varepsilon^2 = \frac{\sum_{i=1}^{n} \sum_{j=1}^{p_i} (Y_{ij} - \bar{Y}_i)^2}{\sum_{i=1}^{n} p_i - n} \qquad (14)$$

are formed. Equation (13) represents the mean square of the residuals, i.e., the mean square deviations from the regression line. Equation (14) represents the mean square deviations within the data sets; it is a measure of the random error which occurs in the experiments at the points x_i.

The hypothesis $\langle Y_i \rangle = \beta'_0 + \beta_1 (x_i - \bar{x})$ is confirmed if the mean squares of the residuals are significantly smaller than the mean deviations within the data sets. The hypothesis can be verified by means of the F-test [4], [5], [20–23]. If, for a preset confidence number $(1 - \alpha)$, for example, 0.95

$$\frac{s_r^2}{s_\varepsilon^2} > F_{(1-\alpha)}^{(n-2),\left(\sum_i p_i - n\right)} \qquad (15)$$

then the mean squares of the residuals are significantly greater than the mean square deviations within one data set, and the hypothesis is not valid. In Equation (15) $F_{(1-\alpha)}^{(n-2),\left(\sum_i p_i - n\right)}$ is the value of the F-distribution for the confidence number $(1 - \alpha)$ and the degrees of freedom $(n-2)$, $\left(\sum_i p_i - n\right)$. In the positive case both s_r^2 and s_ε^2 are good estimates for $\sigma_{\bar{Y}_i}^2$, the variance in $\bar{Y}_i$. If the hypothesis is valid, the variance of $\bar{Y}_i$ is required for the calculation of the variances of the estimated values of the model parameters and their confidence intervals. The variances as given by their definitions are

$$\mathrm{Var}\,(b'_0) = \langle\, (b'_0 - \beta'_0)\, \rangle = \frac{\sigma_{\bar{Y}_i}^2}{\sum_i p_i} \qquad (16)$$

(cf. Eq. 12 a) and

$$\mathrm{Var}\,(b_1) = \langle\, (b_1 - \beta_1)^2\, \rangle = \frac{\sigma_{\bar{Y}_i}^2}{\sum_i p_i (x_i - \bar{x})^2} \qquad (17)$$

(cf. Eq. 12 b). The variance of the model response $\hat{Y}_i$ calculated from the linear model is given by

$$\mathrm{Var}\left(\hat{Y}_i\right) = \mathrm{Var}\left(b_0'\right) + (x_i - \bar{x})^2\,\mathrm{Var}\left(b_1\right) = \sigma^2_{\bar{Y}_i}\left[\frac{1}{\sum\limits_i p_i} + \frac{(x_i - \bar{x})^2}{\sum\limits_i p_i\,(x_i - \bar{x})^2}\right] \tag{18}$$

since b_0' and b_1 are statistically independent. Equation (18) shows that the variance of Y_i is dependent on x and shows a minimum at $\bar{x}$. A first assessment of the linear model defined by the estimated parameter values b_0' and b_1 is achieved by means of the calculated variances.

In addition to the variances, which are a measure of the scatter of the estimated values, the confidence limits of the estimated values are used and can be calculated from the variances for further assessment. Since b_0' and b_1 are linear combinations of Y_i (cf. Eq. 12), they also have a normal distribution about β_0' and β_1 respectively because of the assumed normal distribution of Y_i. Under these conditions the Student t-variable can be formed [4], [5], [21–23]:

$$t = \frac{b_0' - \beta_0'}{s_{b_0'}} = \frac{b_0' - \beta_0'}{\sigma_{\bar{Y}_i} / \left(\sum\limits_i p_i\right)^{1/2}} \tag{19}$$

This has a t-distribution with $\left(\sum\limits_i p_i - 2\right)$ degrees of freedom. Estimates of the confidence interval can be obtained using the t-distribution in the following form:

$$b_0' - t_{(1-\alpha/2)}^{\left(\sum_i p_i - 2\right)}\, s_{b_0'} \le \beta_0' \le b_0' + t_{(1-\alpha/2)}^{\left(\sum_i p_i - 2\right)}\, s_{b_0'} \tag{20}$$

With known $\sigma^2\bar{Y}_i$ the interval in which the expected value β_0' lies is given. Values of the t-distribution can be found in statistics books [4], [5], [21–23].

An analogous procedure is adopted for β_1. The Student t-variable is then

$$t = \frac{b_1 - \beta_1}{s_{b_1}} = \frac{b_1 - \beta_1}{\sigma^2_{\bar{Y}_i} / \left[\sum\limits_i p_i\,(x_i - \bar{x})^2\right]^{1/2}} \tag{21}$$

This quantity also has a t-distribution with $\sum\limits_i p_i - 2$ degrees of freedom. As before, the confidence interval for β_1 becomes:

$$b_1 - t_{(1-\alpha/2)}^{\left(\sum_i p_i - 2\right)}\, s_{b_1} \le \beta_1 \le b_1 + t_{(1-\alpha/2)}^{\left(\sum_i p_i - 2\right)}\, s_{b_1} \tag{22}$$

Confidence intervals for the expected values $\langle \bar{Y}_i \rangle$ are defined in a similar fashion. Since the $\bar{Y}_i$ values are assumed to have a normal distribution around $\langle \bar{Y}_i \rangle$, the t-variable can be expressed in the form

$$t = \frac{\bar{Y}_i - \langle \bar{Y}_i \rangle}{s_{\bar{Y}_i}} = \frac{\bar{Y}_i - \langle \bar{Y}_i \rangle}{\sigma_{\bar{Y}_i} / \left[\frac{1}{\sum\limits_i p_i} + \frac{(x_i - \bar{x})^2}{\sum\limits_i p_i (x_i - \bar{x})^2}\right]^{1/2}} \tag{23}$$

which again has a t-distribution with $\left(\sum\limits_i p_i - 2\right)$ degrees of freedom. In analogy with Equations (20) and (22) the confidence interval for $\langle \bar{Y}_i \rangle$ is as follows:

$$\hat{Y}_i - t_{(1-\alpha/2)}^{\left(\sum_i p_i - 2\right)}\, s_{\bar{Y}_i} \le \langle \bar{Y}_i \rangle \le \hat{Y}_i + t_{(1-\alpha/2)}^{\left(\sum_i p_i - 2\right)}\, s_{\bar{Y}_i} \tag{24}$$

Further discussion and illustration of the method of parameter estimation and assessment are given in Section 3.1.1.4.

3.1.1.3. Sensitivity Analysis

In addition to the analysis of the variances of β_0', β_1, and $\bar{Y}_i$, the sensitivity of the linear empirical model $\langle \bar{Y}_i \rangle = \beta_0' + \beta_1 (x_i - \bar{x})$ can help in interpreting the results obtained from the model as well as in the design of experiments for establishing an empirical model. The sensitivity of the model may be described in various ways. In order to select experiments and methods of measurement for verification of the fundamental variables of the process, it is useful to have a measure of the sensitivity of the sum of the squares of the deviations and of the estimated parameter values with respect to Y_i. Sensitivity can be regarded as the relative change in the sum of squares of the deviations $\partial \boldsymbol{\Phi}_{\min} / \Phi_{\min}$ or the relative change in the estimated parameter values $\partial b_i / b_i$ for a relative change $\partial \bar{Y}_i / \bar{Y}_i$ in $\bar{Y}_i$. Since

$$\boldsymbol{\Phi}_{\min} = \sum_{i=1}^{n} p_i \left(\bar{Y}_i - \hat{Y}_i\right)^2 \tag{25}$$

(cf. Section 3.1.1.1), the sensitivity of $\boldsymbol{\Phi}_{\min}$ with respect to $\bar{Y}_i$ is:

$$\frac{\partial \boldsymbol{\Phi}_{\min}}{\boldsymbol{\Phi}_{\min}} / \frac{\partial \bar{Y}_i}{\bar{Y}_i} = \frac{2\bar{Y}_i p_i \left(\bar{Y}_i - \hat{Y}_i\right)}{\sum\limits_{i=1}^{n} p_i \left(\bar{Y}_i - \hat{Y}_i\right)^2} \tag{26}$$

Hence, relative change in a value $\bar{Y}_i$ causes a change in the minimum of the sum of squares, which increases as the deviation between $\bar{Y}_i$ and the model solution $\hat{Y}_i$ becomes larger and as the minimum becomes smaller.

The sensitivity of the estimated parameter values b_0' and b_1 to changes in $\bar{Y}_i$ can be obtained simply and directly from Equations (12 a) and (12 b):

$$\frac{\partial b_0'}{b_0'} / \frac{\partial \bar{Y}_i}{\bar{Y}_i} = \frac{1}{\sum\limits_i p_i} \frac{p_i \bar{Y}_i}{b_0'} \tag{27}$$

and

$$\frac{\partial b_1}{b_1} / \frac{\partial \bar{Y}_i}{\bar{Y}_i} = \frac{1}{(x_i - \bar{x})^2} \frac{p_i \bar{Y}_i (x_i - \bar{x})}{b_1} \tag{28}$$

The sensitivity obtained from Equations (27) and (28) is equally simple to interpret. The more reliable the measured value is, i.e., the larger p_i becomes in comparison to $\sum\limits_i p_i$, the more rapidly b_0' changes with $\bar{Y}_i$. The same applies to b_1, here the interval $(x_i - \bar{x})$ at which the values are measured has an additional weight.

Finally, an estimate of the sensitivity of the model with respect to its parameters is a sensible measure to evaluate the experimental effort required in the model assessment. This sensitivity is defined by the relative change in the expected value $\langle \bar{Y}_i \rangle$, $\partial \langle \bar{Y}_i \rangle / \langle \bar{Y}_i \rangle$, caused by a change of $\partial \beta_i / \beta_i$ in the model parameter. For the linear model described here formulation of the problem is trivial since the solutions

$$\frac{\partial \langle \bar{Y}_i \rangle}{\langle \bar{Y}_i \rangle} / \frac{\partial \beta_0'}{\beta_0'} = \frac{\beta_0'}{\langle \bar{Y}_i \rangle} \tag{29}$$

and

$$\frac{\partial \langle \bar{Y}_i \rangle}{\langle \bar{Y}_i \rangle} / \frac{\partial \beta_1}{\beta_1} = \frac{\beta_1}{\langle \bar{Y}_i \rangle} (x_i - \bar{x}) \tag{30}$$

are obvious and can be derived directly from Equation (6). For more complex models there are other more subtle relationships for deciding the degree of complexity of the experiments required for the model development. This type of sensitivity analysis will be discussed in detail later (Section 5.3.3.1).

3.1.1.4. Example of a Linear Model

The determination of the model parameters will now be illustrated using a simple example of a linear model with one dependent variable, as defined in Section 3.1.1.1.

As an example we will consider the correlation of the mean heat-transfer coefficients for hot air at the wall of a pipe with turbulent flow (Table 1). The gas velocity (indicated by x in Table 1) is varied and can be precisely adjusted so that it can be regarded as a deterministic variable. The measured values (the mean heat-transfer coefficients denoted by Y) are statistical quantities which satisfy the assumptions presented in Section 3.1.1. Table 1 also contains all conversions necessary for linear regression.

According to Equations (12 a) and (12 b), the estimated values for the parameters of the linear model $\langle \bar{Y}_i \rangle = \beta_0' + \beta_1 (x_i - \bar{x})$ are $b_0' = 15.002$ and $b = 1.282$. To test the hypothesis, the mean square deviations $s^2{}_r$ and s_ε^2 are first obtained from Equations (13) and (14): $s_\mathrm{r}^2 = 7.749/3 = 2.583$ and $s_\varepsilon^2 = 7.613/9 = 0.846$. The ratio of the mean square deviations is then 3.05. The associated F-value [4] is $F_{0.95}^{3.9} = 3.86$, so that the hypothesis is valid for a significance level of 0.05. To obtain an estimated value $s_{\bar{Y}_i}^2$ for $\sigma_{\bar{Y}_i}^2$ it is better to use the combined square deviations from s_r^2 and s_ε^2:

$$s_{\bar{Y}_i}^2 = \frac{\sum\limits_{i=1}^{n} \sum\limits_{j=1}^{p_i} \left(Y_{ij} - \hat{Y}_i\right)^2}{\sum\limits_{i=1}^{n} p_i - 2} = \frac{7.613 + 7.746}{9 + 3} = 1.28$$

The variances of the estimated values of the model parameters can be obtained using Equations (16) and (17):

$$\mathrm{Var}\left(b_0'\right) = \frac{1.28}{14} = 9.14 \times 10^{-2}$$

$$\mathrm{Var}\left(b_1\right) = \frac{1.28}{206.25} = 6.20 \times 10^{-3}$$

and finally the x_i-dependent variance of the model responses $\hat{Y}_i$ are given by Equation (18).

$$\mathrm{Var}\left(\hat{Y}_i\right) = 1.28 \left[\frac{1}{14} + \frac{1}{206.25} (x_i - \bar{x})^2\right]$$
$$= 9.14 \times 10^{-2} + 6.20 \times 10^{-3} (x_i - \bar{x})^2$$

Table 1. Example of linear regression, model with one independent variable

Data set	p_i	x_i	Y	$\bar{Y}_i$	$p_i(x_i-\bar{x})\bar{Y}_i$	$p_i(x_i-\bar{x})^2$	$(Y-\bar{Y}_i)^2$	$(\bar{Y}_i-\hat{Y}_i)^2$	$(x_i-\bar{x})$
1	3	8.00	8.08 9.95 9.26	9.09	−136.35	75.00	1.020 0.739 0.029	0.248	−5.00
2	3	10.50	11.21 12.36 12.98	12.18	−91.35	18.75	0.941 0.032 0.640	0.146	−2.50
3	2	13.00	14.12 15.96	15.04	0.00	0.00	0.846 0.846	0.001	0.00
4	2	15.50	16.80 18.62	17.71	88.55	12.50	0.828 0.828	0.247	2.50
5	4	18.00	19.60 20.08 20.90 20.16	20.18	403.60	100.00	0.336 0.010 0.518 0.004	1.517	5.00
$\sum n=5$	$\sum_i p_i=14$	$\bar{x}=13.00$	$\bar{Y}=15.002$		$\sum_i p_i\bar{Y}_i(x_i-\bar{x})=264.45$	$\sum_i p_i(x_i-\bar{x})^2=206.25$	$\sum_i (Y-\bar{Y}_i)^2=7.613$	$\sum_i p_i\left(\bar{Y}_i-\hat{Y}_i\right)^2=7.746$	

The individual confidence intervals of the expected values of these three quantities can be calculated for a significance level of 0.05 using Equations (20), (22), and (24) ($t_{0.975}^{\left(\sum_i p_i-2\right)}=2.179$, [4]):

$$b_0'-0.658\le\beta_0'\le b_0'+0.658$$

and

$$b_1-0.171\le\beta_1\le b_1+0.171$$

The individual confidence intervals for $\langle\bar{Y}_i\rangle$ are also dependent on x_i because of the dependence of the variance on x_i:

$$\hat{Y}_i-\left[1.99\times10^{-1}+1.35\times10^{-2}\left(x_i-\bar{x}\right)^2\right]^{1/2}\le\langle\bar{Y}_i\rangle\le\hat{Y}_i+\left[1.99\times10^{-1}+1.35\times10^{-2}\left(x_i-\bar{x}\right)^2\right]^{1/2}$$

The results of the regression analysis are illustrated in Figure 6: the regression line, the confidence intervals of the parameters, and the confidence intervals for the expected values of the model responses are shown. In spite of the verification of the hypothesis, the results of the linear regression are only moderately satisfactory due to the relatively large scatter of the measurements. Sensitivity analysis emphasizes this. For example, for data set number 5 from Table 1

$$\frac{\partial\Phi_{\min}}{\Phi_{\min}}/\frac{\partial\bar{Y}_5}{\bar{Y}_5}=\frac{2\times20.28\times4\times1.231}{7.746}=25.66$$

A change in the measured value $\bar{Y}_5$ of 1 % causes a change of 25.55 % in $\Phi_{\min}$. Similarly high sensitivities arise with

$$\frac{\partial b_0'}{b_0'}/\frac{\partial\bar{Y}_5}{\bar{Y}_5}=\frac{1}{14}\cdot\frac{4\times20.18}{15.002}=0.384$$

and

$$\frac{\partial b_1}{b_1}/\frac{\partial\bar{Y}_5}{\bar{Y}_5}=\frac{1}{206.25}\cdot\frac{4\times20.18\times5}{1.282}=1.526$$

so that the experimenter concerned with this task would be well advised to repeat the experiments with a more precise measurement technique. In fact other experiments for this example give the relationship

$$Nu=\frac{\xi/8\,(Re-1000)\,Pr}{1+12.7\sqrt{\xi/8}\,(Pr^{2/3}-1)}\left[1+\left(\frac{D_R}{L}\right)^{2/3}\right]$$

where $\xi = (1.82 \log Re - 1.64)^{-2}$ [24], [25]: this relationship does not produce a linear dependency $Nu = f(Re)$ and hence a linear model appears to be inadequate.

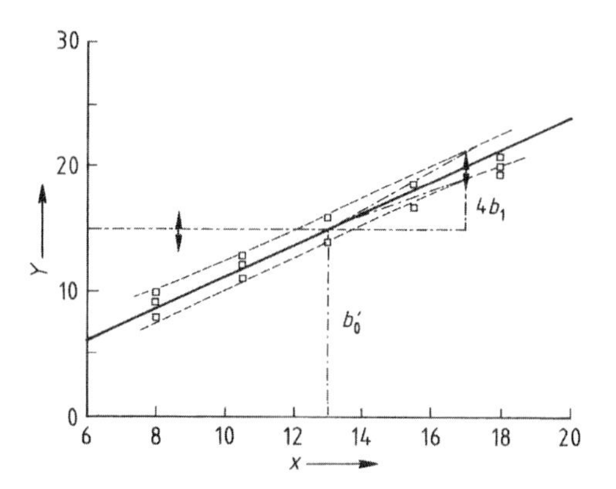

Figure 6. Results of linear regression for the example discussed in Section 3.1.1.4
Arrows indicate the confidence intervals of the parameters, square symbols denote measured values, and the dashed lines the confidence intervals of the expected values.

3.1.1.5. Concluding Remarks

The assumptions stated in Section 3.1.1 are not always appropriate; for example, the random error of the measurements may be a function of the independent variables. In addition, the random errors may be statistically dependent on each other and, in a given process, both the dependent and the independent variables can be statistical quantities. Estimation of the model parameters by linear regression in such cases does not differ in principle from that presented so far: more complex calculations may be required and further hypotheses or statistical rules may be needed. For discussion of these special cases, the reader is referred to the literature [4], [5], [7], [20].

Standardized computer programs are available for linear regression by least squares estimations for models with one dependent variable [26]. The solution of linear equations required for linear regression is also available through program libraries for numerical problems or for problems in linear algebra. Most of these libraries are readily available [26–32], so that special-purpose computer programs can easily be assembled for particular requirements.

3.1.2. Linear Empirical Models with Several Variables

In the case of linear empirical models with several independent variables, similar problems to those in Section 3.1.1 arise. Thus in the following, only the generalized notation will be discussed.

The linear relationship for a model with several independent variables is given by

$$y = \boldsymbol{\beta x} \tag{31}$$

and in extended notation

$$y = \beta_0 + \beta_1 x_1 + \beta_2 x_2 + \ldots + \beta_q x_q \tag{32}$$

so that the problem is extended to q variables and $(q+1)$ parameters. The same assumptions hold as in Section 3.1.1, so that the model given by Equation (31) can be written in the form

$$\langle \bar{Y}_i \rangle = \bar{Y}_i - \varepsilon_i = \beta_0' + \sum_{j=1}^{q} \beta_j \left(x_{ij} - \bar{x}_j \right) \tag{33}$$

The $(q+1)$ model parameters are calculated from the measured values defined at the n data points in the same way as in Section 3.1.1.

In matrix notation Equation (33) reads

$$\langle \bar{\boldsymbol{Y}} \rangle = \bar{\boldsymbol{Y}} - \boldsymbol{E} = \boldsymbol{\beta x}' \tag{34}$$

with the abbreviations

$$\langle \bar{\boldsymbol{Y}} \rangle = \begin{pmatrix} \langle \bar{Y}_i \rangle \\ \langle \bar{Y}_2 \rangle \\ \vdots \\ \langle \bar{Y}_n \rangle \end{pmatrix}, \quad \boldsymbol{\beta} = \begin{pmatrix} \beta_0' \\ \beta_1 \\ \vdots \\ \beta_q \end{pmatrix},$$

and

$$\boldsymbol{x}' = \begin{pmatrix} 1 & (x_{11} - \bar{x}_1) & \cdots & (x_{1q} - \bar{x}_q) \\ 1 & (x_{21} - \bar{x}_1) & \cdots & (x_{2q} - \bar{x}_q) \\ \vdots & \vdots & \ddots & \vdots \\ 1 & (x_{n1} - \bar{x}_1) & \cdots & (x_{nq} - \bar{x}_q) \end{pmatrix} \boldsymbol{p}$$

The first step is the same as for models with only one independent variable – the estimation of model parameters $\boldsymbol{\beta}$.

3.1.2.1. Parameter Estimation

As in Section 3.1.1.1, the estimates of the parameters $\boldsymbol{\beta}$ are found by minimizing the sum of squares of the deviations between the observed values $\bar{Y}_i$ and the expected values $\langle \bar{Y}_i \rangle$ (i.e., the sum of the variances):

$$(\bar{\boldsymbol{Y}} - \langle \bar{\boldsymbol{Y}} \rangle)^T \boldsymbol{p} (\bar{\boldsymbol{Y}} - \langle \bar{\boldsymbol{Y}} \rangle) \stackrel{!}{=} \text{Min} \tag{35}$$

By using Equation (34), the sum of squares of the deviations can be formulated as

$$\boldsymbol{\Phi} = (\bar{\boldsymbol{Y}} - \boldsymbol{\beta x}')^T \boldsymbol{p} (\bar{\boldsymbol{Y}} - \boldsymbol{\beta x}') \tag{36}$$

which has to be differentiated with respect to all the elements of vector $\boldsymbol{\beta}$ and then set to zero. In Equation (36) $\boldsymbol{p}$ once again represents the number of repeated measurements or, in general terms, the statistical weights. In matrix notation

$$\boldsymbol{p} = \begin{pmatrix} p_1 & 0 & \cdots & 0 \\ 0 & p_2 & \cdots & 0 \\ \vdots & \vdots & \ddots & \vdots \\ 0 & 0 & \cdots & p_n \end{pmatrix}$$

Differentiation yields

$$\frac{\partial \boldsymbol{\Phi}}{\partial \boldsymbol{\beta}} = \begin{pmatrix} \frac{\partial \boldsymbol{\Phi}}{\partial \beta_0} \\ \frac{\partial \boldsymbol{\Phi}}{\partial \beta_1} \\ \vdots \\ \frac{\partial \boldsymbol{\Phi}}{\partial \beta_q} \end{pmatrix} = \frac{2\left[\partial (\bar{\boldsymbol{Y}} - \boldsymbol{\beta x}')^T\right] \boldsymbol{p} (\bar{\boldsymbol{Y}} - \boldsymbol{\beta x}')}{\partial \boldsymbol{\beta}} \tag{37}$$

so that for a minimum to occur

$$(\boldsymbol{x}')^T \boldsymbol{p} (\bar{\boldsymbol{Y}} - \boldsymbol{\beta x}') = 0 \tag{38}$$

Further differentiation shows that Equation (38) actually represents a minimum. After rearrangement the normal equation

$$(\boldsymbol{x}')^T \boldsymbol{p x}' \boldsymbol{b} = (\boldsymbol{x}')^T \boldsymbol{p} \bar{\boldsymbol{Y}} \tag{39}$$

is obtained.

In Equation (39) the parameters β_i are replaced by their estimated values. Equation (39) is the matrix equivalent of Equations (9 a, b). Solution of Equation (39) for the parameters gives

$$\boldsymbol{b} = \left[(\boldsymbol{x}')^T \boldsymbol{p x}'\right]^{-1} (\boldsymbol{x}')^T \boldsymbol{p} \bar{\boldsymbol{Y}} \tag{40}$$

Inversion of the matrix $(\boldsymbol{x}')^T \boldsymbol{p x} \equiv \boldsymbol{a}$ is required for the solution. A condition for this is that the determinant is not equal to zero, or in other words that matrix $\boldsymbol{a}$ is nonsingular. Numerical problems related to ill-conditioned matrices $\boldsymbol{a}$ can be avoided by suitably scaling or normalizing the independent variables or by suitable experimental design. This can be illustrated by a simple example. From measurements at

$$\boldsymbol{x} = \begin{pmatrix} 9.5 \\ 10.0 \\ 10.5 \end{pmatrix},$$

a relationship of the form $y = \beta_0 + \beta_1 x$ will be investigated. Estimation of the model parameters involves inversion of the matrix

$$a \equiv \boldsymbol{x}^T \boldsymbol{p x} = \begin{pmatrix} 3.0 & 30.0 \\ 30.0 & 300.5 \end{pmatrix}$$

As can be easily verified,

$$\begin{aligned} \text{Det } (\boldsymbol{a}) &= 3.0 \times 300.5 - 30.0 \times 30.0 \\ &= 901.5 - 900.0 = 1.5 \end{aligned}$$

However, because of the spread of the measured values, the matrix is ill-conditioned for inversion since Det $(\boldsymbol{a}) = 0$ if only the first two digits are significant. If it is scaled instead and we use $y = \beta_0' + \beta_1 (x - \bar{x})$ then

$$\boldsymbol{a} \equiv (\boldsymbol{x}')^T \boldsymbol{p x}' = \begin{pmatrix} 3.0 & 0 \\ 0 & 0.5 \end{pmatrix}$$

or Det $(\boldsymbol{a}) = 3.0 \times 0.5 - 0 = 1.5$

The result is the same. The problem of subtraction of large numbers which arises from the design of the experiments is thereby avoided. As can be seen from the numbers given above, the problem of handling differences between large numbers is aggravated if the empirical model is in the form of a polynomial, i.e.,

$$y = \beta_0 + \beta_1 x_1 + \beta_2 x_2$$

with $x_1 = x$, $x_2 = x^2$ etc.

Numerical problems with the inversion of $\boldsymbol{a}$ are also avoided if the main diagonal of $\boldsymbol{a}$ can be occupied predominantly through proper normalization of the independent variables. This technique can be demonstrated using the simple example of measured values at

$$x = \begin{pmatrix} 5.0 & 10.0 \\ 5.0 & 20.0 \\ 10.0 & 10.0 \\ 10.0 & 20.0 \end{pmatrix}$$

If the variables are normalized in the form

$$\tilde{x}_{i1} = \frac{x_{i1} - 7.5}{2.5}$$

and $\tilde{x}_{i2} = \dfrac{x_{i2} - 15.0}{5.0}$ then

$$\tilde{x}' = \begin{pmatrix} 1.0 & -1.0 & -1.0 \\ 1.0 & -1.0 & 1.0 \\ 1.0 & 1.0 & -1.0 \\ 1.0 & 1.0 & 1.0 \end{pmatrix}$$

With $\boldsymbol{a} \equiv (\tilde{\boldsymbol{x}}')^{\mathrm{T}} \boldsymbol{p} \tilde{\boldsymbol{x}}'$ this leads to the occupation of the main diagonal only with

$$a = \begin{pmatrix} 4.0 & 0 & 0 \\ 0 & 0.4 & 0 \\ 0 & 0 & 4.0 \end{pmatrix},$$

so that inversion does not cause any problems. The variables normalized in this form are orthogonal since $\sum x_0 x_{i1} = \sum x_0 x_{i2} = \sum x_{i1} x_{i2} = 0$. Ill-conditioning of $\boldsymbol{a}$ can therefore be avoided by planning the experiment in a manner which allows simple orthogonalization of the dependent variables.

A brief geometric explanation of the normal equations will now be given. Figure 7 shows the vector $\bar{\boldsymbol{Y}} = (\bar{Y}_1, \bar{Y}_2, \bar{Y}_3)$ in the space of measured values. Since the relationship $y = \beta_0 + \beta_1(x - \bar{x})$ is linear in its parameters, a plane surface for the calculated model solutions that depends on the estimated values b'_0 and b_1 results.

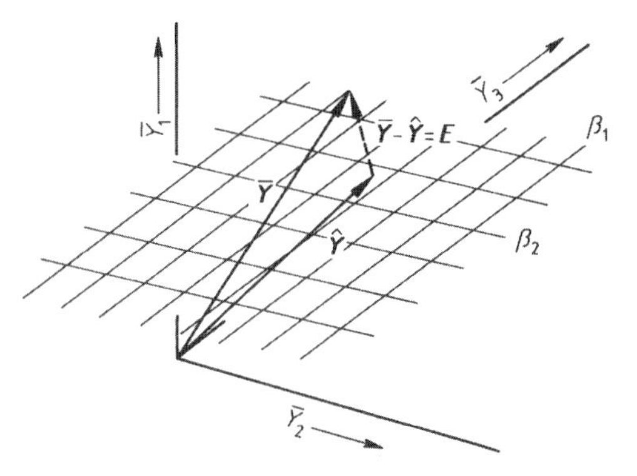

Figure 7. Geometrical illustration of the normal equations for a linear model

The set of estimated values b'_0 and b_1 that gives the shortest distance between the end points of the vectors $\bar{\boldsymbol{Y}}$ and $\hat{\boldsymbol{Y}}$ now has to be found. Obviously the shortest distance $\bar{\boldsymbol{Y}} - \hat{\boldsymbol{Y}}$ is given by the normal from the end of $\bar{\boldsymbol{Y}}$ onto the surface of the calculated model solution. The normal Equations (10) and (39) fix $\boldsymbol{b}$ such that the vector of the residuals $\bar{\boldsymbol{Y}} - \hat{\boldsymbol{Y}}$ passes through the end point of $\bar{\boldsymbol{Y}}$ and is perpendicular to the surface of values calculated from the model.

3.1.2.2. Assessment of Estimated Parameter Values

The approach used to assess the estimated parameter values is the same as that in Section 3.1.1.2, and so will only be briefly discussed. To test the hypothesis $\langle \bar{\boldsymbol{Y}} \rangle = \beta \boldsymbol{x}'$, the mean square deviations from the regression line must be obtained. If $\bar{\boldsymbol{Y}} - \langle \bar{\boldsymbol{Y}} \rangle$ is represented by $\boldsymbol{E}$ then

$$s_{\mathrm{r}}^2 = \frac{(\boldsymbol{E}^T \boldsymbol{p} \boldsymbol{E})}{n - q - 1} \tag{41}$$

where s_{r}^2 is the sum of squares of the residuals divided by the number of degrees of freedom. The number of degrees of freedom is the number of data points minus the number of conditions constraining the minimization, in this case the number of parameters. The mean square deviations for the individual data points are calculated from Equation (19), so that the F-test can be conducted using the ratio $s_{\mathrm{r}}^2/s_{\varepsilon}^2$. The variances of the parameters and the model solutions can be calculated by means of the estimated values $s_{\bar{Y}_i}^2$ of $\sigma_{\bar{Y}_i}^2$ which are obtained from s_{r}^2 and/or s_{ε}^2 in the case of a positive F-test. The covariances of the parameters are

$$\mathrm{Cov}\,(\boldsymbol{b}) = \langle\, (\boldsymbol{b} - \boldsymbol{\beta})^T\, (\boldsymbol{b} - \boldsymbol{\beta})\, \rangle = \sigma_{\bar{Y}_i}^2 \boldsymbol{a}^{-1} \tag{42}$$

If $\boldsymbol{a}^{-1}$ is denoted as $\boldsymbol{c}$, then the variances of the estimated parameter values can be obtained from the diagonal elements of $\boldsymbol{c}$ with estimated values $s_{\bar{Y}_i}^2$ for $\sigma_{\bar{Y}_i}^2$:

$$s_{b_k}^2 = s_{\bar{Y}_i}^2 c_{kk} \tag{43}$$

With

$$\begin{aligned} \mathrm{Var}\left(\hat{Y}_i\right) &= \mathrm{Var}\,(\boldsymbol{x}'\boldsymbol{b}) = \boldsymbol{x}' \mathrm{Var}\,(\boldsymbol{b})\,(\boldsymbol{x}')^T \\ &= \sigma_{\bar{Y}_i}^2 \boldsymbol{x}' \boldsymbol{c}\,(\boldsymbol{x}')^T \end{aligned} \tag{44}$$

and $s^2_{\bar{Y}_i}$ for $\sigma^2_{\bar{Y}_i}$, the following is obtained:

$$s_{\hat{Y}_i} = s_{\bar{Y}_i} \sqrt{\boldsymbol{x}' \boldsymbol{c} (\boldsymbol{x}')^T} \quad (45)$$

so that, as in Section 3.1.1.2 (Eqs. 20, 22, and 24), the confidence intervals for $\boldsymbol{b}$ and $\hat{Y}_i$ are obtained.

More detailed methods for the assessment of the estimated parameter values for models with several independent variables are treated in [4], [5], [20].

3.1.2.3. Concluding Remarks

The sensitivity analysis for empirical models with several independent variables will not be discussed. The relationships presented in Section 3.1.1.3 can be used, and so need not be repeated. The basic rules of linear regression were illustrated using an example with one independent variable, a further example for a model with several independent variables will not be given. Cases which do not comply with the assumptions presented here are described in the literature; for example, processes with random errors which are not normally distributed or with random errors which depend on x are discussed in [4], [7]. Standardized computer programs are available for the treatment of empirical linear models with several independent variables or for multiple linear or polynomial regression [26], [33]. Programs can be easily constructed to fit users' requirements with the aid of modules from extensive program libraries for matrix and linear algebra [27–32].

3.2. Nonlinear Empirical Models

In nonlinear models the relationship from Equation (3) cannot be represented by Equation (4). The derivatives of the model equation with respect to the parameters are themselves functions of the parameters. The nonlinear model $\boldsymbol{y} = f(\boldsymbol{x}, \boldsymbol{\beta})$ is once again based on several assumptions:

1) The dependent variable is a random variable for which n data sets of measured values $\bar{Y}_i$ are available. The model can therefore be represented in the form

$$\langle \bar{\boldsymbol{Y}} \rangle = \bar{\boldsymbol{Y}} - \boldsymbol{E} = f(\boldsymbol{x}, \boldsymbol{\beta}) \quad (46)$$

where $\boldsymbol{E}$ once again represents the random error in $\bar{Y}_i$, so that $\bar{\boldsymbol{Y}} - \langle \bar{\boldsymbol{Y}} \rangle = \boldsymbol{E}$. The expected value of the random error is zero and its variance is constant (independent of x).

2) The random variable Y has a normal distribution about $\bar{Y}_i$ and has the same variance as ε_i.
3) Both the random variable and the random error are statistically independent.
4) The variables $\boldsymbol{x}$ are not random variables.

The empirical nonlinear statistical model can be represented analogously to that in Figure 5, whereby the relationship is nonlinear and is characterized by several independent variables. As with the linear model, the first task consists of estimating the model parameters using the experimental process data.

3.2.1. Parameter Estimation (Nonlinear Regression)

Estimation of the model parameters $\boldsymbol{\beta}$ involves the now familiar task of defining the estimated values $\boldsymbol{b}$ such that the sum of squares of the deviations between the values $\bar{Y}_i$ observed at the n data points and the estimated values $\langle \bar{Y}_i \rangle$ is a minimum. [In the following, the equations are expressed in a simplified form, without repeated measurements, i.e., with unit statistical weights $\boldsymbol{p}$, Det $(\boldsymbol{p}) = 1$]:

$$(\bar{\boldsymbol{Y}} - \langle \bar{\boldsymbol{Y}} \rangle)^T (\bar{\boldsymbol{Y}} - \langle \bar{\boldsymbol{Y}} \rangle) \stackrel{!}{=} \text{Min} \quad (47)$$

A geometric representation of this task is shown in Figure 8. The end point of the vector $\bar{\boldsymbol{Y}} = (\bar{Y}_1, \bar{Y}_2, \bar{Y}_3)$ in the space of measured values should be linked by the shortest possible distance with the end point of the vector $\hat{\boldsymbol{Y}}$. This is once again given by the normal to the surface of model responses in the parameter space, which in this case is curved because of the nonlinearity of the relationship $\boldsymbol{y} = f(\boldsymbol{x}, \boldsymbol{\beta})$.

Partial differentiation of the sum of squares of the deviations

$$\boldsymbol{\Phi} = \left[\bar{\boldsymbol{Y}} - f(x, \boldsymbol{\beta})\right]^T \left[\bar{\boldsymbol{Y}} - f(x, \boldsymbol{\beta})\right] \quad (48)$$

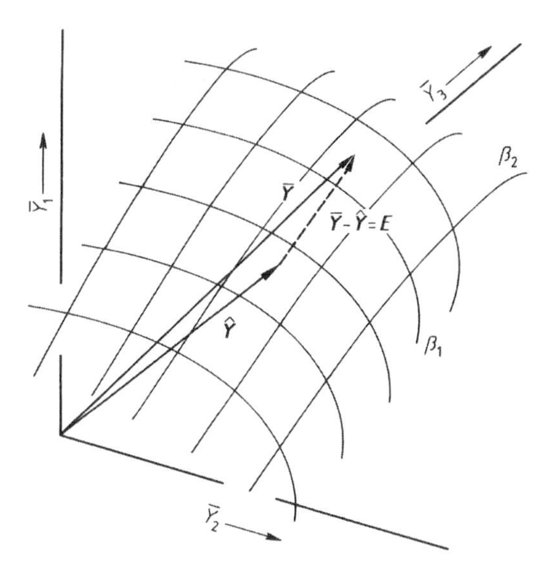

Figure 8. Geometrical illustration of the normal equation for a nonlinear model

with respect to the parameters and then setting the derivatives to zero produces a system of nonlinear normal equations. This is the main difference from linear regression in which a system of linear equations in $\boldsymbol{\beta}$ is obtained for the normal equations. This can be easily verified by analyzing the sums of squares (Eqs. 9, 36, and 48).

Nonlinear regression is therefore the solution of a system of nonlinear equations for $\boldsymbol{\beta}$.

3.2.1.1. Transformation of the Model into Linear Form

The simplest method of estimating parameters in nonlinear regression is transformation of the model equations into linear form to avoid the problem of solving the nonlinear equations described above. Certain types of nonlinear models can be transformed into linear form but this may involve several problems; this will now be briefly illustrated.

The relationship

$$y=\beta_0 x_1^{\beta_1} x_2^{\beta_2} \tag{49}$$

e.g., the rate of a complex chemical reaction, can be transformed into linear form by taking logarithms:

$$\log y=\log\beta_0+\beta_1\log x_1+\beta_2\log x_2 \tag{50}$$

The transformed model is therefore

$$\log\bar{Y}_i-\varepsilon_i=\log\beta_0+\beta_1\log x_{i1}+\beta_2\log x_{i2} \tag{51}$$

The previously presented methods of regression analysis for testing hypotheses and calculating the confidence intervals of the parameters or model solutions presume that the additive errors ε_i and random variables have normal distributions. For the nonlinear model of Equation (49), the additive error must be multiplicative:

$$\bar{Y}_i=\beta_0 x_{i1}^{\beta_1} x_{i2}^{\beta_2}\varepsilon_i' \tag{52}$$

and, like the random variable Y_i, must have a log-normal distribution to satisfy the above-mentioned presumptions. This is not always the case and must be experimentally verified.

Another example is the relationship of the form

$$y=\frac{\beta_1\beta_2 x}{(1+\beta_2 x)^2} \tag{53}$$

which represents, for example, the rate of a heterogeneous catalytic reaction (Hougen – Watson – Langmuir – Hinshelwood kinetics). Transformation into a linear model gives

$$\left(\frac{x}{y}\right)^{1/2}=\frac{1}{(\beta_1\beta_2)^{1/2}}+\frac{\beta_2}{(\beta_1\beta_2)^{1/2}}x \tag{54}$$

The additive statistical error ε_i in the linear model for the estimation of the parameters (Eq. 54) is not identical to the random error of the nonlinear model (Eq. 53). Inverse transformation of Equation (54) gives:

$$y=\frac{\beta_1\beta_2 x}{(1+\beta_2 x)^2}+\frac{\beta_1\beta_2}{(1+\beta_2 x)^2}\left(y\varepsilon'^2-2\sqrt{yx}\varepsilon'\right) \tag{55}$$

which contains an additive random error that is dependent on x, y, β_1, and β_2.

3.2.1.2. Direct Search Methods

Another way of circumventing the problem of solving nonlinear equations is to use direct search methods. For this the normal equations are not derived from the sum of squares of the deviations (Eq. 48), instead the minimum of the function $\boldsymbol{\Phi}$ is sought directly. Vectorial search methods [34], Simplex methods [35], and Simplex methods with variable Simplex geometry [36] are the most commonly used direct search methods. The structure of these methods is presented schematically for a model with two parameters in Figure 9.

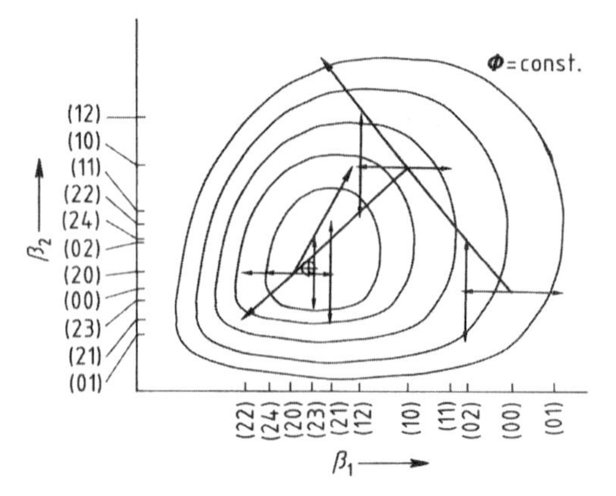

Figure 9. Representation of direct search methods
The first digits in the parentheses indicate the number of iteration. The second digits denote the number of successive variation of β_i. For example, (02) on the β_1 axis means the initial guess of β_1 and the second variation

Vectorial search methods start from estimated values $b_i^{(0)}$ for the parameters β_i. The direction of the search vector is established from successive calculations of Φ by varying the estimated values b_i. The direction of the vector leading to a reduction in Φ is used to calculate improved estimates of b_i. If there is no further reduction in Φ along the search vector, the direction is redefined and the minimum of Φ is located by reducing the step size (Fig. 9).

Simplex methods are more effective, especially with large numbers of parameters. For a twoparameter problem (Fig. 10), a regular Simplex is an equilateral triangle which, starting from estimated values $b_i^{(0)}$, is placed at a point on the surface of Φ in the parameter space. During the search, Φ is evaluated at each of the three vertices, and the triangle is reflected about the side opposite the largest value of Φ. If there is no further reduction in Φ after reflection, then the Simplex is reduced in size to allow more accurate localization of the minimum.

Simplexes with variable geometry are obtained by expanding after successful reflection or contracting after unsuccessful reflection. An example is given in Figure 11 for a twoparameter problem. The reflection algorithm is

$$b_{ir} = (1+\gamma_r)\, b_{ic} - \gamma_r b_{i\max}$$

where b_{ir}, $b_{i\,\max}$, and b_{ic} are the coordinates of the reflected Simplex point, of the Simplex point with the largest sum of squares of deviations, and of the midpoint of the side lying opposite it, respectively; γ_r is the reflection factor. The expansion algorithm is

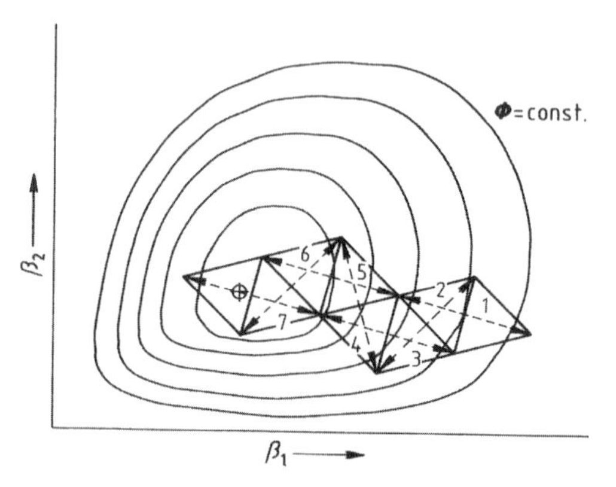

Figure 10. Representation of the Simplex method
The numbers represent the number of reflection

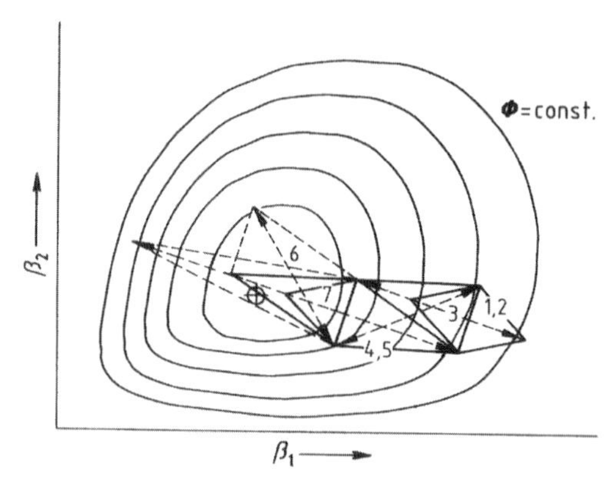

Figure 11. Representation of the Simplex method with variable geometry
The numbers represent the number of reflection, expansion, or contraction

$$b_{ie} = \gamma_e b_{ir} + (1-\gamma_e)\, b_{ic}$$

where b_{ie} are the coordinates of the expanded Simplex point and γ_e is the expansion factor. After unsuccessful expansion or reflection, contraction is achieved with the algorithm

$$b_{ik} = (1-\gamma_k)\, b_{ic} + \gamma_k b_{i\max}$$

where b_{ik} and γ_k have analogous meanings to those in reflection or expansion.

3.2.1.3. Gradient Methods

Discussion of the methods for solving nonlinear systems of equations arising from the minimum conditions for $\boldsymbol{\Phi}$ will be limited to two methods. Further methods are described in the literature on nonlinear optimization [5], [7], [20], [37–39] (see also → Mathematics in Chemical Engineering, Chap. 10.).

Through use of the abbreviation

$$\langle \bar{\boldsymbol{Y}} \rangle \equiv \boldsymbol{\eta} = f(\boldsymbol{x}, \boldsymbol{\beta}) \tag{56}$$

the sum of the squares of the deviations (Eq. 48) can be written as

$$\boldsymbol{\Phi} = (\bar{\boldsymbol{Y}} - \boldsymbol{\eta})^T (\bar{\boldsymbol{Y}} - \boldsymbol{\eta}) \tag{57}$$

The condition for the minimum with respect to the parameters $\boldsymbol{\beta}$ then becomes:

$$\frac{\partial \boldsymbol{\Phi}}{\partial \boldsymbol{\beta}} = \begin{pmatrix} \frac{\partial \boldsymbol{\Phi}}{\partial \beta_1} \\ \frac{\partial \boldsymbol{\Phi}}{\partial \beta_2} \\ \vdots \\ \frac{\partial \boldsymbol{\Phi}}{\partial \beta_q} \end{pmatrix} = \frac{2\left[\partial (\bar{\boldsymbol{Y}} - \boldsymbol{\eta})^T\right] (\bar{\boldsymbol{Y}} - \boldsymbol{\eta})}{\partial \boldsymbol{\beta}} = 0 \tag{58}$$

The first method discussed here is based on the Newton–Raphson method for the solution of nonlinear systems of equations. Here the nonlinear Equation (58) is expanded around an estimated solution into a Taylor series, which is truncated after the first term. For Equation (58) this leads to the expansion of $\boldsymbol{\eta}$ about $\boldsymbol{\eta}^{(0)}$ at $\boldsymbol{b}^{(0)}$, hence

$$\boldsymbol{\eta} = \boldsymbol{\eta}^{(0)} + \left(\frac{\partial \boldsymbol{\eta}}{\partial \boldsymbol{\beta}}\right)^{(0)} \Delta \boldsymbol{b}^{(0)} \tag{59}$$

with $\Delta \boldsymbol{b}^{(0)} = \boldsymbol{\beta} - \boldsymbol{b}^{(0)}$. A system of linear equations for the new variable $\Delta \boldsymbol{b}^{(0)}$ then arises, solution of which leads to an improved estimated value for the parameters:

$$\boldsymbol{b}^{(1)} = \boldsymbol{b}^{(0)} + \Delta \boldsymbol{b}^{(0)} \tag{60}$$

If this procedure is applied to Equation (58), substitution of Equation (59) into Equation (58) gives

$$\frac{2\left[\partial \left(\bar{\boldsymbol{Y}} - \boldsymbol{\eta}^{(0)} - \left(\frac{\partial \eta}{\beta}\right)^{(0)} \Delta \boldsymbol{b}^{(0)}\right)^T\right]\left[\left(\frac{\partial \eta}{\beta}\right)^{(0)} \Delta \boldsymbol{b}^{(0)}\right]}{\partial \Delta \boldsymbol{b}^{(0)}} = 0 \tag{61}$$

This gives the following system of equations for $\Delta \boldsymbol{b}^{(0)}$:

$$\left(\boldsymbol{X}^T \boldsymbol{X}\right)^{(0)} \boldsymbol{B}^{(0)} = \left(\boldsymbol{X}^T \boldsymbol{E}\right)^{(0)} \tag{62}$$

which employs the abbreviations

$$\boldsymbol{X} \equiv \frac{\partial \boldsymbol{\eta}}{\partial \boldsymbol{\beta}}, \quad \boldsymbol{E} \equiv (\bar{\boldsymbol{Y}} - \boldsymbol{\eta}), \text{ and } \boldsymbol{B} \equiv \Delta \boldsymbol{b}$$

Equation (62) is similar to Equation (39). In Equation (62) $\boldsymbol{B}^{(0)}$ is the dependent variable, whereas Equation (39) represents a system of equations for the parameters themselves. (Since there are no repeated measurements according to the assumptions in Section 3.2.1 the statistical weights for all measured points are equal, $\boldsymbol{p}$ is replaced by the unit matrix). The solution of Equation (62) for $\boldsymbol{B}^{(0)}$ gives

$$\boldsymbol{B}^{(0)} = \left[\left(\boldsymbol{X}^T \boldsymbol{X}\right)^{(0)}\right]^{-1} \left(\boldsymbol{X}^T \boldsymbol{E}\right)^{(0)} \tag{63}$$

By using Equation (60) an improved set of parameters is obtained and the method is iterated until a previously defined convergence condition of $\boldsymbol{B}^{(n)}$ or $\boldsymbol{\Phi}$ is satisfied.

With the second method for the estimation of parameters from nonlinear empirical models, the sum of squares itself is linearized. Expansion of Equation (48) around $\boldsymbol{b}^{(0)}$ into a Taylor series truncated after the first term gives

$$\boldsymbol{\Phi} \approx (\boldsymbol{\Phi})^{(0)} + \sum_{j=1}^{q} \left(\frac{\partial \boldsymbol{\Phi}}{\beta_j}\right)^{(0)} \left(\beta_j - b_j^{(0)}\right) \tag{64}$$

The direction of search used to calculate improved estimated values from the relation

$$\boldsymbol{b}^{(n+1)} = \boldsymbol{b}^{(n)} + \boldsymbol{h}^T \Delta \boldsymbol{b}^{(n)} \tag{65}$$

($\boldsymbol{h}$ is a vector of the step size) can be defined by the gradient of $\boldsymbol{\Phi}$; grad $\boldsymbol{\Phi}$ is a vector that is normal to the surface of $\boldsymbol{\Phi}$ in the parameter space and indicates the direction of the steepest ascent in $\boldsymbol{\Phi}$ at the point $\boldsymbol{b}$. Conversely, the vector − grad $\boldsymbol{\Phi}$ shows the direction of the steepest descent of

$\boldsymbol{\Phi}$. For parameter estimation the components of the vectors at $\boldsymbol{b}^{(0)}$

$$-\mathrm{grad}\boldsymbol{\Phi} = -\left(\frac{\partial \boldsymbol{\Phi}}{\partial \beta_1}\right)^{(0)} \delta\beta_1 - \left(\frac{\partial \boldsymbol{\Phi}}{\partial \beta_2}\right)^{(0)} \delta\beta_2 - \ldots - \left(\frac{\partial \boldsymbol{\Phi}}{\partial \beta_q}\right)^{(0)} \delta\beta_q \quad (66)$$

are calculated. For this the vector $-\mathrm{grad}\,\boldsymbol{\Phi}$, normalized by its absolute value is used:

$$\frac{-\mathrm{grad}\boldsymbol{\Phi}}{\|-\mathrm{grad}\boldsymbol{\Phi}\|} = \frac{-\left(\frac{\partial \boldsymbol{\Phi}}{\partial \beta_1}\right)^{(0)} \delta\beta_1 - \left(\frac{\partial \boldsymbol{\Phi}}{\partial \beta_2}\right)^{(0)} \delta\beta_2 - \ldots - \left(\frac{\partial \boldsymbol{\Phi}}{\partial \beta_q}\right)^{(0)} \delta\beta_q}{\sqrt{\left(-\frac{\partial \boldsymbol{\Phi}}{\partial \beta_1}\right)^{(0)2} + \left(-\frac{\partial \boldsymbol{\Phi}}{\partial \beta_2}\right)^{(0)2} + \ldots + \left(-\frac{\partial \boldsymbol{\Phi}}{\partial \beta_q}\right)^{(0)2}}} \quad (67)$$

In Equations (66) and (67), $\delta\beta_i$ denotes the unit vectors in the direction of β_i. The components of the normalized vector (Eq. 67) produce the term $\Delta\boldsymbol{b}$ in Equation (64) from which the iteration formula (Eq. 65) can proceed.

Both methods (solution of the system of nonlinear normal equations and the method of steepest descent) can, under certain conditions of the sum of squares $\boldsymbol{\Phi}$, lead to unfavorable trajectories over the surface of $\boldsymbol{\Phi}$ in the parameter space [40]. To deal with this, according to MARQUARDT a diagonal matrix is added to the coefficient matrix $(\boldsymbol{X}^{\mathrm{T}}\boldsymbol{X})$ in Equation (62):

$$\left(\boldsymbol{X}^T\boldsymbol{X}+\lambda\boldsymbol{I}\right)\boldsymbol{B}=\boldsymbol{X}^T\boldsymbol{E} \quad \lambda\geq 0 \quad (68)$$

where $\boldsymbol{I}$ is the unit matrix and λ is a scaling factor. When $\lambda=0$ Equation (62) is obtained. As $\lambda\to\infty$, then $\boldsymbol{B}\approx\boldsymbol{X}^{\mathrm{T}}\boldsymbol{E}/\lambda$ which, with the identity $\lambda = \frac{\|-\mathrm{grad}\boldsymbol{\Phi}^{(n)}\|}{\boldsymbol{h}^{(n)}}$ leads to the method of steepest descent for calculating $\boldsymbol{B}$. By fitting λ using Marquardt's method, the essential characteristics of both methods can be combined, [4], [7], [40].

Numerical Problems Connected with Gradient Methods. Iterative solution of Equation (62) or the method of steepest descent can give rise to problems, one of them being the initial guesses $\boldsymbol{b}^{(0)}$. Under certain conditions, the selection of $\boldsymbol{b}^{(0)}$ leads to a neighboring minimum in $\boldsymbol{\Phi}$, or is so unfavorable that iteration does not lead to convergence. The first of these problems requires a more precise investigation of the sum of the squares of the deviations or repetition of the calculation with various initial values for $\boldsymbol{b}^{(0)}$. In order to avoid divergence of the iterative method (especially with poorly chosen $\boldsymbol{b}^{(0)}$), only the direction of the vector $\boldsymbol{B}^{(0)}$ should be used; new estimated values for $\boldsymbol{\beta}$ are then calculated using Equation (65) with a reduced step size in place of Equation (60). In Equation (65), $\boldsymbol{h}$ has the character of a user-selected damping factor, which can be adjusted during each iteration in accordance with the evolution of $\boldsymbol{\Phi}$. With favorable conditioning of the nonlinear equation system, this factor can also be used to accelerate convergence.

Another problem lies in the mathematical form of the model. For example, for a model consistent with Equation (53) it is easy to show that $\boldsymbol{\eta}$, as well as the partial derivatives $\partial\eta/\partial\beta_1$ and $\partial\eta/\partial\beta_2$, are unbounded if $\beta_2 x=-1$. The only way to circumvent this problem is to constrain the range of the estimated values of the parameters on the basis of the physicochemical characteristics of the process.

The sum of the squares of the deviations can contain terms of differing orders of magnitude, which lead to very different gradients in the direction of the individual parameters. In such cases scaling of the parameters is always advisable, whereby the individual components in Equation (66) are brought to the same order of magnitude by linear transformation.

Transformations of the independent variables are necessary if interactions exist between individual parameters. This is the case, for example, with a simple Arrhenius type of rate coefficient of a chemical reaction

$$k = k_0\mathrm{e}^{-E_\mathrm{a}/RT}$$

where E_a is the activation energy, R the universal gas constant, and T the temperature.

If the exponential function is expanded so that

$$k \approx k_0 - k_0\frac{E_\mathrm{a}}{RT} + \frac{1}{2}k_0\left(\frac{E_\mathrm{a}}{RT}\right)^2 - \ldots$$

it becomes clear that k_0 has a multiplicative effect on the other parameter E_a. Minima of the sum of squares can thus be found for various combinations of k_0 and E_a that do not make sense in physical terms. In such cases either the range of allowable estimated values of the parameters is restricted on the basis of the physicochemical process characteristics, or the matrix

$(\boldsymbol{X}^{\mathrm{T}}\boldsymbol{X})$ is brought into a form in which the diagonal elements dominate through transformation of the independent variables. The smaller the off-diagonal elements of the matrix $(\boldsymbol{X}^{\mathrm{T}}\boldsymbol{X})$ are compared to the diagonal elements, the smaller the interaction between the parameters becomes. This can be demonstrated for the above example of the simple Arrhenius type of rate coefficient in which T is the independent variable. Since

$$\frac{\partial k}{\partial k_0} = \mathrm{e}^{-\frac{E_\mathrm{a}}{RT}} = \frac{k}{k_0} \text{ and } \frac{\partial k}{\partial (E_\mathrm{a}/R)} = -\frac{k_0}{T}\mathrm{e}^{\frac{-E_\mathrm{a}}{RT}} = \frac{k}{T}$$

then $$\left(\boldsymbol{X}^T\boldsymbol{X}\right) = \begin{pmatrix} \frac{1}{k_0^2}\sum i k_i^2 & -\frac{1}{k_0}\sum i\frac{k_i^2}{T_i} \\ -\frac{1}{k_0}\sum i\frac{k_i^2}{T_i} & \sum i\left(\frac{k_i^2}{T_i}\right)^2 \end{pmatrix}$$

Therefore

$$\mathrm{Det}\left(\boldsymbol{X}^T\boldsymbol{X}\right) = \frac{1}{k_0^2}\left[\sum_i k_i^2 \sum_i \left(\frac{k_i^2}{T_i}\right)^2 - \left(\sum_i \left(\frac{k_i^2}{T_i}\right)\right)^2\right]$$

which can become singular under certain conditions, since all terms in the square brackets are positive for all regions of the measured k_i and T_i.

If the transformation $T^* = (T - \bar{T})/T$ is used, we obtain $k = k_0^*\mathrm{e}^{E^*T^*}$ with $k_0^* = k_0/\mathrm{e}^{E^*}$ and $E^* = E_a/R\bar{T}$. For the transformed model

$$\frac{\partial k}{\partial k_0^*} = \mathrm{e}^{E^*T^*} = \frac{k}{k_0^*}$$

and

$$\frac{\partial k}{\partial E^*} = -k_0^* T^* \mathrm{e}^{E^*T^*} = kT^*$$

Hence

$$\left(\boldsymbol{X}^T\boldsymbol{X}\right) = \begin{pmatrix} \frac{1}{k_0^{*2}}\sum i k_i^2 & -\frac{1}{k_0^*}\sum i k_i^2 T_i^* \\ -\frac{1}{k_0^*}\sum i k_i^2 T_i^* & \sum i\left(k_i T_i^*\right)^2 \end{pmatrix}$$

Since T_i* can be positive or negative, the diagonal elements of the matrix dominate, and in

$$\mathrm{Det}\left(\boldsymbol{X}^T\boldsymbol{X}\right) = \frac{1}{k_0^{*2}}\left[\sum_i k_i^2 \sum_i \left(k_i T_i^*\right)^2 - \left(\sum_i k_i^2 T_i^*\right)^2\right]$$

the first term in the square brackets always dominates. The danger of the matrix $(\boldsymbol{X}^{\mathrm{T}}\boldsymbol{X})$ being singular (which would prevent solution of Eq. 63) is thus overcome.

3.2.2. Assessment of Estimated Parameter Values

Several methods have been developed to assess the estimated parameter values from nonlinear empirical models [20], [41], [42]. A method which is similar to the approach used in Sections 3.1.1.2 and 3.1.2.2 will be presented here.

The nonlinear model is linearized around the estimated values $\boldsymbol{b}$ in the parameter space, i.e., is expanded into a Taylor series truncated after the first term. The approximate variances and covariances of the parameters are then given by (cf. Eqs. 42 and 59):

$$\mathrm{Cov}\,(\boldsymbol{b}) \approx \left(\boldsymbol{X}^T\boldsymbol{X}\right)^{-1}\sigma_{\bar{Y}_i}^2 \tag{69}$$

Each element of $\boldsymbol{X}$ is calculated either analytically or numerically from the value of $\boldsymbol{b}$ indicated by the minimum. If the hypothesis is valid, s_r^2 can once again be used as an estimated value for $\sigma_{\bar{Y}_i}^2$ (Eq. 41). If repeated measurements are available, the hypothesis $\langle\bar{Y}_i\rangle = f(x, \beta)$ can be tested beforehand. The approximate variances of the estimated values of the parameters are obtained from the diagonal elements of $(\boldsymbol{X}^{\mathrm{T}}\boldsymbol{X})^{-1}$:

$$s_{bk}^2 \approx s_{\bar{Y}_i}^2 \cdot c_{kk} \tag{70}$$

using the abbreviation $\boldsymbol{c} = (\boldsymbol{X}^{\mathrm{T}}\boldsymbol{X})^{-1}$. The approximate confidence intervals can be estimated from this as described in Section 3.1.1.2.

The variances of the model solutions are obtained from the linearized model according to

$$s_{\hat{Y}_i}^2 \approx \sum_{j=1}^{q}\left(\frac{\partial\hat{Y}}{\partial b_j}\right) s_{b_j}^2 + \sum_{i=1}^{q}\sum_{j=1, i\neq j}^{q}\left(\frac{\partial\hat{Y}}{\partial b_i}\right)\left(\frac{\partial\hat{Y}}{\partial b_j}\right)\mathrm{Cov}\,(b_i b_j) \approx s_{\hat{Y}_i}\sum_{i=1}^{q}\sum_{j=1, i\neq j}^{q}\left(\frac{\partial\hat{Y}}{\partial b_i}\right)\left(\frac{\partial\hat{Y}}{\partial b_j}\right)c_{ij} \tag{71}$$

so that the approximate confidence intervals of the expected values can also be estimated.

3.2.3. Example of a Nonlinear Model

The estimation of the parameters of a nonlinear empirical model will be demonstrated by further discussion of the process presented in Section 3.1.1.4. Because of the relatively large scatter of the measured values within the individual data sets, the result of the linear regression is merely tolerable in spite of a positive test of the hypothesis $\langle \bar{Y}_i \rangle = \beta_0' + \beta_1 (x_i - \bar{x})$. The assessment of the estimated parameters gives large individual confidence intervals for the parameters and the expected values. Furthermore, the sum of squares and the estimated parameter values show a high sensitivity to the measured values $\bar{Y}_i$.

The same process will now be tested with a second approximation using a model of the form

$$\langle \bar{Y}_i \rangle = \beta_0 x_i^{\beta_1} \qquad (72)$$

The data for nonlinear regression are given in Table 1. The Newton – Raphson method is used to solve the system of nonlinear equations (Eq. 58). The results from the application of the linear model $\langle \bar{Y}_i \rangle = \beta_0' + \beta_1 (x_i - \bar{x})$ are used as an initial guess for b_0 and b_1, hence $b_0^{(0)} = 1.282$, $b_1^{(0)} = 1$. The step size $h_i^{(n)}$ (Eq. 65) is adjusted automatically according to the improvement in $\boldsymbol{\Phi}$ with each iteration. Figure 12 shows the estimated values $b_0^{(n)}$ and $b_1^{(n)}$ and the sum of squares of the deviations $\boldsymbol{\Phi}^{(n)}$ plotted against the number of iterations n. The method is regarded as convergent when $\boldsymbol{\Phi}^{(n-1)} - \boldsymbol{\Phi}^{(n)} \leq 10^{-4}$.

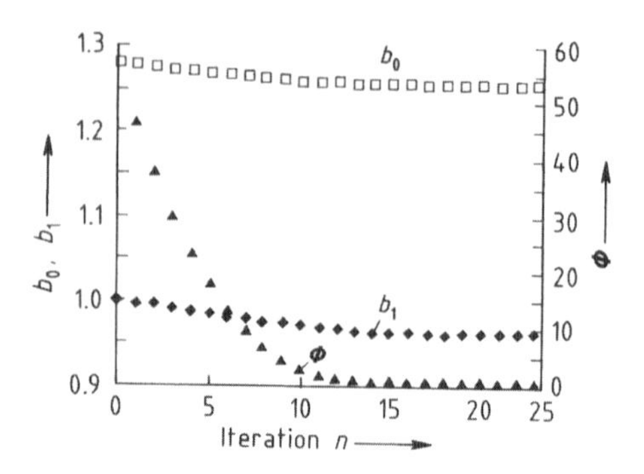

Figure 12. Results of nonlinear regression for the example discussed in Section 3.2.3

For the region of the experimentally adjusted gas velocities, the calculated results $b_0 = 1.257$ and $b_1 = 0.9626$ produce a smaller minimum in $\boldsymbol{\Phi}$, $\boldsymbol{\Phi}_{\min} = 0.332$, than the linear regression with $\boldsymbol{\Phi}_{\min} = 7.746$ (Fig. 6). The improved quality of the regression is obvious from the value of the mean square deviations about the regression function ($s_r^2 = 0.1105$) so that the F-test is fulfilled to a higher level of significance. If the combined squared deviations s_r^2 and s_ε^2 are used to estimate $\sigma_{\bar{Y}_i}^2$, $s_{\bar{Y}_i}^2 = 0.996$ is obtained; the improvement seen in Figure 12 also results in a smaller variance in the estimated parameter values b_0 and b_1 and in the model solution $\bar{Y}_i$. From Equations (70) and (71) the model linearized in the minimum of $\boldsymbol{\Phi}$ gives

$$s_{b_0}^2 \approx 0.996 \times 5.21 \times 10^{-2} = 5.03 \times 10^{-2}$$

$$s_{b_1}^2 \approx 0.996 \times 4.51 \times 10^{-3} = 4.3 \times 10^{-3}$$

$$s_{Y i}^2 \approx 0.996 \left[\left(x_i^{0.9626} \right)^2 \left(5.21 \times 10^{-2} - 3.6 \ln x_i + 7.13 \times 10^{-3} \left(\ln x_i \right)^2 \right) \right]$$

The relation $\langle \bar{Y}_i \rangle = \beta_0 x_i^{\beta_1}$ with $b_1 \leq 1$ reflects the true dependency better (see Section 3.1.1.4).

3.2.4. Concluding Remarks

The treatment of nonlinear empirical models is conducted as in Sections 3.1.1 and 3.1.2 using a relatively simple model to obtain the simplest formulation possible. The principles of nonlinear regression and some of the resulting problems can be clarified using this simple model. For cases in which the assumptions listed in the introduction to Section 3.2 do not hold, more advanced literature should be consulted, [4], [5], [7], [20]. Many computer programs are available for routine problems of parameter estimation with nonlinear empirical models (e.g., for gradient methods [27], for Marquardt's method [43], or for direct search methods [44]). The individual steps of the Newton – Raphson method (see Sections 3.1.1.5 and 3.1.2.3) can be obtained as modules from readily available program libraries [26–32].

3.3. Further Calculation Methods and Model Types

The discussion on the quantitive definition of empirical models (Sections 3.1 and 3.2) referred to the minimization of the sum of squares of the deviations as well as to simple models based on Equations (5) and (46). The section on empirical models will now be concluded with a short discussion of other methods of parameter estimation for empirical models and other types of models.

3.3.1. Further Methods of Parameter Estimation

Further methods of defining the parameters of empirical models include the maximum likelihood method.

A statistical empirical model can be described with the use of the probability density function $P(x, y)$ (Section 2.2 and Fig. 5). If a model for the process can be represented in the form $\langle\bar{Y}_i\rangle = \beta_0' + \beta_1(x_i - \bar{x})$ (Section 3.1.1), the probability density function is defined by the parameters β_0', β_1, and $\sigma^2_{\bar{Y}_i}$, hence $P(\bar{Y}, x; \beta_0', \beta_1, \sigma^2_{\bar{Y}_i})$. The problem of estimating the parameters now consists of defining β_0', β_1, and $\sigma^2_{\bar{Y}_i}$ such that the likelihood function becomes a maximum. With one observation the likelihood function L for the parameters is the probability density function in which the variables $\bar{Y}$ and x are viewed as parameters and the parameters β_0', β_1, and $\sigma^2_{\bar{Y}_i}$ are viewed as variables.

$$L\left(\beta_0',\beta_1,\sigma^2_{\bar{Y}_i};\bar{Y}_1,x_1\right) = P\left(\bar{Y}_1,x_1;\beta_0',\beta_1,\sigma^2_{\bar{Y}_i}\right) \quad (73)$$

With n statistically independent observations the likelihood function is the product of the individual functions:

$$\begin{aligned} &L\left(\beta_0',\beta_1,\sigma^2_{\bar{Y}_i};\bar{Y}_1,x_1\ldots\bar{Y}_n,\bar{x}_n\right) \\ &= \prod_{i=1}^{n} L\left(\beta_0',\beta_1,\sigma^2_{\bar{Y}_i};\bar{Y}_i,x_i\right) \\ &= P\left(\bar{Y}_1,x_1;\beta_0',\beta_1,\sigma^2_{\bar{Y}_i}\right)\cdot P\left(\bar{Y}_2,x_2;\beta_0',\beta_1,\sigma^2_{\bar{Y}_i}\right) \\ &\quad \ldots P\left(\bar{Y}_n,x_n;\beta_0',\beta_1,\sigma^2_{\bar{Y}_i}\right) \end{aligned} \quad (74)$$

To define β_0', β_1, and $\sigma^2_{\bar{Y}_i}$, L must now be maximized. The logarithms of Equation (74) are taken giving the following relations for the parameters:

$$\frac{\partial \ln L}{\partial \beta_0'} = \frac{\partial \sum\limits_{i=1}^{n} P\left(\bar{Y}_i,x_i;\beta_0',\beta_1,\sigma^2_{\bar{Y}_i}\right)}{\partial \beta_0'} = 0 \quad (75a)$$

$$\frac{\partial \ln L}{\partial \beta_1} = \frac{\partial \sum\limits_{i=1}^{n} P\left(\bar{Y}_i,x_i;\beta_0',\beta_1,\sigma^2_{\bar{Y}_i}\right)}{\partial \beta_1} = 0 \quad (75b)$$

$$\frac{\partial \ln L}{\partial \sigma^2_{\bar{Y}_i}} = \frac{\partial \sum\limits_{i=1}^{n} P\left(\bar{Y}_i,x_i;\beta_0',\beta_1,\sigma^2_{\bar{Y}_i}\right)}{\partial \sigma^2_{\bar{Y}_i}} = 0 \quad (75c)$$

For the linear problem $\langle\bar{Y}_i\rangle = \beta_0' + \beta_1(x_i - \bar{x})$ where the random variable Y is assumed to have a normal distribution, it can easily be shown that Equations (9) for defining β_0' and β are obtained from Equation (75) since

$$\begin{aligned} &P\left(\bar{Y}_i,x_i;\beta_0',\beta_1,\sigma^2_{\bar{Y}_i}\right) \\ &= \frac{1}{\sqrt{2\pi\sigma^2_{\bar{Y}_i}}}\cdot\exp\left[-\frac{1}{2\sigma^2_{\bar{Y}_i}}\left(\bar{Y}_i - \langle\bar{Y}_i\rangle\right)^2 p_i\right] \end{aligned} \quad (76)$$

Detailed discussion of this method can be found in [4], [22].

Further search algorithms are discussed in more advanced literature on optimization, → Mathematics in Chemical Engineering, Chap. 10., [5], [7], [38], [39].

3.3.2. Other Types of Models

In addition to the models described by Equations (5) and (46), empirical models in which underlying conditions constrain the parameters are frequently encountered in industrial chemistry. Furthermore the functional relationship may not be in the simple form of algebraic equations (Eqs. 5, 32, and 46) but in the form of differential equations or transfer functions.

3.3.2.1. Models with Constraints on the Parameters

Constraints on parameters are found in many models used in industrial chemistry. A simple example is the reaction rate of a heterogeneously catalyzed reaction according to Equation (53), which only makes sense when

$$\beta_i \geq 0 \quad (77a)$$

Equally common are constraints of the type

$$\beta_i \geq k_i, \text{ where } k_i \text{ is positive} \quad (77b)$$

$$0 \leq \beta_i \leq 1 \quad (77c)$$

$$k_i^* \leq \beta_i \leq k_i \quad (77d)$$

Explicit constraints of this type can easily be removed with appropriate transformations. For example, the general transformation rule for Equation (77 b) is

$$\beta_i = k_i + e^{\beta_i^*} \tag{78a}$$

in which the unknown β_i^* is not affected by any constraints. For Equation (77 d) the general transformation rule

$$\beta_i = k_i^* + (k_i - k_i^*)\sin^2\beta_i^* \tag{78b}$$

can be used, whereas for Equation (77 a)

$$\beta_i = \beta_i^{*2} \text{ or } \beta_i = e^{\beta_i^*} \tag{78c,d}$$

are suitable transformations.

The general constraint

$$g_i(\boldsymbol{Y}, \boldsymbol{x}, \boldsymbol{\beta}) \geq 0, \quad i = 1, \ldots q \tag{79}$$

for which transformations are not applicable, can be added as a penalty function to the sum of squares of the deviations:

$$\boldsymbol{\Phi} + \sum_{i=1}^{q} \lambda_i (g_i)^{\varrho} = \boldsymbol{\Phi}^* \overset{!}{=} \text{Min} \tag{80}$$

For constraints given in Equation (77 b) the penalty function is $g_i = 0$ for $b_j \geq k_j$ and $g_i = (b_j - k_j)^2$ for $b_j \leq k_j$. Here the scaling factors λ and ϱ are set to 1 and 2, respectively.

Similarly constraints in the form

$$g_i(\boldsymbol{Y}, \boldsymbol{x}, \boldsymbol{\beta}) = 0, \quad i = 1, \ldots q \tag{81}$$

can be handled in the form of penalty functions which are added to the sum of squares of the deviations. They disappear when the associated conditions (Eq. 81) are satisfied. Alternatively they are weighted according to the magnitude of the deviation and added to $\boldsymbol{\Phi}$.

A more rigorous method for the treatment of associated constraints in the form of equations is based on their geometrical interpretation. In Figure 13 the sum of squares of the deviations for a two-parameter problem is shown in the parameter space. A constraint of the form

$$\beta_1 + a\beta_2 = c \tag{82}$$

can be represented as a straight line which intersects the lines of constant $\boldsymbol{\Phi}$. The minimum of $\boldsymbol{\Phi}$ is now searched for along this straight line, so that the two-dimensional problem is reduced to a one-dimensional one. This can be easily demonstrated for a linear model of the form according to Equation (5) with the constraint given by Equation (82) because one parameter is directly eliminated by substitution.

In problems involving many parameters or nonlinear models, reduction of a multidimensional problem with one or more constraints by substitution is either impossible or very difficult. In such cases the method of Lagrangian multipliers is used. Figure 13 illustrates that for the constrained minimum, the slopes $\partial\beta_1/\partial\beta_2$ of the two curves $\boldsymbol{\Phi}$ = const. and $\beta_1 + a\beta_2 - c = 0$ are equal. In general these conditions can be expressed as

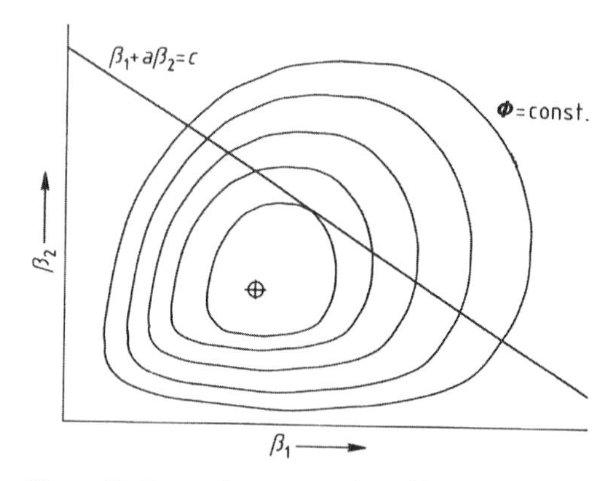

Figure 13. Geometric representation of least squares estimations with constraints on the parameters

$$\frac{\partial\boldsymbol{\Phi}/\partial\beta_1}{\partial\boldsymbol{\Phi}/\partial\beta_2} - \frac{\partial g/\partial\beta_1}{\partial g/\partial\beta_2} = 0 \tag{83}$$

or as the gradient ratio

$$\frac{\partial\boldsymbol{\Phi}/\partial\beta_1}{\partial g/\partial\beta_1} = \frac{\partial\boldsymbol{\Phi}/\partial\beta_2}{\partial g/\partial\beta_2} = \lambda \tag{84}$$

From Equation (84) the system of equations

$$\frac{\partial\boldsymbol{\Phi}}{\partial\beta_1} - \lambda\frac{\partial g}{\partial\beta_1} = 0 \tag{85a}$$

and

$$\frac{\partial\boldsymbol{\Phi}}{\partial\beta_2} - \lambda\frac{\partial g}{\partial\beta_2} = 0 \tag{85b}$$

is obtained. Equations (85 a), (85 b), and the constraint given by Equation (82) may be derived from the extended sum of squares

$$\boldsymbol{\Phi}^* = \boldsymbol{\Phi} + \lambda(\beta_1 + a\beta_2 - c) \tag{86}$$

by partial differentiation with respect to the parameters and the Lagrangian multipliers λ.

In general, therefore, when the associated constraints are in the form of Equation (81), the minimum of the extended sum of squares

$$\boldsymbol{\Phi} + \sum_{i=1}^{q} \lambda_i g_i (\boldsymbol{Y}, \boldsymbol{x}, \boldsymbol{\beta}) = \boldsymbol{\Phi}^* \stackrel{!}{=} \mathrm{Min} \qquad (87)$$

must be found by varying the parameters $\boldsymbol{\beta}$ and the Lagrange multipliers λ_i. This can be carried out using the methods discussed previously (see Section 3.2.1).

3.3.2.2. Models Based on Differential Equations

Chemical processes often give rise to models in the form of differential equations. Examples are models of reactors based on transport equations or the rates of chemical reactions. For differential equations that are simple to integrate the parameters can be determined according to the principles presented in Section 3.2, for example, as with the common first-order differential equation

$$\frac{\mathrm{d}y(t)}{\mathrm{d}t} = \alpha y(t) + x(t), \quad \text{where} \quad y_{t=0} = y_0 \qquad (88)$$

The general solution of Equation (88) is [45]

$$y(t) = y_0 \mathrm{e}^{\alpha t} + \int_0^t x(\tau) \mathrm{e}^{\alpha(t-\tau)} \mathrm{d}\tau \qquad (89)$$

If discrete measured values $Y(t_i)$ are available, the condition for the minimum with respect to α and y_0 is given by

$$\boldsymbol{\Phi} = \sum_{i=1}^{n} \left[Y(t_i) - y_0 \mathrm{e}^{\alpha t_i} + \frac{x_0}{\alpha} \left(1 - \mathrm{e}^{-\alpha t_i}\right)\right]^2 \qquad (90)$$

This leads to the system of nonlinear equations

$$\sum_{i=1}^{n} \left[Y(t_i) - \hat{y}_0 \mathrm{e}^{\hat{\alpha} t_i} + \frac{x_0}{\hat{\alpha}} \left(1 - \mathrm{e}^{-\hat{\alpha} t_i}\right)\right] \mathrm{e}^{\hat{\alpha} t_i} = 0 \qquad (91a)$$

and

$$\sum_{i=1}^{n} \left[Y(t_i) - \hat{y}_0 \mathrm{e}^{\hat{\alpha} t_i} + \frac{x_0}{\hat{\alpha}} \left(1 - \mathrm{e}^{-\hat{\alpha} t_i}\right)\right] \cdot \left[t_i \hat{y}_0 \mathrm{e}^{\hat{\alpha} t_i} - \frac{x_0}{\hat{\alpha}^2} \left(\mathrm{e}^{\hat{\alpha} t_i} (1 - \hat{\alpha} t_i) - 1 \right)\right] = 0 \qquad (91b)$$

which can be solved for $\hat{y}_0$ and $\hat{\alpha}$ by an appropriate method.

In the majority of cases, the differential equations of the model are more complex than Equation (88), so that no analytical solution is available; numerical solution of the differential equations is required to evaluate $\langle Y_i \rangle$. New problems then arise depending on the type of differential equation. Especially with systems of coupled partial differential equations used to model chemical reactors on the basis of transport equations (see Chap. 5), the solution of the differential equations is the real problem. In such cases other methods are often used to determine the model parameters. Changing certain process variables produces a characteristic behavior in some model solutions which is studied in suitably designed experiments.

An example is the spread of the concentration profile of a tracer added in the form of a Dirac function into reactors with axial and/or radial dispersion. By measurement of the concentration at two points in the reactor, the Péclet number *Pe* can be derived in a relatively simple manner. Further examples are discussed in [4], [46]. A feature of these methods is that an analytical or numerical solution of the differential equations is often not necessary because the model parameters are comparatively simple functions of the deterministic moments of the measured variables. Deterministic moments are defined analogously to the moments of probability density functions of statistical variables (Section 2.2) and are related to the distribution functions of the deterministic variables. If, for example, $w(t)$ is the residence time distribution for a chemical reactor, then the first two deterministic moments m_1 and m_2' are defined as

$$m_1 = \bar{t} = \int_0^\infty t \cdot w(t) \, \mathrm{d}t$$

$$m_2' = \int_0^\infty (t - m_1)^2 \, w(t) \, \mathrm{d}t$$

An example is the determination of the Péclet number for a chemical reactor with back mixing on the basis of measurements of the residence-time distribution [46], [47].

Finally, it should be noted that for processes which can be described in the form of systems of common linear differential equations analogous to Equation (88), i.e.,

$$\frac{\mathrm{d}\boldsymbol{y}}{\mathrm{d}t}+\beta \boldsymbol{y}-\boldsymbol{x}\left(t\right), \quad \boldsymbol{y}_{t=0}=0 \tag{92}$$

the parameters can be determined with the Laplace transforms of the transfer functions. The transfer functions can be given in the physical time coordinates as well as in the coordinates of the Laplace transformation. The methods are discussed in detail in [4], [9], [13–15].

4. Models Based on Transport Equations for Probability Density Functions

Probability density functions and the probabilities which can be calculated from them are the most adequate way of describing statistical variables. Examples of statistical variables can be found at the molecular level in the molecular velocity resulting from random thermal movements; the corresponding probability density function is the Maxwell – Boltzmann velocity distribution [48]. The use of this concept in kinetic gas theory and statistical mechanics has been referred to in Section 2.2.

Statistical variables can also be found at the macroscopic level. In the example discussed in Chapter 3, statistical process variables with presumed distribution or probability density functions were examined and their expected values, variances, and functional relationships to other variables were determined.

Another example is the jet stirred fluid-phase reactor with turbulent flow presented schematically in Figure 14. The surrounding fluid entrains the jet by exchange of momentum. The intake volume of the jet is greater than the volume of the surrounding fluid. This leads to recirculation of material from further downstream and mixing of the reactor contents. The fluid in the jet exhibits turbulent flow. The velocity of the fluid particles is so high that they can no longer be held in regular streamlines by the viscous forces, and irregular vortices are formed. Vortices of different sizes with different velocities and different macroscopic properties pass a fixed observer. Therefore, the fixed observer measures statistical, time-dependent variations in the physical quantity Φ.

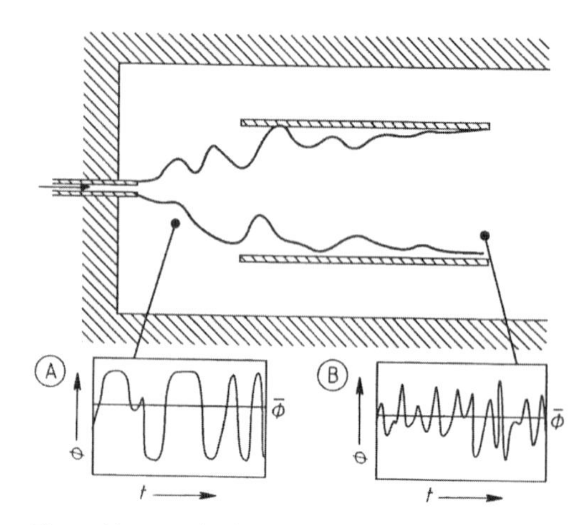

Figure 14. Example of a process with statistical variables: a jet stirred reactor with turbulent flow

In Figure 14 this is shown at two positions in the jet. At the edge of the nozzle (A) either the properties of the jet or the surroundings are measured. Further downstream (B) the macroscopic properties are balanced due to turbulent mixing. All physical parameters are thus characterized by position- and time-dependent distribution or probability density functions. The objective of models based on transport equations for probability density functions is the calculation of functions $P\left(\Phi_1,\ldots,\Phi_n\right)$ in which $\Phi_1,\ldots,\Phi_n$ is the vector of the dependent statistical variables. This has to be specified for each problem. Models based on transport equations for probability density functions generally give a differential equation for $P\left(\Phi_1,\ldots,\Phi_n\right)$.

4.1. Terminology

4.1.1. One-Dimensional Distribution and Probability Density Functions

The example shown in Figure 14 will be used to explain terminology. The temperature of the fluid is treated as a macroscopic physical quantity. Since the temperature varies between two limits $T_b > T > T_a$, it is better to use a normalized variable so that the statistical variable Φ is defined as

$$\Phi=\frac{T-T_b}{T_a-T_b} \tag{93}$$

The sample space for Φ (see Fig. 15) is a line on which only values between 0 and 1 can be realized. An event is defined as $E \equiv \varphi_u \leq \Phi \leq \varphi_o$. The probability of the occurrence of a particular event is between 0 and 1. When $W(E)=0$, $\varphi_u \leq \Phi \leq \Phi_o$ never occurs and when $W(E)=1$, $\varphi_u < \Phi < \varphi_o$ always occurs. Probability defined in this way does not predict E for the next experiment but instead the fraction of occurrences of E over a large number of experiments. Since each event is linked to a defined area of the sample space, the probability of an event is the probability that a random variable lies in this area. The probability of an event can be defined from the distribution function of the random variables:

$$F(\Phi) \equiv W(\Phi \leq \varphi) \tag{94}$$

Since the normalized temperatures defined in Equation (93) can only have values between $0 \leq \Phi \leq 1$, then

$$W(\Phi \leq 0) = F(0) = 0 \quad \text{and} \tag{95a}$$

$$W(\Phi \leq 1) = F(1) = 1 \tag{95b}$$

Any event, i.e., any interval $E \equiv \varphi_u \leq \Phi \leq \varphi_o$ in the sample space can be obtained by subtraction (see Fig. 15):

$$E = (\Phi \leq \varphi_o) - (\Phi \leq \varphi_u) \tag{96}$$

Thus

$$W(E) = W(\varphi_u \leq \Phi \leq \varphi_o) = W(\Phi \leq \varphi_o) - W(\Phi \leq \varphi_u) = F(\varphi_o) - F(\varphi_u) \tag{97}$$

The fundamental characteristics of the distribution function are easily derived by similar reasoning. Since in the general case $W(\Phi \leq -\infty) = F(-\infty) = 0$ and $W(\Phi \leq +\infty) = F(+\infty) = 1$, then

$$W(E) = F(\varphi_o) - F(\varphi_u) \geq 0 \tag{98}$$

and

$$F(\varphi_o) \geq F(\varphi_u) \quad \text{for} \quad \varphi_o \geq \varphi_u \tag{99}$$

The distribution function $F(\Phi)$ is therefore a nondecaying function which grows from 0 at $\varphi = -\infty$ to 1 at $\varphi = +\infty$ (Fig. 15 A).

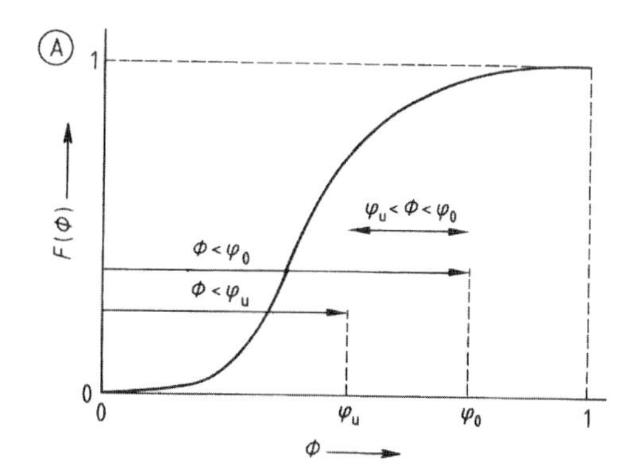

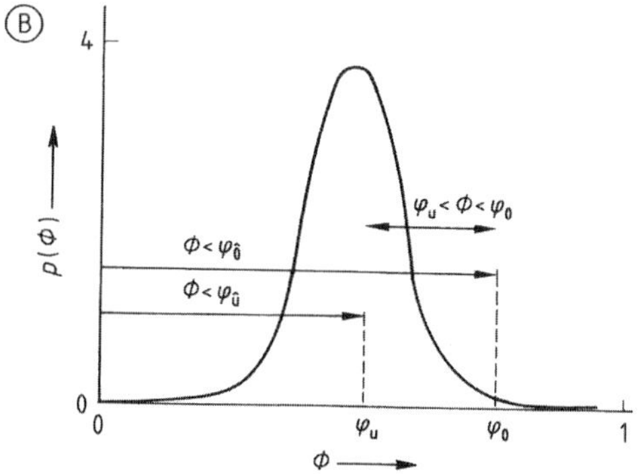

Figure 15. Representation of the distribution function (A) and the probability density function (B) of a statistical variable
The φ axis represents the sample space for the statistical variable Φ. The values φ_o and φ_u confine a finite region within the sample space.

The probability density function of the random variable Φ is the derivative of the distribution function with respect to the random variable (Fig. 15 B).

$$P(\Phi) = \frac{d}{d\Phi}[F(\Phi)] \tag{100}$$

Using this result then

$$\int_{\varphi_u}^{\varphi_o} P(\Phi)\, d\Phi = F(\varphi_o) - F(\varphi_u) = W(\varphi_u \leq \Phi \leq \varphi_o) \tag{101}$$

According to Equation (101) the probability that a random variable falls in a particular area is equal to the integral of the probability density function over that area. For an infinitely small area $d\varphi$ this then becomes

$$W(\varphi \leq \Phi \leq \varphi + d\varphi) = P(\Phi)\, d\Phi \tag{102}$$

(cf. Section 2.2). The fundamental characteristics of $P(\Phi)$ can be derived from Equation (100). Since $F(\Phi)$ is a nondecaying function this gives

$$P(\Phi) \geq 0 \tag{103}$$

Further

$$\int_{-\infty}^{+\infty} P(\Phi)\,\mathrm{d}\Phi = 1 \tag{104}$$

since according to Equation (101) the probability of the occurrence of the random variable in the area from $-\infty$ to $+\infty$ in the sample space must be 1. The third fundamental characteristic of the probability density function follows from the monotonic approximation of the distribution function to 0 or 1 when $|\Phi| \to \infty$. Thus

$$\frac{\mathrm{d}}{\mathrm{d}\Phi}[F(\Phi)]_{-\infty} = P(\Phi)_{-\infty} = \frac{\mathrm{d}}{\mathrm{d}\Phi}[F(\Phi)]_{+\infty} = P(\Phi)_{+\infty} = 0 \tag{105}$$

For discontinuous distribution functions with statistical variables which can only take discrete values (see Chap. 3) the above relationships can be given with additional derivation steps [18], [19], [22]. Continuity of the distribution function is not a necessary condition for the above relationships.

The relationships for transformations of distribution functions and probability density functions are easy to see from Figure 15 and Equations (98)–(100). For example if the relationship $\psi = f(\varphi)$ holds for a given transformation of Φ, then the events $\Phi \leq \varphi$ and $\Psi \leq \psi = f(\varphi)$ are identical. From this $F(\Phi) = F(\Psi)$ and according to Equation (100)

$$P(\Psi) = \frac{\mathrm{d}}{\mathrm{d}\Psi}F(\Psi) = \frac{\mathrm{d}}{\mathrm{d}\Phi}F(\Phi)\cdot\frac{\mathrm{d}\Phi}{\mathrm{d}\Psi} = \mathrm{P}(\Phi)\cdot\frac{\mathrm{d}\Phi}{\mathrm{d}\Psi} \tag{106}$$

Figure 14 shows that the distribution functions and probability density functions can be dependent on position and are thus not always homogeneous. For the area at the edge of the jet close to the nozzle exit an almost bimodal probability density function is produced due to intermittencies of the jet and the surrounding fluid. Further downstream in the middle of the turbulent jet these intermittencies no longer affect the form of the probability density function. Thus probability density functions hold for a given position at a given time; they will be written as $P(\Phi; x, t)$ from now on.

For engineering applications the probability density functions themselves are not very illustrative. Statistical variables are usually described in terms of expected values (see Section 2.2 and Chap. 3). These are easier to handle but have a lower information content. The expected values of statistical variables (synonyms are mean values, mathematical expectations, and first moment) are given by

$$\langle \Phi(\boldsymbol{x},t) \rangle = \int_{-\infty}^{+\infty} \Phi P(\Phi;\boldsymbol{x},t)\,\mathrm{d}\Phi \tag{107}$$

For statistical steady-state processes or for $\mathrm{d}\langle\Phi\rangle/\mathrm{d}t \ll \mathrm{d}\Phi/\mathrm{d}t$ the equivalent form of the averaging is

$$\langle \Phi(\boldsymbol{x},t) \rangle = \frac{1}{\Delta\tau}\int_{t}^{t+\Delta\tau} \Phi(\boldsymbol{x},t)\,\mathrm{d}t \tag{108}$$

For this a continuous function of Φ versus time is necessary as indicated in the plot of Φ against time given in Figure 14.

Functions of statistical variables are themselves statistical values, so that the operations related to Equation (107) for deriving expected values of functions can be used

$$\langle Q(\Phi) \rangle = \int_{-\infty}^{+\infty} Q(\Phi)\,P(\Phi;\boldsymbol{x},t)\,\mathrm{d}\Phi \tag{109}$$

The definitions of Equations (107) and (109) respectively are analogously applicable to higher moments

$$\mu_n \equiv \langle \Phi(\boldsymbol{x},t)^n \rangle = \int_{-\infty}^{+\infty} \Phi^n P(\Phi;\boldsymbol{x},t)\,\mathrm{d}\Phi \tag{110}$$

For example, the second moment around the mean (the variance) is also a characteristic of statistical quantities which is often used in engineering applications (see Chap. 3):

$$\mathrm{Var}(\Phi) \equiv \mu_2' \equiv \left\langle (\Phi - \langle\Phi\rangle)^2 \right\rangle = \int_{-\infty}^{+\infty} (\Phi - \langle\Phi\rangle)^2\,P(\Phi;\boldsymbol{x},t)\,\mathrm{d}\Phi \tag{111}$$

All of these quantities can be determined from the probability density function. However, the probability density function can only be given approximately from a finite number of moments which are relatively easy to measure [49].

If a statistical quantity is split up into the expected value and a fluctuating component Φ^{f}

$$\Phi=\langle\Phi\rangle+\Phi^{\mathrm{f}} \tag{112}$$

then from Equation (111)

$$\operatorname{Var}(\Phi)=\left\langle\left(\Phi^{\mathrm{f}}\right)^{2}\right\rangle \text{ and } \langle\Phi^{\mathrm{f}}\rangle=0 \tag{113a,b}$$

Rules for measuring the statistical moments of statistical steady-state processes are derived from Equation (108). The continuous function $\Phi(\boldsymbol{x},t)$ is necessary for this. However measurements of statistical variables are mostly available in the form of discrete values rather than in continuous form. Therefore all values referred to here must be derived from discrete data. This has been done for the empirical models discussed in Chapter 3 where discrete values of the statistical variables were available.

For n independent repetitions of an experiment under constant conditions the ensemble mean value

$$\langle\Phi\rangle_{n}=\frac{1}{n}\sum_{i=1}^{n}\varphi_{i} \tag{114}$$

gives a good approximation of $\langle\Phi\rangle$ when $n\to\infty$.

In the examples discussed in Chapter 3 the number of occurrences is so small that the mean values from the individual data sets do not represent the expected values but are statistical values. Similarly to Equation (114) the means of functions $Q(\Phi)$ or higher moments are estimated, c.f. Chapter 3 where the sums of squares of the deviations were used as estimates for the variances. The probability density function is measurable through the number of occurrences of φ_i in a given area $\Delta\varphi$ of the sample space. If this number is denoted as $m(\Phi,\Delta\Phi)$, the normalized density of occurrences in the sample space provides a good estimate for the probability density function:

$$P_{n}(\Phi)=\frac{m(\Phi,\Delta\Phi)}{n\Delta\Phi}\approx P(\Phi;\boldsymbol{x},t) \quad \text{for } n\to\infty \tag{115a}$$

The discrete form of the probability density function can also be expressed as

$$P_{n}(\Phi)=\frac{1}{n}\sum_{i=1}^{n}\delta(\Phi-\varphi_{i}) \tag{115b}$$

Here $\delta(a-b)$ is the impulse or Dirac delta function which is defined as the derivative of the Heaviside or step function:

$$\delta(a-b)=\frac{\mathrm{d}H(a-b)}{\mathrm{d}a} \tag{116}$$

with $H(a-b)=0$ for $a\leq b$ and $H(a-b)=1$ for $a>b$.

According to Equation (115 b) the discrete form of the probability density function in the sample space for Φ is the number impulses of discrete occurrences normalized with respect to the total count. The discrete form of the distribution function

$$F_{n}(\Phi)=\frac{1}{n}\sum_{i=1}^{n}H(\Phi-\varphi_{i}) \tag{115c}$$

becomes a stepped curve as shown in Figure 16.

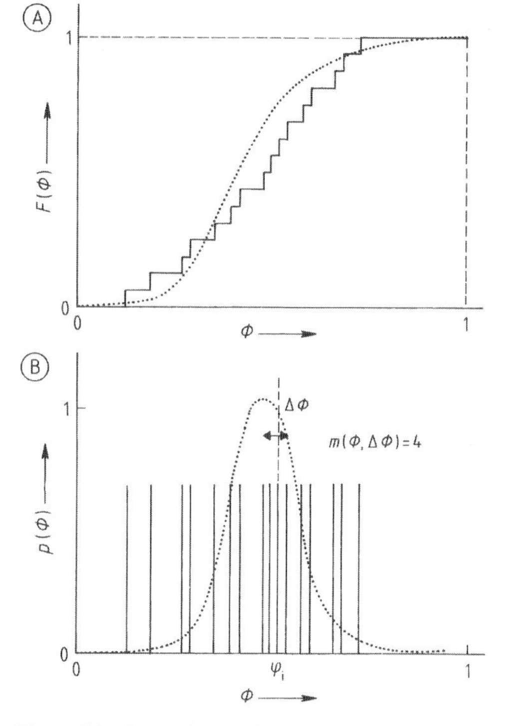

Figure 16. Discrete forms of the distribution function (A) and the probability density function (B)
The continuous functions are shown as dotted lines.

4.1.2. Multidimensional Distribution and Probability Density Functions

For the example shown in Figure 14 the normalized temperature alone is clearly not sufficient to describe the state at a point $\boldsymbol{x}$ and time t. Other variables such as the axial velocity U can be used which is expressed in normalized form as

$$\Psi = \frac{U - U_b}{U_a - U_b} \tag{117}$$

For the statistical variable Ψ the same definitions hold as previously for Φ so that

$$W\,(\varphi \leq \Phi \leq \varphi + \mathrm{d}\varphi) = P\,(\Phi;\boldsymbol{x},t)\,\mathrm{d}\Phi$$

and

$$W\,(\psi \leq \Psi \leq \psi + \mathrm{d}\psi) = P\,(\Psi;\boldsymbol{x},t)\,\mathrm{d}\Psi$$

However, this information is not sufficient. More information is required about the probability of the joint events

$$E_\varphi = (\varphi \leq \Phi \leq \varphi + \mathrm{d}\varphi)$$

and

$$E_\psi = (\psi \leq \Psi \leq \psi + \mathrm{d}\psi)$$

This information can be derived from the joint distribution function $F\,(\Phi,\Psi;\boldsymbol{x},t)$ which has characteristics analogous to those of the simple distribution function:

$$0 \leq F\,(\Phi,\Psi) \leq 1, \quad \text{nondecaying,} \tag{118a}$$

$$F\,(-\infty,\Psi) = F\,(\Phi,-\infty) = 0, \tag{118b}$$

$$F\,(\infty,\Psi) = F\,(\Psi), \quad \text{and} \tag{118c}$$

$$F\,(\Phi,\infty) = F\,(\Phi) \tag{118d}$$

The joint probability density function becomes

$$P\,(\Phi,\Psi;\boldsymbol{x},t) = \frac{\partial^2}{\partial\Phi\partial\Psi}\,[F\,(\Phi,\Psi;\boldsymbol{x},t)] \tag{119}$$

which possesses the same fundamental characteristics as the simple probability density function:

$$P\,(\Phi,\Psi;\boldsymbol{x},t) \geq 0 \tag{120a}$$

$$\int\int P\,(\Phi,\Psi;\boldsymbol{x},t)\,\mathrm{d}\Psi\mathrm{d}\Phi = 1 \tag{120b}$$

The simple probability density function or marginal distributions (the reduced information) can be derived from the joint probability density function:

$$\int P\,(\Phi,\Psi;\boldsymbol{x},t)\,\mathrm{d}\Psi = P\,(\Phi;\boldsymbol{x},t) \tag{121a}$$

$$\int P\,(\Phi,\Psi;\boldsymbol{x},t)\,\mathrm{d}\Phi = P\,(\Psi;\boldsymbol{x},t) \tag{121b}$$

Conversely the joint probability density functions cannot normally be derived from the marginal distributions. Arguments analogous to those above may be used for the functions $Q\,(\Phi,\Psi)$.

Covariances are another quality of joint probability density functions:

$$\mathrm{Cov}\,(\Phi,\Psi) = \langle \Phi^{\mathrm{f}}\Psi^{\mathrm{f}} \rangle = \int_{-\infty}^{+\infty}\int_{-\infty}^{+\infty} (\Phi - \langle\Phi\rangle)\,(\Psi - \langle\Psi\rangle) \cdot P\,(\Phi,\Psi;\boldsymbol{x},t)\,\mathrm{d}\Phi\mathrm{d}\Psi \tag{122}$$

They are mostly given in normalized form as correlation coefficients:

$$\gamma = \frac{\mathrm{Cov}\,(\Phi,\Psi)}{[\mathrm{Var}\,(\Phi)\,\mathrm{Var}\,(\Psi)]^{1/2}} \tag{123}$$

With the help of Equation (122) it is easy to show that $-1 \leq \gamma \leq 1$. If a linear relationship of the form $\psi = \alpha + \beta\,\varphi$ exists, then $|\,\gamma\,| = 1$.

For joint multidimensional probability density functions, transformation rules are also valid. Using the same reasoning as with one-dimensional probability density functions and assuming that $\xi = \xi\,(\varphi,\psi)$ and $\theta = \theta\,(\varphi,\psi)$

$$P\,(\Xi,\Theta) = P\,(\Phi,\Psi) \cdot |\boldsymbol{J}|^{-1} \tag{124}$$

where $\boldsymbol{J}$ is the Jacobi matrix of the system of transformation equations.

4.1.3. Conditional Probability Density Functions

Conditional probability is a prerequisite for the derivation of transport equations for probabiliby density functions. It denotes the probability of an event, for example $\Psi \leq \psi$, subject to the condition that the event $\varphi \leq \Phi \leq \varphi + \mathrm{d}\varphi$ occurs. It is given by the probability for a general event divided by the probability of the condition:

$$W\left[(\Psi \leq \psi) \mid (\varphi \leq \Phi \leq \varphi + \mathrm{d}\varphi)\right] = \frac{W\,(\Psi \leq \psi)\,(\varphi \leq \Phi \leq \varphi + \mathrm{d}\varphi)}{W\,(\varphi \leq \Phi \leq \varphi + \mathrm{d}\varphi)} \tag{125}$$

Consequently

$$W\left[(\Psi \leq \psi) \mid (\varphi \leq \Phi \leq \varphi + \mathrm{d}\varphi)\right] = \frac{\int_{-\infty}^{\psi} \int_{\varphi}^{\varphi+\mathrm{d}\varphi} P\,(\Phi,\Psi)\,\mathrm{d}\Phi \mathrm{d}\Psi}{\int_{\varphi}^{\varphi+\mathrm{d}\varphi} P\,(\Phi)\,\mathrm{d}\Phi} \tag{126}$$

The conditional distribution function is the limit of the conditional probability for $\mathrm{d}\varphi \rightarrow 0$ so that

$$F\,(\Psi|\varphi) = P\left[(\Psi \leq \psi) \mid (\Phi = \varphi)\right] = \int_{-\infty}^{\psi} \frac{P\,(\Psi,\Phi)\,\mathrm{d}\Psi}{P\,(\Phi)} \tag{127}$$

The usual definition of probability density function then gives the following:

$$P\,(\Psi|\varphi) = \frac{\mathrm{d}}{\mathrm{d}\Psi} F\,(\Psi|\varphi) = \frac{P\,(\Psi,\Phi)}{P\,(\Phi)} \tag{128}$$

It is easy to show that the conditional distribution functions and probability density functions exhibit the same characteristics as unconditional ones. By using the same operations as above, the conditional expected value of a function $Q\,(\Psi,\Phi)$ can be derived

$$\langle Q\,(\Psi,\Phi)\,|\Phi = \varphi\rangle = \int_{-\infty}^{+\infty} Q\,(\Psi,\Phi)\,P\,(\Psi|\varphi)\,\mathrm{d}\Psi \tag{129}$$

From Equation (128) it can be seen that the joint probability density function $P\,(\Psi,\Phi)$ can be obtained from the conditional probability density function $P\,(\Psi \mid \varphi)$ and the marginal distribution $P\,(\Phi)$. This important result is used in the determination of expected values and leads to a further important result:

$$\begin{aligned}\langle Q\,(\Psi,\Phi)\,\rangle &= \int_{-\infty}^{+\infty}\int_{-\infty}^{+\infty} Q\,(\Psi,\Phi)\,P\,(\Psi,\Phi)\,\mathrm{d}\Psi \mathrm{d}\Phi \\ &= \int_{-\infty}^{+\infty} P\,(\Phi) \left\{ \int_{-\infty}^{+\infty} Q\,(\Psi,\Phi)\,P\,(\Psi|\varphi)\,\mathrm{d}\Psi \right\} \mathrm{d}\Phi \\ &= \int_{-\infty}^{+\infty} P\,(\Phi) < Q\,(\Psi,\Phi)\,|\Phi = \varphi > \mathrm{d}\Phi \end{aligned} \tag{130}$$

4.2. Transport Equations for Single-Point Probability Density Functions

Before going into some examples, the following sections will briefly discuss transport equations for single-point probability density functions starting from the general conservation equations. Only the important relationships will be described. A more detailed discussion of the general conservation equations is given in [48].

The state of a reacting fluid mixture (e.g., the contents of the reactor shown in Fig. 14) is accurately defined by the velocity $\boldsymbol{U}\,(\boldsymbol{x},t)$, the pressure $p\,(\boldsymbol{x},t)$, the mass fraction of the chemical species present in the mixture $m_i\,(\boldsymbol{x},t)$, $i=1,\ldots,N$, and the specific enthalpy $h\,(\boldsymbol{x},t)$. A complete statistical single-point description of the mixture is given by the joint probability density function $P\,(\boldsymbol{U},\Phi;\boldsymbol{x},t)$ where $\boldsymbol{U}=(U_1,U_2,U_3)$ and $\boldsymbol{\Phi}$ is a vector of length $N+2$ which includes the rest of the scalar quantities. Thus the required function is a surface in the $(N+3+2)$-dimensional Euclidean space.

The relationship between the instantaneous values of the variables stated above is given by the laws of conservation of total mass, momentum, and mass of the individual chemical species, and by the enthalpy. These can be given in the form [48], [50], [51]:

Accumulation + Change due to convective transport = Change due to molecular transport + Source

This leads to

$$\frac{\partial \varrho}{\partial t} + \frac{\partial}{\partial x_k}\,(\varrho U_k) = 0 \tag{131}$$

for the total mass,

$$\varrho \frac{DU_i}{Dt} = \frac{\partial \tau_{ik}}{\partial x_k} - \frac{\partial p}{\partial x_i} + \varrho g_i, \quad i = 1, 2, 3 \tag{132}$$

for the component of the momentum,

$$\varrho \frac{Dm_i}{Dt} = -\frac{\partial J_k^{m_i}}{\partial x_k} + \varrho S_i, \quad i = 1, \ldots, N-1 \tag{133}$$

for the mass of the individual chemical species, and finally

$$\varrho \frac{Dh}{Dt} = -\frac{\partial J_k^h}{\partial x_k} + \varrho S_h \qquad (134)$$

for the enthalpy.

In Equations (132)–(134) the operator D denotes the substantial derivative:

$$\frac{D}{Dt} = \frac{\partial}{\partial t} - U_k \frac{\partial}{\partial x_k} \qquad (135)$$

The Einstein summation convention is also used, for example:

$$\frac{\partial}{\partial x_k}(\varrho U_k) \equiv \frac{\partial}{x_1}(\varrho U_1) + \frac{\partial}{x_2}(\varrho U_2) + \frac{\partial}{x_3}(\varrho U_3)$$

The formulation of Equations (132)–(134) implies the continuity equation; ϱ is the density; m_i is the mass fraction of component i; τ_{ij} is the viscous stress tensor; $\varrho\, g_i$ is the specific mass force in the x_i direction (e.g., $\varrho\, g_1$, where g is the acceleration due to gravity); S_i is the reaction rate and S_h the rate of enthalpy conversion produced by pressure changes, viscosity effects etc.; J^{m} is the mass flux of the chemical species and J^{h} is the enthalpy flux, based on molecular transport mechanisms, like τ_{ij} they are functions of local physical quantities and their gradients.

Equations (131)–(134) can be closed by a thermal equation of state

$$\varrho = \varrho\,(m_i, h, p) \qquad (136)$$

The expressions for the sources S_i and S_h are also generally functions of the scalar quantities:

$$S_i = S_i\,(m_i, h, p) \qquad (137a)$$

$$S_h = S_h\left(m_i, h, p, \frac{\partial p}{\partial t}\right) \qquad (137b)$$

The assumptions and conditions included in this system of equations are discussed fully in [48], [50–52], see also Chapter 5. The system of Equations (131)–(137) is deterministic, this means that a determined solution exists for each set of determined boundary conditions. The solution of this system of equations for deterministic variables is illustrated with some examples in Chapter 5. Problems arise for statistical variables [52] which can only be solved for simple cases [53], [54] or by introducing far-reaching assumptions (see Chap. 5). The system of conservation equations serves as a starting point for the derivation of the transport equations for the probability density functions which are used to overcome the principle problems for solving this set of equations for statistical variables.

4.2.1. General form of the Transport Equations for Probability Density Functions

Equations (131)–(137) show that the state of a chemically reacting system at a position $\boldsymbol{x} = (x_1, x_2, x_3)$ and time t is determined by the velocity vector $\boldsymbol{U} = (U_1, U_2, U_3)$ in the sample space of velocity (Section 3.1.2.1 and Fig. 7) and the vector of scalar quantities $\boldsymbol{\Phi} = (m_1, \ldots, m_N, h)$ of length σ in the sample space of scalar quantities. $\boldsymbol{U}$ and $\boldsymbol{\Phi}$ are vectors of random variables, i.e., random vectors. This formulation holds for low Mach numbers.

The state of the reacting system represents a measured point in the three-dimensional velocity space of the vector $\boldsymbol{V}$ and likewise a measured point in the Euclidean σ-dimensional space of the vector $\boldsymbol{\Phi}$.

For a homogeneous mixture at rest with constant enthalpy, Equation (133) gives

$$\frac{\mathrm{d}m_i}{\mathrm{d}t} = S_i \qquad (138)$$

A chemical reaction can be represented in the composition space as a trajectory between two measured points. A system with two chemical species is shown in Figure 17. The reaction rate is the velocity in the composition space, and the reaction rates of the individual chemical species are given by the tangents to the trajectory in the direction of each species. Since the state of the system can be described by a vector with its end point in the plane of the required probability density function, changes of state due to chemical reactions are therefore trajectories in this plane. Chemical reactions that constitute the source terms in the conservation equations of the chemical species (Eq. 133) are contained in the transport equations for the probability density functions merely as transport in the composition space. Similarly, forces which are the source terms in the momentum conservation equations are expressed in the transport equations for probability density functions as transport in the velocity space. These are the main differences between the simple conservation equations and the transport equations for the probability density functions. The latter do not include forces or chemical reactions as sources.

Both $\boldsymbol{U}$ and $\boldsymbol{\Phi}$ are necessary to complete the information. The complete information at position $\boldsymbol{x}=(x_1,x_2,x_3)$ and time t is represented by a measured point in Euclidean $(\sigma+3)$-dimensional space. Consequently, the objective of the complete description is a $(\sigma+3)$-dimensional distribution function $F(\boldsymbol{U},\Phi;\boldsymbol{x},t)$ or the corresponding probability density function:

$$\frac{\partial^{\sigma+3}}{\partial U_1 \partial U_2 \ldots \partial \Phi_{n+1}}[F(\boldsymbol{U},\boldsymbol{\Phi};\boldsymbol{x},t)] = P(\boldsymbol{U},\boldsymbol{\Phi};\boldsymbol{x},t) \tag{139}$$

The characteristics described in Sections 4.1.1, 4.1.2, 4.1.3 apply to these functions. In the terminology of probabilities these functions predict the fraction of occurrences of discrete values $\boldsymbol{u}$ and φ for a large number of occurrences.

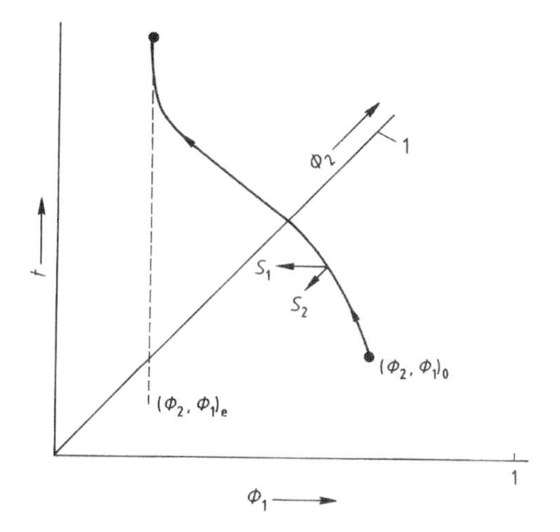

Figure 17. Chemical reactions for a system with two chemical species represented as transport in the sample space of the concentration (Φ_1, Φ_2) of the species
S_1 and S_2 are the partial derivatives of the trajectory in the Φ_1 and Φ_2 direction, the subscripts 0 and e denote initial and final states, respectively.

A transport equation for the joint velocity – composition probability density function can be derived with help of Equations (132) – (134) and the terminology described in Section 4.1. This method is one possibility for deriving the transport equation for probability density functions. This will be used in preference to other methods [18], [55–58] because of its simplicity.

Equations (132) and (133) are rewritten as Equations (140) and (141), respectively:

$$\frac{DU_i}{Dt} = A_i \tag{140}$$

$$\frac{D\Phi_\alpha}{Dt} = \Theta_\alpha \tag{141}$$

where

$$\varrho A_i(\boldsymbol{x},t) = \frac{\partial \tau_{ik}}{\partial x_k} - \frac{\partial p}{\partial x_i} + \varrho g_i \quad \text{and} \tag{142a}$$

$$\varrho \Theta_\alpha(\boldsymbol{x},t) = -\frac{\partial J_k^\alpha}{\partial x_k} + \varrho S_\alpha \tag{142b}$$

For any function $Q(\boldsymbol{U},\boldsymbol{\Phi})$ averaging according to Equation (109) gives

$$\begin{aligned}\langle \varrho \frac{DQ(\boldsymbol{U},\boldsymbol{\Phi})}{Dt}\rangle &= \frac{\partial}{\partial t}\int\int \varrho(\Phi)\, Q(\boldsymbol{U},\boldsymbol{\Phi})\, P \mathrm{d}\boldsymbol{U}\mathrm{d}\boldsymbol{\Phi} \\ &+ \frac{\partial}{\partial x_k}\int\int \varrho(\boldsymbol{\Phi})\, U_k Q(\boldsymbol{U},\boldsymbol{\Phi})\, P\mathrm{d}\boldsymbol{U}\mathrm{d}\boldsymbol{\Phi} \\ &= \int\int Q(\boldsymbol{U},\boldsymbol{\Phi})\left\{\varrho(\boldsymbol{\Phi})\frac{\partial P}{\partial t} + \varrho(\boldsymbol{\Phi})\right. \\ &\left. U_k \frac{\partial P}{\partial x_k}\right\}\mathrm{d}\boldsymbol{U}\mathrm{d}\boldsymbol{\Phi}\end{aligned} \tag{143}$$

In Equation (143) and in the following discussions, P will be used to represent $P(\boldsymbol{U},\Phi;\boldsymbol{x},t)$. In addition, $\mathrm{d}\boldsymbol{U}$ will be used to describe the vector $\mathrm{d}U_1, \mathrm{d}U_2, \mathrm{d}U_3$ and $\mathrm{d}\boldsymbol{\Phi}$ for $\mathrm{d}\Phi_1, \ldots, \mathrm{d}\Phi_{N+1}$. The integral sign in Equation (143) will also be interpreted in this way. The continuity equation (Eq. 131) is used to derive the expression $\langle \varrho \frac{DQ(\boldsymbol{U},\boldsymbol{\Phi})}{Dt}\rangle$ in Equation (143).

A second independent expression for $\langle \varrho \frac{DQ(\boldsymbol{U},\boldsymbol{\Phi})}{Dt}\rangle$ is obtained if $Q(\boldsymbol{U},\Phi)$ is differentiated with respect to $\boldsymbol{U}$ and $\boldsymbol{\Phi}$ (the Einstein summation convention is used for ω in the same way as for k in Eqs. 131 – 135).

$$\frac{DQ(\boldsymbol{U},\boldsymbol{\Phi})}{Dt} = \frac{\partial Q(\boldsymbol{U},\boldsymbol{\Phi})}{\partial U_k}\cdot\frac{DU_k}{Dt} + \frac{\partial Q(\boldsymbol{U},\boldsymbol{\Phi})}{\partial \Phi_\omega}\cdot\frac{D\Phi_\omega}{Dt} \tag{144}$$

If Equations (140) and (141) are used, the substantial derivatives can be substituted with the respective right-hand sides. Thus,

$$\langle \varrho \frac{DQ(\boldsymbol{U},\boldsymbol{\Phi})}{Dt}\rangle = \langle \varrho \frac{\partial Q(\boldsymbol{U},\boldsymbol{\Phi})}{\partial U_k} A_k\rangle + \langle \varrho \frac{\partial Q(\boldsymbol{U},\boldsymbol{\Phi})}{\partial \Phi_\omega}\Theta_\omega\rangle \tag{145}$$

The terms on the right-hand side of Equation (145) can be rewritten as conditional expected values by using Equation (130). For the first term of the right-hand side of Equation (145) this gives

$$\langle \varrho \frac{\partial Q(\boldsymbol{U},\boldsymbol{\Phi})}{\partial U_k} A_k \rangle = \int\int \langle \varrho \frac{\partial Q(\boldsymbol{U},\boldsymbol{\Phi})}{\partial U_k} A_k | \boldsymbol{u}, \varphi \rangle P \mathrm{d}\boldsymbol{U} \mathrm{d}\boldsymbol{\Phi} \quad (146)$$

Since the differential quotient $\partial Q(\boldsymbol{U},\boldsymbol{\Phi})/\partial U_k$ is a known function of $\boldsymbol{u}$ and φ for a given $\boldsymbol{u}$ and φ then from Equation (146)

$$\langle \varrho \frac{\partial Q(\boldsymbol{U},\boldsymbol{\Phi})}{\partial U_k} A_k \rangle = \int\int \varrho \frac{\partial Q(\boldsymbol{U},\boldsymbol{\Phi})}{\partial U_k} \langle A_k | \boldsymbol{u}, \varphi \rangle P \mathrm{d}\boldsymbol{U} \mathrm{d}\boldsymbol{\Phi} \quad (147)$$

If Equation (147) is partially integrated, then

$$\langle \varrho \frac{\partial Q(\boldsymbol{U},\boldsymbol{\Phi})}{\partial U_k} A_k \rangle = I_U - \int\int Q(\boldsymbol{U},\boldsymbol{\Phi}) \frac{\partial}{\partial U_k} \left(\varrho \langle A_k | \boldsymbol{u}, \varphi \rangle P \right) \mathrm{d}\boldsymbol{U} \mathrm{d}\boldsymbol{\Phi} \quad (148)$$

where

$$I_U = \int\int \frac{\partial}{\partial U_k} \left[\varrho(\boldsymbol{\Phi}) Q(\boldsymbol{U},\boldsymbol{\Phi}) \langle A_k | \boldsymbol{u}, \varphi \rangle P \right] \mathrm{d}\boldsymbol{U} \mathrm{d}\boldsymbol{\Phi}$$

An analogous procedure for the second term on the right-hand side of Equation (145) gives

$$\langle \varrho \frac{\partial Q(\boldsymbol{U},\boldsymbol{\Phi})}{\partial \Phi_\omega} \Theta_\omega \rangle = I_{\boldsymbol{\Phi}} - \int\int Q(\boldsymbol{U},\boldsymbol{\Phi}) \frac{\partial}{\partial \Phi_\omega} \left(\varrho \langle \Theta_\omega | \boldsymbol{u}, \varphi \rangle P \right) \mathrm{d}\boldsymbol{U} \mathrm{d}\boldsymbol{\Phi} \quad (149)$$

where

$$I_{\boldsymbol{\Phi}} = \int\int \frac{\partial}{\partial \Phi_\omega} \left[\varrho(\boldsymbol{\Phi}) Q(\boldsymbol{U},\boldsymbol{\Phi}) \langle \Theta_\omega | \boldsymbol{u}, \varphi \rangle P \right] \mathrm{d}\boldsymbol{U} \mathrm{d}\boldsymbol{\Phi}$$

It can be shown that the integrals I_U and I_Φ disappear for a wide class of functions $Q(\boldsymbol{U},\Phi)$. The general condition for this is that the functions are bounded and continuous [19]. With this a second independent expression for $\langle \varrho \frac{DQ(\boldsymbol{U},\boldsymbol{\Phi})}{Dt} \rangle$ is obtained from Equation (145):

$$\langle \varrho \frac{DQ(\boldsymbol{U},\boldsymbol{\Phi})}{Dt} \rangle = -\int\int Q(\boldsymbol{U},\boldsymbol{\Phi}) \frac{\partial}{\partial U_k} \left(\varrho \langle A_k | \boldsymbol{u},\boldsymbol{\varphi} \rangle P \right) \mathrm{d}\boldsymbol{U} \mathrm{d}\boldsymbol{\Phi} - \int\int Q(\boldsymbol{U},\boldsymbol{\Phi}) \frac{\partial}{\partial \Phi_\omega} \left(\varrho \langle \Theta_\omega | \boldsymbol{u},\boldsymbol{\varphi} \rangle P \right) \mathrm{d}\boldsymbol{U} \mathrm{d}\boldsymbol{\Phi} \quad (150)$$

Subtraction of Equation (150) from Equation (143) gives

$$\int\int Q(\boldsymbol{U},\boldsymbol{\Phi}) \left\{ \varrho(\boldsymbol{\Phi}) \frac{\partial P}{\partial t} + \varrho(\boldsymbol{\Phi}) U_k \frac{\partial P}{\partial x_k} + \frac{\partial}{\partial U_k} \left[\varrho(\boldsymbol{\Phi}) \langle A_k | \boldsymbol{u}, \varphi \rangle \mathrm{P} \right] + \frac{\partial}{\partial \Phi_\omega} \left[\varrho(\boldsymbol{\Phi}) \langle \Theta_\omega | \boldsymbol{u}, \varphi \rangle P \right] \right\} \mathrm{d}\boldsymbol{U} \mathrm{d}\boldsymbol{\Phi} = 0 \quad (151)$$

All the terms in the curved brackets are independent of $Q(\boldsymbol{U},\Phi)$. If $Q(\boldsymbol{U},\Phi)$ is the type of function where the integrals I_u and I_Φ disappear, then a sufficient condition to fulfil Equation (151) is that the sums in the curved brackets are zero. With this the transport equation for the joint probability density function $P(\boldsymbol{U},\Phi;\boldsymbol{x},t)$ is

$$\varrho(\boldsymbol{\Phi}) \frac{\partial P}{\partial t} + \varrho(\boldsymbol{\Phi}) U_k \frac{\partial P}{\partial x_k} = -\frac{\partial}{\partial U_k} \left[\varrho(\boldsymbol{\Phi}) \langle A_k | \boldsymbol{u},\boldsymbol{\varphi} \rangle P \right] - \frac{\partial}{\partial \Phi_\omega} \cdot \left[\varrho(\boldsymbol{\Phi}) \langle \Theta_\omega | \boldsymbol{u},\boldsymbol{\varphi} \rangle P \right] \quad (152)$$

For further discussion of this transport equation, A_i and Θ_α are replaced by their respective definition functions:

$$\varrho(\boldsymbol{\Phi}) \langle \Theta_\alpha | \boldsymbol{u},\boldsymbol{\varphi} \rangle = -\langle \frac{\partial J_k^\alpha}{\partial x_k} + \varrho(\boldsymbol{\Phi}) S_\alpha(\boldsymbol{\Phi}) | \boldsymbol{u},\ vec\varphi \rangle = -\langle \frac{\partial J_k^\alpha}{\partial x_k} | \boldsymbol{u},\boldsymbol{\varphi} \rangle + \varrho(\boldsymbol{\Phi}) S_\alpha(\boldsymbol{\Phi}) \quad (153)$$

Using the decomposition $p^{\mathrm{f}} \equiv p - \langle p \rangle$ gives

$$\varrho(\boldsymbol{\Phi}) \langle A_i | \boldsymbol{u},\boldsymbol{\varphi} \rangle = \langle \frac{\partial \tau_{ik}}{\partial x_k} - \frac{\partial \langle p \rangle}{\partial x_i} - \frac{\partial p^{\mathrm{f}}}{\partial x_i} + \varrho g_i | \boldsymbol{u},\boldsymbol{\varphi} \rangle = \langle \frac{\partial \tau_{ik}}{\partial x_k} | \boldsymbol{u},\boldsymbol{\varphi} \rangle - \frac{\partial \langle p \rangle}{\partial x_i} - \langle \frac{\partial p^{\mathrm{f}}}{\partial x_i} | \boldsymbol{u},\boldsymbol{\varphi} \rangle + \langle \varrho g_i | \boldsymbol{u},\boldsymbol{\varphi} \rangle \quad (154)$$

Substitution of these expressions into Equation (152) gives

$$\varrho(\boldsymbol{\Phi}) \frac{\partial P}{\partial t} + \varrho(\boldsymbol{\Phi}) U_k \frac{\partial P}{\partial x_k} + \left[\varrho(\boldsymbol{\Phi}) g_k - \frac{\partial \langle p \rangle}{\partial x_k} \right] \frac{\partial P}{\partial U_k} + \frac{\partial}{\partial \Phi_\omega} \left[\varrho(\boldsymbol{\Phi}) S_\omega(\boldsymbol{\Phi}) P \right] = -\frac{\partial}{\partial U_k} \left[\langle \frac{\partial \tau_{lk}}{\partial x_l} + \frac{\partial p^{\mathrm{f}}}{\partial x_l} | \boldsymbol{u},\boldsymbol{\varphi} \rangle P \right] + \frac{\partial}{\partial \Phi_\omega} \left[\langle \frac{\partial J_k^\omega}{\partial x_k} | \boldsymbol{u},\boldsymbol{\varphi} \rangle P \right] \quad (155)$$

This is the general form of the transport equation for the probability density functions in the combined velocity – composition space (in the first term of the right-hand side k and l must be summed according to the Einstein summation convention). Equation (155) can be solved if the conditional expected values on the right-hand side are known. All physical processes represented by the terms on the left-hand side are described exactly in the context of the method of description given here. These processes are transport in physical space (first and second terms), transport in the velocity space due to gravitational forces and average pressure gradient (third term), and transport in the composition space by chemical reactions (fourth term). All of these processes can be described without any model assumptions.

The expressions on the right-hand side of Equation (155) describe transport in the velocity space due to momentum transfer at the molecular level as well as transport in the composition space due to molecular diffusion. For a turbulent flow the decay of turbulent fluctuations is caused by these mechanisms. Viscous forces are responsible for the dissipation of turbulent (statistical) fluctuations in velocity. Molecular diffusion is responsible for the dissipation of turbulent fluctuations in scalar quantities.

Before Equation (155) can be solved, these terms must be defined or approximated using models. The following sections will describe with some examples how this can be done.

4.2.2. Limitations of Single-Point Probability Density Functions

Solution of Equation (155) gives the joint probability density function $P(\boldsymbol{U}, \Phi; \boldsymbol{x}, t)$ of the random vectors $\boldsymbol{U}$ and $\boldsymbol{\Phi}$ at position $\boldsymbol{x} = (x_1, x_2, x_3)$ and time t. This result contains the complete statistical information at a defined position for a given time. Even this information includes only a partial view of the statistical process being considered. The functionals $P(\boldsymbol{U}, \Phi)$ described in Section 2.2 or multipoint probability density functions, such as the two-point probability density function $P(\boldsymbol{U}_1, \Phi_1; \boldsymbol{x}_1, t_1, \boldsymbol{U}_2, \Phi_2; \boldsymbol{x}_2, t_2)$ which can be derived from them, provide more comprehensive information. The meaning of these multipoint probability density functions can be illustrated by the example shown in Figure 14. The structure of the vortices in the turbulent flow of the jet stirred reactor is dependent on position and time. The "lifetime" and "tracks" of the vortices being transported in the flow are crucial for the chemical reaction. They give an indication of the degree of mixing and the "segregation" of the flow. (Segregation denotes the state of a fluid in which individual fluid parcels maintain their identities when passing through the reactor and do not exchange properties with the surrounding fluid [24], [46].) Second-order moments (the correlation functions) are used to describe mixing and segregation:

$$R_{\boldsymbol{\Phi},\boldsymbol{\Psi}} = \left\langle \left(\boldsymbol{\Phi}(\boldsymbol{x}_1,t_1) - \left\langle \boldsymbol{\Phi}(\boldsymbol{x}_1,t_1)\right\rangle\right)\left(\boldsymbol{\Psi}(\boldsymbol{x}_2,t_2) - \left\langle \boldsymbol{\Psi}(\boldsymbol{x}_2,t_2)\right\rangle\right)\right\rangle$$
$$= \int\int \left(\boldsymbol{\Phi}(\boldsymbol{x}_1,t_1) - \left\langle \boldsymbol{\Phi}(\boldsymbol{x}_1,t_1)\right\rangle\right)\left(\boldsymbol{\Psi}(\boldsymbol{x}_2,t_2) - \left\langle \boldsymbol{\Psi}(\boldsymbol{x}_2,t_2)\right\rangle\right) \cdot P(\boldsymbol{U}_1,\boldsymbol{\Phi}_1;\boldsymbol{x}_1,t_1,\boldsymbol{U}_2,\boldsymbol{\Phi}_2;\boldsymbol{x}_2,t_2)\,\mathrm{d}\boldsymbol{\Phi}\mathrm{d}\boldsymbol{\Psi}$$

For $\boldsymbol{\Phi} = \boldsymbol{\Psi}$ the autocorrelation functions are obtained which provide data about the vortex structure for a physical property of the fluid e.g., the velocity. Time and length scales for the turbulent exchange of the considered property can be derived from the autocorrelation functions [49], [59–61]; these play an important role in the modeling of turbulent processes.

The dimensionality of the transport equations for multipoint probability density functions is proportional to the number of points (Eq. 155). The problem soon becomes unmanageable, even with a small number of scalar quantities and points.

Single-point probability density functions can be derived from probability density functionals or multipoint probability density functions:

$$P(\boldsymbol{\Phi}_1;\boldsymbol{x},t) = \int_{-\infty}^{+\infty} P(\boldsymbol{\Phi}_1;\boldsymbol{x}_1,t_1,\boldsymbol{\Phi}_2;\boldsymbol{x}_2,t_2)\,\mathrm{d}\boldsymbol{\Phi}_2 \qquad (157)$$

Single-point probability density functions only give limited information about the field of random variables. They give complete statistical information at a given site and time but not joint information about two sites and/or two times.

For the example shown in Figure 14 the single-point functions therefore contain no information about the length or time scales of the statistical (turbulent) fluctuations.

This deficiency can be illustrated by a simple example. Suppose that a turbulent fluctuation of a physical variable is in the form of a sine wave of frequency ν and wavelength λ. The statistical character is due to a statistical phase shift ζ which is evenly distributed between 0 and 1. The amplitude of the physical variable is given by

$$\Phi\left(\boldsymbol{x},t\right)=\frac{1}{2}+\frac{1}{2}\sin\left(2\pi\nu t+x_{1}\lambda+\zeta\right) \tag{158}$$

If the amplitude is measured sufficiently often for a given $\boldsymbol{x}, t$ then the probability density function can be obtained from the density of measured points in the sample space (Eq. 115 a). The density of the measured points is given by

$$\left(\frac{\Delta\Phi}{\Delta\zeta}\right)_{\boldsymbol{x},t}\approx\left(\frac{\partial\Phi}{\partial\zeta}\right)_{\boldsymbol{x},t}=\frac{1}{\pi}\left[1-\left(2\Phi-1\right)^{2}\right]^{-1/2}$$

The single-point probability density function is thus independent of the frequency ν and the wavelength λ. In addition, information about values such as $\langle(\partial\Phi/\partial x)^2\rangle$ cannot be derived from the single-point probability density function. This value is, however, dependent on the wavelength, $\langle(\partial\Phi/\partial x)^2\rangle = 1/2(\pi^2/\lambda^2)$, compare Equation (158). The measured single-point probability density function gives no information about the spatial and time propagation of the disturbances.

In spite of these limitations, single-point probability density functions are improvements over the handling of statistical variables with the moments of probability density functions.

4.3. Examples of Calculating Probability Density Functions

For further discussions Equation (155) is rewritten as

$$\begin{aligned}&\varrho\left(\Phi\right)\frac{\partial P\left(\boldsymbol{U},\boldsymbol{\Phi};\boldsymbol{x},t\right)}{\partial t}+\varrho\left(\Phi\right)U_{k}\frac{\partial P\left(\boldsymbol{U},\boldsymbol{\Phi};\boldsymbol{x},t\right)}{\partial x_{k}}\\&+\frac{\partial}{\partial U_{k}}\left[\varrho\left(\boldsymbol{\Phi}\right)\left\langle A_{k}|\boldsymbol{u},\varphi\right\rangle P\left(\boldsymbol{U},\boldsymbol{\Phi};\boldsymbol{x},t\right)\right]\\&+\frac{\partial}{\partial\Phi_{\omega}}\left[\varrho\left(\boldsymbol{\Phi}\right)\left\langle\Theta_{\omega}|\boldsymbol{u},\varphi\right\rangle P\left(\boldsymbol{U},\boldsymbol{\Phi};\boldsymbol{x},t\right)\right]=0\end{aligned} \tag{159}$$

where A_i and Θ_α are given by Equations (142 a, b). Equation (159) is deterministic, i.e., determined solutions exist for given initial and boundary conditions. The conditional expected values $\langle A_i|\,\boldsymbol{u},\,\varphi\rangle$ and $\langle\Theta_\alpha|\,\boldsymbol{u},\,\varphi\rangle$ are not statistical values. In the general case they are unknown functions of the variables $\boldsymbol{u}$ and φ. For the general case (i.e., for fields of statistical variables) model assumptions must be introduced so that Equation (155) or (159) can be solved. For deterministic systems Equations (155) and (159) can be treated without further assumptions because $\langle A_i|\,\boldsymbol{u},\,\varphi\rangle = A_i\,(\boldsymbol{u},\,\varphi)$ and $\langle\Theta_\alpha|\,\boldsymbol{u},\,\varphi\rangle = \Theta_\alpha\,(\boldsymbol{u},\,\varphi)$, respectively.

4.3.1. Solutions for Deterministic Systems

4.3.1.1. Age (Residence Time) Distributions in Chemical Reactors

The individual types of chemical reactors are normally treated as deterministic systems although under industrial conditions flow is not laminar. This simplification is justified if the turbulent time scales for the reactions occurring in the reactor lie several orders of magnitude above that for the turbulent mass or momentum exchange. The reactor contents have then reached the state of molecular mixing, without noticeable chemical conversion having taken place. Distribution functions or probability density functions may be given for the residence time, for example, and can thus indicate the probability for the residence time (the age α) of a volume element in an interval dα around α; residence time distribution functions are easily derived from Equation (159).

Ideally Mixed Stirred-Tank Reactor. In the mathematical sense an ideally mixed reactor is a one-dimensional control volume with homogeneous internal properties. The upstream characteristics are transported into the control volume while the characteristics at the downstream boundary surface relate to the internal properties (Fig. 18). Viscosity, pressure, and gravitational forces and other transport processes (with the exception of convection) are ignored.

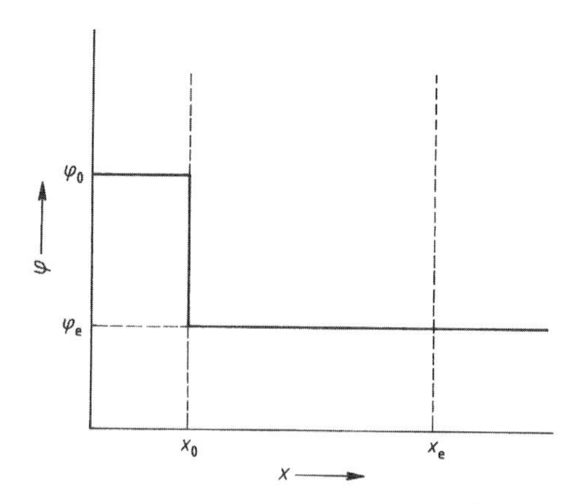

Figure 18. Mathematical definition of an ideally mixed reactor
The subscripts 0 and e denote inlet and outlet, respectively.

For expediency Equation (159) is integrated with respect to the volume:

$$\int_V \varrho(\boldsymbol{\Phi}) \frac{\partial P}{\partial t} \mathrm{d}V + \int_V \varrho(\boldsymbol{\Phi})\, u \frac{\partial P}{\partial x} \mathrm{d}V + \int_V \frac{\partial}{\partial \Phi_\omega} [\varrho(\boldsymbol{\Phi})\, S_\omega(u, \boldsymbol{\Phi})\, P]\, \mathrm{d}V = 0 \tag{160}$$

Using the Gauss – Ostrogradski integrals, Equation (160) can be rewritten as

$$\frac{\partial}{\partial t} \int_V \varrho(\boldsymbol{\Phi})\, P \mathrm{d}V + \oint_F \varrho(\boldsymbol{\Phi})\, u P \mathrm{d}F + \int_V \frac{\partial}{\partial \Phi_\omega} [\varrho(\boldsymbol{\Phi})\, S_\omega(u, \boldsymbol{\Phi})\, P]\, \mathrm{d}V = 0 \tag{161}$$

where integration has to be carried out for the surfaces perpendicular to the direction of the flow.

If the age (residence time) distribution is considered, the variables u, Φ in $P(u, \Phi)$ and $S(u, \Phi)$ must be substituted by the age α. Since intervals in the age space are time intervals, the velocity in the age space is equal to $\partial\alpha/\partial t = S(\alpha) = 1$. With this result and for constant density the following expression can be derived from Equation (161):

$$\frac{\partial}{\partial t} P(\alpha)\, \mathrm{d}V + u P(\alpha)\, \mathrm{d}z \mathrm{d}y - u P_0(\alpha)\, \mathrm{d}z \mathrm{d}y + \frac{\partial}{\partial \alpha} P(\alpha)\, \mathrm{d}V = 0 \tag{162}$$

In the steady state, Equation (162) is easy to solve since the first term on the left-hand side is zero:

$$\frac{\partial}{\partial \alpha} [P(\alpha)] = \frac{1}{\tau} [P_0(\alpha) - P(\alpha)] \tag{163a}$$

where $\tau = u\, \mathrm{d}z\, \mathrm{d}y/\mathrm{d}V$ is the mean hydrodynamic residence time. With the boundary condition $P_0(\alpha) = \delta(\alpha - 0)$ and the condition $\int_0^{+\infty} P(\alpha)\, \mathrm{d}\alpha = 1$ (cf. Eq. 104) then

$$P(\alpha) = \frac{1}{\tau} \exp\left(-\frac{\alpha}{\tau}\right) \tag{164}$$

$\delta(\alpha - 0)$ denotes the impulse function or Dirac delta function (Section 4.1.1).

$$\int_0^{+\infty} \alpha \delta(\alpha - b)\, g(\alpha)\, \mathrm{d}\alpha = a g(b); \quad b \geq 0$$

is used to integrate Equation (163 a). Equation (164) is the familiar residence time distribution function for ideally mixed reactors (denoted as $w(t)$ in Section 3.3.2.2). If it is transformed to normalized age ($\theta = \alpha/\tau$) then

$$P(\theta) \equiv w(\theta) = P(\alpha) \frac{\partial \alpha}{\partial \theta} = \frac{1}{\tau} \exp\left(-\frac{\alpha}{\tau}\right) \tau = \exp(-\theta) \tag{165}$$

In Equation (165) the distinction between age α and time t is no longer maintained. The integral over the residence time distribution

$$W(t) = \int_0^{+\infty} w(t)\, \mathrm{d}t \quad \text{or}$$

$$W(\theta) = \int_0^{+\infty} w(\theta)\, \mathrm{d}\theta$$

is denoted as distribution function. Figure 19 shows the residence time distribution and the distribution function for an ideal stirred-tank reactor.

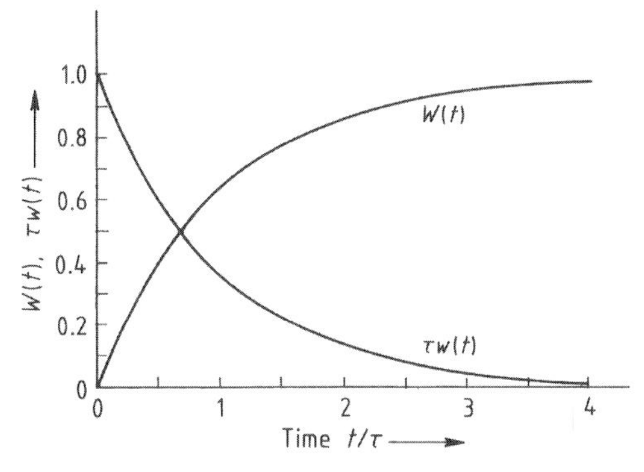

Figure 19. Residence time distribution w (t) and distribution function W (t) for the ideal stirred-tank reactor

Alternative boundary conditions lead to alternative solutions of Equation (163 a). For example, if n ideally mixed reactors are linked in a cascade, then $P_{n-1}(\alpha) = P_{0n}(\alpha)$ etc. With $P_{01}(\alpha) = \delta(\alpha - 0)$ and $\tau_1 = \tau_2 = \cdots \tau_n = \tau/n$, the solution of Equation (163 a) then becomes

$$w_n(t) = \frac{1}{(n-1)!} n^n \frac{t^{(n-1)}}{\tau^n} \exp\left(-n\frac{t}{\tau}\right) \quad (166)$$

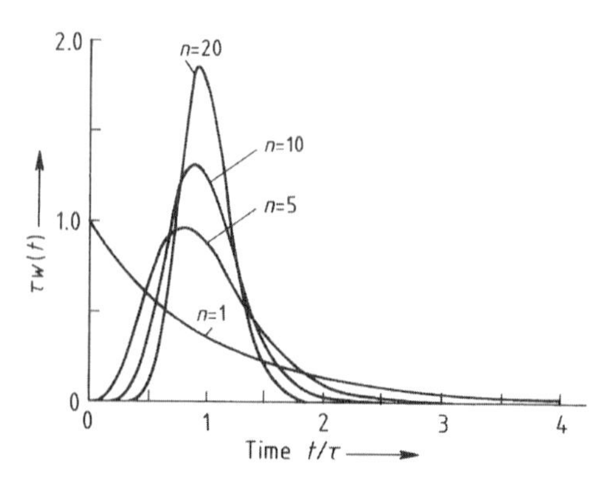

Figure 20. Residence time distribution for a cascade of n ideal stirred-tank reactors

Figure 20 shows residence time distribution functions and distribution functions for a cascade of n ideally mixed reactors for various values of n. It is easy to prove from Equation (166) that simple relationships exist between the deterministic moments

$$m_1 = \int_0^{+\infty} t w(t)\, dt = \bar{t} \quad \text{and}$$

$$m_2' = \int_0^{+\infty} (t - m_1)^2 w(t)\, dt$$

(see Section 3.3.2.2) and the number of elements in the cascade. Thus

$$m_1 = n\tau_i = \tau \quad \text{and} \quad m_2' = n\tau_i^2 = \frac{\tau^2}{n} \quad (167a,b)$$

Given that a reactor can be represented by a cascade of ideal stirred-tank reactors, the number of elements required to represent the reactor can easily be determined with Equation (167) from the first two deterministic moments of the measured residence time distribution, namely from the mean residence time and the variance.

For the nonsteady case, Equation (162) becomes:

$$\frac{\partial}{\partial t}[P(\alpha;t)] + \frac{\partial}{\partial \alpha}[P(\alpha;t)] = \frac{1}{\tau}[P_0(\alpha;t) - P(\alpha;t)] \quad (163b)$$

The general solution for Equation (163 b) can easily be found by using Laplace or Fourier transforms and applying the concept of generalized functions [62], [63]:

$$P(\alpha;t) = \exp\left(-\frac{t}{\tau}\right) P_0(\alpha - t) + \frac{1}{\tau}\exp\left(-\frac{\alpha}{\tau}\right)[H(\alpha) - H(\alpha - t)] \quad (168)$$

The solution holds for the initial conditions $P(\alpha;t)_{t=0} = P(\alpha)$ and the same boundary conditions as above; $H(\alpha)$ denotes the Heaviside or step function (cf. Eq. 116). It can easily be shown that Equation (168) becomes Equation (165) for $t \to \infty$.

Ideally Nonmixed Reactor (Ideal Tubular Reactor). An ideally nonmixed reactor is a onedimensional flow system with inhomogeneous internal properties. Analogously to the ideally mixed reactor, in this one-dimensional model viscosity, pressure, gravitational forces, and other transport processes (excluding convection) are neglected. The changes in the internal properties occur as a result of physical or chemical processes.

The age (residence time) distribution of this model is obtained from Equation (159):

$$\frac{\partial}{\partial t}[P(\alpha;x,t)] + u\frac{\partial}{\partial x}[P(\alpha;x,t)] + \frac{\partial}{\partial \alpha}[P(\alpha;x,t)] = 0 \quad (169a)$$

In the steady state, Equation (169 a) becomes

$$u\frac{\partial}{\partial x}[P(\alpha;x)] + \frac{\partial}{\partial \alpha}[P(\alpha;x)] = 0 \quad (169b)$$

The partial differential Equation (169 b) can be solved in a number of ways. The calculation domain (the tubular reactor) can be divided into a number of finite control volumes, for each of which Equation (169 b) holds. The individual control volumes have the same characteristics as the ideally mixed reactor. The same operations as discussed above can be carried out for each of the finite control volumes and an equation of type (163 b) is obtained for each control volume. Physically this means that an ideally nonmixed reactor is represented by a cascade with an infinite number of elements. Numerically this means that the partial differential Equation (169 b) is solved with an upwind difference scheme. The solution for the boundary condition $P_0(\alpha) = \delta(\alpha - 0)$ is Equation (166) with $n \to \infty$, i.e., an impulse function at $t/n\,\tau_i = 1$.

Equation (169 b) can also be solved analytically, the general solution is

$$P(\alpha;x) = \Omega\left(\alpha - \frac{1}{u}x\right) \tag{170}$$

where Ω is any differentiable function [64]. Equation (170) clarifies the behavior of ideally nonmixed reactors. A fluid flowing into the reactor with a given age distribution retains the form of the age distribution. The age of the individual fluid elements increases uniformly by the time necessary to pass the distance x, $t = x/u$. For example, at the reactor inlet (i.e., $x = 0$)

$$\Omega(\alpha - 0) = \delta(\alpha - 0)$$

at the reactor outlet (i.e., $x = L$)

$$P(\alpha;x) = \delta\left(\alpha - \frac{L}{u}\right) = \delta(\alpha - \tau)$$

This is identical to Equation (166) for $n \to \infty$.

If an ideally mixed reactor with a mean residence time τ_1 and an ideally nonmixed reactor with a mean residence time τ_2 are connected together, then $\Omega(\alpha - 0) = H(\alpha - 0)\exp(-\alpha/\tau_1)$, and from Equation (170)

$$P(\alpha) = H(\alpha - \tau_2) \cdot \frac{1}{\tau_1}\exp\left[-\frac{1}{\tau_1}(\alpha - \tau_2)\right]$$

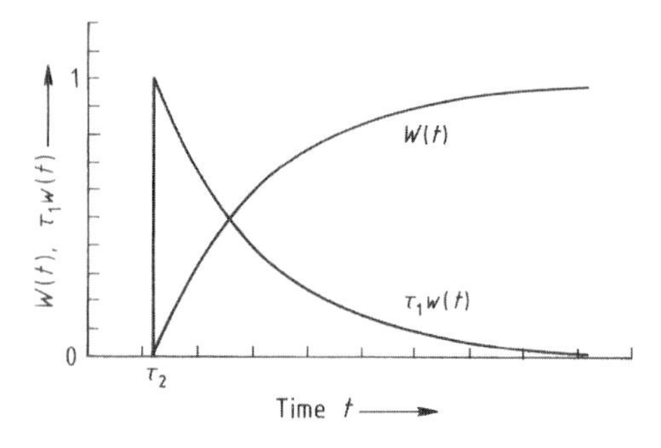

Figure 21. Residence time distribution w (t) and distribution function W (t) for an ideal stirred-tank reactor connected in series with an ideal tubular reactor

It is seen that the age distribution of the ideally mixed reactor is shifted by the mean residence time τ_2 of the nonmixed reactor. Figure 21 shows the residence time distribution and the distribution function for this case.

Reactors with Finite Mixing. Reactors with finite mixing can be thought of as the transport mechanisms discussed in page 242 combined with another transport mechanism. Usually this is treated in the form of a "diffusion" mechanism in which the relevant diffusion coefficient has the character of a model parameter (dispersion model).

Diffusion is expressed in the transport equation for the probability density function as an additional transport term in the sample space of scalar values (see Eq. 155, last term on the right-hand side and Eq. 159). The basic treatment of these terms is discussed in [19], [65–67], see also Section 4.3.2.1. Here, a heuristic method will be used to find a transport equation for the age distribution in reactors where diffusion is superimposed on convective transport.

The problem consists of deriving a term for the additional transport in the age space for a superimposed diffusion process. In chemical reactors with finite mixing, fluid parcels are convected by turbulent motion, i.e., this process consists of the diffusion of fluid elements which can be handled in the same way as molecular diffusion.

The fluid parcels experience a displacement due to the "diffusion" in addition to the displacement resulting from convective transport with velocity u. Analogous to molecular motion, the probability density function for the displacement by diffusion is a normal distribution [68]:

$$P(x)_{\text{diff}} = \frac{1}{2\sqrt{\pi Dt}}\exp\left(-\frac{x^2}{4Dt}\right) \tag{171}$$

where D is the diffusion coefficient. The probability density function for the displacement x of the fluid elements due to diffusion can be rewritten as the probability density function for the age with the help of the transformation rule given by Equation (106) and $\alpha = x/u$:

$$P(\alpha)_{\text{diff}} = P(x)_{\text{diff}} \cdot \frac{\mathrm{d}x}{\mathrm{d}\alpha} = \frac{u}{2\sqrt{\pi D\alpha}} \cdot \exp\left(-\frac{x^2}{4D\alpha}\right) \tag{172}$$

Equation (172) represents the age distribution for a diffusion process that is analogous to molecular diffusion. The term required for the transport equation is the differential of this distribution in the age space, i.e., $\frac{\partial}{\partial\alpha}\left[P\left(\alpha\right)_{\text{diff}}\right]$. Equation (169 b) is now supplemented by the term $\frac{\partial}{\partial\alpha}\left[P\left(\alpha\right)_{\text{diff}}\right]$ so that the transport equation for the age distribution in the steady-state case takes the form

$$\frac{\partial}{\partial\alpha}\left[P\left(\alpha;x\right)\right]+u\frac{\partial}{\partial x}\left[P\left(\alpha;x\right)\right]$$

$$+D\frac{\partial^2}{\partial x^2}\left[P\left(\alpha\right)\right]=0 \qquad (173)$$

Equation (173) can be solved analytically for simple boundary conditions. For the boundary condition

$$\frac{\partial}{\partial x}\left[P\left(\alpha;x\right)\right]=0 \quad \text{for} \quad x=0 \quad \text{and} \quad x=L$$

this gives

$$P\left(\alpha\right)=\frac{1}{2}\sqrt{\frac{Pe}{\pi\tau\alpha}}\exp\left(-\frac{\left(1-\alpha/\tau\right)^2 Pe}{4\alpha/\tau}\right) \qquad (174)$$

where *Pe* is the axial Péclet number, $Pe = u\,L/D$ (Section 3.3.2.2), it is often called the Bodenstein number, *Bo*.

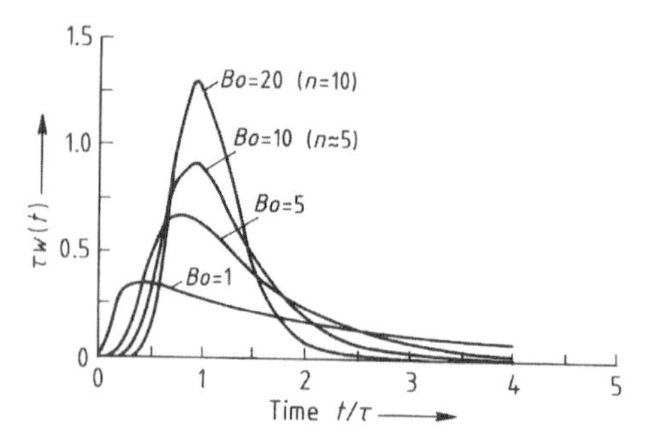

Figure 22. Residence time distribution τ *w* (t) for a reactor with axial mixing
Bo = Bodenstein number; *n* = number of elements of a cascade of ideally mixed reactors to represent the reactor with the corresponding Bodenstein number

Figure 22 shows the residence time distributions (Eq. 174) for different Bodenstein numbers. For large Bodenstein numbers the figure resembles the distributions for the cascade of ideally mixed reactors (Fig. 20). This similarity is inevitable, because the cascade of ideally mixed reactors represents the numerical solution of the transport Equation (169 b) as $n \to \infty$ and Equation (173) will be identical to this for $D \to 0$. The deterministic moments m_1 and m_2' can be calculated from Equation (174):

$$m_1=\left(1+\frac{2}{Bo}\right)\tau \quad \text{and}$$

$$m_2'=\left(\frac{2}{Bo}+\frac{8}{Bo^2}\right)\tau^2 \qquad (175\text{a,b})$$

For $Bo \to \infty$, the first moment corresponds to the mean hydrodynamic residence time (Eq. 167 a). If Equation (175 b) is compared with Equation (167 b) on the basis of the similarity of the solutions of the age distribution for large Bodenstein numbers, then the relationship

$$\frac{1}{n}=\frac{2}{Bo}+\frac{8}{Bo^2}$$

can be derived. For large Bodenstein numbers (i.e., little axial dispersion), $n \approx Bo/2$.

This discussion shows that the definition or experimental determination of model parameters can be relatively simple even for complex models. In the above case the formal diffusion coefficient, the axial mixing coefficient, or the Bodenstein number can be simply determined from the second deterministic moment of the age (residence time) distribution. Furthermore, complex processes may be built up from a number of simple elements. In the case described above, a reactor with finite axial mixing is constructed from a cascade of ideally mixed reactors; a simple relationship exists between the number of elements and the model parameters of this complex one-dimensional process. Further examples for the representation of complex processes by simple elements are found in [4], [11], [12], [24], [46], and text books on chemical reaction engineering, see also Chapter 5.

4.3.1.2. Size Distribution in Continuously Operating Crystallizers

The model used for the representation of a continuously operating crystallizer is an ideally mixed reactor. The supersaturated solution is fed into the crystallizer and the substrate grows on the existing crystals. Using the operations discussed in page 242 this gives the transport equation

$$\frac{\partial}{\partial \Phi}\left[S\left(\Phi\right)P\left(\Phi\right)\right]=\frac{1}{\tau}\left[P_0\left(\Phi\right)-P\left(\Phi\right)\right] \quad (176)$$

The property under consideration is the crystal size L. The rate of crystal growth can be described empirically by $S(L)=S(L,c)$, i.e., it is a function of the crystal size and the concentration of the substrate c. Often $S(L)=S_0(c)L^{\mathrm{b}}$ is used so that Equation (176) becomes

$$\frac{\partial}{\partial L}\left[S_0\left(c\right)L^{b}P\left(L\right)\right]=\frac{1}{\tau}\left[P_0\left(L\right)-P\left(L\right)\right] \quad (177)$$

If the feed does not contain solids, $P_0(L)=\delta(L-0)$. Assuming that the substrate concentration is approximately constant, the solution of Equation (177) is

$$P\left(L\right)=C_1L^{-b}\exp\left[-\frac{L^{1-b}}{S_0\left(c\right)\tau\left(1-b\right)}\right] \quad (178)$$

where C_1 is an integration constant. For $b=0$ (McCabe ΔL law) Equation (178) becomes

$$P\left(L\right)=P\left(L_0\right)\exp\left(-\frac{L}{S_0\left(c\right)\tau}\right) \quad (179)$$

where C_1 is substituted by the initial condition $P(L_0)$, i.e., the number density of the nuclei in the crystallizer. Equation (179) represents the number density distribution of crystal sizes for this simple case ($b=0$).

4.3.2. Solutions for Statistical Systems

The discussion in Section 4.3.1 was confined to deterministic systems for which either the physical quantities are not subject to turbulent (statistical) fluctuations or for which the time scales of the individual physicochemical processes are of vastly different orders of magnitude. In the ideally mixed reactor the time scale for the chemical reactions is so large in comparison with that of the turbulent mass exchange that the reactor contents can be considered as being mixed at the molecular level without having reacted. The point of this discussion was to bring the transport equations for the probability density functions (Eqs. 155, 159) into a simple form which is closed without any model assumptions and is also easy to solve analytically in simple cases.

These simplifications do not apply to statistical systems, such as the chemical reactions in a turbulent single-phase flow shown in Figure 14. The transport equations for the probability density function in the combined velocity – composition space have advantages for describing such processes because, in contrast to the simple equations for turbulent reacting flows (see Section 5.1.2), the most important processes (convection and chemical reaction) can be handled without model assumptions. Assumptions must be introduced for the dissipation of turbulent velocity fluctuations or fluctuations of scalar quantities by molecular mixing. These are manifested in the transport equations for probability density functions as transport in the velocity or composition space.

A method of solving the transport equations must also be found. Analytical solutions can only be obtained for very simple flows, simple models, or deterministic systems under simple conditions (Section 4.3.1). In general, Equation (155) must be solved numerically with the relevant terms on the right-hand side being modeled. Methods using finite differences (see Section 4.3.1.1) can be ruled out for this since $P(\boldsymbol{U},\Phi;\boldsymbol{x},t)$ is a function of $(\sigma+6)$ independent variables and time. If, for example, a case was investigated which required 20 finite volumes in the direction of each of these variables, then the number of grid points for a statistical two-dimensional flow with only two chemical species under isothermal conditions would be $20^7\approx10^9$. Other methods must therefore be used.

4.3.2.1. Closure of the Transport Equation for Probability Density Functions

The equation for the derivation of the joint probability density function for a turbulent reacting flow is given by

$$\begin{aligned}
&\varrho\left(\boldsymbol{\Phi}\right)\frac{\partial P}{\partial t}+\varrho\left(\boldsymbol{\Phi}\right)U_k\frac{\partial P}{\partial x_k}+\left[\varrho\left(\boldsymbol{\Phi}\right)g_k-\frac{\partial\langle p\rangle}{\partial x_k}\right]\\
&\frac{\partial P}{\partial U_k}+\frac{\partial}{\partial \Phi_\omega}\left[\left(\varrho\left(\boldsymbol{\Phi}\right)S_\omega\left(\boldsymbol{\Phi}\right)P\right)\right]=\\
&-\frac{\partial}{\partial U_k}\left[\langle\frac{\partial \tau_{lk}}{\partial x_l}+\frac{\partial p^{\mathrm{f}}}{\partial x_l}|\boldsymbol{u},\boldsymbol{\varphi}\rangle P\right]\\
&+\frac{\partial}{\partial \Phi_\omega}\left[\langle\frac{\partial J_k^\omega}{\partial x_k}|\boldsymbol{u},\boldsymbol{\varphi}\rangle P\right] \quad (155)
\end{aligned}$$

All the terms on the right-hand side of the equation (convection in physical space, transport in the velocity space due to gravitational forces, forces due to the gradient of the average pressure, and transport in the composition space due to chemical reactions) are closed and do not require any model assumptions.

The term $\varrho(\boldsymbol{\Phi})$ is a known function (e.g., a form of thermal equation of state, Eq. 136) and $S_\alpha(\boldsymbol{\Phi})$ is given by Equation (137 a). For example, for a bimolecular reaction A + B → C + D this relationship is $S(A) = r_A = -k m_A m_B$ where the rate coefficient k can often be represented by three parameters in the form $k = k_0 T^\alpha \exp(-E_a/RT)$. If the equations for the reaction rate and the rate coefficients are not considered to be a model, then the representation of the reaction rates in Equation (155) is free of models.

No explicit equation for the mean pressure is contained in the system of Equations (131)–(137). The momentum conservation equations (Eq. 132) contain only the pressure gradient in each of the directions under consideration. Equations (131)–(137) assume a known pressure field which is given implicitly by the conservation equations for the components of momentum and the continuity equation. The mean pressure can be determined from a Poisson equation which is derived from the conservation equations for the momentum components and the continuity equation. Equation (132) can be rewritten as

$$\frac{\partial \varrho U_i}{\partial t} + \frac{\partial}{\partial x_k}(\varrho U_i U_k) = -\frac{\partial \langle p \rangle}{\partial x_i} + \varrho A_i' \qquad (180)$$

with

$$A_i' \equiv \frac{\partial \tau_{ik}}{x_k} - \frac{\partial p^{\mathrm{f}}}{\partial x_i} + \varrho g_i$$

If Equation (180) is averaged and differentiated with respect to x_k and the continuity equation differentiated with respect to t

$$\frac{\partial}{\partial x_k}\frac{\partial}{\partial t}(\langle \varrho U_k \rangle) = -\frac{\partial^2 \langle \varrho \rangle}{\partial t^2} \qquad (181)$$

is subtracted, then the Poisson equation for the mean pressure is obtained which has to be solved simultaneously to the equation for the probability density function:

$$\frac{\partial^2 \langle p \rangle}{\partial x_k \partial x_k} = \frac{\partial^2 \langle \varrho \rangle}{\partial t^2} - \frac{\partial^2 \langle \varrho U_k U_k \rangle}{\partial x_k x_k} + \frac{\partial \langle \varrho A_k' \rangle}{\partial x_k} \qquad (182)$$

In the next step the model assumptions for the conditional expected values on the right-hand side of Equation (155) must be discussed, in particular for

$$\frac{\partial}{\partial U_k}\left[\langle \frac{\partial \tau_{ik}}{\partial x_k} + \frac{\partial p^{\mathrm{f}}}{\partial x_k} | \boldsymbol{u}, \boldsymbol{\varphi} \rangle P\right] \text{ and}$$

$$\frac{\partial}{\partial \Phi_\alpha}\left[\langle \frac{\partial J_k^\alpha}{\partial x_k} | \boldsymbol{u}, \boldsymbol{\varphi} \rangle P\right]$$

Modeling of these expected conditional values is the main problem in developing mathematical models for chemically reacting turbulent flows on the basis of transport equations for probability density functions. The principles used in these methods will be discussed here, for further details see [18], [19], [70], [71], [79], [80].

The probability density functions in the velocity space and the composition space will be treated separately. They can be calculated by integrating Equation (155) over the velocity space and composition space or by using the methods shown in Section 4.2.1 for the momentum components or the scalar quantities. For further simplification a turbulent system is considered with constant density, homogeneous turbulence, and no gradients of the mean velocity.

A uniform fluid flow which flows through a turbulence-generating grid represents a good approximation to these simplified assumptions. If the coordinate system moves with the flow at the mean flow velocity, then no relative convection occurs. The probability density function becomes

$$\frac{\partial P(\boldsymbol{U};t)}{\partial t} = -\frac{\partial}{\partial U_k}\left[\langle \left(\mu \frac{\partial^2 U_k}{\partial x_k^2} - \frac{\partial p^{\mathrm{f}}}{\partial x_k}\right) | \boldsymbol{u} \rangle P(\boldsymbol{U};t)\right] \qquad (183)$$

Equation (183) represents the time development of the probability density function (μ is the dynamic viscosity). The only term to be modeled is

$$\frac{\partial}{\partial U_k}\left[\langle \mu \frac{\partial^2 U_k}{\partial x_k^2} - \frac{\partial p^{\mathrm{f}}}{\partial x_k} | \boldsymbol{u} \rangle P(\boldsymbol{U};t)\right]$$

Experimental observations in a turbulent flow of this type show that the turbulent velocity fluctuations decay with time. Further, if the eddy dissipation model outlined in 4 is used, it can be shown that the turbulent momentum exchange is an isotropic diffusion process of the turbulent eddies [49] that is rate determining for the dissipation of the velocity fluctuations at the molecular level. Similarly to the diffusion process examined in page 245, this model gives a normal distribution for the velocity (a three-dimensional distribution in this case).

A normal distribution is precisely defined by the mean value and the variances of the quantities concerned. Since the normal distribution being discussed is three dimensional, the variances and covariances of the velocities are important and, using Equations (112) and (113), can be expressed as $\langle U^{\mathrm{f}}{}_i U^{\mathrm{f}}{}_j\rangle$ (Reynolds stresses). The problem now is to find models for the Reynolds stresses and their decay over time which will guarantee the required normal distribution as the solution for Equation (183).

A frequently used model for the Reynolds stresses (see also Section 5.1.2.1) which satisfies these conditions, is that developed by LUMLEY [72] and LAUNDER [73] and modified by ROTTA [49], [74]:

$$\frac{\partial}{\partial t}\langle U_i^{\mathrm{f}} U_j^{\mathrm{f}}\rangle = \Big[-\langle U_i^{\mathrm{f}} U_j^{\mathrm{f}}\rangle$$

$$-C_{\langle U_i^{\mathrm{f}} U_j^{\mathrm{f}}\rangle}\left(\langle U_i^{\mathrm{f}} U_j^{\mathrm{f}}\rangle - \frac{2}{3}k\right)\delta_{ij}\Big]/\tau \tag{184}$$

where δ_{ij} is the Kronecker delta, $\delta_{ij}=1$ for $i=j$, $\delta_{ij}=0$ for $i\neq j$; k is the specific kinetic energy of the turbulent velocity fluctuations, $k=1/2\,\langle U_k^{\mathrm{f}} U_k^{\mathrm{f}}\rangle$; and τ is the turbulent time scale, which has to be defined from model assumptions.

The first term on the right-hand side of Equation (184) causes exponential decay of the Reynolds stresses. The second term is responsible for the isotropy of the "diffusion" of the kinetic energy of the turbulent fluctuations.

The second problem for the solution of Equation (183) is the modeling of the conditional expected values

$$\langle \left(\frac{\partial p^{\mathrm{f}}}{\partial x_i}\right) |\boldsymbol{u}\rangle$$

A Poisson equation for the pressure variations can be derived from the conservation equations for the momentum components and the continuity equation in a similar manner to the mean pressure:

$$\frac{\partial^2 p^{\mathrm{f}}}{\partial x_k \partial x_k} = -2\varrho\frac{\partial\langle U_k\rangle}{\partial x_l}\cdot\frac{\partial U_l^{\mathrm{f}}}{\partial x_k} - \varrho\frac{\partial^2 U_k^{\mathrm{f}} U_l^{\mathrm{f}}}{\partial x_k \partial x_l} \tag{185}$$

Thus the pressure fluctuations have two sources. The first is the interactions of the turbulent fluctuations themselves (second term of the right-hand side) and the interactions of the turbulent fluctuations with the gradients of the mean velocities. The second term is described by the equation for the Reynolds stressses (Eq. 184); the first term ("rapid pressure") must be modeled, this can also be carried out with the help of the Reynolds stresses [19], [66].

The third problem is the description of the conditional expected values

$$\frac{\partial}{\partial \Phi_\alpha}\left[\langle\frac{\partial J_k^\alpha}{\partial x_k}|\boldsymbol{u},\boldsymbol{\varphi}\rangle P\right]$$

These terms can be modeled by using the simplifications described above. For the same homogeneous system at constant density with isotropic turbulence and a moving coordinate system, the change in the probability density function over time for a passive scalar quantity, e.g., the mass fraction of a nonreacting chemical species is given by:

$$\frac{\partial P\,(\Phi;t)}{\partial t} = \frac{\partial}{\partial \Phi}\left[\langle\frac{\partial J_k^\Phi}{\partial x_k}|\varphi\rangle P\,(\Phi;t)\right]$$

$$= \frac{\partial}{\partial \Phi}\left[\langle\frac{D}{\varrho}\frac{\partial^2 \Phi}{\partial x_k^2}|\varphi\rangle P\,(\Phi;t)\right] \tag{186}$$

where D is the diffusion coefficient. Equation (186) is obtained for the nonreacting scalar Φ from the general transport equation by integrating over the velocity space or by using the method given in Section 4.2.1. As before with the dissipation of velocity fluctuations, the eddy transport model allows the turbulent exchange of mass to be considered as "eddy diffusion". This "diffusion" is rate determining for the dissipation of the turbulent fluctuations of the scalar at the molecular level. The term

$$\frac{\partial}{\partial \Phi}\left[\langle\frac{D}{\varrho}\frac{\partial^2 \Phi}{\partial x_k^2}|\varphi\rangle P\,(\Phi;t)\right]$$

must be modeled in such a way that the solution of Equation (186) is a normal distribution which decays over time. Thus, the problem consists of modeling the second moment $\langle \Phi^{\mathrm{f}2}\rangle$ (the variance of the scalar quantities).

A simple deterministic model [75] which satisfies the above requirements is

$$\langle \left(\frac{D}{\varrho} \frac{\partial^2 \Phi}{\partial x_k^2} \right) |\varphi\rangle = -\frac{1}{2} C_\Phi \left(\Phi - \langle \Phi \rangle \right) \frac{1}{\tau} \tag{187}$$

in which τ is the turbulent time scale $\tau / C_\Phi = \tau_\Phi$. Since diffusion transport is in the composition space, Equation (187) can be interpreted as transport velocity in the composition space which is proportional to the deviation from the mean value. However, model Equation (187) does not yield a normal distribution from all initial conditions. Instead, the initial conditions must contain information about the form of the distribution [19] as in the applications described in [57], [76], [77]. A general equation can be derived with a statistical particle interaction model, for details refer to [19].

The variances of Φ, $\langle \Phi^{\mathrm{f}2} \rangle$, according to Equation (187) are described by an equation similar to Equation (184):

$$\frac{\partial}{\partial t} \langle \Phi^{\mathrm{f}2} \rangle = -C_{\langle \Phi^{\mathrm{f}2} \rangle} \langle \Phi^{\mathrm{f}2} \rangle / \tau \tag{188}$$

For the combined velocity – composition space of the homogeneous system with isotropic turbulence the same line of reasoning can be applied. The joint probability density function $P(\boldsymbol{U}, \Phi; t)$ is normally distributed so that it is defined by the mean values and the second moments $\langle U_i^{\mathrm{f}} U_j^{\mathrm{f}} \rangle$, $\langle \Phi_\alpha^{\mathrm{f}} \Phi_\beta^{\mathrm{f}} \rangle$, and $\langle U_i^{\mathrm{f}} \Phi_\alpha^{\mathrm{f}} \rangle$. The latter quantities (the Reynolds fluxes) are usually defined by a model equation analogous to Equations (188) and (184):

$$\frac{\partial}{\partial t} \langle U_i^{\mathrm{f}} \Phi_\alpha^{\mathrm{f}} \rangle = -C_{\langle U_i^{\mathrm{f}} \Phi_\alpha^{\mathrm{f}} \rangle} \langle U_i^{\mathrm{f}} \Phi_\alpha^{\mathrm{f}} \rangle / \tau \tag{189}$$

which does not change the character of the solution for the joint probability density function. In the combined velocity – composition space, the diffusion of the Reynolds stresses $\langle U^{\mathrm{f}}{}_i U_j^{\mathrm{f}} \rangle$, the fluctuations of scalar quantities $\langle \Phi_\alpha^{\mathrm{f}2} \rangle$, and the Reynolds fluxes $\langle U_i^{\mathrm{f}} \Phi_\alpha^{\mathrm{f}} \rangle$ are all superimposed on one another. The remaining problem is then the modeling of the turbulent time scales for these "diffusion" processes and the treatment of the "rapid pressure". Some ideas for modeling these are discussed in Sections 5.1 and 5.3.

Under industrial conditions, chemical reactions in turbulent flows occur in nonhomogeneous systems that usually involve gradients of the mean velocities and nonisotropic turbulence. The extension of the simple model described above to such cases is still being developed [18], [19]. The general principle is that the main properties of the model for the second moments are retained. The additional effects are treated as being superimposed on the undisturbed "diffusion". Often "hybrid" models are also used where the expected values of the velocities are computed by other methods (cf. Section 5.1.2.3) and only the joint probability density function for the composition space is evaluated (cf. Section 4.3.2.3).

4.3.2.2. Solution Methods of the Transport Equation for Probability Density Functions

Equation (155) (Section 4.2.1)

$$\begin{aligned}
&\varrho(\boldsymbol{\Phi}) \frac{\partial P}{\partial t} + \varrho(\boldsymbol{\Phi}) U_k \frac{\partial P}{\partial x_k} + \left[\varrho(\boldsymbol{\Phi}) g_k - \frac{\partial \langle p \rangle}{\partial x_k} \right] \\
&\frac{\partial P}{\partial U_k} + \frac{\partial}{\partial \Phi_\omega} \left[\left(\varrho(\boldsymbol{\Phi}) S_\omega(\boldsymbol{\Phi}) P \right) \right] = \\
&-\frac{\partial}{\partial U_k} \left[\langle \left(\frac{\partial \tau_{lk}}{\partial x_l} + \frac{\partial p^{\mathrm{f}}}{\partial x_l} \right) |\boldsymbol{u}, \boldsymbol{\varphi}\rangle P \right] \\
&+\frac{\partial}{\partial \Phi_\omega} \left[\langle \frac{\partial J_k^\omega}{\partial x_k} |\boldsymbol{u}, \boldsymbol{\varphi}\rangle P \right]
\end{aligned} \tag{155}$$

is a partial differential equation in $(\sigma + 6)$ independent variables and time. Finite difference methods for solving this equation fail due to the large number of variables if drastic simplifications cannot be applied. Successful methods of solution have been based on Monte Carlo simulations of the discrete form of the probability density functions in the velocity – composition space [19]. These methods will be described briefly before an example is discussed.

Equation (155) can be rewritten as

$$\frac{\partial P}{\partial t} = (\Omega_1 + \Omega_2 + \Omega_3) P \tag{190}$$

in which Ω_1, Ω_2, and Ω_3 are operators that are obtained by comparison with Equation (155):

$$\Omega_1 \equiv -g_k \frac{\partial}{\partial U_k} - S_\omega \frac{\partial}{\partial \Phi_\omega} - \frac{\partial S_\omega}{\partial \Phi_\omega} I \tag{191a}$$

$$\Omega_2 \equiv [\ldots] \tag{191b}$$

$$\Omega_3 \equiv U_k \frac{\partial}{\partial x_k} + \frac{1}{\varrho(\boldsymbol{\Phi})} \frac{\partial \langle p \rangle}{\partial x_k} \frac{\partial}{\partial U_k} \tag{191c}$$

In Equation (190) P denotes P ($\boldsymbol{U}$, $\boldsymbol{\Phi}$; $\boldsymbol{x}$, t) and in Equation (191 a) I is the identity operator. The terms represented by dots in Equation (191 b) symbolize the models used for each of the "diffusion" processes, i.e., the types of models to describe the conditional expected values on the right-hand side of Equation (155). Equation (190) indicates that all three operators influence the probability density function simultaneously. However, for a solution the operators are separated.

A first-order approximation $P^{(1)}(t)$ for $P(t)$ starting from given initial conditions $P(t_0)$ is estimated by calculating the separate effects of the three operators in the following way:

$$P_1^{(1)}(t) = (I + \Delta t \Omega_1)\, P^{(1)}(t) \tag{192a}$$

$$P_2^{(1)}(t) = (I + \Delta t \Omega_2)\, P_1^{(1)}(t) \tag{192b}$$

$$P^{(1)}(t + \Delta t) = (I + \Delta t \Omega_3)\, P_2^{(1)}(t) \tag{192c}$$

If the fractional changes $P_1^{(1)}(t)$ and $P_2^{(1)}(t)$ are eliminated backwards, then $P^{(1)}(t + \Delta t)$ can be expressed by $P^{(1)}(t)$:

$$\begin{aligned} &P^{(1)}(t + \Delta t) \\ &= (I + \Delta t \Omega_3)(I + \Delta t \Omega_2)(I + \Delta t \Omega_1)\, P^{(1)}(t) \\ &= P^{(1)}(t) + \Delta t\, (\Omega_1 + \Omega_2 + \Omega_3)\, P^{(1)}(t) + O\left(\Delta t^2\right) \end{aligned} \tag{193a}$$

On the other hand if $P^{(1)}(t)$ is developed as a Taylor series, then

$$\begin{aligned} &P^{(1)}(t + \Delta t) = P^{(1)}(t) + \frac{\partial P^{(1)}(t)}{\partial t} \Delta t \\ &+ \frac{\partial^2 P^{(1)}(t)}{\partial t^2} \frac{\Delta t^2}{2!} + \ldots = P^{(1)}(t) \\ &+ \frac{\partial P^{(1)}(t)}{\partial t} \Delta t + O\left(\Delta t^2\right) \end{aligned} \tag{193b}$$

If Equations (193 a) and (193 b) are set equal to each other and divided by Δt then

$$\frac{\partial P^{(1)}(t)}{\partial t} = (\Omega_1 + \Omega_2 + \Omega_3)\, P^{(1)}(t) + O\left(\Delta t^2\right) \tag{194}$$

Equation (194) is identical to Equation (190) apart from the development errors of the order $O(\Delta t^2)$. Thus this procedure of separating the operators gives a first-order approximation for $P(t)$.

The Monte Carlo simulation of the development of the probability density function uses the discrete form of the probability density function with n statistical fluid particles:

$$\begin{aligned} &P_n(\boldsymbol{U}, \boldsymbol{\Phi}; \boldsymbol{x}, t) \\ &= \frac{1}{n} \sum_{i=1}^{n} \delta(\boldsymbol{U} - \boldsymbol{u}_i)\, \delta(\boldsymbol{\Phi} - \boldsymbol{\varphi}_i)\, \delta(\boldsymbol{x} - \boldsymbol{x}_i) \end{aligned} \tag{195}$$

(cf. Eq. 115 b, Section 4.1.1). The discrete form of the probability density function converges with $\sqrt{n}$ towards P ($\boldsymbol{U}$, $\boldsymbol{\Phi}$; $\boldsymbol{x}$, t). The numerical problem is to determine the movement of n statistical fluid particles in the physical, velocity, and composition spaces consistent with the operators Ω_1, Ω_2, and Ω_3. Stated in another way: starting from a given initial distribution P_n ($\boldsymbol{U}$, $\boldsymbol{\Phi}$; $\boldsymbol{x}$, $t = 0$), n fluid particles with the properties $\boldsymbol{U}_i(t_0)$, $\boldsymbol{\Phi}_i(t_0)$, and $\boldsymbol{x}_i(t_0)$ are chosen at random. The fractional changes of P_n are then calculated which lead to P_n ($\boldsymbol{U}$, $\boldsymbol{\Phi}$; $\boldsymbol{x}$, $t = t + \Delta t$) with the particle properties $\boldsymbol{U}_i(t = t_0 + \Delta t)$, $\boldsymbol{\Phi}_i(t = t_0 + \Delta t)$, and $\boldsymbol{x}_i(t = t_0 + \Delta t)$. This gives an algorithm for calculating the development of P_n ($\boldsymbol{U}$, $\boldsymbol{\Phi}$; $\boldsymbol{x}$, t) over time by using forward differences with respect to time. The movement equations consistent with the operators Ω_1, Ω_2, and Ω_3 are for the third fractional change Ω_3

$$\frac{\mathrm{d}\boldsymbol{U}}{\mathrm{d}t} = -\frac{1}{\varrho(\boldsymbol{\Phi}(t))} \left[\frac{\partial \langle p \rangle}{\partial x_k}\right]_{x(t)}, \quad \frac{\mathrm{d}\Phi_\omega}{\mathrm{d}t} = 0, \quad \text{and} \quad \frac{\mathrm{d}\boldsymbol{x}}{\mathrm{d}t} = \boldsymbol{U}(t) \tag{196a,b,c}$$

According to Equation (191 a), for the operator Ω_1 and thus for the first fractional change:

$$\frac{\mathrm{d}\boldsymbol{U}}{\mathrm{d}t} = g_k + [\ldots], \quad \frac{\mathrm{d}\Phi_\omega}{\mathrm{d}t} = S_\omega[\boldsymbol{\Phi}(t)], \quad \text{and} \quad \frac{\mathrm{d}\boldsymbol{x}}{\mathrm{d}t} = 0 \tag{197a,b,c}$$

In Equation (197 a) the contents of the square brackets indicated by dots must be substituted by the model being used for describing the dissipation of the turbulent velocity fluctuations. The second operator only causes a fractional change in the composition space so that the movement equations for the second fractional change are

$$\frac{\mathrm{d}\boldsymbol{U}}{\mathrm{d}t} = 0, \quad \frac{\mathrm{d}\Phi_\omega}{\mathrm{d}t} = [\ldots], \quad \text{and} \quad \frac{\mathrm{d}\boldsymbol{x}}{\mathrm{d}t} = 0 \tag{198a,b,c,}$$

The movement of the statistical fluid particles in the composition space (Eq. 198 b) must again be obtained from the model being used.

After i time steps, each with three fractional changes P_n ($\boldsymbol{U}$, $\boldsymbol{\Phi}$; $\boldsymbol{x}$, $t=t+i\,\Delta t$) is obtained in discrete form and the remaining task is to calculate the mean values, variances, covariances, and all values of interest from the discrete probability density function. This may be achieved by the averaging procedure according to Equation (114) and the analogous rules for higher moments. However, for discrete probability density functions, special interpolation routines are recommended for calculating the moments [19].

4.3.2.3. Example: Combustion of Propane in a Turbulent Diffusion Flame

The solution of Equation (155) for turbulent reacting flows gives the joint probability density function $P(\boldsymbol{U}, \boldsymbol{\Phi}; \boldsymbol{x}, t)$ in discrete form according to the methods outlined in Section 4.3.2.2. All the information required for engineering applications can be obtained from this complete statistical description.

To solve Equation (155) a number of model parameters must be determined (see Eqs. 184, 188, or 189 for simple systems) as well as the turbulent time scale τ. Equation (155) contains $(\sigma+6)$ independent variables so a finite difference method of solution is ruled out. Even if the method of solution shown in Section 4.3.2.2 overcomes these problems, a reduction in the number of variables by a simplification of Equation (155) is worthwhile. In most applications this is achieved by solving Equation (155) for the composition space only [57], [70], [71], [78–80]. The result of this is that, besides the turbulent time scale τ, the expected values of the velocities must also be calculated by other models (Section 5.1.2).

The combustion of propane with air in a turbulent diffusion flame will be described as an example. The propane flows from a simple jet assembly with a Reynolds number of 40 000 related to the nozzle outlet. The gas is mixed with the surrounding air and burns in a turbulent diffusion flame. This example may resemble an industrial flue gas flare.

The numerical calculation is confined to the probability density function in the composition space [70]. Integration of Equation (155) over the velocity space according to Equation (121) gives

$$\frac{\partial\left[\varrho(\boldsymbol{\Phi})\,P(\boldsymbol{\Phi};\boldsymbol{x},t)\right]}{\partial t}+\frac{\partial}{\partial x_k}\left[\varrho(\boldsymbol{\Phi})\,\langle U_k|\varphi\rangle\right.$$
$$\left.\cdot P(\boldsymbol{\Phi};\boldsymbol{x},t)\right]+\frac{\partial}{\partial \Phi_\omega}\left[\left(\varrho(\boldsymbol{\Phi})\,S_\omega(\boldsymbol{\Phi})\,P(\boldsymbol{\Phi};\boldsymbol{x},t)\right)\right]=$$
$$\left|\frac{\partial}{\partial\Phi_\omega}\left[\left\langle\frac{\partial J_k^\omega}{\partial x_k}|\boldsymbol{u},\varphi\right\rangle P(\boldsymbol{\Phi};\boldsymbol{x},t)\right]\right] \qquad (199)$$

By using density-weighted variables (indicated by a tilde) [81]

$$\tilde{P}(\boldsymbol{\Phi};\boldsymbol{x},t)=\frac{1}{\langle\varrho\rangle}\int\varrho(\boldsymbol{\Phi})\,P(\boldsymbol{\Phi};\boldsymbol{x},t)\,\mathrm{d}\varrho \qquad (200)$$

and decomposing the velocities according to Equation (112)

$$U_i|\varphi=\langle\tilde{U}_i\rangle+\langle\tilde{U}_i^{\mathrm{f}}|\varphi\rangle \qquad (201)$$

the following is obtained:

$$\langle\varrho\rangle\frac{\partial\tilde{P}(\boldsymbol{\Phi};\boldsymbol{x},t)}{\partial t}+\frac{\partial}{\partial x_k}\left[\langle\varrho\rangle\langle\tilde{U}_k\rangle\right.$$
$$\left.\cdot P(\boldsymbol{\Phi};\boldsymbol{x},t)\right]+\frac{\partial}{\partial\Phi_\omega}\left[\langle\varrho\rangle S_\omega\tilde{P}(\boldsymbol{\Phi};\boldsymbol{x},t)\right]$$
$$=\frac{\partial}{\partial\Phi_\omega}\left[\left\langle\frac{1}{\varrho}\frac{\partial J_k^\omega}{\partial x_k}|\varphi\right\rangle\tilde{P}(\boldsymbol{\Phi};\boldsymbol{x},t)\right]$$
$$-\frac{\partial}{\partial x_k}\left[\langle\tilde{U}_k^{\mathrm{f}}|\varphi\rangle\langle\varrho\rangle\tilde{P}(\boldsymbol{\Phi};\boldsymbol{x},t)\right] \qquad (202)$$

The problem in using Equation (202) is the unknown velocity $\langle\boldsymbol{U}\rangle$ and the modeling of the terms for the dissipation of turbulent fluctuations of the scalar quantities at the molecular level (first term on the right-hand side of Eq. 202) and transport in the physical space by turbulent velocity fluctuations (second term on the right-hand side of Eq. 202). For the latter term a gradient transport approach is used [70] that describes the transport of higher moments analogously to diffusion. The background of such equations is briefly treated in Section 5.1. In the special case of propane combustion:

$$\langle\tilde{U}_k^{\mathrm{f}}|\varphi\rangle\langle\varrho\rangle\tilde{P}(\boldsymbol{\Phi};\boldsymbol{x},t)$$
$$\approx -C_S\langle\varrho\rangle\frac{\tilde{k}}{\tilde{\varepsilon}}\langle U_k^{\mathrm{f}}U_l^{\mathrm{f}}\rangle\frac{\partial}{\partial x_l}\tilde{P}(\boldsymbol{\Phi};\boldsymbol{x},t) \qquad (203)$$

where C_s is a model parameter; $\tilde{k}$ is the turbulence energy (see Section 4.3.2.1); $\tilde{\varepsilon}=-\,\mathrm{d}\tilde{k}/\mathrm{d}t$ such that $\tilde{k}/\tilde{\varepsilon}\equiv\tau$ the turbulent time scale; and finally $\langle U_i^{\mathrm{f}}U_j^{\mathrm{f}}\rangle$ are the Reynolds stresses. The last three quantities are unknown and must be provided before the solution of Equation (202). In [70] this occurs together with the calculation of the expected velocity values using a "second-order closure model". In this procedure the conservation equations are averaged and the correlations which arise from averaging (the Reynold stresses and Reynolds fluxes) are modeled directly (see also Section 5.1). The remaining problem is the modeling of the dissipation of the turbulent fluctuations of the scalar quantities at the molecular level, this is taken care of with information about the form of the distribution using the method described in Section 4.3.2.1 [57], [71], [76], [78].

The combustion of propane with air in a diffusion flame, comprises ca. 300 elementary reactions between ca. 45 chemical species [82]. The composition space is therefore approximately 45-dimensional and numeric simulation using the Monte Carlo method is prohibitive due to the high dimensionality of the problem. The reaction mechanism for the combustion of propane may be reduced to [70]

$$C_3H_8 + 3/2\,O_2 \rightarrow 3\,CO + 4\,H_2 \qquad \text{(i)}$$

$$CO + OH \rightleftharpoons CO_2 + H \qquad \text{(ii)}$$

and the recombination reactions of the H_2-O_2 system

$$H + O_2 + M \rightleftharpoons HO_2 + M \qquad \text{(iii)}$$

$$H + H + M \rightleftharpoons H_2 + M \qquad \text{(iv)}$$

$$H + OH + M \rightleftharpoons H_2O + M \qquad \text{(v)}$$

$$H + O + M \rightleftharpoons OH + M \qquad \text{(vi)}$$

where M denotes a third body collision partner.

The chemical system is thus described by four scalar variables; the normalized elemental mass fraction of carbon $\Phi_1 \equiv f$, the mixture fraction; the total molar concentration $\Phi_2 \equiv c_{\text{tot}}$; the molar concentration of propane $\Phi_3 \equiv c_{C_3H_8}$; and the molar concentration of carbon monoxide $\Phi_4 \equiv c_{CO}$. This manipulation reduces the number of variables in Equation (202) to four. The reaction rates of these four variables are derived from the reduced mechanism:

$$S(\Phi_1) = 0$$

$$S(\Phi_2) = 4\frac{1}{2} r_{(i)} - r_{(ii)} - r_{(iv)} - r_{(v)} - r_{(vi)}$$

$$S(\Phi_3) = -r_{(i)}$$

$$S(\Phi_4) = 3r_{(i)} - r_{(iv)}$$

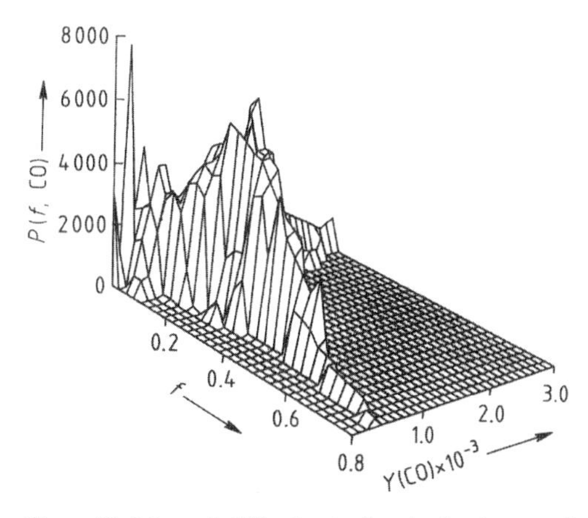

Figure 23. Joint probability density function for the normalized elemental mass fraction (mixture fraction) and mass fraction of CO according to [70] for a propane air diffusion flame at 10 nozzle diameters above the burner nozzle
f = mixture fraction

Figures 23 and 24 show the results of the numeric simulation according to [70]. The results given in Figure 23 are the joint probability density function of f at a distance of 10 jet diameters from the fuel nozzle. There is a peak near the beginning of the coordinates which means that in the region near the fuel nozzle, the air from the surrounding fluid entrains the flow. Figure 24 A gives radial profiles of the mean values of the propane mole fraction and the mixture fraction at a distance of 40 mm from the burner nozzle. These results are important for setting up an exhaust gas flare and can be derived from the general probability density function. There is relatively good correlation between the calculated and measured values. The correlation between the combustion products (CO and CO_2) is not as good (Fig. 24 B). This is due to the models contained in the transport equation for the joint probability density function and the chemical model which contains drastic assumptions about the concentrations of radicals formed during propane combustion (Eqs. i–vi).

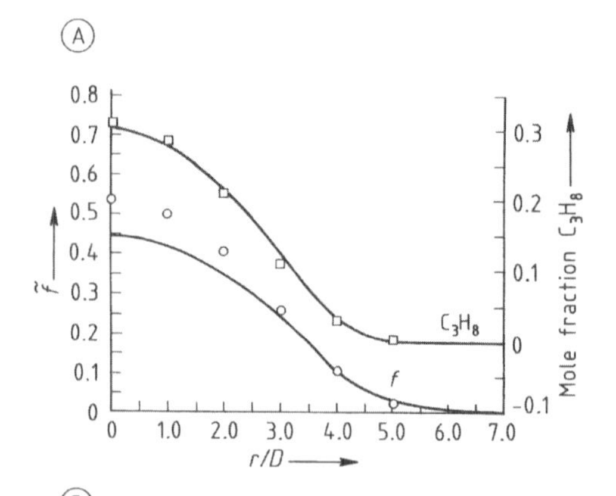

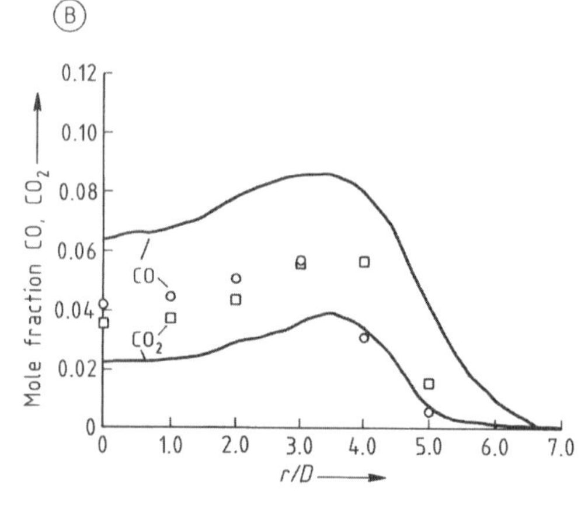

Figure 24. Simulation of a propane air diffusion flame 40 mm above the burner nozzle according to [70]
A) Mean values for the normalized elemental mass fraction (mixture fraction) and the mole fraction of propane; B) Mean values for the mole fraction of carbon monoxide and carbon dioxide
The open squares and circles denote measured values, the curves denote calculated values; r = radial distance; D = nozzle diameter.

5. Models Based on Physicochemical Principles (Transport Phenomena)

Models based on physicochemical principles apply the principles of the transport of momentum, mass, and energy to chemical engineering processes. The representation of processes by this type of model consists of the formulation of the conservation equations for the total mass, the mass of the individual chemical species, components of momentum, and the enthalpy for the system or subsystem under consideration. The conservation equations can be given in the form shown in Section 4.2:

Accumulation + Change due to convective transport =
Change due to molecular transport + Source

The required number, combination, and formulation of the conservation equations and any other physicochemical relationships necessary for closing the set of equations depend on the kind of process and the physical boundary conditions for the system under consideration.

In the following sections, applications of the principles of conservation of mass, momentum, and energy will be discussed. Wherever possible the classification according to Figure 4 will be used.

5.1. Application of the Principle of Conservation of Momentum

(The notation used in this chapter is the same as that employed in Chapter 4.)

The detailed notation for the equation for the conservation of momentum (Eq. 132) is

$$\varrho \frac{\partial u_i}{\partial t} + \varrho u_k \frac{\partial u_i}{\partial x_k} = \frac{\partial \tau_{ik}}{\partial x_k} - \frac{\partial p}{\partial x_i} + \varrho g_i \qquad i = 1, 2, 3 \tag{204}$$

Physically this equation represents the change in momentum flux in a differential control volume caused by volume-specific forces: viscous stress (the first term on the right-hand side of Eq. 204), pressure (the second term on the right-hand side), and body forces, in this case the gravitational forces (the third term on the right-hand side). The assumptions made in the formulation of Equation (204) are discussed in [48], [50], [51].

For Newtonian fluids a gradient expression relates the viscous stresses to the flow velocities. In accordance with Newton's law of friction the shear stress tensor component τ_{ij} can be written as

$$\tau_{ij} = \mu \left(\frac{\partial u_i}{\partial x_j} + \frac{\partial u_j}{\partial x_i} \right) - \frac{2}{3} \mu \frac{\partial u_k}{\partial x_k} \delta_{ij} \tag{205}$$

where δ_{ij} is the Kronecker delta: $\delta_{ij} = 1$ for $i = j$ and $\delta_{ij} = 0$ for $i \neq j$. The assumptions used in the formulation of Equation (205) are discussed in [48], [50], [51]. Except for the pressure gradients, gravitational forces, density ϱ, and the dynamic viscosity of the material μ, Equation

(204) is closed. In the case of gases, for example, μ can be determined from the kinetic theory of gases; other methods of calculating the dynamic viscosity are given in [48], [50]. For systems with constant density (e.g., incompressible fluids) and a known pressure field, Equation (204) is sufficient for calculating the velocity field. This will be demonstrated for the simple example of laminar flow in a tube.

5.1.1. Laminar Tube Flow

Laminar tube flow is illustrated in Figure 25. The flow is assumed to be fully developed and steady state, the accumulation term and the gradients in the direction of flow therefore disappear. There are no components of velocity transverse to the direction of flow. The problem is thus reduced to a one-dimensional problem with one dependent variable.

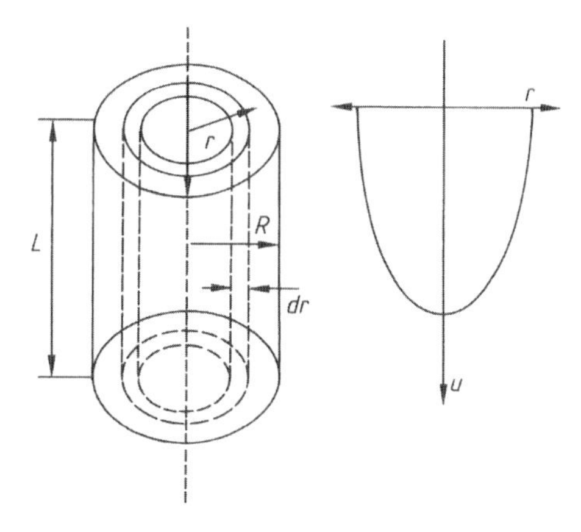

Figure 25. Laminar tube flow and parabolic velocity profile L = length; R = radius; r = radial distance; u = velocity in the axial direction

With the definitions specified in Figure 25 and the appropriate formulation in cylindrical coordinates, Equation (204) can easily be arranged as follows

$$0=\frac{1}{r}\frac{\partial}{\partial r}\left(r\tau_{rz}\right)-\frac{\partial p}{\partial z}+\varrho g_z \tag{206}$$

If the pressure gradient in Equation (206) is replaced by the pressure difference over length L, which is taken as known, Equation (207) is obtained:

$$\frac{\partial}{\partial r}\left(r\tau_{rz}\right)=-\left(\frac{p_0-p_L}{L}+\varrho g\right)r \tag{207}$$

Integration of Equation (207) gives

$$r\tau_{rz}=-\left(\frac{p_0-p_L}{L}+\varrho g\right)\frac{r^2}{2}+C_1 \tag{208}$$

where C_1 is an integration constant. If the term from Newton's law of friction is substituted for the laminar viscous shear stress τ_{rz}, then Equation (208) becomes

$$\frac{\mathrm{d}u}{\mathrm{d}r}=-\left(\frac{p_0-p_L}{L}+\varrho g\right)\frac{r}{2\mu}+\frac{C_1}{\mu r} \tag{209}$$

Integration of Equation (209) results in

$$u=-\left(\frac{p_0-p_L}{L}+\varrho g\right)\frac{r^2}{4\mu}+\frac{C_1}{\mu}\ln r+C_2 \tag{210}$$

Since for $r=0$ the velocity is finite, C_1 must be zero. The constant C_2 can be calculated from a further boundary condition, $u=0$ for $r=R$:

$$C_2=\left(\frac{p_0-p_L}{L}+\varrho g\right)\frac{R^2}{4\mu}$$

Equation (210) thus takes the form

$$u=-\left(\frac{p_0-p_L}{L}+\varrho g\right)\frac{R^2}{4\mu}\left[1-\left(\frac{r}{R}\right)^2\right] \tag{211}$$

This is the well-known parabolic velocity profile for laminar tube flow (Hagen – Poiseuille flow). Further integration of Equation (211) with the relationship $\mathrm{d}q=u\,2\,\pi\,r\,\mathrm{d}r$ provides the Hagen – Poiseuille relationship for the volumetric flow rate q in laminar tube flow

$$q=\left(\frac{p_0-p_L}{L}+\varrho g\right)\frac{\pi R^4}{8\mu} \tag{212}$$

Apart from the assumptions and simplifications associated with the momentum balance itself, the example of laminar flow does not contain any troublesome modeling hypotheses. This is not the case for turbulent flow.

5.1.2. Turbulent, Nonreactive Free Jets

Turbulent flows were described in Chapter 4 as flows with high kinetic energy so that the damping viscous forces can no longer hold the fluid particles on controlled paths due to their high inertial forces. This leads to the development of irregular statistical fluctuations. Turbulent flows are therefore basically unsteady and three-dimensional. This does not, however, represent the principal problem in mathematical modeling. As indicated in Section 2.2, compliance with physicochemical laws does not lose its validity even in statistical processes, such as turbulent flows. Direct numerical simulation of such flows in the nonsteady state with threedimensionally formulated conservation equations is possible in principle.

However, systematic difficulties arise which can only be solved for simple cases [49], [53], [54]. These problems include the small measurements of characteristic turbulent length scales (Kolmogorov length) and the extreme sensitivity of the nonsteady, three-dimensional conservation equations to changes in the boundary conditions. Other options for describing statistical processes are the use of probability density functions (Sections 4.2 and 4.3.2) or the moments of probability density functions. Although the latter do not provide complete statistical information, time- or ensemble-averaged values are often adequate for engineering purposes.

The most important features of models of statistical processes in the form of conservation equations for the moments of probability density functions will be outlined below for turbulent nonreactive flow. The aim is to derive transport equations for the mean values of the velocities starting from the conservation Equation (204) for the compounds of momentum which form a system of differential equations for the instantaneous values of the velocity components.

The example used is based on a jet stirred reactor, similar to that depicted in Figure 14.

The first step is easily carried out by using the decomposition Equation (112) and is demonstrated on a convective term in Equation (204). Equation (204) is first converted into

$$\frac{\partial}{\partial t}(\varrho U_i)+\frac{\partial}{\partial x_k}(\varrho U_k U_i)=\frac{\partial \tau_{ik}}{\partial x_k}-\frac{\partial p}{\partial x_i}+\varrho g_i \quad i=1,2,3 \tag{213}$$

using the continuity Equation (131). From the convective term in one direction and Equation (112) the following is obtained:

$$\begin{aligned}\varrho U_i U_j &= \left(\langle\varrho\rangle+\varrho^{\mathrm f}\right)\left(\langle U_i\rangle+U_i^{\mathrm f}\right)\left(\langle U_j\rangle+U_j^{\mathrm f}\right)\\ &=\langle\varrho\rangle\langle U_i\rangle\langle U_j\rangle+\varrho\langle U_i\rangle U_j^{\mathrm f}\\ &+\langle\varrho\rangle U_i^{\mathrm f}\langle U_j\rangle+\langle\varrho\rangle U_i^{\mathrm f}U_j^{\mathrm f}\\ &+\varrho^{\mathrm f}\langle U_i\rangle\langle U_j\rangle+\varrho^{\mathrm f}\langle U_i\rangle U_j^{\mathrm f}\\ &+\varrho^{\mathrm f}U_i^{\mathrm f}\langle U_j\rangle+\varrho^{\mathrm f}U_i^{\mathrm f}U_j^{\mathrm f}\end{aligned} \tag{214}$$

If Equation (214) is averaged, then by taking Equation (113 b) into account we get

$$\begin{aligned}\langle\varrho U_i U_j\rangle&=\langle\varrho\rangle\langle U_i\rangle\langle U_j\rangle+\langle\varrho\rangle\langle U_i^{\mathrm f}U_j^{\mathrm f}\rangle\\ &+\langle U_i\rangle\langle\varrho^{\mathrm f}U_j^{\mathrm f}\rangle+\langle U_j\rangle\langle\varrho^{\mathrm f}U_i^{\mathrm f}\rangle\\ &+\langle\varrho^{\mathrm f}U_i^{\mathrm f}U_j^{\mathrm f}\rangle\end{aligned} \tag{215}$$

This shows the basic problem in the derivation of conservation equations for the expected values: in addition to the expected values, the convective terms contain the covariances of the velocity components and density–velocity covariances. These terms must be modeled or replaced by assumptions. This is the major difference to dealing with the problem in terms of probability density functions, where convection is not modeled in the transport equations (cf. Section 4.2).

If Equation (215) is substituted into Equation (213), taking the average of the other terms and neglecting fluctuations in density, then

$$\begin{aligned}&\frac{\partial}{\partial t}(\langle\varrho\rangle\langle U_i\rangle)+\frac{\partial}{\partial x_k}(\langle\varrho\rangle\langle U_k\rangle\langle U_i\rangle)\\ &=\frac{\partial}{\partial x_k}\left(\langle\tau_{ik}\rangle-\langle\varrho\rangle\langle U_i^{\mathrm f}U_k^{\mathrm f}\rangle\right)\\ &-\frac{\partial\langle p\rangle}{\partial x_i}+\langle\varrho\rangle\langle g_i\rangle,\quad i=1,2,3\end{aligned} \tag{216}$$

Neglection of the density fluctuations in Equation (216) represents an important modeling assumption. Further model formation concerns the terms $\langle\varrho\rangle\,\langle U_i^{\mathrm f} U_j^{\mathrm f}\rangle$ which have the units of stresses and are thus called Reynolds stresses.

5.1.2.1. Models for Reynolds Stresses

The modeling of Reynolds stresses has already been referred to in Section 4.3.2.1. Discussion is continued here for the closure of the averaged transport equations for the momentum components. More detailed information is given in [49], [51], [52], [83–86].

The most commonly used model in engineering is based on the analogy between laminar and turbulent transfer of momentum (Chap. 4), established by BOUSSINESQ [87] and expanded by PRANDTL [88], [89]. Analogous to the gradient expression for laminar viscous shear stresses (Eq. 205) the apparent turbulent viscous stresses

(Reynolds stresses) are coupled with the gradients of the mean velocity.

A tensor for the turbulent shear stresses is obtained:

$$\tau_{t_{ij}}=\mu_t\left(\frac{\partial\langle U_i\rangle}{\partial x_j}+\frac{\partial\langle U_j\rangle}{\partial x_i}\right)-\frac{2}{3}\mu_t\frac{\partial\langle U_k\rangle}{\partial x_k}\delta_{ij} \quad (217)$$

in which δ_{ij} is the Kronecker delta and μ_t is the apparent turbulent viscosity.

The problem is now to determine the apparent turbulent viscosity, which is not (as in the laminar case) a propensity of the material, but a characteristic generated by the flow. When determining this characteristic the analogy between molecular and turbulent transfer of momentum is retained.

In the laminar case the viscosity can be given as

$$\mu=\frac{1}{2}\varrho\lambda\bar{u} \quad (218)$$

from the kinetic gas theory, where $\bar{u}$ is the mean velocity and λ the mean free path of the molecules. For turbulent transfer of momentum the same argument leads to

$$\mu_t\sim\bar{\varrho}L^*u^* \quad (219)$$

where $L*$ is the analogous term for the mean free path of the molecules and $u*$ the analogue for the mean velocity of the molecules. This corresponds with Prandtl's theory of mixing lengths [60]. In the turbulent case, not molecules but turbulent eddies are the elements which transfer momentum. Both of the quantities $L*$ and $u*$ representing the eddy viscosity hypothesis must therefore be determined by plausible arguments [84]. The characteristic velocity for the turbulent eddies is proportional to the fluctuations of the velocities $u*\sim(2/3\,\langle U_k^f U_k^f\rangle)^{1/2}$ and the characteristic length is the distance which the turbulent eddies travel perpendicularly to the direction of flow, before they lose their identity through dissipation at a molecular level.

This can be expressed as $L*\sim u*\,\tau$, where τ is the turbulent time scale; using the approximation $\tau=\langle k\rangle/\langle\varepsilon\rangle$ (Section 4.3.2.3) gives $L*\sim\langle U_k^f U_k^f\rangle^{3/2}/\langle\varepsilon\rangle$. The eddy viscosity hypothesis is thus

$$\mu_t=C_\mu\langle\varrho\rangle\frac{\langle k\rangle^2}{\langle\varepsilon\rangle} \quad (220)$$

where C_μ is a modeling parameter. As a result of measurements on turbulent boundary layers its value is given as $C_\mu=0.09$ [60].

Many of the problems involved in applying the eddy viscosity hypothesis (e.g., the physical requirements for gradient transport approaches) are discussed in the literature [84–86]. A mathematical model of this simplicity entails a series of other problems which are physically awkward to deal with. For the model with gradient transport approach and the eddy viscosity hypothesis ($k-\varepsilon$ turbulence model) these problems lie in the determination of the turbulence energy $\langle k\rangle$ and its dissipation rate $\langle\varepsilon\rangle$ and require the derivation of corresponding transport equations [84].

The description of turbulent flows in terms of the mean values of the velocities thus requires the calculation of second moments, in this case of the variances $\langle U_i^f U_j^f\rangle$, from which the turbulence energy and its dissipation rate are derived. An equation for the turbulence energy may be derived by multiplying Equation (213) formulated for the i-direction by U_j^f and formulated for the j-direction by U_i^f. Addition of the two equations followed by use of Equation (112) for the components of the velocity, rearrangement of differential quotients, consideration of the continuity equation, and averaging finally give

$$\begin{aligned}&\frac{\partial}{\partial t}\left(\langle\varrho\rangle\langle U_i^f U_j^f\rangle\right)+\frac{\partial}{\partial x_k}\left(\langle\varrho\rangle\langle U_k\rangle\langle U_i^f U_j^f\rangle\right)\\&=-\frac{\partial}{\partial x_k}\left(\langle\varrho\rangle\langle U_k^f U_i^f U_j^f\rangle\right)\\&\quad-\langle\varrho\rangle\langle U_i^f U_k^f\rangle\frac{\partial\langle U_j\rangle}{\partial x_k}\\&\quad-\langle\varrho\rangle\langle U_j^f U_k^f\rangle\frac{\partial\langle U_i\rangle}{\partial x_k}\\&+\langle U_i^f\frac{\partial\tau_{jk}}{\partial x_j}\rangle+\langle U_j^f\frac{\partial\tau_{ik}}{\partial x_i}\rangle\\&\quad-\langle U_i^f\frac{\partial p}{\partial x_j}\rangle-\langle U_j^f\frac{\partial p}{\partial x_i}\rangle\end{aligned} \quad (221)$$

gravitational forces being ignored. Summation over the diagonal elements from Equation (221) gives half the trace of the tensor for the Reynolds stresses $\langle U_l^f U_l^f\rangle$, $\langle k\rangle=1/2\,\langle U_l^f U_l^f\rangle$, and hence the equation for turbulence energy

$$\begin{aligned}&\frac{\partial}{\partial t}\left(\langle\varrho\rangle\langle k\rangle\right)+\frac{\partial}{\partial x_k}\left(\langle\varrho\rangle\langle U_k\rangle\langle k\rangle\right)\\&=-\frac{\partial}{\partial x_k}\left(\langle\varrho\rangle\langle U_k^f k\rangle\right)-\langle\varrho\rangle\langle U_l^f U_k^f\rangle\frac{\partial\langle U_l\rangle}{\partial x_k}\\&+\langle U_l^f\frac{\partial\tau_{lk}}{\partial x_l}\rangle-\langle U_l^f\frac{\partial p}{\partial x_l}\rangle\end{aligned} \quad (222)$$

The description of statistical flows using mean values (first moments) additionally requires the calculation of second moments, in this case those of the velocity variances in the form of turbulence energy. Transport equations can be simply derived for this (see Eq. 222), but they contain terms with moments of the next higher order so that this procedure never leads to a closed equation system. The first expression on the right-hand side of Equation (222), for example, contains moments of the third order. To close the transport equations at a certain order, model assumptions have to be introduced for the higher order moments. These are often based on the same assumptions as the simple gradient transport assumptions and the eddy viscosity hypothesis. The physical basis and the larger number of model parameters arising in Reynolds stress models are the reasons that further development of these [83], [90], [91] is only gradually gaining acceptance in engineering.

The generally accepted form of the transport equation for turbulence energy (Eq. 223) is obtained from Equation (222) by using the definition for turbulent shear stresses, gradient transport approaches for third-order moments, the dissipation hypothesis [49], and other model assumptions [92]:

$$\frac{\partial}{\partial t}\left(\langle\varrho\rangle\langle k\rangle\right)+\frac{\partial}{\partial x_k}\left(\langle\varrho\rangle\langle U_k\rangle\langle k\rangle\right)=\frac{\partial}{\partial x_k}\left[\frac{\mu_{\mathrm{t}}}{\sigma_{k\mathrm{eff}}}\frac{\partial\langle k\rangle}{\partial x_k}\right]+G_k-\langle\varrho\rangle\langle\varepsilon\rangle \tag{223}$$

with the abbreviation

$$G_k=\mu_{\mathrm{t}}\left[\left(\frac{\partial\langle U_l\rangle}{\partial x_k}+\frac{\partial\langle U_k\rangle}{\partial x_l}\right)-\frac{2}{3}\left(\frac{\partial\langle U_k\rangle}{\partial x_k}+\langle\varrho\rangle\langle k\rangle\right)\delta_{ik}\right]\frac{\partial\langle U_k\rangle}{\partial x_k} \tag{224}$$

The individual steps in model formation for Equation (123) are described in [48], [51], [86], [92] and Section 4.3.2.1. For the simplifying homogeneous cases described in Section 4.3.2.1, Equation (223) becomes Equation (184) where $\langle\varepsilon\rangle=\langle k\rangle/\tau$.

For complete definition one more equation is needed for the dissipation rate $\langle\varepsilon\rangle=\mathrm{d}\langle k\rangle/\mathrm{d}t$ which occurs both in Equation (222) and also in the eddy viscosity hypothesis (Eq. 220). The derivation takes a similar path to that for the equation for turbulence energy and is described in [49], [92]. The accepted form is

$$\frac{\partial}{\partial t}\left(\langle\varrho\rangle\langle\varepsilon\rangle\right)+\frac{\partial}{\partial x_k}\left(\langle\varrho\rangle\langle U_k\rangle\langle\varepsilon\rangle\right)=\frac{\partial}{\partial x_k}\left[\frac{\mu_{\mathrm{t}}}{\sigma_{\varepsilon\mathrm{eff}}}\frac{\partial\langle\varepsilon\rangle}{\partial x_k}\right]+\left[C_{\varepsilon1}G_k-2C_{\varepsilon2}\langle\varrho\rangle\langle\varepsilon\rangle\right]\frac{\langle\varepsilon\rangle}{\langle k\rangle} \tag{225}$$

In both Equations (223) and (225) $\sigma_{\varphi\,\mathrm{eff}}$ is the turbulent Schmidt number, which compares the mixing length for the velocity fluctuations to the mixing length for the Reynolds stresses or the dissipation rate respectively; $C_{\varepsilon1}$ and $C_{\varepsilon2}$ are further model parameters. From the solution of Equation (225), the turbulent time scale can be determined; this is a precondition for the closure of the models discussed in Sections 4.3.2.1 and 4.3.2.3.

Describing a nonreactive, isothermal turbulent flow with the model discussed above in terms of the expected velocity values leads to the averaged conservation equations of momentum (Eq. 216), as well as the equations for turbulence energy (Eq. 223) and the dissipation rate (Eq. 225). This system of simultaneous partial differential equations has to be solved by a suitable numerical procedure (see Section 5.1.2.2).

5.1.2.2. Solution Method of the Resulting System of Partial Differential Equations

The partial, simultaneous nonlinear differential Equations (216), (223), and (225) can be put into the form

$$\operatorname{div}\boldsymbol{J}_\varphi=\langle S_\varphi\rangle$$

$$\langle\Phi\rangle=\langle U\rangle,\langle V\rangle,\langle W\rangle,\langle k\rangle,\langle\varepsilon\rangle \tag{226}$$

for the steady-state problem being considered here. In Equation (226) $\boldsymbol{J}_\varphi$ is

$$\boldsymbol{J}_\varphi=\langle\varrho\rangle\langle\Phi\rangle\langle U_k\rangle-\frac{\mu_{\mathrm{t}}}{\sigma_{\varphi\mathrm{eff}}}\frac{\partial\langle\Phi\rangle}{\partial x_k} \tag{227}$$

and $\langle S_\varphi\rangle$ accomodates all other expressions not included by Equation (227). For the momentum equation $\langle S_\varphi\rangle$ also contains specific terms for the turbulent shear stress tensor from Equation (217) and the right-hand sides of the equation for

$\langle k\rangle$ and $\langle\varepsilon\rangle$ can be seen from Equations (223) and (225).

The laminar transport of momentum is generally ignored compared with the turbulent shear stresses. To develop a numerical solution procedure, Equation (226) is integrated over a control volume (page 242). Using the Gauss–Ostrogradski integration principle this gives

$$\int_V \operatorname{div} \boldsymbol{J}_\varphi = \oint_F J_{\varphi k} = \int_V \langle S_\varphi\rangle \tag{228}$$

The integration has to be carried out over the surfaces of the control volume which are perpendicular to the components $J_{\varphi i}$, so that for a cartesian system of coordinates the relationship

$$\Delta J_{\varphi k} \mathrm{d}x_i \mathrm{d}x_{j,i,j\neq k} = \int_V \langle S_\varphi\rangle$$

$$\langle\Phi\rangle = \langle U\rangle,\langle V\rangle,\langle W\rangle,\langle k\rangle,\langle\varepsilon\rangle \tag{229}$$

results, in which the operator Δ denotes the difference at two opposite surfaces of the control volume. Equation (229) is the basis for setting up difference equations which are substituted for the continuous differential operators in Equations (216), (223), and (225) to allow numerical integration.

The conservation equations for the momentum components contain the simple pressure gradients as sources; a reduction in pressure in the direction of the ith coordinate thus leads to an increase of momentum in this direction. Equations (216), (223), and (225) do not, however, contain a pressure term so that the pressure $\langle p\rangle$ or the pressure gradients $\partial\langle p\rangle/\partial x_i$ cannot be directly calculated from them. The same problem occurs in closing the transport equation for the probability density functions (Section 4.3.2.1). Solution of these equations assumes a known pressure field, which is implicitly given through the conservation equations for the momentum components and the continuity equation. If an estimate for the pressure field is assumed for a numerical solution, then the continuity equation (Eq. 131) is usually not complied with. Consequently the numerical solution in an iterative procedure may get further and further away from physically meaningful results because the equation system does not contain correction factors to ensure compliance with the continuity equation.

In the numerical simulation of turbulent steady flow or when using the Navier–Stokes equations, this problem is frequently solved by replacing the primitive variables $\langle \boldsymbol{U}\rangle$ and $\langle p\rangle$ with stream functions and vorticity [94], [95] through the von Mises transformation [93]. The stream function is defined by the von Mises transformation such that the continuity equation is complied with (see also Section 5.2.2.1). In the procedure outlined below another possibility is suggested, in which the pressure correction is obtained from the continuity equation and a simplified form of the momentum balance (the SIMPLE algorithm) [96], [97].

For the example considered here of a turbulent free jet the formulation in cylinder coordinates is used as in Section 5.1.1. Due to the rotational symmetry Equation (226) can be expressed as

$$L(\Phi) = \langle S_\varphi\rangle,\quad \Phi = \langle U\rangle,\langle V\rangle,\langle k\rangle,\langle\varepsilon\rangle \tag{230}$$

The turbulent transport operator $L(\Phi)$ is given by

$$L(\Phi) \equiv \frac{\partial(\langle\varrho\rangle\langle U\rangle\langle\Phi\rangle)}{\partial x} + \frac{1}{r}\frac{\partial(r\langle\varrho\rangle\langle V\rangle\langle\Phi\rangle)}{\partial r} - \frac{\partial}{\partial x}\left(\frac{\mu_t}{\sigma_{\varphi\mathrm{eff}}}\frac{\partial\langle\Phi\rangle}{\partial x}\right) - \frac{1}{r}\frac{\partial}{\partial r}\left(r\frac{\mu_t}{\sigma_{\varphi\mathrm{eff}}}\frac{\partial\langle\Phi\rangle}{\partial r}\right)$$

$$\langle\Phi\rangle = \langle U\rangle,\langle V\rangle,\langle k\rangle,\langle\varepsilon\rangle \tag{231}$$

Since an axially symmetrical system is invariant with respect to rotation around the axis of symmetry, then a two-dimensional grid with L nodal points in the x-direction and M nodal points in the r-direction is used for replacing the differential operators in Equation (231) by differences (Fig. 26). The spatial structure of the grid arises from rotation about the axis of symmetry, and the control volumes are defined by revolving the grid through the unit angle. The control volume is indicated in Figure 26 by a shaded area, the relevant flows, for example $J_{\varphi 1_\mathrm{w}}$, can be expressed as

$$J_{\varphi 1_w} = -\Big(\mathrm{Max}\left\{+\left(\langle\varrho\rangle\langle U\rangle\right)_\mathrm{w}, 0\right\}\langle\Phi\rangle_{l-1} - \mathrm{Max}\left\{-\left(\langle\varrho\rangle\langle U\rangle\right)_\mathrm{w}, 0\right\}\langle\Phi\rangle_l - \frac{\mu_t}{\sigma_{\varphi\mathrm{eff}}}\frac{\langle\Phi\rangle_l - \langle\Phi\rangle_{l-1}}{x_l - x_{l-1}}\Big)F_\mathrm{w} \tag{232}$$

The formulation given in Equation (232) contains an “up-wind” difference scheme

[Max $\{+(\langle\varrho\rangle\langle U\rangle)_{\mathrm{w}},0\}$ $\langle\Phi\rangle_{l-1}-$ Max $\{-(\langle\varrho\rangle\cdot\langle U\rangle)_{\mathrm{w}},0\}\langle\Phi\rangle_l$] for the convective terms (analogous to the cascade of ideally mixed stirred reactors discussed in Section 4.3.1.1) and a central difference scheme for the diffusive terms. As with the model for ideally mixed reactors, the up-wind difference scheme is based on the assumption that the upstream value of a physical quantity is transported over the control surfaces.

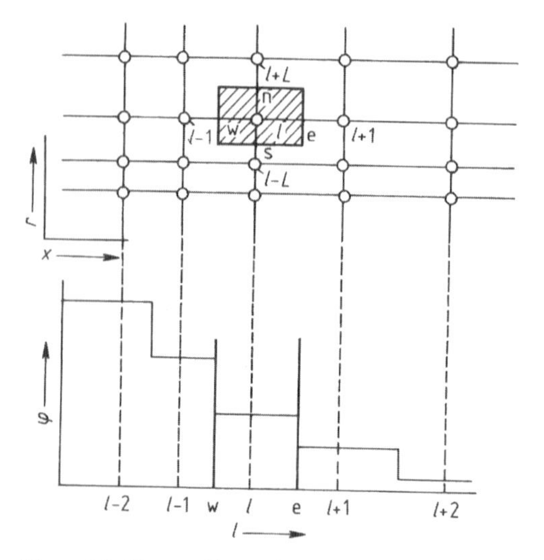

Figure 26. Diagram illustrating the "up-wind" difference scheme for elliptical differential equations
l = number of grid point; L = number of grid points in n direction; n = north; e = east; s = south; w = west; φ = any variable

The profiles of the variables brought into discrete form in the x-direction in accordance with the "up-wind" difference scheme are also shown in Figure 26. Problems associated with this difference scheme (e.g., "numerical diffusion") and measures to limit them are described in [97–101].

The grid with L nodal points in the x-direction and M nodal points in the r-direction and the difference scheme can be used to develop the difference equations from Equation (229); the fluxes through the remaining surfaces of the control volume are defined analogously to Equation (232). Summation of the fluxes gives

$$
\begin{aligned}
&-\Big(\mathrm{Max}\left\{+\left(\langle\varrho\rangle\langle U\rangle\right)_{\mathrm{w}},0\right\}\langle\Phi\rangle_{l-1}\\
&-\mathrm{Max}\left\{-\left(\langle\varrho\rangle\langle U\rangle\right)_{\mathrm{w}},0\right\}\langle\Phi\rangle_{l}\\
&-\frac{\mu_{\mathrm{t}}}{\sigma_{\varphi\mathrm{eff}}}\frac{\langle\Phi\rangle_{l}-\langle\Phi\rangle_{l-1}}{x_{l}-x_{l-1}}\Big)F_{\mathrm{w}}\\
&+\Big(\mathrm{Max}\left\{+\left(\langle\varrho\rangle\langle U\rangle\right)_{\mathrm{e}},0\right\}\langle\Phi\rangle_{l}\\
&-\mathrm{Max}\left\{-\left(\langle\varrho\rangle\langle U\rangle\right)_{\mathrm{e}},0\right\}\langle\Phi\rangle_{l+1}\\
&-\frac{\mu_{\mathrm{t}}}{\sigma_{\varphi\mathrm{eff}}}\frac{\langle\Phi\rangle_{l+1}-\langle\Phi\rangle_{l}}{x_{l+1}-x_{l}}\Big)F_{\mathrm{e}}\\
&-\Big(\mathrm{Max}\left\{+\left(\langle\varrho\rangle\langle V\rangle\right)_{\mathrm{s}},0\right\}\langle\Phi\rangle_{l-L}\\
&-\mathrm{Max}\left\{-\left(\langle\varrho\rangle\langle V\rangle\right)_{\mathrm{s}},0\right\}\langle\Phi\rangle_{l}\\
&-\frac{\mu_{\mathrm{t}}}{\sigma_{\varphi\mathrm{eff}}}\frac{\langle\Phi\rangle_{l}-\langle\Phi\rangle_{l-L}}{x_{l}-x_{l-L}}\Big)F_{\mathrm{s}}\\
&+\Big(\mathrm{Max}\left\{+\left(\langle\varrho\rangle\langle V\rangle\right)_{\mathrm{n}},0\right\}\langle\Phi\rangle_{l}\\
&-\mathrm{Max}\left\{-\left(\langle\varrho\rangle\langle V\rangle\right)_{\mathrm{n}},0\right\}\langle\Phi\rangle_{l+L}\\
&-\frac{\mu_{\mathrm{t}}}{\sigma_{\varphi\mathrm{eff}}}\frac{\langle\Phi\rangle_{l+L}-\langle\Phi\rangle_{l}}{x_{l+L}-x_{l}}\Big)F_{\mathrm{n}}=\langle S\rangle_{\varphi l}
\end{aligned}
\tag{233}
$$

In Equation (233) $\langle S\rangle_{\varphi l}$ represents the source terms from Equation (230) integrated over the control volume. If the continuity equation in the corresponding discrete form is taken into account

$$
\begin{aligned}
&\left(\langle\varrho\rangle\langle U\rangle\right)_{\mathrm{w}}F_{\mathrm{w}}-\left(\langle\varrho\rangle\langle U\rangle\right)_{\mathrm{e}}F_{\mathrm{e}}+\left(\langle\varrho\rangle\langle V\rangle\right)_{\mathrm{s}}F_{\mathrm{s}}\\
&-\left(\langle\varrho\rangle\langle V\rangle\right)_{\mathrm{n}}F_{\mathrm{n}}=0
\end{aligned}
\tag{234}
$$

then Equation (233), after addition of Equation (234) multiplied by $\langle\Phi\rangle_l$, can be transformed into

$$
\begin{aligned}
&-a_{l-L}\langle\Phi\rangle_{l-L}-a_{l-1}\langle\Phi\rangle_{l-1}+b_{l}\langle\Phi\rangle_{l}\\
&-a_{l+1}\langle\Phi\rangle_{l+1}-a_{l+L}\langle\Phi\rangle_{l+L}-\langle S\rangle_{\varphi l}=0
\end{aligned}
\tag{235}
$$

In accordance with Equation (233) the coefficients for Equation (235) are

$$a_{l-L} = \left[\text{Max}\left\{+\left(\langle\varrho\rangle\langle V\rangle\right)_{\text{s}}, 0\right\} + \frac{\mu_{\text{t}}}{\sigma_{\varphi\text{eff}}} \frac{1}{x_l - x_{l-L}}\right] F_{\text{s}} \quad (236a)$$

$$a_{l-1} = \left[\text{Max}\left\{+\left(\langle\varrho\rangle\langle U\rangle\right)_{\text{w}}, 0\right\} + \frac{\mu_{\text{t}}}{\sigma_{\varphi\text{eff}}} \frac{1}{x_l - x_{l-1}}\right] F_{\text{w}} \quad (236b)$$

$$a_{l+1} = \left[\text{Max}\left\{-\left(\langle\varrho\rangle\langle U\rangle\right)_{\text{e}}, 0\right\} + \frac{\mu_{\text{t}}}{\sigma_{\varphi\text{eff}}} \frac{1}{x_{l+1} - x_l}\right] F_{\text{e}} \quad (236c)$$

$$a_{l+L} = \left[\text{Max}\left\{-\left(\langle\varrho\rangle\langle V\rangle\right)_{\text{n}}, 0\right\} + \frac{\mu_{\text{t}}}{\sigma_{\varphi\text{eff}}} \frac{1}{x_{l+L} - x_l}\right] F_{\text{n}} \quad (236d)$$

$$b_l = a_{l-L} + a_{l-1} + a_{l+1} + a_{l+L} \quad (236e)$$

Difference equations corresponding to Equation (235) can be formulated for the $(L-2)$, $(M-2)$ inner grid points of the calculation domain for all variables $\langle \Phi \rangle = \langle U \rangle, \langle V \rangle, \langle k \rangle, \langle \varepsilon \rangle$. Similar relationships are derived for the grid points lying on the boundaries of the calculation domain. The resulting system of equations is nonlinear and coupled via $\langle S \rangle_{\varphi l}$ and the coefficients a and b, which are not constant but depend on the components of the velocity and turbulent viscosity. Because the numerical procedure for solving Equation (235) is an iterative one these dependencies are not resolved.

An analogous equation for pressure corrections can be obtained with the procedure suggested in [96], [97]:

$$-a_{pl-L}\Delta\langle p\rangle_{l-L} - a_{pl-1}\Delta\langle p\rangle_{l-1} + b_{pl}\Delta\langle p\rangle_l - a_{pl+1}\Delta\langle p\rangle_{l+1} - a_{pl+L}\Delta\langle p\rangle_{l+L} - Rs_l = 0 \quad (237)$$

where $\Delta\langle p\rangle$ represents the corrections for the given pressure field, and $R\,s_l$ the "continuity error", which is obtained from the residuals of the continuity equation with the calculated velocity field. The coefficients of Equation (237) are obtained from a simplified form of the momentum balance [96], [97]. To avoid numerical problems which originate from the structure of the averaged momentum equations, Equations (235) and (237) are best solved with a "staggered grid" arrangement for the velocities [98].

Equations (235) and (237) constitute a system of nonlinear equations with $5\,(L-2)\,(M-2)$ unknowns for the example considered. The values of the variables at the boundaries of the calculation domain are given by plausible boundary conditions. The system matrix for this system of nonlinear equations has a block pentadiagonal structure due to the elliptical formulation of the problem (Eq. 231), and the difference scheme used. The system of equations can now be solved, for example with the Newton – Raphson procedure outlined in Section 3.2.1.3 or other methods.

The model for the Reynolds stresses discussed in Section 5.1.2.1 as well as the numerical procedure given in this section are examples from a rapidly developing field of engineering. Other models for the Reynolds stresses represent the phenomena observed in turbulent flows better than the $k-\varepsilon$ turbulence model [52], [84–86], [90–92]. Difference schemes other than the combination of "up-wind" and central differences are also used for partial elliptical differential equations and other solution procedures [101–104] are suggested.

5.1.2.3. Example: Turbulent Flow of Nitrogen in Air

In the example discussed here nitrogen is discharged at room temperature from a tube of 10 mm diameter with an exit velocity of 85 m/s into air which is almost at rest (0.3 m/s) (Fig. 14). The system of nonlinear Equations (235) and (237) is solved for a tensor product grid by an alternating direction implicit (ADI) procedure [101]. In this procedure the block structure is first resolved by solving the system of equations one after the other for the individual variables. The estimated values from the previous iteration are used for the variables contained in the couplings. The system is converted into a quasi-one-dimensional one by including the elements of the two outer diagonals in the right-hand sides.

The system of equations is solved alternately along a line of the grid in Figure 26 in the x-

or r-direction. The coefficients situated on the neighboring lines are calculated with the estimated values from the previous iteration.

To increase the velocity of convergence, linear relaxation is used (Section 3.2.1.3) and the source terms are linearized in the form $\langle S_\varphi \rangle = \langle S_{\mathrm{lin}\varphi} \rangle + \langle S_{\mathrm{prop}\varphi} \rangle \ \langle \Phi \rangle$. The latter measure is particularly useful for stiff differential equation systems, for which the source terms are highly nonlinear and coupled. The modeling parameters [91] used for calculation are $C_\mu = 0.09$, $C_{\varepsilon 1} = 1.4$, $C_{\varepsilon 2} = 0.925$, $\sigma_{k\ \mathrm{eff}} = 1.0$, and $\sigma_{\varepsilon\ \mathrm{eff}} = 1.3$. Figure 27 shows the predicted values for the mean velocity in the x-direction for the turbulent flow of nitrogen at two distances from the nozzle outlet in comparison with the similarity solutions from [105]. The latter give the flow velocity as

$$\langle U \rangle_m = \langle U \rangle_0 / \left(0.16 \frac{x}{D_\mathrm{R}} - 1.5 \right) \tag{238}$$

and

$$\langle U \rangle = \langle U \rangle_m \exp \left[-C_u \left(\frac{r}{x} \right)^2 \right] \tag{239}$$

where $C_u = 90$.

5.2. Applications of the Principle of Conservation of Enthalpy

The conservation equation of enthalpy (Eq. 134) is written in detail as

$$\varrho \frac{\partial h}{\partial t} + \varrho u_k \frac{\partial h}{\partial x_k} = - \frac{\partial J_k^h}{\partial x_k} + \varrho S_h \tag{240a}$$

In accordance with Equation (240 a) enthalpy transport occurs due to convection and molecular transport. The term S_h contains all the mechanisms for conversion of enthalpy (e.g., volumetric work due to a change in pressure, work against physical forces or radiation).

Fourier's simple gradient transport approach can be used to represent molecular transport:

$$J_i^h = - \lambda \frac{\partial T}{\partial x_i} \tag{241}$$

where λ is the thermal conductivity. If a Lewis number, $Le = Pr/Sc$, of one is assumed and other transport effects are neglected, the simple form of the conservation of energy equation is obtained

$$\varrho \frac{\partial h}{\partial t} + \varrho u_k \frac{\partial h}{\partial x_k} = \frac{\partial}{\partial x_k} \lambda \frac{\partial T}{\partial x_k} + \varrho S_h \tag{240b}$$

Detailed discussion of the assumptions and simplifications for this formulation are found in [48], [50], [51].

The conservation equation of enthalpy also contains the density and the components of the velocity. The enthalpy and momentum equations are thus coupled and have to be solved simultaneously. To avoid this a simple heat conduction problem is discussed first.

5.2.1. Heat Conduction

A simple heat conduction problem is shown in Figure 28. A pipe consisting of two different materials (pipe and insulation) has an internal temperature T_{Wi}, the external surface has a temperature of T_{Wo}. If only molecular conduction is considered, Equation (240 a) is reduced to

$$- \frac{1}{r} \frac{\mathrm{d} \left(r J_r^h \right)}{\mathrm{d}r} = 0 \tag{242}$$

Integration gives

$$- r J_r^h = C_1 \tag{243a}$$

Applying Equation (241) gives

$$r \lambda \frac{\mathrm{d}T}{\mathrm{d}r} = C_1 \tag{243b}$$

which after integration results in the relationship

$$T = \frac{C_1}{\lambda} \ln r + C_2 \tag{244a}$$

for the temperature profile. For the inner pipe surface, the boundary condition $T = T_{\mathrm{Wi}}$ applies for $r = r_{\mathrm{Wi}}$. Thus $C_2 = T_{\mathrm{Wi}} - (C_1 / \lambda_\mathrm{i}) \ln r_{\mathrm{Wi}}$ and Equation (244 a) becomes

$$T = T_{\mathrm{Wi}} - \frac{C_1}{\lambda_i} - \ln \frac{r}{r_{\mathrm{Wi}}} \tag{244b}$$

Correspondingly for the outer pipe surface the boundary condition is

$$T_{\mathrm{Wi/o}} = T_{\mathrm{Wi}} - \frac{C_1}{\lambda_i} \ln \frac{r_{\mathrm{Wi/o}}}{r_{\mathrm{Wi}}}$$

and

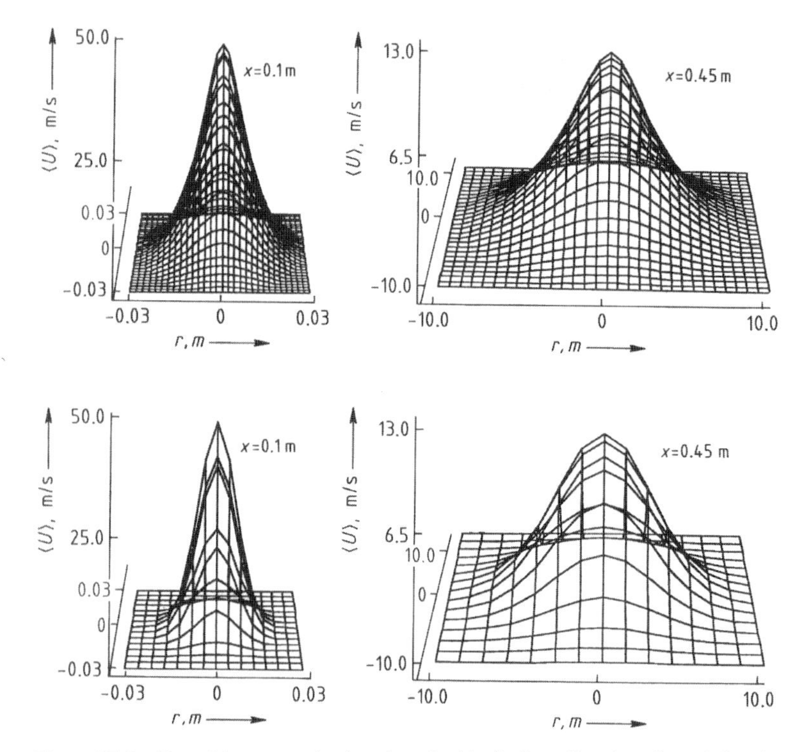

Figure 27. Profiles of the expected values for velocities in the x-direction of a turbulent inert flow from a free turbulent jet
The upper diagrams show numerical simulations, the lower diagrams show similarity solutions. The radial profiles at two distances from the nozzle are given.

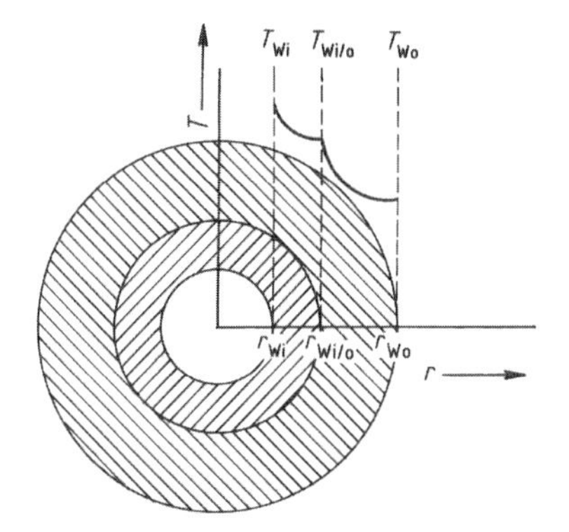

Figure 28. Arrangement and temperature profile for a simple heat conduction problem
The subscripts Wi and Wo denote inner wall and outer wall, respectively.

$$T = T_{\mathrm{Wi/o}} - \frac{C_1}{\lambda_{\mathrm{o}}} \ln \frac{r}{r_{\mathrm{Wi/o}}} \qquad (244c)$$

In Equations (244 a – c), C_1 is given by the total heat flux (Eq. 243 a).

5.2.2. Heat Transfer

5.2.2.1. Exact Solution for a Boundary Layer Problem

An exact solution for the heat transfer from a fluid to a solid wall is possible for laminar flow over a flat plate (Fig. 29). A fluid flows over a flat plate of infinitely small thickness and of length L at a velocity u_∞. Due to the viscous stresses at the wall the fluid particles slow down and a boundary layer of increasing thickness forms along the plate in which the velocity increases from zero to u_∞. This gives the boundary conditions $u = 0$ for $y = 0$ and $u_{\delta_u}(x) = u_\infty$. For the sake of simplicity, constant pressure is assumed, $p_{\delta_u}(x) = p_\infty = \text{const}$. The flow is steady-state and the plate extends infinitely in one direction so that the problem can be formulated twodimensionally. The boundary conditions for temperature are $T = T_\mathrm{W}$ for $y = 0$ and $T_{\delta_T}(x) = T_\infty$. An equation for the temperature can be obtained from Equation (240 b) by neglecting sources such as the conversion of mechanical energy or viscous dissipation and by using the equation for the definition of enthalpy $dh = c_p dT$ and the usual boundary layer approximations [59], [60]:

$$c_p \varrho u \frac{\partial T}{\partial x} + c_p \varrho v \frac{\partial T}{\partial y} = \lambda \frac{\partial^2 T}{\partial y^2} \quad (245a)$$

At constant density and specific heat

$$u \frac{\partial T}{\partial x} + v \frac{\partial T}{\partial y} = a \frac{\partial^2 T}{\partial y^2} \quad (245b)$$

where $a = \lambda / \varrho c_p$.

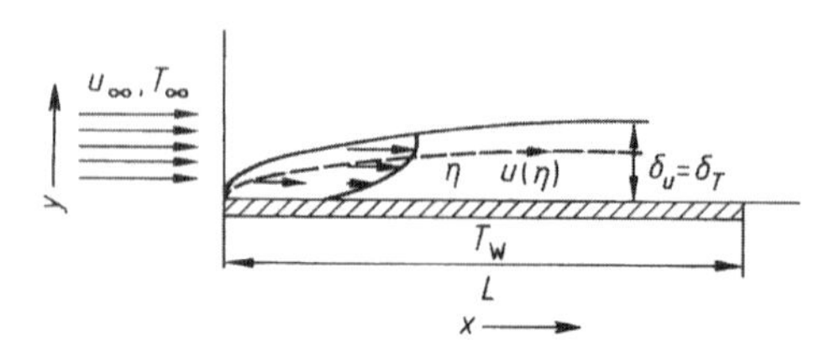

Figure 29. Development of a velocity (δ_u) and temperature (δ_T) boundary layer for laminar incompressible flow over a flat plate
$Pr = 1$; u_∞ = velocity in the x-direction outside the boundary layer; T_∞ = temperature outside the boundary layer; η = similarity coordinate; T_W = wall temperature; L = plate length

The differential equation for the temperature still contains the unknown velocities u and v. Equations for these variables are obtained from the continuity equation (Eq. 131) and the momentum equation (Eq. 204) in the x-direction

$$\frac{\partial u}{\partial x} + \frac{\partial v}{\partial y} = 0 \quad (246)$$

$$u \frac{\partial u}{\partial x} + v \frac{\partial u}{\partial y} = \nu \frac{\partial^2 u}{\partial y^2} \quad (247)$$

where $\nu = \mu / \varrho$ is the kinematic viscosity. For $Pr = \nu / a = 1$ and hence for $\nu = a$, the enthalpy equation is identical with the momentum equation, so that the solution of the latter produces a solution to the enthalpy equation. Furthermore, the coupling of the momentum and enthalpy equations are removed by these assumptions.

The general expression for temperature is

$$\frac{T}{T_\infty} = A \frac{u}{u_\infty} + B$$

$$\text{where } A = \frac{T_\infty - T_W}{T_\infty} \text{ and } B = \frac{T_W}{T_\infty} \quad (248)$$

The heat flux to the wall is then

$$J_W^h = -\lambda \left(\frac{\partial T}{\partial y} \right)_W = -\lambda \frac{T_\infty - T_W}{u_\infty} \left(\frac{\partial u}{\partial y} \right)_W \quad (249)$$

Determination of the heat flux to the wall thus depends on determination of the velocity gradient at the wall, which is a function of x and can be obtained from the solution of Equations (246) and (247).

For the solution, the stream function Ψ is used to eliminate the velocities u and v [59], [60]:

$$u = \frac{\partial \Psi}{\partial y} \text{ and } v = -\frac{\partial \Psi}{\partial x} \quad (250a,b)$$

The stream function invariably satisfies the continuity equation, as can be seen by suitable differentiation. With this transformation, Equation (247) becomes a partial differental equation of the third order

$$\Psi_y \Psi_{yx} - \Psi_x \Psi_{yy} = \nu \Psi_{yyy} \quad (251)$$

in which Ψ_x and the appropriate symbols for each of the differential operators are used.

In the boundary layer formed in laminar flow locations of constant velocity u lie on a parabola $\eta \sim y/\sqrt{x}$ (Fig. 29). If the equation system is therefore transformed to this similarity coordinate η, then Equations (247) and (251) can be reduced to ordinary differential equations. The exact transformation is

$$\eta(x, y) = y \sqrt{\frac{u_\infty}{\nu x}} \quad (252)$$

in which u_∞ / ν represents a scale factor that makes η dimensionless. As u is only a function of η, the general expression $u/u_\infty = g(\eta)$ can be used for making the stream function dimensionless:

$$\Psi(x, y) = \int u \mathrm{d}y = u_\infty \int g(\eta) \mathrm{d}\eta$$

$$= u_\infty \sqrt{\frac{u_\infty}{\nu}} \int g(\eta) \mathrm{d}\eta \quad (253)$$

where $\int g(\eta) \mathrm{d}\eta \equiv f(\eta)$ is called the dimensionless stream function. Equations (250 a, b), (252), and (253) provide all the transformations and conversions that are needed to convert the partial differential Equation (251) into an ordinary differential equation of dimensionless quantities.

$$\Psi_y = u_\infty f_\eta, \quad \Psi_x = \frac{1}{2}\sqrt{\frac{u_\infty}{\nu x}}\,[\eta f_\eta - f(\eta)],$$

$$\Psi_{yx} = -\frac{1}{2}\frac{u_\infty}{x}\eta f_{\eta\eta},$$

$$\Psi_{yy} = u_\infty\sqrt{\frac{u_\infty}{\nu x}}\,f_{\eta\eta},$$

$$\Psi_{yyy} = u_\infty \frac{u_\infty}{\nu x} f_{\eta\eta\eta} \qquad (254\text{a,b,c,d,e})$$

In Equations (254) f_η and the corresponding symbols denote the derivatives of f with respect to η. The ordinary differential equation

$$f_{\eta\eta\eta} + \frac{1}{2} f \cdot f_{\eta\eta} = 0 \qquad (255)$$

is thus obtained from Equation (251) which has to be solved for the boundary conditions $f(\eta) = 0$, $f_\eta = 0$ for $y = 0$ or $\eta = 0$ and also for $f_\eta = 1$ for $y \to \infty$ or $\eta \to \infty$.

Equation (255) can be solved numerically but not analytically. The essential quantities for the boundary layer problem, $f(\eta) = \int (u/u_\infty)\,\mathrm{d}\eta$, $f_\eta = u/u_\infty$, and $f_{\eta\eta} = u^{-1}{}_\infty\,(\partial u/\partial \eta)$, can be found in the literature, tabulated as a function of η [60], [106]. According to [106], $u/u_\infty = 0.99$ for $\eta = 5.0$ and $f_{\eta\eta} = 0.332$ for $\eta = 0$.

The boundary layer thickness is calculated from the first of these two values and is defined here as the distance δ_u at which the velocity reaches 99 % of the flow velocity outside the boundary layer. It then follows from Equation (252) that

$$\delta_u = 5\sqrt{\frac{\nu x}{u_\infty}} \quad \text{or} \quad \frac{\delta_u}{x} = \frac{5}{\sqrt{Re_x}} \qquad (256\text{a,b})$$

The essential information provided by Equation (256) is that the development of the boundary layer thickness for laminar flow over a flat plate is proportional to $\sqrt{x}$ and $1/\sqrt{u_\infty}$, and inversely proportional to the square root of the Reynolds number $Re_x = u_\infty x/\nu$. The solutions of the momentum and enthalpy equations are similar because $Pr = 1$; as a result $\delta_u = \delta_T$. A detailed discussion of further implications of the solution of the boundary layer equations is given in [59], [60].

Using the value

$$0.332 = f_{\eta\eta} = \frac{1}{u_\infty}\frac{\partial u}{\partial \eta} = \frac{1}{u_\infty}\frac{\partial u}{\partial y}\frac{\partial y}{\partial \eta} = \frac{1}{u_\infty}\frac{1}{\sqrt{u_\infty/\nu x}}\frac{\partial u}{\partial y}$$

the following relation is obtained:

$$\left(\frac{\partial u}{\partial y}\right)_\mathrm{W} = 0.332 u_\infty \sqrt{\frac{u_\infty}{\nu x}} \qquad (257)$$

Substitution of this term in Equation (249) gives

$$J_\mathrm{W}^h = -0.332\lambda\sqrt{\frac{u_\infty}{\nu x}}\,(T_\infty - T_\mathrm{W}) \qquad (258)$$

If the definition for the heat-transfer coefficient $J_\mathrm{W}^h \equiv \alpha\,\Delta T$ where $\Delta T = (T_\mathrm{W} - T_\infty)$ is now substituted in Equation (258), then

$$\alpha = 0.332\lambda\sqrt{\frac{u_\infty}{\nu x}} = 0.332\lambda\frac{1}{x}\sqrt{Re_x} \qquad (259)$$

Rearrangement of this equation in dimensionless groups using $Nu_x = \alpha\,x/\lambda$ yields the solution for the heat-transfer problem for laminar flow over a flat plate:

$$\frac{Nu_x}{\sqrt{Re_x}} = 0.332 \qquad (260)$$

The physical background of Equations (259) and (260) is that the local heat-transfer coefficient, like the boundary layer thickness, is inversely proportional to the distance in the x-direction and the velocity of flow. This is easy to understand because the temperature and velocity gradients level out at the wall with the formation of the boundary layer.

In technical problems the local heat-transfer coefficient is often not as important as its mean value α_m from the beginning of the plate to a distance x or the total length L. Equation (259) can be written as $\alpha = \text{const.}\; x^{-1/2}$:

$$\alpha_m = \frac{1}{L}\int_0^L \alpha(x)\,\mathrm{d}x = \frac{\text{const.}}{L}\int_0^L \frac{\mathrm{d}x}{\sqrt{x}}$$

$$= 2\frac{\text{const.}}{L}\sqrt{x}\Big|_0^L = 2\alpha_L \qquad (261\text{a})$$

The mean heat-transfer coefficient for a plate with length L is thus double the local heat-transfer coefficient at the distance L. This can also be formulated with Equation (260):

$$Nu_m = 0.664\sqrt{Re_L} = 2Nu_L \qquad (261\text{b})$$

The analysis described here only applies under the highly simplified assumptions of incompressible flow and constant density and also $Pr = \nu/a = 1$. Nevertheless this allows the general principles of modeling for heat transfer to be described.

5.2.2.2. General Principles of Modeling Heat Transfer

Under industrial conditions the simplifications introduced in the previous section become void. The consequence of $Pr \neq 1$ and variable density is that Equations (245), (213), and (131) are coupled through the density. Moreover, the solutions for the momentum balances and the enthalpy balances are no longer identical. The dimensions of the temperature boundary layer δ_T and the hydrodynamic boundary layer δ_u differ. In addition, flow is mostly turbulent sometimes with very complicated geometries.

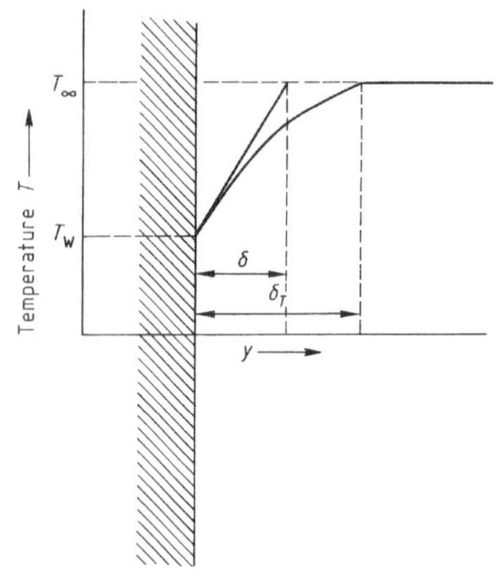

Figure 30. Representation of a general model for heat transfer
δ = film thickness; δ_T = thickness of temperature boundary layer; T_W = wall temperature; T_∞ = temperature outside the boundary layer.

As the heat transfer under such conditions can still be defined as

$$J^h_W = -\lambda \left(\frac{\partial T}{\partial y}\right)_W \equiv \alpha \Delta T \qquad (262)$$

or from the appropriate mean value over a surface, the problem is thus to determine the temperature gradients at the wall. Heat transfer at the wall is primarily a heat conduction problem. The temperature profile in the vicinity of the wall is however determined by simultaneous transfer of momentum, heat, and mass. Analytical or numerical solutions for the system of Equations (131), (213), and (245) will only succeed in a few cases with a justifiable amount of effort [25], [59], [107], [108], see also the solution of the momentum equations for the turbulent inert jet of gas in Section 5.1.2.2.

In most engineering applications semi-empirical methods are therefore used which assume a linearized temperature profile at the wall. Figure 30 shows the temperature profile for laminar flow over a plate.

If the temperature is developed at the wall, $y = 0$ in a Taylor series that is truncated after the first member, then

$$T = T_W + \left(\frac{\partial T}{\partial y}\right)_W \mathrm{d}y \text{ or } T_\infty = T_W + \left(\frac{\partial T}{\partial y}\right)_W \delta \qquad (263)$$

If this is substituted into Equation (262) then

$$\lambda \frac{T_\infty - T_W}{\delta} = \alpha (T_\infty - T_W) \quad \text{or} \quad \alpha = \frac{\lambda}{\delta} \qquad (264a,b)$$

Here δ is the film thickness produced by linearization of the temperature profile as shown in Figure 30. This is not identical with the boundary layer thickness δ_u or δ_T. As can be seen from the solutions for the laminar boundary layer

$$\left(\frac{\partial T}{\partial y}\right)_W = 0.332\,(T_\infty - T_W)\sqrt{Re_x}/x$$

whereas

$$\left(\frac{\partial T}{\partial y}\right)^*_W = \frac{1}{5}(T_\infty - T_W)\sqrt{Re_x}/x$$

when the thickness δ_T of the temperature boundary layer is used for the difference quotients. This latter quantity is thus only ca. 60 % of the true temperature gradient. The determination of heat transfer thus depends on determination of the film thickness as in Equation (264). This problem has been solved experimentally in most technical applications, where the mean heat-transfer number depending on the various variables is represented in the form $Nu = f\,(Re, Pr, L/D \ldots)$, see also Equations (260) and (261).

These relationships for industrial heat exchangers are not treated here. They are described in [25]. An example of such a relationship for the transfer of heat for turbulent flow was discussed in Sections 3.1.1.4 and 3.2.3.

With the establishment of the mean heat-transfer coefficients α and the relevant temperature difference, heat transfer (Eq. 262) is clearly defined. Arrangements such as that shown in Figure 28 can be calculated completely with regard to temperature profiles or heat losses. Summation of the temperature differences for the individual "phases" of the arrangement allows heat transfer to be related to an "overall resistance"

$$\frac{1}{k}=\frac{1}{\alpha_{\mathrm{Wi}}}+\sum\frac{1}{\lambda_{\mathrm{i}}/\left(r_{\mathrm{Wo}}-r_{\mathrm{Wi}}\right)}+\frac{1}{\alpha_{\mathrm{Wo}}}$$

and to the total temperature difference. The equation for heat transfer analogous to Equation (262) will then be used depending on the statement of the question. The calculation procedure will be shown in an analogous problem of mass transfer in Section 5.3.1.3.

5.3. Applications of the Law of Conservation of Mass

The law of conservation of mass is formulated through the continuity equation (Eq. 131) and the conservation equations for the mass of the individual chemical species, m_i

$$\varrho\frac{\partial m_i}{\partial t}+\varrho u_k\frac{\partial m_i}{\partial x_k}=-\frac{\partial J_k^{m_i}}{\partial x_k}+\varrho S_i \quad i=1,\ldots,N-1 \qquad (265\text{a,b})$$

A further equation results from the condition $\sum_1^N m_i=1$. Analogous to the formulations for the conservation of momentum and enthalpy, the transport of chemical species results from convection and molecular transport. The term S_i contains sources or sinks based on all possible forms of transformation, for example due to chemical reactions or due to phase changes in multiphase systems. If molecular transport is considered to be adequately described by Fick's law (gradient transport approach)

$$J_j^{m_i}=-\varrho D_i\frac{\partial m_i}{\partial x_j} \qquad (241\text{a})$$

then a simple form of the conservation equations is obtained that is analogous to Equation (240 b)

$$\varrho\frac{\partial m_i}{\partial t}+\varrho u_k\frac{\partial m_i}{\partial x_k}=\frac{\partial}{\partial x_k}\left[\varrho D_i\frac{\partial m_i}{\partial x_k}\right]+\varrho S_i \quad i=1,\ldots,N-1 \qquad (265\text{b})$$

In Equation (265 b) thermal and pressure diffusion are neglected. Furthermore in general $S_i=S_i\,(m_i,\,h,\,p)$ (Eq. 137 a) is to be set in general for the transformation term so that Equation (265 b) is coupled with the conservation equation of momentum through the density and velocities and coupled with the conservation equation of enthalpy through the transformation term. The coupled system of equations given in Section 5.2 is thus expanded by one equation.

5.3.1. Mass Transfer without Chemical Reaction

5.3.1.1. Exact Solution for a Boundary Layer Problem

The flat plate in Figure 29 is replaced by a water surface (Fig. 31) from which the water evaporates into the higher temperature air flow. Under the same assumptions as in Section 5.2.2.1 and with the boundary layer approximations for Equation (265 b), the following system of equations is obtained for this steady-state problem:

$$\frac{\partial u}{\partial x}+\frac{\partial v}{\partial y}=0 \qquad (246)$$

$$u\frac{\partial u}{\partial x}+v\frac{\partial u}{\partial y}=\nu\frac{\partial^2 u}{\partial y^2},\quad \nu=\frac{\mu}{\varrho} \qquad (247)$$

$$u\frac{\partial T}{\partial x}+v\frac{\partial T}{\partial y}=a\frac{\partial^2 T}{\partial y^2},\quad a=\frac{\lambda}{\varrho c_p} \qquad (245\text{b})$$

$$u\frac{\partial m}{\partial x}+v\frac{\partial m}{\partial y}=D\frac{\partial^2 m}{\partial y^2} \qquad (266)$$

where $m=m_{\mathrm{H_2O}}$. The equation system is to be solved for the boundary conditions $u=0$, $v=v_{\mathrm{W}}$, $T=T_{\mathrm{W}}$, and also $m=m_{\mathrm{W}}$ for $y=0$ and $u=u_\infty$, and $m=m_\infty$ for $y\to\infty$.

Since the inclusion of the conservation equation of mass has not altered the nature of the equations, it is again logical to transform the equations into the similarity coordinates

$$\eta\,(x,y)=y\sqrt{\frac{u_\infty}{\nu x}}=\frac{y}{x}\sqrt{Re_x}$$

(see Eq. 252), and to express them in the dimensionless streams function $f\,(\eta)=\int\,(u/u_\infty)\,\mathrm{d}\eta$. Due to the different boundary conditions for temperature and mass fraction and in analogy with $f_\eta=u/u_\infty$, the term

$\theta = (T - T_W)/(T_\infty - T_W)$ is used for the temperature and $\varphi = (m - m_W)/(m_\infty - m_W)$ for the mass fraction. With the aid of the relationships given in Section 5.2.2.1, these transformations yield the equation system

$$f_{\eta\eta\eta} + \frac{1}{2} f f_{\eta\eta} = 0 \tag{255}$$

$$\theta_{\eta\eta} + \frac{1}{2} Pr f \theta_\eta = 0 \tag{267}$$

$$\varphi_{\eta\eta} + \frac{1}{2} Sc f \varphi_\eta = 0 \quad \text{where} \quad Sc = \nu/D \tag{268}$$

that has to be solved simultaneously.

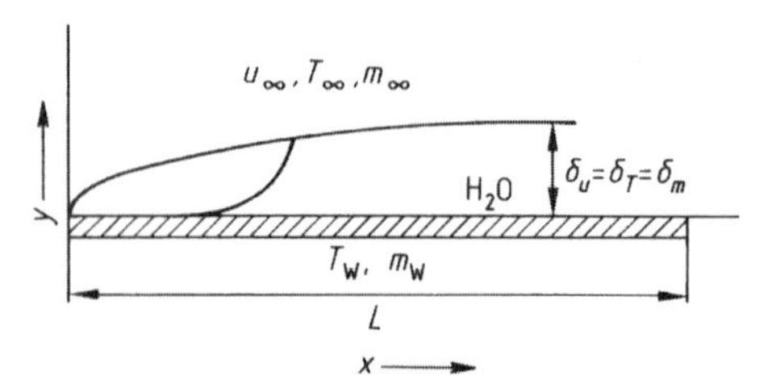

Figure 31. Development of a velocity (δ_u), temperature (δ_T), and concentration (δ_m) boundary layer for laminar incompressible flow over a flat water surface
$Pr = Sc = 1$
The subscripts ∞ and W indicate values outside the boundary layer and at the wall, respectively.

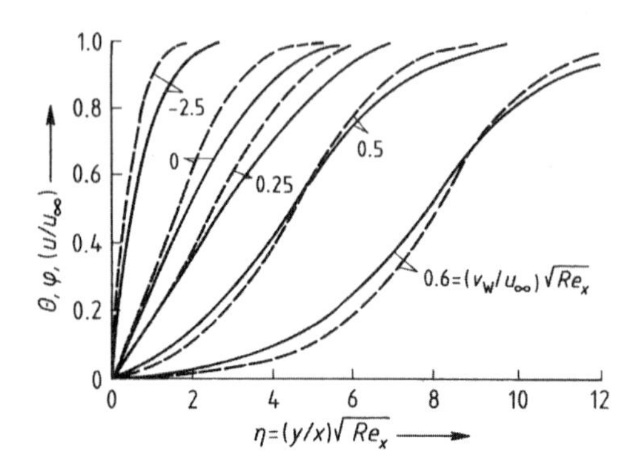

Figure 32. Concentration and temperature distribution with mass transfer for laminar flow over a flat surface
The curves for $Pr = Sc = 1$ indicated by the dashed lines also give the velocity profiles. The curves for $Pr = Sc = 0.7$ indicated by the solid lines show temperature and concentration distributions.
$\theta = (T - T_W)/(T_\infty - T_W)$; $\varphi = (m - m_W)/(m_\infty - m_W)$;
η = similarity coordinate $= (y/x)\, Re_x \sqrt{Re_x}$;
v = velocity in y-direction; u = velocity in x-direction; Re_x = Reynolds number defined with u_∞ and axial distance x; the subscripts W and ∞ indicate values at the wall and outside the boundary layer, respectively.

The boundary conditions with the transformed variables are $f_\eta = 0$, $\theta = 0$, $\varphi = 0$ and also $(v_W/u_\infty)\sqrt{Re_x} = \text{const.}$ for $\eta = 0$ and $f_\eta = 1$, $\theta = 1$, $\varphi = 1$ for $\eta \to \infty$.

The system of Equations (255), (267), (268) cannot be solved analytically, but for $Le = Pr/Sc = 1$ all profiles are identical and thus result in identical values for δ_u, δ_T, and δ_m. All transport mechanisms exhibit the same development of the boundary layer. For $Pr = Sc = 1$ and $Pr = Sc = 0.7$ and also some values of $(v_W/u_\infty)\sqrt{Re_x}$, Figure 32 shows the profiles $f_\eta = u/u_\infty$, θ, and φ [59]. The curve for $Pr = Sc = 1$ and $(v_W/u_\infty)\sqrt{Re_x} = 0$ corresponds to the profile discussed in the previous sections and is called the Blasius profile. The value of $\eta = 5$ for $\theta = \varphi = u/u_\infty \approx 0.99$ can be read off from the curve for $(v_W/u_\infty)\sqrt{Re_x} = 0$. Figure 32 shows that, depending on the magnitude of the "blow out parameter" $(v_W/u_\infty)\sqrt{Re_x}$, the profiles of the flat plate will be either blown off or sucked on. For many technical applications this is taken advantage of for stabilization or removal of boundary layers.

In the case of the evaporation example shown in Figure 31, $(v_W/u_\infty)\sqrt{Re_x}$ can be neglected so that the mass transfer, which is to be defined analogously to the heat transfer, is written in the form

$$J_W^m = -\varrho D \left(\frac{\partial m}{\partial y}\right)_W = -\varrho D \,(m_\infty - m_W) \sqrt{\frac{u_\infty}{\nu x}} \left(\frac{\partial \varphi}{\partial \eta}\right)_W \tag{269}$$

If the Blasius solution $f_{\eta\eta} = \varphi_\eta = \theta_\eta = 0.332$ is used for $Pr = Sc = 1$ and, analogous to the heat-transfer coefficient, a mass-transfer coefficient is defined as $J_W^m = \varrho\, \beta\, \Delta m$ where $\Delta m = (m_W - m_\infty)$, it follows that

$$\beta = 0.332 D \sqrt{\frac{u_\infty}{\nu x}} = 0.332 D \frac{1}{x} \sqrt{Re_x} \tag{270}$$

In order to complete the analogy with heat transfer the Sherwood number $Sh_x = \beta x/D$ should be substituted. Thus for $Pr = Sc = 1$

$$\frac{Sh_x}{\sqrt{Re_x}} = 0.332 \tag{271}$$

Further treatment of mass-transfer coefficients – for example averaging over the length – follows the scheme given in Section 5.2.2.1 and need not be repeated here.

5.3.1.2. General Principles for Modeling Mass Transfer

As mentioned in Section 5.2.2.2 for heat transfer, only in very few cases do the Prandtl or Schmidt numbers have the value one; in addition the geometric or hydrodynamic assumptions listed in Section 5.3.1.1 seldom apply. Even if the system of Equations (204), (240 b), and (265 b) can be solved under such conditions, the resulting solutions deviate from the complete analogy.

For engineering applications mass transfer is therefore also treated by semi-empirical models which assume a linearization of the profile of the concentrations at the phase boundary. Because of the analogy with heat transfer (Fig. 30) discussion of this topic can be kept short. If the mass fraction profile is linearized at the phase boundary

$$m = m_\mathrm{W} + \left(\frac{\partial m}{\partial y}\right)_\mathrm{W} \mathrm{d}y$$

or $m_\infty = m_\mathrm{W}$

$$+ \left(\frac{\partial m}{\partial y}\right)_\mathrm{W} \delta \qquad (272)$$

the equation defining the mass-transfer coefficient results in

$$D\frac{m_\infty - m_\mathrm{W}}{\delta} = \beta\,(m_\infty - m_\mathrm{W}) \quad \text{or} \quad \beta = \frac{D}{\delta} \qquad (273\mathrm{a,b})$$

where δ again represents the film thickness produced by linearization of the mass fraction profile. Determination of the mass-transfer coefficient thus depends on the determination of the film thickness, which is carried out experimentally for many engineering systems. The results for the mean mass-transfer coefficients in the form of the relationships *Sh* = *f* (*Re, Sc, L/D*) are described in [25], [59], [109]. For evaluation of industrial mass transfer equipment there remains the problem, analogous to heat transfer, of the addition of the "individual resistances" for mass transfer and reference to the total concentration difference.

5.3.1.3. Examples of Simultaneous Mass and Heat Transfer: Dynamic Models

The system used to illustrate the general principles of the modeling of simultaneous heat and mass transfer (Sections 5.2.2.2 and 5.3.1.2) is based on the boundary layer example shown in Figure 31. Figure 33 shows an evaporation cooler which is used in an air-conditioning device. A liquid (water) trickles in a thin film over suitable packings and comes into contact with a gas (air) flowing in the same direction. The liquid evaporates and is thus cooled, so that heat flow occurs from gas to liquid thus lowering the gas temperature.

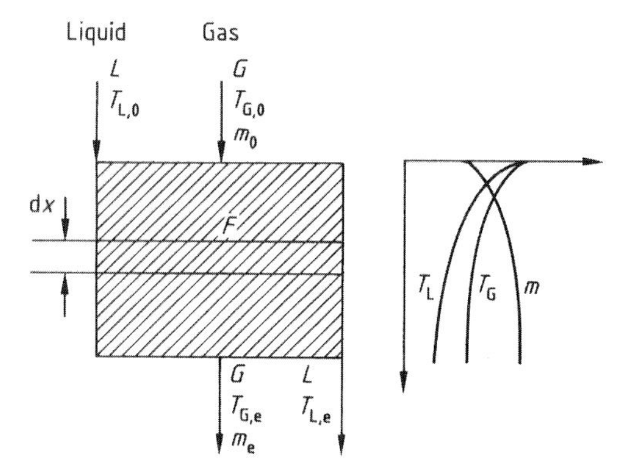

Figure 33. Example of simultaneous mass and heat transfer in an evaporation cooler
L = mass flow of liquid; G = mass flow of gas; F = cross sectional area; x = axial distance
The subscripts L, G, 0, and e denote liquid, gas, initial, and final, respectively.

The final steady-state temperature of the gas must be determined for a given area, $S = s\,F\,H$, where s is the specific transfer area. All terms are defined in Figure 33. The thickness of the layer of liquid flowing over the packings is assumed to be so small that the temperature in the liquid phase is uniform. The mean heat- and mass-transfer coefficients for this problem are known. The temperature changes lie in a range which allows the material values to be regarded as constants. The functions for the Nusselt and Sherwood numbers generally contain terms which are functions of the properties of the material. The assumption of approximately constant material properties avoids recalculation of the example with improved Nusselt and Sherwood numbers. The mass transfer is so small compared with the mass flow of gas and liquid (G and L, respectively) that the latter can likewise be regarded as nearly constant. The required quantities only change in the x-direction so that the problem can be formulated one-dimensionally.

The heat balance for the gas phase gives the temperature change in the gas based on the heat transfer to the liquid

$$\frac{dT_G}{dx} = -\alpha \frac{sF}{Gc_{pG}} (T_G - T_L) \tag{274}$$

where c_p is the specific heat. The mass transfer to the gas phase gives the change in mass fraction in the gas phase

$$\frac{dm}{dx} = \beta \frac{\varrho sF}{G} (m_{Ph} - m) \tag{275}$$

where β is the mass-transfer coefficient and m_{Ph} the mass fraction of liquid in the phase boundary. The temperature change in the liquid phase is finally obtained from the heat balance for the liquid phase

$$\frac{dT_L}{dx} = \alpha \frac{sF}{L\,c_{pL}} (T_G - T_L)$$

$$-\beta \frac{\Delta h_v \varrho sF}{L\,c_{pL}} (m_{Ph} - m) \tag{276}$$

where Δh_v is the specific latent heat of vaporization of the liquid.

Equations (274) – (276) represent a system of simultaneous ordinary differential equations for T_G, T_L, and m. In order to close these equations a relationship has to be found for m_{Ph}. This can be obtained from the Clausius–Clapeyron equation for vapor pressure $\ln p = -A'/T + B'$. Since the changes are assumed to be very small, a linearized expression can be used in the temperature interval being considered, so that $m_{Ph} = A\,T_L + B$ or $dm_{Ph}/dx = A\,(dT_L/dx)$. By substituting the transformations $\theta = T_G - T_L$ and $\varphi = m_{Ph} - m$, the number of equations can be reduced by one:

$$\frac{d\theta}{dx} = -\alpha \frac{sF}{Gc_{pG}\,L\,c_{pL}} (Gc_{pG} + L\,c_{pL})\,\theta$$

$$+\beta \frac{\Delta h_v \varrho sF}{L\,c_{pL}} \varphi \tag{277a}$$

$$\frac{d\varphi}{dx} = \alpha \frac{AsF}{L\,c_{pL}} \theta + \left[\beta \frac{A \Delta h_v \varrho sF}{L\,c_{pL}} + \beta \frac{\varrho sF}{G}\right] \varphi \tag{277b}$$

Equations (277 a) and (277 b) can also be abbreviated as

$$\theta_x = a_1 \theta + b_1 \varphi \tag{278a}$$

$$\varphi_x = a_2 \theta - b_2 \varphi \tag{278b}$$

where θ_x and φ_x are the derivatives of θ and φ with respect to x. Differentiation of Equation (278 a) and elimination of φ_x and φ give the ordinary homogeneous second-order differential equation

$$\theta_{xx} + (a_1 + b_2)\,\theta_x + (a_1 b_2 - a_2 b_1)\,\theta = 0 \tag{279}$$

to which the system of coupled equations is now reduced. Since $(a_1 + b_2)^2 > 4\,(a_1\,b_2 - a_2\,b_1)$, the characteristic equation of differential Equation (279) $\lambda^2 + (a_1 + b_2)\,\lambda + (a_1\,b_2 - a_2\,b_1) = 0$ has two real roots. The general solution for Equation (279) is thus

$$\theta = C_1 e^{\lambda_1 x} + C_2 e^{\lambda_2 x} \tag{280a}$$

The integration constants C_1 and C_2 are determined from the boundary conditions, which are simplified here as $T_{L0} = T_{G0}$, since water and air are both assumed to have the same (room) temperature. This leads to $\theta = 0$ and $\theta_x = b_1\,\varphi_0$ for $x = 0$. These conditions give $C_1 = -\,C_2$ and $\lambda_1 C_1 + \lambda_2 C_2 = b_1\,\varphi_0$. Thus the solution of Equation (279) is

$$\theta = \frac{b_1 \varphi_0}{\lambda_1 - \lambda_2} \left[e^{\lambda_1 x} - e^{\lambda_2 x}\right] \tag{280b}$$

where φ_0 is to be determined from the known initial temperature of the water and the initial mass fraction of water in the air; λ_1 and λ_2 have to be calculated from the characteristic equation. Equation (280 b) shows that cooling is initiated entirely by mass transfer into the gas phase. The gas temperature can now be determined with the known solution for θ by integration of Equation (274):

$$T_G = T_{G0} - a_3 \frac{b_1 \varphi_0}{(\lambda_1 - \lambda_2)} \left[\frac{1}{\lambda_1} e^{\lambda_1 x} - \frac{1}{\lambda_2} e^{\lambda_2 x}\right] \tag{281}$$

where $a_3 = \alpha sF / Gc_{pG}$

5.3.1.4. Examples of Simultaneous Mass and Heat Transfer: Static Models

In the examples outlined in Section 5.3.1.3, the changes in temperature and mass fractions were calculated from the rates of transport of enthalpy and mass between the individual phases. The concept used in this model is shown in Figure 34 A for the mass transfer of a component from the gas phase into a counterflowing liquid phase. At the phase boundary the mass fraction profile shows a jump caused by the relevant phase equilibrium which may be expressed by the relationship e.g., $m_{\mathrm{PhG}}=Am_{\mathrm{PhL}}$. In contrast to the cases discussed in Sections 5.3.1.1 and 5.3.1.3 the component being transferred is dissolved in the liquid phase. The mass fraction differences between the bulk phases and the phase boundary (Fig. 34 A) can be expressed as

$$(m_{\mathrm{G}}-m_{\mathrm{PhG}})=\frac{j^{m}}{\varrho_{\mathrm{G}}\beta_{\mathrm{G}}sF\mathrm{d}x} \tag{282a}$$

and

$$(m_{\mathrm{PhL}}-m_{\mathrm{L}})=\frac{j^{m}}{\varrho_{\mathrm{L}}\beta_{\mathrm{L}}sF\mathrm{d}x} \tag{282b}$$

where j^{m} is the mass flux of the transferred component. If Equations (282 a) and (282 b) are added and the relationship for the phase equilibrium is used to eliminate m_{PhL}, then the following equation is obtained for the steady-state case, taking into consideration the mass balance for the gas phase:

$$j^{m}=\frac{sF}{\frac{1}{\varrho_{\mathrm{G}}\beta_{\mathrm{G}}}+\frac{A}{\varrho_{\mathrm{L}}\beta_{\mathrm{L}}}}(m_{\mathrm{G}}-Am_{\mathrm{L}})\,\mathrm{d}x$$
$$=-\mathrm{d}m_{\mathrm{G}}G \tag{283}$$

Rearrangement gives

$$-\frac{\mathrm{d}m_{\mathrm{G}}}{(m_{\mathrm{G}}-Am_{\mathrm{L}})}=\frac{sF}{\left(\frac{1}{\varrho_{\mathrm{G}}\beta_{\mathrm{G}}}+\frac{A}{\varrho_{\mathrm{L}}\beta_{\mathrm{L}}}\right)G}\mathrm{d}x \tag{284}$$

and after integration

$$\int_{m_{\mathrm{Ge}}}^{m_{\mathrm{G0}}}\frac{\mathrm{d}m_{\mathrm{G}}}{(m_{\mathrm{G}}-Am_{\mathrm{L}})}=$$
$$\frac{sF}{\left(\frac{1}{\varrho_{\mathrm{G}}\beta_{\mathrm{G}}}+\frac{A}{\varrho_{\mathrm{L}}\beta_{\mathrm{L}}}\right)G}\int_{0}^{H}\mathrm{d}x \tag{285a}$$

which can be abbreviated as

$$\mathrm{NTU}=\frac{H}{\mathrm{HTU}} \tag{285b}$$

where NTU denotes the number of transfer units and HTU the height of one transfer unit.

The notation and assumptions for the derivation of Equation (285 b) are the same as those in Section 5.3.1.3. The same procedure as indicated in Section 5.2.2.2 for the heat transfer leads

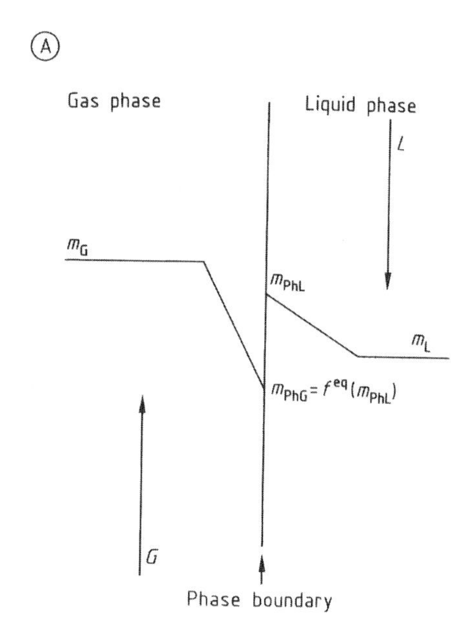

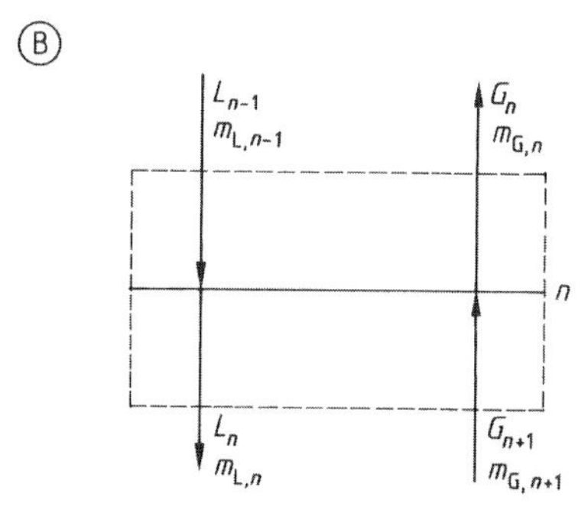

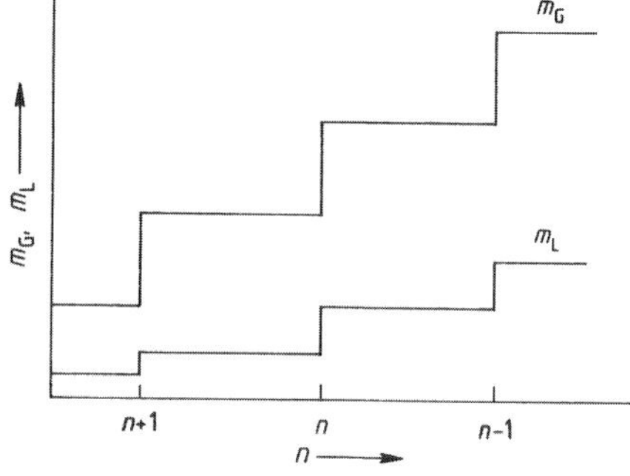

Figure 34. Representation of the dynamic (A) and static (B) models for mass transfer
L = mass flow of liquid; G = mass flow of gas
The subscripts L, G, Ph, and n denote liquid, gas, phase boundary, and the number of theoretical plates, respectively. The superscript eq denotes equilibrium.

to Equation (285 b). A similar relationship is often used for total heat transfer in heat exchangers [108].

Equation (285 b) is used to determine the height of a separation column with a known specific area and cross-sectional area or conversely to determine the concentration difference for a given height of the separation column. The quantity Am_{L} contained in the integral N T U still has to be expressed in terms of m_{G} and the boundary conditions $m_{\mathrm{G},0}$ and $m_{\mathrm{L},0}$ or $m_{\mathrm{L},e}$. All quantities in the H T U expression are known.

A completely different model for mass and heat transfer in mass separation processes is shown in Figure 34 B. Here the mass transfer device consists of a series of separation stages, in which phase equilibrium occurs between the counterflowing phases. Mass transfer is assumed to occur infinitely quickly and the separation stage is homogeneous with regard to the variables of state. The mass balance in the steady-state case for the components transferred is then

$$Lm_{\mathrm{L},n-1}+Gm_{\mathrm{G},n+1}=Lm_{\mathrm{L},n}+Gm_{\mathrm{G},n} \quad (286)$$

Since phase equilibrium should exist in the separation stage, then $m_{\mathrm{G},n}$ can be expressed by $m_{\mathrm{L},n}$. Due to the modeling assumptions, the mass balance does not contain any transport terms. If the relationship for the phase equilibrium is expressed in the form $m_{\mathrm{G}}=A\,m_{\mathrm{L}}+B$, Equation (286) becomes

$$\begin{aligned} &Lm_{\mathrm{L},n-1}+G\left(Am_{\mathrm{L},n+1}+B\right)\\ &\quad=Lm_{\mathrm{L},n}+G\left(Am_{\mathrm{L},n}+B\right) \end{aligned} \quad (287a)$$

or

$$Lm_{\mathrm{L},n-1}-(L+AG)\,m_{\mathrm{L},n}+AGm_{\mathrm{L},n+1}=0 \quad (287b)$$

For a separation column with N separation stages (Fig. 35) for given mass fractions $m_{\mathrm{G},0}$ and $m_{\mathrm{L},0}$, the relevant linear equation system is

$$-(L+AG)\,m_{\mathrm{L},1}+AGm_{\mathrm{L},2}=-Lm_{\mathrm{L},0} \quad (288,1)$$

$$\vdots$$

$$Lm_{\mathrm{L},n-1}-(L+AG)\,m_{\mathrm{L},n}+AGm_{\mathrm{L},n+1}=0 \quad (288,n)$$

$$\vdots$$

$$Lm_{\mathrm{L},N-1}-(L+AG)\,m_{\mathrm{L},N}=-G\left(m_{\mathrm{G},0}-B\right) \quad (288,N)$$

In matrix form this can be written as

$$\begin{pmatrix} -(L+AG) & AG & & \cdots \\ L & -(L+AG)\;AG & & \cdots \\ \vdots & \ddots\quad\ddots & & \vdots \\ \cdots & L & & -(L+AG) \end{pmatrix}$$

$$\times\begin{pmatrix} m_{\mathrm{L},1}\\ \vdots \\ m_{\mathrm{L},N}\end{pmatrix}=\begin{pmatrix} -Lm_{\mathrm{L},0}\\ \vdots \\ -G\left(m_{\mathrm{G},0}-B\right)\end{pmatrix}$$

or

$$\boldsymbol{Bm}=\boldsymbol{c} \quad (288b)$$

The solution of this equation with one of the methods discussed in Section 3.1.2 is

$$\boldsymbol{m}=\boldsymbol{B}^{-1}\boldsymbol{c} \quad (289)$$

and provides the mass fraction profile in discrete form for the component being transferred to the liquid phase over the entire length of the column. Using the appropriate relationships for phase equilibrium the corresponding profile can easily be determined for the gas phase. The system of equations could equally well be derived for the mass fractions in the gas phase.

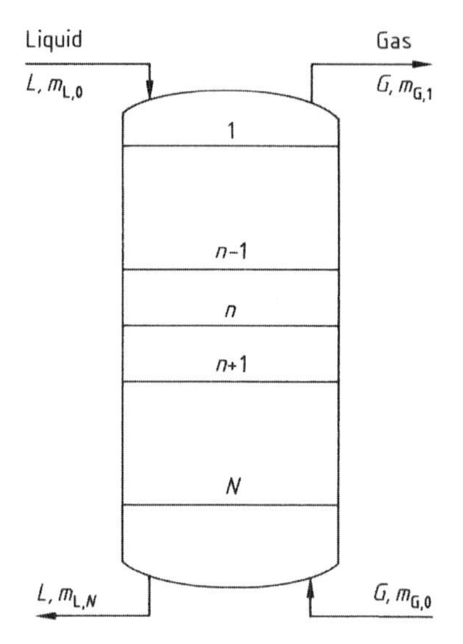

Figure 35. Schematic representation of a counterflow separation column
For explanation of symbols see Figure 34.

The above problem can easily be solved in another way. Transformation of Equation (286) gives

$$m_{L,1} - m_{L,0} = \frac{G}{L}\left(m_{G,2} - m_{G,1}\right) \quad \text{or} \tag{290a}$$

$$\Delta m_L = \frac{G}{L}\Delta m_G \tag{290b}$$

for the first separation stage in the column shown in Figure 35. The geometric interpretation of this equation is to find the point with coordinates ($m_{L,\,1}$, $m_{G,\,2}$) on a straight line of slope (L/G) starting from the point with the coordinates ($m_{L,0}$, $m_{G,\,1}$). As the first step, a point ($m_{L,\,1}$, $m_{G,\,1}$) has to be found that satisfies the phase equilibrium relationship $m_G = Am_L + B$. This is done by inserting $m_{G,\,1}$ into the phase equilibrium relationship as shown by the horizontal line in Figure 36. Working from this point, point ($m_{L,\,1}$, $m_{G,\,2}$) can be found by inserting $m_{L,1}$ into the mass balance equation (Eq. 290 a); this is shown in Figure 36 by the vertical line. In this way Equation (288) can easily be solved graphically by completing the construction for all separation stages step by step. The graphical solution method is the basis for the McCabe – Thiele method and similar procedures for calculating mass separation equipment

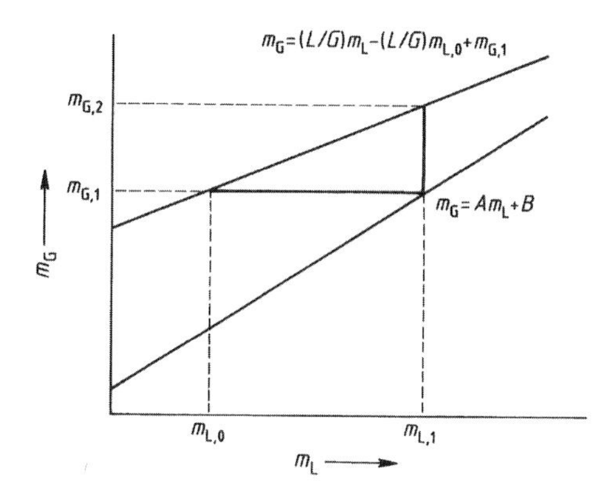

Figure 36. Graphical solution of the equation system (288) for a counterflow separation column: McCabe – Thiele diagram
A and B are coefficients of the relation for thermodynamic phase equilibrium $m_G = f^{\,eq}(m_L)$. For explanation of other symbols see Figure 34.

Static models can also be used for simultaneous heat and mass transfer in mass separation processes. This will be described briefly for the separation stage of the rectification column shown in Figure 37. Phase equilibrium is assumed in the separation stage and the separation stage is regarded as being homogeneous with respect to the variables of state. Phase equilibria relationships exist both for the mass fractions and the enthalpies. With these preconditions, balances for the total mass, the mass of the individual components of the mixture, and the total enthalpy can be given.

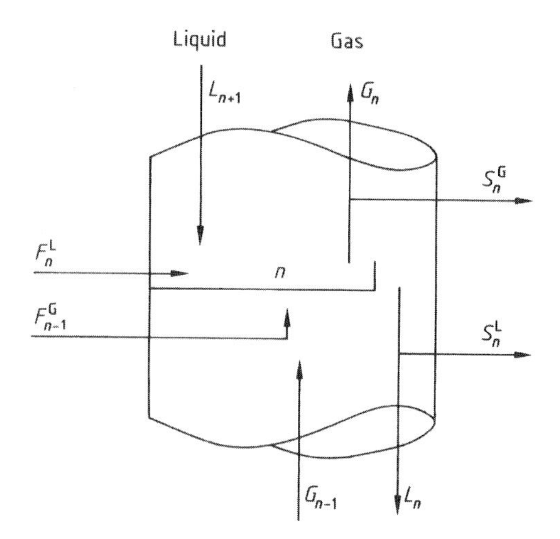

Figure 37. Mass flows in a separation stage of a rectification column
F = liquid or gaseous feed; S = liquid or gaseous sink
The superscripts L and G denote liquid and gas phase, respectively. For explanation of other symbols see Figure 34.

For the sake of simplicity the mixture is taken to be a binary mixture of components A and B. The balances are considered for the steady-state case. (The mass and enthalpy balances can, however, also be formulated without difficulty for a multicomponent mixture or for the nonsteady state, for details see [3], [8], [9].) The equation for the total mass is (notation is defined in Fig. 37)

$$L_{n+1} + G_{n-1} + F_n^L + F_{n-1}^G = L_n + G_n + S_n^G + S_n^L \tag{291}$$

Analogously, the equation for the mass of a component is

$$L_{n+1} m_{L,n+1} + G_{n-1} m_{G,n-1} + F_n^L m_{L,n} + F_{n-1}^G m_{G,n-1} = L_n m_{L,n} + G_n m_{G,n} + S_n^G m_{G,n} + S_n^L m_{L,n} \tag{292}$$

and finally that for the total enthalpy is

$$h_{L,n+1}+h_{G,n-1}+h_{L,F_n^L}+h_{G,F_{n-1}^G} = h_{L,n}+h_{G,n}+h_{G,S_n^G}+h_{L,S_n^L} \quad (293)$$

For a rectification column of N separation stages, for example, the mass fraction in the liquid phase, the mass flows of the liquid phase, and the temperature have to be calculated. To do this Equations (291) – (293) are simplified by setting all flows to and from the external surroundings to zero (see also Fig. 37). This simplification does not change the character of the resulting equations.

The expansions of the equations for inputs or outputs at the individual separation stages, and likewise the formulations for the condensor at the top of the column and the reboiler at the bottom of the column, offer no major difficulties. To reduce Equations (291) – (293) to a system of equations in L, m_L, and T_L further relationships are needed to eliminate m_G and G. Relationships for the phase equilibrium are used for this: $m_G = f^{eq}(m_L)$, and $G_n/G_{n-1} = R_n$.

Here f^{eq} is the relation for phase equilibrium and R_n is the ratio of the evaporation enthalpies $R_n = \Delta h_{v,n}/\Delta h_{v,n-1}$. If mixing effects are neglected the enthalpies are additive, so that for example $h_L = (h_L^A - h_L^B)\, m_L + h_L^B$. Finally, neglecting the mixing effects, $p_{tot} = [p^A(T) - p^B(T)]\, m_L + p^B(T)$ is used to obtain a correlation between the temperature and the composition of the liquid phase. Using $h = h_0 + c_p(T - T_0)$, the enthalpies can finally be eliminated. The resulting system of equations is

$$\left[p^A(T_n) - p^B(T_n)\right] m_{L,n} + p^B(T_n) = p_{tot} \quad (294a)$$

$$\left[m_{L,n} - \left(\frac{1}{1+R_n}\right) f^{eq}(m_{L,n}) + \left(\frac{R_n}{1-R_n}\right) f^{eq}(m_{L,n-1})\right] L_n + \left[\left(\frac{1}{1+R_n}\right) f^{eq}(m_{L,n}) - \left(\frac{R_n}{1-R_n}\right) f^{eq}(m_{L,n-1}) - m_{L,n-1}\right] L_{n+1} = 0 \quad (294b)$$

$$\left[\left(\frac{R_n}{1-R_n}\right)(L_{n+1}-L_n)\right] \cdot \left\{\left[\left(h_L^A+\Delta h_v^A\right) - \left(h_L^B+\Delta h_v^B\right)\right] m_{L,n-1} + \left(h_L^B+\Delta h_v^B\right)\right\} - L_n\left[\left(h_L^A-h_L^B\right) m_{L,n}+h_L^B\right]$$
$$- \left[\left(\frac{1}{1-R_n}\right)(L_{n+1}-L_n)\right] \cdot \left\{\left[\left(h_L^A+\Delta h_v^A\right) - \left(h_L^B+\Delta h_v^B\right)\right] m_{L,n} + \left(h_L^B+\Delta h_v^B\right)\right\}$$
$$+ L_{n+1}\left[\left(h_L^A-h_L^B\right) m_{L,n+1}+h_L^B\right] = 0 \quad (294c)$$

In Equations (294) the relationships for $m_G = f^{eq}(m_L), p^i(T), h_L^i(T)$, and R_n are not explicitly formulated. In this form Equation (294) should be adequate to point to the similarity to Equation (288). A system of N equations is obtained whose system matrix exhibits a distorted tridiagonal structure. Each element consists of a block with 3×3 elements. In contrast to Equation (288) the sytem is nonlinear and coupled.

Consideration of output and input at the individual separation stages or at the top and bottom of the column does not alter the structure of this system of equations; further terms are simply added to the right-hand side and the dimensions of the blocks of the system matrix are increased on account of consideration of a multicomponent mixture. Equation (294) is easy to solve because of the conditioning of the system matrix. Equation (294 a) contains entries only in the main diagonals, and Equation (294 b) on two diagonals. In most applications the system is not therefore solved simultaneously, but is solved successively starting with Equation (294 a) and predefined boundary conditions. Examples of computer programs for the solution of these problems with multicomponent formulations and nonsteady-state conditions are found in [8], [9].

5.3.2. Mass Transfer with Chemical Reactions

In the dynamic and static models for calculating simultaneous mass and heat transfer, chemical reactions have not yet been considered. In many industrial mass separation processes, mass transfer is associated with chemical reactions. The effect of chemical reactions on mass transfer is now discussed using a simple example (Fig. 38) based on the simple boundary layer problem given in Sections 5.2.2.1 and 5.3.1.1. The plate is replaced by a soluble component A which passes into the fluid phase where it reacts with a second component B. The reaction rate is first order with respect to the individual components, the reaction rate is thus

$$\frac{\mathrm{d}m_{\mathrm{A}}}{\mathrm{d}t} = -k'(T)\, m_{\mathrm{A}} m_{\mathrm{B}} \tag{295}$$

The component B is assumed to be present in excess such that $m_{\mathrm{B}} \approx$ const. If $m = m_{\mathrm{A}}$, then Equation (295) can be simplified as

$$\frac{\mathrm{d}m}{\mathrm{d}t} = -k\,(T, m_{\mathrm{B}})\, m \tag{296}$$

Using the same assumptions as in Sections 5.2.2.1 and 5.3.1.1 (the boundary layer approximations and negligible reaction enthalpy), the following system of equations is obtained

$$\frac{\partial u}{\partial x} + \frac{\partial v}{\partial y} = 0 \tag{246}$$

$$u\frac{\partial u}{\partial x} + v\frac{\partial u}{\partial y} = \nu\frac{\partial^2 u}{\partial y^2}, \tag{247}$$

$$u\frac{\partial T}{\partial x} + v\frac{\partial T}{\partial y} = a\frac{\partial^2 T}{\partial y^2} \tag{245b}$$

From the addition of a source term S_{A} to Equation (266) and with $S_A = \mathrm{d}m/\mathrm{d}t$:

$$u\frac{\partial m}{\partial x} + v\frac{\partial m}{\partial y} = D\frac{\partial^2 m}{\partial y^2} - k\,(T, m_{\mathrm{B}})\, m \tag{297}$$

Due to the simplification used in the formulation of Equation (296), the conservation equations for the chemical species are no longer coupled.

Equation system (246), (247), (245 b), and (297) cannot be solved analytically. Numerical solutions for this problem will not be discussed here. However, also for $Pr = Sc = 1$ the similarity of the equations is lost due to the source term $-k\,(T, m_{\mathrm{B}})\, m$ in Equation (297) originating from the chemical reaction. Thus the boundary layer δ_m develops differently from δ_u and δ_T. Since the source term in Equation (297) is negative, a larger gradient of the mass fraction of A occurs and this leads to a thinner boundary layer δ_m in comparison with δ_u, δ_T. From the definition of the mass-transfer number (see Section 5.3.1.1) it can be seen that the local mass-transfer coefficient increases due to the larger gradients. Mass transfer is accelerated by the chemical reaction.

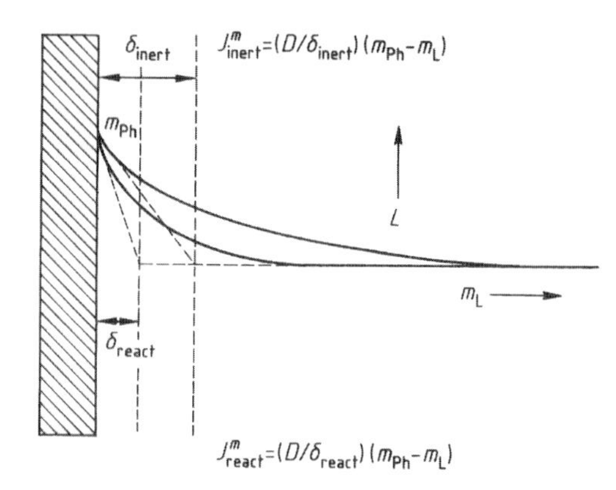

Figure 38. Acceleration of mass transfer by chemical reactions
δ = film thickness; J^{m} = mass fluxes; D = diffusion coefficient
For explanation of other symbols see Figure 34.

For a more thorough discussion of this acceleration effect, the boundary layer problem is further simplified by considering the situation at a large distance from the edge of the plate. Since $\delta_u \sim \sqrt{x}$, the gradients of u, T, and m in the x-direction become extremely small, and v approaches zero. Thus for $x \to \infty$ Equation (297) is simplified to

$$D\frac{\mathrm{d}^2 m}{\mathrm{d}y^2} - k\,(T, m_{\mathrm{B}})\, m \approx 0 \tag{298}$$

If y is transformed into $\eta = y/\delta_m$, then

$$D\frac{\mathrm{d}^2 m}{\mathrm{d}\eta^2} - Ha^2 m = 0 \tag{299}$$

where

$$Ha = \delta_m\sqrt{\frac{k}{D}} = \frac{\delta_m}{D}\sqrt{Dk} = \frac{1}{\beta}\sqrt{Dk}$$

is the Hatta number. The boundary conditions given in Figure 38 are $m = m_{\mathrm{Ph}}$ for $\eta = 0$ and $m = m_{\mathrm{L}}$ for $\eta = 1$. The characteristic equation of the ordinary homogeneous differential Equation (299), $\lambda^2 - Ha^2 = 0$ has the solutions $\lambda_1 = Ha$ and $\lambda_2 = -Ha$ and thus the general solution for Equation (299) is

$$m = C_1 \mathrm{e}^{Ha\eta} + C_2 \mathrm{e}^{-Ha\eta} \tag{300}$$

From the above boundary conditions

$$C_1 = \frac{1}{\mathrm{e}^{Ha} - \mathrm{e}^{-Ha}}\left[m_{\mathrm{L}} - m_{\mathrm{Ph}}\mathrm{e}^{-Ha}\right]$$

and

$$C_2 = -C_1 = \frac{1}{e^{Ha} - e^{-Ha}} \left[m_{Ph} e^{-Ha} - m_L \right]$$

so that the solution becomes

$$m = \frac{1}{\sinh Ha} \left[m_{Ph} \sinh \left(Ha \left(1 - \eta \right) \right) \right.$$

$$\left. + m_L \sinh \left(Ha \eta \right) \right] \tag{301}$$

Mass transfer is defined as usual by the condition

$$J_W^m = -\varrho D \left(\frac{dm}{dy} \right)_{y=0} = -\varrho D \frac{1}{\delta_m} \left(\frac{dm}{d\eta} \right)_{\eta=0}$$

From Equation (301) this condition gives

$$\left(J_W^m \right)_{react} = \varrho \frac{D}{\delta_m} \frac{Ha}{\sinh Ha} \left[m_{Ph} \cosh Ha - m_L \right] \tag{302}$$

If the mass-transfer rates from Equation (302) are compared with the condition without reaction (indicated by the subscript inert)

$$\left(J_W^m \right)_{inert} = \varrho \frac{D}{\delta_m} \left(m_{Ph} - m_L \right) \tag{303}$$

the accelerating effect of the chemical reaction is immediately obvious. The ratio of Equations (302) and (303) gives the "acceleration factor" β_r, i.e., the factor by which mass transfer is accelerated as a result of the chemical reaction:

$$\beta_r = \frac{Ha}{\tanh Ha} \frac{m_{Ph}}{m_{Ph} - m_L} \left[1 - \frac{m_L}{m_{Ph} \cosh Ha} \right] \tag{304a}$$

For $m_L = 0$

$$\beta_r = \frac{Ha}{\tanh Ha} \tag{304b}$$

which in turn, for $Ha > 3$, gives

$$\beta_r = Ha \tag{304c}$$

The acceleration of mass transfer is therefore proportional to the square root of the rate coefficient of the reaction rate. This only applies, however, if the assumptions made in deriving the equation for the expression of the reaction rate (Eq. 296) are fulfilled. For very fast reactions, i.e., for very high Hatta numbers, this is no longer the case.

The concentration of component B is then no longer "unlimited". On the contrary, transport of B into the boundary layer becomes rate-determining. The simultaneous solution for the conservation equations for A and B produces a result which is given in Figure 39 [110]. For $Ha >> 3$

$$\beta_r = 1 + \frac{\nu_A m_{BL} D_B}{\nu_B m_{APh} D_A}$$

The accelerating effect of the reaction is controlled by the limited availability of B.

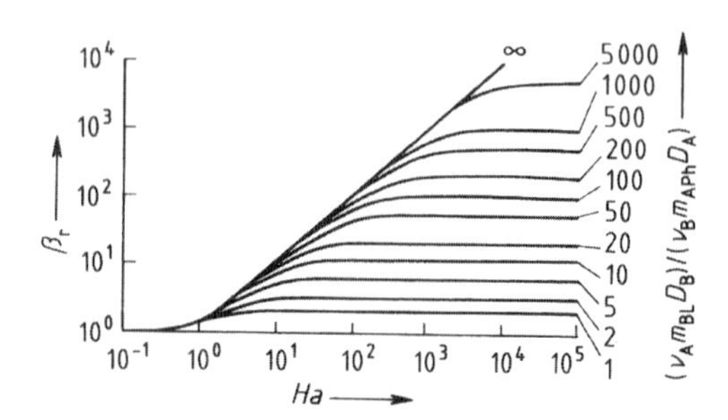

Figure 39. Acceleration factor β_r as a function of the Hatta number *Ha* for various stoichiometric characteristic values [110]
ν = stoichiometric coefficient; D = diffusion coefficient
The subscripts A and B denote components, L and Ph denote liquid and phase boundary, respectively.

Introduction of the acceleration factor β_r means that the effect of the chemical reactions is incorporated into the mass transfer; the chemical reactions are no longer coupled with the mass transfer. Further aspects of mass transfer with chemical reactions are discussed in Section 5.3.3.1.

5.3.3. Chemical Reactions in the Homogeneous Phase

Many chemical reactors are single phase. The term single phase should be understood here to mean that mass transfer from one phase to the other is not rate determining. Many heterogeneous systems also come within the term "single phase", for example heterogeneous catalytic reactions, which are dealt with as quasi-single phase. The mathematical treatment of single-phase reactors differs from the problems discussed in the earlier chapters due to the boundary conditions and the models for the processes occurring at and in the phase boundary layers.

Definition of the state of the reactive mixture in chemically reactive flows with the fields $\boldsymbol{u}(\boldsymbol{x}, t)$, $p(\boldsymbol{x}, t)$, $m_i(\boldsymbol{x}, t)$, $i = 1, \ldots, N$ and $h(\boldsymbol{x}, t)$ requires the solution of the conservation equations for the total mass, the components of the momentum, the mass of the individual chemical species, and the enthalpy. For closure of the system of equations other fundamental equations may have to be incorporated. With the simplifications introduced in the earlier sections for the molecular transport processes this system of equations can be given in the form

$$\frac{\partial \varrho}{\partial t} + \frac{\partial}{\partial x_k}(\varrho u_k) = 0 \tag{131}$$

for the total mass,

$$\varrho \frac{\partial u_i}{\partial t} + \varrho u_k \frac{\partial u_i}{\partial x_k} = \frac{\partial \tau_{ik}}{\partial x_k} - \frac{\partial p}{\partial x_i} + \varrho g_i \quad i = 1, 2, 3 \tag{204}$$

for the components of the momentum,

$$\varrho \frac{\partial h}{\partial t} + \varrho u_k \frac{\partial h}{\partial x_k} = \frac{\partial}{\partial x_k} \lambda \frac{\partial T}{\partial x_k} + \varrho S_h \tag{240b}$$

for the enthalpy, and

$$\varrho \frac{\partial m_i}{\partial t} + \varrho u_k \frac{\partial m_i}{\partial x_k} = \frac{\partial}{\partial x_k}\left[\varrho D_i \frac{\partial m_i}{\partial x_k}\right] + \varrho S_i \quad i = 1, \ldots, N-1 \tag{265b}$$

for the chemical species. Since the system of Equations (131), (204), (240 b), and (265 b) is coupled, simultaneous solution is necessary. This is usually done numerically.

Before discussing a numerical procedure, some fundamental characteristics of the modeling of reactors will be shown by a simplified version of the system of Equations (131), (204), (240 b), and (265 b). The first simplification is the assumption of constant density, isothermal conditions, chemical reactions that do not involve volume changes, and frictionless flow.

5.3.3.1. Isothermal Reactors with Frictionless Flow, Constant Density, and Reactions Without Volume Changes

Under the above conditions it is easy to show that the equation system

$$\varrho \frac{\partial m_i}{\partial t} + \varrho u \frac{\partial m_i}{\partial x} = \frac{\partial}{\partial x}\left[\varrho D_i \frac{\partial m_i}{\partial x}\right] + \varrho S_i \quad i = 1, \ldots, N-1 \tag{305}$$

is adequate for the definition of the problem provided that pressure effects and physical forces are neglected and the assumption is made that $v = w = 0$ (one-dimensional formulation).

If only two chemical species A and B are considered which react in a first-order reaction $A + B \rightarrow$ products, Equation (305) for the steady-state case and with $S_A = -k m_A$ is

$$\varrho u \frac{dm}{dx} - \varrho D \frac{d^2 m}{dx^2} + \varrho k m = 0 \tag{306}$$

In Equation (306) $m = m_A$, and thus also $S_A = S = -k m$. Analytical solutions can be found for this equation for various boundary conditions. If the equation is, for example, normalized after dividing by the density ϱ in the form $\eta = x/L$, then

$$\frac{d^2 m}{d\eta^2} - Pe \frac{dm}{d\eta} - Da_I Pe m = 0$$

Here $Pe = u L/D$, and Da_I is the Damköhler number of the first kind, $Da_I = k\tau$, where $\tau = L/u$ is the mean hydrodynamic residence time. For $Da_I < Pe$, the characteristic equation of the above differential equation has two real roots

$$\lambda_{1,2} = \frac{1}{2} Pe \pm \sqrt{\frac{Pe^2}{4} + Da_I Pe}$$

The general solution is thus

$$m = C_1 e^{\lambda_1 x} + C_2 e^{\lambda_2 x}$$

where C_1 and C_2 are found from the boundary conditions $m = m_0$ for $x = 0$ and $dm/dx = 0$ for $x = L$ as

$$C_1 = -m_0 \frac{\lambda_2 e^{\lambda_2 L}}{\lambda_1 e^{\lambda_1 L} - \lambda_2 e^{\lambda_2 L}} \quad \text{and}$$

$$C_2 = m_0 - C_1 = m_0 \left[1 + \frac{\lambda_2 e^{\lambda_2 L}}{\lambda_1 e^{\lambda_1 L} - \lambda_2 e^{\lambda_2 L}}\right]$$

Here however the numerical procedure analogous to Section 5.1.2.2 is discussed in order to compare various concepts of mathematical modeling.

To convert the differential operators into differences, Equation (306) is subjected to the same manipulations as Equation (226) and brought into discrete form with an "up-wind" difference scheme. For a one-dimensional problem a line along the x-direction of the grid in Figure 26 is adequate (Fig. 40). In Figure 40 the profile for m is also shown. When the operations carried out in Section 5.1.2.2 are applied to Equation (306) with the grid given in Figure 40, then the system of equations

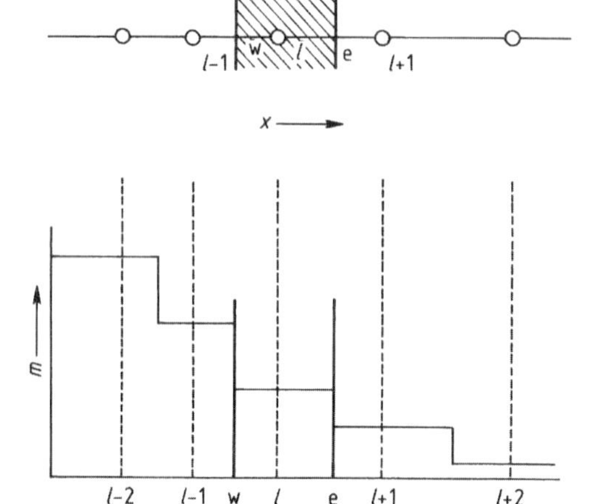

Figure 40. "Up-wind" difference scheme for a second-order differential equation
For explanation of symbols see Figure 26.

$$-a_{l-1}m_{l-1}+b_l m_l - a_{l+1}m_{l+1} - \varrho S \Delta V_l = 0 \qquad (307)$$

is obtained. The coefficients of this tridiagonal system of equations, with the notation from Figure 40 are given by

$$a_{l-1} = \left[+ (\varrho u)_{\mathrm{w}} + \varrho D \frac{1}{x_l - x_{l-1}} \right] F_{\mathrm{w}} \qquad (308a)$$

$$a_{l+1} = \left[+ \varrho D \frac{1}{x_{l+1} - x_l} \right] F_{\mathrm{e}} \qquad (308b)$$

$$b_l = a_{l-1} + a_{l+1} \qquad (308c)$$

Under the conditions given above $u > 0$. Assuming that $(u F) = \Delta V_l / \tau_l$ where τ_l is the hydrodynamic residence time over a distance l, and Pe_l is a local Péclet number $Pe_l = u(x_l - x_{l-1})/D$, the coefficients can be converted into

$$a_{l-1} = \frac{\varrho}{\tau_l} \Delta V_l + \frac{\varrho}{\tau_l Pe_l} \Delta V_l \qquad (308d)$$

$$a_{l+1} = \frac{\varrho}{\tau_l Pe_{l+1}} \Delta V_l \qquad (308e)$$

$$b_l = \frac{\varrho}{\tau_l} \Delta V_l + \frac{\varrho}{\tau_l Pe_l} \Delta V_l + \frac{\varrho}{\tau_l Pe_{l+1}} \Delta V_l \qquad (308f)$$

From Equations (307) and (308 d – f) the following tridiagonal equation system

$$\frac{\varrho}{\tau_l}\left[1 + \frac{1}{Pe_l}\right] m_{l-1} - \frac{\varrho}{\tau_l}\left[1 + \frac{1}{Pe_l} + \frac{1}{Pe_{l+1}}\right] m_l + \frac{\varrho}{\tau_l}\frac{1}{Pe_{l+1}} m_{l+1} + \varrho S(m_l) = 0 \qquad (309)$$

is obtained. If the case of convectively dominated flows is considered where $Pe \to \infty$, Equation (309) becomes

$$\frac{\varrho}{\tau_l}(m_{l-1} - m_l) + \varrho S(m_l) = 0 \qquad (310)$$

This is the mass balance for a cascade of ideally mixed reactors. This analysis again shows that reactors with very large Bodenstein or Péclet numbers (ideally unmixed reactors) can be represented by a cascade of ideally mixed reactors (see Fig. 41 and Section 4.3.1.1).

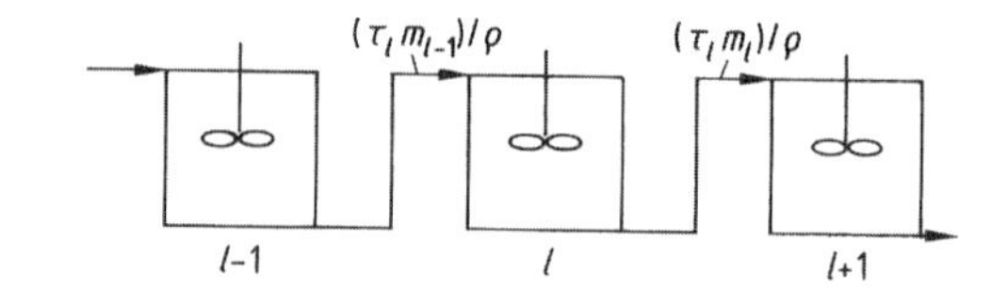

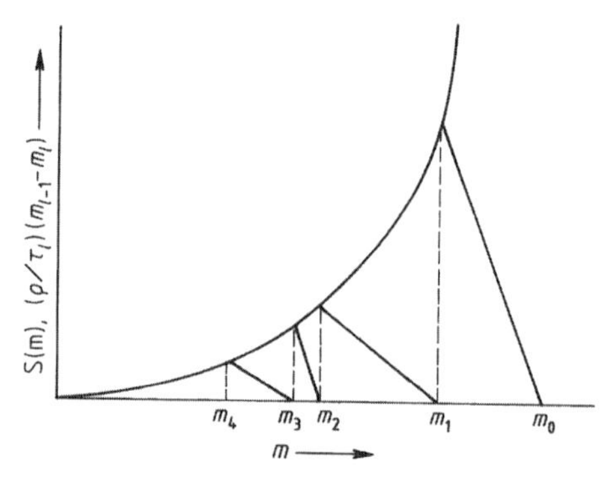

Figure 41. Cascade of ideally mixed reactors equivalent to an ideal tubular reactor
τ = residence time; ϱ = density; l = number of reactor.

Numerically this physical model finds its equivalent in the discrete form of differential Equation (306) using an "up-wind" difference scheme for the convectively dominated case. Equation (310) can easily be solved either by elimination or graphically by plotting both terms of the equation against m_l. The graphical solution is shown in Figure 41 for a nonequidistant grid, i.e., for ideally mixed reactors with different hydrodynamic residence times. It is important that there is no feedback from elements further downstream.

If the flow becomes predominantly diffusive, then *Pe* is very small and Equation (309) is written as

$$\frac{\varrho}{\tau_l}\frac{1}{Pe_l}m_{l-1}-\frac{\varrho}{\tau_l}\left[\frac{1}{Pe_l}+\frac{1}{Pe_{l+1}}\right]m_l$$
$$+\frac{\varrho}{\tau_l}\frac{1}{Pe_{l+1}}m_{l+1}+\varrho S\left(m_l\right)=0 \tag{311}$$

Equation (311) has the same structure as Equation (310) but the feed for the ideally mixed reactor with the number *l* consists of a flow with the composition m_{l-1} and a flow with the composition m_{l+1} (Fig. 42). There is therefore a feedback from the elements further downstream, and the volumetric mass flows are increased by the factor $1/Pe_l$ etc., as compared with the convective mass flow $\varrho m_l/\tau_l$. Algorithms for the solution of the tridiagonal Equation (311) on the basis of LU decomposition of the system matrix can be found in [29–32].

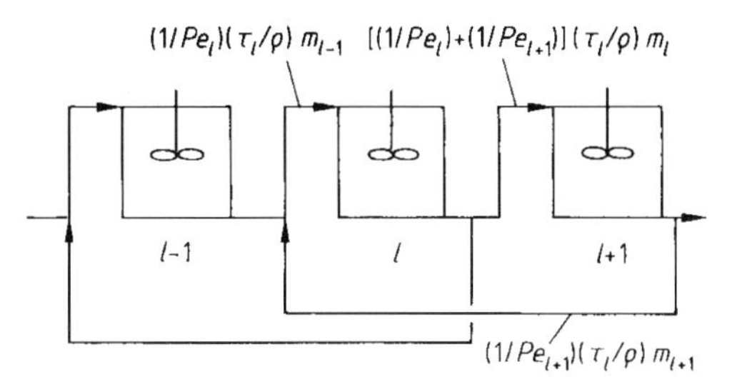

Figure 42. Cascade of ideally mixed reactors with feedback, equivalent to a tubular reactor with strong back-mixing
Pe = Péclet number. For explanation of other symbols see Figure 41.

The general case where $1/Pe_l = O(1)$ has already been formulated with Equation (309). Feedback dependent on the local Péclet number exists from the element $l+1$ and the volumetric mass flows are increased as against the convective flows by the factor $[1+(1/Pe_l)]$ etc. (Fig. 43). For solution, the modeling parameter *Pe* is required, which contains the molecular diffusion coefficients or experimentally determined effective mixing coefficients (see page 245).

In the foregoing analysis real reactors were described by numerical solution of differential Equation (306). This is equivalent to a representation by elements of ideally mixed reactors. Alternative structures of matrices of elements of ideal reactors can be derived by physical analysis of complex reactors [4], [11], [12], [24], [111], [112]. An example which is of some importance in chemical engineering will be described here.

Example of Combinations of Elements of Ideal Reactors: the Loop Reactor. The layout of an ideal plug flow reactor with recirculation (loop reactor) is shown in Figure 44, part of the product flow is added to the feed. For analysis the same assumptions apply as in Section 5.3.3.1.

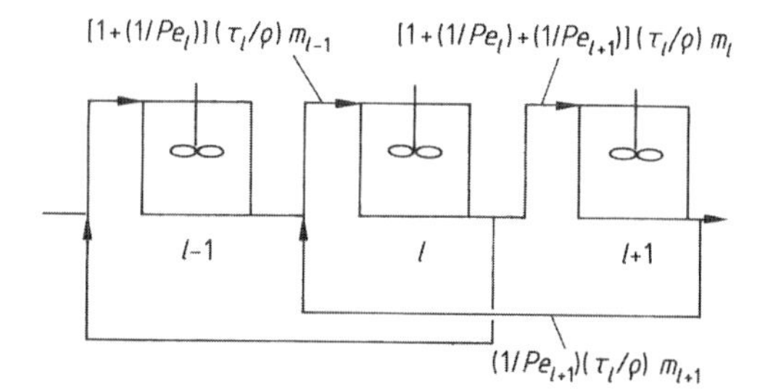

Figure 43. Cascade of ideally mixed reactors with feedback, equivalent to a tubular reactor with back-mixing, general case
Pe = Péclet number. For explanation of symbols see Figure 41.

With the definitions given in Figure 44, the mass fraction m_e at the outlet of the reactor can be obtained from Equation (310)

$$m_e=m_0'\left[\frac{1}{(1+k\tau_l)^n}\right] \tag{312}$$

For the sake of simplicity the ideal plug flow reactor in *n* equidistant sections so that $\tau_l = \tau = \tau_{tot}/n$. Thus from Equation (312)

$$m_e = m'_0 \left[\frac{1}{\left(1 + k\tau_{tot} + \frac{n-1}{n2!}(k\tau_{tot})^2 + \frac{(n-1)(n-2)}{n^2 3!}(k\tau_{tot})^3 + \ldots\right)} \right] \quad (313)$$

For $n \gg 1$ and $k\tau_{tot} \ll 1$

$$m_e \approx m'_0 \left[\frac{1}{\left(1 + k\tau_{tot} + \frac{1}{2!}(k\tau_{tot})^2 + \frac{1}{3!}(k\tau_{tot})^3 + \ldots\right)} \right] = m'_0 e^{-k\tau_{tot}} \quad (314)$$

where m'_0 and τ_{tot} are unknown and depend on the magnitude of the recirculated volume q_r.

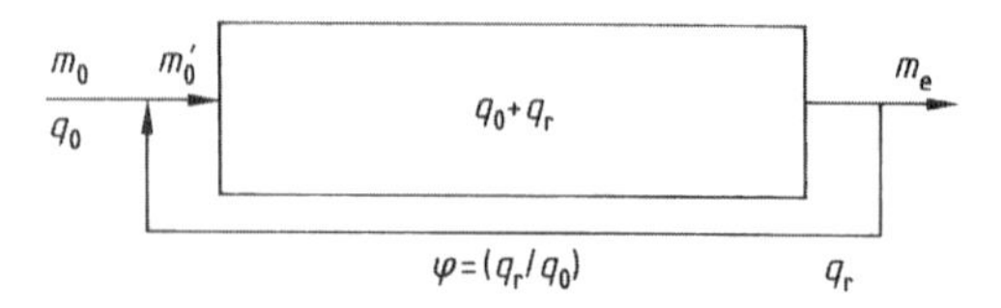

Figure 44. Schematic representation of a loop reactor
q = flow rate; m = mass fraction
The subscripts 0, r, and e denote inlet, loop (recirculated), and final, respectively; the superscript ($'$) denotes conditions after mixing of inlet and backflow.

The term m'_0 results from the combined mass fractions of the flows entering the plug flow reactor and τ_{tot} from the volumes of the flow reactor and the total volumetric flow. If $\varphi = q_r/q_0$ (the recirculation ratio), then

$$m'_0 = m_0 \frac{1}{1+\varphi} + m_e \frac{\varphi}{1+\varphi} \quad (315a)$$

and

$$\tau_{tot} = \tau_0 \frac{1}{1+\varphi} \quad (315b)$$

where τ_0 is the hydrodynamic residence time without recirculation. Substitution into Equation (314) gives

$$m_e = m_0 \frac{e^{-k\tau_0/(1+\varphi)}}{1+\varphi\left(1 + e^{-k\tau_0/(1+\varphi)}\right)} \quad (316)$$

For $\varphi \rightarrow 0$ the loop reactor takes on the character of the ideal plug flow reactor and $m_e = m_0\, e^{-k\tau_0}$. For $k\tau_0 = Da_I = 0.1$, $m_e/m_0 \approx 90\,\%$. For $\varphi \rightarrow \infty$, $e^{-k\tau_0/\varphi} = 1 - (k\tau_0/\varphi)$ and $m_e = m_0/(1 + k\tau_0)$. The loop reactor is similar to an ideally stirred reactor with correspondingly lower conversions.

The adjustable recirculation ratio of the loop reactor is fully exploited in important engineering applications. When measuring kinetic data in laboratory reactors the lowest possible conversion is required in the reactor so that the measured reaction rates can be assigned to specific temperatures and concentrations. Thus $Da_I \ll 1$. On the other hand the concentration differences measured for evaluating the reaction rates should be as large as possible to minimize statistical errors. As may be calculated from Equations (314) and (316), for $Da_I = 0.1$ and $\varphi = 10$ the conversion in the plug flow reactor is $1 - (m_e/m'_0) \approx 1\,\%$. The measured difference in the mass fractions is, however, $m_e/m_0 \approx 90\,\%$.

For reactions with positive reaction order relative to the reactants, the reaction rate falls off as the concentration decreases. Recirculation of the product lowers the concentration in the feed due to dilution. There is however a class of reactions in which the reaction rate first increases with decreasing concentration and then decreases after passing through a maximum (e.g., in heterogeneous catalysis, enzymatic catalysis, and autocatalysis). Recirculation has a positive effect on conversion in these cases.

5.3.3.1.1. Stability of Isothermal Reactors

Steady-State Cases. The importance of the stability of chemical reactors for safety and economic reasons is obvious. The description of stability analysis given here refers to the description of processes in physical space, location $\boldsymbol{x} = (x_1, x_2, x_3)$ and time t. Stability analysis and theories often refer to the description of chemical processes in the form of transfer functions in the space of Laplace-transformed variables or to the frequency behavior. Detailed discussion of these relationships can be found in [3], [9], [113–116].

To illustrate stability analysis in isothermal, steady-state reactors one element of the ideal plug flow reactor will be used (Fig. 40).

This can be regarded as an ideally mixed reactor where the essentials of stability analysis can be shown. Three types of chemical reactions will be considered:

1) Normal reactions with declining reaction rate
$A + B \rightarrow$ products
where $dm_A/dt = -k_2 m_A m_B$ and $m_B = m_{B_0} - (m_{A_0} - m_A)$
2) Autocatalysis
$A + 2B \rightarrow 3B$
where $dm_A/dt = -k_3 m_A m_B^2$ and $m_B = m_{B_0} + (m_{A_0} - m_A)$
3) Self poisoning
$A \rightarrow B$
where $dm_A/dt = -k_1 m_A/(1 + K m_A)^2$ (Section 3.2.1.1, Eq. 53).

Along with the assumptions from Section 5.3.3.1, equal molar masses of A and B are also assumed. Using the abbreviations $\alpha = m_A/m_{A_0}$ and $\beta_0 = m_{B_0}/m_{A_0}$ and Equation (310), the following equations are obtained for the three cases

$$\frac{1-\alpha}{\tau k_2 m_{A_0}} = \alpha^2 + (\beta_0 - 1)\,\alpha \tag{317a}$$

$$\frac{1-\alpha}{\tau k_3 m_{A_0}^2} = \alpha\,[(\beta_0 + 1) - \alpha]^2 \tag{317b}$$

$$\frac{1-\alpha}{\tau k_1} = \alpha\left[\frac{1}{1+\alpha m_{A_0} K\,(2+\alpha m_{A_0} K)}\right] \tag{317c}$$

where $1-\alpha$ is the conversion. The term $\tau k_2 m_{A_0}$, $\tau k_3 m_{A_0}^2$, or τk_1 can be regarded as the ratio of the hydrodynamic residence time to a characteristic chemical time and is denoted as the Damköhler number of the first kind Da_I. Equations (317) can conveniently be solved graphically by plotting both sides against $1-\alpha$. Figures 45, 46, 47 show such plots for the three cases above. For the normal case (Fig. 45 A), decreasing reaction rate with decreasing reactant concentration, an unambiguous steady-state solution exists for each Damköhler number. With increasing Damköhler number (i.e., increasing hydrodynamic residence time), the steady-state solutions give higher conversions. All steady-state solutions, which are plotted against the Damköhler number in Figure 45 B, are stable. If a small disturbance occurs to the left or right on the conversion axis in the steady-state solution, opposing changes are produced in the right-hand side (RHS) or left-hand side (LHS) respectively of Equation (317 a).

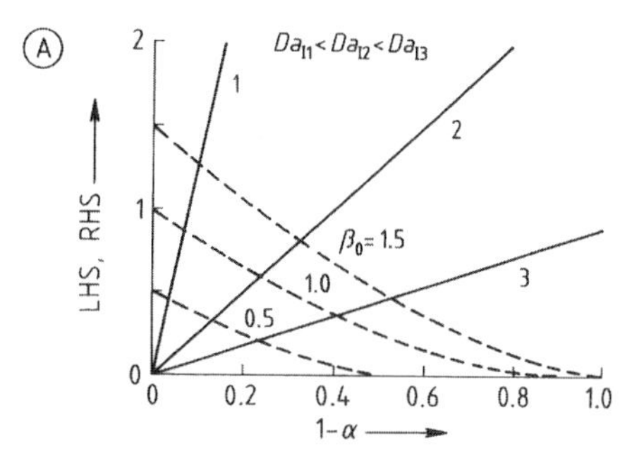

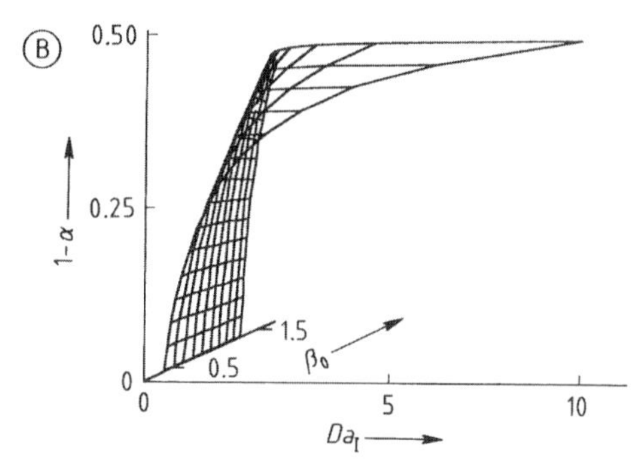

Figure 45. Mono-steady-state conversion points for "normal" chemical reactions (A) and bifurcation diagram (B) for isothermal conditions in a well-stirred reactor
LHS = left-hand side (Eq. 317 a); RHS = right-hand side (Eq. 317 a); $1-\alpha$ = conversion; Da_I = Damköhler number of first kind; $\beta_0 = m_{B_0}/m_{A_0}$

The "convective term" on the LHS decreases (increases), whilst the "reaction term" on the RHS increases (decreases). A lesser (greater) availability of A due to convection therefore counterbalances a greater (lesser) consumption of A due to reaction, so that the conversion is shifted to the right or to the left. The general condition for stability can be given in the terminology of Equation (317) as

$$\frac{\mathrm{d}}{\mathrm{d}\,(1-\alpha)}\,[\mathrm{LHS}] > \frac{\mathrm{d}}{\mathrm{d}\,(1-\alpha)}\,[\mathrm{RHS}] \quad \text{at LHS} = \text{RHS} \tag{318a}$$

For Equation (317 a), $\frac{\mathrm{d}}{\mathrm{d}(1-\alpha)}\,[\mathrm{LHS}] = 1/Da_I$ and $\frac{\mathrm{d}}{\mathrm{d}(1-\alpha)}\,[\mathrm{RHS}] = 2\,(1-\alpha) - \beta - 1$; this satisfies the condition for stability for all possible values of $1-\alpha$ based on the inlet condition β_0.

A completely different picture results for autocatalysis (Fig. 46 A) or self poisoning (Fig. 47 A). In these cases multiple steady-state solutions occur for a particular range of Damköhler numbers. Some of the solutions do not comply with the conditions for stability (Eq. 318). For Equation (317 b) $\frac{d}{d(1-\alpha)}$ [RHS] $= [\beta_0 + (1-\alpha)]\,[2 - 3\,(1-\alpha) - \beta_0]$, so that the stability criterion is not met for small values of $(1-\alpha)$.

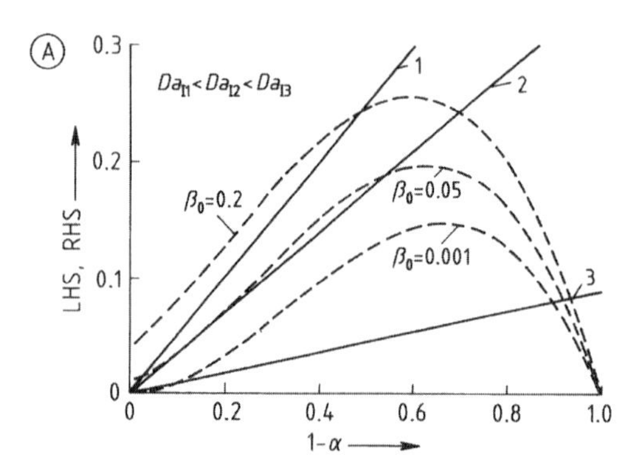

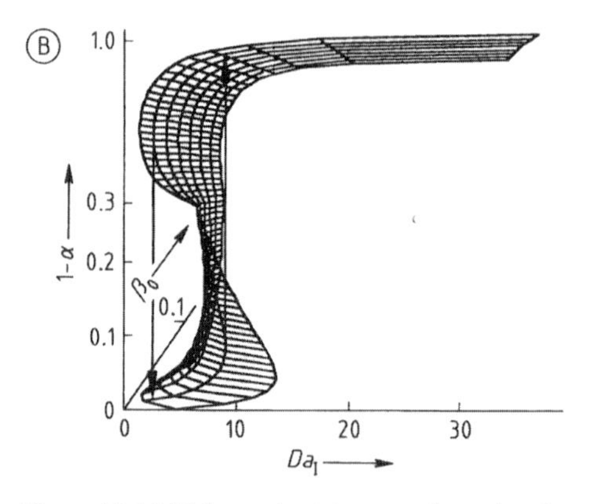

Figure 46. Multiple steady-state conversion points for autocatalytic chemical reactions (A) and bifurcation diagram (B) for isothermal conditions in a well-stirred reactor
LHS = left-hand side (Eq. 317 b); RHS = right-hand side (Eq. 317 b)
For explanation of other symbols see Figure 45.

In this range a small disturbance of $(1-\alpha)$ to the right leads to a larger increase in the "reaction term" than in the availability due to convection; conversion therefore increases. The reaction system thus always heads for the nearest adjacent stable operating point. In the bifurcation diagrams shown in Figures 46 B and 47 B,

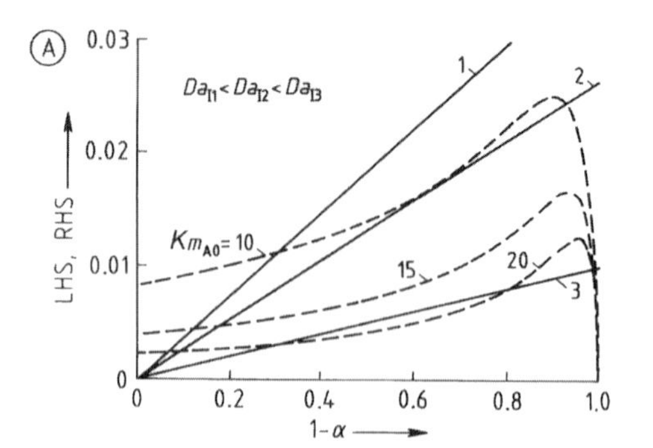

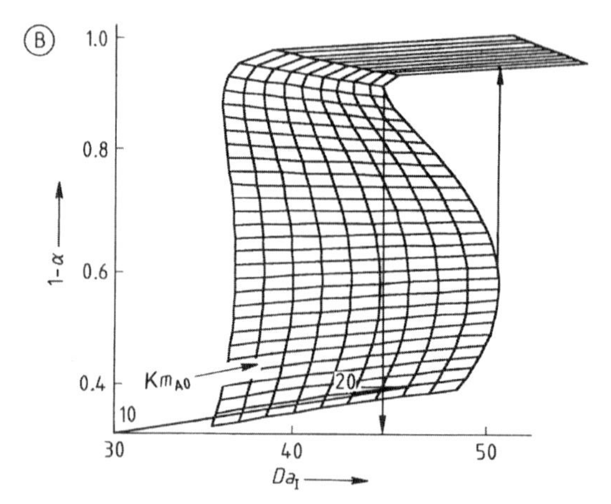

Figure 47. Multiple steady-state conversion points for chemical reactions with self poisoning (A) and bifurcation diagram (B) for isothermal conditions in a well-stirred reactor
K = adsorption coefficient; m_{A_0} = inlet mass fraction of component A; LHS = left-hand side (Eq. 317 c), RHS = right-hand side (Eq. 317 c)
For explanation of other symbols see Figure 45.

these unstable steady-state solutions lie on an S-shaped branch. With a continuous increase in the Damköhler number (residence time) however the system cannot pass through this S-shaped branch, but jumps from a condition with low conversion to one with high conversion. The bifurcation diagrams show that in both cases with multiple steady-state solutions the Damköhler number for "igniting" the reaction is higher than that for "extinguishing" it. Thus hysteresis occurs. The jumping points are given by the condition for the coalescence of an unstable with a stable steady-state solution:

$$\frac{d}{d(1-\alpha)}\,[\text{LHS}] = \frac{d}{d(1-\alpha)}\,[\text{RHS}]$$
$$\text{at LHS} = \text{RHS} \qquad (318\text{b})$$

In the case of autocatalysis this condition occurs for

$$\frac{8}{Da_{\mathrm{I}}}=1+20\beta_0-8\beta_0^2\pm(1-4\beta_0)\sqrt{(1-8\beta_0)}$$

The positions of the two jumping points move closer together and, with an increase in the inlet concentration of the catalyst β_0, the S-shaped curve unfolds. Increasing the concentration of the catalyst at the reactor inlet results in stable reactor behavior, shown by the disappearance of multiple steady-state solutions.

An analogous picture results for self poisoning. The appropriate conditions can easily be calculated from Equation (317 c) and the stability condition. Here the parameter controlling the nonambiguity of the solutions is Km_{A_0} which can be regarded as a measure of the poisoning. A decrease in this quantity (lower pressure or lower inlet concentration of A in heterogeneous catalytic reactions) leads to a range of conversions that exhibits unambiguous, stable, steady-state solutions (compare Figs. 45 B, 46 B, 47 B).

Nonsteady-State Cases. The stability analysis of steady-state, isothermal, ideally mixed reactors has shown that the phenomenon of multiple steady states with hysteresis always occurs when there is feedback. A feedback system will now be considered for the stability analysis of nonsteady-state reactions. The chemical reaction is formulated as

$$\mathrm{P}\xrightarrow{k_0}\mathrm{A}\xrightarrow{k_u}\mathrm{B}\xrightarrow{k_2}\mathrm{C}\quad(k_1:\ \mathrm{A}\to\mathrm{B})$$

where the step A → B is a noncatalytic first-order reaction and a cubic (third-order) reaction catalyzed with B. The feedback mechanism in this reaction is clear: the more B is formed, the faster A is consumed. Since the supply of A from P remains limited, at high concentrations of catalyst B the reaction A → B breaks down until adequate A is again supplied.

Analysis will be performed on a batch reactor so that the system of Equations (305) can be written in the form

$$\frac{dm_{\mathrm{P}}}{dt}=-k_0m_{\mathrm{P}} \tag{319a}$$

$$\frac{dm_{\mathrm{A}}}{dt}=k_0m_{\mathrm{P}}-k_1m_{\mathrm{A}}m_{\mathrm{B}}^2-k_{\mathrm{u}}m_{\mathrm{A}} \tag{319b}$$

$$\frac{dm_{\mathrm{B}}}{dt}=k_1m_{\mathrm{A}}m_{\mathrm{B}}^2+k_{\mathrm{u}}m_{\mathrm{A}}-k_2m_{\mathrm{B}} \tag{319c}$$

A condition for m_{C} results from the mass balance of the system. The quantity $(k_2/k_1)^{1/2}$ is used to normalize the mass fraction, and the normalization of the time is achieved through $(1/k_2)$. The system of equations can then be written as

$$\frac{\mathrm{d}\pi}{\mathrm{d}\tau}=-\varepsilon\pi \tag{320a}$$

where $\pi=m_{\mathrm{P}}(k_2/k_1)^{1/2}$, $\varepsilon=(k_0/k_2)$, and $\tau=t/k_2$

$$\frac{\mathrm{d}\alpha}{\mathrm{d}\tau}=\varepsilon\pi-\alpha\beta^2-\kappa\alpha \tag{320b}$$

where $\alpha=m_{\mathrm{A}}(k_2/k_1)^{1/2}$, $\beta=m_{\mathrm{B}}(k_2k_1)^{1/2}$, and $\kappa=(k_{\mathrm{u}}/k_2)$, and finally

$$\frac{\mathrm{d}\beta}{\mathrm{d}\tau}=\alpha\beta^2+\kappa\alpha-\beta \tag{320c}$$

The initial conditions are $\pi=\pi_0$, $\alpha=\beta=0$ for $\tau=0$. Thus Equation (320 a) can easily be solved. The result

$$\pi=\pi_0\mathrm{e}^{-\varepsilon\tau} \tag{321}$$

is used in the system of Equations (320) which thus become a system of coupled first-order differential equations

$$\frac{\mathrm{d}\alpha}{\mathrm{d}\tau}=\mu-\alpha\beta^2-\kappa\alpha=f(\alpha,\beta) \tag{322a}$$

$$\frac{\mathrm{d}\beta}{\mathrm{d}\tau}=\alpha\beta^2+\kappa\alpha-\beta=g(\alpha,\beta) \tag{322b}$$

where the abbreviation μ denotes $\mu=\varepsilon\,\pi_0\,\mathrm{e}^{-\varepsilon\tau}=\mu_0\mathrm{e}^{-\varepsilon\tau}$. For $k_0\ll k_1,\ k_{\mathrm{u}},\ k_2$ the principle of quasi-steady state is applicable for the mass fractions α and β. Thus Equations (322 a, b) are solved without integration; approximate solutions for α and β can be obtained from the conditions

$$\frac{\mathrm{d}\alpha}{\mathrm{d}\tau}=\frac{\mathrm{d}\beta}{\mathrm{d}\tau}=0$$

From

$$\mu-\alpha\beta^2-\kappa\alpha=0 \tag{323a}$$

and

$$\alpha\beta^2+\kappa\alpha-\beta=0 \tag{323b}$$

the quasi-steady-state mass fractions α_{qs}, β_{qs} are given by

$$\alpha_{qs}=\mu/\left(\mu^2+\kappa\right) \text{ and } \beta_{qs}=\mu \tag{324a,b}$$

For the stability analysis α and β are regarded as being subject to small disturbances $\Delta\alpha$ and $\Delta\beta$, respectively. The time evolution of these disturbances are then considered. The perturbed quantities thus become

$$\alpha = \alpha_0+\Delta\alpha, f(\alpha,\beta)=f(\alpha,\beta)_0+\left(\frac{\partial f(\alpha,\beta)}{\partial\alpha}\right)_0\Delta\alpha + \left(\frac{\partial f(\alpha,\beta)}{\partial\beta}\right)_0\Delta\beta+... \tag{325a}$$

and

$$\beta = \beta_0+\Delta\beta, g(\alpha,\beta)=g(\alpha,\beta)_0+\left(\frac{\partial g(\alpha,\beta)}{\partial\alpha}\right)_0\Delta\alpha + \left(\frac{\partial g(\alpha,\beta)}{\partial\beta}\right)_0\Delta\beta+... \tag{325b}$$

The functions $f(\alpha, \beta)$ and $g(\alpha, \beta)$ in Equations (325) are developed in Taylor series at α_0 and β_0 that are truncated after the first term. This is a reasonable approximation for small disturbances. If Equation (325) is substituted into Equation (322), the time evolution of the disturbances can be expressed as

$$\frac{\partial\alpha_0}{\partial\tau}+\frac{\partial\Delta\alpha}{\partial\tau}=f(\alpha,\beta)_0+\left(\frac{\partial f(\alpha,\beta)}{\partial\alpha}\right)_0\Delta\alpha + \left(\frac{\partial f(\alpha,\beta)}{\partial\beta}\right)_0\Delta\beta \tag{326a}$$

and

$$\frac{\partial\beta_0}{\partial\tau}+\frac{\partial\Delta\beta}{\partial\tau}=g(\alpha,\beta)_0+\left(\frac{\partial g(\alpha,\beta)}{\partial\alpha}\right)_0\Delta\alpha + \left(\frac{\partial g(\alpha,\beta)}{\partial\beta}\right)_0\Delta\beta \tag{326b}$$

If the quasi-steady-state solutions are considered, then Equation (326) becomes the system of coupled ordinary differential equations

$$\begin{pmatrix}\frac{d\Delta\alpha}{d\tau}\\ \frac{d\Delta\beta}{d\tau}\end{pmatrix} - \boldsymbol{J}_{qs}\begin{pmatrix}\Delta\alpha\\ \Delta\beta\end{pmatrix}=0 \tag{327a}$$

where $\boldsymbol{J}_{qs}$ is the Jacobi matrix of Equation (323)

$$\boldsymbol{J}_{qs}=\begin{pmatrix}\frac{\partial f(\alpha,\beta)}{\partial\alpha} & \frac{\partial f(\alpha,\beta)}{\partial\beta}\\ \frac{\partial g(\alpha,\beta)}{\partial\alpha} & \frac{\partial g(\alpha,\beta)}{\partial\beta}\end{pmatrix}_{qs} \tag{327b}$$

Equation (327) can easily be solved analytically after being transformed into a second-order differential equation (see Section 5.3.1.3). The general solution is

$$\Delta\alpha(\tau)=C_1 e^{\lambda_1\tau}+C_2 e^{\lambda_2\tau} \tag{328a}$$

$$\Delta\beta(\tau)=C_3 e^{\lambda_1\tau}+C_4 e^{\lambda_2\tau} \tag{328b}$$

where the eigenvalues λ_1 and λ_2 are determined from the characteristic equation

$$\lambda^2-\mathrm{Tr}(\boldsymbol{J}_{qs})\lambda+\mathrm{Det}(\boldsymbol{J}_{qs})=0 \tag{329}$$

This expression has already been used in Sections 5.3.1.3, 5.3.2, and 5.3.3.1. The trace of the Jacobi matrix is the sum of the diagonal elements

$$\mathrm{Tr}(\boldsymbol{J}_{qs})=\left(\frac{\partial f(\alpha,\beta)}{\partial\alpha}\right)_{qs}+\left(\frac{\partial g(\alpha,\beta)}{\partial\beta}\right)_{qs}$$

and its determinant is

$$\mathrm{Det}(\boldsymbol{J}_{qs})=\left(\frac{\partial f(\alpha,\beta)}{\partial\alpha}\right)_{qs}\left(\frac{\partial g(\alpha,\beta)}{\partial\beta}\right)_{qs}-\left(\frac{\partial g(\alpha,\beta)}{\partial\alpha}\right)_{qs}\left(\frac{\partial f(\alpha,\beta)}{\partial\beta}\right)_{qs}$$

so that the solution of the characteristic equation is

$$\lambda_{1,2}=\frac{1}{2}\left[\mathrm{Tr}(\boldsymbol{J}_{qs})\pm\left\{\mathrm{Tr}(\boldsymbol{J}_{qs})^2-4\mathrm{Det}(\boldsymbol{J}_{qs})\right\}^{1/2}\right] \tag{330}$$

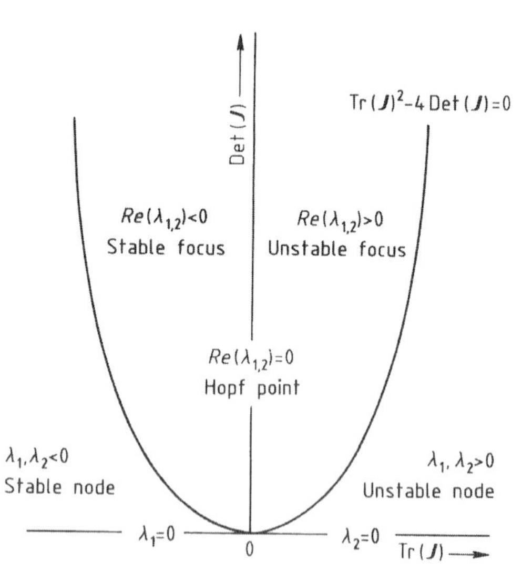

Figure 48. Stability diagram and nature of the steady-state solutions for the example in Section 5.3.3.1
$\lambda_{1,2}$ = eigenvalues of Equations (328).

The time evolution of the disturbances $\Delta\alpha$ and $\Delta\beta$ is affected by the eigenvalues of $\lambda_{1,2}$, which depend on the values of Tr ($\boldsymbol{J}_{qs}$) and Det ($\boldsymbol{J}_{qs}$). The different domains of the solutions of Equations (327) are shown in Figure 48 in a stability diagram.

To demonstrate some of the possible cases the simplifying assumption is made that the first-order noncatalytic reaction (A → B) proceeds very slowly, i.e., $\kappa \ll 1$. Equations (323) and (324) can then be further simplified

$$\mathrm{Tr}\,(\boldsymbol{J}_{qs}) = (1-\mu^2) \tag{331a}$$

$$\mathrm{Det}\,(\boldsymbol{J}_{qs}) = \mu^2 \tag{331b}$$

$$\mathrm{Tr}\,(\boldsymbol{J}_{qs})^2 - 4\mathrm{Det}\,(\boldsymbol{J}_{qs}) = \mu^4 - 6\mu^2 + 1 \tag{331c}$$

As μ is a function of time, it is to be expected that during the course of the reaction with the depletion of the reservoir P, the discriminants, Equation (331 c), as well as Tr ($\boldsymbol{J}_{qs}$) and Det ($\boldsymbol{J}_{qs}$) will change their values and signs, so that $\lambda_{1,2}$ take all possible combinations

1) Initial condition, $\mu > 1+\sqrt{2}$
   Initially only P is present so μ is very large. Then Tr ($\boldsymbol{J}_{qs}$) is negative, Det ($\boldsymbol{J}_{qs}$) is positive, and the discriminant is also positive. Thus $\lambda_{1,2}$ become both real and negative. Small disturbances $\Delta\alpha$ and $\Delta\beta$ decay monotonically, and the quasi-steady-state mass fractions represent a stable node of the system. The terms node (and focus) are derived from the phase diagrams of $\Delta\alpha$ and $\Delta\beta$, see [114]. In Figure 49 this range is shown in the plot of α_{qs} and β_{qs} against μ.
2) Progressive reaction, $1 < \mu < 1+\sqrt{2}$
   In this range of conversion of P, Tr ($\boldsymbol{J}_{qs}$) is negative and Det ($\boldsymbol{J}_{qs}$) positive. The discriminant has now changed its sign however. Due to this $\lambda_{1,2}$ become conjugate complex with a negative real part. In this case the disturbances $\Delta\alpha$ and $\Delta\beta$ behave as damped oscillations and the quasi-steady-state solutions represent a stable focus of the system.
3) Further progress of the reaction, $\sqrt{2}-1 < \mu < 1$
   In this concentration range of P, Tr ($\boldsymbol{J}_{qs}$) is positive and the discriminant negative. Here too $\lambda_{1,2}$ are conjugate complex but with a positive real part. The disturbances $\Delta\alpha$ and $\Delta\beta$ increase in the form of divergent oscillations. The quasi-steady-state solutions are an unstable focus of the system.
4) Depletion of P, $0 < \mu < \sqrt{2}-1$
   Towards the end of the reaction the trace of the Jacobi matrix is always positive; the discriminant has changed its sign and is also positive again. Now $\lambda_{1,2}$ are both real and positive so that the disturbances grow exponentially. The state of the system is described as an unstable node.

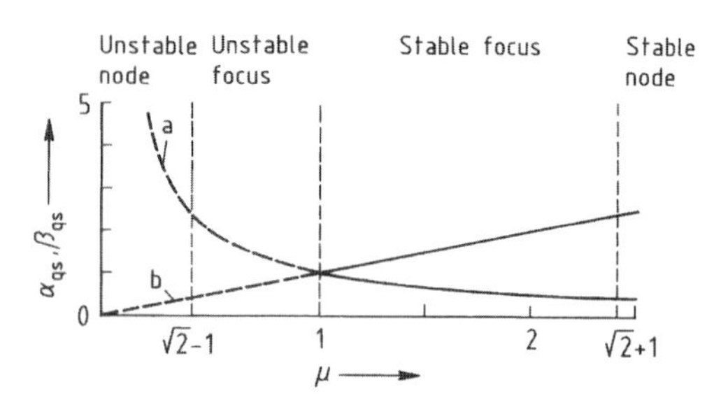

Figure 49. Quasi-steady-state solutions (a) $\alpha_{qs}(\mu)$ and (b) $\beta_{qs}(\mu)$ for an isothermal nonsteady-state system with cubic autocatalysis
For further explanation of symbols see text.

Between the conditions (2) and (3), a state exists at which $\mu = 1$. At this point Tr ($\boldsymbol{J}_{qs}$) = 0 and Det ($\boldsymbol{J}_{qs}$) > 0 so that $\lambda_{1,2}$ become imaginary. Here the disturbances $\Delta\alpha$ and $\Delta\beta$ behave as undamped sinusoidal oscillations and the system forms stable oscillations around the quasi-steady-state solutions. This point is designated the Hopf bifurcation point.

For the simplified system without the noncatalytic reaction, the trace of the Jacobi matrix is given by Equation (331 a). Thus the condition for the change of sign, i.e., Tr ($\boldsymbol{J}_{qs}$) = 0 and dTr ($\boldsymbol{J}_{qs}$)/d$\mu \neq 0$, only occurs when $\mu = 1$. If the noncatalytic reaction is also considered, then

$$\mathrm{Tr}\,(\boldsymbol{J}_{qs}) = \frac{\mu^4 - (1-2\kappa)\,\mu^2 + \kappa\,(1+\kappa)}{\mu^2 + \kappa}$$

(see also Eq. 323). Two Hopf bifurcation points are then obtained from

$$(\mu_{1,2\mathrm{Bf}})^2 = \frac{1}{2}\left[(1-2\kappa) \pm (1-8\kappa)^{1/2}\right]$$

For $\kappa \ll 1$ the upper equation may be expanded into a series and the two bifurcation points are

$$\mu_{1\mathrm{Bf}} = 1 - \frac{3}{2}\kappa + \ldots \quad \text{and}$$

$$\mu_{2\mathrm{Bf}} = \kappa^{1/2}\,(1 + 2\kappa) + \ldots$$

Since Det ($\boldsymbol{J}_{qs}$) is always positive, then with consumption of P the system passes through two Hopf bifurcation points between which solutions for $\lambda_{1,\,2}$ occur with a predominantly imaginary part. In this range instabilities can develop in the form of stable oscillations, the amplitude and frequency of which alter with the decreasing concentration of P [113].

In chemical engineering practice instabilities can only be avoided in systems with the properties quoted above for the range $\mu > 1$. Details of the Hopf bifurcation analysis and the detailed calculation of the oscillation properties (frequency, amplitude, and stability) are given in [113–116], see also → Mathematics in Chemical Engineering, Chap. 6.7.

5.3.3.1.2. Sensitivity Analysis

In Section 3.1.1.3 sensitivity analysis was described as the change in the dependent variables of a model with a change of its modeling parameters. If the dependent variables $\boldsymbol{\Phi}$ are developed as a Taylor series at a specific solution $\boldsymbol{\Phi}_0$, the change is given by

$$\boldsymbol{\Phi} = \boldsymbol{\Phi}_0 + \left(\frac{\partial \boldsymbol{\Phi}}{\partial \boldsymbol{\beta}}\right)_0 \Delta\boldsymbol{\beta} + \left(\frac{\partial^2 \boldsymbol{\Phi}}{\partial \boldsymbol{\beta}^2}\right)_0 \frac{\Delta\boldsymbol{\beta}^2}{2!} + \ldots \tag{332}$$

where $\boldsymbol{\beta}$ is the vector of the parameters.

The simplest measure of the system's sensitivity are the gradients ($\partial\boldsymbol{\Phi}/\partial\boldsymbol{\beta}$) with which the change in the dependent variables is calculated with the aid of a Taylor series truncated after the first term. The gradients ($\partial\boldsymbol{\Phi}/\partial\boldsymbol{\beta}$) are denoted as a first-order sensitivity coefficients, the curve and the higher derivatives of Equation (332) as sensitivity coefficients of the second and correspondingly higher orders.

If the solution $\boldsymbol{\Phi}$ is available in explicit form (e.g., Eq. 323), calculation of the sensitivity coefficients is simple. For very many models of chemical engineering processes the differential equations which constitute the model cannot, however, be solved analytically or with the help of simple assumptions. The sensitivity coefficients must then be determined from the differential equations or from the numerical solutions [117], [118]. The latter method will now be described.

Numerical solutions provide the dependent variables not in the form of continuous functions but in discrete form (discrete values), calculation of the gradients is therefore not trivial. The numerical solution of a differential equation is (e.g., for the example in Section 5.3.3.1) given implicitly through Equation (309) in the form of a system of equations.

$$F\,(\boldsymbol{\Phi}) = 0 \tag{333}$$

The first-order sensitivity coefficients are obtained from Equation (333) by the differentiation rule

$$\frac{\partial F\,(\boldsymbol{\Phi})}{\partial\,(\boldsymbol{\Phi})}\frac{\partial\,(\boldsymbol{\Phi})}{\partial \boldsymbol{\beta}} + \frac{\partial F\,(\boldsymbol{\Phi})}{\partial \boldsymbol{\beta}} = 0 \tag{334}$$

Thus the gradients are given by

$$\frac{\partial\,(\boldsymbol{\Phi})}{\partial \boldsymbol{\beta}} = -\boldsymbol{J}^{-1}\frac{\partial F\,(\boldsymbol{\Phi})}{\partial \boldsymbol{\beta}} \tag{335}$$

In the example from Section 5.3.3.1 the corresponding relations are easily verified: the Jacobi matrix is given by the coefficients a_{l-1}, $-[b_l - \varrho\,(\partial S\,(m_l)/\partial m_l)]$, and a_{l+1} and is identical to the coefficient matrix for the equation system (Eq. 309). For $\beta = k$ the derivatives with respect to the parameters are simply $\varrho\,(\partial S\,(m_l)/\partial k)$. For a first-order reaction $\varrho\,(\partial S\,(m_l)/\partial k) = -\,\varrho\, m_l$. If we simplify in Equation (309) $Pe_l = Pe_{l+1} = Pe$ then

$$\frac{\partial F}{\partial Pe} = -\frac{\varrho}{\tau_l\, Pe^2}\,(m_{l-1} - 2m_l + m_{l+1})$$

so for the sensitivity coefficients the equation system (336) has to be solved. Equation system (336) for the gradients $\partial(\boldsymbol{\Phi})/\partial\beta$ has the same structure as the original equation system (Eq. 309) and can consequently be solved by similar methods.

For the calculation and discussion of the sensitivity coefficients of higher order, reference should be made to the literature [117], [118]. An example of calculating first-order sensitivity coefficients for a complex model follows in Section 5.3.3.2.

$$\begin{pmatrix} \frac{\partial m_1}{\partial k} & \frac{\partial m_1}{\partial Pe} \\ \vdots & \vdots \\ \frac{\partial m_n}{\partial k} & \frac{\partial m_n}{\partial Pe} \end{pmatrix} = - \begin{pmatrix} -(b_1+\varrho k) & a_2 & & & \\ & a_1 & -(b_2+\varrho k) & a_3 & \\ & & & \ddots & \\ & & & a_{n-1} & -(b_n+\varrho k) \end{pmatrix}^{-1} \cdot \begin{pmatrix} -\varrho m_1 & -\frac{\varrho}{\tau_1 Pe^2}(m_0-2m_1+m_2) \\ \vdots & \vdots \\ -\varrho m_n & -\frac{\varrho}{\tau_n Pe^2}(m_{n-1}-2m_n) \end{pmatrix} \quad (336)$$

5.3.3.2. Nonisothermal Reactors

In nonisothermal reactors the density does not remain constant. The couplings in the system of Equations (131), (204), (240 b), and (265 b) therefore cannot be resolved and the equations must be solved simultaneously.

5.3.3.2.1. Heterogeneous Catalytic Reactions

Example: Heterogeneous Catalytic Dehydration of Ethylbenzene to Styrene. The heterogeneous catalytic dehydration of ethylbenzene to styrene is outlined in Figure 50. Ethylbenzene is preheated and mixed with superheated steam. The mixture is passed through a cylindrical reactor packed with an iron oxide catalyst. Dehydration is endothermic. The steam, used as a heat carrier, causes some secondary reactions. A suitable reaction scheme is [11], [119], [120]:

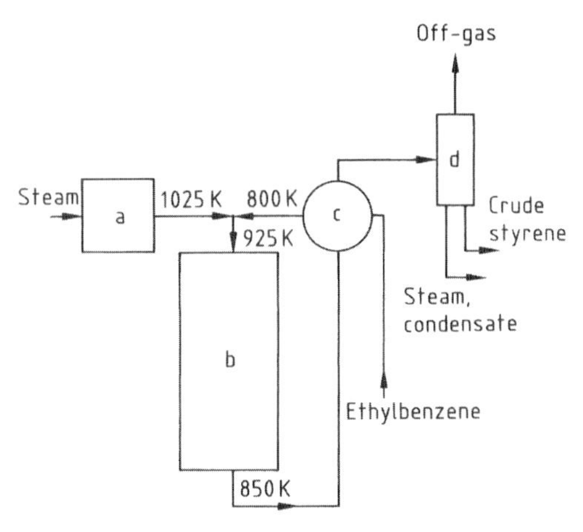

Figure 50. Schematic of a pilot plant for catalytic dehydration of ethylbenzene
a) Superheater; b) Catalytic reactor; c) Heat exchanger; d) Separator.

$$C_6H_5{-}C_2H_5(\mathbf{1}) \rightarrow C_6H_5{-}C_2H_3(\mathbf{2}) + H_2(\mathbf{3}) \quad \text{(i)}$$

$$C_6H_5{-}C_2H_5 \rightarrow C_6H_6(\mathbf{4}) + C_2\,H_4(\mathbf{7}) \quad \text{(ii)}$$

$$C_6H_5{-}C_2H_5 + H_2 \rightarrow C_6H_5{-}CH_3(\mathbf{5}) + CH_4(\mathbf{6}) \quad \text{(iii)}$$

$$1/2C_2H_4 + H_2O\,(\mathbf{10}) \rightarrow CO\,(\mathbf{8}) + 2\,H_2 \quad \text{(iv)}$$

$$CH_4 + H_2\,O \rightarrow CO + 3\,H_2 \quad \text{(v)}$$

$$CO + H_2O \rightarrow CO_2\,(\mathbf{9}) + H_2 \quad \text{(vi)}$$

The formulation of this set of reactions is necessary because they all have an appreciable reaction enthalpy so that conversion into the various byproducts also affects the temperature profile. Furthermore the reaction rate of reaction (i) is given by $r_{(i)} = k_1(m_1 - m_2\, m_3/K)$ [8], [108] which indicates poisoning due to the products styrene and hydrogen, and thus the mass fraction m_1 is coupled with m_2, m_3.

The appropriate formulation of Equations (131), (204), (240 b), and (265 b) is in cylindrical coordinates for this axially symmetrical problem. The continuity equation in this form is

$$\frac{\partial \varrho}{\partial t} + \varrho \frac{\partial}{\partial x}(u) + \varrho \frac{1}{r}\frac{\partial}{\partial r}(rv) = 0 \quad (337a)$$

If we use the normalized variables $u* = u/u_0$, $v* = v/u_0$, $\varrho* = \varrho/\varrho_0$, $t* = t/\tau$, $x* = x/L$, and $r* = r/R$, then from Equation (337 a) it follows that

$$\frac{\varrho_0}{\tau}\frac{\partial \varrho^*}{\partial t^*} + \frac{\varrho_0 u_0}{L}\varrho^* \frac{\partial}{\partial x^*}(u^*) + \frac{\varrho_0 u_0}{R}\varrho^* \frac{1}{r^*}\frac{\partial}{\partial r^*}(r^* v^*) = 0 \quad (337b)$$

This can be transformed into

$$\frac{\partial \varrho^*}{\partial t^*}+\varrho^*\frac{\partial}{\partial x^*}(u^*)+\varrho^*\frac{1}{r^*}\frac{\partial}{\partial r^*}\left(r^*v^*\frac{R}{L}\right)=0 \tag{337c}$$

Since $u*$, $x*$, and $r*$ are all of the order of magnitude $O(1)$, then $v*(L/R)=O(1)$ or $v*\approx(R/L)$, and thus for $(R/L)\ll 1$ the third term in Equation (337 c) can be neglected. In dimensional quantities the continuity equation for the steady-state case thus has the form

$$\varrho\frac{\mathrm{d}u}{\mathrm{d}x}=0 \tag{337d}$$

The balance of momentum (Eq. 204) with the above result in cylindrical coordinates is

$$\varrho\frac{\partial u}{\partial t}+\varrho u\frac{\partial u}{\partial x}=\left[\frac{\partial \tau_{xx}}{\partial x}+\frac{1}{r}\frac{\partial(r\tau_{rx})}{\partial r}+\frac{1}{r}\frac{\partial \tau_{\theta x}}{\partial \theta}\right]-\frac{\partial p}{\partial x} \tag{338a}$$

For flows through packed catalyst beds the viscous stress term can be expressed by the permeability κ according to Darcy's Law [8], [121]. The term in the square brackets in Equation (338 a) then becomes

$$\left[\frac{\partial \tau_{xx}}{\partial x}+\frac{1}{r}\frac{\partial(r\tau_{rx})}{\partial r}+\frac{1}{r}\frac{\partial \tau_{\theta x}}{\partial \theta}\right]=\frac{\mu u}{\kappa}$$

If the same normalized variables are used as in Equation (337) together with the normalized pressure $p*=p/p_0$, Equation (338 a) becomes

$$\frac{R}{\tau u_0}\frac{\partial u^*}{\partial t^*}+\frac{R}{L}u^*\frac{\partial u^*}{\partial x^*}=-\frac{\mu R}{\kappa u_0\varrho_0}\frac{u^*}{\varrho^*}-\frac{Rp_0}{L\,\varrho_0u_0^2}\frac{1}{\varrho^*}\frac{\partial p^*}{\partial x^*} \tag{338b}$$

Estimates of the orders of magnitude of the factors in front of the differential quotients in Equation (338 b) under pilot plant conditions ($\tau\approx 1$ s, $u_0\approx 0.2$ m/s, $R/L\approx 5\times10^{-2}$, $\varrho_0\approx 0.5$ kg/m^3, $p_0\approx 1.5\times10^5$ Pa, $\mu\approx 10^{-5}$ kg m^{-1} s^{-1}, and finally $\kappa\approx 1.5\times10^{-8}$ m^2) leads to

$$\frac{R}{L}<\frac{R}{\tau u_0}<1<\frac{\mu R}{\kappa u_0\varrho_0}\ll\frac{Rp_0}{L\,\varrho_0u_0^2}$$

Thus, for the given conditions, the pressure term predominates in the steady-state balance of momentum which in dimensional quantities simplifies to

$$\frac{\mathrm{d}p}{\mathrm{d}x}\approx 0 \quad\text{or}\quad p\approx\text{const.} \tag{338c}$$

Consequently the continuity equation and the momentum equation are no longer coupled with the other transport equations and pressure p and velocity u are simple to estimate.

Using the definition equation for enthalpy, the conservation equation for the enthalpy (Eq. 240 b) can be transformed into a differential equation for temperature (see Section 5.2.2.1). For a multicomponent system the conservation equations for the mass of the chemical species must be used for this. The result, also in cylindrical coordinates, is

$$\varrho c_p\frac{\partial T}{\partial t}+\varrho c_p u\frac{\partial T}{\partial x}+\lambda\left[\frac{\partial^2 T}{\partial x^2}+\frac{1}{r}\frac{\partial}{\partial r}\left(r\frac{\partial T}{\partial r}\right)\right]=\sum r_i(-\Delta h_{r_i})+S_\mathrm{C} \tag{339a}$$

Equation (339 a) contains the assumptions introduced in Section 5.2. In addition since $p\approx$ const., S_C only includes a term for heat transfer from the gas phase to the catalyst. Due to the transformation of the conservation equation of enthalpy into a differential equation for temperature, the term $\sum r_i(-\Delta h_{r_i})$ occurs which describes the release of heat due to chemical reactions. As in Equations (337) and (338), all quantities are normalized to the conditions at the inlet of the reactor. For Equation (339 a) this results in

$$\begin{aligned}&\frac{L}{\tau u_0}\varrho^*c_p^*\frac{\partial T^*}{\partial t^*}+\varrho^*c_p^*u^*\frac{\partial T^*}{\partial x^*}\\&=\frac{\lambda_0}{L\,\varrho_0c_{p0}u_0}\lambda^*\frac{\partial^2 T^*}{\partial x^{*2}}\\&+\frac{\lambda_0 L}{R^2\varrho_0c_{p0}u_0}\frac{\lambda^*}{r^*}\frac{\partial}{\partial r^*}\left(r^*\frac{\partial T^*}{\partial r^*}\right)\\&+\frac{L\alpha s}{\varepsilon\varrho_0c_{p0}u_0}(T^*_\mathrm{Cat}-T^*)\\&+\frac{L}{u_0T_0\varrho_0c_{p0}}\sum r_i(-\Delta h_{r_i})\end{aligned} \tag{339b}$$

where α is the heat-transfer coefficient, s the specific surface area, ε the intergranular volume, and T_Cat the catalyst temperature. Estimation of the factors preceding the differential quotients in Equation (339 b) ($c_{p0}\approx 2.5$ kJ kg^{-1} K^{-1}, $\lambda_0\approx 4\times10^{-5}$ kJ m^{-1} K^{-1} s^{-1}, $\alpha\approx 4\times10^{-2}$ kW $\cdot$ m^{-2} K^{-1}, $\varepsilon\approx 0.36$, and $s\approx 1.5\times10^3$ m^{-1}) results in

$$\frac{\lambda_0 L}{R^2\varrho_0c_{p0}u_0}<\frac{\lambda_0}{L\,\varrho_0c_{p0}u_0}\ll\frac{L}{\tau u_0}\approx 1\ll\frac{L\alpha s}{\varepsilon\varrho_0c_{p0}u_0}$$

The fundamental mechanisms for energy transport are therefore expressed in the one-dimensional form of the temperature equation, which in steady-state formulation with dimensional quantities becomes

$$\varrho c_p u \frac{\mathrm{d}T}{\mathrm{d}x} = \frac{\alpha s}{\varepsilon}\left(T_{\mathrm{Cat}} - T\right) + \sum r_i \left(-\Delta h_{r_i}\right) \quad (339\mathrm{c})$$

The heat-transfer term in Equation (339 c) couples the temperature of the gas phase T with the catalyst temperature T_{Cat}. This coupling requires an equation for determining the catalyst temperature, which is derived from the heat balance for the catalyst bed:

$$\varrho_{\mathrm{Cat}} c_{p\mathrm{Cat}} \frac{\partial T_{\mathrm{Cat}}}{\partial t} = \lambda_{\mathrm{eff}} \frac{\partial^2 T_{\mathrm{Cat}}}{\partial x^2} + \frac{\lambda_{\mathrm{eff}}}{r} \frac{\partial}{\partial r}\left(r \frac{\partial T_{\mathrm{Cat}}}{\partial r}\right) - \left(\frac{1}{1-\varepsilon}\right) \alpha s \left(T_{\mathrm{Cat}} - T\right) \quad (340\mathrm{a})$$

Normalizing the equation in similar fashion to Equations (337), (338), or (339) results in

$$\varrho_{\mathrm{Cat}} c_{p\mathrm{Cat}} \left(\frac{T_0}{\tau}\right) \frac{\partial T^*_{\mathrm{Cat}}}{\partial t^*} = \frac{\lambda_{\mathrm{eff}} T_{\mathrm{Cat0}}}{L^2} \frac{\partial^2 T^*_{\mathrm{Cat}}}{\partial x^{*2}} + \lambda_{\mathrm{eff}} \frac{T_{\mathrm{Cat0}}}{R^2} \frac{1}{r^*} \frac{\partial}{\partial r^*}\left(r^* \frac{\partial T^*_{\mathrm{Cat}}}{\partial r^*}\right) - \left(\frac{1}{1-\varepsilon}\right) \alpha s T_{\mathrm{Cat0}} \left(T^*_{\mathrm{Cat}} - T^*\right) \quad (340\mathrm{b})$$

Estimation of the orders of magnitude of the factors preceding the differential quotients gives

$$\frac{\lambda_{\mathrm{eff}} T_{\mathrm{Cat0}}}{L^2} \ll \lambda_{\mathrm{eff}} \frac{T_{\mathrm{Cat0}}}{R^2}$$

and thus the fundamental mechanism for the transport of enthalpy in the catalyst bed in the steady state is represented by an equation which is one-dimensional in the radial direction

$$0 = \frac{\lambda_{\mathrm{eff}}}{r} \frac{\mathrm{d}}{\mathrm{d}r}\left(r \frac{\mathrm{d}T_{\mathrm{Cat}}}{\mathrm{d}r}\right) - \left(\frac{1}{1-\varepsilon}\right) \alpha s \left(T_{\mathrm{Cat}} - T\right) \quad (340\mathrm{c})$$

The remaining problem is the expression of the conservation equations for the chemical species by means of similar estimates of orders of magnitudes. The conservation equation for the chemical species with the simplified continuity equation (Eq. 337 d)

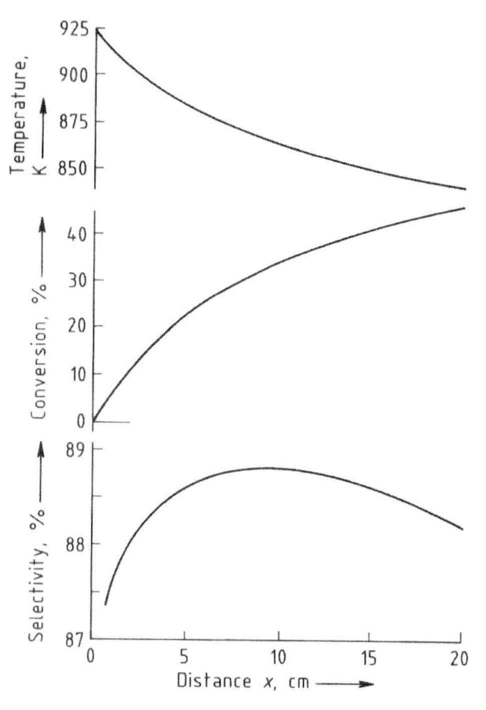

Figure 51. Calculated temperature, conversion, and selectivity profiles for the catalytic dehydration of ethylbenzene [119]

$$\varrho \frac{\partial m_i}{\partial t} + \varrho u \frac{\partial m_i}{\partial x} = \frac{\partial}{\partial x}\left[\varrho D_i \frac{\partial m_i}{\partial x}\right] + \frac{1}{r} \frac{\partial}{\partial r}\left[r \varrho D_i \frac{\partial m_i}{\partial r}\right] + \varrho S_i \qquad i = 1, \ldots, 8 \quad (341\mathrm{a})$$

is first formulated with the variables normalized to the state at the reactor inlet.

$$\frac{\varrho_0 m_{i0}}{\tau} \varrho^* \frac{\partial m_i^*}{\partial t^*} + \frac{\varrho_0 m_{i0} u_0}{L} \varrho^* u^* \frac{\partial m_i^*}{\partial x^*} = \frac{\varrho_0 D_{i0} m_{i0}}{L^2} \varrho^* D_i^* \frac{\partial^2 m_i^*}{\partial x^{*2}} + \frac{\varrho_0 D_{i0} m_{i0}}{R^2} \frac{1}{r^*} \frac{\partial}{\partial r^*}\left(r^* \varrho^* D_i^* \frac{\partial m_i^*}{\partial r^*}\right) + \varrho_0 \varrho^* S_i, \quad i = 1, \ldots, 8 \quad (341\mathrm{b})$$

With $D_{i0} \approx 10^{-3}\,\mathrm{m^2\,s^{-1}}$

$$\frac{\varrho_0 D_{i0} m_{i0}}{L^2} \ll \frac{\varrho_0 D_{i0} m_{i0}}{R^2} \ll \frac{\varrho_0 m_{i0} u_0}{L}$$

and for the pilot plant conditions assumed here, obviously convective transport in the axial direction is dominant. The following one-dimensional equation results for the steady-state case

$$\varrho u \frac{\mathrm{d}m_i}{\mathrm{d}x} = \varrho S_i = \varrho r_i, \quad i = 1, \ldots, 8 \quad (341\mathrm{c})$$

The above estimates have resulted in a reduction of the model for the fixed bed catalyst reactor: convective transport of mass in the axial direction is the main transport mechanism for the gas phase, whilst for the catalyst phase conductive transport of heat in the radial direction along with heat transfer predominates. Consequently the couplings in Equations (131), (204), (240 b), and (265 b) are largely reduced.

To solve the system of Equations (339 c), (340 c), and (341 c), Equation (340 c) for the catalyst temperature is transformed into normalized variables with the transformations

$$r^* = \frac{r}{R}, \quad \theta = \frac{T_{\mathrm{Cat}} - T}{T_{\mathrm{W}} - t}$$

(T_{W} is the wall temperature) and

$$\beta^2 = \frac{R^2 \alpha s}{\lambda_{\mathrm{eff}}\,(1-\varepsilon)}$$

This gives

$$\frac{\mathrm{d}^2\theta}{\mathrm{d}r^{*2}} + \frac{1}{r^*}\frac{\mathrm{d}\theta}{\mathrm{d}r^*} - \beta^2\theta = 0 \tag{342}$$

with the boundary condition $(\mathrm{d}\theta/\mathrm{d}r*) = 0$ at $r* = 0$. This is the modified Bessel differential equation with variable coefficients, the general solution of which is given by [122]:

$$\theta = C_1 I_0\,(\beta r^*) + C_2 K_0\,(\beta r^*) \tag{343}$$

where

$$I_p\,(\beta r^*) = \sum_{n=0}^{\infty} \left(\frac{\beta r^*}{2}\right)^{2n+p} \frac{1}{n!\,(n+p)!} \quad \text{and}$$

$$K_p\,(\beta r^*) = I_p\,(\beta r^*) \int \frac{\mathrm{d}\,(\beta r^*)}{\beta r^*\, I_p^2\,(\beta r^*)}$$

From the above boundary condition and the properties of the Bessel functions $\mathrm{d}I_0/\mathrm{d}(\beta r*) = I_{-1}(\beta r*)$ and $\mathrm{d}K_0/\mathrm{d}(\beta r*) = K_{-1}(\beta r*)$, then C_2 is zero. With a further boundary condition $\theta = 1$ for $r* = 1$, it follows that $C_1 = 1/I_0(\beta)$, so the general solution changes to

$$\theta = \frac{I_0\,(\beta r^*)}{I_0\,(\beta)} \quad \text{or} \quad (T_{\mathrm{C}} - T) = \frac{I_0\,(\beta r^*)}{I_0\,(\beta)}\,(T_{\mathrm{W}} - T) \tag{344}$$

Equations (339 c) and (341 c) can only be solved numerically. Equation (339 c) has to be transformed into a difference equation in the axial direction. With the solution of Equation (344) for the catalyst temperature the cross-section-averaged heat flow to the catalyst bed can be calculated for each control volume. For this Equation (339 c) with solution Equation (344) is integrated over the cross section of the reactor:

$$0 = -\int_0^1 \varepsilon 2\pi r^* \varrho c_p u \frac{\mathrm{d}T}{\mathrm{d}x}\mathrm{d}r^* + \int_0^1 \varepsilon 2\pi r^* \frac{\alpha s}{\varepsilon}\,(T_{\mathrm{C}} - T)\,\mathrm{d}r^* + \int_0^1 2\pi r^* \sum r_i\,(-\Delta h_{r_i})\,\mathrm{d}r^* \tag{345}$$

The solution of the integral $\int_0^1 (T_{\mathrm{C}} - T)\, r^* \mathrm{d}r^*$ also results from the general properties of the Bessel functions

$$\int_0^1 (T_{\mathrm{C}} - T)\, r^* \mathrm{d}r^* = \frac{R^{3/2}}{\sqrt{\beta}} \frac{I_1\,(\beta)}{I_0\,(\beta)}\,(T_{\mathrm{W}} - T) \tag{346}$$

This expression can be inserted into Equation (339 c) which is then solved numerically and simultaneously with Equation (341 c).

Numerical solutions for pilot plant conditions with integration routines from [28] are taken from [119] and shown in Figure 51. Kinetic data for reactions (i) – (vi) (see above) are given in [8], [11], [119]. A drop in temperature of ca. 75 K and a continuous increase in conversion occur over the length of the reactor. At higher conversions the reactions forming the byproducts gain greater importance.

The estimates used to obtain the model in the form of Equations (339 c), (340 c), and (341 c) are undoubtedly drastic. They should therefore be regarded as an example of how a two-dimensional problem (a reactor model with axial and radial dispersion) can be reduced to a one-dimensional problem and of how the system of model differential equations is decoupled. Nevertheless the model successfully reproduces experimental temperature and concentration profiles measured in a pilot plant [11]. Another special feature of the above model is the treatment of the heterogeneous catalytic reactions. Mass transport processes in the catalyst phase were assumed to have no effect on the reaction rates. The heterogeneous system is thereby regarded as being quasi-homogeneous and all couplings between mass fractions in the gas phase and catalyst phase are neglected. This assumption is not always true. The effect of transport processes on the reaction rates and the repercussion on models for reactors with heterogeneous catalysis are discussed below.

Some Particulars of Heterogeneous Catalytic Reactions. The effect of transport processes on reaction rates in heterogeneous catalysis can be treated analogously to the acceleration of mass transport by chemical reactions described in Section 5.3.2 (Fig. 38). The core of the liquid phase is replaced by a porous catalyst sphere that is one-dimensional in spherical coordinates (Fig. 52). Furthermore if a simple reaction A $\rightarrow$ products is considered, the heterogeneous reaction rate of which is assumed to follow a formal first-order rate law $r = -k' m_A = -k' m$, where k' is a rate coefficient related to the surface area of the catalyst.

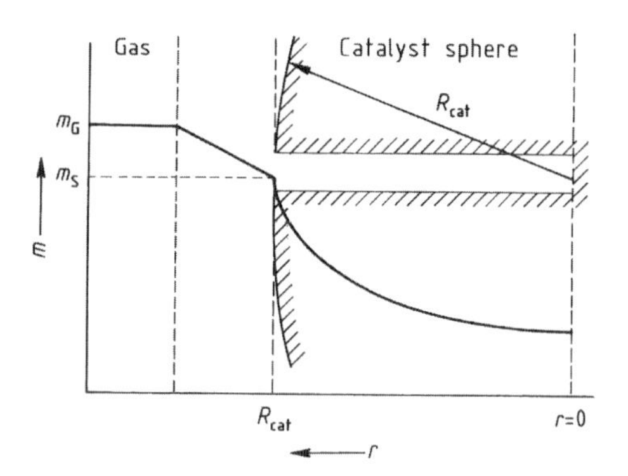

Figure 52. The effect of mass transport in heterogeneous catalytic reactions
m = mass fraction; r = radial distance; R_{cat} = radius of catalyst sphere
The subscripts G and S denote gas phase and surface of catalyst, respectively.

The steady-state mass flux to the surface of the catalyst (see also Section 5.3.1) is given by

$$j^m = \varrho \beta s (m_G - m_S) \tag{347}$$

This flow is compensated for by the consumption of component A by the chemical reaction. If an "effective" reaction rate $r_S = -k' s m_S$ with the mass fraction at the surface of the catalyst is formulated, then the mass fraction m_S for steady-state conditions is given by

$$m_S = \frac{m_G}{\left(1 + \frac{k}{\beta s}\right)} \tag{348}$$

where $k = k' s$. In Equation (347) m_S can now be eliminated so that

$$j^m = \varrho \beta s m_G \left(1 - \left(1 + \frac{k}{\beta s}\right)^{-1}\right) \tag{349}$$

Equation (349) also gives the effective reaction rate for steady-state conditions. As in Section 5.3.2, Equation (349) illustrates the acceleration of mass transport with an increasing ratio $Da_{II} = k/\beta s$, i.e., with increasing reaction rate where Da_{II} is the Damköhler number of the second kind. The degree of external utilization (i.e., effectiveness) of the catalyst η_{ext} is defined as the ratio of the surface reaction rate r_S to the reaction rate for $Da_{II} \rightarrow 0$, i.e., $m_S \rightarrow m_G$. It thus follows that (cf. Eq. 348)

$$\eta_{ext} = \frac{k m_G}{k m_G \left(1 + \frac{k}{\beta s}\right)} = \frac{1}{1 + Da_{II}} \tag{350}$$

For nonisothermal conditions the temperature dependence of the reaction rate must be considered. If Da_{II} is defined with the conditions in the gas phase and constant density is assumed, it follows from Equation (350) that

$$\eta_{ext} = \frac{k(T_S)}{k(T_G)} \left(1 + \frac{k(T_S)}{k(T_G)} Da_{II}\right)^{-1} \tag{351}$$

Given the temperature dependence of the rate coefficient in the form of an Arrhenius equation $k = k_0 e^{-E_a/RT}$, the temperature at the surface T_S must be eliminated from Equation (351) to calculate η_{ext}. The temperature at the surface T_S may be calculated in a similar way as m_S. In the steady-state case

$$j^h = \alpha s (T_S - T_G) \tag{352}$$

The heat transferred to the surroundings is released by chemical reactions occurring at the catalyst. Thus from the steady-state condition $j^h = r_S(-\Delta h_r)$

$$\frac{T_G}{T_S} = 1 + \left[\frac{\varrho(-\Delta h_r) m_G}{\frac{\lambda}{D_{eff}} Le T_G} \frac{\eta_{ext} k(T_G)}{\beta s}\right]^{-1} \tag{353}$$

To derive Equation (353) full analogy for heat and mass transfer, i.e., $Pr = Le \cdot Sc$, is assumed.

Using the abbreviations $\gamma = E_a/RT_G$ (Arrhenius number), and

$$\beta_H = \frac{\varrho\,(-\Delta h_r)\,m_G}{\frac{\lambda}{D_{\text{eff}}} Le T_G} \quad \text{(Prater number)}$$

Equations (351) and (353) are combined to give

$$\frac{\eta_{\text{ext}}}{\exp\left[-\gamma\,(1+\beta_H \eta_{\text{ext}} Da_{II})^{-1}\right]} = \left\{1+Da_{II}\exp\left[-\gamma\,(1+\beta_H\eta_{\text{ext}} Da_{II})^{-1}\right]\right\}^{-1} \quad (354)$$

η_{ext} can also take values greater than 1 in the nonisothermal case if $\beta_H > 0$. In this case the reaction at the surface of the catalyst occurs at a higher temperature than the gas-phase temperature T_G. Since all quantities in Equation (354) are related to T_G, then values of $\eta_{\text{ext}} > 1$ are also consistent with the definition of η_{ext}.

Conditions inside the catalyst can be illustrated in a similar fashion. For spherical geometry (Fig. 52) diffusive mass flux inside the catalyst is given by

$$j^m = -\varrho D_{\text{eff}} 4\pi r^2 \frac{\mathrm{d}m}{\mathrm{d}r} \quad (355)$$

Changes in the diffusive mass flux are caused by chemical reactions at the internal surface of the catalyst, so that under steady-state conditions the mass balance is

$$\frac{\mathrm{d}j^m}{\mathrm{d}r} = \varrho k' 4\pi r^2 m \quad (356)$$

Using Equation (355)

$$\frac{\mathrm{d}^2\varphi}{\mathrm{d}\eta^2} + \frac{2}{\eta}\frac{\mathrm{d}\varphi}{\mathrm{d}\eta} - \Phi_K^2\varphi = 0 \quad (357)$$

is obtained where the normalized quantities $\varphi = m/m_S$ and $\eta = r/R_{\text{Cat}}$ and the abbreviation $\Phi_K = R_{\text{Cat}}\sqrt{k's/D_{\text{eff}}}$ are used (R_{Cat} is the radius of the catalyst sphere). The solution of Equation (357) with the boundary conditions $\varphi = 1$ for $\eta = 1$ and $\mathrm{d}\varphi/\mathrm{d}\eta = 0$ for $\eta = 0$ is

$$\varphi = \frac{\sinh(\Phi_K\eta)}{\eta\sinh(\Phi_K)} \quad (358)$$

Completely analogous to the external mass transfer, a degree of internal utilization of the catalyst η_{int} (internal effectiveness factor) can be defined as the ratio of the mean reaction rate in the catalyst,

$$r_{\text{eff}} = \int_0^{R_{\text{Cat}}} k' sm 4\pi r^2 \mathrm{d}r$$

and reaction rate for $m \to m_S$. Integration of Equation (358) gives

$$\eta_{\text{int}} = \frac{3}{\Phi_K}\left(\tanh\Phi_K - \frac{1}{\Phi_K}\right)^{-1} \quad (359)$$

For $\Phi_K > 3$, $\eta_{\text{int}} \approx 3/\Phi_K$, for $\Phi_K < 0.3$, $\eta_{\text{int}} \approx 1$. The result is similar to the acceleration of mass transfer by chemical reactions, cf. Section 3.3.2. The calculation of η_{int} for other catalyst geometries including more realistic pore structures and other classes of reaction are described in [24], [123].

For nonisothermal conditions the temperature dependence of the reaction rate must once again be taken into consideration. If the simple Arrhenius expression is substituted for k' in Equation (356), then assuming constant density and neglecting the temperature dependence of the effective diffusion coefficient, the result is

$$\frac{\mathrm{d}^2 m}{\mathrm{d}r^2} + \frac{2}{r}\frac{\mathrm{d}m}{\mathrm{d}r} = \frac{k_0'}{D_{\text{eff}}} s\exp\left(-\frac{E_a}{RT}\right) m \quad (360)$$

Inside the catalyst heat is transported by conduction. The conductive heat flow is given by

$$j^h = -\lambda 4\pi r^2 \frac{\mathrm{d}T}{\mathrm{d}r} \quad (361)$$

Changes in heat flow are caused by heat release due to chemical reactions. Under steady-state conditions

$$\frac{\mathrm{d}j^h}{\mathrm{d}r} = \varrho k' s 4\pi r^2 m\,(-\Delta h_r) \quad (362)$$

From Equations (361) and (362) a differential equation for the temperature is obtained

$$\frac{\mathrm{d}^2 T}{\mathrm{d}r^2} + \frac{2}{r}\frac{\mathrm{d}T}{\mathrm{d}r} = \frac{\varrho\,(-\Delta h_r)\,k_0' s\exp\left(-\frac{E_a}{RT}\right)}{D_{\text{eff}}\frac{\lambda_{\text{eff}}}{D_{\text{eff}}}} m \quad (363)$$

Subtracting Equation (363) from Equation (360) gives

$$\frac{\mathrm{d}^2 m}{\mathrm{d}r^2} + \frac{2}{r}\frac{\mathrm{d}m}{\mathrm{d}r} - \frac{\frac{\lambda_{\text{eff}}}{D_{\text{eff}}}}{\varrho\,(-\Delta h_r)}\left[\frac{\mathrm{d}^2 T}{\mathrm{d}r^2} + \frac{2}{r}\frac{\mathrm{d}T}{\mathrm{d}r}\right] = 0 \quad (364)$$

which, after integrating twice with the boundary conditions $m = m_S$ and $T = T_S$ for $r = R_{\text{Cat}}$, and $\mathrm{d}m/\mathrm{d}r = 0$, and $\mathrm{d}T/\mathrm{d}r = 0$ for $r = 0$, leads to the result

$$T - T_S = \frac{\varrho D_{eff}(-\Delta h_r)}{\lambda_{eff}}(m_S - m) \qquad (365)$$

With this the temperature can now be eliminated from Equation (360). If the Prater number, Arrhenius number, and Thiele modulus Φ_K are defined with the conditions at the outer surface of the catalyst, i.e.,

$$\beta_H = \frac{\varrho(-\Delta h_r)m_S}{\frac{\lambda}{D_{eff}}T_S}, \quad \gamma = \frac{E_a}{RT_S}, \quad \text{and}$$

$$\Phi_K^2 = R_{Cat}^2 \frac{k_0' s}{D_{eff}} \exp\left(-\frac{E_a}{RT_S}\right)$$

Equation (364) becomes

$$\frac{d^2\varphi}{d\eta^2} + \frac{2}{\eta}\frac{d\varphi}{d\eta} - \Phi_K^2 \varphi \exp\left[\gamma\beta_H \frac{(1-\varphi)}{1+\beta_H(1-\varphi)}\right] = 0 \qquad (366)$$

Equation (366) cannot be solved analytically, numerical solutions are given in [124]. If the internal effectiveness factor η_{int} is calculated from the numerical solutions [124] the results shown in Figure 53 are obtained. For $\gamma = 20$, multiple solutions exist for η_{int} at high values of β_H and low values of the Thiele modulus. The middle solutions are unstable so that ignition and extinction of the exothermic chemical reaction in the catalyst can occur for certain values of Thiele moduli. As all parameters are related to the state at the outer surface of the catalyst, values of $\eta_{int} > 1$ are consistent with the definition of the internal effectiveness factor. Due to the exothermic reaction the temperature inside the catalyst can be higher than temperature on the outer surface.

By introducing an effective reaction rate $r_{eff} = \eta_{eff}\, r_G$, the rate of heterogeneous catalytic reaction is related to the conditions in the gas phase and is thus no longer coupled with the internal transport in the catalyst. This permits the description of reactors with heterogeneous catalysis as quasi-single-phase systems. The effectiveness factors η_{int} or η_{ext} can be calculated from the variables of state at the surface of the catalyst or in the gas phase, respectively. For the isothermal case and spherical geometry

$$\eta_{eff} = \left(\frac{1}{\eta_{int}} + \frac{\Phi_K^2}{3Bi}\right)^{-1}$$

where the Biot number $Bi = \beta R_{Cat}/D_{eff}$.

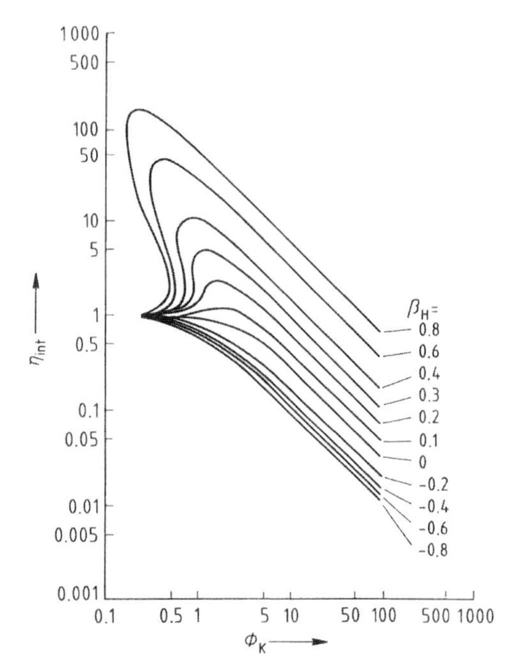

Figure 53. Effectiveness factor of a catalyst as a function of the Thiele modulus Φ_K and the Prater number β_H for spherical catalyst geometry (Arrhenius number $\gamma = 20$) [124] γ = Arrhenius number; η_{int} = internal effectiveness factor.

Nonisothermal conditions and other geometries are discussed in [24], [123], [124].

The discussion in this section was based on first-order reactions with no changes in the mole number. Very few reactions in heterogeneous catalysis can be described with these expressions, higher order reactions are much more frequent. Furthermore the catalysts may be "poisoned" by adsorption of the reactants or products (see page 287) yielding nonintegral orders of reaction. Inhibition by mass transport influences the order of the catalytic reaction. In the extreme case $r_{eff} = \beta\, s\,(m_G - m_S)$. This rate expression is of the first order with respect to the mass fraction of A. Inhibition by mass transport shifts the order of reaction in the direction of first order. Selectivity in parallel or secondary reactions will be similarly affected.

The temperature dependence of the reaction rate is also influenced by inhibition by mass transport. If the expression $r_{eff} = -\eta_{eff}\, k\, m_G = -k{*}\, m_G$ is used for a first-order reaction, and the temperature dependence is given as usual by $k{*} = k_0^* e^{-E_a^*/RT}$, then

$$\ln k^* = \ln k_0^* - \frac{E_a^*}{RT} \qquad (367)$$

For $\eta_{\rm eff}=1$ there is chemical control, $k_0^*=k_0$ and $E_a^*=E_a$. If diffusion in the catalyst pores is rate controlling, $\eta \approx 3/\Phi_K$. Then $\ln k*= \ln k_0+\ln 3-\ln \Phi_K - E_a/RT$. Substitution of $\Phi_K=R_{\rm Cat}\sqrt{k/D_{\rm eff}}$ gives

$$\ln k^*=\frac{1}{2}\ln k_0+\ln 3+\frac{1}{2}\ln D_{\rm eff}-\frac{1}{2}\frac{E_a}{RT}$$

thus $k_0^*=3\sqrt{k_0 D_{\rm eff}}$ and $E_a^*=\frac{1}{2}E_a$. If the temperature dependence of the effective diffusion coefficients is neglected, the activation energy for pore diffusion-controlled reaction is about half that for chemical control. If control is exclusively due to mass transfer, then $r_{\rm eff}\approx -\beta s m_G$ and $k*=\beta s \sim D_{\rm eff}$. Taking into account the temperature dependence of $D_{\rm eff}$, then $\ln k^*=\ln k_0^*+\frac{3}{2}\ln T$ which leads to an activation energy E_a^* of ca. 5 kJ/mol for this case. Thus the experimental determination of the effective formal activation energy for a heterogeneous catalytic reaction clarifies the extent of control by different transport processes. For further discussion of these relationships see [24], [123], [124].

In addition to the formulation of conservation equations for reactor models for heterogeneous catalytic reactions, combinations of simple reactor elements are frequently used for modeling these systems. For further information, see [12], [125]. Other complicated reactors (e.g., fluidized bed reactors) for heterogeneous catalytic reactions can be similarly dealt with [24], [123].

5.3.3.2.2. Stability Analysis of Nonisothermal Reactors

Steady-State Cases. The stability analysis of isothermal reactors discussed in Section 5.3.3.1 demonstrated the close connection of the phenomenon of multiple steady states in steady-state systems and the instability of nonsteady-state systems to a feedback mechanism. For the isothermal cases a chemical feedback mechanism was discussed in the form of autocatalysis and self-poisoning. Another feedback mechanism is the heat release from a chemical reaction. This will be discussed for a simple first-order reaction, A → B + heat, the temperature dependence of the rate coefficient is of the Arrhenius form, $k=k_0 e^{-E^a/RT}$. With increasing conversion of A heat is released. The released heat increases the temperature of the reaction mixture and the reaction rate. This in turn increases the rate of heat release. The reaction rate is an exponential function of the temperature. Therefore the feedback mechanism is nonlinear in temperature.

Stability analysis for nonisothermal steady-state reactors will again be carried out for a well-stirred reactor (one element of a reactor cascade of ideally mixed reactors which is used to represent a flow reactor with convectively dominated flow). In analogy with Equation (310), the mass balance for A gives

$$\frac{\varrho}{\tau}(m_0-m)-\varrho m k_0 e^{-E_a/RT}=0 \tag{368}$$

From the enthalpy balance and transformation of Equation (240 b) into an equation for temperature, the following equation is obtained:

$$\frac{c_p \varrho}{\tau}(T_0-T)+(-\Delta h_r)\varrho m k_0 e^{-E_a/RT} - \alpha s(T-T_A)=0 \tag{369}$$

where the specific surface area s is the effective surface for heat transfer to the surroundings divided by the total volume of the ideally mixed reactor. The source term in Equation (240 b) is thus represented by a heat-transfer term. As in Section 5.3.3.1, Equations (368) and (369) are transformed into normalized variables: $\alpha=m_A/m_{A_0}$, $\theta=(T-T_0)E_a/RT_0^2$, $\theta_C=(T_A-T_0)E_a/RT_0^2$, and $Da_I=\tau k_0$. The Damköhler number (the ratio of the mean residence time to a characteristic chemical time, the time required until $m/m_0=e^{-1}$) is defined with the inlet conditions. If the adiabatic temperature difference $\Delta T_{ad}=(-\Delta h_r)m_0/c_p$ is also defined with E_a/RT_0^2, then $\theta_{ad}=(-\Delta h_r)m_0 E_a/c_p RT_0^2$. Finally a "cooling time"

$$\tau_N=\frac{c_p \varrho}{\alpha s t_{\rm chem}}=\frac{c_p \varrho k_0}{\alpha s}$$

can be defined from the heat transfer. A reciprocal Arrhenius number $\gamma_r=RT_A/E_a$ is also introduced. With these abbreviations Equations (368) and (369) become

$$\frac{1-\alpha}{Da_I}-\alpha\exp\left(\frac{\theta}{1+\gamma_r\theta}\right)=0 \tag{370a}$$

$$\theta_{ad}\alpha\exp\left(\frac{\theta}{1+\gamma_r\theta}\right)-\left(\frac{1}{Da_I}+\frac{1}{\tau_N}\right)\theta+\frac{\theta_C}{\tau_N}=0 \tag{370b}$$

Equations (370 a, b) contain the variables α and θ as a function of six parameters. To reduce the number of parameters, the case of very high activation energy, $\gamma_r \ll 1$, $T_A = T_0$, and $\theta_C = 0$ is considered. With these assumptions the following simplified equations are obtained

$$\frac{1-\alpha}{Da_I} - \alpha e^{\theta} = 0 \tag{371a}$$

$$\theta_{ad}\alpha e^{\theta} - \left(\frac{1}{Da_I} + \frac{1}{\tau_N}\right)\theta = 0 \tag{371b}$$

In order to remain as close as possible to the notation used in Section 5.3.3.2, θ in Equation (371 a) is replaced by Equation (371 b) so that

$$\frac{1-\alpha}{Da_I} = \alpha \exp\left[\frac{(1-\alpha)\,\theta_{ad}}{1+\frac{Da_I}{\tau_N}}\right] \tag{372a}$$

The temperature increase can easily be attributed to α from Equations (371 a, b):

$$\theta = \left[\frac{(1-\alpha)\,\theta_{ad}}{1+\frac{Da_I}{\tau_N}}\right] \tag{372b}$$

Equation (372 a) can be solved graphically by plotting both sides against the progress of the reaction $1-\alpha$ (conversion). In order to further reduce the number of parameters, ideal adiabatic conditions are assumed, i.e., $\tau_N \to \infty$. Thus the left-hand side (LHS) of Equation (372 a) is only dependent on Da_I, whilst the right-hand side (RHS) is only dependent on the adiabatic temperature increase. The solutions for the case discussed here (Fig. 54) are qualitatively similar to the solutions for the isothermal case with self poisoning (see page 280). Here too for a given residence time (Damköhler number) multiple steady states occur for particular areas of the adiabatic temperature increase θ_{ad}. Using the same argumentation as in Section 5.3.3.2, the stability condition Equation (318 a)

$$\frac{d}{d(1-\alpha)}(\text{LHS}) > \frac{d}{d(1-\alpha)}(\text{RHS}) \quad \text{at LHS} = \text{RHS} \tag{318a}$$

is not complied with for the middle one of the multiple solutions. In this state a small perturbation of the conversion causes an increase or decrease in the "convective term" (LHS). The "reaction term" (RHS) however reacts in the same way so that a greater (smaller) availability of A due to convection is followed by a greater (smaller) consumption due to reaction. The system therefore tends towards one of the external stable steady-state solutions. As can be seen from Figure 54 A, the shape of the curve depends on the adiabatic temperature increase θ_{ad}. Figure 54 B shows that as θ_{ad} becomes smaller the S-shaped profile vanishes and the surface of solutions unfolds. Figure 54 B also demonstrates that the unstable branch of the solutions is not passed. The system jumps from low conversion to high conversion with increasing residence time Da_I. On igniting and extinguishing the reaction, hysteresis occurs and the jumping points can be calculated from Equation (318 b)

$$\frac{d}{d(1-\alpha)}(\text{LHS}) = \frac{d}{d(1-\alpha)}(\text{RHS}) \quad \text{at LHS} = \text{RHS} \tag{318b}$$

For the example discussed here the jumping points are given by

$$(1-\alpha) = \frac{1}{2}\left[1 \pm \sqrt{\left(1-\frac{4}{\theta_{ad}}\right)}\right] \tag{373}$$

Since θ_{ad} is always positive for exothermic reactions, then for the point at which the discriminant in Equation (373) disappears, $\theta_{ad} = 4$. If $\theta_{ad} > 4$, multiple steady states are observed.

If the simplification $\gamma_r \to 0$ is discarded, the following equation is obtained for the steady-state solutions

$$\frac{1-\alpha}{Da_I} = \alpha \exp\left[\frac{(1-\alpha)\,\theta_{ad}}{1+\gamma_r \theta_{ad}(1-\alpha)}\right] \tag{374a}$$

The right-hand side of Equation (374 a) now depends on two parameters. In general it can be written in the form

$$F(x, y; p, q, r, s, \ldots) = 0 \tag{374b}$$

where x is the variable for describing the steady state, y the bifurcation parameter which controls the multiple steady state, and p, q, r, s, ... are additional parameters which produce folding or unfolding of the surface of x in space y and p, q, r, s, ...

One or more steady-state solutions exist for the condition given by Equation (374 b). The steady-state solution is stable for

$$F(x, y; p, q, r, s, \ldots) = 0 \text{ and } F_x > 0 \tag{374c}$$

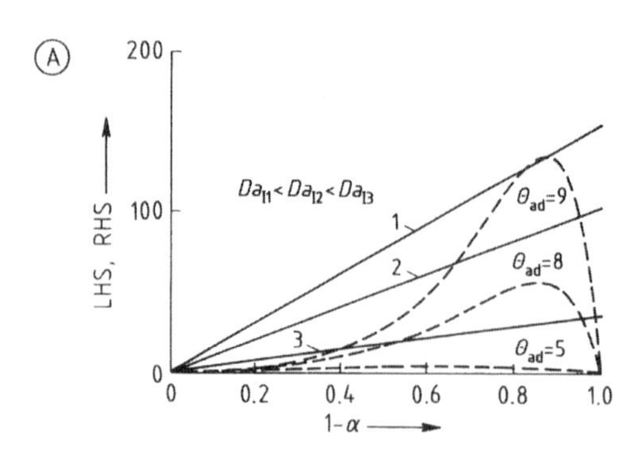

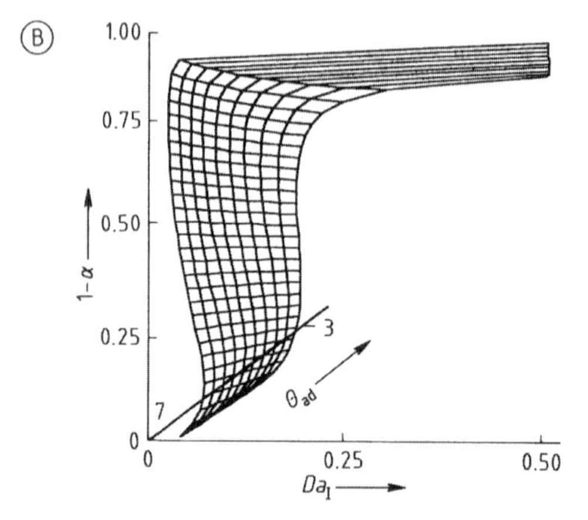

Figure 54. Multiple steady-state conversion points for "normal" exothermic chemical reactions (A) and bifurcation diagram (B) for nonisothermal conditions in a well-stirred reactor
θ_{ad} = adiabatic temperature increase; for explanation of other symbols see Figure 45.

For the unfolding of the hysteresis loop the following applies

$$F(x, y; p, q, r, s, \ldots) = 0, \quad F_x > 0,$$
$$\text{and} \quad F_{xx} = 0 \tag{374d}$$

and "islands" finally occur in the bifurcation diagram which grow into "mushrooms" when

$$F(x, y; p, q, r, s, \ldots) = 0, \quad F_x = 0, \quad F_y = 0,$$
$$F_{xy} \neq 0, \quad F_{xx} \neq 0, \quad \text{and} \quad F_{yy} \neq 0 \tag{374e}$$

Depending on the process under consideration these conditions apply for various parameter combinations. For Equation (374 a) the condition for unfolding occurs when $\theta_{ad}(4\gamma_r - 1) + 4 = 0$. In the system of Equations (374), F_x, F_y, etc. denote the derivatives according to the variable in the index.

Nonsteady-State Cases. As in the stability analysis of isothermal reactors, the stability of nonsteady-state nonisothermal reactors is investigated for a consecutive reaction P → A → B where A is produced from a large reservoir of P. The second reaction A → B is exothermic and its rate coefficient k_1 is temperature dependent. For the reaction sequence P → A → B where $k_1 = k_{01}\,e^{-E_a/RT}$, the mass and enthalpy balances for a batchwise operating ideally mixed reactor are given by

$$\frac{dm_P}{dt} = -k_0 m_P \tag{375}$$

$$\frac{dm_A}{dt} = k_0 m_P - k_1 m_A \tag{376a}$$

$$\varrho c_p \frac{dT}{dt} = (-\Delta h_r)\, k_1 m_A - \alpha s\,(T - T_A) \tag{376b}$$

(see page 283). The feedback in this system results from the temperature dependence of the second reaction rate which increases the reaction rate with increasing conversion. As previously, normalized variables are used: $\alpha = m_A/m_{ref}$, $\pi = m_P/m_{ref}$, $\theta = (T - T_A)\,E_a/RT_A^2$ and $\gamma_r = RT_A/E_a$. For the temperature-dependent reaction rate coefficient it thus follows that

$$k_1 = k_{01} e^{-E_a/RT_A} \exp\left[\theta/(1+\gamma_r\theta)\right]$$

The cooling time is not related to a characteristic chemical time $\tau_N = \varrho c_p/\alpha s$, so that a normalized time can be given as $\tau = t/\tau_N$. Thus the rate coefficients k_0 and k_1 are normalized so that $\varepsilon = k_0\tau_N$ and $\kappa = \tau_N k_{01}\,e^{-E_a/RT_A}$. Finally m_{ref} is established from

$$m_{ref} = \frac{\alpha s R T_A^2}{\varrho E_a\,(-\Delta h_r)\, k_{01} e^{-E_a/RT_A}}$$

There is no feedback to reservoir P thus Equation (375) can be integrated separately:

$$m_P = m_{0P} e^{-k_0 t} \quad \text{or} \quad \pi = \pi_0 e^{-\varepsilon\tau} \tag{377}$$

Using this Equations (376 a) and (377 b) give

$$\frac{d\alpha}{d\tau} = \varepsilon\pi_0 e^{-\varepsilon\tau} - \kappa\alpha \exp\left[\frac{\theta}{(1+\gamma_r\theta)}\right]$$
$$= \mu - \kappa\alpha \exp\left[\frac{\theta}{(1+\gamma_r\theta)}\right] \tag{378a}$$

$$\frac{d\theta}{d\tau}=\alpha\exp\left[\frac{\theta}{(1+\gamma_r\theta)}\right]-\theta \tag{378b}$$

where $\mu=\varepsilon\,\pi_0 e^{-\varepsilon\tau}=\mu_0 e^{-\varepsilon\tau}$. For the sake of simplicity the case of high activation energy (i.e., $\gamma_r \ll 1$) will be investigated. Equations (378 a, b) are thus again simplified into

$$\frac{d\alpha}{d\tau}=\mu-\kappa\alpha e^{\theta} \tag{379a}$$

$$\frac{d\theta}{d\tau}=\alpha e^{\theta}-\theta \tag{379b}$$

For very small values of ε (i.e., for a very much quicker time scale for the changes in α and θ than for the consumption of P) the principle of quasi-steady-state can again be used, and hence Equations (379 a, b) do not have to be integrated. The quasi-steady-state solutions for α and θ result from $d\alpha/d\tau=0$ and $d\theta/d\tau=0$, hence

$$\mu-\kappa\alpha e^{\theta}=0 \tag{380a}$$

$$\alpha e^{\theta}-\theta=0 \tag{380b}$$

giving

$$\theta_{qs}=\frac{\mu}{\kappa} \tag{381a}$$

$$\alpha_{qs}=\frac{\mu}{\kappa}e^{-\mu/\kappa} \tag{381b}$$

The approximate solutions α_{qs} and θ_{qs} are a function of the ratio μ/κ, and also of time since μ is a function of time. With these operations and simplifications the quasi-steady-state solutions α_{qs} and θ_{qs} in Equation (381) have the same form as for the isothermal case (Eq. 324). The stability analysis which now follows is carried out in the same way as for the isothermal case (Section 5.3.3.1).

The development of Equation (380 a) into a Taylor series around the quasi-steady-state solutions provides a system of coupled differential equations for the disturbances $\Delta\alpha$ and $\Delta\theta$, the general solution of which is given by

$$\Delta\alpha(\tau)=C_1 e^{\lambda_1\tau}+C_2 e^{\lambda_2\tau} \tag{382a}$$

$$\Delta\theta(\tau)=C_3 e^{\lambda_1\tau}+C_4 e^{\lambda_2\tau} \tag{382b}$$

The eigenvalues λ_1, λ_2 once again determine the characteristics of the time evolution of the disturbances. They are given by the solutions to the characteristic Equation (329)

$$\lambda^2-\mathrm{Tr}(\boldsymbol{J}_{qs})\lambda+\mathrm{Det}(\boldsymbol{J}_{qs})=0 \tag{329}$$

i.e.,

$$\lambda_{1,2}=\frac{1}{2}\left[\mathrm{Tr}(\boldsymbol{J}_{qs})\pm\left\{\mathrm{Tr}(\boldsymbol{J}_{qs})^2-4\mathrm{Det}(\boldsymbol{J}_{qs})\right\}^{1/2}\right] \tag{330}$$

In this case the Jacobi matrix takes the form

$$\boldsymbol{J}_{qs}=\begin{pmatrix}\frac{\partial f(\alpha,\theta)}{\partial\alpha} & \frac{\partial f(\alpha,\theta)}{\partial\theta}\\ \frac{\partial g(\alpha,\theta)}{\partial\alpha} & \frac{\partial g(\alpha,\theta)}{\partial\theta}\end{pmatrix}_{qs} \tag{383}$$

The three quantities which decide the nature of the solutions are the discriminant of the characteristic equation $\mathrm{Tr}(\boldsymbol{J}_{qs})^2-4\,\mathrm{Det}(\boldsymbol{J}_{qs})$, the trace of the Jacobi matrix

$$\mathrm{Tr}(\boldsymbol{J}_{qs})=\left(\frac{\mu}{\kappa}\right)-1-\kappa e^{\mu/\kappa} \tag{384a}$$

and also their determinant

$$\mathrm{Det}(\boldsymbol{J}_{qs})=\kappa e^{\mu/\kappa} \tag{384b}$$

As μ/κ is always positive, not all the combinations of λ_1, λ_2 given in the stability diagram (Fig. 48) will be passed through with increasing conversion of P.

Two states will now be discussed that are of practical importance. The first case is the transition of the solutions from a node of the system into a focus. The condition for this is that λ_1, λ_2 become conjugate complex, the discriminant thus passes through zero. If μ in Equations (384 a, b) is replaced by Equation (381 a), the result is the following equation for the discriminant

$$\kappa^2 e^{2\theta_{qs}}-2\kappa(1+\theta_{qs})e^{\theta_{qs}}+(\theta_{qs}-1)^2=0 \tag{385}$$

For each positive value of θ_{qs}, there are thus two values of κ

$$\kappa_{1,2}=\left(\sqrt{\theta_{qs}}\pm1\right)^2 e^{-\theta_{qs}} \tag{386a}$$

which satisfy the above condition for the discriminant to be zero. The appropriate values for μ are

$$\mu_{1,2}=\kappa_{1,2}\theta_{qs} \tag{386b}$$

The stability diagram is best displayed in the form given in Figure 55 with the two parameters μ and κ. Two closed loops are obtained, the outer of which corresponds to the larger root from

Equation (386 a) and the inner to the smaller. The area outside the outer curve represents stable nodes, stable or unstable foci lie between the two curves, and the area within the inner curve gives unstable nodes.

The second case is the loss of stability of the system given by the condition $\mathrm{Tr}\,(\boldsymbol{J}_{\mathrm{qs}})=0$. This is also a condition for Hopf bifurcation. If μ in Equation (384 a) is replaced by Equation (381 a), this condition is given as

$$\kappa=(\theta_{\mathrm{qs}}-1)\,\mathrm{e}^{-\theta_{\mathrm{qs}}} \quad \text{and} \quad \mu=\kappa\theta_{\mathrm{qs}} \tag{387a,b}$$

The pertinent curve in the stability diagram is a closed loop which lies between the two loops for the transition from nodes to foci. The region within this curve denotes unstable foci, see the enlarged section of Figure 55 in the area of the origin. The maximum of this curve is at e^{-2}.

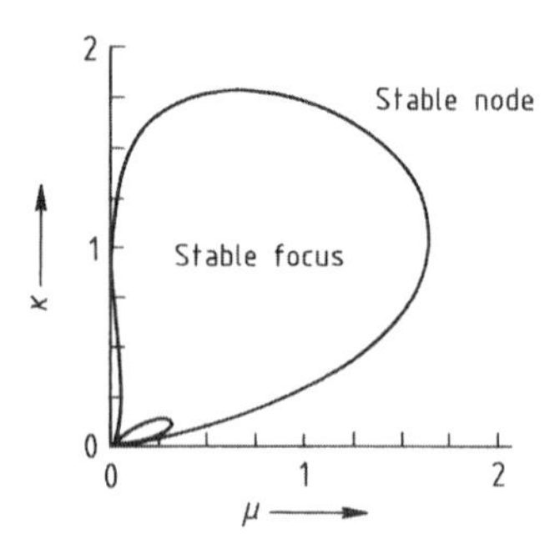

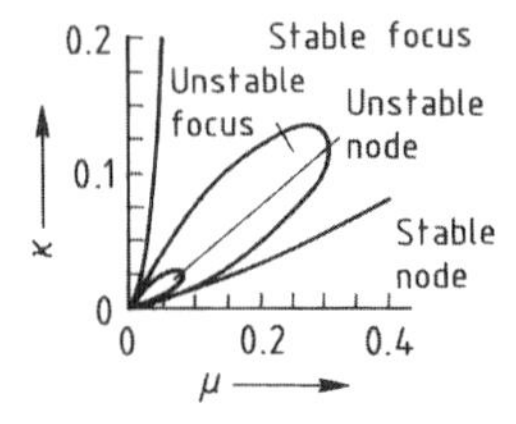

Figure 55. Stability diagram for exothermic consecutive reactions for quasi-steady-state nonisothermal conditions [113]
For explanation of symbols see Figure 49 and text.

Stability analysis for nonisothermal nonsteady-state batchwise operating models exhibits the following important results: for each combination of the experimental conditions μ and κ there is an unambiguous solution for the quasi-steady-state concentration of the intermediate α_{qs} and the normalized temperature increase θ_{qs}. If the normalized reaction rate coefficient is $\kappa>\mathrm{e}^{-2}$, then the quasi-steady-state solutions are always stable. For $\kappa\approx 1.78$ transition from stable nodes to stable foci occurs. For $\kappa<\mathrm{e}^{-2}$ the quasi-steady-state solutions are unstable. At the location of the Hopf bifurcations two solutions occur (Fig. 55). Between the two bifurcation points for α_{qs} and θ_{qs} as functions of the conversion (i.e., as functions of μ), oscillations of varying frequency and amplitude occur.

For engineering purposes it is important that, for the simplified case discussed here, stability can be controlled through the parameter κ. For $\kappa>\mathrm{e}^{-2}$ the system remains stable. If the properties of the chemical reaction are given by $k_1=k_{01}\,\mathrm{e}^{-\mathrm{E_a}/\mathrm{RT}}$, stability can only be achieved by altering the cooling time by means of increased heat transfer to the surroundings. Detailed discussions on the stability of nonisothermal reactors can be found in [113], [114].

5.3.3.2.3. Use on Statistical Processes

The description of statistical processes by averaging the conservation equations of momentum is discussed in Section 5.1.2 for isothermal inert systems. The averaged equations however contain terms with higher moments which must be modeled; in Equation (216) these are the Reynolds stresses $\langle\varrho\rangle\,\langle U_i^{\mathrm{f}}U_j^{\mathrm{f}}\rangle$.

A model for the Reynolds stresses which shows them in analogy to the laminar viscous stresses as an isotropic tensor of turbulent viscous stresses

$$\tau_{t_{ij}}=\mu_t\left(\frac{\partial\langle U_i\rangle}{\partial x_j}+\frac{\partial\langle U_j\rangle}{\partial x_i}\right)-\frac{2}{3}\mu_t\frac{\partial\langle U_k\rangle}{\partial x_k}\partial_{ij}$$

is given in Section 5.1.2.1. The averaged equations then take the form

$$\frac{\partial}{\partial t}\left(\langle\varrho\rangle\langle U_i\rangle\right)+\frac{\partial}{\partial x_k}\left(\langle\varrho\rangle\langle U_k\rangle\langle U_i\rangle\right)=\frac{\partial}{\partial x_k}\left(\langle\tau_{t_{ik}}\rangle\right)-\frac{\partial\langle p\rangle}{\partial x_i}+\langle\varrho\rangle\langle g_i\rangle,\quad i=1,2,3 \tag{388}$$

which is a transport equation for the first moment of the velocities. The accompanying problem of describing the turbulent viscosity was solved using the eddy viscosity hypothesis, $\mu_t=C_\mu\langle\varrho\rangle\left(\langle k\rangle^2/\langle\varepsilon\rangle\right)$.

As well as the solution of Equation (388), the model requires the solution of the equations for the turbulence energy, $\langle k \rangle$ (Eq. 222), and its dissipation rate, $\langle \varepsilon \rangle$ (Eq. 225), thereby defining a turbulent time scale according to $\tau = \langle k \rangle / \langle \varepsilon \rangle$. A numerical solution that takes into account the special nature of the averaged conservation equations of momentum is demonstrated in Section 5.1.2.2.

For nonisothermal reacting systems this concept has to be expanded to the equations for the conservation of enthalpy (Eq. 134 or 240 a) and for the conservation of mass of the chemical species (Eq. 133 or 265 a). Averaging Equations (240 a) and (265 a) using the same assumptions as in Section 5.1.2.1 results in the equations

$$\frac{\partial}{\partial t}\left(\langle \varrho \rangle \langle h \rangle\right) + \frac{\partial}{\partial x_k}\left(\langle \varrho \rangle \langle U_k \rangle \langle h \rangle\right) = -\frac{\partial}{\partial x_k}\left(\langle J^h \rangle + \langle \varrho \rangle \langle h^{\mathrm{f}} U_k^{\mathrm{f}} \rangle\right) + \langle \varrho S_h \rangle \tag{389}$$

and

$$\frac{\partial}{\partial t}\left(\langle \varrho \rangle \langle m_i \rangle\right) + \frac{\partial}{\partial x_k}\left(\langle \varrho \rangle \langle U_k \rangle \langle m_i \rangle\right) = -\frac{\partial}{\partial x_k}\left(\langle J^{m_i} \rangle + \langle \varrho \rangle \langle m_i^{\mathrm{f}} U_k^{\mathrm{f}} \rangle\right) + \langle \varrho S_i \rangle \quad i=1,\ldots,N-1 \tag{390}$$

The problem now lies in modeling the Reynolds fluxes $\langle \varrho \rangle \langle h^{\mathrm{f}} U_j^{\mathrm{f}} \rangle$ and $\langle \varrho \rangle \langle m_i^{\mathrm{f}} U_j^{\mathrm{f}} \rangle$, and also the expected values of the source term $\langle \varrho S_i \rangle$ and $\langle \varrho S_h \rangle$.

Modeling the Reynolds Fluxes. The problem of modeling the Reynolds fluxes $\langle \varrho \rangle \langle h^{\mathrm{f}} U^{\mathrm{f}}{}_j \rangle$ and $\langle \varrho \rangle \langle m^{\mathrm{f}}{}_i U^{\mathrm{f}}{}_j \rangle$ can be solved analogously to the modeling of the Reynolds stresses. As with the transport equations for the Reynolds stresses (see Section 5.1.2.1, Eq. 221), the analogous transport equation for the Reynolds fluxes is not closed however. The same reasons as were discussed in Section 5.1.2.1 for the Reynolds stresses, lie behind the fact that transport equations for Reynolds fluxes are usually not solved in engineering but are modeled directly.

The direct modeling of Reynolds fluxes is based on the Boussinesq hypothesis and assumes that there is a similarity between the turbulent transport of momentum, and enthalpy, and mass. This principle has already been used in Sections 5.2.2.1 and 5.3.1.1 for deterministic systems. The turbulent flux of a scalar quantity is thus related to the gradients of the expected values of the scalar. The proportionality number is a turbulent transport coefficient which is obtained from the turbulent viscosity:

$$-\langle \varrho \rangle \langle \Phi^{\mathrm{f}} U_k^{\mathrm{f}} \rangle = \frac{\mu_t}{\sigma_\varphi} \frac{\partial \langle \Phi \rangle}{\partial x_k} \tag{391}$$

This expression has already been used in the equation for turbulence energy (Eq. 223); σ_φ is a turbulent Prandtl or Schmidt number, which relates the turbulent transfer of the scalar in question and the turbulent transfer of momentum. With the help of the eddy viscosity hypothesis, σ_φ can be interpreted as the ratio of the mixing length of the velocity fluctuations to the mixing length of the fluctuations in the scalar.

With this hypothesis and neglecting the proportion of laminar flux to turbulent flux, Equations (389) and (390) become

$$\frac{\partial}{\partial t}\left(\langle \varrho \rangle \langle h \rangle\right) + \frac{\partial}{\partial x_k}\left(\langle \varrho \rangle \langle U_k \rangle \langle h \rangle\right) = \frac{\partial}{\partial x_k} \frac{\mu_t}{\sigma_h} \frac{\partial \langle h \rangle}{\partial x_k} + \langle \varrho S_h \rangle \tag{392}$$

and

$$\frac{\partial}{\partial t}\left(\langle \varrho \rangle \langle m_i \rangle\right) + \frac{\partial}{\partial x_k}\left(\langle \varrho \rangle \langle U_k \rangle \langle m_i \rangle\right) = \frac{\partial}{\partial x_k} \frac{\mu_t}{\sigma_m} \frac{\partial \langle m_i \rangle}{\partial x_k} + \langle \varrho S_i \rangle \quad i = 1,\ldots,N-1 \tag{393}$$

The remaining problem is now the modeling of the terms $\langle \varrho S_h \rangle$ and $\langle \varrho S_i \rangle$. If the change in enthalpy due to mechanical work and radiation are neglected in Equation (392), then the term $\langle \varrho S_h \rangle$ vanishes. A modeling expression then only needs to be found for $\langle \varrho S_i \rangle$.

Modeling of the Mean Reaction Rates. The discussion of models for the mean reaction rates in turbulent (statistical) flows is of similar importance in engineering to the modeling of Reynolds stresses and Reynolds fluxes. Reviews are given in [126–128].

The reaction rates are generally a function of mass fractions and temperature (Eq. 137 a). The expected values $\langle \varrho S_i \rangle$ are obtained from Equation (109) since the averaging rule (Eq. 107) is applicable to functions. If changes in density are neglected, then Equation (137 a) gives

$$\langle S_i\rangle=\int_{T_u}^{T_0}\int_0^1\ldots\int_0^1 P\left(m_l,\ldots,m_o,T\right)$$
$$\cdot S_i\left(m_l,\ldots,m_o,T\right)\mathrm{d}m_l\ldots\mathrm{d}m_o\mathrm{d}T \qquad (394)$$

In Equation (394) $P(m_l,\ldots,m_o,T)$ is a joint probability density function of temperature and the mass fractions of the chemical species that are contained in the expression for the reaction rates. $S_i(m_l,\ldots,m_o,T)$ are usually known functions and hence to determine $\langle S_i\rangle$, the joint probability density functions have to be established. The example shown in Figure 14 demonstrates that a priori determinations for this are difficult; there are intermittencies in the region at the jet boundary so that $P(m_l,\ldots,m_o,T)$ becomes "bimodal". The bimodal structure is not dominant further downstream and within the turbulent jet. An acceptable assumption for the form of $P(m_l,\ldots,m_o,T)$ is a multidimensional normal distribution, which must be suitably clipped due to the restricted domain of definition of temperature and mass fractions. Although this form does not cover the intermittencies at the boundary of the turbulent jet, it is a good approximation for the inside of the turbulent jet and further downstream. For many applications the deficiencies in representation by a multidimensional normal distribution is of no consequence because the mean reaction rates in the areas with strong intermittencies are mostly small compared with those inside the turbulent jet. If the approximation of $P(m_l,\ldots,m_o,T)$ by means of a multidimensional normal distribution is adequate, the form of $P(m_l,\ldots,m_o,T)$ must be established quantitatively. As in the one-dimensional case, this is achieved through the expected values and the second moments. Consequently in order to determine $P(m_l,\ldots,m_o,T)$ calculation of $\langle m_l\rangle,\ldots,\langle m_o\rangle$, $\langle T\rangle$, $\langle m_i^{\mathrm{f}}m_j^{\mathrm{f}}\rangle$, $i,j=l,\ o$, and $\langle m_i^{\mathrm{f}}T^{\mathrm{f}}\rangle$, $i=l,\ o$ is necessary.

Equation (394) can be presented in the form

$$\langle S_i\rangle=S_i\left(\langle m_l\rangle,\ldots,\langle m_o\rangle,\langle T\rangle\right)$$
$$\cdot F\left(\langle m_i^{\mathrm{f}}m_j^{\mathrm{f}}\rangle,\langle m_i^{\mathrm{f}}T^{\mathrm{f}}\rangle,i,j=l,o\right) \qquad (395)$$

in which $F(\langle m_i^{\mathrm{f}}m_i^{\mathrm{f}}\rangle, \langle m_i^{\mathrm{f}}T^{\mathrm{f}}\rangle, i,j=l,o)$ are correction functions for reproducing the effect of turbulent fluctuations in temperature and mass fractions in the averaging procedure of weighting with the joint probability density functions.

These correction functions can be given as polynomials of the second moments $\langle m_i^{\mathrm{f}}m_j^{\mathrm{f}}\rangle$, $\langle m_i^{\mathrm{f}}T^{\mathrm{f}}\rangle$ (or more precisely of the turbulence intensities and correlation coefficients) and of the activation energies for the reactions in question [129], [130]. The coefficients of these polynomials can be obtained by applying the averaging procedure (Eq. 394) for a number of predefined values of $\langle m_i^{\mathrm{f}}m_j^{\mathrm{f}}\rangle$ $\langle m_i^{\mathrm{f}}T^{\mathrm{f}}\rangle$ and E_{a}; $\langle S_i\rangle$ is then represented as a function of these quantities by an empirical polynomial expression [129]. Thus integration of Equation (394) is unnecessary for the numerical solution of the resulting system of equations.

Calculation of the expected values $\langle S_i\rangle$ of the reaction rates from Equation (395) assumes knowledge of the second moments $\langle m_i^{\mathrm{f}}m_j^{\mathrm{f}}\rangle$, $i,j=1,\ldots,N$ and $\langle m_i^{\mathrm{f}}T^{\mathrm{f}}\rangle$, $i=1,\ldots,N$. In this case too closure of the equation system at the level of the expected values necessitates calculation of the second moments. The relevant transport equations can be derived by similar considerations and modeling assumptions used for deriving the equation for other second moments, see Section 5.1.2.1 [52], [126–128], [130].

The numerical work involved in calculating a statistical reacting nonisothermal flow is considerable, despite the relatively simple structure of the modeling components. In addition to the solution of the balances of momentum, direct modeling of Reynolds stresses requires the solution of two further equations for turbulence energy and the turbulent time scale. The Reynolds fluxes are modeled on the same basis as the Reynolds stresses, so that apart from the enthalpy balance (Eq. 392) and the transport equations for the mass of the chemical species (Eq. 393), no other equations originate from this part of the model. Modeling of the mean chemical reaction rates through the simple expression (Eq. 394) with presumed shape joint probability density functions requires the solution of equations for the variances and covariances $\langle m_i^{\mathrm{f}}m_j^{\mathrm{f}}\rangle$, $i,j=1,\ldots,N$, and $\langle m_i^{\mathrm{f}}T^{\mathrm{f}}\rangle$, $i=1,\ldots,N$. In addition the density has to be calculated from the averaged form of the thermal equation of state

$$\langle \varrho\rangle = \frac{\langle p\rangle}{R}\sum_{i=1}^{N}\frac{m_i}{\langle T\rangle\langle m_i\rangle+\langle m_i^{\mathrm{f}}T^{\mathrm{f}}\rangle} \qquad (396)$$

which also contains the second moments $\langle m_i^{\mathrm{f}} T^{\mathrm{f}} \rangle$, $i = 1, \ldots, N$. Equation (396) can easily be derived from the ideal gas equation. The temperature must be determined from the expected value of the enthalpy of the mixture through the definition of the enthalpy:

$$\langle h \rangle = \sum_{i=1}^{N} \langle m_i h_i \rangle =$$

$$\sum_{i=1}^{N} \langle m_i \left(h_{0i} + \int c_{pi} \mathrm{d}T \right) \rangle \qquad (397)$$

Equation (397) is not solved here for the temperature, for further details see [129–131].

Example: Combustion of Hydrogen in a Turbulent Diffusion Flame. The simulation of a turbulent nonisothermal reacting flow by the system of equations outlined above will be demonstrated briefly on an example. A circular jet of hydrogen with a Reynolds number of 12 000 relative to the nozzle outlet is injected into virtually stationary air and is burned in a horizontal combustion chamber. In the experiments described in [132–134], velocities, temperatures, and concentrations of stable chemical species and of OH radicals are measured downstream of the burner nozzle. The system is axially symmetrical.

The numerical solution contains the solutions of Equations (223), (225), (388), (392), (393) and also of the equations for the variances and covariances $\langle m_i^{\mathrm{f}} m_j^{\mathrm{f}} \rangle$, $i, j = 1, \ldots, N$ and $\langle m^{\mathrm{f}}{}_i T^{\mathrm{f}} \rangle$, $i = 1, \ldots, N$ in axially symmetrical formulation. The numerical procedure is the same as that outlined for the isothermal case in Section 5.1.2.3. The chemical reactions are described by 44 elementary reactions between eleven chemical species: H_2, O_2, H_2O, N_2, H, O, OH, HO_2, H_2O_2 and also N and NO [135]. The reaction rates of each of these components consists of the summation of reaction rates of all the elementary reactions *j* in which the component in question *i* occurs, i.e., $r_i = \sum_{j=1}^{R} r_{ij}$. The rate coefficients of the reactions are expressed in the form $k_j = k_{0j}\, T^{\alpha j}\, \mathrm{e}^{-E_{\mathrm{a}}/RT}$ and are thus not freely selectable modeling parameters. For details of the reaction rate coefficients of the H_2–O_2 system see [130], [135]. The modeling parameters used are the coefficients $C_\mu = 0.09$, $C_{\varepsilon 1} = 1.4$, $C_{\varepsilon 2} = 0.925$, $\sigma_{k\,\mathrm{eff}} = 1.0$, $\sigma_{\varepsilon\,\mathrm{eff}} = 1.3$; the turbulent Prandtl numbers in Equations (392) and (393); and also the equations for the second moments which are defined as $\sigma_\varphi = 0.7$ [91].

Figure 56 shows the result of the simulations from the model compared with the measured results given in [132–134]. The correspondence between the measured results and the model is satisfactory. In particular the penetration of oxygen to the axis of the turbulent jet close to the nozzle is well predicted. Along with the calculation of the OH radical concentrations, this is a particular potential of the model that is based on an elementary reaction scheme for describing the chemical reactions for the H_2–O_2–N_2 system.

Another aspect of the model presented here can be shown with the results of sensitivity analysis. The first-order sensitivity coefficients are calculated as described in Section 5.3.3.1.2. Due to the elliptical form of the conservation equations and the difference form which follows from this (see Section 5.1.2.2, Equation 232), the resulting system of equations has a block pentadiagonal structure and is solved by an ADI procedure.

Figure 57 A shows the results in the form of the gradients of the axial velocity $\langle U \rangle$ with respect to the model parameter C_μ, which controls the turbulent viscosity. The sensitivity coefficients are given in relative form $(\partial \langle U \rangle / \langle U \rangle) / (\partial \langle C_\mu \rangle / C_\mu)$ as in Equations (29) and (30). The sensitivity with respect to a change in the parameter C_μ is particularly large in areas with high velocity gradients. The sensitivity analysis thus reflects the physical background and the limitations of the model used: since the turbulent stresses are described by an approach which relates them to the gradients of the mean velocities, large sensitivities inevitably result in response to changes in the apparent turbulent viscosity in areas with high velocity gradients.

Figure 57 B shows the gradients of the molar fractions for oxygen with respect to the rate coefficient for the reaction $O_2 + H \rightarrow OH + O$, this is one of the most important oxygen-consuming reactions. Here to, the sensitivity coefficients are given in relative form $(\partial \langle X_{O_2} \rangle / \langle X_{O_2} \rangle) / (\partial k_{01} / k_{01})$. Sensitivity is only obvious in the main reaction zones of the tur-

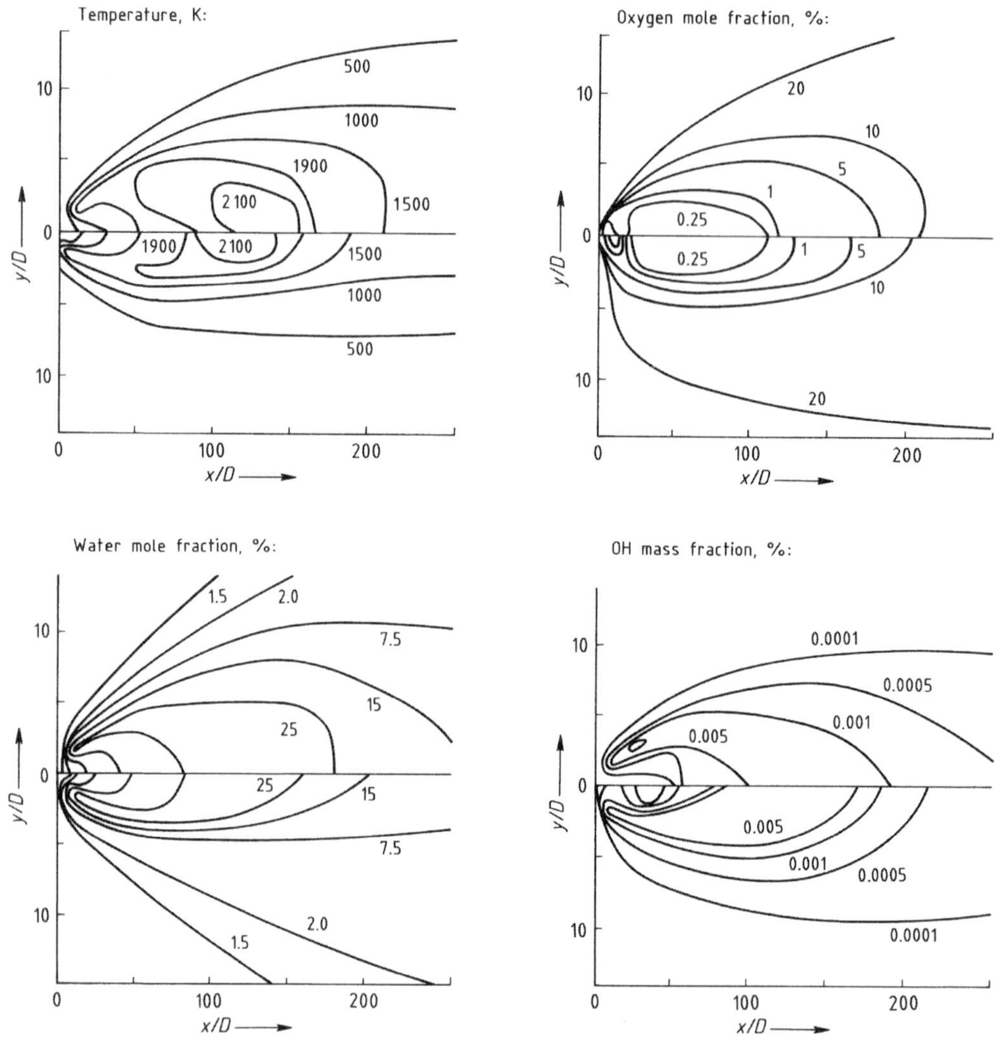

Figure 56. Measured (upper section) and calculated (lower section) profiles for temperature, mole fraction of oxygen and water, and mass fraction of OH radicals for a turbulent hydrogen – air diffusion flame [130], [131]
x = axial distance; y = radial distance; D = nozzle diameter

bulent hydrogen flame. Thus, the molar fraction of oxygen only reacts sensitively to the change in these rate coefficients in the main reaction zone of the flame.

The result of averaging the reaction rates with Equations (394) and (395) can be interpreted as the change in the rate coefficients due to the correction functions. The unsatisfactory representation of the joint probability density function $P(m_l, \ldots, m_o, T)$ by a multidimensional normal distribution used to calculate these correction functions is therefore no longer of importance at the edge of the jet for the prediction of the oxygen molar fraction. This is due to the low reaction rates for the consumption of oxygen in this region.

For further discussion of sensitivity analysis, especially for discussion of the possibility of reduction of the chemical models, see [131].

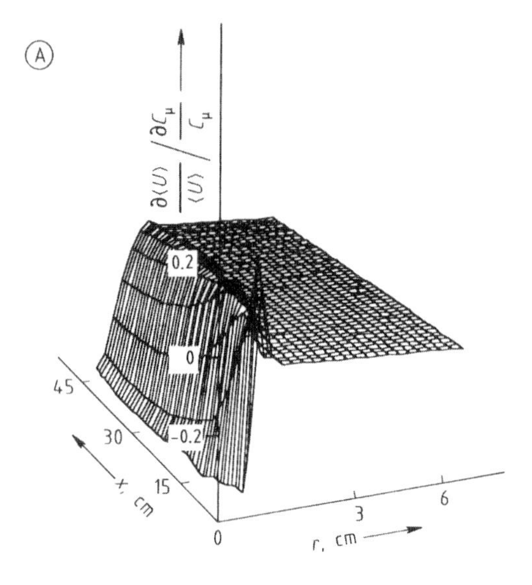

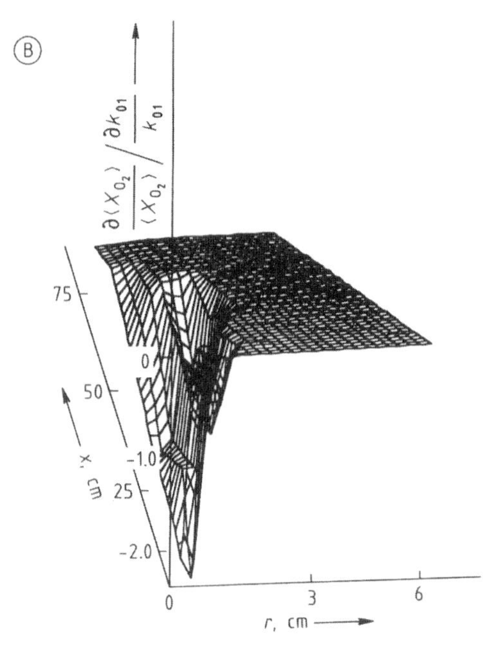

Figure 57. Relative sensitivity coefficient for a H_2–O_2 diffusion flame
A) Coefficients for the axial velocity $\langle U \rangle$ with respect to the modeling parameter C_μ; B) Coefficients for the mole fraction of oxygen $\langle X_{O_2} \rangle$ with respect to the rate coefficient k_{01} x = axial distance; r = radial distance.

6. References

1. G. Klaus: *Wörterbuch der Kybernetik,* Fischer Verlag, Frankfurt 1969.
2. A. I. Bojarinow, W. W. Kafarow: *Optimierungsmethoden in der chemischen Technologie,* Verlag Chemie, Weinheim 1972.
3. D. M. Himmelblau, K. B. Bischoff: *Process Analysis and Simulation: Deterministic Systems,* J. Wiley and Sons, New York 1968.
4. D. M. Himmelblau: *Process Analysis by Statistical Methods,* J. Wiley and Sons, New York 1970.
5. U. Hoffmann, H. Hofmann: *Einführung in die Optimierung mit Beispielen aus dem Chemie-Ingenieur-Wesen,* Verlag Chemie, Weinheim 1971.
6. D. M. Himmelblau in A. Bisio, R. Kabel (eds.): *Scaleup of Chemical Processes,* J. Wiley and Sons, New York 1985.
7. T. E. Edgar, D. M. Himmelblau: *Optimization of Chemical Processes,* McGraw-Hill, New York 1988.
8. F. W. Ramirez: *Computational Methods for Process Simulation,* Butterworth Publishers, Stoneham 1989.
9. W. F. Luyben: *Process Modeling, Simulation and Control for Chemical Engineers,* McGraw-Hill, New York 1990.
10. *Ullmann,* 4th ed., **4,** 451.
11. G. Emig, *Ber. Bunsenges. Phys. Chem.* **90** (1986) 986.
12. H. Hofmann: "Future Trends in Chemical Engineering Modelling," *Proceedings of the 18th EFCE Congress,* Apr. 26 – 30, 1987, p. 579.
13. W. W. Kafarow: *Kybernetische Methoden in der Chemie und chemischen Technologie,* Verlag Chemie, Weinheim 1971.
14. L. A. Gould: *Chemical Process Control: Theory and Applications,* Addison-Wesley Publishing Company, Reading 1969.
15. P. S. Buckley: *Techniques of Process Control,* Robert E. Krieger Publishing Company, Huntington 1979.
16. P. V. Danckwerts, *Chem. Eng. Sci.* **2** (1953) 1.
17. A. D. Randolph, *Can. J. Chem. Eng.* **42** (1964) 280.
18. E. E. O'Brien in P. A. Libby, F. A. Williams (eds.): *Turbulent Reacting Flows,* Springer Verlag, Berlin-Heidelberg 1980, p. 191.
19. S. B. Pope, *Prog. Energy Combust. Sci.* **11** (1985) 119.

20. G. Emig, U. Hoffmann, H. Hofmann: *DECHEMA-Kurs: Planung und Auswertung von Versuchen zur Erstellung mathematischer Modelle,* **parts I and II,** DECHEMA, Frankfurt-Main 1974.
21. E. Kreyszig: *Statistische Methoden und ihre Anwendung,* Verlag Vandenhoeck und Ruprecht, Göttingen 1972.
22. M. Fisz: *Wahrscheinlichkeitsrechnung und mathematische Statistik,* VEB Deutscher Verlag der Wissenschaften, Berlin 1980.
23. A. Lindner: *Statistische Methoden,* Birkhäuser, Stuttgart 1964.
24. M. Baerns, H. Hofmann, A. Renken: *Chemische Reaktionstechnik, Lehrbuch der Technischen Chemie 1,* Thieme Verlag, Stuttgart 1987.
25. VDI-Gesellschaft Verfahrenstechnik und Chemieingenieurwesen: "Berechnungsblätter für den Wärmeübergang," *VDI-Wärmeatlas,* VDI-Verlag, Düsseldorf 1984.
26. IBM – Scientific Subroutine Package – System 360, IBM Corporation, White Plains 1972.
27. NAG Fortran Library-Mark 8, Numerical Algorithms Group Inc., Downers Grove 1981.
28. IMSL Library Reference Manual, IMSL Inc., Houston 1980.
29. W. H. Press, B. B. Flannery, S. A. Teukolsky, W. T. Vetterling: *Numerical Recipes. The Art of Scientific Computing (Fortran Version),* Cambridge University Press, Cambridge 1989 (also available in Pascal and C).
30. W. H. Press, B. B. Flannery, S. A. Teukolsky, W. T. Vetterling: *Numerical Recipes. Example Book (Fortran),* Cambridge University Press, Cambridge 1985 (also available in Pascal and C).
31. G. Engeln-Müllges, F. Reutter: *Numerische Mathematik für Ingenieure,* B.-I. Wissenschaftsverlag, Mannheim 1987.
32. G. Engeln-Müllges, F. Reutter: *Formelsammlung zur Numerischen Mathematik mit Standard-Fortran 77-Programmen,* B.-I. Wissenschaftsverlag, Mannheim 1988.
33. J. Hildenrath et al.: "OMNITAB, A Computer Program for Statistical and Numerical Analysis," *Handbook 101,* Nat. Bur. of Standards, U.S. Government Printing Office, Washington, D.C., 1968.
34. R. Hooke, T. A. Jeeves, *J. Assoc. Compt. Mach.* **8** (1961) 212.
35. N. Spendley, G. R. Hext, F. R. Himsworth, *Technometrics* **4** (1962) 441.
36. J. A. Nelder, R. Mead, *Comput. J.* **7** (1965) 308.
37. H. H. Rosenbrock, C. Storey: *Computational Techniques for Chemical Engineers,* Pergamon Press, Oxford 1966.
38. C. S. Beightler, D. T. Phillips, D. G. Wilde: *Foundations of Optimization,* Prentice Hall, Englewood Cliffs, N.Y. 1979.
39. P. E. Gill, W. Murray, M. H. Wright: *Practical Optimization,* Academic Press, New York 1981.
40. D. W. Marquardt, *J. Soc. Ind. Appl. Math.* **11** (1963) 431.
41. H. O. Hartley, A. Booker, *Ann. Math. Stat.* **36** (1965) 638.
42. E. J. Williams, *J. Royal Stat. Soc. B* **24** (1962) 125.
43. J. J. More in G. A. Watson (ed.): *Numerical Analysis, Lecture Notes in Mathematics,* **vol. 630,** Springer Verlag, Berlin 1977, p. 105.
44. Z. E. Beisinger, S. Bell: H2 SAND MIN, Sandia Corporation.
45. I. N. Bronstein, K. H. Semendjajew: *Taschenbuch der Mathematik,* Verlag Harry Deutsch, Thun 1983.
46. O. Levenspiel: *Chemical Reaction Engineering,* J. Wiley and Sons, New York 1972.
47. O. Levenspiel, W. K. Smith, *Chem. Eng. Sci.* **6** (1975) 227.
48. J. O. Hirschfelder, C. F. Curtiss, R. B. Bird: *Molecular Theory of Gases and Liquids,* J. Wiley and Sons, New York 1954.
49. J. C. Rotta: *Turbulente Strömungen,* B. G. Teubner, Stuttgart 1972.
50. R. B. Bird, W. E. Stewart, E. N. Lightfoot: *Transport Phenomena,* J. Wiley and Sons, New York 1960.
51. F. A. Williams: *Combustion Theory,* The Benjamin/Cummings Publishing Company, Menlo Park, Calif. 1985.
52. P. A. Libby, F. A. Wiliams in P. A. Libby, F. A. Williams (eds.): *Turbulent Reacting Flows,* Springer Verlag, Berlin – Heidelberg 1980, p. 1.
53. P. Givi, *Prog. Energy Combust. Sci.* **15** (1989) 121.
54. M. Lesieur: *Turbulence in Fluids,* Martinus Nijhoff Publishers, Dordrecht 1987.
55. T. S. Lundgreen, *Phys. Fluids* **12** (1969) 485.
56. S. B. Pope, *Combust. Flame* **27** (1976) 294.
57. J. Janicka, W. Kolbe, W. Kollmann in C. T. Crowe, W. L. Grosshandler (eds.): *Proc. of the 1978 Heat Transf. Fluid Mech. Inst.,* Standford University Press, Stanford 1978, p. 296.

58. S. B. Pope, *Philos. Trans. R. Soc. Lond. Ser. A* **291** (1979) 529.
59. M. Jischa: *Konvektiver Impuls-, Wärme- und Stoffaustausch,* Vieweg und Sohn, Braunschweig 1982.
60. H. Schlichting: *Grenzschichttheorie,* Verlag G. Braun, Karlsruhe 1982.
61. N. Peters in E. S. Oran, J. P. Boris (eds.): *Approaches to Combustion Modeling,*
62. M. J. Lighthill: *Introduction to Fourier Analysis and Generalized Functions,* Cambridge University Press, Cambridge 1958.
63. H. M. Hulburt, S. Katz, *Chem. Eng. Sci.* **19** (1964) 555.
64. E. Kamke: *Differentialgleichungen, Lösungsmethoden und Lösungen II,* B. G. Teubner Verlag, Stuttgart 1979.
65. P. Langevin, *C. R. Acad. Sci. Paris* **146** (1908) 530.
66. S. B. Pope, *Phys. Fluids* **24** (1981) 588.
67. L. Arnold: *Stochastic Differential Equations: Theory and Applications,* J. Wiley and Sons, New York 1974.
68. W. Jost: *Diffusion in Solids, Liquids and Gases,* Academic Press, New York 1952.
69. O. Levenspiel, K. B. Bischoff: *Adv. Chem. Eng.* **4** (1963) 95.
70. J.-Y. Chen, W. Kollmann: *Twenty-Second Symposium (International) on Combustion,* The Combustion Institute, Pittsburgh 1989, p. 645.
71. J.-Y. Chen, W. Kollmann, *Combust. Flame* **79** (1990) 75.
72. J. L. Lumley, *Adv. Appl. Mech.* **18** (1978) 123.
73. B. E. Launder, C. J. Reece, W. Rodi, *J. Fluid. Mech.* **68** (1975) 573.
74. J. C. Rotta, *Z. Phys.* **129** (1951) 547.
75. C. Dopazo, *Phys. Fluids* **18** (1975) 397.
76. J. Janicka, W. Kolbe, W. Kollman, *J. Non-Equilib. Thermodyn.* **4** (1978) 47.
77. C. Dopazo, *Phys. Fluids* **22** (1979) 20.
78. J. Janicka, W. Kollman in L. J. S. Bradbury et al. (eds.): *Turbulent Shear Flows 4,* Selected Papers from the Fourth International Symposium on Turbulent Shear Flows, Springer Verlag, Berlin – Heidelberg 1985, p. 73.
79. S. B. Pope, *Combust. Sci. Technol.* **25** (1981) 159.
80. T. V. Nguyen, S. B. Pope, *Combust. Sci. Technol.* **42** (1984) 13.
81. A. Favre in Society for Industrial and Applied Mathematics (ed.): *Problems of Hydrodynamics and Continuous Mechanics,* Philadelphia 1969, p. 231.
82. G. Stahl, J. Warnatz, B. Rogg in AIAA (ed.): *Progr. Aeronaut. Astronaut.,* **vol. 113,** Washington 1988, p. 195.
83. P. Bradshaw in P. Bradshaw (ed.): *Turbulence,* Springer Verlag, Berlin – Heidelberg 1978, p. 1.
84. B. E. Launder, D. B. Spalding: *Lectures in Mathematical Models of Turbulence,* Academic Press, London – New York 1972.
85. K. K. Kuo: *Principles of Combustion,* J. Wiley and Sons, New York 1986.
86. J. O. Hinze: *Turbulence,* McGraw-Hill, New York 1975.
87. J. Boussinesq, *Mém. prés. Acad. Sci.* **XXIII, 46,** Paris1877.
88. L. Prandtl, *Z. Angew. Math. Mech.* **22** (1942) 241.
89. L. Prandtl, *Nachr. Akad. Wiss. Göttingen, Math. Phys. Kl. 2* (1945) 6 – 19.
90. J. Janicka: *Twenty-First Symposium (International) on Combustion,* The Combustion Institute, Pittsburgh 1988, p. 1409.
91. S. M. Correa, W. Shyy, *Prog. Energy Combust. Sci.* **13** (1987) 249.
92. H. Tennekes, J. L. Lumley: *A First Course in Turbulence,* MIT-Press, Boston 1972.
93. W. F. Ames: *Nonlinear Partial Differential Equations in Engineering,* Academic Press, New York – London 1965.
94. A. D. Gosman et al.: *Heat and Mass Transfer in Recirculating Flows,* Academic Press, London 1969.
95. M. D. Smooke, A. A. Turnbull, R. E. Mitchell, D. E. Keyes in C.-M. Brauner, C. Schmidt-Lainé (eds.): *Mathematical Modeling in Combustion and Related Topics,* Martinus Nijhoff Publishers, Dordrecht 1988, p. 261.
96. D. B. Spalding: *Basic Equations of Fluid Mechanics and Heat and Mass Transfer,* Imperial College London, Mechanical Engineering Department, Report HTS/76/6, London 1976.
97. S. V. Patankar: *Numerical Heat Transfer and Fluid Flow,* McGraw-Hill, New York 1980.
98. F. H. Harlow, J. E. Welch, *Phys. Fluids* **8** (1965) 2182.
99. A. O. Demurren, *Comput. Fluids* **13** (1985) 411.
100. P. J. Roach: *Computational Fluid Dynamics,* Hermosa Publishers, Albuquerque 1976.
101. E. S. Oran, J. P. Boris: *Numerical Simulation of Reactive Flow,* Elsevier Science Publishing Company, New York 1987.

102. C. Hirsch: *Numerical Computation of Internal and External Flows,* vol. 1, J. Wiley and Sons, New York 1988.
103. C. Hirsch: *Numerical Computation of Internal and External Flows,* vol. 2, J. Wiley and Sons, New York 1990.
104. A. J. Baker: *Finite Element Computational Fluid Mechanics,* McGraw-Hill, New York 1985.
105. J. M. Beer, N. A. Chigier: *Combustion Aerodynamics,* Applied Science Publishers Ltd., London 1972.
106. L. Horvarth, *Proc. R. Soc. London Ser. A* **164** (1938) 547.
107. E. R. G. Eckert, R. M. Drake: *Analysis of Heat and Mass Transfer,* McGraw-Hill, New York 1972.
108. M. N. Özisik: *Basic Heat Transfer,* McGraw-Hill, New York 1985.
109. H. Brauer: *Stoffaustausch einschließlich chemischer Reaktionen,* Verlag Sauerländer, Aarau 1971.
110. P. Trambouze in A. E. Rodriguez, J. M. Calo, N. H. S. Sweed (eds.): *Multiphase Chemical Reactors, Fundamentals* **(vol. I),** Sijthoff and Noordhoff, Alphen 1981.
111. J. A. van der Vusse, *Chem. Eng. Sci.* **17** (1962) 507.
112. L. Cloutier, *Can. J. Chem. Eng.* **37** (1959) 105.
113. P. Gray, S. K. Scott: *Chemical Oscillations and Instabilities: Nonlinear Chemical Kinetics,* Clarendon Press, Oxford 1990.
114. G. Iooss, D. D. Joseph: *Elementary Stability and Bifurcation Theory,* Springer Verlag, New York 1980.
115. J. H. Merkin, D. J. Needham, S. K. Scott, *Proc. R. Soc. London Ser. A* **406** (1986) 299.
116. B. F. Gray, M. J. Roberts, *Proc. R. Soc. London Ser. A* **416** (1988) 391.
117. P. M. Frank: *Introduction to System Sensitivity Theory,* Academic Press, New York 1978.
118. T. P. Coffee, J. M. Heimerl, *Combust. Flame* **50** (1983) 323.
119. D. E. Clough, W. F. Ramirez, *AIChE J.* **22** (1976) no. 4, 1097.
120. J. C. P. Sheel, C. M. Crowe, *Can. J. Chem. Eng.* **47** (1969) 183.
121. R. A. Greenkorn: *Flow Phenomena in Porous Media,* Marcel Dekker, New York 1983.
122. E. Kreyszig: *Advanced Engineering Mathematics,* J. Wiley and Sons, New York 1988.
123. G. F. Froment, K. B. Bishoff: *Chemical Reactor Analysis and Design,* J. Wiley and Sons, New York 1990.
124. P. B. Weisz, J. S. Hicks, *Chem. Eng. Sci.* **17** (1962) 265.
125. J. Ganoulis, F. Durst: "Finite Elements in Water Resources," *Proceedings VI Int. Conf.,* Lisbon 1986, p. 655.
126. S. N. B. Murthy (ed.): *Turbulent Mixing in Nonreactive and Reactive Flows,* Plenum Press, New York – London 1975.
127. R. W. Bilger in P. A. Libby, F. A. Williams (eds.): *Turbulent Reacting Flows,* Springer Verlag, Berlin – Heidelberg 1980, p. 65.
128. R. Borghi, *Prog. Energy Combust. Sci.* **14** (1988) 245.
129. H. Bockhorn in C.-M. Brauner, C. Schmidt-Lainé (eds.): *Mathematical Modeling in Combustion and Related Topics,* Martinus Nijhoff Publishers, Dordrecht 1988, p. 411.
130. H. Bockhorn: *Twenty-Second Symposium (International) on Combustion,* The Combustion Institute, Pittsburgh 1988, p. 665.
131. H. Bockhorn: *Twenty-Third Symposium (International) on Combustion,* The Combustion Institute, Pittsburgh.
132. M. C. Drake et al.: *Twentieth Symposium (International) on Combustion,* The Combustion Institute, Pittsburgh 1984, p. 327.
133. M. C. Drake, R. W. Pitz, M. Lapp: *AIAA Pap.* **84-0544** (1984).
134. G. M. Faeth, G. S. Samuelson, *Prog. Energy Combust. Sci.* **12** (1986) 305.
135. J. Warnatz in W. C. Gardiner, Jr. (ed.): *Combustion Chemistry,* Springer Verlag, New York 1984, p. 196.

Molecular Modeling

DONALD B. BOYD, Indiana University-Purdue University at Indianapolis (IUPUI), Indianapolis, IN, United States

See also: Molecular Dynamics Simulations

1. **Definitions** 307
2. **Purpose of Modeling** 307
3. **Brief History** 308
4. **Methodologies of Molecular Modeling** 309
4.1. **Computer Graphics** 309
4.2. **Databases** 310
4.3. **Conformational Modeling** 310
4.4. **Quantum Mechanical Modeling** 311
4.5. **Force Field Modeling** 312
4.6. **Statistical Modeling** 313
4.7. **Homology Modeling** 314
4.8. **Drug Design Modeling** 314
5. **Economic Aspects** 315
6. **References** 317

Molecular modeling is used to understand the characteristics of molecules and to design new molecules in industry and elsewhere. The computer-based techniques of molecular modeling include molecular mechanics, molecular graphics, and many other methodologies of computational chemistry. Most large research-based companies involved in molecular science, particularly pharmaceutical companies, use molecular modeling daily.

1. Definitions

In the narrowest sense used by some chemists, molecular modeling refers to conformational energy calculations and interactive manipulation of three-dimensional (3D) molecular structures on a computer screen. For instance, an early definition of molecular modeling was offered by TROST: "Molecular mechanics, [which is] the empirical computation of conformational energy, combined with computer graphics" [2]. This definition reflects the perspective of many organic chemists who want the computer to help them better understand the 3D shapes of molecules. HOPFINGER, a computational chemist, suggested a somewhat broader definition: "The generation, manipulation, and/or representation of molecular structures and associated physicochemical properties" [3]. From the perspective of a chemist in industry (DuPont), an even broader definition was set forth: "Virtually anything which is done to depict, describe, or evaluate any aspect of the properties or structure of a molecule that requires the use of a computer" [4].

The terms computational chemistry and molecular modeling are intertwined and are often used interchangeably in industry. Computational chemistry has been broadly defined as "those aspects of molecular research that are expedited or rendered practical by computers" [5]. Definitions for molecular modeling and computational chemistry have been discussed on the Internet (see, e.g., [6]). Deliberations of an IUPAC working party [7] resulted in a recommendation to define computational chemistry loosely as "a discipline using mathematical methods for the calculation of molecular properties or for the simulation of molecular behavior. It also includes, e.g., synthesis planning, database searching, and combinatorial library manipulation". The IUPAC group defined molecular modeling as "the investigation of molecular structures and properties using computational chemistry and graphical visualization techniques in order to provide a plausible three-dimensional representation under a given set of circumstances".

2. Purpose of Modeling

Modeling involves working with what is usually a simplified description of a system. Despite the simplifications, the model should capture the essence of a particular object, property, or process of interest. The models help scientists sort out complex relationships between variables. Being able to devise and use good

Ullmann's Modeling and Simulation

ISBN: 978-3-527-31605-2

models can help rationalize experimental data. Modeling can be used to estimate values of molecular properties in cases where experiment is difficult or impossible [8].

It is important to keep in mind that because of the mathematics involved, modeling requires attention to quantitation and detail. When molecules are being generated, manipulated, represented, depicted, etc., it is important to strive for realistic molecular structures and associated physicochemical properties. Although it is fairly easy to display a molecular structure on a computer screen, the model should closely represent an experimentally determined structure, so that any conclusions that are derived from the model are useful. State-of-the-art hardware and software make it easy for even the novice to create a structure on a computer screen, but does it mean anything? It may not be immediately apparent from just looking at a structural depiction that some bond lengths are too short or too long, for instance. A bench chemist may be used to writing chemical structures without the hydrogen atoms, but to minimize the quantum mechanical energy of a molecular structure leaving off hydrogen atoms will lead to useless results. Proper use of the models requires an appreciation of quantitative details and the underlying theory.

3. Brief History

Models of molecules have been around for a long time. Chemists have used hand-held molecular models to help them visualize conformation and understand stereochemistry. These models were built of cork, wood, metal, or plastic. Well-respected types, such as Dreiding or CPK, were once familiar in most chemistry laboratories. Because of the prevalence of computers today, some young students may not have ever heard of these models unless they happen to look in some old books and journals. Today computers perform many of the functions of the hand-held models and can do much more.

Theoretical chemistry had its origins in the quantum mechanics developed in the 1920s. However, it was not till the late 1950s that the early slow computers became available to research chemists. Two camps of quantum chemists developed [9]. One camp was interested in rigorous applications of the theory, which of necessity were confined to small molecules of only a few atoms, too small to be of interest to most industrial chemists. This ab initio (from first principles) research was mainly of interest to theorists, spectroscopists, and physicists. The other camp was willing to forego theoretical purity to render the methods applicable to molecules large enough to be of interest to a wide body of chemists. The latter, so-called semiempirical approach began to be applied to molecules of biological and commercial interest.

Simultaneously, organic chemists, starting with Westheimer, Allinger [10], and others, recognized the importance of understanding the conformations of organic molecules and developed empirical methods for computing the relative energies of molecular structures. Likewise, protein chemists, such as Scheraga [11], became interested in empirical modeling because it allowed study of the conformations of polypeptides and proteins. The empirical or molecular mechanics methods were faster than either ab initio or semiempirical quantum mechanical methods and opened the door to computer simulations of large biomolecules, including enzymes, nucleic acids, and materials. The use of molecular mechanics spread in the 1970s and continues to grow [12, 13].

Other physical organic chemists were interested in the lipophilicity of molecules because of the relevance of this property to drug action. Starting in the 1960s, computational methods were developed for estimating the partitioning of solutes between aqueous and nonpolar solvents [14]. These methods have been used frequently to quantitate the relationship of biological activity and chemical structure.

Visionary scientists in the pharmaceutical industry recognized in the 1960s that theoretical modeling could add value to their company's research programmes and convinced their managers to hire one or two computational chemists in the 1960s and 1970s. However, it was not until the 1980s that industry broadly recognized the need to hire scientists to do molecular modeling. The field blossomed because of four factors:

1) Mini- and supermini-computers, particularly the VAX series from Digital Equipment Corporation.

2) High-resolution computer graphics terminals and workstations, such as the Evans and Sutherland Picture System.
3) User-friendly, multifunctional computational chemistry software packages.
4) The promising results of the initial modeling efforts.

Today the use of molecular modeling continues to grow due to:

1) Ever more powerful workstations (running UNIX) and personal computers (running Windows or Linux) at low relative cost.
2) Modern computational chemistry software with graphical user interfaces (GUI).

All major research-based pharmaceutical companies and many biotechnology companies employ computational chemists for molecular modeling. Large companies may have hundreds of scientists using molecular modeling. For a more detailed history of the use of computers and software in the pharmaceutical industry, see [15]. In addition, molecular modeling groups can be found in the agrochemical, chemical, and other industries.

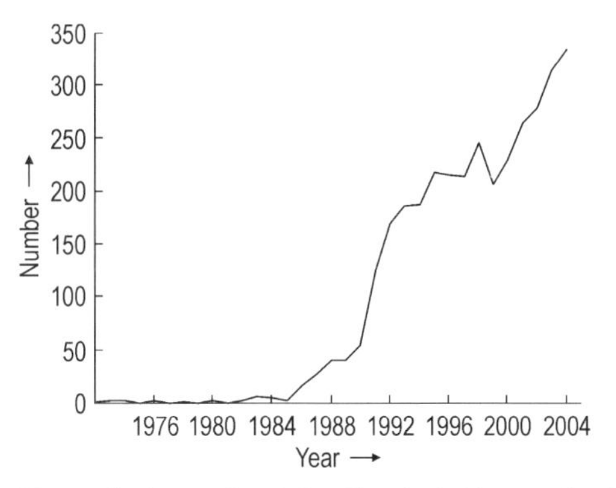

Figure 1. A search of the Chemical Abstract Service CAPLUS database indicates that as of late 2005 more than 3100 original research papers and nearly 400 book chapters and review articles are indexed as relevant to the two terms molecular modeling and computer. The number published annually is plotted.

Thousands of papers have been written about the methods and uses of molecular modeling. The growth in the annual number of publications can be seen in Figure 1. Use of the term molecular modeling with respect to computers increased in the 1980s, accelerated in the 1990s, and continues to grow. The most popular venues for these publications have been the Journal of Medicinal Chemistry, the Journal of Molecular Graphics and Modelling, and the Journal of Computer-Aided Molecular Design. Many books and reviews of molecular modeling continue to appear, see, e.g., [16 – 28].

4. Methodologies of Molecular Modeling

Molecular modeling entails a wide variety of complex, computer-based methods. These methods include ones for determining properties, accessing prior knowledge about them (databases), and analyzing data. Brief qualitative descriptions of some of these methods are presented.

Standard molecular modeling programs [29] are in use at many companies and universities, whereas other software packages, especially those specialized for molecular structure database management and drug design, are found mainly in the pharmaceutical industry because of their expense. Compilations [30, 31] have presented brief descriptions of programs, information about developers and vendors, and resources on the World Wide Web (WWW). More recent information about software can be found with the Google search engine on the WWW. A number of databases (e.g., PubChem) and software tools are in the public domain. Again, a search of the WWW will reveal both free and commercial resources.

4.1. Computer Graphics

The most important component of molecular modeling is computer graphics because this is the interface between the user and the computer. Molecular structure being the universal language of chemists, a user sitting at a computer can draw a molecule to be modeled on the screen using a "point and click" paradigm. Program options can be set up through a graphical user interface (GUI). After a calculation (such as described in the sections below) is complete, the results may be examined visually on the computer screen.

The complexity of molecular models makes high resolution, color, and stereo graphics highly desirable, if not essential [32]. The simplest graphical depiction of molecules has lines representing bonds connecting points representing atoms. More complex graphics representing molecules as solid objects with colored or translucent surfaces help to convey molecular properties that pertain to each region of the molecule. Examples of computer graphics are shown in Figures 2 [33] and 3 [34 – 37].

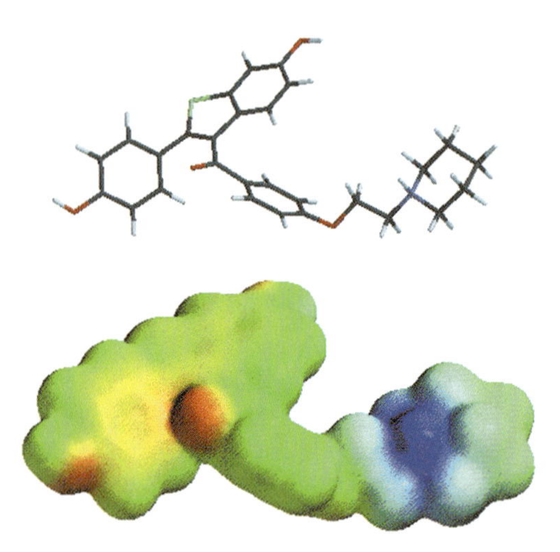

Figure 2. Molecular graphics of raloxifene (Evista), a drug for preventing osteoporosis and other health problems in women. Atoms are color-coded by element: carbons in black, oxygens in red, hydrogens in light blue, etc. The upper image is a stick figure; the lower image shows the electrostatic potential color-coded on an outer contour of the electron density distribution. Red denotes regions with high concentration of negative charge that would strongly attract a positive probe point charge; regions with a slight excess of electron density are in yellow; regions with a slight deficiency of electron density are in green; regions that would strongly repel a positive probe point charge are in blue.

Computer graphics is also used to visualize the results of statistical analyses. Complex response surfaces can be color-coded to show how several variables relate to each other [38]. The quantum chemical and molecular simulations methods (see below) are computationally intense and generate large quantities of data; computer visualization turns the data into something meaningful to chemists.

4.2. Databases

Almost all pharmaceutical and chemical companies have their own proprietary databases of compounds that have been synthesized or purchased for testing in various biological assays. In addition, there are commercial databases of compounds, such as the Chemical Abstracts Service's Registry database, and MDL Information Systems' Available Chemicals Directory (ACD). Industrial chemists use these databases to answer questions such as where can a compound be purchased, or has a chemical structure already been synthesized, or what compounds exist that are similar to a structure of interest?

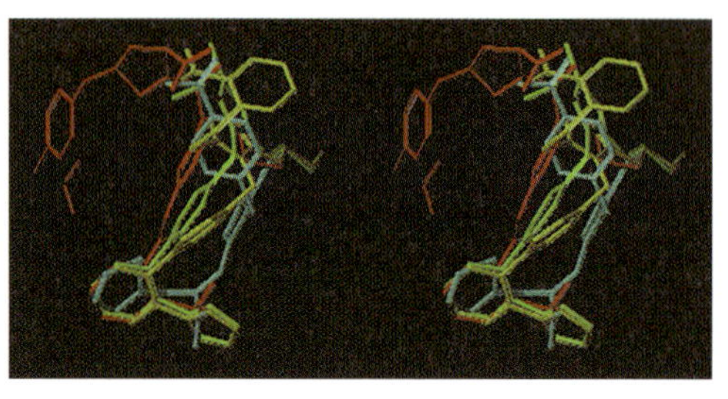

Figure 3. Stereo image of a possible superposition of four nonpeptide angiotensin II receptor antagonists.
a) losartan (green); b) eprosartan (light blue); c) Lilly's LY301875 (red); d) 3-*n*-butyl-5-[(2-carboxybenzyl)thio]-4-[[2′(1*H*-tetrazol-5-yl)biphenyl-4-yl]methyl]-4*H*-1,2,4-triazole (yellow). This overlay is based on alignments of the alkyl chains (right), aromatic rings (bottom), acidic groups (tetrazole or sulfonic acid) (bottom), and hydrogen bonding groups (top).

Database management tools specifically designed to handle chemical skeletons are used for rapid storage and retrieval. There are two types of molecular structure database: two-dimensional and three-dimensional. The 2D type stores atoms (chemical elements) and connectivity information (which atoms are bonded in a molecule). The 3D type stores, in addition, the x, y, z Cartesian coordinates of each atom in a molecule.

Structures determined by X-ray crystallography [39] or NMR spectroscopy [40] are stored in 3D databases. International repositories of the atomic coordinate data are available [41], such as the Cambridge Structural Database (CSD) for small organic and organometallic compounds, and the Protein Data Bank (PDB) and the Nucleic Acid Database for biomacromolecules. The structures can be used as starting points for molecular modeling experiments.

4.3. Conformational Modeling

Shape and energy are two molecular properties by which chemists understand how compounds interact and react. The properties are thus important in the pharmaceutical and agrochemical industries, where compounds must be found that will bind specifically to a receptor mediating a biological effect. To bind to a receptor, a molecule must have the shape — or the ability to adopt the shape — required for fitting into the receptor site. Most molecules are flexible, and at room temperature they are constantly undergoing conformational changes. The lowest energy structure is the global minimum one, but it may have no relevance to the conformation adopted upon receptor binding. As more ligand – receptor structures become solved by X-ray crystallography, it has become clear that is it not uncommon for the bioactive conformation of a ligand to be several kJ/mol higher in energy than the global minimum conformation. This situation arises because, when a ligand interacts with a receptor site, additional interatomic forces, not present in aqueous solution, occur that can induce the ligand to adapt its shape for better fit. The other forces arise from electrostatics, hydrogen bonding, van der Waals interactions, and hydrophobic effects. However, if too much energy is required to force a molecule to adopt a conformation required for receptor recognition, it is unlikely that the molecule will bind well.

The process of exploring all the conformations a molecule can adopt is conformational searching [42]. This involves molecular modeling techniques to map out the potential energy surface of a molecule, which describes the energy of the molecule as a function of the spatial location of its atoms. One type of conformational searching is systematic searching, where, as the name implies, the torsional angles at the rotatable bonds are varied systematically in increments. Another type is the Monte Carlo method, in which the geometrical variables are varied randomly, and the energy of each random conformation is calculated [43]. These and other methods for conformational searching tend to be computationally intensive and become prohibitively lengthy for molecules with more than about ten rotatable bonds. Energy evaluation during a search can be done with a force field or quantum chemical method (see below), although the force field methods are preferred because of their much greater speed and comparable accuracy.

Once the conformations have been determined, then molecular shape comparisons are possible. Molecular modeling offers several ways to compare and superpose molecular structures: alignments can be made by least-squares fitting of atoms or of projections to hypothetical hydrogen bonding points. Molecules can also be aligned based on their "surfaces" (e.g., the van der Waals surface) [44] or a computed field (measuring, e.g., steric, electrostatic, or lipophilic effects). Overlaying molecules is easy with computer models and is one of the most common applications of modeling.

4.4. Quantum Mechanical Modeling

Quantum mechanics (QM) is embodied in the Schrödinger equation, $H\Psi = E\Psi$, where H is a mathematical operator of kinetic and potential energy terms, Ψ is the wavefunction describing the positions of particles (electrons and nuclei) in a system, and E is the energy of the system [45, 46]. In molecular orbital (MO) theory, which has been the most common implementation of quantum mechanics used by chemists, electrons are distributed around the atomic nuclei until they reach a so-called self-consistent field (SCF), that is, until the attractive and repulsive forces between all the particles (electrons and nuclei) are in a steady state, and the energy is at a minimum. By computing the energy of different nuclear configurations, the lowest energy structure can be found. Such geometry optimization is useful because the bond lengths and bond angles that give the lowest energy are usually closest to what is observed experimentally. It has been pointed out that reliable reproductions of potential energy surfaces of flexible molecules sometimes need more stringent SCF and optimization convergence criteria during the calculations [47].

In ab initio MO calculations, very few approximations are made [48], and in principle they have the best ability to give agreement with experimental molecular properties. The electron density is represented by distributing it among certain standard sets of mathematical functions, called basis sets [49]. The different levels of ab initio theory are referred to as "theoretical models" [50]. Unless the basis set is adequate, there is no great advantage of doing a calculation ab initio. Sometimes electron correlation (the phenomenon of electrons avoiding each other) and relativistic effects must be included in a calculation to get good results. Of the various quantum methods available, ab initio calculations are preferred, but unfortunately these calculations are sometimes too slow for molecular modeling applications of interest to industrial chemists. In certain situations, however, a molecule with several hundred atoms can be treated by an ab initio method [51, 52]. Large molecular systems are best left to one of the force field approaches (see Section 4.5).

In semiempirical MO calculations, additional approximations are made, such as treating only the outermost (i.e., valence) electrons while assuming the core (inner) electrons neutralize part of the nuclear charge [53 – 55]. Parameters in semiempirical theories are set to reproduce as closely as possible selected experimental molecular properties of well-studied compounds. However, parameters that work well at reproducing a small set of data are usually not robust enough to reproduce a broad range of properties and compounds. Semiempirical methods are sometimes used for "intermediate-sized" molecules (20 - 200 atoms). Semiempirical methods can sometimes predict molecular properties adequately, at least for similar molecules in which the effects of systematic errors are minimized.

A third implementation of QM is density functional theory (DFT) [56 – 59]. It is particularly useful in materials applications and is increasingly used for determining molecular properties. DFT is popular because it offers a compromise: it is faster than ab initio methods and more accurate than semiempirical methods. In DFT, the energy of a molecule is determined from the electron density, not from an MO wavefunction. Whereas ab initio theory includes the kinetic energy of the electrons and the potential energies of electron-electron repulsion and electron-nuclear attraction, DFT also includes an energy term called exchange-correlation that improves the result.

Electronic structure calculations are suited for predicting a variety of molecular systems (ground states, excited states, and transition states) and properties (optimum bond lengths and bond angles, conformational preferences, ionization potentials, electron affinities, charge distributions, hydrogen bonding, etc. [60, 61]). Molecular similarity can be calculated based on electronic structure [62]. Many types of reactions have been modeled (see, e.g., [63]). For modeling an enzymatic reaction, a chemically reasonable subset of atoms in the active site can be treated quantum mechanically and the rest of the system can be treated empirically [64]. This is the so-called quantum mechanics/molecular mechanics (QM/MM) approach.

Although the charge on an atom is not an experimental observable, charge distributions are of great interest to chemists [65, 66]. From a wavefunction, one can calculate the molecular electrostatic potential (MEP), which is an energy of attraction or repulsion experienced by a hypothetical unit charge as it moves in the vicinity of a molecule [67]. The MEP provides a model of how one molecule looks to another as they approach, such as when a ligand is approaching a receptor.

4.5. Force Field Modeling

In molecular mechanics (MM), as well as in molecular dynamics (MD) and Monte Carlo (MC) simulations, an empirical force field (FF) describes how the energy of a molecule changes as a function of bond lengths l, bond angles θ, and torsional angles ϕ. Atoms are treated as points in space except for the van der Waals term, which represents the atoms as spheres with soft surfaces. A set of energy terms and associated parameters is called a force field (FF) [68]. Simpler FFs assume harmonic force constants. The energy is a combination of bond stretching/compression, bond angle bending, torsional, van der Waals, and electrostatic forces that tend to hold the atoms at their equilibrium positions [69 – 73].

$$E_{\text{total}}=E_b+E_\theta+E_\phi+E_\tau+E_{el}+E_{vdw}$$

$$E_b=\sum_{\text{bonds}} k_b(l-l_0)^2$$

$$E_\theta=\sum_{\text{bond angles}} k_\theta(\theta-\theta_0)^2$$

$$E_\phi=\sum_{\text{dihedral angles}} k_\phi[1+\cos(n\phi-\delta)]$$

$$E_{\tau}=\sum\nolimits_{\text{improper angles}} k_{\tau}(\tau-\tau_0)^2$$

$$E_{el}=\sum\nolimits_{i,j>i}\frac{q_iq_j}{\varepsilon r_{ij}}$$

$$E_{vdw}=\sum\nolimits_{i,j>i}\frac{A_{ij}}{r_{ij}^{12}}-\frac{B_{ij}}{r_{ij}^{6}}$$

Here the force constants are the various parameters *k*, the "natural" bond lengths and angles are indicated by the subscript zero, ε is the dielectric constant, r_{ij} is the interatomic distance between atoms *i* and *j*, and "improper" angles τ refer to a mathematical trick to keep sp^2 atoms from deviating too far from planarity. Partial charges are denoted by q. n is 1, 2, or 3 depending on the periodicity of bond rotation, and δ determines where the minima occur. A_{ij} and B_{ij} are parameters determining the depth and distance of the energy minimum in a van der Waals interaction, which is attractive at long range and repulsive at short distances. The van der Waals and Coulombic interactions are summed between nonbonded atoms, whereas the other FF terms involve atoms bonded to each other. In better force fields, additional terms take into account anharmonicity and special electronic effects. The FF parameters are calibrated to reproduce experimental geometries and relative energies.

In molecular mechanics, the objective is to compute the energy of a molecular structure and/or to minimize the energy as a function of geometrical degrees of freedom, i.e., the potential energy surface. The computed energy reflects how far a geometrical arrangement around each atom deviates from an idealized bonding situation. A "better" geometry has a lower energy. Optimization methods [74] are used for energy minimization.

The MM model is conceptually simple, gives quite accurate molecular structures, and provides its information at least an order of magnitude faster than do QM methods. These strengths make MM well suited for treating both small and large molecules. Limitations of MM are that it requires empirical parameters to describe each "spring", and predictions depend on the care with which parameterization has been done, just as in quantum calculations where the results depend on the adequacy of the basis set and theoretical model.

Molecular dynamics (MD) simulations (→ Molecular Dynamics Simulations) give information about the variation in structure and energy of a molecule over an interval of time [75, 76]. In MD, atoms move according to Newton's equations of motion (force = mass × acceleration, and force = negative gradient of potential energy). Thus, atoms under the most strain move fastest and farthest. To satisfactorily solve Newton's equations requires extremely short time steps, typically only 1 fs (10^{-15} s). If the time increment is too long, atoms can come too close in each step, and the energy of interaction from the FF become so high that the system becomes erratic in subsequent steps [77].

MD simulations are useful in computing accurate relative free energies. Free energy perturbation (FEP) or thermodynamic cycle calculations simulate the mathematical conversion of one molecule to another [78 – 80]. By changing the FF parameters in tiny increments from those of one chemical structure to those for another similar structure, the change in free energy can be accurately simulated. Assuming the incremental changes are small enough that the molecular system stays in equilibrium, FEP calculations can predict relative binding energies of two similar ligands associating with a receptor. Although in principle FEP calculations should be useful for testing molecular design ideas, in practice the calculations may be too slow for many industrial applications [81, 82].

Monte Carlo (MC) simulations are like MD calculations, but instead of proceeding along a time course, the system follows a random walk through configuration space [83]. Huge numbers of more or less randomly chosen Cartesian coordinates of the atoms of the system are considered. An ensemble average of the computed energies is used to calculate thermodynamic properties via statistical mechanics.

4.6. Statistical Modeling

By performing regression analysis and other statistical techniques, it is possible to quantitate the relation between molecular descriptors and an observed molecular property for a related set of compounds. The molecular property can be, for instance, the wavelength of an electronic transition, bioactivity, boiling point, chemical stability, etc. Descriptors can be almost any quantity (calculated or experimental) that helps describe a molecule, such as molecular weight, atom counts, bond topological indices [84], molecular shape [85], or lipophilicity [14]. Descriptors related to substituents on an aromatic or aliphatic template have been assembled into large collections [86]. If a quantitative structure – property relationship (QSPR) can be found, then it can be used to predict (preferably by interpolation) the property of new structures. With the proliferation of descriptors, machine-learning techniques (see, e.g., [87]) have been developed to help choose the best ones [88].

4.7. Homology Modeling

Because of advances in the human genome project and in molecular biology in general, amino acid sequences of proteins are becoming known experimentally at a high rate, tens of thousands per year. Unfortunately, structural chemists cannot solve the 3D structures of these at the same rate, and it is very difficult to predict reliably how an amino acid sequence will fold.

One approach to obtaining a new 3D structure of a protein is to use homology modeling [89 – 92]. Here, the 3D structure has to be known for a sequence homologous to the new protein. The new sequence is threaded onto the backbone of the known structure, taking into account insertions and deletions in the sequence. The new model is then refined by energy minimization and an MD technique called simulated annealing. Thereby the theoretical temperature of the protein (determined by the kinetic energy of the atoms in the system) is gradually lowered in the hope that the structure will settle into a low minimum on the potential energy surface.

4.8. Drug Design Modeling

Some uses of the above methods have already been mentioned in regard to computer-aided drug design. In addition, a panoply of methods has been created specifically for applying molecular modeling to the design of chemical structures of interest to the pharmaceutical industry [93 – 98]. The vast majority of medicines perform their effect by binding to an enzyme or other protein. The rest target nucleic acids or other natural chemicals in the human body. Generally, synthetic molecules that bind to a biomacromolecule are called ligands.

Software tools [99, 100] have been developed to convert 2D chemical structures (as typically found in corporate and commercial databases) to 3D for further analysis by molecular modeling. Proprietary and commercial 3D databases are used by computational chemists in the pharmaceutical industry to search for structures matching a pharmacophore (the smallest set of atoms in a specific geometrical arrangement that will allow a molecule to be recognized by a biological receptor and bind to it). If appropriate hits from a 3D search are obtained, samples of the compounds can be submitted for testing in biological screens.

By combining quantitative structure – activity relationships (QSAR) and database tools, molecular modeling comes into play as a way to measure and maximize structural diversity in selecting compounds from databases for screening or to focus new compound libraries around a known lead structure. The ability of molecular modeling to provide molecular descriptors is important in database mining [101, 102] and combinatorial chemistry [103].

Comparative molecular field analysis (CoMFA) is a QSAR method that was developed in an attempt to model how ligands may be perceived in 3D by a receptor [104, 105]. The nonbonded van der Waals and electrostatic energies of a probe atom interacting at grid points surrounding a set of ligands are used as descriptors. The resulting data set has far more columns (descriptors) than rows (compounds). This situation is handled by the statistical method of partial least squares (PLS) of latent variables. If the ligands can be aligned in a strictly comparable way, then CoMFA can sometimes reveal which spatial regions surrounding a set of ligands enhance a property and which detract from it.

Some software tools combine conformational analysis and statistical modeling. These programs are designed to analyze the shapes of a series of compounds that are known to bind to the same receptor [106]. The algorithm tries to find conformations that let the more active compounds present a pharmacophore that the less active ones cannot. A statistical relationship is sought between activity and features in the molecules, such as hydrophobic regions, hydrogen bonding sites, and ionized acidic or basic groups with formal positive or negative charge.

The 3D structure of many drug targets has not been determined. In such cases, methods for QSAR and modeling known ligands can be employed to elucidate the requirements for biological activity. In an increasing number of cases, a 3D structure of the drug target is known from experiment or can be built by molecular modeling. With a 3D structure in hand (structure-based drug design, SBDD), ligands can be docked in the receptor site [107 – 110]. Visual inspection of the docked molecules may reveal where adding or subtracting atoms/groups to/from the ligands would increase binding affinity. Usually, the better a drug binds to its intended target, the lower the required dose, and hence the lower the risk of side effects. The process of estimating the binding affinity is called scoring. A ligand may be able to orient in more than one way in a receptor; each orientation is called a pose. Computers may be used to automate the process in docking and scoring; this is called virtual screening [111].

A 3D model of a receptor can be further exploited with de novo design modeling programs [112, 113]. These programs generally have a library of molecular fragments, such as peptides or organic subunits, which they automatically assemble into possible 3D ligands. The hypothetical structures are evaluated in terms of the computed energy of interaction with the target receptor or other means. A weakness of early de novo design efforts was that they assumed that the target receptor structure is frozen. However, it is known both experimentally and computationally that some receptor sites change shape significantly when a ligand binds. Moreover, X-ray crystallography on ligand – receptor complexes shows that even similar ligands can occupy the receptor in different ways. Nevertheless, the de novo programs may suggest designs not previously considered, which can be followed up if they are synthetically feasible.

Pharmaceutical and biopharmaceutical companies currently use synthetic strategies, collectively called combinatorial chemistry, to produce thousands of compounds for biological screening. These sets of compounds are called libraries. Molecular modeling techniques are employed to increase the likelihood that bioactive compounds will be in the libraries [114, 115].

Genetic algorithms (GA) is a computational technique for solving optimization problems [116, 117]. The properties or constituents of a ligand can be modeled mathematically by a string of binary bits. Algorithms recombine and mutate these bit strings and then measure the resulting predicted property until the optimum combination of bits is found. GA can be used, for instance, to optimize the substituents on a molecular framework to fill a receptor cavity.

Computational drug design is extending beyond simply designing new ligands. Some of the new directions include studying the absorption, distribution, metabolism, and elimination (ADME) properties of compounds being considered for pharmaceutical development [118 – 121]. In addition, modeling is done to study potential toxicity of compounds. The preferred route of administration for most medicines is by mouth. Molecular modeling is used to study oral bioavailability [122]. One of the characteristics a compound must have to become a medicine is sufficient stability. Even this property is being scrutinized with molecular modeling [123].

The new field of cheminformatics is evolving mainly in regard to drug design as applied in industry [124, 125]. Cheminformatics arises from the cross-pollination of computer science, information science, and computational chemistry.

5. Economic Aspects

Most of the applications of molecular modeling alluded to in this article have dealt with compounds of interest to the pharmaceutical industry. This is because this area has to date been the largest area of applicability of the techniques. The total economic impact of molecular modeling is unknown, but the impact can be partially appreciated from the following data for the pharmaceutical industry. The major pharmaceutical companies operating in the United States, for instance, invested steadily greater amounts of their income in research and development: \$ 2×10^9 in 1980, \$ 15.2×10^9 in 1995, and \$ 39.4×10^9 in 2005. Much of the money goes for clinical testing, and only a very tiny fraction is invested

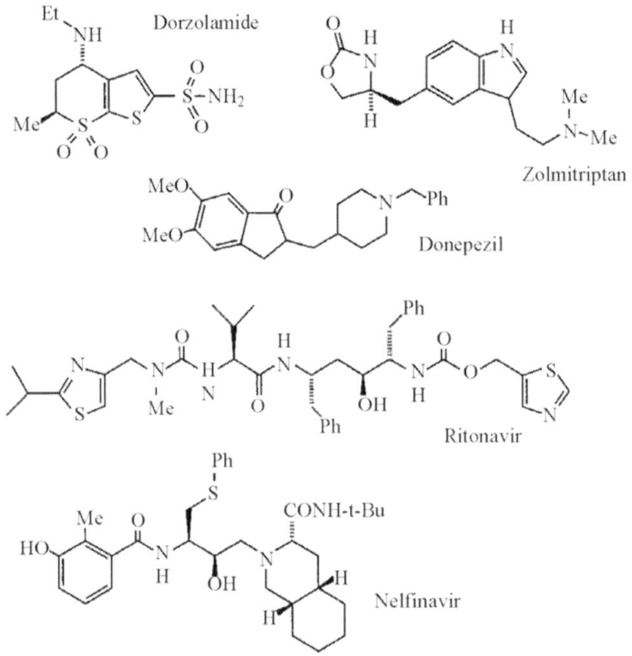

Figure 4. Examples of drugs discovered with some help of computer modeling: dorzolamide (Trusopt), an agent for treating glaucoma; zolmitriptan (Zomig), a serotonin-1D agonist; donepezil (Aricept), an acetylcholinesterase inhibitor; ritonavir (Norvir) and nelfinavir (Viracept), both HIV protease inhibitors.

in modeling. Because the number of new drugs reaching the market has been relatively flat (due to scientific difficulty and regulatory factors), it is not surprising that the average cost of discovering and developing a new pharmaceutical product is huge and has grown from \$ 597×10^6 in 1994 to about \$ 10^9 ten years later. Almost the entire burden of the R& D costs must be born by sales of existing pharmaceuticals. The incentive to innovate and discover new and better medicines is made possible by the high cost of prescription drugs. Innovative medicines save lives, improve the quality of lives, and in general help the economy of a nation by increasing productivity.

A large company may spend \$ 10×10^6 per year to support molecular modeling; this amount includes the investment in the computational chemists and computer support staff, the cost of new computers to stay current with technology, the purchase of new software and databases, and the licenses to maintain existing software and databases. However, a small company can launch an entry-level molecular modeling effort for under \$ 50 000.

Because of the extreme difficulty of finding viable new pharmaceutical products, researchers must use every effective technology to help them discover the few new compounds that meet medical and market needs. Molecular modeling has proven to be one such technology. Rarely, if ever, can it be claimed that a marketed product was designed purely by molecular modeling. Usually modeling plays a supportive role to the other scientific disciplines of a research organization. Sometimes modeling supplies information that answers specific research questions. Sometimes modeling can nudge a project in a direction that eventually leads to a new product. Cases of the discovery of marketed pharmaceuticals which were influenced or assisted by some type of computer modeling have been documented [126 – 135]. Structure-based drug design efforts have been particularly productive [136]. The chemical structures of some of these compounds are shown in Figure 4. In addition, modeling has proved useful for finding marketed agrochemicals, such as those depicted in Figure 5 [137 – 139].

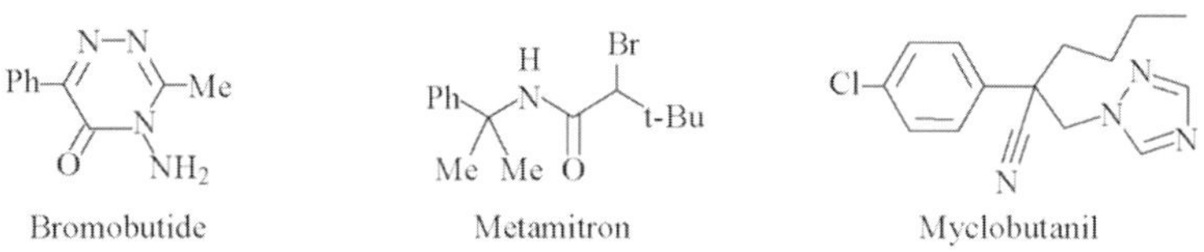

Figure 5. Examples of agrochemicals supported by computer modeling: bromobutide, a herbicide; metamitron, a photosystem II herbicide; myclobutanil (Systhane, Rally), a cytochrome P450 oxidase inhibitor.

Molecular modeling applications extend well beyond the pharmaceutical and agrochemical industries. Today, the methods are being used in industries commercializing adhesives, carbohydrates, catalysts, chromatographic stationary phases, clays [140], cleansing agents, corrosion inhibitors, crystals [141], explosives, flavorants, food components, fullerenes [142], glasses [143], hair colorings, lubricants and other petroleum products, materials [144], nanomaterials [145], nonlinear optical materials [146], nutrients, paint pigments, pesticides [147], photographic dyes, polymers, polysaccharides, semiconductors, soils [148], surfactants, water [149, 150], zeolites [151], and a host of other interesting materials (see, e.g., [152]). Uses of molecular modeling continue to grow because the computer methodologies help researchers gain a better appreciation of substances at a fundamental level.

6. References

General Reference

1. A. R. Leach: *Molecular Modeling: Principles and Applications*, 2nd ed., Prentice Hall, Harlow 2001.

Specific References

2. B. M. Trost, *Science* **227** (1985) 908–916.
3. A. J. Hopfinger, *J. Med. Chem.* **28** (1985) 1133–1139.
4. D. A. Pensak, *Pure Appl. Chem.* **61** (1989) 601–603.
5. K. B. Lipkowitz, D. B. Boyd (eds.): *Reviews in Computational Chemistry*, vol. 1, VCH Publ., New York 1990.
6. Archives of messages posted on the Computational Chemistry List (CCL) are available on the World Wide Web at http://www.ccl.net. See, e.g., the postings of A. Hocquet on 16 June 1997 and 4 July 1997, and A. Hocquet, L'actualite Chimique **7** (1997) 27–33.
7. H. Van de Waterbeemd et al., *Pure Appl. Chem.* **69** (1997) 1137–1152.
8. D. B. Boyd, J. D. Snoddy in D. Shugar, W. Rode, E. Borowski (eds.): *Molecular Aspects of Chemotherapy*, Springer-Verlag, New York 1992, p. 1–22.
9. J. D. Bolcer, R. B. Hermann in K. B. Lipkowitz, D. B. Boyd (eds.): *Reviews in Computational Chemistry*, vol. 5, VCH Publ., New York 1994, p. 1–63.
10. U. Burkert, N. L. Allinger: *Molecular Mechanics*, Am. Chem. Soc., Washington DC 1990.
11. H. A. Scheraga in K. B. Lipkowitz, D. B. Boyd (eds.): *Reviews in Computational Chemistry*, vol. 3, VCH Publ., New York 1992, p. 73–142.
12. D. B. Boyd, K. B. Lipkowitz in K. B. Lipkowitz, D. B. Boyd (eds.): *Reviews in Computational Chemistry*, vol. 6, VCH Publ., New York 1995, p. 317–354.
13. D. B. Boyd, *J. Mol. Struct.: THEOCHEM* **369** (1997) 219–225.
14. P.-A. Carrupt, B. Testa, P. Gaillard in K. B. Lipkowitz, D. B. Boyd (eds.): *Reviews in Computational Chemistry*, vol. 11, Wiley-VCH, New York 1997, p. 241–315.
15. D. B. Boyd, M. M. Marsh in S. Ekins (ed.): *Computer Applications in Pharmaceutical Research and Development*, Wiley, Hoboken 2006, p. 1–50.
16. P. Kollman, *Annu. Rev. Phys. Chem.* **38** (1987) 303–316.
17. A. D. French, J. W. Brady (eds.): *Computer Modeling of Carbohydrate Molecules*, American Chemical Society, Washington, DC, 1990.
18. G. Barnickel, *Pharm. Z.* **138** (1993) no. 45, 9–15.
19. G. Barnickel, *Chem. unserer Zeit* **29** (1995) no. 4, 176–185.
20. P. Comba, T. W. Hambley: *Molecular Modeling of Inorganic Compounds*, VCH Verlagsgesellschaft., Weinheim 1995.
21. B. J. Howlin, *Annu. Rep. Progr. Chem., Sect. C: Phys. Chem.* **92** (1996) 75–95.

22. W. B. Smith: *Introduction to Theoretical Organic Chemistry and Molecular Modeling*, VCH Publ., New York 1996.
23. H. D. Höltje, G. Folkers: *Molecular Modeling — Basic Principles and Applications*, VCH Verlagsgesellschaft., Weinheim 1996.
24. W. Gans, A. Amann, J. C. A. Boeyens (eds.): *Principles of Molecular Modeling*, Plenum Press, New York 1996.
25. N. B. Leontis, J. SantaLucia Jr. (eds.): *Molecular Modeling of Nucleic Acids*, Am. Chem. Soc., Washington DC 1998.
26. J. M. Goodman: *Chemical Applications of Molecular Modeling*, Royal Society of Chemistry, Cambridge 1998.
27. M. F. Schlecht: *Molecular Modeling on the PC*, Wiley-VCH, New York 1998.
28. H.-D. Höltje, W. Sippl, D. Rognan, G. Folkers: *Molecular Modeling — Basic Principles and Applications*, 2nd ed., Wiley-VCH, Weinheim 2003.
29. N. C. Cohen et al., *J. Med. Chem.* **33** (1990) 883–894.
30. D. B. Boyd in K. B. Lipkowitz, D. B. Boyd (eds.): *Reviews in Computational Chemistry*, vol. 7, VCH Publ., New York 1995, p. 303–380.
31. D. B. Boyd in K. B. Lipkowitz, D. B. Boyd (eds.): *Reviews in Computational Chemistry*, vol. 11, Wiley-VCH, New York 1997, p. 373–399.
32. T. E. Ferrin et al., *J. Mol. Graphics* **9** (1991) 27–32, 37–38.
33. D. B. Boyd, R. D. Coner, *J. Mol. Struct.: THEOCHEM* **368** (1996) 7–15.
34. J. V. Duncia et al., *J. Med. Chem.* **33** (1990) 1312–1329.
35. W. T. Ashton et al., *J. Med. Chem.* **36** (1993) 591–609.
36. R. M. Keenan et al., *J. Med. Chem.* **36** (1993) 1880–1892.
37. D. B. Boyd et al. in C. H. Reynolds, M. K. Holloway, H. K. Cox (eds.): *Computer-Aided Molecular Design: Applications in Agrochemicals, Materials, and Pharmaceuticals*, Am. Chem. Soc. Symp. Series 589, Washington DC 1995, p. 14–35.
38. D. B. Boyd, *J. Med. Chem.* **36** (1993) 1443–1449.
39. K. Müller in B. Jensen, F. S. Jorgensen, H. Kofod (eds.): *Frontiers in Drug Research — Crystallographic and Computational Methods*, Munksgaard, Copenhagen 1990, p. 210–221.
40. A. E. Torda, W. F. van Gunsteren in K. B. Lipkowitz, D. B. Boyd (eds.): *Reviews in Computational Chemistry*, vol. 3, VCH Publ., New York 1992, p. 143–172.
41. P. E. Bourne, H. Weissig (eds.): *Structural Bioinformatics*, Wiley-Liss, Hoboken 2003.
42. A. R. Leach in K. B. Lipkowitz, D. B. Boyd (eds.): *Reviews in Computational Chemistry*, vol. 2, VCH Publ., New York 1991, p. 1–55.
43. M. Saunders et al., *J. Am. Chem. Soc.* **112** (1990) 1419–1427.
44. P. G. Mezey in K. B. Lipkowitz, D. B. Boyd (eds.): *Reviews in Computational Chemistry*, vol. 1, VCH Publ., New York 1990, p. 265–294.
45. D. B. Boyd in K. B. Lipkowitz, D. B. Boyd (eds.): *Reviews in Computational Chemistry*, vol. 1, VCH Publ., New York 1990, p. 321–354.
46. C. J. Cramer: *Essentials of Computational Chemistry: Theories and Models*, Wiley, Hoboken 2002.
47. D. B. Boyd et al., *J. Comput. Chem.* **9** (1988) 387–398.
48. R. J. Bartlett, J. F. Stanton in K. B. Lipkowitz, D. B. Boyd (eds.): *Reviews in Computational Chemistry*, vol. 5, VCH Publ., New York 1994, p. 65–169.
49. D. Feller, E. R. Davidson in K. B. Lipkowitz, D. B. Boyd (eds.): *Reviews in Computational Chemistry*, vol. 1, VCH Publ., New York 1990, p. 1–43.
50. W. J. Hehre et al.: *Ab Initio Molecular Orbital Methods*, Wiley, New York 1986.
51. E. R. Davidson in K. B. Lipkowitz, D. B. Boyd (eds.): *Reviews in Computational Chemistry*, vol. 1, VCH Publ., New York 1990, p. 373–382.
52. J. Cioslowski in K. B. Lipkowitz, D. B. Boyd (eds.): *Reviews in Computational Chemistry*, vol. 4, VCH Publ., New York 1993, p. 1–33.
53. J. J. P. Stewart in K. B. Lipkowitz, D. B. Boyd (eds.): *Reviews in Computational Chemistry*, vol. 1, VCH Publ., New York 1990, p. 45–81.
54. M. C. Zerner in K. B. Lipkowitz, D. B. Boyd (eds.): *Reviews in Computational Chemistry*, vol. 2, VCH Publ., New York 1991, p. 313–365.
55. C. J. Cramer, D. G. Truhlar in K. B. Lipkowitz, D. B. Boyd (eds.): *Reviews in Computational Chemistry*, vol. 6, VCH Publ., New York 1995, p. 1–72.
56. L. J. Bartolotti, K. Flurchick in K. B. Lipkowitz, D. B. Boyd (eds.): *Reviews in Computational Chemistry*, vol. 7, VCH Publ., New York 1995, p. 187–216.

57. A. St.-Amant in K. B. Lipkowitz, D. B. Boyd (eds.): *Reviews in Computational Chemistry*, vol. 7, VCH Publ., New York 1995, p. 217–259.
58. E. Wimmer, *Analysis* **24** (1996) no. 1, M37–M42.
59. F. M. Bickelhaupt, E. J. Baerends in K. B. Lipkowitz, D. B. Boyd (eds.): *Reviews in Computational Chemistry*, vol. 15, Wiley-VCH, New York 2000, p. 1–86.
60. S. Scheiner in K. B. Lipkowitz, D. B. Boyd (eds.): *Reviews in Computational Chemistry*, vol. 2, VCH Publ., New York 1991, p. 165–218.
61. C. E. Dykstra et al. in K. B. Lipkowitz, D. B. Boyd (eds.): *Reviews in Computational Chemistry*, vol. 1, VCH Publ., New York 1990, p. 83–118.
62. P. Bultinck, X. Gironés, R. Carbó-Dorca in K. B. Lipkowitz, R. Larter, T. R. Cundari (eds.): *Reviews in Computational Chemistry*, vol. 21, Wiley-VCH, Hoboken 2005, p. 127–207.
63. N. Matsunaga, S. Koseki in K. B. Lipkowitz, R. Larter, T. R. Cundari (eds.): *Reviews in Computational Chemistry*, vol. 20, Wiley-VCH, Hoboken 2004, p. 101–152.
64. J. Gao in K. B. Lipkowitz, D. B. Boyd (eds.): *Reviews in Computational Chemistry*, vol. 7, VCH Publ., New York 1995, p. 119–185.
65. D. E. Williams in K. B. Lipkowitz, D. B. Boyd (eds.): *Reviews in Computational Chemistry*, vol. 2, VCH Publ., New York 1991, p. 219–271.
66. S. M. Bachrach in K. B. Lipkowitz, D. B. Boyd (eds.): *Reviews in Computational Chemistry*, vol. 5, VCH Publ., New York 1994, p. 171–227.
67. P. Politzer, J. S. Murray in K. B. Lipkowitz, D. B. Boyd (eds.): *Reviews in Computational Chemistry*, vol. 2, VCH Publ., New York 1991, p. 273–312.
68. D. B. Boyd, K. B. Lipkowitz, *J. Chem. Educ.*, **59** (1982) 269–274.
69. U. Dinur, A. T. Hagler in K. B. Lipkowitz, D. B. Boyd (eds.): *Reviews in Computational Chemistry*, vol. 2, VCH Publ., New York 1991, p. 99–164.
70. R. L. DeKock et al. in K. B. Lipkowitz, D. B. Boyd (eds.): *Reviews in Computational Chemistry*, vol. 4, VCH Publ., New York 1993, p. 149–228.
71. K. B. Lipkowitz, M. A. Peterson, *Chem. Rev.* **93** (1993) 2463–2486.
72. C. R. Landis, D. M. Root, T. Cleveland in K. B. Lipkowitz, D. B. Boyd (eds.): *Reviews in Computational Chemistry*, vol. 6, VCH Publ., New York 1995, p. 73–148.
73. I. Pettersson, T. Liljefors in K. B. Lipkowitz, D. B. Boyd (eds.): *Reviews in Computational Chemistry*, vol. 9, VCH Publ., New York 1996, p. 167–189.
74. T. Schlick in K. B. Lipkowitz, D. B. Boyd (eds.): *Reviews in Computational Chemistry*, vol. 3, VCH Publ., New York 1992, p. 1–71.
75. W. F. van Gunsteren, H. J. C. Berendsen, *Angew. Chem. Int. Ed. Engl.* **29** (1990) 992–1023.
76. K. E. Gubbins, N. Quirke: *Molecular Simulation and Industrial Applications: Methods, Examples and Prospects*, Gordon and Breach, Amsterdam 1996.
77. D. B. Boyd in A. Kent, J. G. Williams (eds.): *Encyclopedia of Computer Science and Technology*, vol. 33, Marcel Dekker, New York 1995, p. 41–71.
78. T. P. Lybrand in K. B. Lipkowitz, D. B. Boyd (eds.): *Reviews in Computational Chemistry*, vol. 1, VCH Publ., New York 1990, p. 295–320.
79. T. P. Straatsma in K. B. Lipkowitz, D. B. Boyd (eds.): *Reviews in Computational Chemistry*, 9, VCH Publ., New York 1996, p. 81–127.
80. H. Meirovitch in K. B. Lipkowitz, D. B. Boyd (eds.): *Reviews in Computational Chemistry*, 12, Wiley-VCH, New York 1998, p. 1–74.
81. L. M. Balbes, S. W. Mascarella, D. B. Boyd in K. B. Lipkowitz, D. B. Boyd (eds.): *Reviews in Computational Chemistry*, vol. 5, VCH Publ., New York 1994, p. 337–379.
82. M. R. Reddy, M. D. Erion, A. Agarwal in K. B. Lipkowitz, D. B. Boyd (eds.): *Reviews in Computational Chemistry*, 16, Wiley-VCH, New York 2000, p. 217–304.
83. R. J. Woods in K. B. Lipkowitz, D. B. Boyd (eds.): *Reviews in Computational Chemistry*, vol. 9 VCH Publ., 1996, p. 129–165.
84. L. H. Hall, L. B. Kier in K. B. Lipkowitz, D. B. Boyd (eds.): *Reviews in Computational Chemistry*, vol. 2, VCH Publ., New York 1991, p. 367–422.
85. G. A. Arteca in K. B. Lipkowitz, D. B. Boyd (eds.): *Reviews in Computational Chemistry*, 9, VCH Publ., New York 1996, p. 191–253.
86. C. Hansch, A. Leo, D. Hoekman: *Exploring QSAR; Hydrophobic, Electronic, and Steric Constants*, Am. Chem. Soc., Washington, DC, 1995.
87. K. L. Peterson in K. B. Lipkowitz, D. B. Boyd (eds.): *Reviews in Computational Chemistry*, 16, Wiley-VCH, New York 2000, p. 53–140.

88. D. J. Livingstone, D. W. Salt in K. B. Lipkowitz, R. Larter, T. R. Cundari (eds.): *Reviews in Computational Chemistry*, 21, Wiley-VCH, Hoboken 2005, p. 287–348.
89. J. M. Troyer, F. E. Cohen in K. B. Lipkowitz, D. B. Boyd (eds.): *Reviews in Computational Chemistry*, vol. 2, VCH Publ., New York 1991, p. 57–80.
90. M. F. Hibert in C. G. Wermuth (ed.): *Practical Medicinal Chemistry*, Academic Press, London 1996, p. 523–546.
91. H. Cid, M. Bunster, *Biol. Res.* **29** (1996) 77–100.
92. N. Eswar et al., *Nucleic Acids Res.* **31** (2003) 3375–3380.
93. P. Gund et al., *Science* **208** (1980) 1425–1431.
94. A. S. V. Burgen, G. C. K. Roberts, M. S. Tute (eds.): *Molecular Graphics and Drug Design*, Elsevier, Amsterdam 1986.
95. J. G. Vinter, M. Gardner (eds.): *Molecular Modelling and Drug Design*, CRC, Boca Raton 1994.
96. N. C. Cohen (ed.): *Guidebook to Molecular Modeling and Drug Design*, Academic Press, San Diego 1997.
97. A. Tropsha, W. Zheng in O. M. Becker (ed.): *Computational Biochemistry and Biophysics*, Dekker, New York, 2001 p. 351–369.
98. P. Bultinck, H. De Winter, W. Langenaeker, J. P. Tollenaere (eds.): *Computational Medicinal Chemistry for Drug Discovery*, Dekker, New York 2004.
99. J. Sadowski, J. Gasteiger, *Chem. Rev.* **93** (1993) 2567–2581.
100. J. Sadowski, C. H. Schwab, J. Gasteiger in P. Bultinck, H. De Winter, W. Langenaeker, J. P. Tollenaere (eds.): *Computational Medicinal Chemistry for Drug Discovery*, Dekker, New York 2004, p. 151–212.
101. G. M. Downs, P. Willett in K. B. Lipkowitz, D. B. Boyd (eds.): *Reviews in Computational Chemistry*, vol. 7, VCH Publ., New York 1995, 1–66.
102. A. C. Good, J. S. Mason in K. B. Lipkowitz, D. B. Boyd (eds.): *Reviews in Computational Chemistry*, vol. 7, VCH Publ., New York 1995, 67–117.
103. E. J. Martin et al. in K. B. Lipkowitz, D. B. Boyd (eds.): *Reviews in Computational Chemistry*, vol. 10, VCH Publ., New York 1996, p. 75–100.
104. T. I. Oprea, C. L. Waller in K. B. Lipkowitz, D. B. Boyd (eds.): *Reviews in Computational Chemistry*, vol. 11, Wiley-VCH, New York 1997, p. 127–182.
105. G. Greco, E. Novellino, Y. C. Martin in K. B. Lipkowitz, D. B. Boyd (eds.): *Reviews in Computational Chemistry*, vol. 11, Wiley-VCH, New York 1997, p. 183–240.
106. A. Smellie, S. D. Kahn, S. L. Teig, *J. Chem. Inf. Comput. Sci.* **35** (1995) 285–304.
107. J. P. Tollenaere, *Pharm. World Sci.* **18** (1996) no. 2, 56–62.
108. J. M. Blaney, J. S. Dixon in K. B. Lipkowitz, D. B. Boyd (eds.): *Reviews in Computational Chemistry*, vol. 5, VCH Publ., New York 1994, p. 299–335.
109. I. Muegge, M. Rarcy in K. B. Lipkowitz, D. B. Boyd (eds.): *Reviews in Computational Chemistry*, vol. 17, Wiley-VCH, New York 2001, p. 1–60.
110. H.-J. Böhm, M. Stahl in K. B. Lipkowitz, D. B. Boyd (eds.): *Reviews in Computational Chemistry*, vol. 18, Wiley-VCH, New York 2002, p. 41–87.
111. P. Kolb et al. in J. Alvarez, B. Schoichet (eds.): *Virtual Screening in Drug Discovery*, CRC Press, Boca Raton 2005, p. 349–378.
112. M. A. Murcko in K. B. Lipkowitz, D. B. Boyd (eds.): *Reviews in Computational Chemistry*, vol. 11, Wiley-VCH, New York 1997, p. 1–66.
113. D. E. Clark, C. W. Murray, J. Li in K. B. Lipkowitz, D. B. Boyd (eds.): *Reviews in Computational Chemistry*, vol. 11, Wiley-VCH, New York 1997, p. 67–125.
114. R. A. Lewis, S. D. Pickett, D. E. Clark in K. B. Lipkowitz, D. B. Boyd (eds.): *Reviews in Computational Chemistry*, vol. 16, Wiley-VCH, New York 2000, p. 1–51.
115. A. Dominik in W. Bannwarth, B. Hinzen (eds.): *Combinatorial Chemistry — From Theory to Application*, 2nd ed., Wiley-VCH, Weinheim 2006, p. 559–613.
116. R. Judson in K. B. Lipkowitz, D. B. Boyd (eds.): *Reviews in Computational Chemistry*, vol. 10, Wiley-VCH, New York 1997, p. 1–73.
117. D. E. Clark (ed.): *Evolutionary Algorithms in Molecular Design*, Wiley-VCH, Weinheim 2000.
118. A. M. ter Laak, N. P. E. Vermeulen in B. Testa (ed.): *Pharmocokinetic Optimization in Drug Research: Biological, Physicochemical, and Computational Strategies*, Verlag Helvetica Chimica Acta, Zurich 2001, p. 551–588.
119. D. Butina, M. D. Segall, K. Frankcombe, *Drug Discovery Today* **7** (2002) S83–S88.
120. S. Ekins, P. W. Swaan in K. B. Lipkowitz, R. Larter, T. R. Cundari (eds.): *Reviews in Computational Chemistry*, vol. 20, Wiley-VCH, Hoboken 2004, p. 333–415.

121. M. P. Payne in M. T. Cronin, D. J. Livingstone (eds.): *Predicting Chemical Toxicity and Fate*, CRC Press, Boca Raton 2004, p. 205–227.
122. H. van de Waterbeemd, H. Lennernäs, P. Artursson, *Drug Bioavailability: Estimation of Solubility, Permeability, Absorption and Bioavailability*, Wiley VCH, Weinheim 2003.
123. S. W. Baertschi (ed.), *Pharmaceutical Stress Testing: Predicting Drug Degradation*, Taylor and Francis, Boca Raton 2005.
124. A. R. Leach, V. J. Gillet, *An Introduction to Chemoinformatics*, Kluwer, Dordrecht 2003.
125. T. I. Oprea (ed.): *Chemoinformatics in Drug Discovery*, Wiley-VCH, Weinheim 2005.
126. D. B. Boyd in K. B. Lipkowitz, D. B. Boyd (eds.): *Reviews in Computational Chemistry*, vol. 1, VCH Publ., New York 1990, p. 355–371.
127. J. Greer et al., *J. Med. Chem.* **37** (1994) 1035–1054.
128. D. J. Kempf et al., *Proc. Natl. Acad. Sci. USA* **92** (1995) 2484–2488.
129. R. C. Glen et al., *J. Med. Chem.* **38** (1995) 3566–3580.
130. A. K. Patick et al., *Antimicrob. Agents Chemother.* **40** (1996) 292–297.
131. Y. Kawakami et al., *Bioorg. Med. Chem.* **4** (1996) 1429–1446.
132. J. R. Damewood, Jr. in K. B. Lipkowitz, D. B. Boyd (eds.): *Reviews in Computational Chemistry*, vol. 9, VCH Publ., New York 1996, p. 1–79.
133. D. B. Boyd in T. Liljefors, F. S. Jorgensen, P. Krogsgaard-Larsen (eds.): *Rational Molecular Design in Drug Research*, vol. 42, Munksgaard, Copenhagen 1998.
134. D. B. Boyd in A. L. Parrill and M. R. Reddy (eds.): *Rational Drug Design: Novel Methodology and Practical Applications*, American Chemical Society, Washington, DC, 1999, p. 346–356.
135. D. B. Boyd in P. v. R. Schleyer, N. L. Allinger, T. Clark, J. Gasteiger, P. Kollman, H. F. Schaefer III (eds.): *Encyclopedia of Computational Chemistry*, vol. 1, Wiley, Chichester 1998, p. 795–804.
136. M. Congreve, C. W. Murray, T. L. Blundell, *Drug Discovery Today* **10** (2005) 895–907.
137. O. Kirino, *Nippon Noyaku Gakkaishi* **9** (1984) 571–579.
138. W. Draber, *Z. Naturforsch. Teil C* **42** (1987) 713–717.
139. T. T. Fujimoto et al., *Pesticide Biochem. Physiol.* **30** (1988) 199–213.
140. S.-H. Park, G. Sposito in S. M. Auerbach, K. A. Carrado, P. K. Dutta (eds.): *Handbook of Layered Materials*, Marcel Dekker, New York 2004, p. 39–89.
141. P. Verwer, F. J. J. Leusen in K. B. Lipkowitz, D. B. Boyd (eds.): *Reviews in Computational Chemistry*, vol. 12, Wiley-VCH, New York 1998, p. 327–365.
142. Z. Slanina, S.-L. Lee, C.-H. Yu in K. B. Lipkowitz, D. B. Boyd (eds.): *Reviews in Computational Chemistry*, vol. 8, VCH Publ., New York 1996, p. 1–62.
143. A. N. Cormack, J. Du, T. Zeitler in *Proceedings of the 19th International Congress on Glass*, Society of Glass Technology, Sheffield 2001, p. 584–588.
144. J.-R. Hill, C. M. Freeman, L. Subramanian in K. B. Lipkowitz, D. B. Boyd (eds.): *Reviews in Computational Chemistry*, vol. 16, Wiley-VCH, New York 2000, p. 141–216.
145. R. Q. Topper et al. in K. B. Lipkowitz, R. Larter, T. R. Cundari (eds.): *Reviews in Computational Chemistry*, vol. 19, Wiley-VCH, Hoboken, 2003, p. 1–41.
146. H. A. Kurtz, D. S. Dudis in K. B. Lipkowitz, D. B. Boyd (eds.): *Reviews in Computational Chemistry*, vol. 12, Wiley-VCH, New York 1998, p. 241–279.
147. E. L. Plummer in K. B. Lipkowitz, D. B. Boyd (eds.): *Reviews in Computational Chemistry*, vol. 1, VCH Publ., New York 1990, p. 119–168.
148. M. H. Gerzabek et al., *Bodenkultur* **52** (2001) 133–146.
149. J. C. Shelley, D. R. Bérard in K. B. Lipkowitz, D. B. Boyd (eds.): *Reviews in Computational Chemistry*, vol. 12, Wiley-VCH, New York 1998, p. 137–205.
150. A. Wallqvist, R. D. Mountain in K. B. Lipkowitz, D. B. Boyd (eds.): *Reviews in Computational Chemistry*, vol. 13, Wiley-VCH, New York 1999, p. 183–247.
151. B. van de Graff, S. L. Njo, K. S. Smirnov in K. B. Lipkowitz, D. B. Boyd (eds.): *Reviews in Computational Chemistry*, vol. 14, Wiley-VCH, New York 2000, p. 137–223.
152. C. H. Reynolds, M. K. Holloway, H. K. Cox (eds.): *Computer-Aided Molecular Design: Applications in Agrochemicals, Materials, and Pharmaceuticals*, Am. Chem. Soc. Symp. Series 589, Washington, DC, 1995.

Molecular Dynamics Simulation

PHILIPPE A. BOPP, Université Bordeaux I, France

JÖRN B. BUHN, Technische Universität Darmstadt, Darmstadt, Germany

MANFRED J. HAMPE, Technische Universität Darmstadt, Darmstadt, Germany

See also: Molecular Modeling

1. **Introduction** . . . 324
2. **Fundamental Approximations** . . . 324
3. **Interaction Models** . . . 326
3.1. **Intramolecular Interactions** . . . 326
3.2. **Intermolecular Interactions** . . . 327
3.2.1. Pair Potentials . . . 327
3.2.2. Three-Body and Higher Order Intermolecular Potentials . . . 328
3.3. **Example: Intermolecular Potentials for an Aqueous Salt Solution** . . . 329
3.4. **Data for Interaction Models** . . . 330
4. **Approximations for the Simulations** 331
5. **Constructing the Ensemble** . . . 332
5.1. **Method** . . . 332
5.2. **Setting up a Simulation** . . . 333
6. **Obtaining Results** . . . 334
6.1. **Simple Averages** . . . 335
6.2. **Time-Correlation Functions** . . . 336
7. **Examples** . . . 336
8. **Applications** . . . 337
9. **References** . . . 339

Symbols and Abbreviations

A, B, C state functions, i.e., functions that can be computed knowing a configuration
A_{ij},B_{ij} parameters of the Lennard–Jones (12–6) potential
$\overleftrightarrow{\alpha}$ molecular polarizability tensor
$c_{BC}(t)$ time correlation function between state functions B and C
d interparticle distance
D depth of Morse potential
D_s self-diffusion coefficient
δt time step size
$|e|$ elementary electric charge
$\overrightarrow{E}$ electric field
E total energy of the system
E^{kin} kinetic energy
ε parameter of the Lennard–Jones (12–6) potential and of potentials derived from the LJ potential
$\overrightarrow{F}_i$ force acting on particle i
γ constant of the Axilrod–Teller potential
h Planck's constant, $h = 6.62608 \times 10^{-34}$ J s
$\overrightarrow{k}$ wave vector
k_{B} Boltzmann's constant, k_{B}= 1.38066 10^{-23} J/K
k_{lm},k_{lmn},k_{lmno} (generalized) force constants
Λ De Broglie wavelength
m_i mass of particle i
M number of events or occurrences
M_{dof} degrees of freedom
N total number of particles in a simulated system
ω angular frequency
p pressure
$\overrightarrow{p}$ momentum
$\overrightarrow{P}$ molecular polarization
q electric (partial) charge, i.e. multiples of the elementary charge $|e|$
$\Re$ configuration
$\overrightarrow{r}_i$ vector x_i, y_i, z_i of particle coordinates
$\ddot{\overrightarrow{r}}_i$ acceleration of particle i
$\overrightarrow{r}_{ij}$ position of site i relative to j, $(\overrightarrow{r}_i - \overrightarrow{r}_j)$
r_{cut} cut-off radius
ρ internal coordinate
S length of the periodic box
σ parameter of the Lennard–Jones (12–6) potential
t, τ time
T absolute temperature
T transport property
θ angle
U potential energy
V volume
$\overrightarrow{V}$ vector of velocities

Ullmann's Modeling and Simulation

ISBN: 978-3-527-31605-2

$\vec{v}_i$ vector $V_{x_i}, V_{y_i}, V_{z_i}$ of particle velocities, $\vec{v}_i = \dot{\vec{r}}_i$
$\langle \rangle$ average over configurations
CPMD Car–Parrinello molecular dynamics
DFT density functional theory
DPD dissipative particle dynamics
LJ Lennard–Jones
MD molecular dynamics
MC Monte Carlo
NpT isothermal–isobaric ensemble
NVE microcanonical ensemble
NVT canonical ensemble
NEMD nonequilibrium molecular dynamics
RDF radial distribution function $g(r)$
VACF velocity autocorrelation function

1. Introduction

More than a century ago LUDWIG BOLTZMANN (1844–1906) elaborated a procedure allowing the macroscopic properties of matter to be linked with the molecular details and interactions of its constituent particles. This approach is known today as statistical thermodynamics or, more generally, as statistical mechanics. However, up to the arrival of sufficiently cheap and accessible numerical computers in the last third of the twentieth century, practical results could be obtained only for the simplest systems, in some limiting cases, or under drastic assumptions. This has proven intellectually challenging and has yielded many insights, but the urge has remained to overcome the simplifications and to look at specific systems of interest in various disciplines such as physics, chemistry, biology, geoscience, and engineering. The limitations have been pushed back in the last few decades, and many systems can now be investigated by following Boltzmann's principles. Nevertheless, it is still impossible to carry out this method for any real system of even the smallest macroscopic size. Approximations, although much weaker than before, must still be made, and the task remains daunting in many cases.

The computational methods used today to follow Boltzmann's lead, i.e., to cross the bridge between the microscopic, molecular world and the macroscopic observables measured in the laboratory, are known as molecular simulations. They come in many variants, the two major families of which are Monte-Carlo (MC) simulations [1] and the molecular dynamics (MD) simulations. The basic idea underlying all molecular simulations is simple and follows directly from Boltzmann's considerations: Construct a sufficient number of microscopic configurations, or states, compatible with (1) knowledge about the intermolecular (or interatomic) interactions in the system and (2) the macroscopic thermodynamic constraints (temperature, density, etc.) of the system under consideration. This could be called the "Gallup poll" [2] approach to the thermodynamic ensemble: Since this powerful abstract notion upon which the Boltzmann approach rests cannot be practically realized, one must be satisfied with a representative sample. Like in the polls, the trouble lies, of course, in the representativeness. However, once the sample has been constructed, it can be used to do statistics, i.e. to compute average values of quantities that can be calculated from the configurations in these samples.

Example 1. Assume that the positions of all atoms in a sufficiently large sample of water are known over a sufficient length of time under the external constraints (e.g., at a given total energy, temperature, or...) imposed on the system. We want to know the thermodynamic average of all distances between neighboring pairs of oxygen atoms, d_{O-O}, under these conditions. Since the positions of the atoms have been constructed by means of the simulation so as to be compatible with the external conditions, and there are enough of them, according to Boltzmann the thermodynamic average, denoted here $\langle d_{O-O} \rangle$, is simply the usual average over all values d_{O-O} computed from the atomic positions in the simulated system. If a total of M such values d^i_{O-O} can be computed from the configurations, Equation (1) applies.

$$\langle d_{o-o} \rangle = \frac{1}{M} \sum_i^M d^i_{o-o} \tag{1}$$

2. Fundamental Approximations

At the energies relevant for most of biology, chemistry, and engineering, matter can be considered as consisting of molecules which can undergo changes (chemical reactions). The molecules themselves consist of atoms, which in turn consist of nuclei and electrons. Such particles of molecular, atomic, or subatomic dimensions are, in principle, subject to quantum mechanics, as expressed, e.g., through the Schrödinger equation for the wavefunction, rather than to classical mechanics, i.e., the Newton equation of motion. In most cases of interest here, however, approximations can be used to simplify the problem. In the Born – Oppenheimer approximation [3], it is recognized that the mass of the electrons is much smaller than the mass of the nuclei. The problem concerning the electrons and that concerning the nuclei can thus be treated separately. The energies holding together the atoms to form molecules as well as the energies holding together molecules, e.g., in liquids or solids, can thus be computed by solving quantum mechanical Schrödinger equations only for the electrons. It is one of the main purposes of quantum chemistry to devise methods, usually numerical, to tackle this complex problem for molecular or supramolecular systems [3 – 5]. Unfortunately, even with the largest computers available in the foreseeable future, it will not be possible to elaborate these energies for extended systems consisting of large disordered assemblies of molecules. Yet, according to what has been said above, in order to do statistical mechanics, as many molecular arrangements as possible compatible with a given set of constraints must be constructed, and to do this, the energies are needed.

The interaction energies between particles (individual atoms, atoms in molecules, groups of atoms, molecules themselves), which can be obtained, in principle, from the solution of the Schrödinger equations for the electrons, can then be expressed as potential functions, or potentials. They are usually written as functions of the positions (and, if needed, orientations) of the particles. Simplifying assumptions as to the mathematical form of these potentials are introduced at this level (see Chap. 3).

The second major approximation is that classical dynamics is used for the nuclei themselves, i.e., the atoms and molecules, rather than quantum mechanics, to construct the ensemble of configurations. The heavier the masses involved, the weaker the interactions, and the higher the temperature, the better this approximation will be. Equivalently, the smaller the so-called thermal De Broglie wavelength $\Lambda = \sqrt{\frac{h^2}{2\pi m k_B T}}$ associated with a particle of mass m at temperature T compared to a characteristic length (in a gas, e.g., on the order of the mean free path), the better the classical approximation [6]. At room temperature, the classical dynamics become a limitation only for the lightest atoms (e.g. hydrogen, helium) when they are strongly bonded.

Before we proceed we should note here that in selected cases, essentially for small systems for which not too large a number of configurations is sufficient, the quantum mechanical problem for the electrons can be solved (usually in a simplified approach) simultaneously with the generation of the configurations. Thus, no explicit interaction model must be specified, although the quantum mechanical calculations usually involve a number of adjustable quantities which are empirically chosen. One popular approach is Car – Parrinello MD (CPMD) [7]. Here, electrons belonging to all particles present in the system are treated, in the density-functional theory (DFT) approach, through equations of motion. This method is not free of arbitrariness, e.g., in the choice of the so-called functional; it is extremely CPU-intensive and is thus restricted to very small systems (a few tens of small molecules) which can be explored in a short time (a few picoseconds). In other approaches, the quantum mechanical equations are retained only for a small part of the system, while the larger part is described classically. The system is usually divided up geometrically into a smaller region which is treated quantum mechanically (by the ab initio, DFT, or, mostly, semi-empirical approach) and a larger one in which the interactions are described by an interaction model (force field). Coupling these two regions together is problematic and again involves choices and approximations. These so-called quantum-classical simulations are mostly outside the scope of this article, but we stress that if systems with real chemical reactions are to be studied approaches like the ones sketched here must be used.

Besides the two fundamental approximations (classical mechanics and interactions via potentials), several more are needed in order to study most systems of interest, mainly in two areas: The way the interactions are described, and the approximations needed to mimic extended systems with finite computational resources. The following sections are devoted to these aspects. In the generally available simulation packages (freeware, shareware, commercial) these approximations are often dealt with in a way somewhat hidden from the user. It is not the purpose of this article to evaluate the various approaches proposed in these packages and to give recommendations. We believe, however, that users of such software should be aware of the approximations involved and of their consequences: All results should always be critically evaluated in the light of the approximations made.

3. Interaction Models

The interaction model, often called force field, is, beyond all further "technical" approximations to be discussed below, the crucial input to a classical MD simulation. Within the restrictions in time and space inherent to the technique, the simulation is supposed to yield the "exact" results associated with a given interaction model. In other words, failure of the simulation to reproduce a given observable in a given system should be ultimately ascribable to deficiencies in the interaction model.

Two different philosophies can be distinguished: In the first, MD simulations are used to explore systematically the statistical mechanics of certain types of interaction models, usually highly idealized simple ones (e.g., hard spheres or Lennard–Jones spheres [8], see Eq. 7). Such simulations thus yield in some sense the "exact" result for the idealized model. The second approach attempts to reproduce as precisely as possible the properties of specific chemical systems. Water has been, for obvious reasons, a frequent target of such studies [9]. In the latter case, it is a reasonable and efficient approximation to write the total potential energy of a molecular system U^{total} as a sum of terms (Eq. 2)

$$U^{\text{total}} = \sum_{\substack{\text{all molecules}}} U^{(1)} + \sum_{\substack{\text{all pairs}\\\text{of molecules}}} U^{(2)} + \sum_{\substack{\text{all triples}\\\text{of molecules}}} U^{(3)} + \cdots \tag{2}$$

where $U^{(1)}$ is an energy associated with a single molecule, e.g., its intramolecular energy (see below) or its energy in an external field. For small molecules, this term is often nearly constant and thus omitted in the simulations. $U^{(2)}$ is the interaction energy between pairs of molecules. In most cases, only terms of this type are used in the simulations. $U^{(3)}$ is an additional interaction energy that depends simultaneously on (e.g., the positions of) three molecules. Figure 1 illustrates the second and third terms of this series of approximations.

Figure 1. Left: Interaction $U^{(2)}$ (solid lines) between two molecules represented as circles. Middle: Interactions $U^{(2)}$ between three molecules; the total energy is the sum of the three $U^{(2)}$ terms, each of which is the same as in the two-body case, i.e., each interaction is deemed to be independent of the presence or absence of the third molecule. Right: Interactions between three molecules as the sum of the two-body terms $U^{(2)}$ plus a modifying three-body term $U^{(3)}$ (dashed circle).

In a system of N molecules, there are N terms of the first kind; the number of terms of the second kind increases as $N(N-1)/2 \propto N^2$, and the number of terms of the third kind increases proportionally to $N(N-1)(N-2)/6 \propto N^3$. Even though not all terms need to be evaluated, these proportionalities govern the amount of computer time needed to evaluate the interactions in a simulation.

3.1. Intramolecular Interactions

Intramolecular interactions are usually stronger than intermolecular interactions. If this is the case, the molecular geometry can be assumed not to change much when a molecule interacts with its neighbors. Since the intramolecular energy will then be constant, it can be neglected, even though it may be large: A rigid body is then a suitable molecular model. This works in some instances for small molecules. In many cases, however, and in particular for large molecules, the energetics involved with changes in the molecular geometry must be included in the simulation. Since it is chemically intuitive, the intramolecular energy is often expressed as a power series in internal displacement coordinates ρ [10] (Eq. 3).

$$U^{(1)} = \sum_{\substack{\text{all pairs }(l,m)\\ \text{of coordinates}}} k_{lm} \cdot \rho_l \cdot \rho_m + \sum_{\substack{\text{all triples}(l,m,n)\\ \text{of coordinates}}} k_{lmn} \cdot \rho_l \cdot \rho_m \cdot \rho_n + \sum_{\substack{\text{all quadruples}(l,m,n,o)\\ \text{of coordinates}}} k_{lmno} \cdot \rho_l \cdot \rho_m \cdot \rho_n \cdot \rho_o + \cdots \quad (3)$$

Examples of such internal displacement coordinates are the bond stretch $\rho = r_{ij} - r_{ij}^0$, where r_{ij} denotes the module of the distance between two atoms in a molecule and r_{ij}^0 its equilibrium value, and the angle bend $\rho = \alpha_{ijk} - \alpha_{ijk}^0$, where α_{ijk} denotes the angle spanned by particles *i* and *k* at the position of particle *j*. The constants k_{lm} are called the force constants; their definition in the literature varies due to the inclusion or omission of the factor of 1/2 originating from the definition of the power series. The same holds for the higher constants k_{lmn} and k_{lmno}. Note that this definition of $U^{(1)}$ rests upon the notion of equilibrium geometry, i.e., it is suitable for molecules for which only one conformer must be considered.

In the first term of Equation 3, the terms with $\rho_l = \rho_m$ are called diagonal terms. Nondiagonal terms ($l \neq m$) are usually included only if the corresponding coordinates ρ_l, ρ_m share a common particle (*i* or *j*). The higher terms in Equation 3, usually called (mechanical) anharmonicities, are also not frequently used. In some cases, however, they cannot be omitted since they play an important role in the redistribution of energy in the system. Another frequently used approximation for anharmonic intramolecular bond stretches is the Morse potential (Eq. 4) [11]

$$U^{\text{Morse}}(r_{ij}) = D \cdot \left(1 - e^{-a\left(r_{ij} - r_{ij}^o\right)}\right)^2 \quad (4)$$

where *D* is the depth of the potential well, and *a* a constant.

We note again that the number of operations needed in a simulation to evaluate Equation 3 is proportional to *N*, the number of molecules. Thus, the computational weight of these terms is usually not very large, in contrast to the intermolecular terms (see Section 3.2). More details about the modeling of intramolecular interactions, e.g., further definitions of internal coordinates, information on the quantum mechanical background, and potential functions suitable for molecules with several conformers can be found in → Molecular Modeling.

3.2. Intermolecular Interactions

3.2.1. Pair Potentials

In most cases the intermolecular interactions are expressed in terms of potentials of type $U^{(2)}$ only (see Eq. 2). A frequently used approach consists of writing this interaction potential as a sum of so-called site–site pair potentials (Eq. 5)

$$U^{(2)} = \sum_{\substack{\text{all sites } i\\ \text{on molecule } \alpha}} \sum_{\substack{\text{all sites } j\\ \text{on molecule } \beta}} U_{ij}(r_{ij}) \quad (5)$$

In many instances, "site" is identified with "atom position", but if molecules are described as rigid entities, other points on the molecular frame can be used as sites. Examples of such site–site pair potentials are the electrostatic (Coulomb) interaction between two partial electric charges *q* (Eq. 6)

$$U^{\text{Coulomb}}(r_{ij}) = C \cdot \frac{q_i \cdot q_j}{r_{ij}} \quad (6)$$

where *C* is a constant depending on the energy unit used. We note already here that special techniques such as Ewald summations or reaction field techniques [12], which are computationally expensive, must be used to evaluate Coulomb terms in the periodic arrangements that are mostly used to represent extended systems. This is due to the long range ($U \propto r^{-1}$) of this interaction.

Shorter range repulsive and attractive (van der Waals-type) interactions are often approximated by the Lennard–Jones (LJ or 12-6) potential, which contains two constants ε and σ (or *A* and *B*; Eq. 7).

$$U^{\text{Lennard-Jones}}(r_{ij}) = 4 \cdot \varepsilon_{ij} \cdot \left[\left(\frac{\sigma_{ij}}{r_{ij}}\right)^{12} - \left(\frac{\sigma_{ij}}{r_{ij}}\right)^{6}\right] = \frac{A_{ij}}{r_{ij}^{12}} - \frac{B_{ij}}{r_{ij}^{6}} \quad (7)$$

The constants ε and σ used in the first notation represent the depth of the potential well and the value of the argument *r* for which $U^{\text{Lennard-Jones}} = 0$, respectively. As an illustration of the typical order of magnitude for the two interaction types mentioned above, Figure 2 shows a Lennard–Jones potential and a Coulomb potential between two small partial charges of opposite sign, typical for those on nonpolar molecular groups. The examples given below will show that about ten times higher partial charges are used to represent polar groups, and even higher ones for ions.

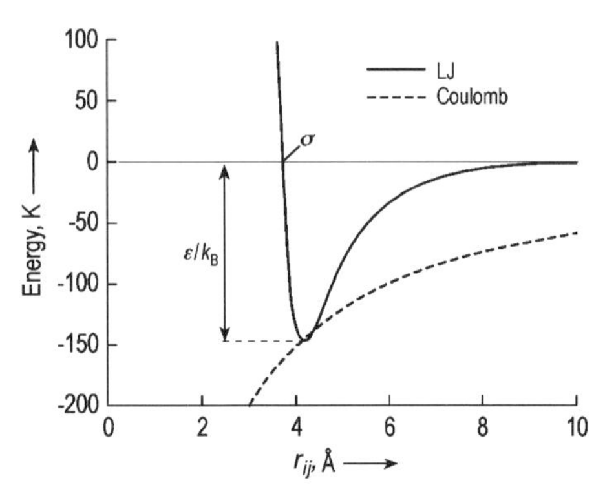

Figure 2. Typical interaction potentials as a function of distance. Lennard–Jones (LJ) potential between two methane (CH_4) molecules carrying one site each (ε/k_B = 148.6 K, σ = 3.728 Å, 1 Å = 10^{-10} m) (effective potential in the united-atom approximation). For comparison: Electrostatic Coulomb potential between two opposite charges of $q = 0.06\ |e|$, where $|e|$ is the elementary charge. The energy is given in Kelvin, i.e., divided by k_B, since the thermal energy is in some way the measure of all things in a thermodynamic system.

Since the number of parameters needed to specify the interactions (Eqs. 6 and 7) increases rapidly when several types of molecules must be described (see page 329 in Section 3.2), empirical rules, called combination rules, are often used to construct the LJ parameters ε and σ for interactions between unlike sites (*ij*) from those for interactions between like sites (*ii* and *jj*). There are several sets of such rules, the most frequently used of which are due to Kong [13] and Lorentz–Berthelot [14]. The Lorentz – Berthelot rule is given here as an example (Eq. 8).

$$\sigma_{ij} = \frac{1}{2} \cdot (\sigma_{ii} + \sigma_{jj}), \quad \varepsilon_{ij} = \sqrt{\varepsilon_{ii} \cdot \varepsilon_{jj}} \tag{8}$$

Another way to simplify the expressions for the intermolecular energies is to attempt to reduce the number of sites used to represent a group of atoms or a molecule. The CH_n-groups in organic molecules are thus often represented by just one site instead of several (e.g., one site on each atom). This is known as the united-atom approximation. It may work if the group is not too polar. Figure 2 shows the interaction potential resulting between two CH_4 molecules, each of which is represented by just one site. We note that this potential is spherically symmetric, and it thus neglects the exact "shape" of the molecule. Typical simulation results for this system are discussed in Examples.

Many other functional forms have been proposed besides the Lennard – Jones ansatz to describe intermolecular interactions of type $U^{(2)}$. The powers 12 and 6 in the LJ-ansatz (Eq. 7) have thus been modified and/or written as functions of *r*. In other instances (e.g., in the Buckingham potential) the inverse power of *r* describing the repulsive part of the potential has been replaced by an exponential function. Alternative forms of molecule–molecule potentials may involve functions of relative molecular orientations instead of sums of site–site potentials. The Gay–Berne model [15], frequently used in simulations of liquid-crystalline systems, again uses a LJ-type form, but with parameters dependent on the relative orientation of two molecules. A detailed discussion of intermolecular interactions in simple systems can be found in, e.g. [16].

3.2.2. Three-Body and Higher Order Intermolecular Potentials

Since they are difficult to establish and computationally expensive, three-body ($U^{(3)}$) and higher terms are not often used in simulations. This is an area of active research, and we mention here only two examples among the many approaches.

The three-body Axilrod – Teller interactions [17] (Eq. 9)

$$U^{\text{Axilrod-Teller}} = \frac{\gamma \cdot (1 + 3 \cdot \cos\theta_{ij} \cdot \cos\theta_{jk} \cdot \cos\theta_{ki})}{r_{ij} \cdot r_{jk} \cdot r_{ki}} \tag{9}$$

where *i*, *j*, and *k* are three particles, similar to Figure 1, *r* their distances, and θ the angles. This model potential contains only one constant, γ, which can be related to the molecular polarizability [3].

The Sutton – Chen [18] model, which has been used successfully, e.g., in studies of metals and surfaces, is formally similar to an LJ-ansatz with generalized exponents *n* and *m*. The second term, however, is replaced by an average over all interaction partners j. This is thus a true many-body potential, and the total potential energy of the N-particle system is written as Equation 10

$$U^{\text{Sutton-Chen}} = \varepsilon \cdot \sum_{i}^{N} \left(\frac{1}{2} \sum_{i \neq j} \left(\frac{a}{r_{ij}} \right)^{n} - b \cdot \sqrt{\sum_{i \neq j} \left(\frac{a}{r_{ij}} \right)^{m}} \right) \tag{10}$$

Five constants, ε, a, b, n, and m must be specified here.

An important physical effect neglected in the pair potential approach discussed up to here is the mutual electronic polarization of molecules. There have been many attempts to devise workable schemes allowing to treat the polarization $\vec{P}$ of a molecule efficiently and as exactly as possible in a simulation. $\vec{P}$ can be seen as the additional dipole moment created on a molecule by the electric field $\vec{E}$ generated by its neighbors. In the picture of Figure 1 this would mean that the intermolecular interaction between two molecules, denoted there $U^{(2)}$, is modified by the presence of a third, fourth, fifth molecule in the neighborhood. In the notation of electrodynamics [19], one has $\vec{P} = \overleftrightarrow{\alpha} \cdot \vec{E}$, where $\overleftrightarrow{\alpha}$ is the molecular polarizability tensor. Many difficulties arise: $\overleftrightarrow{\alpha}$ is often not well known; if it is, it is not clear with which site of a molecule it should be identified or how it should be distributed among the possible locations on the molecular frame. Furthermore, if there is more than one polarizable species in the system, an iterative self-consistent procedure is in principle needed to evaluate the energies resulting from this effect: The field generated by the neighborhood polarizes molecule i, say, but the changing electric moments (usually only the dipole moment is considered) of molecule i will change the electric field experienced by the neighbors, hence their polarization, hence the field acting on i, etc., etc. Several sophisticated approximate methods have been proposed to treat this problem (see, e.g., [20, 21]); they cannot be described here. We note, however, that if intramolecular potentials are specified (flexible molecules) some effective mutual polarization will arise from the mutual geometrical distortions of the molecules, e.g., if they lead to distance changes between point charges. In another particularly simple approach, called the fluctuating charge model, the partial electric charges q_{i} are allowed to vary, e.g., as explicit functions of some distances or angles in the system.

Finally, we note that in certain instances the interaction energies and/or their derivatives, the forces, need not be expressed explicitly as mathematical functions, but can simply be tabulated in one- or multidimensional arrays. Interpolation schemes or splines are then used in the simulations to compute the relevant quantities.

3.3. Example: Intermolecular Potentials for an Aqueous Salt Solution

We stress again that the choice of the interaction model is crucial. It involves the choice of a proper mathematical representation and the choice of all the required parameters. In particular, the choice of the number of sites used to express the intermolecular interactions is crucial. All these choices are largely based on experience and result from a compromise between a representation of the interactions that is as accurate as possible and the resulting computational expenditure required to generate a sufficient number of configurations (see Section 5.1). An example will illustrate this:

Example 2. A simulation of a solution of a 1:1 salt, say NaCl, in water. The site–site pair potential approach is used for all intermolecular interactions. Rigid molecules (i.e. no intramolecular potentials need to be specified) are assumed and there is no external field.

We consider a system consisting of three species, the solvent (H_2O), cations (Na^+), and anions (Cl^-). The simplest reasonable representation of the water molecules needs three sites (say the atoms O and H), while one site suffices for these simple ions. There are thus four types of sites, leading to $(4 \times 5)/2 = 10$ site–site intermolecular pair potentials $U^{(2)}$ (see Equation 2), which can be arranged in an upper triangular matrix for better visibility:

$$\begin{matrix} U^{(2)\mathrm{O-O}} & U^{(2)\mathrm{O-H}} & U^{(2)\mathrm{O-cation}} & U^{(2)\mathrm{O-anion}} \\ & U^{(2)\mathrm{H-H}} & U^{(2)\mathrm{H-cation}} & U^{(2)\mathrm{H-anion}} \\ & & U^{(2)\mathrm{cation\text{-}cation}} & U^{(2)\mathrm{cation\text{-}anion}} \\ & & & U^{(2)\mathrm{anion\text{-}anion}} \end{matrix} \tag{11}$$

To write each of these 10 potentials as a sum of a Coulomb interaction (Eq. 6) between electric charges and a Lennard – Jones term (Eq. 7), 20 Lennard – Jones constants (ε and σ, or A and B) plus four electric partial charges are needed. In this case, it can be assumed that the entire ionic charge of $\pm|e|$ resides on the ions. The water molecules being neutral, one has $q_{\mathrm{O}} = -2q_{\mathrm{H}}$; there is thus only one parameter to determine, say q_{O}, which is of the order of magnitude of $q_{\mathrm{O}} \approx -2/3|e|$. As far as the Lennard–Jones parameters are concerned, assuming combination rules (e.g., those in Eq. 8) would reduce the number of parameters to $2 \times 4 = 8$ for the potentials $U^{(2)}$ in the diagonal of the matrix above, the parameters of the off-diagonal $U^{(2)}$ resulting from those of the diagonal terms. We note in passing that the notion of "hydrogen bond" does not appear explicitly in the model. The typical features associated with these bonds in terms of structural and dynamical properties of the liquid arise from the interplay between the pair potentials. In any case, it is easily seen that the number of parameters which must be known increases very quickly with increasing complexity of the system, i.e., with the number of species in the system and with the sophistication (or complexity) of the description.

3.4. Data for Interaction Models

Details of the various approaches used to determine consistent sets of parameters for interaction models are beyond the scope of this article. There are essentially three sources of information used for this purpose:

1) Quantum chemistry: This is the "in principle" approach outlined in Chapter 2. For practical aspects (→ Molecular Modeling, Chap. 4.4). It will suffice here to say that quantum mechanics, at various levels of approximation (e.g. ab initio, semi-empirical, etc.), can be used to determine, as a function of the relative arrangements of molecules, intermolecular interaction energies in ways very similar to the determination of intramolecular ones. One then endeavors to fit these energies with expressions of type $U^{(2)}$, $U^{(3)}$, etc. We note, in particular, that although the notion of partial charge is not inherent to quantum chemistry, most quantum chemical program packages provide algorithms which help in the determination of these quantities by suitable integrations over the nuclear charges and the wave functions or by fitting procedures.
2) Experiments: Thermodynamic data have frequently been used to gauge the comparative depths of potential wells. Very careful work has, e.g., been carried out to reproduce IR and Raman spectra of many molecules by "harmonic force fields" (essentially the first term in Eq. 3) or higher order expansions. These provide valuable input for intramolecular model potentials. Results from scattering experiments can provide insight into molecular shapes, i.e., essentially the repulsive part of the interaction potential.
3) Simulations themselves: Simulations are often used in a sort of trial and error procedure to refine model potentials, i.e., to adjust potential parameters. Again, this is often done with thermodynamic data. Model potentials have thus been adjusted to yield the proper pressure–density dependence. Self-diffusion coefficients have also been used for such procedures. Some systematic procedures have been proposed recently [22]. Note that there is in general no guarantee that the solution thus found for the interaction model is unique.

In all cases, "true" pair potentials, as constructed, e.g., from quantum mechanical calculations of pairs of molecules, must be distinguished from "effective" potentials, in which averages over the higher order terms are in some way worked into the $U^{(2)}$ term of Eq. 2. Many models, e.g., have a distribution of electric charges (Eq. 6) leading to enhanced dipole moments compared to the gas-phase value in order to account for the average polarization of molecules by their neighbors in a condensed phase. Consequently, such models are in principle valid only for a restricted range of densities and temperatures.

4. Approximations for the Simulations

Molecular dynamics simulations are mostly used to study condensed phases, homogeneous or inhomogeneous. They are also very well adapted to studies of interfaces. For the sake of simplicity we shall, however, focus here on simulations of simple homogeneous systems, say a liquid under ambient conditions. The question arises: How many particles does one need for the typical properties of a liquid to be observable? Or, in other words, how many particles/molecules need to be included in the simulation so that the properties determined from this simulation reflect those of a bulk liquid, and not of some sort of cluster, aggregate, or droplet?

Assume for the present qualitative argument the particles to be arranged in a cube with an edge length equal to a multiple of the particle dimension (not worrying what exactly this dimension might be). We consider cubes of $3 \times 3 \times 3 = 27, 4 \times 4 \times 4 = 64, 125, 216, 343, \ldots, 15 \times 15 \times 15 = 3375, \ldots$, particles (the number 3375 being already relatively high for a simulation with anything but the simplest interaction model, although simulations with many more particles have been reported in special cases). Assuming furthermore that the influence of the surface extends about three molecular layers deep into the system, at best $(15-6)(15-6)(15-6) = 729$ particles would display bulklike properties (i.e., unperturbed by the surface). Thus, in this example, only 729/3375 or 21 % of the particles included in the simulation would contribute to the result. Placing the particles in a sphere would of course improve this ratio, but not decisively.

To avoid this unaffordable inefficiency, the system to be investigated is replaced by a periodic one through the introduction of periodic boundary conditions. The simplest case of cubic periodic boundary conditions is illustrated in Figure 3. The construction is equivalent to replacing the N-particle system by a system of N cubic primitive lattices: Each particle i forms with its image particles i' in the neighboring image boxes a cubic primitive lattice. Together with the minimum-distance convention, which stipulates that the interaction is computed between the two closest partners, be they two particles in the basic box or a particle in the basic box and one in a neighboring image box, two advantages are gained:

1) There are no "surfaces" in the system, since each particle is, owing to this construction, at the center of all its interaction partners (green box in Figure 3).

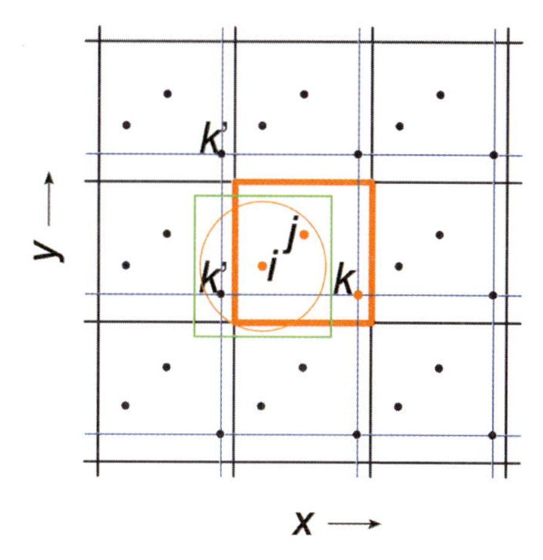

Figure 3. A system with cubic periodic boundary conditions containing three particles i, j, and k (in red). The basic cube, usually called box, i.e., the volume containing the N (here 3) particles of the simulation, is highlighted in red. It is surrounded by 26 (in 2 dimensions 8) "image boxes" (in black), which themselves are surrounded by further boxes. Each image box contains periodic images (in black) of all particles in the basic box. The green box is centered on particle i. It can be seen that it contains particle j but instead of particle k it contains an image k'. Under the minimum-distance convention particles j and k' are the interaction partners of particle i. The orange sphere (circle in 2 dimensions) centered on particle i indicates that under certain conditions the interactions of particle i with particles inside the green box, but outside of this sphere, can be neglected. The radius of the sphere is known as the cutoff radius r_{cut}. If applicable, this procedure can save a lot of computer time, the neglected volume (cube vs sphere) being roughly 50 %.

2) The density in the basic box (and thus in the system) remains constant. If a particle leaves the basic box on one side during a simulation, it becomes an image particle. An image particle enters simultaneously on the other side of the box and becomes the particle.

These procedures are not without incidence on some properties computed from an ensemble simulated in this way. In some instances, the correction is easy (maybe even trivial), in others, it is difficult even to estimate its influence. As a rule of thumb, the size of the simulation box and, related through the density of the system, the number of particles, should be chosen in such a way that spatial correlations (expressed, e.g., through the radial pair distribution function; see Chap. 8) decay over less than half the box length (i.e., from a central particle to the edge). Furthermore, in systems where collective motions (e.g., phonons) play an important role, one should not forget that the box will limit the wavelengths associated with the phenomenon. In mixtures, e.g., solutions, the desired concentration may also set limits as to the minimum number of particles that must be included.

As indicated in Figure 3, the calculations can, under certain circumstances (i.e., when the range of the interactions is short) be accelerated even more by truncating the interactions beyond a cutoff radius r_{cut}, i.e., the potential used in the simulation is given by Equation 12.

$$\mathrm{U}^{\mathrm{simulation}} = \begin{cases} U^{\mathrm{model}} & \text{if } r < r_{\mathrm{cut}} \\ 0 & \text{else} \end{cases} \quad (12)$$

There are several ways to implement this truncation scheme [12]. Note that truncation schemes can generate discontinuities in the forces which must be dealt with.

5. Constructing the Ensemble

5.1. Method

The task is now to construct a sufficient number of microscopic configurations, or microstates, compatible with the interactions and the macroscopic constraints. We consider only systems set up as described above, i.e., with a fixed number of particles N in a fixed (through the periodic boundary conditions) volume V. We assume that it is possible to construct one such microstate, i.e., to find positions and velocities for all particles such that the total energy of the system E (Eq. 13)

$$E = U^{\mathrm{total}} + E^{\mathrm{kin}} \quad (13)$$

where E^{kin} is the kinetic energy, has a desired value. A series of new microstates, constrained by N = const., V = const., and E = const., can be generated by solving the Newton equation (or any equivalent set of equations) for all particles under the influence of their mutual forces. We shall call this an NVE simulation. Written here, for simplicity, for N point masses m_{i}, i.e., without rotations, these are Equations 14.

$$\begin{array}{lcl} m_1 \ddot{\vec{r_1}} & = & \vec{F_1}(\vec{r_1},\vec{r_2},\vec{r_3},\ldots,\vec{r_N}) \\ m_2 \ddot{\vec{r_2}} & = & \vec{F_2}(\vec{r_1},\vec{r_2},\vec{r_3},\ldots,\vec{r_N}) \\ \vdots & & \vdots \\ m_N \ddot{\vec{r_N}} & = & \vec{F_N}(\vec{r_1},\vec{r_2},\vec{r_3},\ldots,\vec{r_N}) \end{array} \quad (14)$$

They are coupled through their right-hand sides, i.e., through the forces. The force acting on mass i depends, in general, on the position of mass i itself and on the positions of all other masses j through Equation 15.

$$\vec{F}_i = -\nabla_i U^{\mathrm{total}} \quad (15)$$

Note that U^{total} (Eq. 2) is given as a function of the site positions, which may or may not coincide with the positions of the masses. In molecules or groups of atoms treated as rigid bodies, the equations become more complex [12], but the principle remains the same.

Additionally to the constraints (NVE), the microstates constructed in this way fulfill the additional constraint of constant total momentum (Eq. 16).

$$\vec{p} = \sum (m_i \vec{v}_i) = \text{const.} \quad (16)$$

This constant is usually set equal to zero, which facilitates the subsequent evaluation procedures (see Section 6.1). The configurations thus generated constitute a particular sample of the microcanonical ensemble of statistical mechanics (see, e.g., [6] or any other book on statistical mechanics) and averages can be computed over this sample, as exemplified in Equation 1.

Except for very few specific cases it is not possible to find a closed, analytical solution of Equations 14. They are thus solved numerically, and several numerical schemes have been devised for this purpose. Without going into details, all these integration schemes can be thought of as procedures in which the following steps are iterated:

1) Knowing the positions and velocities of all particles (sites), calculate all forces (and torques if rigid groups are considered) acting in the system
2) For a short time step δt, allow the particles (a) to move according to their velocities and (b) accelerate according to the forces acting on them
3) Determine new positions and new velocities;
4) Go to step 1

The microstates, or configurations (which we shall denote as $\Re$; $\Re$ = all positions and velocities of all particles) are thus points on the trajectory of the system evolving with time according to the Newton equation. The generated ensemble can thus be written as a series of configurations ordered according to time and starting at a time origin t_0:

$$\{\Re(t_0), \Re(t_0+\delta t), \Re(t_0+2\delta t), \ldots\ldots\ldots, \Re(t_0+M\delta t)\} \quad (17)$$

if M integration steps have been carried out. In simulation parlance the term "history" of the system is often used for the trajectory. The availability of the time evolution of the system is the main difference between MD and MC simulations.

The length of the time step δt must be chosen carefully. It should be as large as possible in order to sample the system efficiently (minimize computational costs) but short enough to ensure correct integration of the equations of motion. In NVE simulations, the choice of δt can be checked by making sure that the fluctuations in total energy (which should be zero since E = const., Eq. 13) are small compared to real fluctuations in the system, e.g., between kinetic and potential energy. In terms of dynamics, the time-step size must be at least one order of magnitude smaller (say a factor of 20–50) than the period of the fastest motion occurring in the system (e.g., molecular O–H or C–H vibrations, Debye frequency in a solid, etc.). If there are very different motional regimes in the system, it is sometimes possible to use a multiple-time-step method in which the interactions of the slowly fluctuating parts are evaluated less often than those of the fast ones.

We have dealt here explicitly only with the generation of the NVE ensemble. Most simulation packages offer the possibility to generate other ensembles, mostly NVT or NpT, i.e., with constant temperature T and/or constant pressure p instead of constant energy E and/or constant volume V. There are various ways to proceed, e.g., by solving additional differential equations associated with "thermostats" and "barostats" or by adding "virtual particles" that act as heat reservoirs; see [23] for details. Both methods may occasionally lead to artefacts [24]. We note in particular that in both cases the dynamics are no longer strictly Newtonian, which may have consequences for the dynamical properties computed from such simulations. However, using these methods for equilibration can facilitate this step, and so an equilibration procedure (see below) using these methods is often combined with a final production run in the NVE ensemble.

5.2. Setting up a Simulation

It was assumed above that a suitable configuration is available to start the simulation. It is by no means obvious how to generate such a configuration but a procedure similar to that outlined in Figure 4 often leads to success.

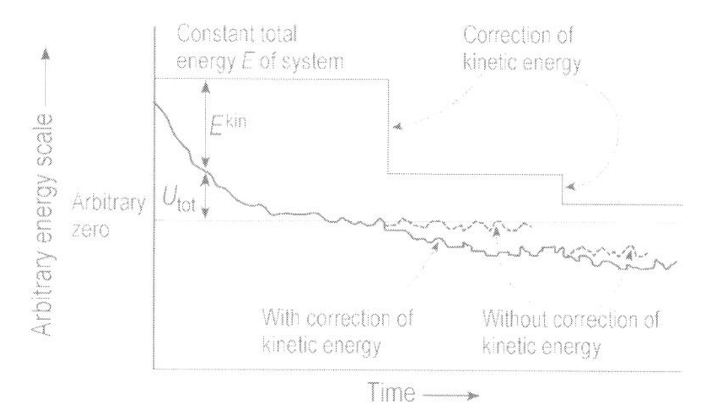

Figure 4. Schematic equilibration procedure to generate a starting configuration for an equilibrium simulation under NVE conditions.

The particles are placed in the simulation box as "reasonably" as possible, e.g., randomly but with reasonable distances and orientations, or on a lattice or distorted lattice. Velocities are randomly assigned or simply set equal to zero. The integration is started. Assuming the numerical integration works perfectly, the total energy of the system will remain constant. Since the selected configuration will most probably have a surplus of potential energy (a slightly too short interparticle distance on a repulsive part of a Lennard–Jones potential (Eq. 7) suffices) conversion of potential energy to kinetic energy will occur, and the temperature of the system will rise. If the interaction potentials are sufficiently anharmonic, the surplus energy will be redistributed among all degrees of freedom, and the system will converge toward a regime where there are only fluctuations between the two energies, but no systematic drift. (In contrast, a system exhibiting motions of the normal-mode type will not show this behavior.) However, the average kinetic energy (i.e., the system temperature) is now usually much higher than desired. Kinetic energy must thus be removed, either "instantaneously" (as in Figure 4) by scaling the velocities, or more gradually through thermostatting procedures. In any case, since average kinetic and potential energies of a system are interrelated, there will be a relaxation following this adjustment. However, through a series of such adjustments the system can be brought to a state with a total energy compatible with the desired average temperature. All configurations generated during this equilibration procedure are then discarded, except the final one, which is the true initial configuration of the simulation.

It is generally not easy to determine whether one has really reached equilibrium. Necessary conditions are that there should be no drift with time of the kinetic or potential energies, that the particle velocities should follow approximately the Maxwell–Boltzmann distribution, and that the time derivatives of various quantities computed from the ensemble should be zero. However, all this does not guarantee that the system is not trapped in a metastable state.

Equilibration is usually not a problem for homogeneous liquids consisting of small molecules under conditions of temperature and density where the molecular mobilities are not too low (say compared to the time accessible in a simulation, see Figure 5). In mixtures, the equilibration may be more difficult. It becomes a major problem in simulations of systems of large molecules such as polymers or biomolecules.

Another test is to use different starting configurations and equilibration procedures. Once equilibrium has been reached, properties computed over the equilibrium ensemble should of course not depend on the way the equilibrium was reached.

Many simulation programs report observables (such as the system temperature, kinetic energy, potential energy, momenta, etc., and also partial and total averages of these quantities) "on the fly", i.e., while the simulation is still in progress. While this information is very useful for monitoring the advancement of a simulation and detecting problems, one will eventually want to extract much more information from the simulated ensemble. It is thus necessary to store the configurations on some storage device at time intervals short enough for a subsequent statistical evaluation. This can result in large files. As an example, for a system of 1000 argon atoms, storing just 10 000 configurations (positions and velocities) in single precision (unformatted) would take 1000 (particles) × 10 000 (configurations) × 6 ($x/y/z/v_x/v_y/v_z$) × 4 bytes = 240 Mbyte of storage space.

6. Obtaining Results

Once the sample of the thermodynamic ensemble has been constructed, statistical methods can be used to extract the macroscopic observables of the system. This is where a very critical assessment is needed whether the ensemble of simulated configurations is a sufficiently representative sample of the thermodynamic ensemble. Questions to be answered include: Has sufficient decorrelation from the initial configuration been achieved? Could the system be trapped in a metastable state? There are no rigorous methods to tackle these questions, and trial and error methods as well as experience are most often called upon here. Analytical methods [25] can be used in some cases to gauge the reliability of the computed averages in a strictly statistical sense.

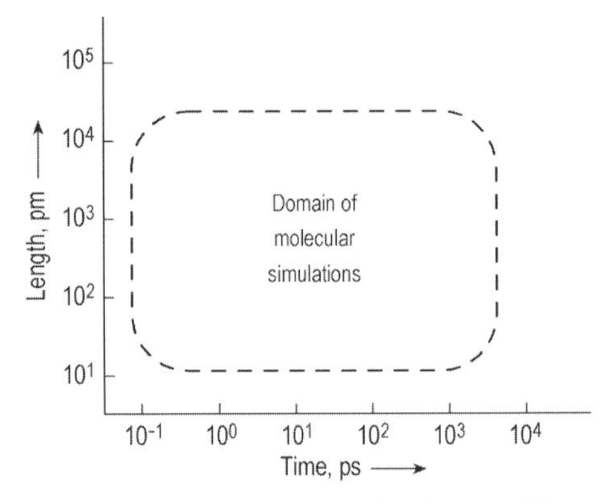

Figure 5. Typical space and time domain accessible to molecular dynamics; see text for typical physical phenomena occurring in this domain.

In summary, taking into account the approximations and limitations described above, Figure 5 shows roughly the space–time window accessible to MD computer simulations with today's (ca. 2004) programs and computers. The lower left corner of the graph illustrates the time–space domain of typical fast (in a chemical sense) phenomena such as the electronic transitions observed in optical spectroscopy. The time domain associated with molecular vibrations ranges from femto- to picoseconds; molecular reorientations in liquids have correlation times of several picoseconds. These phenomena can be probed, e.g., by IR and Raman spectroscopy in classical and time-resolved modes, as well as by NMR spectroscopy and dielectric techniques. Longer times are associated with motions of large molecules and with collective motions of groups of smaller ones. The length scale extends from the Ångström range, typical of interatomic distances, accessible experimentally by scattering methods (X-rays, neutrons, etc.), to the nanometer range, which characterizes medium-sized assemblies of molecules. The hydrodynamic regime starts at the top right corner of the graph, it is accessible through other types of simulations (see Chap. 8).

Two kinds of averages are usually distinguished: Simple averages over all available configurations $\Re$, and averages in which the order of the configurations, i.e., the time evolution of the system, is taken into account. In the equilibrium ensemble, this is most often done by computing so-called time-correlation functions. We note here, without going into detail, that these time-correlation functions and the spatial correlations briefly discussed below can be seen as limiting cases of a general space–time function [6].

6.1. Simple Averages

The average value of a quantity A, denoted $\langle A \rangle$, is computed as Equation 18

$$\langle A \rangle = \frac{1}{M} \cdot \sum_{i}^{M} A\left(\Re_i\right) \tag{18}$$

where A is a quantity that can be determined from the particle positions, velocities, (and possibly forces) contained in the system configurations R_i. Equation 1 is an example of this procedure. Another example follows from the equipartition theorem of statistical mechanics, written in Equation 19 for a system of N masses

$$\frac{3N-3}{2} \cdot k_{\mathrm{B}} \cdot \langle T \rangle = \langle E^{\mathrm{kin}} \rangle = \frac{1}{2} \cdot \sum_{i}^{N} m_i v_i^2 \tag{19}$$

where k_{B} is the Boltzmann constant, $\langle E^{\mathrm{kin}} \rangle$ is the average kinetic energy and $\langle T \rangle$ the average temperature. The term $3N-3$ is the number of degrees of freedom in this system. The three degrees of freedom corresponding to the movement of the center of mass must be removed by virtue of Equation 16. In general, for systems with constraints or with molecules represented by rigid or semirigid entities, the real total number of degrees of freedom must be entered into Equation 19 instead of $3N-3$. Examples of such averages are, besides $\langle T \rangle$ in Equation 19, the average potential energy $\langle U \rangle$, the average pressure $\langle p \rangle$, and many other thermodynamic quantities which can be directly related to experimental values. Thermodynamic quantities involving derivatives can be obtained from fluctuations (e.g., heat capacities, for which quantum corrections may have to be included) or by more advanced statistical mechanical treatments [26 – 28]. Spatial correlations like the radial pair distribution function (RDF) $g(r)$ or its analogue, the radial pair correlation function $h(r) = g(r)-1$, are routinely computed. These functions are suitably normalized histograms over interparticle distances between pairs of molecules; they are related to the so-called potential of mean forces (PMF) and are often used in thermodynamics calculations [6]. They are furthermore related through Fourier transformations to the scattering functions obtained in X-ray or neutron scattering experiments [29].

We note that in Equation 18 the result will not depend on the order of the configurations $\Re_i$. The fact that the $\Re_i$ were constructed using the Newton equation is thus of no consequence. Quantities like $\langle A \rangle$ are thus often termed "static" or "structural". If they depend only on the positions they can also be obtained from Monte-Carlo (MC) simulations, in which the ensemble is constructed through a random-walk procedure.

6.2. Time-Correlation Functions

The time-correlation function $c_{\mathrm{BC}}(t)$ between two quantities B and C is defined in the equilibrium ensemble as Equation 20

$$c_{BC}\left(t\right) = \left\langle B\left(\Re\left(\tau\right)\right) \otimes C\left(\Re\left(t+\tau\right)\right)\right\rangle_{\tau} \tag{20}$$

where $B(\Re)$ and $C(\Re)$ are quantities like $A(\Re)$ above, $\otimes$ denotes a general operator (most often a product or scalar product, sometimes also a logical AND or OR operator) and $\langle\,\rangle_\tau$ denotes an average over all times τ. If $C = B$, $c_{\mathrm{BB}}(t)$ is called an autocorrelation function. Otherwise, it is called a cross-correlation function.

Example 3. One of the simplest and most widely studied autocorrelation functions is the velocity autocorrelation function, i.e., $B(\Re) = \vec{V}$, where $\vec{V}$ is the vector of the velocities of N equivalent particles (e.g., the same type of atom, or a given type of atom fulfilling some condition) at a given time (Eq. 21).

$$\vec{V}\left(\Re\left(t\right)\right) = \begin{pmatrix} \vec{\vartheta}_1 \\ \vec{\vartheta}_2 \\ \vdots \\ \vec{\vartheta}_N \end{pmatrix} \tag{21}$$

The velocity autocorrelation function then becomes in terms of the $\vec{\vartheta}_i$ Equation 22

$$\begin{aligned} c_{\vec{V}\vec{V}}\left(t\right) &= \langle \vec{V}\left(\Re\left(\tau\right)\right) \cdot \vec{V}\left(\Re\left(t+\tau\right)\right)\rangle_\tau \\ &= \frac{1}{N \cdot M}\sum_{j=0}^{M}\sum_{i=1}^{N}\left(\vec{\vartheta}_i\left(\tau_j\right)\cdot\vec{\vartheta}_i\left(\tau_j+t\right)\right) \\ &= \langle c_{\vec{\vartheta}_i\vec{\vartheta}_i}\left(t\right)\rangle \end{aligned} \tag{22}$$

Equation 22 can be easily evaluated from the trajectory; an example is given in Sect. 7

Integrals of correlation functions of the type Equation 23

$$\mathcal{T} = \lim_{t\to\infty}\int_0^t c_{AA}\left(\tau\right)\mathrm{d}\tau \tag{23}$$

are related to transport properties, in the case of the velocity autocorrelation function, to the self-diffusion coefficient D_{S}. For the integral to converge, the correlation function must decay within the simulation time. Actually, it must decay much faster so that a sufficient number of averages (the sum over j in Eq. 22) can be carried out. If the expectation value of the autocorrelated quantity is not equal to zero, the definition of the function is usually modified so that it indeed decays to zero at long times.

Classical time correlations are even in time. It is thus sufficient to consider their Fourier–Laplace transforms. For a function $c_{\mathrm{AA}}(\tau)$, the transform (Eq. 24)

$$c_{AA}\left(\omega\right) \propto \int_0^\infty c_{AA}\left(\tau\right)\cos\left(\omega\tau\right)\mathrm{d}\tau \tag{24}$$

is called the spectral density, or the simply the spectrum. Depending on which quantity A is correlated, the spectra can be observed experimentally (e.g. in IR spectroscopy, corresponding to the correlation of the dipole moment or derivative of the dipole moment of the system). The velocity autocorrelation function is again used as an example in Chapter 7.

7. Examples

The first example is a simulation of a system consisting of 864 Lennard–Jones particles (Eq. 7) parameterized to represent methane (CH_4, united-atom approximation, effective potential from Figure 2) in an NVE simulation with volume and energy chosen to yield an average temperature (Eq. 19) of $\langle T\rangle = 130$ K and a density of 0.3952 g/cm^3. Figure 6 shows the simulated radial pair distribution function (RDF) $g(r)$, which is a measure of the microscopic structure in a liquid [29]. It represents the relative density, compared to the overall bulk density, as a function of distance from a fixed particle at $r = 0$. In the region of the first peak, located roughly where the Lennard–Jones potential has its minimum, the local density is thus enhanced by a factor of approximately 2.6. These density fluctuations extend to further "shells" of neighbors, and finally the $g(r)$ function converges toward a value of 1 (bulk density). RDFs can be determined experimentally, under certain approximations, from X-ray, neutron, or other scattering methods; they are also an important tool for the computation of thermodynamics quantities in statistical mechanics.

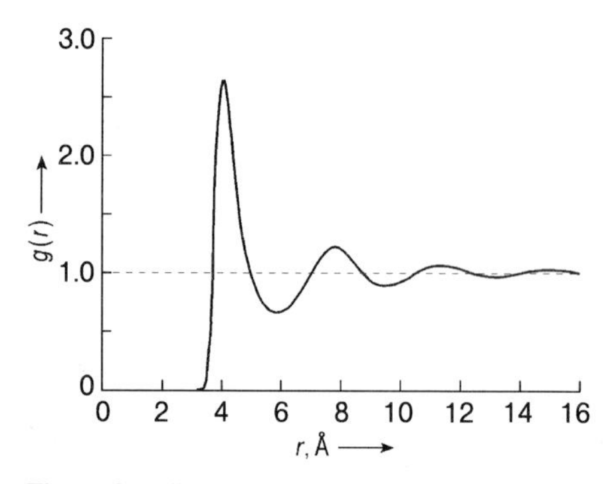

Figure 6. Radial distribution function $g(r)$ of liquid methane at $\langle T\rangle = 130$ K and a density of 0.3952 g/cm^3.

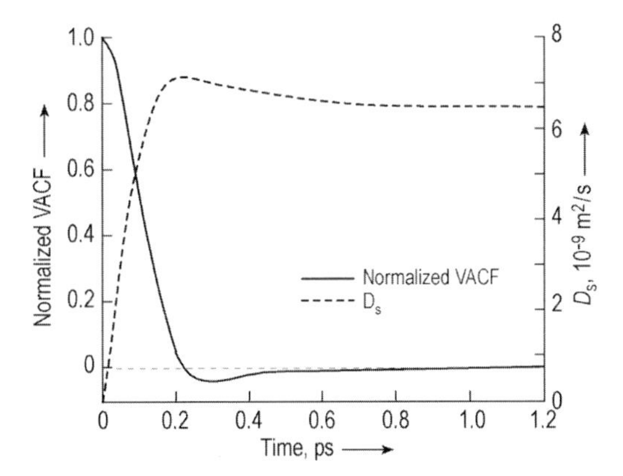

Figure 7. Normalized VACF of liquid methane and integral of the unnormalized function, yielding at long times a self-diffusion coefficient of $D_S \approx 6.66 \times 10^{-9}$ m^2/s.

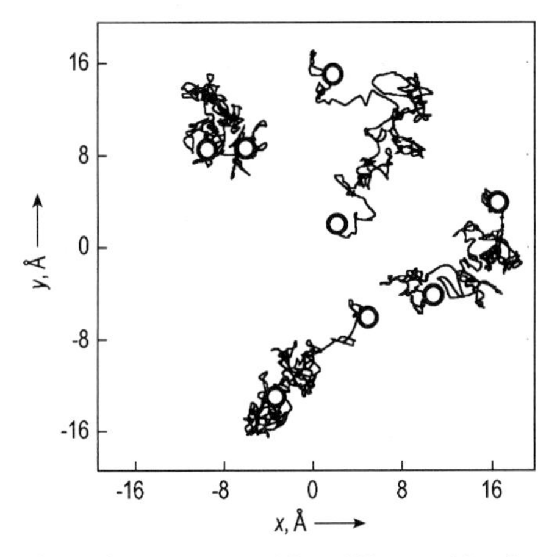

Figure 8. Trajectories of four different arbitrarily chosen methane molecules over 25 ps of simulation. The frame represents the quadratic simulation box with a side length of 38.8 Å. Initial and end points are marked by a bold circle.

As outlined in Section 6.2, dynamical quantities are accessible by computing suitable time-correlation functions. The self-diffusion coefficient D_S of liquid methane is thus obtained from the molecular velocity autocorrelation function (VACF) using Equations 22 and 23. The underlying simulation was run under the condition of zero total momentum $\vec{p} = 0$ (Eq. 16), i.e., the expectation value of a particle velocity is zero (see above). Figure 7, left axis, shows this VACF, normalized to unity at $t = 0$. The right-hand scale in Figure 7 refers to an integral like in Equation 23, suitably normalized to yield at long times the self-diffusion coefficient D_S.

To illustrate the real movements of some methane molecules during a simulation, Figure 8 displays the projections onto a plane of four particle trajectories generated during 25 ps of simulation time. We note that this time is much longer than the correlation time of the velocities in Figure 7. Such plots are useful to gauge how extensively the thermodynamic ensemble has been sampled by the simulation.

8. Applications

The applications of MD simulations are by now almost innumerable, and only a few examples can be given here. The first examples, discussed in the previous sections, were taken from studies of liquids such as liquid methane or aqueous ionic solutions at ambient conditions, for which much of the basic methodology was developed. However, one of the main advantages of simulation methods is that no assumptions have to be made concerning the state of the system: gases, liquids, and solids, as well as inhomogeneous systems containing all kinds of interfaces, are thus, in principle, amenable to MD simulations. As a caveat, we note, however, that in some cases this method will be quite inefficient. In gases at low densities, for instance, very much time is spent integrating the motions of molecules with almost negligible mutual interactions.

It is easy in MD simulations to study systems under extreme conditions of density (pressure) and temperature; such systems are usually difficult to study experimentally. Thus, molten salts were among the first systems studied by simulation techniques. Studies of liquids of geological interest, i.e., usually at pressures and temperatures difficult to maintain in a laboratory experiment, are other examples for this kind of approach. Supercritical fluids in thermodynamic conditions not too close to the critical point can be investigated in the same fashion. Studies on highly complex liquids such as polymer melts or liquid crystalline systems are becoming more and more numerous as the available computational resources increase and models and methods for their simulations are developed.

Besides neat liquids and solutions, the structure and dynamics of liquid mixtures has become a very active field of investigation. We note here, however, that studies of certain thermodynamic properties, in particular of the phase equilibria, are most often undertaken by using specific Monte-Carlo (MC) techniques rather than MD simulations. The development of adequate interaction models, i.e., ones able to represent both the neat liquids and the mixtures, is one of the main challenges when mixtures are to be studied.

Among the many applications concerning interfaces (liquid–liquid, liquid–gas, and solid–liquid) the free surfaces of pure liquids and solutions have been studied. Furthermore, liquids near smooth hydrophilic or hydrophobic walls with or without corrugation have been extensively simulated. Systems of electrochemical interest, e.g., solid metal electrodes in contact with aqueous ionic solutions, but also confined liquids such as liquids in pores or in clays, have been of particular interest. The dissolution and crystallization processes of salts have been investigated in simulations of salt solutions in contact with the corresponding crystals. Layers of amphiphilic molecules, soap films, micelles, and micellar solutions of various kinds have been investigated in very extensive MD simulations, as have polymer layers and model biomembranes. In some of the studies on interfacial or porous systems it is sufficient to consider the solid phase as an immobile "frozen" obstacle. However, if the dynamics of such crystalline solids are to be simulated, which is often necessary, the restrictions imposed by the periodic boundary conditions must be considered. Multiples of the elementary cell are used as simulation boxes, and special techniques are available to adjust the box, e.g., if structural phase transitions are to be studied. Multiple studies on inclusion compounds such as clathrate hydrates and zeolites, with or without guest molecules, have been carried out in this way.

The biosciences are by now possibly the largest application domain for MD simulations. The main problems here are the sizes of the systems and the consequent complexity of the interaction models. Furthermore, in many instances these systems are characterized by correlation times that are long compared to the times usually accessible in MD simulations (see Figure 5). Extremely long simulation runs are thus necessary, and present-day resources often do not allow a satisfactory sampling. Yet events like the penetration of small solvated molecules, or indeed water itself, into membranes or trans-membrane ion channels have been studied by simulation techniques. Derived techniques are used in drug design. Understanding protein folding remains one of the challenges in this area, and large-scale efforts are underway. This is, however, beyond the scope of this article.

As discussed above, transport coefficients can be obtained, within the framework of the fluctuation-dissipation theorem [6], from correlation functions computed for the system in equilibrium (see Section 6.2). However, it is often more efficient to perform nonequilibrium molecular dynamics (NEMD). This can be done both for stationary (e.g., in a thermal gradient) and nonstationary (e.g. for relaxation phenomena) conditions. Many methods have been proposed [12, 30 – 32]. Essentially, in the first instance, gradients are maintained across the system during the simulation in a way compatible with the periodic boundary conditions. In the second instance, an equilibrium simulation is performed and configurations are periodically extracted, perturbed, and their evolution is compared with the ongoing equilibrium run. The NEMD simulation allows one to go beyond the linear response and to study nonlinear transport phenomena.

In a coarse-graining approach, macroscopic quantities obtained from MD simulations are often used as input parameters to other types of simulations, e.g. Brownian Dynamics [12] or dissipative particle dynamics (DPD) [33]. In these approximations, much larger systems can be studied over longer periods of time, going well beyond the domain sketched in Figure 5.

Finally, we note that more and more approaches combining MD simulations (in thermodynamic equilibrium, as discussed here, or out of equilibrium) with other approaches are being proposed for various purposes. Chemical reactions are thus tackled by combining MD with quantum chemistry calculations. Other efforts aim at extending the space and time domains accessible to the simulations by using suitable coarse-graining techniques. These methods are mostly still in the stage of methodic developments. Like the now classical methods discussed here, which were developed in the last third of the 20th century, they will one day become tools for research in all kinds of disciplines.

9. References

General textbooks

1. M. P. Allen, D. J. Tildesley, *Computer Simulation of Liquids*, Clarendon Press, Oxford, 1990.
2. K. Binder, *The Monte Carlo Method in Condensed Matter Physics*, Springer-Verlag, Berlin, 1992.
3. G. Ciccotti, D. Frenkel, I. R. McDonald, *Simulation of Liquids and Solids: Molecular Dynamics and Monte Carlo Methods in Statistical Mechanics*, North-Holland, Amsterdam, 1987.
4. D. J. Evans, G. P. Morriss, *Statistical Mechanics of Non-Equilibrium Liquids*, Academic Press, London, 1990.
5. D. Frenkel, B. Smit, *Understanding Molecular Simulation — From Algorithms to Application*, Academic Press, San Diego, 2002.
6. K. E. Gubbins, N. Quirke (Eds), *Molecular Simulation and Industrial Applications. Methods, Examples and Properties*, Gordon and Breach, Amsterdam, 1996.
7. R. Haberlandt, S. Fritzsche, G. Peinel, K. Heinzinger, *Molekulardynamik — Grundlagen und Anwendungen*, Vieweg Verlag, Braunschweig, 1995.
8. J. M. Haile, *Molecular Dynamics Simulation: Elementary Methods*, Wiley, New York, 1992.
9. D. M. Heyes, *The Liquid State: Applications of Molecular Simulation*, Wiley, Chichester, 1998.
10. W. G. Hoover, *Computational Statistical Mechanics*, Elsevier, Amsterdam, 1991.
11. G. C. Maitland, M. Rigby, E. B. Smith, W. A. Wakeham, *Intermolecular Forces: Their Origin and Determination*, Clarendon Press, Oxford, 1981.
12. D. C. Rapaport, *The Art of Molecular Dynamics Simulation*, Cambridge University Press, Cambridge, 1995.
13. R. J. Sadus, *Molecular Simulation of Fluids — Theory, Algorithms and Object-Orientation*, Elsevier, Amsterdam, 1999.
14. M. Schoen, *Computer Simulation of Condensed Phases in Complex Geometries*, Springer-Verlag, Berlin, 1993.
15. F. Vesely, *Computerexperimente an Flüssigkeitsmodellen*, Physik Verlag, Weinheim, Germany, 1978.
16. F. Vesely, *Computational Physics. An Introduction*, Kluwer Academic/Plenum Publishers, New York – London, 2001.

Selected Articles

17. K. Binder, J. Horbach, W. Kob, W. Paul, F. Varnik, "Molecular Dynamics Simulations", *J. Phys. Condens. Matter* **16** (2004) 429–453.
18. G. Ciccotti, G. Jacucci, I. R. McDonald, "Thought-Experiments by Molecular Dynamics", *J. Stat. Phys.* **21** (1979) no. 1, 1–22.
19. W. G. Hoover, W. T. Ashurst, "Nonequilibrium Molecular Dynamics" in *Theoretical Chemistry. Advances and Perspectives*, **Vol. 1**, Academic Press, San Diego, 1975, pp. 2–51.
20. W. Smith, C. W. Young, P. M. Rodger, "DL_POLY: Application to Molecular Simulation", *Mol. Sim.* **28** (1975) no. 5, 385–471.
21. W. F. van Gunsteren, H. J. C. Berendsen, "Computer Simulation of Molecular Dynamics: Methodology, Applications and Perspectives in Chemistry", *Angew. Chem. Int. Ed. Engl.* **29** (1990) 992–1023.

Web resources (as of 2004):

22. A glossary of terms: http://xeon.concord.org:8080/modeler1.2/guide/html/glossary/index.html
23. A. S. Côté, B. Smith, P. J. D. Lindau, *Democritus — A Molecular Dynamics Tutorial*: http://www.compsoc.man.ac.uk/~lucky/Democritus/Basic/Democritus.html, Daresbury Laboratory, Daresbury, UK, 2001.
24. P. Cummings, D. A. Kofke, A. Panagiotopolous, J. de Pablo, R. Rowley, H. Cochran, D. Evans, *Molecular Simulation. A World Wide Web Book*, Written by a task force and published under the auspices of the National Science Foundation: http://boltzmann.vuse.vanderbilt.edu/~w3press/.
25. F. Ercolessi, A Molecular Dynamics Primer: http://www.fisica.uniud.it/~ercolessi/md/md/, June 1997. University of Udine, Italy.
26. D. A. Kofke, Molecular Simulation Lectures: http://www.cheme.buffalo.edu/courses/ce530/index.html, Department of Chemical Engineering, SUNY Buffalo.
27. F. Vesely. Computational Physics. Course material: http://www.ap.univie.ac.at/users/ves/cp0102/dx/index.html, Institute of Experimental Physics, University of Vienna, 2002.
28. F. Vesely. Statistical Physics. Course material with JAVA applets: http://www.ap.univie.ac.at/users/ves/sp_english/sp/sp.html, Institute of Experimental Physics, University of Vienna, 2002.

Specific references

29. K. Binder, D. W. Heermann, *Monte Carlo Simulation in Statistical Physics*, Springer-Verlag, Berlin, 1997.
30. The Gallup Organization, http://www.gallup.com, Princeton, NJ, accessed 2004.
31. P. W. Atkins, J. de Paula, *Atkins' Physical Chemistry*, Oxford University Press, Oxford, 2002.
32. J. B. Foresman, A. Frisch, *Exploring Chemistry with Electronic Structure Methods*, Gaussian, Inc., Pittsburgh, PA, 1993.
33. A. Hinchliffe, *Computational Quantum Chemistry*, Wiley, New York, 1988.
34. D. A. McQuarrie, *Statistical Mechanics*, Harper & Row, New York, London, 1976.
35. R. Car, M. Parrinello, "Unified Approach for Molecular Dynamics and Density-Functional Theory", *Phys. Rev. Lett.* **55** (1985) 2471–2474.
36. J. M. Haile, *Molecular Dynamics Simulation: Elementary Methods*, Wiley, New York, 1992.
37. B. Guillot, "A Reappraisal of What We Have Learnt During Three Decades of Computer Simulations on Water", *J. Mol. Liq.* **101** (2002) nos. 1–3, 219–260.
38. E. B. Wilson, J. C. Decius, P. C. Cross, *Molecular Vibrations*, McGraw-Hill, New York, 1955.
39. G. Herzberg, *Molecular Spectra and Molecular Structure I*, Van Nostrand, New York, 1950.
40. M. P. Allen, D. J. Tildesley, *Computer Simulation of Liquids*, Clarendon Press, Oxford, 1990.
41. C. L. Kong, "Combining Rules for Intermolecular Potential Parameters. II. Rules for the Lennard-Jones (12-6) Potential and the Morse Potential", *J. Chem. Phys.* **59** (1973) 2464.
42. J. O. Hirschfelder, C. F. Curtiss, R. B. Bird, *Molecular Theory of Gases and Liquids*, Wiley, New York, 1954.
43. J. G. Gay, B. J. Berne, "Modification of the Overlap Potential to Mimic a Linear Site-Site Potential", *J. Chem. Phys.* **74** (1981) 3316.
44. G. C. Maitland, M. Rigby, E. B. Smith, W. A. Wakeham, *Intermolecular Forces: Their Origin and Determination*, Clarendon Press, Oxford, 1981.
45. B. M. Axilrod, E. Teller, "Interactions of the van der Waals' Type between Three Atoms", *J. Chem. Phys.* **11** (1943) 299–300.
46. A. P. Sutton, J. Chen, "Long-Range Finnis-Sinclair Potentials", *Philos. Mag. Lett.* **61** (1990) no. 3, 139–146.
47. J. D. Jackson, *Classical Electrodynamics*, Wiley, New York, 1975.
48. L. X. Dang, "A Mechanism for Ion Transport across the Water/Dichloromethane Interface: A Molecular Dynamics Study Using Polarizable Potential Models", *J. Phys. Chem. B* **105** (2001) 804–809.
49. J.-C. Soetens, G. Jansen, C. Millot, "Molecular Dynamics Simulation of Liquid CCl_4 with a New Polarizable Model", *Mol. Phys.* **96** (1999) 1003.
50. H. Meyer, O. Biermann, R. Faller, D. Reith, F. Müller-Plathe, "Coarse Graining of Nonbonded Inter-particle Potentials Using Automatic Simplex Optimization to Fit Structural Properties", *J. Chem. Phys.* **113** (2000) 6264–6275.
51. W. G. Hoover, *Computational Statistical Mechanics*, Elsevier, Amsterdam, 1991.
52. S. Toxvaerd, O. H. Olsen, "Canonical Molecular Dynamics of Molecules with Internal Degrees of Freedom", *Ber. Bunsenges. Phys. Chem.* **94** (1990) 274.
53. H. Flyvberg, H. G. Petersen, "Error Estimates on Averages of Correlated Data", *J. Chem. Phys.* **91** (1989) 461.
54. R. Lustig, "Statistical Thermodynamics in the Classical Molecular Dynamics Ensemble. I. Fundamentals", *J. Chem. Phys.* **100** (1994) 3048–3059.
55. R. Lustig, "Statistical Thermodynamics in the Classical Molecular Dynamics Ensemble. II. Application to Computer Simulation", *J. Chem. Phys.* **100** (1994) 3060–3067.
56. R. Lustig, "Statistical Thermodynamics in the Classical Molecular Dynamics Ensemble. III. Numerical Results", *J. Chem. Phys.* **100** (1994) 3068–3078.
57. J. P. Hansen, I. R. McDonald, *Theory of Simple Liquids*, Academic Press, New York, 1986.
58. D. J. Evans, G. P. Morriss, *Statistical Mechanics of Non-Equilibrium Liquids*, Academic Press, London, 1990.
59. W. G. Hoover, "Nonequilibrium Molecular Dynamics", *Ann. Rev. Phys. Chem.* **34** (1983) 103–127.
60. W. G. Hoover, W.T. Ashurst, "Nonequilibrium Molecular Dynamics" in *Theoretical Chemistry. Advances and Perspectives, Vol. 1*, Academic Press, San Diego, 1975, pp. 2–51.
61. D. Frenkel, B. Smit, *Understanding Molecular Simulation — From Algorithms to Application*, Academic Press, San Diego, 2002.

Computational Fluid Dynamics

ANJA R. PASCHEDAG, University of Technology Berlin, Berlin, Germany (Chap. 1)

1.	**Introduction**	342
2.	**Procedure**	342
3.	**Modeling**	343
3.1.	**Transport Equations**	343
3.2.	**Initial and Boundary Conditions**	345
3.3.	**Turbulent Flow**	346
3.4.	**Multiphase Approaches**	349
4.	**Numerics**	352
4.1.	**Basics**	352
4.2.	**Finite Volume Method**	354
4.3.	**Pressure Correction Methods**	357
4.4.	**Lattice Boltzmann Method**	357
5.	**Interpretation**	358
6.	**Industrial Application**	359
7.	**References**	361

Computational Fluid Dynamics (CFD) is a numerically based tool for the prediction of flow field, concentration and temperature distribution. Its main parts are mathematical modeling, discretization, numerical solution of the discretized equations and the interpretation of numerical results.Basic equations in all mathematical models for CFD are balances for momentum and total mass determining velocity, pressure and density field. Depending on the case considered they are supplemented by mass balances for single species and a heat balance. Additional models are required to describe, e.g. turbulence, multiphase flows, chemically reactive systems and other special cases.Basis for the discretization of the balance equation is the discretization of a space — the grid generation. Most codes can handle unstructured grids. Nevertheless, certain requirements concerning grid structure have to be fulfilled to get stable convergence and an accurate solution. Traditionally, most CFD codes use finite volume discretization for the balance equations, even if finite element algorithms are of increasing relevance for simulations with adaptively moving grids and for coupling CFD with structural dynamics simulations. A new approach for simulations with high resolution in space and time is the lattice Boltzmann method.Finally, the numerical results have to be graphically presented and interpreted. Because of the huge amount of numerical data provided by each simulation this cannot be done with a general method. It must always be analysed in reference to a certain research question. Errors caused by the model formulation and by the numerical scheme have to be analyzed in order to judge the accuracy of a simulation. Quantitative estimates are required for an adequate interpretation of the results.

Symbols

A	m^2	cell surface (FVM)
$C_{BC1}...C_{BC4}$		constants of mathematical expressions for boundary conditions
$C_\mu, C_{\varepsilon 1}, C_{\varepsilon 2}$		constants in k—ε model
c	mol/m^3	concentration
c_p	$J\,kg^{-1}K^{-1}$	specific heat capacity
C_Δ		Courant number
D	m^2/s	diffusion coefficient
$\boldsymbol{F}$	N	force
$\boldsymbol{F}$		general function
f		weighting factor in time discretization
G		filter function (LES)
$\boldsymbol{g}$	m/s^2	gravitational acceleration
I	$kg{\cdot}m^2$	moment of inertia
J_ϕ		molecular flux
k	m^2/s^2	turbulent kinetic energy
m	kg	mass
$\boldsymbol{n}$	m	normal vector
N		number of nodes
$\hat{P}$	m^2/s^3	source term in transport equation for k
p	N/m^2	pressure
$\dot{q}$	$J\,m^{-2}s^{-1}$	heat flux density
r		residual
Re		Reynolds number
S_c		source in concentration equation
S_{hr}		source in temperature equation
S_u		source in energy equation
S_ϕ		source in general transport equation
$\boldsymbol{T}$	N·m	torque
T	K	temperature
t	s	time
u	J/m^3	specific internal energy

Ullmann's Modeling and Simulation

ISBN: 978-3-528-31605-2

V	m^3	volume
v	m/s	velocity
$\mathbf{x}$	m	position vector
α_i		volume fraction of phase i
Γ	m^2/s	molecular transport coefficient
$\boldsymbol{\delta}$		unit tensor
ε	m^2/s^3	energy dissipation rate
κ	Pa·s	dilatational viscosity
λ	J $(m^{-1}s^{-1}K^{-1})$	heat conductivity
μ	Pa·s	dynamic viscosity
μ_t	Pa·s	turbulent viscosity (RANS)
ρ	kg/m^3	density
$\boldsymbol{\tau}$	N/m^2	Newtonian shear stress
$\boldsymbol{\tau}^s$	N/m^2	turbulent stress tensor (LES)
$\boldsymbol{\tau}_t$	N/m^2	turbulent stress tensor (RANS)
ϕ		general transport quantity
Ω	m/s	rotational speed of particles
Indices		
B,E,N,P,S,T,W		nodes of the FVM grid
b,e,n,s,t,w		cell faces of the FVM grid
Pi		i^{th} particle parcel

1. Introduction

The increasing pressure on development time and product quality in industrial process development stimulates the application of numerical methods in addition to traditional experimental ones. In many cases the numerical tools can save time and costs if they are used complementary to experimental methods. The simulation tool for modeling fluid dynamic processes is called computational fluid dynamics (CFD). Highly complex mathematical models are needed for a sufficient description of the relevant interactions in reactive flows occurring on different timescales. Even if not all aspects can be considered in detail to keep the equation system in a manageable frame, the complex mathematical models are far from being solvable analytically. Therefore, numerical methods play an important role in CFD, and many users associate CFD mainly with its numerical aspects. But the subject is defined much more widely: formulation of the mathematical model, visualization, and interpretation are all essential parts of the tool.

2. Procedure

This chapter sketches the general steps of a CFD simulation. It gives an overview of the procedure and the relation of the individual parts. The points mentioned here are described in more detail in the following chapters.

As with all investigation methods, the formulation of the exact question is a crucial point for the entire CFD procedure. The question includes the target value and its accuracy, the dependencies of importance and possible approximations and assumptions which can be made. Also the range in space and time to be considered is a part of it. All significant physical parameters or the range of their variation must be specified together with the physical state at the boundaries and at the initial time. The formulation of the question determines the effort and the time required to solve the problem and should be therefore as strict as necessary and as weak as possible.

The mathematical model can be formulated on the basis of the question posed. Its central elements are balance equations for mass, momentum and energy. These are supplemented by additional model equations if necessary. All partial differential equations require initial and boundary conditions for their solution. Parameters and constants in the equations have to be specified in a way that the model is mathematically closed. For this purpose, additional laws, like an equation of state or temperature dependencies of physical properties (e.g., density, viscosity) may be required. Not always can the most exact mathematical formulation be solved with reasonable effort. In such cases, additional model assumptions such as averaging procedures, reduction of the number of variables or restriction of the range in space and time considered must be introduced.

The resulting system of differential, integral and algebraic equations is too complex to be solved analytically. Therefore, numerical algorithms are applied to derive a system of linear algebraic equations which provides an approximative solution of the original mathematical model. In most cases this discretization is performed by using the finite volume method or the finite element method. Specialized methods exist for certain applications. A relatively new alternative to classic discretization algorithms are statistical methods like the lattice Boltzmann method.

As a result of the great complexity of the mathematical model, the numerical solution of the governing equations must be postprocessed to extract the information necessary to answer the original question. Most often this information can best be presented in a graphical plot, but the data to be plotted are usually not the direct solution of the model. Relevant data must be selected, linked, transformed, and in some cases statistically treated and normalized to create a data set which allows the user to answer the question.

Last but not least, the accuracy of the numerical result must be judged, even if the simulation is not erroneous in an obvious sense. Starting from the setup of the mathematical model over all numerical steps, up to postprocessing, inaccuracies are introduced or at least accepted. Only a rough quantitative error estimation can prove whether the solution provided is accurate enough for the question considered.

3. Modeling

The central CFD modeling approach is to balance momentum, mass and heat. This approach is based on the continuity hypothesis which states that all balance quantities and physical parameters change continuously within each phase. Consequently, limiting considerations for infinitesimally small volumes never consider changes at the molecular level. Discontinuous jumps are found only at interfaces. Therefore, in some points, modeling of multiphase systems requires different approaches than modeling of single-phase systems.

For an easier understanding of the following discussion a few terms will be defined:

- Balance quantities: Quantities governed by laws of conservation can be balanced. In general, mass, momentum and energy are such balance quantities. For detailed considerations masses of different chemical species or different kinds of energy can be balanced separately. In such balances source terms for possible conversions appear. Balance quantities are extensive values.
- Transport quantities: While the basic step of balancing is an integral consideration of the balance quantities, differential transport equations can be derived from the original balances. Transport equations describe the distribution and transport of intensive values, either mass-based (mass fraction, specific energy) or volume-based (density, concentration), which are the so-called transport quantities.
- Independent variables: Independent variables in transport equations are normally the three spacial directions and time. In special cases symmetry can be used to reduce the number of spacial directions; in the case of stationary considerations, time is not considered. For certain model approaches additional independent variables must be included. In probability density models, e.g. for micromixing, the species composition is an additional independent variable, in population balances particle properties such as the characteristic diameter take on this functionality.
- Parameters: All known values in the transport equations — besides transport quantities and independent variables — are called parameters. “Known” must be understood in a mathematical sense: the solution of the equations is only possible if values for the parameters are given. In the physical sense the determination of these values might be problematic or only possible with the help of an additional equation.

3.1. Transport Equations

Transport equations are derived on the basis of balances. For the general transport quantity ϕ the transport equation reads:

$$\begin{aligned} \frac{\partial \phi}{\partial t} &= -\nabla(\boldsymbol{v}\phi) - \nabla \boldsymbol{J}_\phi + S_\phi \\ (a) &= (b) \qquad + (c) \quad + (d) \end{aligned} \tag{1}$$

Term (a), the time derivative, codes transient changes. Term (b) models the convective transport with $\boldsymbol{v}$ being the convective velocity. Term (c) in this general form is the molecular transport term including the molecular flux vector $\boldsymbol{J}_\phi$. In certain cases, in particular in turbulence modeling, structurally similar terms can occur which are numerically treated in the same way but have different physical meanings. Finally, term (d) is the source term. The kinds of sources which occur depend strongly on the transport quantity ϕ. In the momentum balance these are forces, in the energy balance these are exchange rates to other types of energy, and in mass balances for single species these are conversion rates due to chemical reactions.

Detailed derivations and discussions of the transport equations can be found in [1] (see also Transport Phenomena). Only a short overview is presented here.

Continuity Equation. A basic physical law is the conservation of mass. From this the continuity equation can be derived:

$$\frac{\partial \rho}{\partial t} = -\nabla\, (\rho \boldsymbol{v}) \tag{2}$$

As there is no molecular mass transfer relative to the convective velocity (with its mass-based definition) and no source of mass in a nonrelativistic system, this equation contains only terms (*a*) and (*b*) of Equation 1 with density ρ being the transport value.

Equation of Motion. CFD considers transport phenomena in fluids, so the convective velocity is a central quantity. This becomes obvious also from Equation 1, where velocity is the only general, ϕ-independent value. The velocity is determined by the equation of motion:

$$\frac{\partial \rho \boldsymbol{v}}{\partial t} = -\nabla\, (\rho \boldsymbol{v} \boldsymbol{v}) - \nabla\, \boldsymbol{\tau} - \nabla p + \rho \boldsymbol{g} \tag{3}$$

The velocity itself is not the transport quantity here but rather the volume-based momentum $\rho\boldsymbol{v}$. The stress tensor $\boldsymbol{\tau}$ plays the part of the molecular flux. It is a function of the velocity gradient, but this functionality can not be expressed in a general form because it depends on the rheologic properties of the fluid. In many applications a Newtonian fluid can be assumed, and $\boldsymbol{\tau}$ can be expressed in the following way:

$$\boldsymbol{\tau} = \mu[\nabla \boldsymbol{v} + (\nabla \boldsymbol{v})^T] + (\kappa - \frac{2}{3}\mu)(\nabla \cdot \boldsymbol{v})\boldsymbol{\delta} \tag{4}$$

where μ is the viscosity, which is independent of shear stress and time in Newtonian fluids, and κ is the dilatational viscosity, which in most real cases has no influence on the flow.

The last two terms of Equation 3 are the source terms to be considered in most cases. The pressure term is of greatest importance. It has a dual role in the structure of the equation. As mentioned, it can be seen as a source term, but it also could be counted as a special type of stress and included together with $\boldsymbol{\tau}$ in a general stress term. The first concept is applied here because it provides a more suitable structure for the discussion of numerical methods.

Gravity acts in all systems and is therefore also written here. In special cases other forces acting on a fluid like electromagnetic forces or centrifugal forces also must be considered.

Two things complicate the numerical handling of the equation of motion. The first is the nonlinear structure of the convective terms. While in other transport equations the velocity can be handled like a parameter, in the equation of motion it is the unknown. The second problem results from the pressure term. There is no transport equation available for pressure. In compressible fluids an equation of state is used additionally which allows one to calculate the pressure if the density field is known from the equation of continuity. The solution of these three coupled equations (one of which has three components) is numerically not handsome, but possible without significant problems. In incompressible systems the density is constant and an equation of state cannot be used to determine the pressure. The remaining equation for the pressure is the equation of continuity, but it does not contain the pressure. The relation can be described such that only the velocities which satisfy the equation of motion with the right pressure gradient also satisfy the equation of continuity.

Navier–Stokes Equations. A special and commonly used version of the equation of continuity and the equation of motion are obtained for Newtonian fluids with constant density and constant viscosity. The equations simplify to:

Equation of continuity:

$$\nabla \boldsymbol{v} = 0 \tag{5}$$

Equation of motion:

$$\rho \frac{\partial \boldsymbol{v}}{\partial t} = -\rho \boldsymbol{v} \nabla \boldsymbol{v} + \mu \nabla^2 \boldsymbol{v} - \nabla p + \rho \boldsymbol{g} \tag{6}$$

This equations are called Navier–Stokes equations. They are of restricted validity because changes in temperature as well as changes in the chemical composition effect density and velocity. Nevertheless, they are a reasonable approach in many cases.

Concentration Equation. The equation of continuity results from the overall mass balance. If a multicomponent system is considered and the concentration of the different species is of interest for the simulation, separate transport equations must be solved for them. As the global mass balance is solved in all cases, the number of additional equations needed for a complete description is one less than the total number of species in the system. The transport equation for a concentration c reads:

$$\frac{\partial c}{\partial t} = -\boldsymbol{v}\nabla c + \nabla(D\nabla c) + S_c \tag{7}$$

where D is the molecular diffusion coefficient and S_c is the source term resulting from a chemical reaction. It is expressed via the chemical reaction kinetics, which in many cases include nonlinear expressions and dependencies between concentrations of different chemical species.

Energy Equation. Formulating energy equations is a large area because there are many different forms of energy which can be converted into each other. For CFD the equation of the internal energy is usually considered to obtain an equation for the temperature profile. For the application of turbulence modeling, a certain part of the kinetic energy is of interest. This second case is discussed in Section 3.3 only the internal energy is considered here. Internal energy covers the kinetic energy of atomic and molecular motion, potential energy of intermolecular interactions, energy of chemical bonds, and nuclear energy. Not all aspects of this energy can be determined absolutely. Therefore, only differences in the internal energy between two states are discussed. Balancing of the internal energy gives:

$$\begin{array}{llllll} \frac{\partial \rho u}{\partial t} & = -\nabla(\rho u \boldsymbol{v}) & - \nabla \dot{\boldsymbol{q}} & - (\boldsymbol{\tau}:\nabla \boldsymbol{v}) & - p(\nabla \cdot \boldsymbol{v}) & + S_u \\ (a) & = (b) & + (c) & + (d) & + (e) & + (f) \end{array} \tag{8}$$

with

(a) = temporal change
(b) = convective transport
(c) = heat conduction
(d) = increase by viscous dissipation
(e) = increase by compression
(f) = conversion from other types of energy

If an incompressible fluid is considered, in addition viscous dissipation, radiation and other kinds of energy conversion can be neglected, and if the mixing enthalpy is also negligible, the only potential source is the enthalpy of a chemical reaction S_{hr}. Density, thermal conductivity, and heat capacity values are calculated for the mixture in the case of a multicomponent system. With these assumptions the temperature equation can be derived in the form of a commonly used equation:

$$\frac{\partial \rho c_p T}{\partial t} = -\boldsymbol{v}\nabla(\rho c_p T) - \nabla(\lambda \nabla \rho c_p T) + S_{hr} \tag{9}$$

where c_{p} is the heat capacity and λ the thermal conductivity. If the assumptions do not hold as described above, extended forms of the temperature equations must be derived.

3.2. Initial and Boundary Conditions

Partial differential equations cannot be solved without initial and boundary conditions. In a mathematical sense this can be interpreted such that the solving of a differential equation requires its integration, and to perform this the integration constants must be determined. Consequently, the number of additional conditions needed for each independent variable is equal to the number of derivatives. Therefore, in time only one condition is needed, which in most cases is given at the initial time, while for the space coordinates two conditions are required, which are realized normally by one condition at each boundary point.

For the initial conditions values of the transport quantities ϕ must always be given. These values may vary in space but cannot be a function of ϕ.

Three different types of boundary conditions are possible:

Dirichlet condition

$\phi|_{\mathrm{BC}} = C_{\mathrm{BC1}}$ — fixed value of ϕ given at the boundary

Neumann condition

$\left.\frac{\partial \phi}{\partial n}\right|_{\mathrm{BC}} = C_{\mathrm{BC2}}$ — derivative of ϕ normal to the boundary is given

Cauchy condition

$\left.\frac{\partial \phi}{\partial n}\right|_{\mathrm{BC}} + C_{\mathrm{BC3}}\,\phi|_{\mathrm{BC}} = C_{\mathrm{BC4}}$ — combination of value and derivative of ϕ is given

In a finite region of the boundary a Dirichlet condition is required. The use of these boundary conditions can be segregated in relation to the physical nature of boundaries. The ones most often used are presented here, but for special flow situations other types can also be defined. It is not always easy to find the expression required, and in some cases a setup is more easily obtained by moving the boundary location than by constructing a complex formulation at the original location. An example is shown in Figure 1.

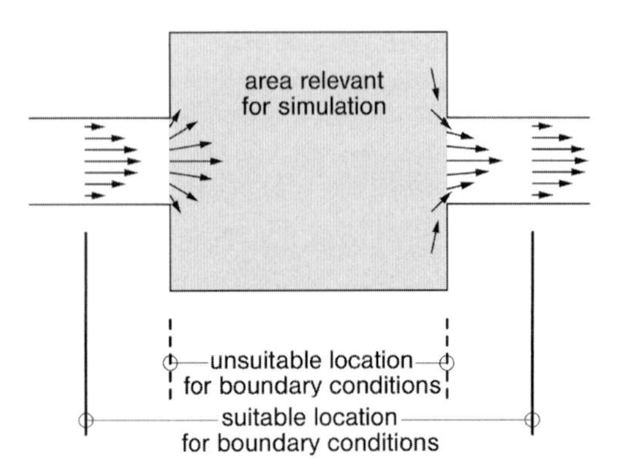

Figure 1. Location of boundary conditions

- Inlet: It is assumed that the conditions at the inlet are known. Therefore, fixed values are given at the inlet for all transport quantities. These fixed values may vary along the inlet, e.g., a parabolic velocity profile can be given for a laminar flow in a tube.
- Wall: Most model systems are bounded at least partly by solid walls. They can be fixed or moving (e.g., wall and stirrer of a stirred vessel). Walls considered here are impermeable for mass fluxes but may possibly conduct heat. To prevent convective fluxes through the wall the normal velocity component is set to zero or equal to the normal velocity of the moving wall. Assuming no slip conditions, the tangential velocity is equal to the wall velocity, which is zero with fixed walls. If slip is considered, the gradient of the tangential velocity components normal to the wall is zero. To avoid diffusive fluxes through the wall the concentration gradient normal to the wall is set to zero.
  For heat transfer different types of walls must be considered. The definition of temperature boundary conditions at walls is an approximation in most cases, because neither the physical conditions are well known there nor are they easy to control. Adiabatic walls are modeled by using a zero temperature gradient normal to the wall. At walls of fixed temperature this temperature is given as boundary condition. Heat fluxes across a wall can be described by using Neumann or Cauchy conditions.
- Outlet: No assumptions about the quantities of the transport values can be made for outlet conditions. However, it is assumed that the flow is uniform and no further sources occur. In that case velocities tangential to the outlet are zero. For all other unknowns (normal velocity, concentrations, temperature) zero gradients normal to the outlet are used. Furthermore, for incompressible flows the mass flow at each outlet must be specified to preserve the total mass.
  It is not always easy to find a location at which outlet boundary conditions can be specified with sufficient accuracy. Special problems occur in turbulent flows if vertex structures are not averaged but resolved and the assumptions for the velocity field do not hold. In such a case, boundary conditions for the pressure are stated instead of velocity conditions. The strong coupling between pressure and velocity described in context of the equation of motion allows one to solve these equations with this type of boundary conditions.
- Symmetry: Some systems show symmetries which allow one to reduce the model space and consequently the numerical effort. Symmetry can be utilized if it holds not only for the geometry, but also for the flow and for all transport variables.
  Symmetry is given when all gradients normal to the symmetry plane are zero and the velocity component normal to the symmetry plane is zero, too.

3.3. Turbulent Flow

While laminar flow is characterized by parallel stream lines, in a turbulent flow vortex structures of a large size range cause highly frequent fluctuations in all transport variables. All the equations introduced up to now are valid in laminar as well as in turbulent flows, as long as the conditions mentioned hold. The greatest problem in simulations of turbulent flows is that the resolution of the high-frequency fluctuations requires a fine numerical grid and small time steps. The number of grid cells N needed for a three-dimensional simulation can be estimated by

$$N=5^3 Re^{9/4} \tag{10}$$

where Re is the Reynolds number of the flow. Such simulations of turbulent flows, called *direct numerical simulations* (DNS) are possible only for small academic cases and relatively low Reynolds numbers with current (2004) computing power.

In most cases the exact structure of the turbulent flow is of minor interest and information on local averaged values is sufficient. Unfortunately, such an averaging can not be done in a simple way because the course-scale structure of the flow interacts strongly with the small-scale structure. Special attention must be given to chemically reactive flows. Chemical reactions can take place only if the species are mixed on the molecular scale. A model averaging this scale may overestimate the rate of fast chemical reactions.

Two general modeling approaches for turbulent flows are used nowadays: *Reynolds averaging* (Reynolds-averaged Navier–Stokes RANS) and *large-eddy simulations* (LES). RANS modeling is based on averaging all transport quantities over all scales of turbulent fluctuations. This allows one to use relatively coarse grids and can be applied to geometrically large systems with reasonable effort. The influence of turbulent structures is included in additional models. Large-eddy simulations work with a grid size which allows direct resolution of the large-scale part of the turbulence structures. Since the large eddies contain the majority of the turbulent energy, the return in accuracy is superproportional to the increase in effort with this approach. The filter width is a function of the grid spacing and can therefore be influenced by the user. Nevertheless, subgrid-scale models (SGS models) are required also for LES. The numerical effort for LES is significantly larger than for RANS simulations and can be handled at the moment only on high-performance parallel computers. Figure 2 gives an impression of the resolution of turbulent eddy structures by the different methods.

LES and RANS are illustrated in more detail for the equation of motion. Concentration and energy equations can be handled in a similar way, but all turbulence models are developed in the first step for momentum transport. Thus the mass and energy transport models use proportionality approaches, in most cases based on the parameters computed for momentum transport.

Reynolds Averaging. Reynolds averaging is based on the assumption that scales of the main flow and of turbulent fluctuations differ significantly. Therefore, each transport quantity can be split up into a time-averaged value $\overline{\phi}$ and a fluctuating value $\phi\prime$ in such a way that possible macroscale fluctuations of $\phi\prime$ are included in $\overline{\phi}$ while turbulent fluctuations are covered by $\phi\prime$:

$$\phi=\overline{\phi}+\phi' \tag{11}$$

In accordance with the goal of RANS to obtain averaged values, the transport equations are averaged. This yields for the Navier–Stokes equations:

$$\begin{aligned} \nabla\,\overline{\boldsymbol{v}} &= \quad 0 \\ \rho\frac{\partial\overline{\boldsymbol{v}}}{\partial t} &= -\rho\overline{\boldsymbol{v}}\nabla\overline{\boldsymbol{v}}\; -\nabla\boldsymbol{\tau}_t+\mu\nabla^2\overline{\boldsymbol{v}}-\nabla\overline{p}+\rho\boldsymbol{g} \end{aligned} \tag{12}$$

where an additional term including the *Reynolds stress tensor* $\boldsymbol{\tau}_{\mathrm{t}}$ occurs.

$$\boldsymbol{\tau}_t=\rho\boldsymbol{v}'\times\boldsymbol{v}'=\begin{pmatrix} \overline{\rho v'_x v'_x} & \overline{\rho v'_x v'_y} & \overline{\rho v'_x v'_z} \\ \overline{\rho v'_y v'_x} & \overline{\rho v'_y v'_y} & \overline{\rho v'_y v'_z} \\ \overline{\rho v'_z v'_x} & \overline{\rho v'_z v'_y} & \overline{\rho v'_z v_z} \end{pmatrix} \tag{13}$$

With this term the averaged equation of motion is not closed because the velocity fluctuations are unknowns.

There are a large number of approaches to model the Reynolds stress tensor. To explain all of them is beyond the scope of this article. A detailed presentation is given in [2]. Therefore, only the most commonly used concept of the eddy viscosity approach together with the k—ε model will be sketched. The effect of the Reynolds stress tensor can be interpreted in analogy to the Newton stress tensor [3]. Therefore, it can be expressed proportionally to the gradient of the averaged velocity with a proportionality factor μ_{t} called *eddy viscosity*:

$$\boldsymbol{\tau}_t=\mu_t\left[\nabla\overline{\boldsymbol{v}}+(\nabla\overline{\boldsymbol{v}})^T\right]-\frac{2}{3}\rho k\delta \tag{14}$$

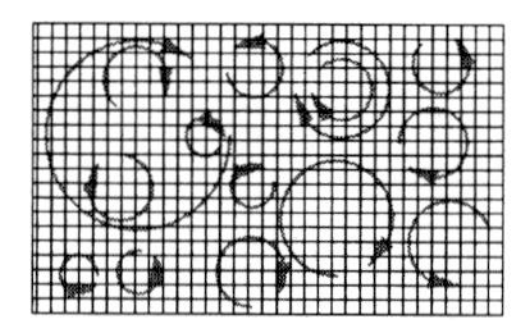
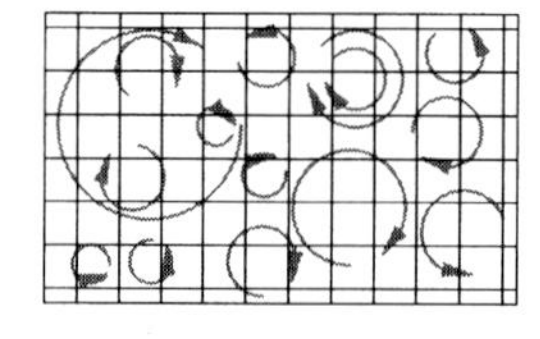
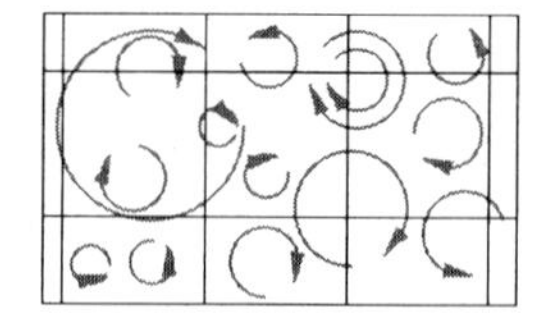

Figure 2. Resolution of turbulent eddy structures by different modeling approaches
a) Direct numerical simulation; b) Large-eddy simulation; c) Reynolds-averaged simulation

where δ is the unit tensor and k is the turbulent kinetic energy defined by

$$k=\frac{1}{2}\overline{\boldsymbol{v}'\cdot\boldsymbol{v}'} \tag{15}$$

In contrast to the molecular viscosity, μ_t is the *eddy viscosity*, which is not a material constant but depends on the flow structure. The k—ε model expresses this dependence as a function of the turbulent kinetic energy k and its dissipation rate ε:

$$\mu_t=C_\mu\rho\frac{k^2}{\varepsilon} \tag{16}$$

with

$$\varepsilon=\frac{\mu}{2\rho}\overline{|\nabla\boldsymbol{v}'+\nabla\boldsymbol{v}'^{T}|^2} \tag{17}$$

and k and ε computed by their own transport equations. The original model [4] uses the following form:

$$\begin{aligned}\frac{\partial\rho k}{\partial t} &= -\nabla\left(\overline{\boldsymbol{v}}\rho k\right)+\nabla\left(\tfrac{\mu_t}{\sigma_k}\nabla k\right)+\quad\hat{P}\quad-\quad\rho\varepsilon\\ \frac{\partial\rho\varepsilon}{\partial t} &= -\nabla\left(\overline{\boldsymbol{v}}\rho\varepsilon\right)+\nabla\left(\tfrac{\mu_t}{\sigma_\varepsilon}\nabla\varepsilon\right)+C_{\varepsilon1}\tfrac{\varepsilon}{k}\hat{P}-C_{\varepsilon2}\rho\frac{\varepsilon^2}{k}\end{aligned} \tag{18}$$

with the production term $\hat{P}$ defined as

$$\hat{P}=-\frac{\mu_t}{2}\,|\nabla\overline{\boldsymbol{v}}+\nabla\overline{\boldsymbol{v}^{T}}|^2 \tag{19}$$

The constants of the model are C_μ = 0.09, $C_{\varepsilon1}$ = 1.44, $C_{\varepsilon2}$ = 4.92, σ_k = 4.0, σ_ε = 1.3

The k—ε model is valid for fully developed isotropic turbulence. If this condition is not given, either extended RANS models or LES are required for an acceptable flow prediction.

Large-Eddy Simulations. A high-accuracy alternative to Reynolds averaging is given by large-eddy simulations. In this approach the transport values are filtered with a filter function $G(\mathbf{x},\mathbf{x}')$:

$$\hat{\phi}\left(\mathbf{x}\right)=\int G(\mathbf{x},\boldsymbol{x}')\phi\left(\mathbf{x}'\right)\mathrm{d}\mathbf{x}' \tag{20}$$

where G describes the influence of the value of ϕ in position $\mathbf{x}'$ on the filtered value $\hat{\phi}$ at position $\mathbf{x}$. Several forms of G are possible, but it is always a function of the grid size.

Filtering the equation of motion gives:

$$\rho\frac{\partial\hat{\boldsymbol{v}}}{\partial t}=-\rho\hat{\boldsymbol{v}}\nabla\hat{\boldsymbol{v}}-\nabla\boldsymbol{\tau}^{s}+\mu\nabla^2\hat{\boldsymbol{v}}-\nabla\hat{p}+\rho\boldsymbol{g} \tag{21}$$

In this equation the subgrid-scale Reynolds stress tensor $\boldsymbol{\tau}^{\mathrm{s}}$ occurs. Its meaning is analogous to the Reynolds stress tensor in RANS but it considers only the effect of unresolved subgrid-scale structures.

Equation 21 can be solved only if $\boldsymbol{\tau}^{\mathrm{s}}$ is expressed as a function of filtered values. For this purpose so-called subgrid-scale models (SGS models) are used.

SGS models of the Smagorinsky type are widespread [5]. They are similar to the eddy viscosity models in RANS, apart from the fact that they consider only sub-grid scale effects. In their structure they do not use information about the eddy structure of the resolved scales. Similarity models like that in [6] are based on the assumption that there is a similarity in the eddy structures of different scales. Therefore, the analysis of the resolved structure can be used to model the subgrid-scale. These models fail to describe dissipation at the smallest eddy scales, because this effect is not found at larger scales. So they have to be combined with other models to cover this aspect. Dynamic models, e.g., [7, 8] are based on this idea and extend upon it. For this approach two simulations on different grids are carried out. The SGS model is adapted to the differences in the predictions of both models. In this way parameters of the SGS scale model can be varied in time and space and dissipation effects are described. Such models are interesting for cases in which constant parameters are only

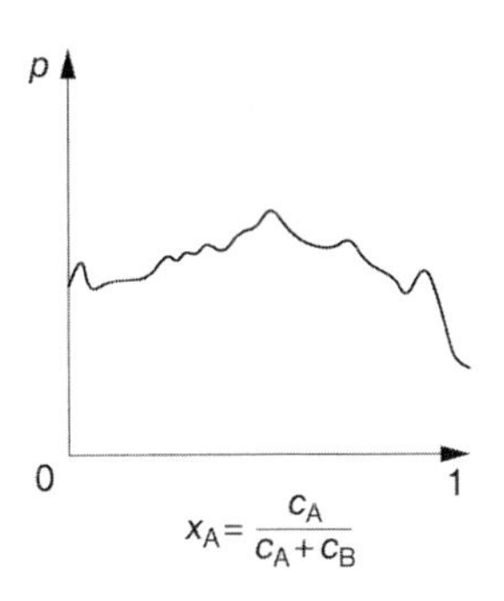

Figure 3. Modeling micromixing in a computational cell with probability density function approach
a) Distribution of species A (black) and B (white) in a turbulent flow with incomplete micromixing; b) Probability density function of mixing between A and B

a rough approximation. The major drawback of this method is its numerical effort, which is significantly higher than for the other methods.

Micromixing Models. If systems with fast chemical reactions are to be simulated, a description of the mixing state on the microscopic level is required. Neither Reynolds averaging nor large-eddy simulation provides such an description because they average or filter small-scale fluctuations. A remedy is the use of micromixing models which provide information about the micromixing quality as a function of turbulence quantities, like eddy viscosity or turbulent kinetic energy dissipation.

A widespread micromixing model is the probability density function approach (PDF) [9, 10]. It considers the possible mixing states of a fluid of certain composition and gives for each computational cell a probability density function for each state in such a away that the integral composition in this cell is satisfied. For a mixture of only two species A and B this is shown in Figure 3. The local probability density functions are computed from transport equations for the PDF.

Each mixing state is related to a certain chemical reaction rate. The integral of the reaction rates over the whole composition space gives the total reaction rate in the appropriate numerical cell.

Alternative micromixing models are

- Eddy break up model [11]: This is a simple empirical model which assumes binary chemical reaction with only one product. It is mainly applied for the modeling of combustion.
- Flamelet model [12, 13]: This model assumes chemical reaction in a two-dimensional laminar layer. It was developed for premixed and nonpremixed flames with a different definition of the reaction layer for the two cases.
- Engulfment model [14]: This model assumes that a fluid is mixed into another of different composition by formation of a rotational cylinder. Within the cylinder diffusion is the dominant mixing process and the composition changes proportionally to the change in cylinder size.

3.4. Multiphase Approaches

Multiphase systems are characterized by interfaces at which the fluid properties change discontinuously. This requires extensions of the model approaches introduced up to now for their application to multiphase systems. At the beginning of this chapter multiphase systems will be classified to describe the appropriate model for each type.

One type of classification refers to the geometric structure of the system (see Figure 4). If one phase consists of solid or fluid particles with a small size compared to the total system the system is called a *disperse system*. The particles form the disperse phase while the other phase is called the continuous one. For modeling disperse systems, an extended form of the continuum hypothesis is applied. Properties of both phases are considered to be continuously distributed over the whole system; specific properties of single particles are not relevant. The

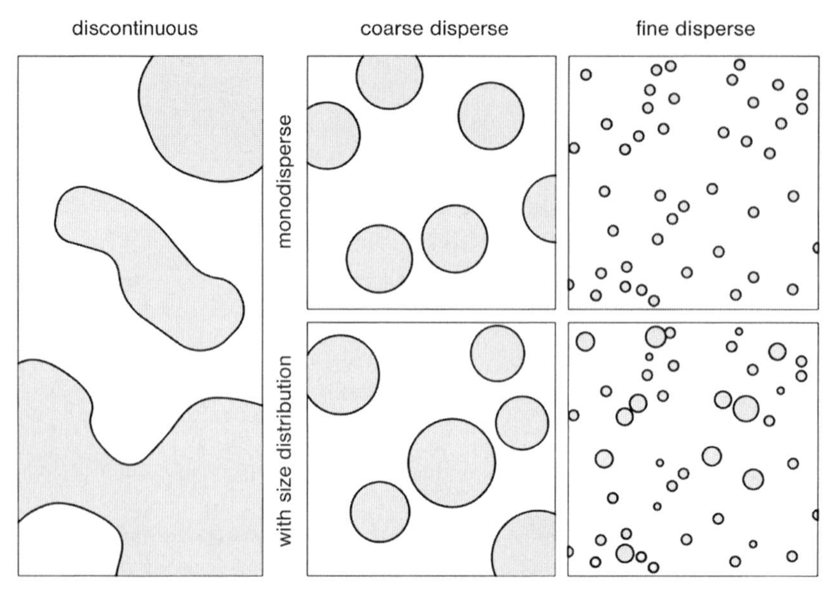

Figure 4. Geometric structure of multiphase systems

central property in that case is the volume fraction α which for phase i is defined as the ratio of the volume V_i of this phase and the total volume V:

$$\alpha_i = \frac{V_i}{V} \tag{22}$$

The sum of all volume fractions must be unity:

$$\sum_i \alpha_i = 1 \tag{23}$$

The assumption that the continuum hypotheses holds for disperse systems implies that changes of properties in a particle are not resolved. This means, e.g., in a temperature distribution that each particle is characterized by a mean temperature, but the different mean temperatures of all particles determine a particle temperature field of the system. If this approach is insufficient for the description of a system, e.g., because the temperature gradient at the interface is needed to determine the heat transfer between the phases exactly, it cannot be modeled with the approach for disperse systems.

An important property of disperse systems is the size distribution. Many phase interactions like buoyancy force or heat transfer rate depend on particle size, and therefore the particle size distribution is needed to determine these interactions exactly. The easiest case from the point of modeling are monodisperse systems; which all particles have the same size. For some applications the differences in the interactions do not need to be resolved, and this approach can be used. If the size distribution has to be considered the particle size becomes an additional independent variable in the system, which increases the complexity of the mathematical treatment significantly. Size distributions are relevant for all cases where the size range of the particles is very wide or the size distribution changes in the course of the process, e.g., by nucleation, evaporation, agglomeration, or similar processes. The computation of particle size distributions is based on the solution of population balances. This model approach is not discussed here further, but details can be found, e.g., in [15 – 17].

Another characteristic of a disperse system is the relative velocity between the phases. If the differences in gravitational forces are small compared to the momentum transfer between the phases, both phases have nearly the same velocity. This is the case for phases with similar densities or for systems with very small particles.

Another description refers to the interactions during momentum transfer. If particles move with the continuous phase but do not significantly influence the flow field of this phase it is called a *one-way interaction*. Systems with one-way interactions do not have to be modeled with multiphase approaches. The flow field is modeled with the single-phase equations and the volume fraction can be handled like a concentration. The term *two-way interaction* is used, if there is a relative velocity between the two phases which causes a reciprocal momentum transfer. If, in addition, the volume fraction of the disperse phase is so high that interactions between the particles and also between particles and walls have to be considered, one speaks of *four-way* interactions.

If in a multiphase system a disperse phase cannot be defined clearly because the extension of all phases is too large to neglect gradients of the transport values in them, the system is called a *discontinuous multiphase system*. The way to model such a system is to use the approaches for single-phase systems and additionally to describe the condition at the interface. This works well if the extent of each single-phase region covers a significant part of the total domain.

The conditions at the interface act as boundary conditions for the balances in both phases. Therefore, two conditions must be defined for each transport variable in each point. One is the equilibrium of fluxes, and the other is the relation between the values of the transport quantity. As neither momentum nor mass or energy can be accumulated at the interface, the fluxes must be equal on both sides. The ratio of the values depends on the quantity considered. As a result of the nonslip condition the tangential velocities are equal. Unless there is a significant mass transfer over the interface (e.g., melting or evaporation) this holds also for the normal velocity; otherwise source terms must be considered in the stress balance at the interface. Thermal equilibrium at the interface states that the interfacial temperature for both phases is the same. The distribution of chemical species is not equal in equilibrium but described by a distribution coefficient. Detailed explanations and equations for such systems can be found, e.g., in [18, 19].

Euler–Euler Method. The Euler–Euler method is used for modeling disperse systems. It utilizes continuous transport equations for each phase. These equations are based on single-phase equations but weighted with the volume fraction and appended with source terms describing the phase interaction [20].

For the equation of motion the interactions between phases are forces. In all systems with a relative velocity between the phases the drag force plays a major role. It is the main force which prevents a continuous increase in the relative velocity between the phases. Further forces which have to be considered depending on the properties of the phases are the virtual mass force, the Basset force, and other forces caused by different relative velocities at different points of the particle surface, such as the Saffman force and the Magnus force.

Mass and heat transfer between the phases determine the interaction terms in the concentration and temperature equations. For systems with a significant mass transfer over the interface it must be considered that heat and momentum are also transferred together with that mass.

The Euler–Euler approach is preferably used for systems with high volume fractions of the disperse phase or for relatively large particle sizes, because there are no formal restrictions for these properties. Models for the changes in turbulence caused by phase interaction are not well developed yet. Size distributions can be described only with a high numerical effort because most forces change as a function of the particle size, and therefore the size distribution must be divided into size classes, whereby each class is handled as an individual phase.

Euler–Lagrange Method. Alternatively, the Euler–Lagrange approach can be used for disperse systems. In this approach the continuous transport equations are also used for the continuous phase and extended by interaction terms. The disperse phase is modeled differently: particle parcels are defined and distributed in such a way that they represent the particle phase with its volume fraction and properties. For each of these parcels Pi the path and the velocities (convective and rotational) are determined by ordinary differential equations:

$$\begin{aligned} \frac{d\boldsymbol{x}_{Pi}}{dt} &= \boldsymbol{v}_{Pi} \\ m_{Pi}\frac{d\boldsymbol{v}_{Pi}}{dt} &= \sum \boldsymbol{F}_i \\ I_{Pi}\frac{d\omega_{Pi}}{dt} &= \boldsymbol{T}_i \end{aligned} \tag{24}$$

where $\boldsymbol{x}_{Pi}$ is the position of the parcel, $\boldsymbol{v}_{Pi}$ and $\boldsymbol{\omega}_{Pi}$ are its convective and rotational velocity, $\boldsymbol{F}_i$ the forces acting between the faces, $\boldsymbol{I}_{Pi}$ is the moment of inertia, and $\boldsymbol{T}_i$ the torque.

The velocity of the continuous phase and the particle phase are corrected in an iterative procedure considering interactions based on current particle paths. The particle parcels only give a statistical representation of the particle phase if their characteristics are averaged over a reasonable time interval. The time interval for averaging has no limits in the stationary case aside from giving statistically reliable results. For transient investigations the time interval of averaging determines the time resolution of the simulation. If statistical reliability is insufficient, the time steps of the particle simulation must be reduced.

The forces for phase interaction in the second of Equations 24 are the same as those mentioned for the Euler–Euler approach. Mass and energy transfer also have to be considered along the particle paths and statistically evaluated.

Consideration of the particle paths reduces the volume fraction and particle sizes which can be handled by this method. The volume of a parcel presents can not be larger than the smallest grid cell. For high volume fractions the danger of having more than one parcel in a cell at a time further restricts the volume ratio between a cell and a parcel. Reasonable results are therefore only obtained for very low volume fractions, usually in the range of 1%, exceptionally up to 5%. The numerical effort of the method increases with the volume fraction, because additional parcels have to be introduced.

On the other hand, this method provides better models of the influence of particles on turbulence than the Euler–Euler approach. Defining parcels with different properties allows one to model size distributions with only a small additional effort. A detailed description of the Euler–Lagrange method is given in [21, 22].

Grid Adaptation to Interface. The best way to model a discontinuous multiphase system is to define the reference system in such a way that the interface has a fixed position in space. In that case the grid can be constructed so that the interface is located at grid lines. The interfacial conditions can then be applied easily.

In most cases interfaces move, and the reference system cannot be defined as described above. In this case the location of the interface is a part of the solution of the model. To incorporate the interfacial conditions as with boundary conditions the grid can be adapted to the new location of the interface. As long as the dislocation is small, this affects only the cell layer close to the interface and the numerical error introduced is small. For large dislocations the cells near the interface would be deformed severely and it is necessary to add or delete cells. Detailed descriptions of adaptive methods are given in [23].

Volume of Fluid. If there is a strong movement of the interface, the grid adaptation method is numerically laborious and erroneous. An alternative is to locate the interface at an arbitrary position of the grid. Different methods act in this way (segment method [24], marker and cell method [25]). The most popular method is the volume of fluid method [26]. A volume fraction is defined and a transport equation is solved for it similarly to the Euler–Euler method. The volume fraction in most cells is either zero or one and the interface is located in the cells where the volume fraction lies between these values. The exact location of the interface is determined by the volume fraction in the cell. The direction of the interface is normal to the gradient of the volume fraction.

4. Numerics

4.1. Basics

In chapter 3 mathematical models for the description of transport processes in fluids have been introduced. Clearly, the coupled system of integrodifferential equations and algebraic equations can be solved analytically only for a very small number of limiting cases. For all practically relevant situations only a numerical, computer-based solution is possible.

The basic idea of a numerical solution is to replace the partial differential equation which is continuously defined over the range in space and time with a system of algebraic equations which gives the solution only at defined discrete points or for discrete intervals. For the finite difference method (FDM), values at single points are computed; for the finite volume method (FVM), characteristic values are determined, which are constant for a numerical cell; and for the finite element method (FEM), parameters of a polynomial function over a cell yield the solution. All these are available for separate times. A completely different approach, used with the lattice Boltzmann method, is described separately.

As the solution of the algebraic equations, called the discretized equations, is fixed to certain points or elements in space and time, the discretization of space and time is the first step to be performed. Time is a one-dimensional coordinate, and so it can simply be split into intervals, i.e., time steps. The time step size can vary over the total time period considered. Stability and accuracy of the procedure on the one hand and computing time on the other depend on the time-step size chosen in relation to the space discretization (see Section 4.2).

The discretization of the three-dimensional space offers much more possibilities. The two discretization methods mainly used in CFD, FVM and FEM, require the subdivision of the total model space into non overlapping cells. In the ideal case the grids formed from these cells meet the following requirements:

- Geometrical requirements: The outer border of the grid should fit as closely as possible to the physical borders of the system to introduce the boundary conditions at the correct location. In the case of simple geometrical structures this is not a problem but it might become one for jagged structures. The more variable the cell shapes a CFD code accepts, the more easily such structures can be described without extensive refinement close to the boundaries.
- Physical requirements: To achieve sufficient accuracy in the solution the grid size should increase in areas of steep gradients and especially where they are varying. This criterion has the major problem that in can be fulfilled precisely only if the solution is known. In many cases general physical understanding of the flow investigated is sufficient to estimate where local grid refinement is necessary. In other situations preliminary computations on a rough grid give an orientation for grid improvements.
  The simulation of turbulent flows requires a few grid cell layers parallel to the boundary for an appropriate modeling of the laminar sublayer.
- Numerical requirements: The numerical requirements depend mainly on the numerical method used and the way it is implemented in the code. The crux of the matter is whether a structured grid is needed or an unstructured one is allowed. Especially for unstructured grids a wide variety of cell shapes can be imagined but not every one is supported by every code. Nearly all unstructured solvers can handle tetrahedral and hexahedral cells separately or in combination in one grid. In some cases more general formulations of polyhedral cells are available but are not standard.
  From a numerical point of view serious jumps in the size of adjacent grid cells should be avoided. Furthermore, cells should not tend to degenerate, that is, the ratio of volume to surface should not become too small.

In many cases very large numbers of cells would be needed to meet all requirements sufficiently. This might push the computational effort to an unrealistic level. Therefore, a reasonable compromise has to be found.

The major distinction of grid types is made between structured and unstructured ones. A structured grid is defined by three bands of grid faces, where the faces of one band do not intersect with each other but with the faces from the other bands. There is no restriction on the shape of the faces or the coordinate system in which they are defined. The cell shapes in structured grids are hexahedral, sometimes with one collapsed cell face. Structured grids have only restricted applicability to complex geometries but for simple geometries they can be generated easily.

In unstructured grids no global grid faces or lines are defined, but the domain may be divided in any way. This offers many more possibilities for grid fitting, but it normally cannot be done without an automated grid generator.

For some simulations it is necessary to perform grid adaptation in the course of the simulation. Reasons might be, e.g., moving boundaries (as in stirred vessels), movements of interfaces which have to be tracked, or strong transient changes in the solution which require local changes in refinement. If the new structure of the grid depends on the numerical solution, the numerical effort for the adaptation strategies is relatively high. Nevertheless, in many cases with grid adaptation a significant improvement in accuracy and efficiency of a simulation can be gained.

Discretization of the model equations can be based on the discretization of space and time. The different methods that are available vary in their mathematical approach and range of application. An insight is given here for two methods: the finite volume method, and the lattice Boltzmann method, with stress on the forme.

A third widespread method is the finite element method. It is extensively explained in → Mathematics in Chemical Engineering, Chap. 7.6 and therefore not discussed here.

4.2. Finite Volume Method

Historically, the most common numerical method in CFD is the finite volume method (FVM). It is also used by the presently most widest distributed commercial CFD codes FLUENT, CFX, and Star-CD. Its success in CFD is related to physical equivalents in the approach.

The fundamental concept is the consideration of grid cells (also called *control volumes*) in which all physical parameters and system characteristics are assumed to be constant. The balance equations are integrated over each of this cells. For volume sources these integrals can be solved directly by considering these sources as constant over the cell. The volume integrals of the transport terms are transformed into surface integrals by using the theorem of Gauss. As the physical properties are constant within each cell, they can also considered to be constant along each cell face, which permits the surface integral to be resolved. The value or the derivative of the solution at the cell face is contained in the expressions gained. While the values are constant within each cell, they jump unsteadily at the cell faces, and the values and derivatives at the faces have to be expressed by the values in the cells, which are fixed at certain reference points, i.e., the nodes.

This approach is, in general terms, the reverse of the derivation of the transport equations, where a finite model volume is considered initially and fluxes over the boundaries and volume sources are balanced. In the second step the limiting expression for the model volume approaching zero is determined. This analogy reveals another advantage of the method in a physical sense: as the surface integral of the transport terms expresses the fluxes over the cell surface, a numerical error in the determination of these fluxes does not violate the integral conservation of the balanced property. The laws of conservation of mass, momentum, and energy are the fundamentals of modeling for CFD, and therefore a method upholding these laws inherently has been most widely accepted by engineers.

Nowadays, this advantage of FVM is no longer significant, firstly, because the accuracy of all methods has been increased, so that for well-designed grids the error in the computation of fluxes is small compared to other errors in the course of simulations, and, secondly, because advanced grid techniques, like certain types of unstructured grids or sliding grids, do not uphold this advantage exactly. Nevertheless, it is still the most widely distributed CFD method and is therefore described here as an example. More details can be found, e.g., in [27 – 29].

Discretization in Space. We first consider the general shape of a transport equation for the quantity ϕ. For purposes of simplicity only the two-dimensional version is used, but the extension to three dimensions is straightforward. Discretization in time as discussed later, so we consider here the stationary form. The velocity field is assumed to be known as well as the molecular transport coefficient Γ, which might vary in a known way over the analyzed space.

$$
\begin{aligned}
0 = & -v_x \frac{\partial \phi}{\partial x} - v_y \frac{\partial \phi}{\partial y} \\
& + \frac{\partial}{\partial x}\left(\Gamma \frac{\partial \phi}{\partial x}\right) + \frac{\partial}{\partial y}\left(\Gamma \frac{\partial \phi}{\partial y}\right) + S_\phi
\end{aligned}
\qquad (25)
$$

For reasons of simplicity the integration is shown here for an equidistantly structured grid. But there is no general restriction — finite volume discretization can also be carried out for nonequidistant and unstructured grids. Note that in the two-dimensional case the "volume" for integration is a face and the "volume surface" is a line.

For the grid the compass notation shown in Figure 5 is normally used [27]. The nodes are marked with capital letters, *P* being that of the cell considered. The points at cell faces are marked by lower-case letters. The final linear equation system can contain only values with

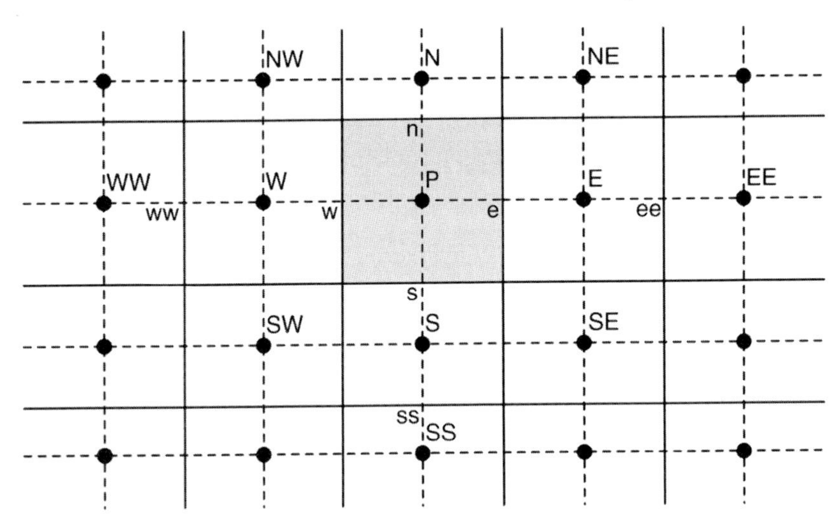

Figure 5. Nomenclature in a two-dimensional grid for finite volume method

capital letter indices, values with lower case indices are not defined and must be replaced by the former.

Integration of Equation 25 over the control volume gives

$$
\begin{aligned}
0 = &-\int\limits_V v_x \frac{\partial \phi}{\partial x} dV - \int\limits_V v_y \frac{\partial \phi}{\partial y} dV \\
&+ \int\limits_V \frac{\partial}{\partial x}\left(\Gamma \frac{\partial \phi}{\partial x}\right) dV \\
&+ \int\limits_V \frac{\partial}{\partial y}\left(\Gamma \frac{\partial \phi}{\partial y}\right) dV \\
&+ \int\limits_V S_\phi dV
\end{aligned} \qquad (26)
$$

Conversion of volume integrals of the transport terms and resolution lead to:

$$
\begin{aligned}
0 = &-\int\limits_A \boldsymbol{n} v_x \phi dA - \int\limits_A \boldsymbol{n} v_y \phi dA \\
&+ \int\limits_A \boldsymbol{n}\left(\Gamma \frac{\partial \phi}{\partial x}\right) dA + \int\limits_A \boldsymbol{n}\left(\Gamma \frac{\partial \phi}{\partial y}\right) dA \\
&+ \int\limits_V S_\phi dV \\
0 = &-(v_x A\phi)_{\mathrm{e}} + (v_x A\phi)_{\mathrm{w}} - (v_y A\phi)_{\mathrm{n}} + (v_y A\phi)_{\mathrm{s}} \\
&+ \left(\Gamma A \frac{\partial \phi}{\partial x}\right)_{\mathrm{e}} - \left(\Gamma A \frac{\partial \phi}{\partial x}\right)_{\mathrm{w}} \\
&+ \left(\Gamma A \frac{\partial \phi}{\partial y}\right)_{\mathrm{n}} - \left(\Gamma A \frac{\partial \phi}{\partial y}\right)_{\mathrm{s}} \overline{S} V
\end{aligned} \qquad (27)
$$

A_{e}, A_{w}, A_{n} and A_{s} are the cell faces crossing e, w, n and s and $\boldsymbol{n}$ is the face normal.

Finally, the values and the derivatives of ϕ at the cell faces must be replaced as a function of the values at the nodes. The derivatives are approximated by using the central difference, which is accurate to the second order, e.g.

$$\left(\Gamma A \frac{\partial \phi}{\partial x}\right)_w = \Gamma_w A_w \left(\frac{\phi_P - \phi_W}{\delta x_{WP}}\right) \qquad (28)$$

where δx_{wp} is the distance between the points W and P.

To express the values of ϕ various approaches are used which differ in their stability and accuracy. Most of them are asymmetric and depend therefore on the flow direction. We discuss them for cell face w and a velocity $\nu_{\mathrm{x,w}}$ from W to P.

- *Upwind Differencing Scheme (UD)*

$$\phi_w = \phi_W \qquad (29)$$

This scheme is unconditionally stable, but it is only accurate for the first order. It is the only scheme which does not produce any overshots or undershots without additional limiters and therefore guarantees the solution to be in the physically meaningful range (no negative concentrations, etc.). On the other hand, the numerical error is relatively large. It has the effect of smoothing strong changes in the gradients—a property called *numerical diffusion*. The error can be minimized if the local velocity vector is nearly parallel to the face normal vector.

- *Central Differencing Scheme (CD)*

$$\phi_w = \frac{\delta x_{Ww}\phi_W + \delta x_{wP}\phi_P}{\delta x_{WP}} \quad (30)$$

This scheme is only conditionally stable but it is second-order. Central schemes normally give a better approximation of steep slopes, but they show significant nonphysical oscillations of the solution. In the worst case these oscillations can build up and lead to a diverging solution. CD schemes are stable and advisable for diffusion-dominated processes.

- *Quadratic Upstream Interpolation for Convective Kinetics (QUICK)*

$$\phi_w = -\frac{1}{8}\phi_{WW} + \frac{6}{8}\phi_W + \frac{3}{8}\phi_P \quad (31)$$

QUICK is the basis for most third-order schemes used today. With a combination of central and upwind aspects it joints properties of both schemes discuss above. Oscillations occur, but to a lesser extent than with the CD scheme. QUICK has a much wider range of stability than CD but restrictions remain for systems without significant diffusion terms. To improve this third-order scheme various advancements have been developed. The main advantage is gained from the combination with so-called *limiters*. Limiters are extensions of the numerical scheme which prevent the formation and pronunciation of local extrema but keep the numerical accuracy of a scheme.

For the source term a linear approach in ϕ is used:

$$\overline{S}V = S_0 + S_1\phi \quad (32)$$

Discretization in Time. For transient cases a time derivative of the transport quantity also occurs in the equation. If the two-dimensional Equation 25 is extended in this way one obtains:

$$\frac{\partial\phi}{\partial t} = -v_x\frac{\partial\phi}{\partial x} - v_y\frac{\partial\phi}{\partial y} + \frac{\partial}{\partial x}\left(\Gamma\frac{\partial\phi}{\partial x}\right) + \frac{\partial}{\partial y}\left(\Gamma\frac{\partial\phi}{\partial y}\right) + S_\phi \quad (33)$$

The philosophy for the time discretization is the same as for the discretization in space: The equation is integrated over the grid cell as well as over the time interval. The integration of the left-hand side can be performed easily:

$$\int_V \int_t^{t+\Delta t} \frac{\partial\phi}{\partial t} dt dV = V(\phi_P^{t+\Delta t} - \phi_P^t) \quad (34)$$

For the time integral of the right-hand side a linear relation to the values of ϕ at the old and the new time step is assumed:

$$\int_t^{t+\Delta t} \phi dt = \left[f\phi^{t+\Delta t} + (1-f)\phi^t\right]\Delta t \quad (35)$$

where f is a weighting factor between zero and one. For a space discretization which needs only the values of ϕ in P and its direct neighbor points the discretized equation reads:

$$\begin{aligned} a_P^{t+\Delta t}\phi_P^{t+\Delta t} &= a_E\left[f\phi_E^{t+\Delta t} + (1-f)\phi_E^t\right] \\ &+ a_W\left[f\phi_W^{t+\Delta t} + (1-f)\phi_W^t\right] \\ &+ a_N\left[f\phi_N^{t+\Delta t} + (1-f)\phi_N^t\right] \\ &+ a_S\left[f\phi_S^{t+\Delta t} + (1-f)\phi_S^t\right] \\ &+ a_P^t\phi_P^t \end{aligned} \quad (36)$$

in which the coefficients $a_{\mathrm{P}}^{t+\Delta t}$ and a_{P}^t depend on f linearly, and a_{i} are the coefficients resulting from space discretization and from Equation 34.

For different values of f different methods can be derived. The most common ones are shown in Figure 6 .

- *f=0 (explicit method)*
  This method is numerically simple to use because only independent linear equations are derived. It is of first-order accuracy. On the other hand, it is only conditionally stable. For applications in which the flux is dominated by the convective part (and this is the major part of all CFD applications) the *Courant–Friedrich–Lewy criterion* determines the stability range:

$$1 \geq C_\Delta = \frac{|\boldsymbol{v}|\,\Delta t}{\Delta x} \quad (37)$$

  C_Δ is the so-called *Courant number.*

- *f=1 (implicit method)*
  This method is unconditionally stable and also first-order accurate. If a final state of a process is more interesting than the transient behavior, this method can be used with relatively large time steps.

- *f=0.5 (Crank–Nicolson method)*
  This method is second-order accurate. It is

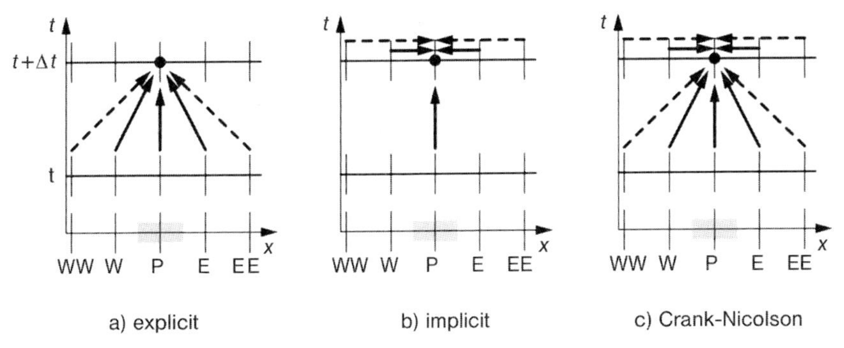

Figure 6. Discretization in time – coupling of values according to different methods
a) Explicit; b) Implicit; c) Crank-Nicolson

conditionally stable but larger Courant numbers than in the explicit method can be handled. It is used if a high accuracy in time is required.

4.3. Pressure Correction Methods

One of the crucial points of fluid dynamic modeling are the nonlinearities in the equation of motion and the lack of an explicit pressure equation for incompressible flow. This was outlined in connection with the equation of motion.

There are different approaches to deal with this problem. The directly coupled solution of the equation of motion and the equation of continuity is possible, but is mathematically unstable, especially for the finite volume method, and is seldom used. Also, the approach of an artificial compressibility is only successful and efficient in a few cases.

The most commonly used alternatives are pressure correction schemes. They break up the velocity and pressure into an estimated value and a corrective part. Starting with an estimated pressure field, they compute an estimated velocity field, or vice versa, using the equation of motion. For the velocities in front of the derivatives in the convective terms (see Equation 6) estimates are used. Reorganization of the discretized momentum balance gives an equation for the pressure correction as a function of velocity correction. With the help of this equation the discretized continuity equation can be transformed into a pressure correction equation. Iterative solution of all these equations lead to a stepwise approach of velocity and pressure. It is important to update the "parameter velocity" in the convective term and other dependent parameters in each iterative cycle. If necessary, coupled equations (e.g., turbulence models) must be solved at the end of each step.

One of the most basic pressure correction algorithms is the SIMPLE algorithm [27]. Because it does not converge stably, a significant under-relaxation has to be used. Reversed algorithms have been developed since then, but the underlying basic principles are the same.

4.4. Lattice Boltzmann Method

The lattice Boltzmann method (LBM) differs greatly from FVM because it is not a discretization of the transport equations but a statistical method which gives an average result for the solution of the equations. To provide sufficient statistics the method requires a relatively fine grid and small time steps, but it performs much faster on a fine grid than FVM does on the same grid. The grid used is completely regularly structured, but the location of wall boundaries is not restricted to the grid lines. Therefore, the method is suited for geometries with sophisticated wall shapes like porous media. While some years ago the grid requirements caused an impassable numerical effort, such a method can be handled today on parallel computers. It is especially suited for simulations which require a fine grid to handle the models, like large-eddy simulations of turbulent flows.

For lattice Boltzmann simulations a number of numerical particles with properties like velocity, temperature, and concentration are distributed over the grid in such a way that averaging of their properties gives the initial conditions of the flow. The simulation consists of a first step in which the particles move due to their individual velocities. In a second step collisions between particles and also between particles and walls are interpreted according to collision rules. The collision rules describe the change in the properties of each of the particles in the collision. They must be formulated in such a way that they cause statistical changes so that the averaged property field fulfills the transport equations. [30, 31]

As the method is still relatively new, not all models can be used yet, but the current development is rapid.

5. Interpretation

As mentioned in Chapter 2, the volume of computed data of a CFD simulation is too large to interpret the numbers directly. In a graphic visualization the data are much easier to comprehend.

A basic step of a suitable visualization is an appropriate reduction of the data. Three-dimensional transient data cannot be visualized at once, and the visualization of the velocity vector field can not be realized effectively in a two-dimensional plane. But not all aspects of the data field are relevant to answering the initial question. Relevant data, however, can be extracted, and data can be averaged or combined to derive meaningful results. To transfer information into a graphical presentation requires that the figure be clearly readable and quantitative. Aspects of data reduction (location of a slice, chosen velocity components) have to be marked, and a legend must be given which allows quantitative interpretation of colored figures.

Visualization is not only useful for finding the result of the simulation. It also helps to detect errors. For this purpose the whole data field should at least be scanned roughly. As the aim of this procedure is to find irregularities in the solution, the layout of the presentation is of minor interest, but a quantitative interpretation of the data must be possible.

To judge the results all errors of the simulation must be analyzed as well as possible and quantified. The main categories are:

- *Modeling errors* refer to the differences between the values of all quantities in a real physical system and the values resulting from an exact solution of the mathematical model. Not all physical relations can be reproduced exactly in a mathematical formulation. Furthermore, not all mathematical models can be solved numerically with reasonable effort. Therefore, additional assumptions and restrictions are introduced which increase the modeling error. Examples are the application of turbulence models, the neglect of temperature dependencies of parameters or simplified approaches for boundary conditions.
  A quantitative estimate of the modeling errors is very difficult. If the mathematical model can be solved analytically (at least for a limiting case) and the corresponding quantities can be measured, a direct comparison is possible. Even in this case the measurement error from the experiment must be considered. In most cases such a direct comparison is not possible and one is restricted to experience from similar cases or to comparison of different model approaches for the same situation.
- *Discretization errors* refer to the difference between the solution of the continuous mathematical model and the discretized equations. The discrete solution is computed only for the nodes (which are representative for a certain volume according to FVM) and for time steps. For the locations between these points in space and time only approximations based on the values at the nodes can be used.
  The solution of the discrete equations at the nodes differs from the solution of the continuous equations at the same points. This is caused by the assumptions and approaches used in the course of discretization. While this is valid for all discretization schemes, it can be most easily explained for the finite difference method (FDM) (see → Mathematics in Chemical Engineering, Chap. 7.3 or [32]).
  According to this method, the approximation is derived from the Taylor series where only the first terms are considered. The neglected terms give the discretization error. The procedure is based on the assumption that the following terms in the Taylor series are decreasing. Therefore, the discretization error can be approximated by the first neglected term. This term is proportional to the grid size with a certain power. This power is the so-called *order*

of accuracy. The more terms considered the higher the order of the scheme. The order of a scheme is also used for other discretization methods. It states by which power the discretization error is reduced in relation to the reduction of the grid size. This means that a scheme of higher order is not always more accurate than a scheme of lower order if the schemes are derived in completely different manners. But the influence of grid size reduction is greater for the scheme of higher order.

- *Truncation of iterative procedures*: Iterative procedures are used on different levels for the solution of the discrete equations. Normally the linear algebraic equation systems are solved iteratively. Also, the coupling between the equations is often handled iteratively. The most important example of this are the pressure correction schemes (see Section 4.3). Iterative procedures approach the solution asymptotically. They are terminated if a certain accuracy is reached. Criterion for termination is the residual. It can be defined in two different ways. The first alternative is the difference in the solution between two iterations. This value is easily determined, but it is dangerous, because for slow convergence the differences between two iterations is small, even far away from the solution. The second alternative can be used if the problem can be formulated as:

$$\boldsymbol{F}(x) = 0 \tag{38}$$

The residual of the nth iteration $\boldsymbol{r}^n$ is then defined as:

$$\boldsymbol{F}(\boldsymbol{x}^n) = \boldsymbol{r}^n \tag{39}$$

Both definitions of the residual give values for each node. For a truncation criterion different statistical methods are used to derive a characteristic value. This can be either the maximum, the average, or a weighted average of the value itself or of a normalized value.
- *Representation errors*: In the computer all numbers are represented by a limited number of digits. This might result in significant errors, e.g., if a small difference of large numbers is computed. Normally, this type of error is of minor importance in CFD simulations.

The different types of errors are usually additive. Therefore, it does not help to improve the accuracy of one aspect if the error in this regard is smaller than other errors. In most cases for simulation the limiting factor for error reduction is the modeling error. All other errors are usually small by comparison. A quantitative judgment has a higher priority than indifferent formulations such as "grid independence".

6. Industrial Application

CFD is a tool of increasing importance in industrial applications. The expertise of industrial users is mainly in the field of fluid dynamics or chemical engineering, not in numerics or programming. Therefore, industrial problems are solved with commercial codes which are available at high level with regard to modeling and numerics. Furthermore, such codes have a user-friendly interface and support which allows even inexperienced users to obtain first results quite fast. Commercial codes are continuously being developed and achieve more or less the state of the art. Nevertheless, the different codes are specialized in different fields of applications and therefore it is necessary to check the available models before a new kind of application is tackled. Often additional models can be implemented by user coding, but the effort required to do so is great and it should be checked before beginning whether another code provides a better foundation for the case considered.

To illustrate different modeling requirements, some examples for applications in chemical engineering are presented. This cannot give a complete overview of the most current usage of CFD in chemical engineering but gives an impression of the variety of applications.

- Tubular reactor: Tubes without internals have a simple geometric structure. They are a part of nearly all technical apparatuses as inlet and outlet pipes and in some cases they are used as reactors. Problems in modeling occur due to the large ratio between length and diameter, which requires a compromise in meshing between number of cells and shape of cells. In some cases the number of cells can be reduced by utilizing symmetry properties of the system.
  Tubular reactors are often used to mix non-Newtonian fluids or, in a jet configuration, to

mix Newtonian fluids very fast down to microscopic levels. For non-Newtonian fluids, e.g., in polymerization processes, chaotic mixers [33] consisting of tubular loops or static mixers [34, 35] within the tubes are used. In both cases the geometry of the flow becomes much more complicated. Especially for static mixers sufficient flexibility is only provided by unstructured grids.
In jet flows the modeling of turbulence is of great importance. As tubular reactors are relatively narrow, wall effects have a significant influence on the flow. Therefore, in most cases special wall models have to be used, and this increases the modeling and numerical effort. Furthermore, turbulence in tubes is often nonisotropic. This reduces the applicability of RANS simulations.

- Stirred vessel: Stirred vessels are widely used in chemical engineering for single phase and multiphase flows. The flow field is markedly inhomogeneous and transient. Characteristic of all stirred vessel configurations is the combination of static and moving elements. In the simplest case the static parts are rotationally symmetric. Then they can be modeled as a moving wall in a rotating reference frame [36]. If there are baffles, the rotational symmetry is broken and extended approaches are used. A good overview on these approaches is given by in [37]. If only the stationary solution averaged over a large number of stirrer revolutions is at interest, the power input from the stirrer can be included in source terms. This system can be solved for the stationary case on a fixed grid. For more detailed simulations the grid is divided in two overlapping or nonoverlapping parts. One of them moves with the stirrer; the other is static. As the stirrer motion has to be resolved sufficiently in time, only small time steps can be used, and the simulation becomes relatively extensive.
  RANS simulations often do not give a sufficient description, especially in presence of chemical reactions, because the flow field is inhomogeneous. As the fine resolution required for large-eddy simulations is numerically extensive, a remedy is to apply the lattice Boltzmann method [38].
- Precipitation reactor: Precipitation is usually simulated to predict the particle size distribution in the system [39, 40]. The initial supersaturation in precipitation processes is high, and therefore nucleation and particle growth start at a high rate. This requires appropriate turbulence modeling, including micromixing models [41]. The particles are very small, so only one-way coupling is relevant and the simulation can be carried out as pseudo-single phase. On the other hand, the solution of the population balance must be integrated into the simulation [15].
- Bubble column: Bubble columns are a typical example for two-phase flows. The difference in density between gas and liquid causes a relevant relative velocity between the phases. Depending on bubble size and volume fraction, either Euler–Euler [42, 43] or Euler–Lagrange [44, 45] approaches are used. In most practical applications the requirements for the Lagrangian approach are not met because the volume fraction is too high. But to describe the nonuniform bubble size population, balances have to be solved. This is much easier in connection with the Lagrangian approach than with the Eulerian approach.
- Membrane module: A newer apparatus used in chemical engineering is the membrane module. Its numerical simulation is at the very beginning because of the variety of aspects which have to be considered. These are, for instance:
  Presently the modeling approaches do not consider all of these aspects. The main stress is on modeling the two flow regions divided by the porous membrane under certain pressure conditions [46, 47].
  - Flow through a porous medium
  - Coupling of two flow regions
  - Multiphase flow
  - Non-Newtonian fluids
  - Fouling

It is obvious from this listing that the qualified application of CFD can not be performed without well-founded knowledge of applicable models and some insight into the numerical procedures.

The development of refined products like fine chemicals and active agents will be of growing importance in the chemical industry, and CFD will play an important part in the design of the necessary apparatus. This requires the consideration of parallel and consecutive chemical reactions of different rates and simulations of multiphase systems including the prediction of particle size distributions. Consequently, emphasis in CFD development for this field will be on models and solution methods for fine spatial and temporal resolution and for a detailed description of the physics.

The advances in CFD interact strongly with hardware development. On the one hand the available computing power determines which mathematical models and which numerical methods can be used with reasonable numerical effort, and on the other it affects the direction of development of new models and methods. Due to the fast pace at which new powerful computer hardware evolves and becomes available, model development does not restrict itself to quantitative refinement but regularly includes new fields of research. More recent improvements are large-eddy simulations of turbulent flows instead of Reynolds-averaged approaches, and models for multiphase systems. A newer trend in the field of numerical methods is the application of the lattice Boltzmann method. Further innovations can be expected.

7. References

1. R.B. Bird, W.E. Steward, E.N. Lightfoot: *Transport Phenomena*, 2nd ed., John Wiley & Sons, New York 2002.
2. S. Pope: *Turbulent Flows*, Cambridge University Press, Cambridge 2000.
3. J.V. Boussinesq: "Essa sur la theories des eaux courantes", *Mém. prés. par div. savants à l'acad. sci. de Paris* **23** (1877) 1.
4. B.E. Launder, D.B. Spalding: "The Numerical Computation of Turbulent Flows", *Comput. Methods Appl. Mech. Eng.* **3** (1974) 269.
5. J. Smagorinsky: "General Circulation Experiments with the Primitive Equations, Part I: The Basic Experiment", *Monthly Weather Rev.* **91** (1963) 99.
6. J. Bardina, J.H. Ferziger, W.C. Reynolds: Improved Turbulence Models Based on Large Eddy Simulation of Homogeneous, Incompressible, Turbulent Flows, Technical Report TF-19, Thermal Sciences Div., Dept. of Mech. Eng., Stanford University, Stanford, CA. 1980.
7. M. Germano, U. Piomelli, P. Moin, W.H. Cabot: "A Dynamic Subgrid Scale Eddy Viscosity Model", in Proc. Summer Workshop, Center for Turbulent Research, Stanford CA 1990.
8. C. Meneveau, T.S. Lund, W.H. Cabot: "A Lagrangian Dynamic Subgrid-Scale Model of Turbulence", *J. Fluid Mech.* **319** (1996) 353.
9. J. Baldyga: "A Closure Model for Homogeneous Chemical Reactions", *Chem. Eng. Sci.* **49** (1994) 1985.
10. R.O. Fox: "On the Relationship between Lagrangian Micromixing Models and Computational Fluid Dynamics", *Chem. Eng. Process* **37** (1998) 521.
11. B.F. Magnussen, B.W. Hjertager: "On the Structure of Turbulence and a Generalised Eddy Dissipation Concept for Chemical Reaction in Turbulent Flow" in *19th AIAA Aerospace Meeting*, St. Louis, USA, 1981.
12. A. Linan: On the Internal Structure of Laminar Diffusion Flames, Technical Note, Inst. nac. de tec. aeron., Esteban Terradas, Madrid, Spain, 1961.
13. N. Peters: "A Spectral Closure for Premixed Turbulent Combustion in the Flamelet Regime", *J. Fluid Mech.* **242** (1992) 611.
14. J. Baldyga, J.R. Bourne: *Turbulent Mixing and Chemical Reactions*, John Wiley & Sons Ltd., Chichester 1999.
15. A.R. Paschedag: *CFD in der Verfahrenstechnik: Allgemeine Grundlagen und mehrphasige Anwendungen*, Wiley-VCH, Weinheim, Germany 2004.
16. D. Ramkrishna: *Population Balances. Theory and Application to Particulate Systems in Engineering*, Academic Press, San Diego 2000.
17. A. Gerstlauer, A. Mitrovic, S. Motz, E.-D. Gilles: "A Population Balance Model for Crystallization Processes using two Independent Particle Properties", *Chem. Eng. Sci.* **56** (2001) 2553.
18. W.M. Deen: *Analysis of Transport Phenomena*, Oxford University Press, New York 1998.
19. J.C. Slattery: *Advanced Transport Phenomena*, Cambridge University Press, New York 1999.
20. D.A. Drew: "Mathematical Modeling of Two-Phase Flow", *Ann. Rev. Fluid Mech.* **15** (1983) 261.
21. C.T. Crowe., M. Sommerfeld, Y. Tsuji: *Multiphase Flows with Droplets and Particles*, CRC Press, Boca Raton 1998.
22. M. Sommerfeld, W. Krebs: "Particle Dispersion in a Swirling Confined Jet Flow", *Part. Part. Syst. Characterization* **7** (1990) 16.

23. J.H. Ferziger, M. Peric: *Computational Methods for Fluid Dynamics*, 2nd rev. ed., Springer, Berlin 1999.
24. B.D. Nichols, C.W. Hirt: "Calculating Three-Dimensional Free Surface Flows in the Vicinity of Submerged and Exposed Structures", *J. Comput. Phys.* **8** (1971) 434.
25. F.H. Harlow, J.E. Welch: "Numerical Calculation of Time-Dependent Viscous Incompressible Flow of Fluid with Free Surface", *Phys. Fluids* **8** (1965) 2182.
26. C.W. Hirt, B.D. Nichols: "Volume of Fluid (VOF) Method for the Dynamics of Free Boundaries", *J. Comput. Phys.* **39** (1981) 201.
27. S.V. Patankar: *Numerical Heat Transfer and Fluid Flow*, Hemisphere Publishing Co., Washington 1980.
28. H.K. Versteeg, W. Malalasekera: *An Intr. to Computational Fluid Dynamics. The Finite Volume Method*, Longman, Harlow 1995.
29. C.A.J. Fletcher: *Computational Techniques for Fluid Dynamics*, vols. **1** and **2**, Springer, New York 1988.
30. U. Frisch, D. d'Humieres, B. Hasslacher: "Lattice Gas Hydrodynamics in Two and Three Dimensions", *Complex Syst.* **1** (1987) 649.
31. Succi S.: *The Lattice Boltzmann Equation for Fluid Dynamics and Beyond*, Oxford University Press, Oxford 2001.
32. J.D. Anderson Jr.: *Computational Fluid Dynamics: the Basics with Applications*, McGraw-Hill, New York 1998.
33. A. Birtigh, G. Lauschke, W.F. Schierholz, D. Beck, Ch. Maul, N. Gilbert, H.-G. Wagner, C.Y. Werninger: "CFD in der chemischen Verfahrenstechnik aus industrieller Sicht", *Chem. Ing. Tech.* **72/3** (2000) 175.
34. Th. Avalosse, M.J. Crochet: "Finite Element Simulation of Mixing: 2. Three-Dimensional Flow through a Kenics Mixer", *AIChE J.* **43/3** (1997) 563.
35. E.S. Mickaily-Huber, F. Bertrand, P. Tanguy, T. Meyer, A. Renken, F.S. Rys, M. Wehrli: "Numerical Simulations of Mixing in an SMRX Static Mixer", *Chem. Eng. J.* **63** (1996) 117.
36. F. Bertrand, P.A. Tanguy, E. Brito de la Fuente, P. Carreau: "Numerical Modelling of the Mixing Flow of Second-Order Fluids with Helical Ribbon Impellers", *Comput. Meth. Appl. Mech. Eng.* **180** (1999) 267.
37. A. Brucato, M. Ciofalo, F. Grisafi, G. Micale: "Numerical Prediction of Flow Fields in Baffled Stirred Vessels: A Comparison of Alternative Modelling Approaches", *Chem. Eng. Sci.* **53/21** (1998) 3653.
38. J. Derksen, H.E.A. Van den Akker: "Large Eddy Simulations on the Flow Driven by a Rushton Turbine", *AIChE J.* **45/2** (1999) 209.
39. D.L. Marchisio, A.A. Barresi, R.O. Fox: "Simulation of Turbulent Precipitation in a Semi-batch Taylor-Couette Reactor Using CFD", *AIChE J.* **47/3** (2001) 664.
40. A.R. Paschedag: "Modelling of Mixing and Precipitation Using CFD and Population Balances", *Chem. Eng. Tech.* **27/3** (2004) 232.
41. D.L. Marchisio: Precipitation in Turbulent Fluids, PhD thesis, Politecnico di Torino, 2002.
42. A. Sokolichin, G. Eigenberger: "Applicability of the Standard $k-\varepsilon$ Turbulence Model on the Dynamic Simulation of Bubble Columns: Part I Detailed Numerical Simulations", *Chem. Eng. Sci.* **54** (1999) 2273.
43. L.I. Zaichik, V.M. Alipchenkov: "A Kinetic Model for the Transport of Arbitrary Density Particles in Turbulent Shear Flows", in: Proc. Turbulence and Shear Flow Phenomena 1, Santa Barbara 1999.
44. A. Lapin, A. Lübbert: "Numerical Simulation of the Dynamics of Two-Phase Gas-Liquid Flows in Bubble Columns", *Chem. Eng. Sci.* **49** (1994) 3661.
45. L. Sanyal, S.V. Squez, S. Roy, M.P. Dudukovic: "Numerical Simulation of Gas-Liquid Dynamics in Cylindrical Bubble Column Reactors", *Chem. Eng. Sci.* **54** (1999) 5071.
46. D.E. Wiley, D.F. Fletcher: "Techniques for Computational Fluid Dynamics Modelling of Flow in Membrane Channels", *J. Membrane Sci.* **211** (2003) 127.
47. C.A. Serra, M.R. Wiesner: "A comparison of Rotating and Stationary Membrane Disk Filters using Computational Fluid Dynamics", *J. Membrane Sci.* **165** (2000) 19.

Design of Experiments

Sergio Soravia, Process Technology, Degussa AG, Hanau, Germany (Chap. 1, 2, 3 and 8)

Andreas Orth, University of Applied Sciences, Frankfurt am Main, Germany (Chap. 4, 5, 6, 7 and 8)

1. **Introduction** . . . 363
1.1. **General Remarks** . . . 363
1.2. **Application in Industry** . . . 364
1.3. **Historical Sidelights** . . . 365
1.4. **Aim and Scope** . . . 365
2. **Procedure for Conducting Experimental Investigations: Basic Principles** . . . 365
2.1. **System Analysis and Clear Definition of Objectives** . . . 366
2.2. **Response Variables and Experimental Factors** . . . 366
2.3. **Replication, Blocking, and Randomization** . . . 367
2.4. **Interactions** . . . 368
2.5. **Different Experimental Strategies** . . . 368
2.6. **Drawback of the One-Factor-at-a-Time Method** . . . 369
3. **Factorial Designs** . . . 370
3.1. **Basic Concepts** . . . 370
3.2. **The 2^2 Factorial Design** . . . 371
3.3. **The 2^3 Factorial Design** . . . 374
3.4. **Fractional Factorial Designs** . . . 377
4. **Response Surface Designs** . . . 380
4.1. **The Idea of Using Basic Empirical Models** . . . 381
4.2. **The Class of Models Used in DoE** . . . 381
4.3. **Standard DoE Models and Corresponding Designs** . . . 382
4.4. **Using Regression Analysis to Fit Models to Experimental Data** . . . 384
5. **Methods for Assessing, Improving, and Visualizing Models** . . . 384
5.1. **R^2 Regression Measure and Q^2 Prediction Measure** . . . 385
5.2. **ANOVA (Analysis of Variance) and Lack-of-Fit Test** . . . 386
5.3. **Analysis of Observations and Residuals** . . . 388
5.4. **Heuristics for Improving Model Performance** . . . 388
5.5. **Graphical Visualization of Response Surfaces** . . . 389
6. **Optimization Methods** . . . 390
6.1. **Basic EVOP Approach Using Factorial Designs** . . . 390
6.2. **Model-Based Approach** . . . 391
6.3. **Multi-Response Optimization with Desirability Functions** . . . 391
6.4. **Validation of Predicted Optima** . . . 392
7. **Designs for Special Purposes** . . . 393
7.1. **Mixture Designs** . . . 393
7.2. **Designs for Categorical Factors** . . . 395
7.3. **Optimal Designs** . . . 397
7.4. **Robust Design as a Tool for Quality Engineering** . . . 398
8. **Software** . . . 399
9. **References** . . . 399

1. Introduction

1.1. General Remarks

Research and development in the academic or industrial context makes extensive use of experimentation to gain a better understanding of a process or system under study. The methodology of *Design of Experiments (DoE)* provides proven strategies and methods of experimental design for performing and analyzing test series in a systematic and efficient way. All experimental parameters are varied in an intelligent and balanced fashion so that a maximum of information is gained from the analysis of the experimental results. In most cases, the time and money spent on the experimental investigation will be greatly reduced. In all cases, an optimal ratio between the number of experimental trials and the information content of the results will be achieved.

DoE is a powerful target-oriented tool. If it is properly employed, creative minds with a scientific and technical background will best deploy their resources to reach a well-defined

Ullmann's Modeling and Simulation

ISBN: 978-3-527-31605-2

goal of their studies. In contrast to what researchers sometimes fear, experimenters will not be hampered in their creativity, but will be empowered for structuring their innovative ideas. Of course, adopting DoE requires discipline from the user, and it has proved very helpful to take the initial steps together with an expert with experience in the field. The rewards of this systematic approach are useful, reliable, and well-documented results in a clear time and cost frame. A comprehensible presentation and documentation of experimental investigations is gratefully acknowledged by colleagues or successors in research and development teams. The application of DoE is particularly essential and indispensable when processes involving many factors or parameters are the subject of empirical investigations of cause – effect relationships.

DoE is a scientific approach to experimentation which incorporates statistical principles. This ensures an objective investigation, so that valid and convincing conclusions can be drawn from an experimental study. In particular, an honest approach to dealing with process and measurement errors is encouraged, since experiments that are repeated under identical conditions will seldom lead to the same results. This may be caused by the measuring equipment, the experimenter, changes in ambient conditions, or the natural variability of the object under study. Note that this inherent experimental error, in general, comprises more than the bare repeatability and reproducibility of a measurement system. DoE provides basic principles to distinguish between experimental error and a real effect caused by consciously changing experimental conditions. This prevents experimenters from drawing erroneous conclusions and, as a consequence, from making wrong decisions.

1.2. Application in Industry

In industry, increasingly harsh market conditions force companies to make every effort to reach and maintain their competitive edge. This applies, in particular, in view of the following goals:

- Quality of products and services in conformance to market requirements
- Low costs to ensure adequate profits
- Short development periods for new or improved products and production processes (time to market)

Quality engineering techniques are powerful elements of modern quality management systems and make it possible to reach these goals. One important challenge in this context is not to ensure quality downstream at the end of the production line, but to ensure product quality by a stable and capable production process which is under control. By this means, ongoing tests and checks to prove that the product conforms to specification requirements are avoided. This can be realized by knowing the important and critical parameters or factors governing the system and through the implementation of intelligent process management strategies.

A methodical approach, sometimes referred to as off-line quality engineering [1, 2], focuses even further upstream. By considering quality-relevant aspects in the early stages of product and process development, quality is ensured preventively in terms of fault prevention [3, 4]. Naturally, there is considerable cost-saving potential in the early stages of product and process design, where manufacturing costs are fixed to a large extent. The losses incurred for each design modification of a product or process gradually increase with time. In addition, design errors with the most serious consequences are known to be committed in these early stages.

As an outstanding quality engineering tool, DoE occupies a key position in this context. The emphasis is on engineering quality into the products and processes. At the same time, DoE opens up great economic potential during the entire development and improvement period. It is well-known that the implementation and use of corresponding methods increases competitiveness [5 – 9]. DoE is applied successfully in all high-technology branches of industry. In the process industry, it makes essential contributions to optimizing a large variety of procedures and covers the entire lifecycle of a product, starting from product design in chemical research (e.g., screening of raw materials, finding the best mixture or formulation, optimizing a chemical reaction), via process development in process engineering (e.g., test novel technological solutions, determine best operating conditions, optimize the performance of processes), up to production (e.g., start up a plant smoothly, find an operating window which meets customer requirements at low cost, high capacity, and under stable conditions) and application technology (give competent advice concerning the application of the product, customize the properties of products to specific customer needs). In particular, technologies like high-throughput screening or combinatorial synthesis with automated workstations require DoE to employ resources reasonably.

1.3. Historical Sidelights

The foundations of modern statistical experimental design methods were laid in the 1920s by R. A. FISHER [10] in Great Britain and were first used in areas such as agricultural science, biology, and medicine. By the 1950s and 1960s, some of these methods had already spread into chemical research and process development where they were successfully employed [11 – 13]. During this period and later, G. E. P. BOX et al. made essential contributions to the advancement and application of this methodology [5, 14 – 18].

In 1960, J. KIEFER and J. WOLFOWITZ initiated a profound research of the mathematical theory behind optimal designs, and in the early 1970s the first efficient algorithms for so-called D-optimal designs were developed.

Around the same time, G. TAGUCHI integrated DoE methods into the product and process development of numerous Japanese companies [19]. One of his key ideas is the concept of robust design [1, 2, 4]. It involves designing products and processes with suitable parameters and parameter settings so that their functionality remains as insensitive as possible to unavoidable disturbing influences. Taguchi's ideas were discussed fruitfully in the United States during the 1980s [1, 5, 20, 21] and caused an increasing interest in this subject in Western industries. At the same time, another set of tools, which in general is not suitable for chemical or chemical engineering applications, became popular under the name of D. SHAININ [22, 23].

The 1980s also saw the advent and spread of software for DoE. Various powerful software tools have been commercially available for several years now. They essentially support the generation, analysis, and documentation of experimental designs. Experience has shown that such software can be used by experimenters once the basic principles of the methods applied have been learned. A list of software tools is given in Chapter 8.

Despite these stimulating developments, the majority of scientists and engineers in industry and academic research have still not yet used DoE.

1.4. Aim and Scope

The target group of this article consists of scientists and engineers in industry or academia. The intention is

- To give an insight into the basic principles and strategies of experimental design
- To convey an understanding of the important design and analysis methods
- To give an overview of optimization techniques and some advanced methods
- To demonstrate basically the power of proven methods of DoE and to encourage their use in practice

References for further reading and detailed study are given on all subjects. In particular, [16, 24], and [25] may be regarded as standard monographs on DoE in their respective languages.

2. Procedure for Conducting Experimental Investigations: Basic Principles

DoE should become an integral part of the regular working tools of all experimenters. The methods provided by this approach may be used in experimental investigations on a smaller scale as well as in large projects which, depending on their importance and scope, may involve putting together an interdisciplinary team of appropriate size and composition. It proved to be advantageous to also include staff who take care of the equipment or facility on-site. They often bring in aspects and experiences with which the decision-makers involved are largely unfamiliar. Moreover, involving, for instance, laboratory staff in the planning phase of experiments has a positive effect on their motivation. A DoE

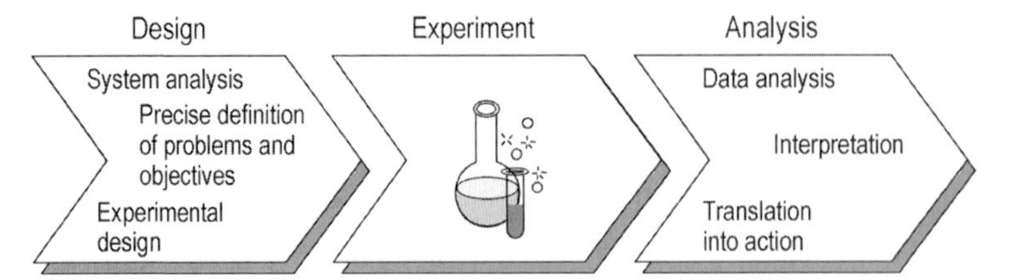

Figure 1. Three essential phases characterize the basic structure of a DoE project

project essentially subdivides into three characteristic phases: the design or planning phase, the actual experimental or realization phase, and the analysis or evaluation phase (Fig. 1). It should be stressed that the design phase is decisive because it is this phase that determines the level of information attainable by the analysis of the experimental results. For more information on this subject and on basic principles of DoE, see [26 – 28].

2.1. System Analysis and Clear Definition of Objectives

A system analysis involves collecting all existing information about the system to be examined and describing the current situation. A precise formulation of the problem and a clear definition of the objectives are very important prerequisites for a successful procedure and for attaining useful results from an experimental investigation. Experiments should never be conducted for their own sake but are to provide objective and reliable information, particularly, as a sound basis for decisions to be made. A clear statement of objectives is crucial, since the experimental strategy and hence the design of the experiments is essentially influenced by the goals to be reached (see Section 2.5). This sounds trivial but is frequently not handled carefully enough in practice. The actual planning or even the performance of experimental trials should not start until all of the aforementioned points have been settled satisfactorily.

2.2. Response Variables and Experimental Factors

Each experiment can be regarded as an inquiry addressed to a process or system (see Fig. 2). Naturally, it must be possible to record its answer or result in terms of measurable or quantifiable response variables (dependent variables). Response variables must be selected such that the characterization of the interesting properties of the system is as complete and simple as possible. Let us consider a batch reaction, for instance. In this case it is certainly not only the conversion of starting material that is of interest as a test result; other potential response variables could be the concentration of undesired byproducts or the duration of the reaction (capacity). For some system characteristics, such as foaming during stirring operations or the visual impression of a pigment, there are often no measurable quantities. In these cases, it may be helpful to assess the results by a subjective rating (e.g., 0: no foam or very beautiful pigment, 1: some foam or beautiful pigment, up to about 7: very much foam or very ugly pigment). When the response variables are selected, it is important to ask questions about the reliability of the corresponding values: How much do the results vary if the experimental runs are conducted under conditions that are as identical as possible? Ideally, a detailed analysis of the measurement system is available which, in particular, provides information on repeatability and reproducibility.

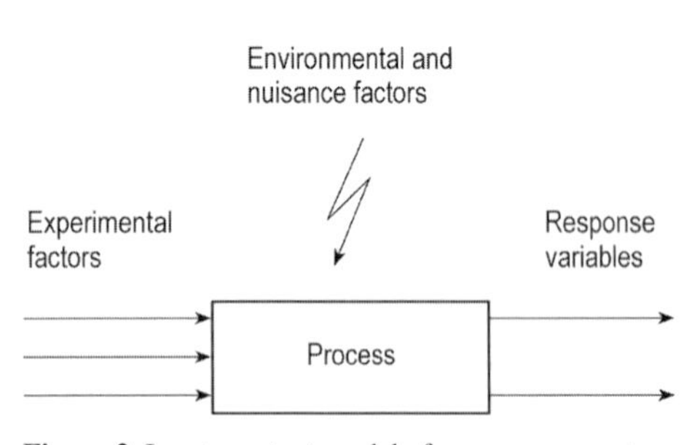

Figure 2. Input – output model of a process or system

The results of experimental runs, i.e., the values of response variables, are affected by various factors. In practically all applications there are disturbing environmental factors, which may cause undesired variations in the response variables. Some of them are hard to control or uncontrollable, others simply may not even be known. Examples of such variables are different batches of a starting material, various items of equipment with the same functionality, a change of experimenters, atmospheric conditions, and — last but not least — time-related trends such as warming-up of a machine, fouling of a heat exchanger, clogging of a filter, or drifting of a measuring instrument.

However, besides collecting and discussing the uncontrollable and disturbing variables, it is essential to collect and weigh up those factors or parameters that can be controlled or adjusted (independent variables), such as temperature, pressure, or the amount of catalyst used. The decision as to which of these experimental factors are to be kept constant during a test series and which are to be purposefully and systematically varied must also be carefully weighed up. The entire scientific and technical system know-how available by then from literature and experience, as well as intuition, must decisively influence not only the choice of the factors to be varied (one should focus on the important ones here according to the latest knowledge) but also the determination of the experimental region, i.e., of the specific range over which each factor will be varied (e.g., temperature between 120 and 180 °C and pressure between 1200 and 1800 mbar). A good experimental design will efficiently cover this experimental domain such that the questions related to the objectives may be answered when the experimental results are analyzed.

2.3. Replication, Blocking, and Randomization

To take the effects of disturbing environmental variables into account and decrease their impact to a large extent, the principles of *replication*, *blocking*, and *randomization* are employed. Moreover, by considering these principles, the risk of misleading interpretation of the results is minimized.

Replicates serve to make results more reliable and to obtain information about their variability. A genuine replicate should consider all possible environmental influences leading to variations within the response variables, i.e., the whole experiment with its corresponding factor-level combination should be repeated from the beginning and with some time delay in between. The results of data analyses must always be seen in the light of this inherent experimental error. Hence, replication does not mean, for instance, analyzing a sample of a product several times. This variability of a response variable is solely a measure of the precision of the analytical test procedure (laboratory assistant, measuring instrument).

It is of crucial importance that environmental factors and experimental factors of interest do not vary together such that changes in a response variable cannot be unambiguously attributed to the factors varied. For example, if two catalyst types were to be compared at various reaction temperatures and if two differently sized reactors were available for this purpose, it would be unwise to conduct all experiments with one catalyst in the smaller reactor and all experiments with the other catalyst in the larger reactor. If, in this case, the results differed from each other, it would be impossible to decide whether the catalyst type or the reactor type or both caused the deviations. The objective of blocking is to predetermine relatively similar blocks — in this case, the two reactors — in which test conditions are more homogeneous and which allow a more detailed study of the experimental factors of interest. Regarding the selection of the catalyst type and reaction temperature, the experiments to be conducted in each of the two reactors must be similar in the sense that variations in the values of a response variable can be interpreted correctly.

It is not always possible, however, to clearly identify unwanted influences and to take them into account, as is the case when blocking is used. Yet these side effects can be counterbalanced by a general use of randomization. Here, in contrast to systematically determining *which* experiments are to be conducted, *the order* of the experiments is randomized. In particular, false assignments of time-related trends are avoided. Let us consider a rectifying column, for instance, in which the effects of operating pressure, reflux ratio, and reboiler duty on the purity of the top product are to be examined. Let us assume that the unit is started up in the morning and that the whole test series could be realized within one day. Now, if one conducted all experiments involving a low reflux ratio before those involving a high reflux ratio, the effect of the reflux ratio could be falsified more or less by the unit's warming up, depending on how strong this influence is (poor design in Fig. 3). Such an uncontrolled mixing of effects is prevented by choosing the order of the experimental runs at random (good design in Fig. 3).

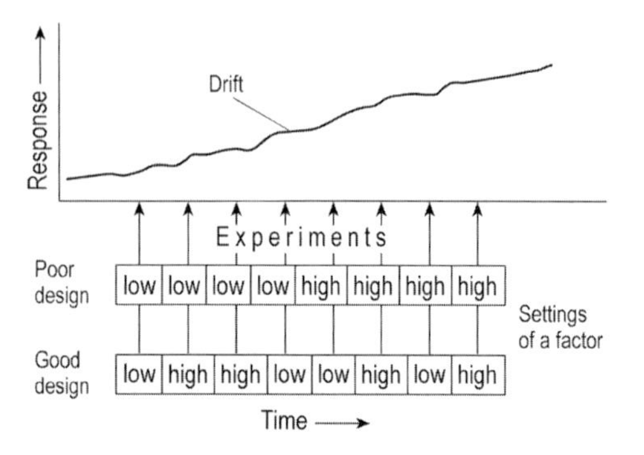

Figure 3. Time-related effects caused, e.g., by instrument drifts may falsify the analysis of the results when factor settings are not changed randomly (poor design). If environmental conditions vary during the course of an experiment, their effect will be damped or averaged out by randomizing the sequence of the experiments (good design)

The decisive reason for employing the principles of replication, blocking, and randomization is therefore to prevent the analysis of systematically varied experimental factors from being unnecessarily contaminated by the influences of unwanted and often hidden factors. While blocking and randomization basically do not involve additional experimental runs, each replicate is a completely new realization of a combination of factor settings. An appropriate relation between the number of experimental runs and the reliability of its results must be established here on an individual basis.

2.4. Interactions

To avoid an overly limited view on the behavior of systems, it is of great importance to know about the joint effects of experimental factors, that is, their *interactions*. Two variables are said to interact if, by changing one, the extent of impact on a third, namely, a response variable, depends on the setting of the other variable. In other words, interaction between two experimental factors measures how much the effect of a factor variation on a response variable depends on the level of the other factor. Interactions are often not heeded in practice, or they are studied at the price of spending large amounts of time and money on the associated experimental investigation. In addition, what interaction actually means is often not clearly understood. In particular, interaction is not to be confused with correlation. Two variables are said to be correlated if an increase of one variable tends to be associated with an increase or decrease of the other. Especially factorial experimental designs (see Chap. 3) allow, among other things, a quantitative determination of interactions between varied experimental factors.

2.5. Different Experimental Strategies

When processes are to be improved or novel technical solutions are to be tested, but also when plants are started up, several factors are often varied, in the hope of meeting with short-term success, by using an unsystematic iterative trial-and-error approach until satisfactory results are eventually produced. The expenditure involved quickly takes on unforeseeable proportions without affording important insights into the cause-and-effect relationships of the system. The result of this procedure is that, in the end, a comprehensible documentation is not available, and objective, reliable reasons for process operations or factor settings are missing. Furthermore, very little is known in most cases about the impact of factor variations. Experimenting in this way might be acceptable for orientation purposes in a kind of pre-experimental phase. However, one should switch to a judicious program as soon as possible.

To study causal relationships systematically, the experimental factors or parameters are usually varied separately and successively, and the values of a response variable (product, process, or quality characteristic), such as the yield of a chemical product, are shown in a diagram (see Fig. 4). This one-factor-at-a-time method (see

Section 2.6), however, provides only few insights into the subject under study because the effect of a particular factor is only known at a single factor-level combination of the other factors. The response variable may have quite another shape if the levels of the remaining factors are set differently. If the experimental factors in their effect on a response variable do not act additively according to the superposition principle, i.e., if the factors influence each other in their effect on the response variable by existing interactions, a misinterpretation of the results is easily possible, particularly when optimum conditions are to be attained.

When statistical experimental design methods are used, all considered factors are varied in a systematic and balanced way so that a maximum of information is gained from the analysis of the corresponding experiments. This may comprise the statistically sound quantitative determination of the effects of factor variations on one or several response variables (see Chap. 3) or a systematic optimization of factor settings (see Chap. 6). Depending on the experimenter's intention, the following questions can be answered:

- What are the most important factors of the system under investigation?
- To what extent and in which direction does a response variable of interest change when an experimental factor is varied?
- To what extent is the size and direction of the effect of a factor variation dependent on the settings of other experimental factors (interactions)?
- With which factor settings does one obtain a desired state of a response variable (maximum, minimum, nominal value)?
- How can this state be made insensitive to disturbing environmental factors or how can an undesired variability of a response variable be reduced (robust design)?

The question of which experimental strategy should be chosen from a comprehensive range of methods will be governed by the objectives to be achieved in each individual case, taking, e.g., system-inherent, financial, and time-related boundary conditions into account. Every project has its peculiarities. Carefully planned experiments cover the experimental region to be investigated as evenly as possible, while ensuring to the largest possible extent that changing values in a response variable can be attributed unambiguously to the right causes. Information of crucial importance is frequently obtained by a simple graphical analysis of the data without having to employ sophisticated statistical analysis methods, such as variance analysis or regression analysis. On the other hand, the best statistical analysis is not capable of retrieving useful information from a badly designed series of experiments. It is therefore decisive to consider basic DoE principles right from the beginning, above all, however, before conducting any experiments.

2.6. Drawback of the One-Factor-at-a-Time Method

A crystallization process is used in the following to illustrate the deficiency of the frequently used one-factor-at-a-time method. Factors influencing this system are, for instance, crystallization conditions such as geometry of the crystallizer, type and speed of the agitator, temperature, residence time, and concentrations of additives like crystallization and filter aids, as well as of two presumed additives A and B. Possible response variables may be bulk density, abrasion, hardness, and pourability of the crystallization product. Let us assume the simple case that the effects of the two experimental factors — additive A and additive B — on the material's bulk density are to be systematically examined with the aim of obtaining a maximum bulk density. As mentioned before, the experimental factors are usually examined and/or optimized separately and successively. In the example considered here, one would therefore begin by keeping factor B constant and varying A over a certain range until A has been optimally adjusted in terms of a maximum bulk density and enter the result in a diagram (see Fig. 4). The optimal value for A would then be selected and kept constant. The same procedure would then be employed for B. The result of this is a presumably optimal setting for A and B, and hasty experimenters would jump to the conclusion that, in this case, after varying A between 15 and 40 g/L and B between 5 and 17.5 g/L, the highest bulk density is obtained by setting A to 33 g/L and B to 8.5 g/L and that its value is approximately 825

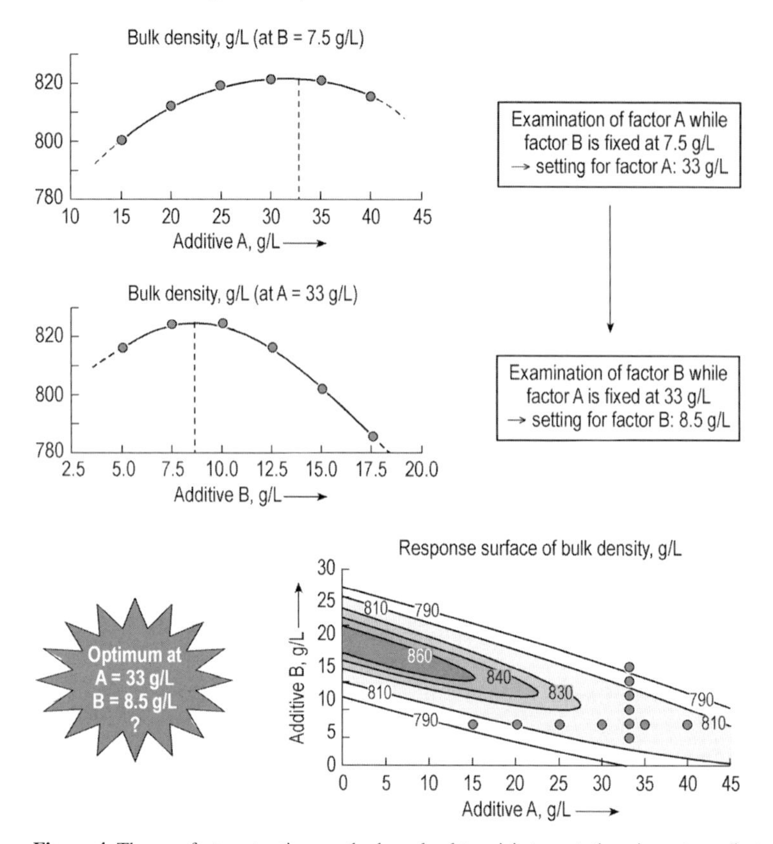

Figure 4. The one-factor-at-a-time method can lead to misinterpretations in systems that are subject to interactions

g/L. However, the response surface in Figure 4, which shows the complete relationship between both experimental factors and the response variable, reveals how misleading such a conclusion can be. This drastic misinterpretation is based on the, in this case, false assumption that the effect of varying one factor is independent of the settings of the other factor. For instance, by using the response surface, one can show that the bulk density values take on a decidedly different shape compared to the first diagram when varying A for B = 15 g/L. The following should be noted: If there are interactions between experimental factors, the one-factor-at-a-time method is an unsuitable tool for a systematic analysis of these factors, which holds true in particular when factor settings are to be optimized. If such interactions can definitely be ruled out, it might well be used. In chemistry, however, it is rather the rule that interactions occur.

3. Factorial Designs

3.1. Basic Concepts

The statistical experimental designs most frequently used in practice are the *two-level factorial designs*. These designs are called two-level because there are only two levels of settings, a lower (−) and an upper level (+), for each of the experimental factors. A full two-level factorial design specifies all combinations of the lower and upper levels of the factors as settings of the experimental runs (2^n design, where n denotes the number of factors). Their principle is illustrated by a simple example of a chemical reaction for which the influence of 2 (2^2 design) and 3 (2^3 design) experimental factors on the product yield is to be examined (Sections 3.2 and 3.3). For a growing number of factors, the number of runs of a full factorial design increases exponentially, and it provides much more information than is generally needed. Particularly for $n > 4$, the number of experimental settings can be reduced by selecting a well-defined subgroup of all 2^n possible settings of the factors without losing important information. This leads to the fractional factorial designs (Section 3.4).

The restriction of initially using just two levels for each experimental factor often causes some uneasiness for experimenters using this method for the first time. But by using two-level factorial designs, a balanced coverage of the interesting experimental region is achieved very economically. Moreover — owing to the special combination of factor levels — it is also possible to gain deeper insights from the associated individual values of the response variables. A decisive advantage of the two-level factorial designs is that they allow the effects of factor variations to be systematically and reliably analyzed and quantified and that they provide information on how these effects depend on the settings of the other experimental factors. These insights are gained by calculating so-called *main effects* and *interaction effects*. In the calculation of these effects, all experimental results can be used and are included to form well-defined differences of corresponding averages (see Figs. 6 and 9), thereby increasing the degree of reliability. The essential results of this effect analysis can be visualized by simple diagrams (see, e.g., Fig. 7).

In a factorial design, not only continuous experimental factors, such as temperature, pressure, and concentration, which can be set to any intermediate value, but also discrete or categorical factors, such as equipment or solvent type, may be involved. If at least one categorical variable with more than two levels is involved or if curvatures in the response variables are expected and to be explored, factorial designs with more than two levels may be used, e.g., 3^n designs, in which all factors are studied at three levels each, or hybrid factorial designs with mixed factor levels like the 2×3^2 design, in which one factor is varied at two levels, and two factors at three levels [24]. However, especially in the case of continuous factors, other so-called response surface designs are more efficient (see Chap. 4). In the following, the expression "factorial designs" always refers to two-level factorial designs.

For the sake of simplicity, replicates are neglected in the following examples, and variability in the process and in measurement are assumed to be very small. Note, however, that being aware of the impact of experimental error on the reliability or significance of calculated effects is an essential principle of DoE and crucial to drawing valid conclusions. Variability within individual runs having the same settings of the experimental factors will propagate and cause variability in each calculated variable, e.g., main effect or interaction effect, deduced from these single results. Experimental designs, particularly the factorial designs, minimize error propagation.

Factorial designs are treated in most textbooks on DoE, e.g., [16, 24, 25, 29, 30].

3.2. The 2^2 Factorial Design

Let us suppose the influence of two factors — catalyst quantity A and temperature B — on yield y of a product in a stirred tank reactor is to be examined. Figure 5 shows the two levels of both factors involved, as well as a tabular and a graphical representation of the associated experimental 2^2 factorial design. The — in this case two — columns which contain the settings of the experimental factors form the so-called design matrix. The resultant values of the response variable y obtained for the settings A−B−, A+B−, A−B+, and A+B+ are referred to as y_{A-B-}, y_{A+B-}, y_{A-B+}, and y_{A+B+} respectively. They are entered in the column of the response variable and in the corresponding positions of the graph.

Due to the special constellation of the experimental runs, it is possible to see how y changes when factor A is varied at the two levels of B and what happens when factor B is varied at the two levels of A. Figure 5 reveals that, at the lower temperature of 70 °C (B−), an increase of the catalyst quantity from 100 g (A−) to 150 g (A+) increases the response variable yield by 2 %, while at the higher temperature of 90 °C (B+), increasing the catalyst quantity enlarges the value of the response variable by 18 %, i.e.,

$$\begin{aligned}&\text{Effect of A for B}-\\&=E_{B-}(A)\\&=y_{A+B-}-y_{A-B-}\\&=+2\%\end{aligned}$$

$$\begin{aligned}&\text{Effect of A for B}+\\&=E_{B+}(A)\\&=y_{A+B+}-y_{A-B+}\\&=+18\%\end{aligned}$$

ID	Factor	Unit	Lower level (-)	Upper level (+)
A	Catalyst	g	100	150
B	Temperature	°C	70	90

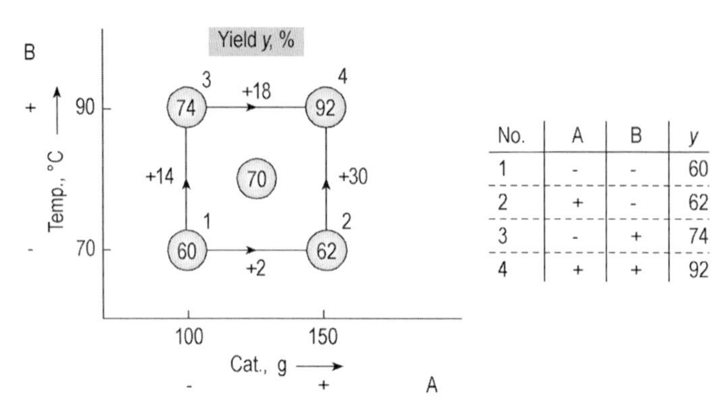

No.	A	B	y
1	-	-	60
2	+	-	62
3	-	+	74
4	+	+	92

Figure 5. Example of a 2^2 factorial design, including a graphical and a tabular representation of the experimental settings and results

Accordingly, the individual effects of a temperature increase can be determined for the different catalyst quantities:

Effect of B for A−

$$= \mathrm{E}_{A-}(\mathrm{B}) = y_{A-B+} - y_{A-B-} = +14\%$$

Effect of B for A+

$$= \mathrm{E}_{A+}(\mathrm{B}) = y_{A+B+} - y_{A+B-} = +30\%$$

As mentioned before, key results of a factorial design are obtained by the calculation of main effects and interaction effects (Fig. 6). The main effect of a factor is a measure of the extent to which a response variable changes on average when this factor is varied from its lower to its upper level. To calculate this main effect, all experimental results obtained at the upper (+) and lower (−) level of the factor are averaged, and the result at the lower setting is subsequently subtracted from that at the upper. This leads to the following equations:

Main effect of A

$$= \mathrm{ME}(\mathrm{A}) = \frac{y_{A+B-} + y_{A+B+}}{2} - \frac{y_{A-B-} + y_{A-B+}}{2} = \frac{(y_{A+B-} - y_{A-B-}) + (y_{A+B+} - y_{A-B+})}{2} = \frac{\mathrm{E}_{B-}(\mathrm{A}) + \mathrm{E}_{B+}(\mathrm{A})}{2} = 10\%$$

Main effect of B

$$= \mathrm{ME}(\mathrm{B}) = \frac{y_{A-B+} + y_{A+B+}}{2} - \frac{y_{A-B-} + y_{A+B-}}{2} = \frac{(y_{A-B+} - y_{A-B-}) + (y_{A+B+} - y_{A+B-})}{2} = \frac{\mathrm{E}_{A-}(\mathrm{B}) + \mathrm{E}_{A+}(\mathrm{B})}{2} = 22\%$$

The expressions preceding the numerical results in the calculations above show that the main effect of a factor can also be calculated from the mean of the individual effects involved. The two calculated results can now be read as follows: When the catalyst quantity A is increased from 100 to 150 g, the yield increases on average by 10 % from $\frac{60+74}{2} = 67\%$ to $\frac{62+92}{2} = 77\%$. The temperature B has a stronger impact on the response variable yield within the range of 70 – 90 °C. An increase in temperature leads to an average increase in yield of 22 %. The main-effect diagrams in Figure 7 illustrate these relations graphically.

The effect of a factor variation is strongly dependent on the respective setting of the other factor. This was already shown in Figure 5 and is further illustrated by the two interaction diagrams in Figure 7. The four lines in these diagrams correspond to the four edges of the response surface, also shown in Figure 7. This interaction is a typical case of a synergetic interaction. A simultaneous increase of A and B has a clearly higher impact on the response variable than the additive superposition of the individual effects, $E_{B-}(A)$ and $E_{A-}(B)$, would lead one to expect. If the value of y_{A+B+} were not 92 but 76 %, there would be no interaction between A and B. The effect of a variation of one of the two factors on the response variable would be independent of the adjustment of the other factor. In this case, the two individual effects would be identical for each factor, i.e., $E_{B-}(A) = E_{B+}(A) = +2\,\%$ and $E_{A-}(B) = E_{A+}(B) = +14\,\%$. The corresponding lines in the interaction diagrams would then be parallel. It seems reasonable now to determine a quantitative measure for the interaction of two factors as the difference of these individual effects, i.e.,

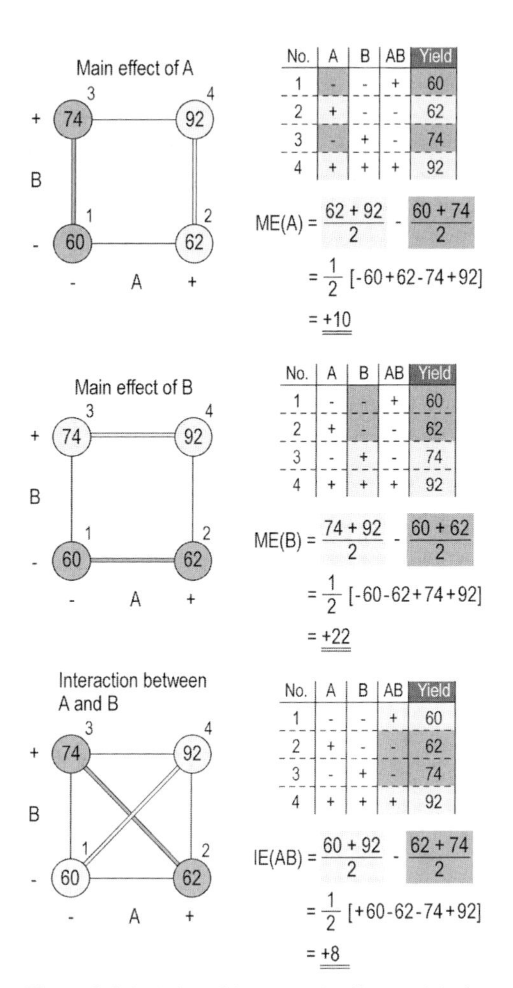

Figure 6. Calculation of the two main effects and the interaction effect in a 2^2 factorial design

Interaction effect between A and B

$$
\begin{aligned}
&= \mathrm{IE}\,(AB) \\
&= \tfrac{1}{2}\,[E_{B+}(A) - E_{B-}(A)] \\
&= \frac{(y_{A+B+} - y_{A-B+}) - (y_{A+B-} - y_{A-B-})}{2} \\
&= \frac{y_{A+B+} + y_{A-B-}}{2} - \frac{y_{A-B+} + y_{A+B-}}{2} \\
&= \frac{(y_{A+B+} - y_{A+B-}) - (y_{A-B+} - y_{A-B-})}{2} \\
&= \tfrac{1}{2}\,[E_{A+}(B) - E_{A-}(B)] \\
&= \mathrm{IE}\,(BA) \\
&= +8\%
\end{aligned}
$$

From these sequences of expressions it can be seen that the interaction between A and B could be equally defined through the difference of the individual effects of A or the individual effects of B, giving the same numerical result each time. Moreover, the expression in the middle shows that the interaction between A and B — as is the case for the main effects — is nothing but the difference of two averages (this is basically the reason why a factor of $\frac{1}{2}$ is introduced in the definition). This corresponds to the difference of the averages of the results located on the diagonals in Figure 6.

The calculation of the two main effects and of the interaction effect is also represented geometrically in Figure 6. The corresponding analysis table contains columns of signs, which allow calculation of these effects. Each effect may be computed as the sum of signed response values divided by half the number of experiments, where the signs are taken from the column of the desired effect. Note that the signs of the interaction column AB can be generated by a row-wise multiplication of the main-effect columns A and B. Today, however, the effects do not need to be computed like this anymore. Specific DoE software tools (see Chap. 8) use methods such as those described in Section 4.4 and yield numerical and graphical results more easily.

A final interpretation of the experiments could read as follows: The catalyst shows a not yet satisfactory activity at the lower temperature (70 °C). An increase of the catalyst quantity from 100 to 150 g gives a slight improvement but does not yet yield satisfactory values. The situation is different at the higher temperature (90 °C), where the catalyst clearly performs better. In addition, an increase in the catalyst quantity at this temperature has a strong impact on yield.

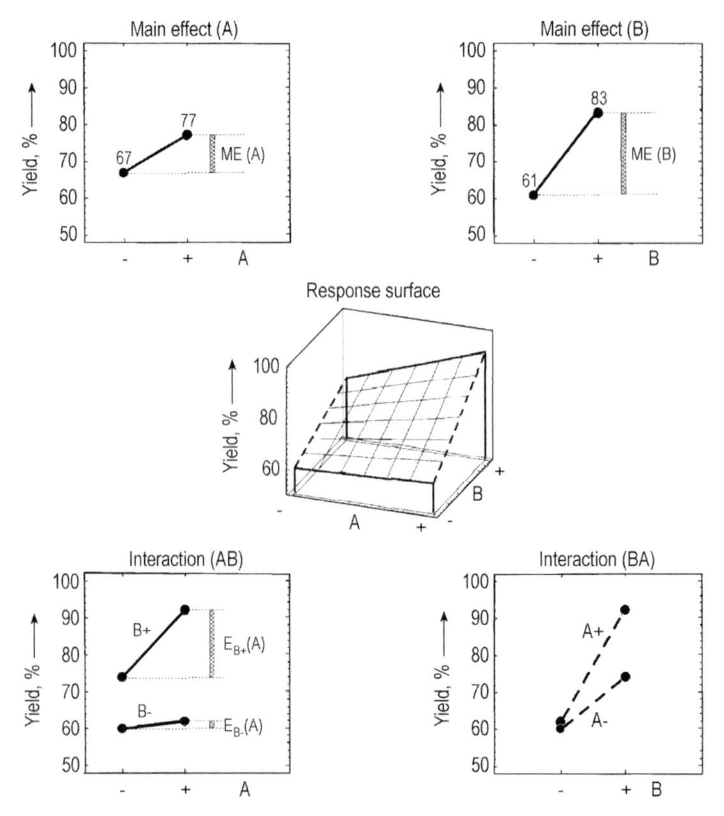

Figure 7. Diagrams of main effects, interactions, and response surfaces illustrate conspicuously important relations governing the system

If all experimental factors are continuous, it will be possible and useful to perform experimental runs at the center point. It is obtained by setting each factor to the midpoint of its factor range, in our example, to 125 g of catalyst and 80 °C. This isolated additional experimental point gives a rough impression of the behavior of response variables inside the experimental region of interest. If the result in the center point does not correspond to the mean of the results obtained in the corner points of the factorial design, then the response surface of the response variable will have a more or less pronounced curvature that depends on the magnitude of this deviation. The graph in Figure 5 shows a result of 70 % at the center point, which is slightly below the 72 % obtained by averaging the four results of the factorial design. Thus, the response surface must be imagined as slightly sagging in its middle. Of course, with this single additional experiment, it is impossible to determine which of the experimental factors is (are) ultimately responsible for the curvature. This question can only be settled by a response surface design (see Chap. 4).

The analysis shown for one response variable is performed for each response variable so that the effects of factor variations on every response variable are finally known.

3.3. The 2^3 Factorial Design

The concepts and notions introduced in the Section 3.2 can be generalized to three or more factors. Let us suppose that, in addition to the example in Section 3.2, not only the effects of temperature and catalyst quantity but also the impact of changing the agitator type on the yield are to be examined. In analogy to Figure 5, the experimental factors with their respective two settings as well as the 2^3 factorial design, which is obtained by realizing all possible combinations of factor settings, are represented in graphical and tabular form in Figure 8.

ID	Factor	Unit	Lower level (-)	Upper level (+)
A	Catalyst	g	100	150
B	Temperature	°C	70	90
C	Agitator type		current	new

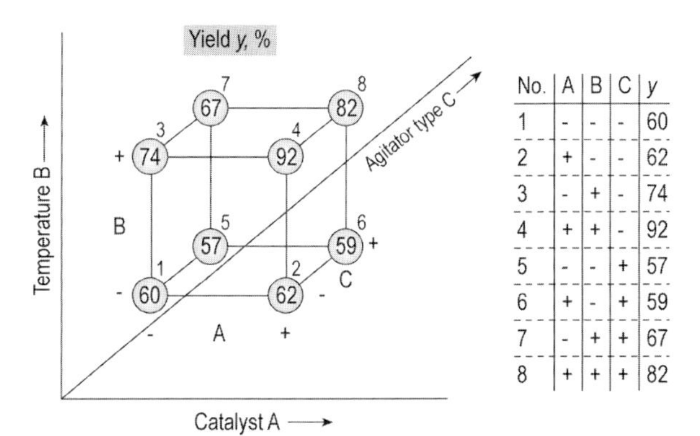

No.	A	B	C	y
1	-	-	-	60
2	+	-	-	62
3	-	+	-	74
4	+	+	-	92
5	-	-	+	57
6	+	-	+	59
7	-	+	+	67
8	+	+	+	82

Figure 8. Example of a 2^3 factorial design, including a graphical and a tabular representation of the experimental settings and results. The experimental settings for agitator type "current" correspond to the 2^2 factorial design in Figure 5

By comparing the values at the ends of the various edges of the cube in Figure 8, it is possible to perform a very elementary analysis. For each factor, the effect of its variation can be studied for the four different constellations of the other two factors. For example, the change from the current to the new agitator type does not lead to the presumed improvement in yield. This may be verified by looking at the four edges going from the front face to the back face of the cube. The deterioration is particularly severe at high temperature where the yield decreases from 74 to 67 % (for the lower level of catalyst) and from 92 to 82 % (for the upper level of catalyst).

More detailed information about the specific and joint effects of the factors is obtained by calculation of the main effects and interaction effects introduced in Section 3.2.

The main effect thereby indicates how much a response variable is affected on average by a variation of a factor and is measured as the difference in the average response for the two factor levels. Figure 9 illustrates this for the factors A (catalyst) and B (temperature). The main effect for C (agitator type) is obtained analogously by calculating the difference of the two averages at the back face and the front face of the cube.

The interaction effect of two factors, more precisely, the two-factor interaction was introduced in Section 3.2. Generally, there will be

$$\binom{n}{2} = \frac{n \cdot (n-1)}{2}$$

two-factor interactions, where n denotes the number of factors. For three factors there are

$$\binom{3}{2} = 3$$

two-factor interactions, namely, AB, AC, and BC. The calculation of a two-factor interaction in a 2^3 factorial design will be demonstrated for AB. This interaction is obtained by calculating the two-factor interaction $\mathrm{IE}_{C-}(\mathrm{AB})$ at the lower level of factor C and the two-factor interaction $\mathrm{IE}_{C+}(\mathrm{AB})$ at the upper level of factor C, and then by averaging these two values:

Interaction effect between A and B

$$
\begin{aligned}
&= \mathrm{IE}\,(\mathrm{AB}) \\
&= \frac{\mathrm{IE}_{C-}(\mathrm{AB}) + \mathrm{IE}_{C+}(\mathrm{AB})}{2} \\
&= \frac{\left(\frac{60+92}{2} - \frac{62+74}{2}\right) + \left(\frac{57+82}{2} - \frac{59+67}{2}\right)}{2} \\
&= \frac{60+92+57+82}{4} - \frac{62+74+59+67}{4} \\
&= +7.25\%
\end{aligned}
$$

A two-factor interaction in a factorial design with more than two factors is obtained by taking the average of all individual two-factor interactions at the different constellations of the other factors. The calculation of IE(AB) is also illustrated in Figure 9, where it is seen to be the difference of averages between results on two

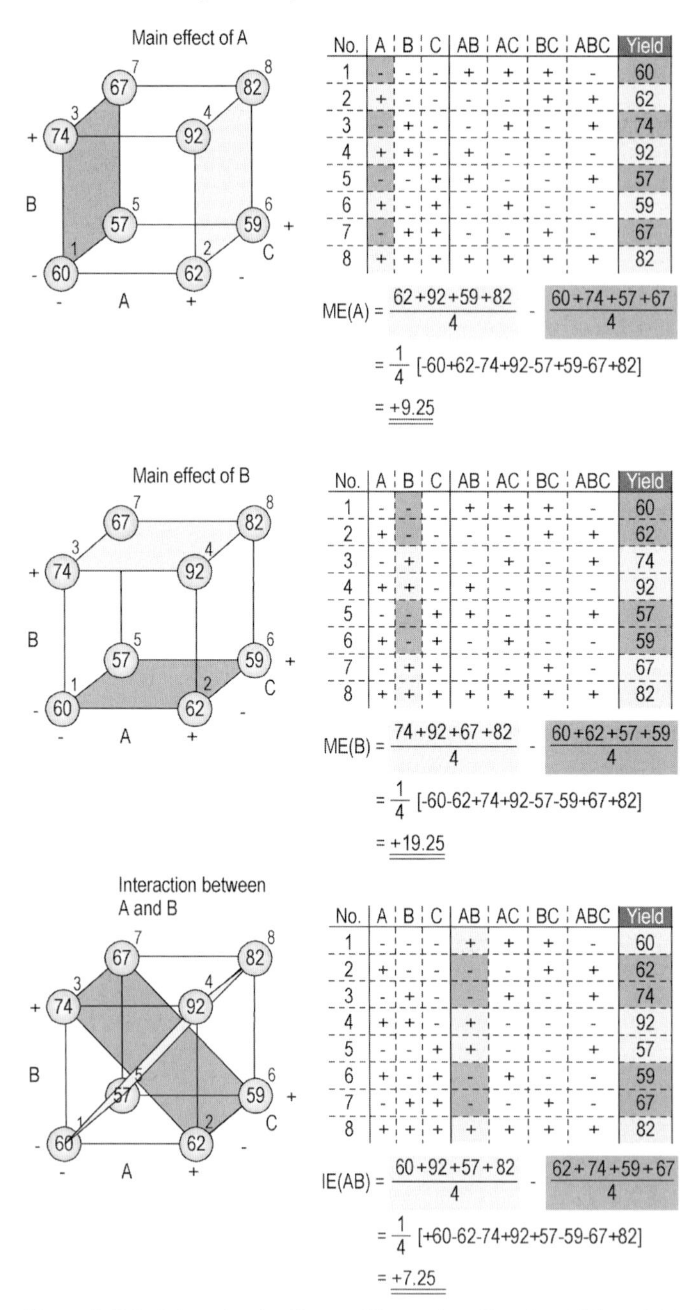

Figure 9. Calculation of main effects and interaction effects in a 2^3 factorial design shown for the main effects of A and B and their two-factor interaction AB

diagonal planes. Due to the inherent symmetry of the design, the calculation of the other two-factor interactions, IE(AC) and IE(BC), can be performed in a similar way. This leads to the respective columns of signs within the analysis table in Figure 9 and to the corresponding diagonal planes within the cube.

Now, if the interaction between A and B at the lower level of C differs from their interaction at the upper level of C, there will be a three-factor interaction, which is defined by the difference of these two individual two-factor interactions:

$$\begin{aligned}
&\text{IE (ABC)}\\
&=\tfrac{1}{2}\ [\text{IE}_{\text{C}+}\ (\text{AB}) - \text{IE}_{\text{C}-}\ (\text{AB})]\\
&=\tfrac{1}{2}\ [(\tfrac{57+82}{2} - \tfrac{59+67}{2}) - (\tfrac{60+92}{2} - \tfrac{62+74}{2})]\\
&=\tfrac{62+74+57+82}{4} - \tfrac{60+92+59+67}{4}\\
&= -0.75\%
\end{aligned}$$

Note that this result can also be obtained by using the column of signs corresponding to the three-factor interaction ABC in Figure 9. As in the case of two-factor interactions, the signs within this column are obtained by a row-wise multiplication of the signs of the main effect columns A, B, and C. Obviously, the three-factor interaction is also the difference of two averages (as in the case of the two-factor interaction presented in Section 3.2, this is the reason why the factor of $\frac{1}{2}$ is introduced in the definition). They are obtained by averaging the results located at the vertices of the respective two tetrahedra which make up the cube.

The numerical results of all effects obtainable from the 2^3 factorial design are summarized in the following:

ME(A)	=	9.25 %
ME(B)	=	19.25 %
ME(C)	=	– 5.75 %
IE(AB)	=	7.25 %
IE(AC)	=	– 0.75 %
IE(BC)	=	– 2.75 %
IE(ABC)	=	– 0.75 %

Note again that numerical results like these and graphical results like those of Figure 7 are easily obtained by using specific DoE software (see Chap. 8). Diagrams of main effects and interaction effects may also be generated for factorial designs with more than two factors.

Even though three-factor interactions may really exist in some cases, it is seldom that they play an essential role. So, if their absolute value is clearly higher than most of the main effects and two-factor interactions, it seems reasonable to conclude that the experimental error is of a magnitude that does not allow the reliable estimation of most of the effects (perhaps because they are very small) and/or that at least one response value has been corrupted by a gross systematic error.

If, on the whole, systematic errors can be excluded, it is a legitimate practice to neglect higher-order interactions, such as three- and four-factor interactions, because main effects tend to be larger than two-factor interactions, which in turn tend to be larger than three-factor interactions, and so on. Moreover, if no information about the magnitude of the experimental error is available, it will be possible to obtain a rough estimate of this error by using higher-order interactions, like five-, four-, and even three-factor interactions.

3.4. Fractional Factorial Designs

For a growing number n of experimental factors, the number of experimental settings 2^n increases exponentially in a full factorial design. Simultaneously, the proportion of higher-order interactions increases rapidly. For instance, if $n = 5$, there are 5 main effects, 10 two-factor interactions, and 16 interactions of higher order. Obviously, if n is not small, there is some redundancy in a full factorial design, since higher-order interactions are not likely to have appreciable magnitudes. At this point, the following questions arise:

- Is it possible to reduce the amount of experimental effort in a sophisticated way so that the most important information can still be obtained by analysis of the data?
- Is it possible to study more experimental factors instead of higher-order interactions with the same number of experimental settings?

The answer in both cases is "yes". It leads to the 2^{n-k} fractional factorial designs, where 2^{n-k} denotes the number of experimental settings, n the number of experimental factors, and k the number of times by which the number of settings has been halved compared to the corresponding complete 2^n design with the same number of experimental factors ($1/2^k \cdot 2^n = 2^{n-k}$). The application of fractional factorial designs yields a reduction in experimental effort, which is adapted to the complexity of the system under investigation and to the information required.

Once the experimental runs have been performed, fractional factorial designs are analyzed like full factorial designs, except that fewer values are available. As introduced in Sections 3.1, 3.2, and 3.3, effects are obtained by calculating corresponding differences of averages. However, by doing so, one will discover that effects

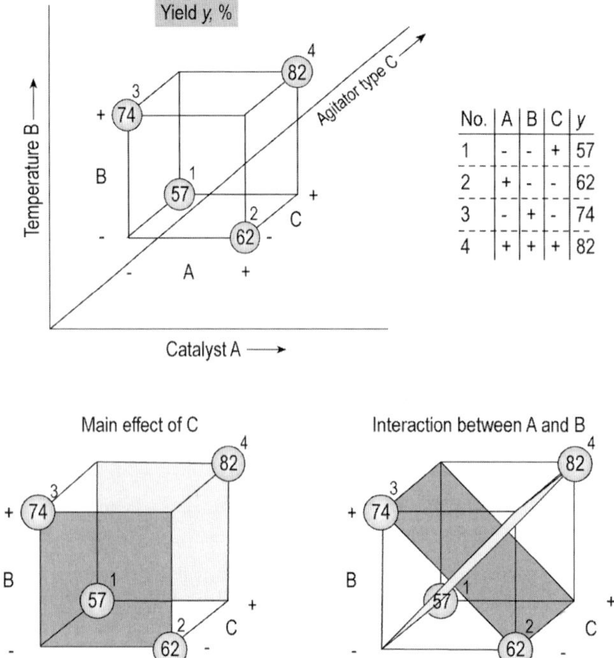

No.	A	B	C	y
1	-	-	+	57
2	+	-	-	62
3	-	+	-	74
4	+	+	+	82

Main effect of C

Interaction between A and B

$$ME(C) = \frac{57 + 82}{2} - \frac{62 + 74}{2}$$

$$IE(AB) = \frac{57 + 82}{2} - \frac{62 + 74}{2}$$

Confounding !

Figure 10. Example of a 2^{3-1} fractional factorial design. In the top section the experimental settings and results are shown in geometrical and tabular form. In the bottom section the problem of confounding is illustrated: Calculating ME(C) and IE(AB) leads to the same expression, which is actually the sum of both. It is not possible to determine whether the calculated value is caused by the main effect of C, by the interaction effect of A and B, or by both

are confounded. *Confounding* is defined as a situation where an effect cannot unambiguously be attributed to a single main effect or interaction.

Let us consider the 2^{3-1} design, in which the effects of three factors are studied with four experimental settings (see Figure 10). This is an example of a half-fraction factorial design (see Figure 11). By calculating the main effect of C [ME(C) on the left-hand side of Fig. 10] the outcome is not the difference between the averages of four (as in the case of the 2^3 design) but the averages of only two results which are calculated. A similar situation occurs when calculating the interaction between A and B [IE(AB) on the right-hand side of Fig. 10]. Moreover, calculating these effects leads to the same expression. It is actually the sum ME(C)+IE(AB) of both effects. It can also be verified that ME(A) is confounded with IE(BC), and ME(B) with IE(AC). This is an example of a so-called resolution III design.

A slightly different situation occurs when considering the 2^{4-1} design (see Fig. 11). Here the main effects are confounded with three-factor interactions, e.g. ME(A) with IE(BCD), and the two-factor interactions are confounded with each other, e.g., IE(AC) with IE(BD). This is an example of a resolution IV design. The 2^{5-1} design also shown in Figure 11 is an example of a very efficient resolution V design.

The *resolution* of a fractional factorial design largely characterizes the information content obtainable by analyzing the results of a fractional factorial design:

- A fractional factorial design of resolution III does not confound main effects with each other but does confound main effects with two-factor interactions.

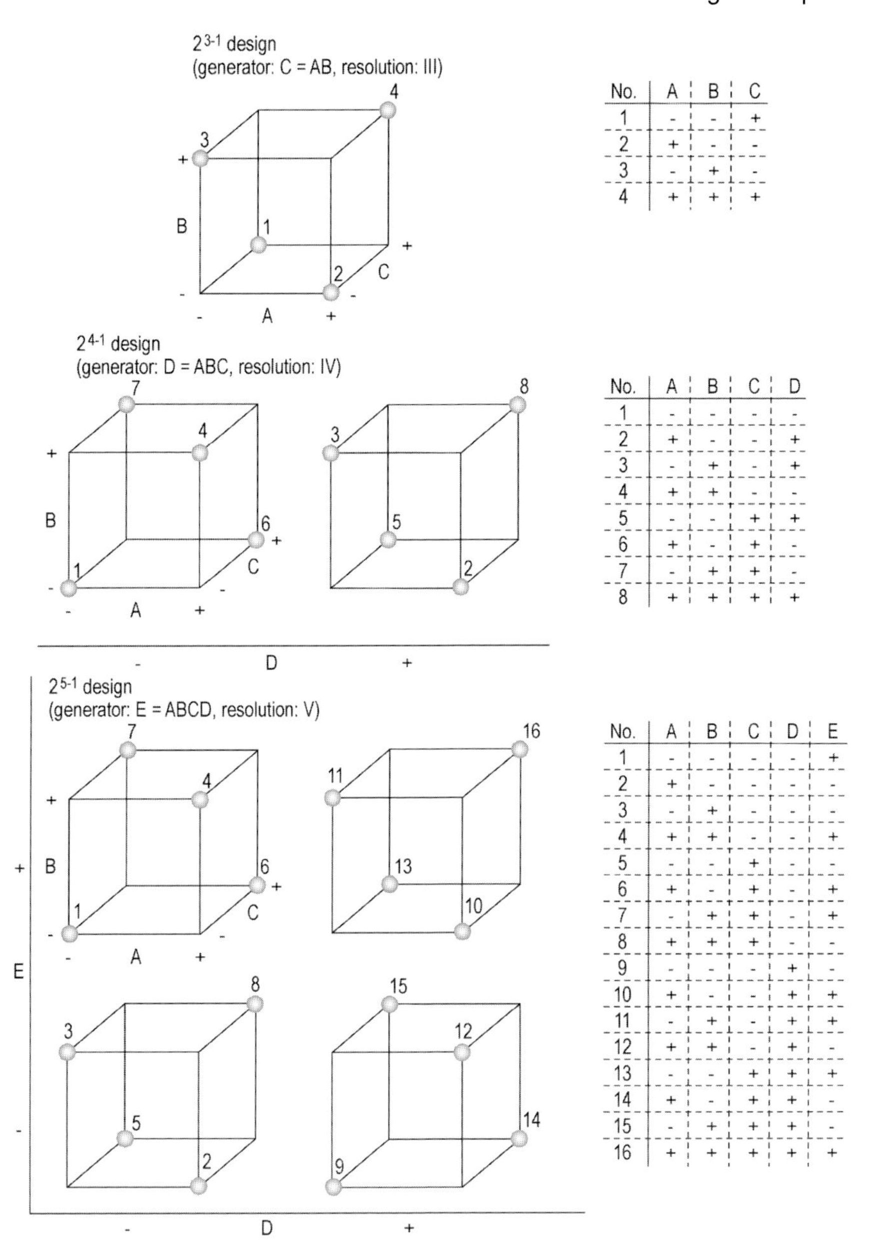

No.	A	B	C
1	-	-	+
2	+	-	-
3	-	+	-
4	+	+	+

No.	A	B	C	D
1	-	-	-	-
2	+	-	-	+
3	-	+	-	+
4	+	+	-	-
5	-	-	+	+
6	+	-	+	-
7	-	+	+	-
8	+	+	+	+

No.	A	B	C	D	E
1	-	-	-	-	+
2	+	-	-	-	-
3	-	+	-	-	-
4	+	+	-	-	+
5	-	-	+	-	-
6	+	-	+	-	+
7	-	+	+	-	+
8	+	+	+	-	-
9	-	-	-	+	-
10	+	-	-	+	+
11	-	+	-	+	+
12	+	+	-	+	-
13	-	-	+	+	+
14	+	-	+	+	-
15	-	+	+	+	-
16	+	+	+	+	+

Figure 11. Geometric representation of the most important half-fraction factorial designs and their respective tabular representation, i.e., their design matrices

- A fractional factorial design of resolution IV does not confound main effects with two-factor interactions but does confound two-factor interactions with other two-factor interactions.
- A fractional factorial design of resolution V does not confound main effects and two-factor interactions with each other but does confound two-factor interactions with three-factor interactions.

The resolution of the most important fractional factorial designs can be seen in Table 1.

Designs of resolution III are used in the early stages of experimental investigations to gain a first insight into the possibilities and behaviors of systems. Particularly beneficial are *saturated*

Table 1. The 2^{n-k} (fractional) factorial designs with a maximum of 32 experimental settings and their resolution. 2^{n-k} denotes the number of experimental settings, n the number of experimental factors, and $1/2^k$ the factor by which the number of settings has been reduced compared to the corresponding full factorial design with the same number of experimental factors.

Resolution	Number of experimental settings			
	4	8	16	32
III	2^{3-1}	2^{5-2} – 2^{7-4}	2^{9-5} – 2^{15-11}	2^{17-12} – 2^{31-26}
IV		2^{4-1}	2^{6-2} – 2^{8-4}	2^{7-2} – 2^{16-11}
V			2^{5-1}	
VI				2^{6-1}
Full	2^2	2^3	2^4	2^5

designs for $2^n - 1$ factors in which all degrees of freedom of a 2^n design are exploited. The above-mentioned 2^{3-1} design is an example of such a design in which three factors are studied with four experimental settings. It is "generated" by replacing all columns of signs in the analysis table of Figure 6 by experimental factors, i.e., the settings of factor C at the top of Figure 11 are determined by the column of the two-factor interaction AB in the analysis table of Figure 6. A further important example of a saturated design is the 2^{7-4} design, which allows seven factors to be studied with eight experimental settings. It is generated by replacing AB by D, AC by E, BC by F, and ABC by G in the heading of the columns of signs used for the calculation of effects in Figure 9. With resolution III designs special care must be taken when interpreting the analysis results. Calculated main effects may also be two-factor interactions of other factors.

Resolution IV designs are employed to gain unambiguous information about the individual impact of the experimental factors, while unambiguous information about their two-factor interactions is not yet required. Designs of resolution III and IV are mostly used to find out which factors play an important role. This technique of isolating the important factors is sometimes referred to as *screening*.

Using designs of resolution V or higher does not lead to any loss of decisive information, since the main effects and two-factor interactions are only confounded with higher-order interactions, which in most cases can be neglected. Particularly when $n > 4$, experimental settings can be halved to achieve designs of at least resolution V.

Fractional factorial designs support the iterative nature of experimentation. The experimental runs of two or more fractional factorial designs conducted sequentially may be combined to form a larger design with higher resolution. In this way it is possible to resolve ambiguities by the addition of further experimental runs.

Half-fraction designs possess an interesting and useful projection property: the omission of one arbitrary column in the designs always leads to a full factorial design with respect to the remaining columns or factors. So, if a factor proves to have no significant effect on a response variable, the remaining factors can be analyzed as in the full factorial design. For example, the omission of factor B in the 2^{4-1} design of Figure 11 leads to a 2^3 design for the factors A, C, and D. This may be verified by examining the three remaining columns in the table but also by pushing the upper faces of the two cubes into the lower faces.

4. Response Surface Designs

In some situations factorial designs in which all factors are varied by only using two settings are not adequate for describing the behavior of an experimental system, because a more detailed insight is needed to predict its responses or to find optimal factor settings. In this case, it is often necessary to extend factorial designs and to do additional experiments at other points in the experimental domain. To decide which points to use, *response surface designs* are used [14, 31]. These designs are based on mathematical models that describe how responses depend on experimental factors.

4.1. The Idea of Using Basic Empirical Models

A model is a way to describe a part of reality; ideally it is much simpler and more manageable than that which it describes, but it should nevertheless adequately fulfill a predefined modeling purpose. Since a model can only be an approximate description of the original, it is important to be aware of this purpose when constructing models (see Chap. 2 for typical goals in conjunction with DoE). Models used in DoE are polynomials in several variables, in which the y variable is a response, and the x variables are experimental factors. A set of coefficients is used to describe how the y variables depend on the x variables. Often a process or experimental system can only be adequately described by more than one response, in which case there will be one model for each response, and each model will have its own set of coefficients.

Coefficients are estimated from experimental data which are collected in a corresponding experimental design. Estimating model coefficients from experimental data is called model fitting and represents the principle task of statistical analysis to be performed as soon as response values from experiments are available. An equally important task consists of assessing the quality of a fitted model, as an important step towards qualifying it for use in prediction and optimization or whatever other purpose. In fact it is often possible to improve the performance of a model by taking small corrective measures such as those described in Section 5.4.

The important ideas behind empirical modeling are:

- The model must describe the entire behavior of the process or experimental system that is relevant to answering the questions of the experimenter.
- The experimental design is based on the model and determines which information can be extracted from the experimental results. Statistical analysis is only the tool for extracting this information.
- Neglecting experimental design essentially means missing out on finding all the relevant answers. Picking the wrong design also means missing out on relevant answers.
- Enlarging the scope of the questions always means extending the model and adding experiments to the design.

Using models and setting up designs in this fashion requires that one proceed in the systematic way that has been described in Chapter 2. There are several additional aspects that play an important role in modeling:

- Definition of the experimental domain in which the model should be useful for prediction
- Selection of the correct model type
- Choice of the experimental design that corresponds optimally to the experimental domain and to the model type chosen
- Estimation of coefficients by regression analysis
- Qualification and refinement of the model by continued statistical analysis
- Validation of model predictions by confirmatory experiments
- Use of the model for the purpose of finding optimal factor settings

4.2. The Class of Models Used in DoE

The class of models that is normally used in DoE contains only models which are linear with regard to the unknown coefficients. This is why, in order to estimate coefficients, linear regression methods can efficiently be used [32]. Nonlinear or first-principle models, such as mechanical models, reaction-kinetic models, and more general dynamic models are only rarely used directly when designing experiments; they are commonly approximated by simple polynomial models at the cost of restricting the domain of validity of the model. A direct generation of optimal designs for nonlinear models, i.e., models that are nonlinear in the parameters that are to be estimated, is sometimes possible. However, it is particularly important that such models are very accurate, that experimental errors are small and that initial estimates of the parameters are already available. It is this last point that often makes setting up the correct design very difficult [17].

Polynomial models used in DoE are built up as a sum of so-called model terms:

$$y=b_0+b_A x_A+b_B x_B+b_{AB} x_A x_B+b_{AA} x_A^2+b_{BB} x_b^2+\varepsilon$$

This is an example of a quadratic model for two factors A and B, containing a constant term b_0, linear terms $b_A x_A$ and $b_B x_B$, an interaction term $b_{AB} x_A x_B$, quadratic terms $b_{AA} x_A^2$ and $b_{BB} x_B^2$, and an error term ϵ.

The x represent the settings of the factors in the experimental domain. In factorial designs, they are coded as − 1 and + 1. When factors have continuous scales, like temperature or pressure, this coding can be understood as a simple linear transformation of the factor range onto the interval [− 1, + 1]. This transformation is called scaling and centering; the corresponding equation is:

$$x_{\text{centered \& scaled}}=2\ (x-x_{\text{center}})\ /\ (x_{\max}-x_{\min})\ .$$

Since $x_{center} = (x_{max} + x_{min})/2$, x_{max} transforms into + 1 and x_{min} transforms into − 1 (or simply + and − respectively, when using factorial designs). Centering and scaling allows the influences of different factors with different scales to be compared. In the following discussion it is assumed that all factors are either coded or scaled and centered in this fashion.

The b are the coefficients that are estimated by regression after the experiments have been completed. A question that arises is: How do calculated effects in a factorial design as described in Chapter 3 compare to estimated coefficients of linear or interaction terms in a fitted model?

The answer to this question is quite interesting. Estimating coefficients by multiple linear regression and calculating effects for factorial designs as described in Chapter 3 are mathematically equivalent. In fact, coefficients are simply half of the corresponding effects: Calculating a main effect of a factor ME(A) means estimating the difference Δy in the response that has been provoked by changing the factor A from its lower to its upper level. In contrast, the corresponding coefficient, b_A, is the geometrical slope of the curve describing the dependency of y upon x, i.e., $\Delta y/\Delta x$. Since DoE is based on scaled and centered variables, Δx is exactly 2. So b_A can be estimated by ME(A)/2. This is also true for interaction effects: b_{AB} can be estimated by IE(AB)/2. Estimators are often denoted by $\hat{}$, so $\hat{b}_A=\text{IE (AB)}/2$. The constant b_0 can be estimated by calculating the mean of all response values.

For the example that was discussed in Section 3.2, the model equation is $y = b_0 + b_A x_A + b_B x_B + b_{AB} x_A x_B$ and the estimated coefficients are:

$$\hat{b}_0=\tfrac{1}{5}\ (74+92+70+60+62)=71.6\%$$

$$\hat{b}_A=\text{ME (A)}\ /2=5\%$$

$$\hat{b}_b=\text{ME (B)}\ /2=11\%$$

$$\hat{b}_{AB}=\text{IE (AB)}\ /2=4\%$$

The benefit of using coefficients lies in the greater generality of the response surface models. These allow:

- Prediction of response variables within the experimental region
- Use of quadratic models for modeling maxima and minima
- Use of mixture models for modeling formulations (see Section 7.1)
- Use of dummy variables for modeling categorical factors (see Section 7.2)
- Correcting factor settings when prescribed settings cannot be exactly met in the experiment
- Nonstandard domains for the model (and the design) that are subject to additional constraints

4.3. Standard DoE Models and Corresponding Designs

For standard DoE models the designs can be chosen off the peg. This means that the design structure is predefined and available in the form of a design table (see, e.g., Fig. 11 for half-fraction factorial designs). For nonstandard models an optimal design has to be generated by using a mathematical algorithm (see Section 7.3).

Standard DoE models are

- Linear models (i.e., linear with regard to the factor variables), containing the constant term and linear terms for all factors involved
- Interaction models, which additionally contain interaction terms of the factors involved
- Quadratic models, which, in addition to all interaction terms, contain quadratic terms

Standard models for two and three factors are shown in Table 2, examples of response surfaces are shown in Figure 12.

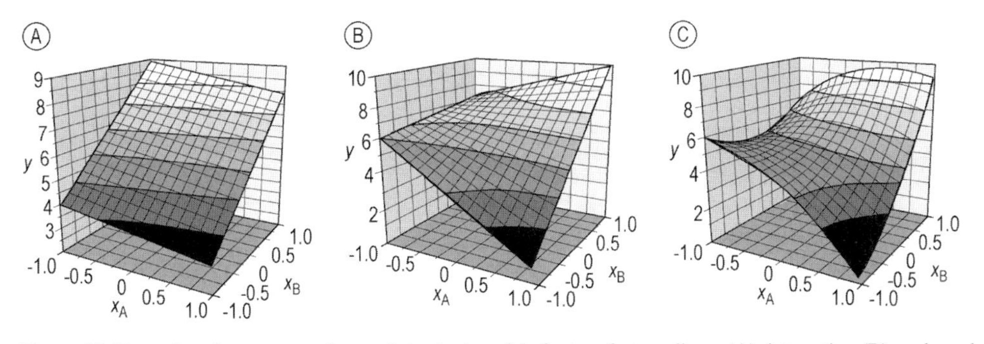

Figure 12. Examples of response surfaces of standard models for two factors: linear (A), interaction (B), and quadratic (C) models

Table 2. Standard models for two and three factors

	Linear model	Interaction model	Quadratic model
Two factors	$y = b_0 + b_A x_A + b_B x_B$	$\ldots + b_{AB} x_A x_B$	$\ldots + b_{AA} x_A^2 + b_{BB} x_B^2$
Three factors	$y = b_0 + b_A x_A + b_B x_B + b_C x_C$	$\ldots + b_{AB} x_A x_B + b_{AC} x_A x_C + b_{BC} x_B x_C$	$\ldots + b_{AA} x_A^2 + b_{BB} x_B^2 + b_{CC} x_C^2$

Adding isolated interaction terms to linear models, taking away interaction terms from interaction models, taking away square terms from quadratic models, or even adding cubic terms like $b_{AAB} x_A^2 x_B$, $b_{ABC} x_A x_B x_C$, or $b_{AAA} x_A^3$, to quadratic models gives nonstandard models which can also be used for DoE. Nonstandard models are also obtained when mixture components are investigated together with normal factors, or when so-called dummy variables (indicator variables) are used to code categorical factors that have three or more settings.

Optimal designs for standard and nonstandard models are:

- Linear models: resolution III factorial designs or so-called Plackett – Burman designs (which are very similar to factorial designs) if no interactions are present
- Interaction models with only some interaction terms: resolution IV factorial designs or so-called D-optimal designs (see Section 7.3); resolution IV designs are also used for linear models when interactions may be present but are assumed to be unimportant
- Interaction models with all interaction terms: resolution V (or higher) factorial designs or D-optimal designs
- Quadratic models: central composite designs (CCD; see below) or so-called Box – Behnken designs
- All nonstandard models: D-optimal designs (see Section 7.3)

Table 3. Design matrix of a CCD for two factors. The run Nos. 1 to 4 form the factorial design, Nos. 5 to 8 the star points, and No. 9 the center point.

No.	A	B	y
1	− 1	− 1	
2	1	− 1	
3	− 1	1	
4	1	1	
5	− 1.41	0	
6	1.41	0	
7	0	− 1.41	
8	0	1.41	
9	0	0	

Factorial designs are discussed in detail in Chapter 3. *Central composite designs* (CCD) are extensions of factorial designs, in which so-called star points and additional replicates at the center points are added to allow estimation of quadratic coefficients (see Fig. 13 and Table 3). The number of star points is simply twice the number of factors and, ideally, the number of replicates at the center is roughly equal to the number of factors. The distance α from the center point to the star points should be greater than one, i.e., the star points should be outside the domain defined by the factorial design. A star distance $\alpha = 1$ is sometimes used; then star points lie in the faces of the factorial design. A good alternative to a CCD is the Box – Behnken design, in which no design points leave the factorial domain [14]. Box – Behnken designs do not contain a classical factorial design as a basis.

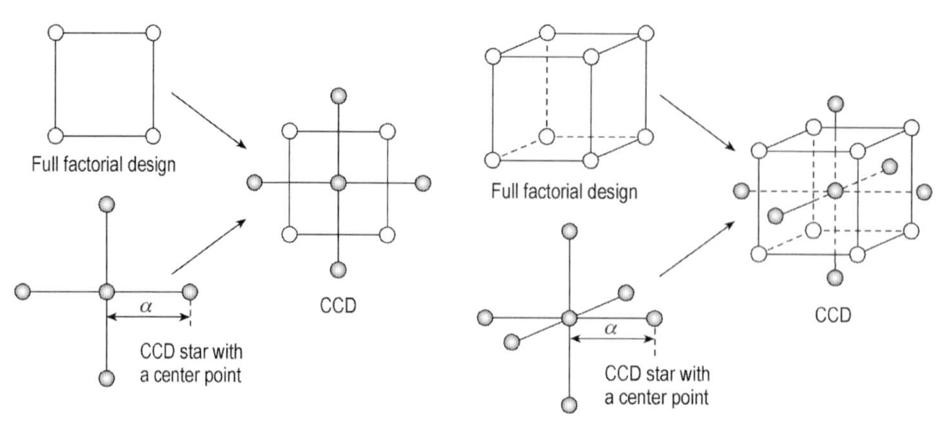

Figure 13. Central composite design for two factors (left) and three factors (right)

CCDs are normally used with full factorial designs, and sometimes with resolution V designs. Another class of designs, called Hartley designs [33], are similar to CCDs and are based on resolution III factorial designs. When, as is sometimes done in industrial practice, only some quadratic terms are added to an existing interaction model, it is useful to include star points only for those factors for which quadratic terms have been added.

4.4. Using Regression Analysis to Fit Models to Experimental Data

Factorial designs are orthogonal for linear and interaction models which means that coefficients can be estimated independently of each other. This is why calculating effects as described in Chapter 3 is so easy. For more complex models and designs, the analysis will be based on multiple linear regression (MLR) and its variants, such as stepwise regression, variable subset selection (VSS), ridge regression (RR), and partial least squares (PLS) [34]. All of these methods represent ways of fitting the model to the data, in the sense of minimizing the sum of squared distances from measured response values to model values (Fig. 14). They differ in that the minimization procedure is subject to different constraints, and their performance differs only in the case of badly conditioned designs, i.e., when the design is not really adequate for estimating all model coefficients.

If y_i stands for the observed response value at experiment i and $\hat{y}_i$ represents the predicted value at that point, then the least-squares estimates for the coefficients $(b_0, b_A, b_B, b_{AB}, \ldots)$ are those for which $\Sigma(y_i - \hat{y}_i)^2$ is minimized (remember: $\hat{y}_i$ depends on the $\hat{b}$ values).

5. Methods for Assessing, Improving, and Visualizing Models

There are many statistical tools that allow a basic judgement of whether a fitted model is sound. A useful selection of these is:

- The regression measure R^2 to check the quality of fit
- The prediction measure Q^2 to check the potential for prediction and to prevent so-called over-fit, which means that the model is so close to the data that it models experimental errors
- Analysis of variance (ANOVA), to compare the variance explained by the model with variance attributed to experimental errors and to check for significance
- Lack-of-fit test (LoF) to assess the adequacy of the model
- Analysis of the residuals to find structural weaknesses in the model and outliers in the collected data

These methods are used both to qualify models for prediction and optimization purposes and also to find indications of how to improve the models in the sense of increasing their reliability. The different statistical methods are explained and ways to interpret and use them toward model improvement are discussed.

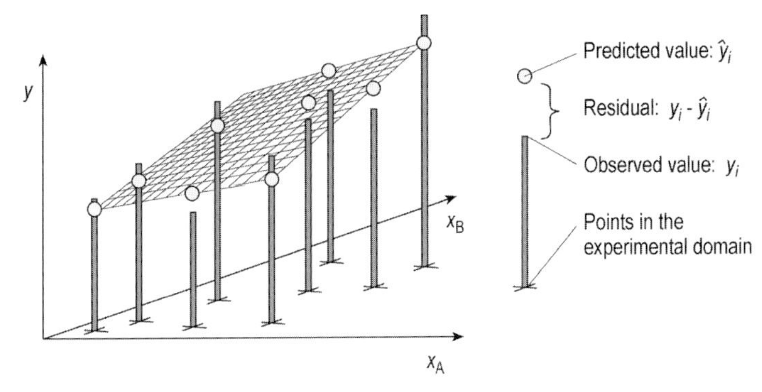

Figure 14. How least-squares regression works: squared distances from the model to the observed values are minimized

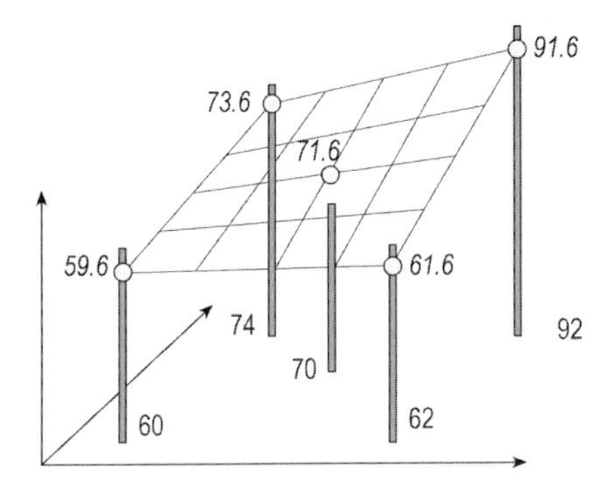

Figure 15. From the example in Section 3.2: observed values y_i and predicted values $\hat{y}_i$ from a model with coefficients $\hat{b}_0 = \bar{y} = 71.6$, $\hat{b}_A = 5$, $\hat{b}_B = 11$, $\hat{b}_{AB} = 4$ (see Section 4.2), so that $\hat{y}_i$, $y_i - \hat{y}_i$ and $y_i - \bar{y}$ can be calculated

5.1. R^2 Regression Measure and Q^2 Prediction Measure

The regression measure R^2 is the quotient of squared deviations due to the model, $SS_{reg} = \Sigma(\hat{y}_i - \bar{y})^2$, and the total sum of squared deviations of the measured data about the mean value $\bar{y}$ $SS_{tot} = \Sigma(y_i - \bar{y})^2$, i.e.,

$$R^2 = SS_{reg}/SS_{tot}.$$

If ordinary least squares is used for fitting the model, $R^2 = 1 - SS_{res}/SS_{tot}$, where $SS_{res} = \Sigma(y_i - \hat{y}_i)^2$, because $SS_{res} = SS_{tot} - SS_{reg}$, the sum of squared residuals. An example for the calculation of R^2 is given in Figure 15 and Table 4. R^2 is always between 0 and 1 and should be as close as possible to 1; how close it should be depends upon the context of the application. When a measuring device is calibrated, R^2 should be above 0.99, whereas when the output of a chemical reaction with several influencing factors is examined, R^2 may well be around 0.7 and still belong to a very useful model.

R^2 is a deceptive measure because it is prone to manipulation: by adding terms to a model, it will always be possible to get a value of R^2 that is almost one, without really improving the quality of the model. On the contrary, many models with a very high R^2 tend to "over-fit" the data, i.e., they model the experimental errors. This is a common and unwanted phenomenon because it decreases the prediction strength of a model. Especially when a design is not orthogonal, one should be wary of over-fit.

To counteract overfit and improve the reliability of model predictions it is useful to consider the prediction measure Q^2. The calculation of Q^2 is similar to that of R^2, except that in the second equation, the prediction error sum of squares (PRESS) is used instead of SS_{res}:

$$Q^2 = 1 - PRESS/SS_{tot}$$

Here, $PRESS = \Sigma\left(y_i - \hat{\hat{y}}_i\right)^2$, where $\hat{\hat{y}}_i$ is the prediction for the *i*th experiment from a model that has been fitted by using all experiments except this *i*th one. In a sense $y_i - \hat{\hat{y}}_i$ is a fair measure of prediction errors. PRESS is always greater then SS_{res}, and therefore Q^2 is always smaller then R^2. The relationship between Q^2 and R^2 and an example for the calculation of Q^2 are given in Figure 16 and Table 5.

In good models, Q^2 and R^2 lie close together, and Q^2 should at least be greater than 0.5. This may not be the case if the design is saturated or almost saturated, which means that the number

Table 4. Calculation of R^2 for the example above: $SS_{res} = 3.2$, $SS_{reg} = 648$, $SS_{tot} = 651.2$, hence $R^2 = 1 - SS_{res}/SS_{tot} = 0.995$. This is a very good value for the regression measure R^2.

	A	B	y_i	$\hat{y}_i$	$y_i - \hat{y}_i$	$y_i - \bar{y}$
1	–	–	60	59.6	0.4	– 11.6
2	+	–	62	61.6	0.4	– 9.6
3	–	+	74	73.6	0.4	2.4
4	+	+	92	91.6	0.4	20.4
5	0	0	70	71.6	– 1.6	– 1.6
Squared sum					$SS_{res} = 3.2$	$SS_{tot} = 651.2$

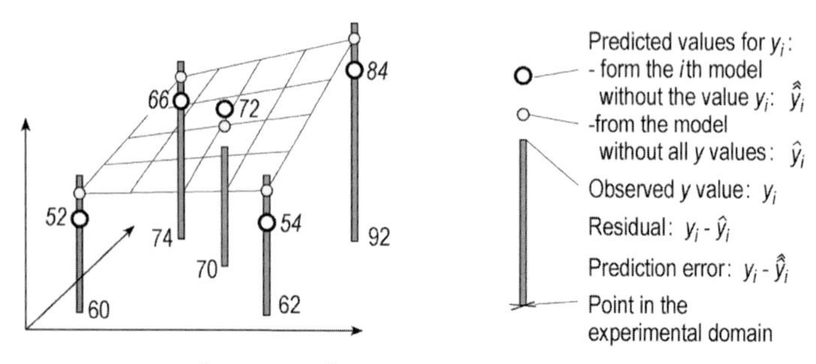

Figure 16. How Q^2 relates to R^2: PRESS $\geq SS_{res}$ and $Q^2 \leq R^2$; $\hat{y}_i$ is usually between y_i and $\hat{\hat{y}}_i$

Table 5. Calculation of Q^2 for the example above: PRESS = 260 and $Q^2 = 1 - 260/651.2 = 0.601$. This is quite a reasonable Q^2 value. Hence, there is no indication of over-fit.

	A	B	y_i	$\hat{\hat{y}}_i$	$y_i - \hat{\hat{y}}_i$	$y_i - \bar{y}$
1	–	–	60	52	8	– 11.6
2	+	–	62	54	8	– 9.6
3	–	+	74	66	8	2.4
4	+	+	92	84	8	20.4
5	0	0	70	72	– 2	– 1.6
Squared sum					260	651.2

of experiments that have been carried out equals or only slightly exceeds the number of terms in the model. In this case, Q^2 may underestimate the quality of the model, because leaving out single measurements may destroy the structure of the design.

It is more dangerous, however, to overestimate the quality of a model, which may happen if many experiments are replicated. In these experiments $\hat{\hat{y}}_i$ will be very close to $\hat{y}_i$ because only one of the measurements is left out in the calculation of $\hat{\hat{y}}_i$. This means that PRESS may be unduly close to SS_{res}, and Q^2 unduly close to R^2. Nevertheless, Q^2 is normally quite a useful measure to prevent overfit.

5.2. ANOVA (Analysis of Variance) and Lack-of-Fit Test

Analysis of variance, usually abbreviated as ANOVA, is a general statistical tool which is used to analyze and compare variability in different data sets. It becomes a powerful tool that can be used for significance testing when assumptions about the error structure underlying the data can be made. In the context of DoE, ANOVA can be used to complement regression analysis and to compare the variability caused by the factors with the variability due to experimental error.

Strictly speaking, "analysis of variance" should be referred to as "analysis of sum of squares" or, even more correctly, "analysis of the sum of squared deviations", because it is actually this sum of squares that is decomposed (Fig. 17).

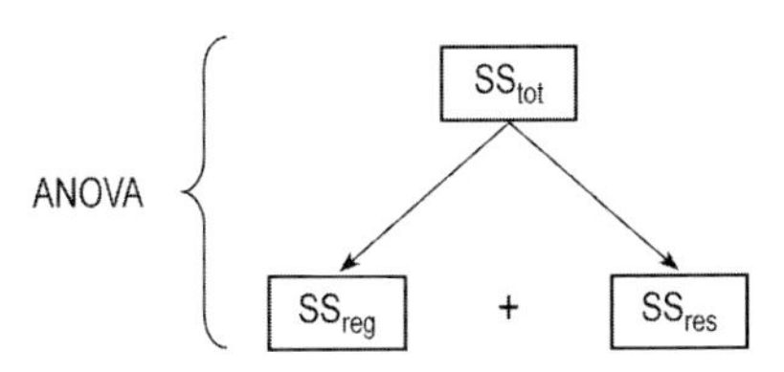

Figure 17. Decomposition of SS_{tot} into SS_{reg} and SS_{res}

However, the aim of ANOVA is to see to what extent the variability in the measured data is explainable by the model and to judge whether the model is statistically significant. To make this comparison, SS_{reg} and SS_{res} must first be made comparable by considering the degrees of freedom.

Already when R^2 is calculated, it is arguable that the number of terms in the model p and the number of runs in the design N should be considered in the calculation, and that $R^2_{adjusted} = 1 - [(N-1)SS_{res}]/[(N-p)SS_{tot}]$ should be used instead of $R^2 = 1 - SS_{res}/SS_{tot}$. But since $R^2_{adjusted}$ normally lies somewhere between R^2 and Q^2, these two are quite sufficient for a first model assessment. In any case, $N-1$ is the total number of degrees of freedom (of variation with respect to the mean value), $p-1$ is the number of degrees of freedom of the model (not counting the constant), and $N-p$ is the so-called residual number of degrees of freedom.

ANOVA compares the model to the residuals and tells us whether the model is statistically significant under the assumption that the residuals can be used to estimate the size of random experimental error. $MS_{reg} = SS_{reg}/(p-1)$ is compared to $MS_{res} = SS_{res}/(N-p)$, also called the mean square error (MSE), by subjecting them to a so-called F-test for significance. If the quotient, $F_{emp} = MS_{reg}/MS_{res}$, is greater than 1, then there is reason to suspect that the model is needed in order to explain the variability in the experimental data. If this has to be proven "beyond a reasonable doubt" (i.e., for statistical significance), F_{emp} must be greater than the theoretical value of $F_{crit}(p-1, N-p, \gamma)$, which is always greater than 1, where γ is the desired level of confidence.

Taking the square root of MS_{res} furnishes a reasonable estimate of the standard deviation of a single measurement, provided $N-p$ is not too small and MS_{res} does not change dramatically when terms with small coefficients are added to or removed from the model. This number is called the residual standard deviation (RSD or SD_{res}).

In the example that has been used $SS_{reg} = 648$, $p-1=3$, hence $MS_{reg} = 216$. $SS_{res} = 3.2$, $n-p=1$, hence $MS_{res} = 3.2$. F_{emp} can be calculated as $F_{emp} = \frac{MS_{reg}}{MS_{res}}$, but a comparison to F_{crit} (3, 1, 95 %) is not meaningful, because there is just one degree of freedom. Also the estimate $RSD = \sqrt{MS_{res}} = 1.789$ should not be taken too seriously in this example.

The lack-of-fit test addresses the question of whether the model may have missed out on some of the systematic variability in the data. This test can only be performed if some of the experimental runs have been replicated: When replicates are present, SS_{res} can be further decomposed into a pure error part $SS_{p.e.}$ and a remaining lack-of-fit part SS_{lof} (Fig. 18).

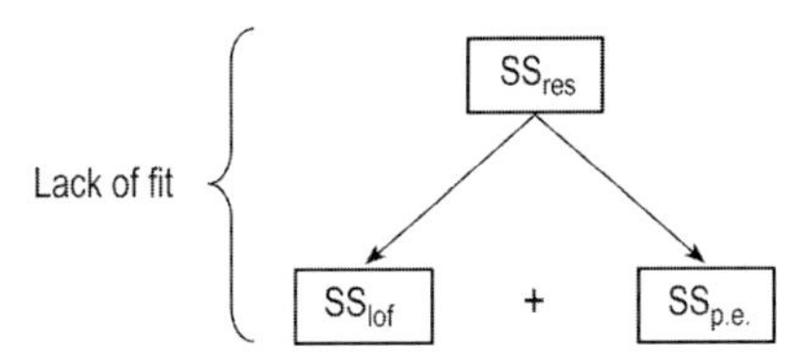

Figure 18. Decomposition of SS_{res} into SS_{lof} and $SS_{p.e.}$

The corresponding significance test compares $MS_{lof} = SS_{lof}/(N-p-r)$ to $MS_{p.e.} = SS_{p.e.}/r$, where r is the total number of replicates of experimental runs in the design. (A replicate count does not include the original run: if there are five runs at the center point, then one of these runs is counted as the original and the four others as replicates; hence, $r = 4$; if a design consisting of eight runs is completely replicated, there are eight runs that have each been replicated once, hence $r = 8$.)

As in ANOVA, a quotient $MS_{lof}/MS_{p.e.}$ greater than 1 means that there may be a lack of fit although the evidence is still weak, whereas a quotient greater than $F_{crit}(N-p-r, r, 1-\alpha) > 1$ means a significant lack of fit at the error probability level α.

The lack-of-fit test method can be regarded as a good complement to the Q^2 prediction measure because the former works well when many replicates are involved, while the latter makes sense when there are only few replicates. Of course, the latter situation is more common in DoE.

5.3. Analysis of Observations and Residuals

Ideally, when models are fitted to data, the model describes all of the deterministic part of the measured values and the part due to experimental error is reflected by the residuals. It is a presumption in linear modeling, i.e., when using least squares fitting as a criterion, that the experimental error is

- Identically (i.e., evenly) distributed over all measurements
- Statistically independent of the measured value, the order of the experiments, the preceding measurement, the settings of the factor variables, etc.

To verify this assumption and to detect outliers, a rudimentary examination of residuals should always be performed as a step toward qualifying the model. The most important tests for residual structure are:

- Test for influence of single observations on the model
- Test for normal distribution and test for outliers
- Test for uniform variance
- Test for independency of measurement errors

These tests can be performed formally as described above for ANOVA and lack-of-fit testing. However, formal testing of hypotheses of the type necessary here usually requires a very high number of residual degrees of freedom $N - p$. Since this number is actively and consciously kept low in DoE, these tests do not often yield useful results. This is why it is common practice to plot observations and residuals in different types of graphs in order to detect structural weaknesses and to find hints on what the problem might be and on how to avoid it.

Common plots are:

- Observed values versus predicted values to see whether there are observations that had undue influence on the model
- Ordered residuals in a normal probability plot to detect outliers and indications of a possibly nonnormal distribution of residuals
- Residuals versus predicted values in order to spot inhomogeneities in variance
- Residuals versus different factor variables to detect weaknesses in the model
- Residuals versus run order to check latent influences of time and autocorrelation

There is a further systematic approach to coping with possible inhomogeneities in variance that was proposed by G. E. P. Box and D. R. Cox in 1964 and that can be summarized in the so-called Box – Cox plot: In addition to fitting a model to the response y, they suggest fitting models to the transformed response $(y^\lambda - 1)/\lambda$ for different λ and then to choose $\lambda_{\max}$ such that residuals are closest to normally distributed. The Box – Cox plot, displays performance of a transformation against λ. It is a practical way of finding indications of what type of transformation to the response data may be useful.

5.4. Heuristics for Improving Model Performance

An interesting although somewhat dangerous aspect of DoE and statistical analysis is that of "pruning" models. What is meant by this is the following: When a model displays some weaknesses in the analysis phase explained above, there is often the possibility to improve the performance of a model by simple measures that do not require additional experiments:

- Excluding model terms with insignificant coefficients
- Introducing a transformation
- Excluding observations that seem to unduly dominate the model or that lie far away from the model
- Include one or two terms in the model without overstressing the design, i.e., without having to perform further experiments

Of course there are always situations in which a model remains inadequate and further experiments must be performed to reach the objectives. The following is a set of heuristics, which may help to improve a model in one of several possible situations that may arise during the analysis phase:

- R^2 and Q^2 are both small (i.e., there is just bad fit): check for outliers using a normal probability plot; check that the response values correspond to the factor variables; check pure error if experiments have been repeated; check that no important factors and interactions are missing in the model.
- R^2 is high but Q^2 is very low (below 0.4, i.e., tendency for over-fit): remove very small and insignificant terms from the model (they may reduce the predictive power of the model); check for dominating outliers (by comparing observed and predicted values) and try fitting the model without them (remember, however, that for screening designs with few residual degrees of freedom, Q^2 may be low although the model is good).
- There are clear outliers in the normal probability plot: usually these outliers have not had much influence on the model (otherwise they would be seen when comparing observed and predicted values), however, they often lead to low R^2 values; check what happens when they are removed from the model; check the records, repeat the experiment; mistrust predictions of the model in the vicinity of such outliers; consider that the outlier may contain important information (maybe this is a new and better product).
- There is some structure in a plot showing residuals versus a factor variable (Fig. 19): this is a sign that the model is too weak and should be expanded; this can usually not be done without enlarging the design.

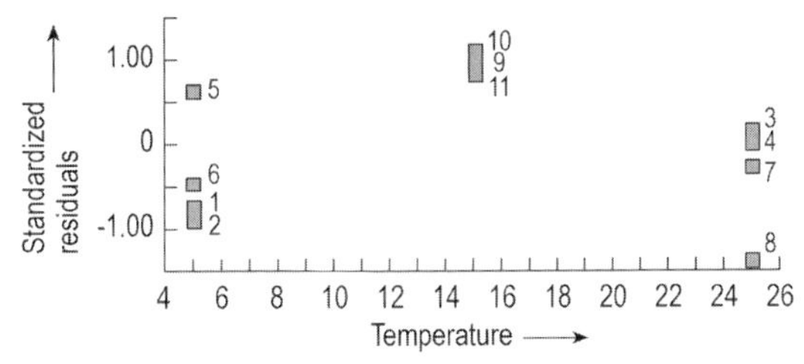

Figure 19. Standardized residuals are residuals divided by RSD = SD_{res} (see Section 5.2). When plotted against a factor variable they may give an indication of how a model can be improved

- In rare cases of analysis it may be observed in a plot displaying residuals versus predicted values that residuals are not homogeneous; this may be the case when a response varies over several orders of magnitude and the size of errors is either proportional to the response values or satisfies some other relation. Indications of this can also be seen in the Box – Cox plot if the optimal exponent λ_{max} deviates significantly from unity. A transformation of the response, e.g., taking logarithms or square roots, may be the correct measure.

Sometimes, the measures suggested above do not lead to a stable model; outliers seem to be present in spite of all attempts to improve the situation, there are bends and curves in the normal probability plot, coefficients are not significant but also not negligible, and so on. This is usually an indication that not enough experiments have been done. It is the authors' recommendation in this case to either

- Reduce all "manipulation" to a minimum (eliminate only obvious outliers from the data and very small coefficients from the model), and use the model knowing that it is weak but may still be useful, or to
- Strip the model down to linear, find the optimal conditions and repeat a larger design here.

In any case the heuristics above cannot repair a model that is simply incorrect, and care must be taken not to succumb to the temptation of systematically perfecting such an incorrect model. This is usually not the purpose of modeling.

5.5. Graphical Visualization of Response Surfaces

When statistical tests such as those described above detect no serious flaws in the model, it is useful to plot the model by reducing it to two or three dimensions and by representing it as a contour diagram (see Fig. 20) or a response surface plot (see Fig. 12).

When the model consists of more than two factors, surplus factors are set constant, usually to their optimum level, i.e., where responses are optimal. Setting a factor constant corresponds to slicing through the experimental domain and reducing its dimension by one. An interesting possibility is setting a third factor constant at three levels, and placing the three contours side by side.

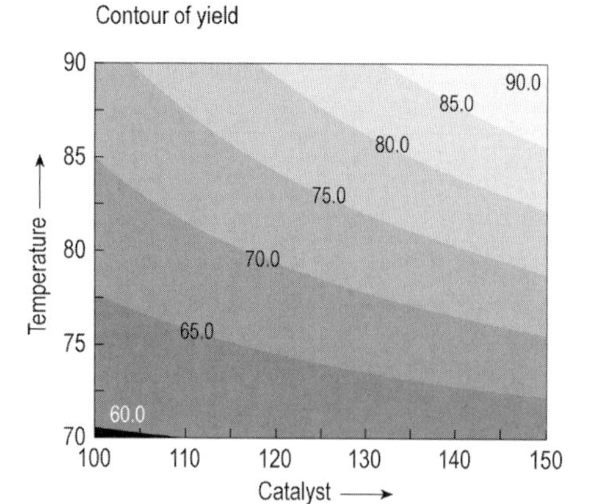

Figure 20. Contour plot of yield, showing the dependency on catalyst and temperature (for the data from Fig. 6 and Fig. 15). It can be seen that there is an interaction between the two factors

Visualizing models is particularly interesting when the modeling involves several responses, because contour diagrams can be used to find specification domains for the factors when specification ranges for the response variables have been imposed (e.g., by the customer of the chemical product to be produced; see Fig. 21).

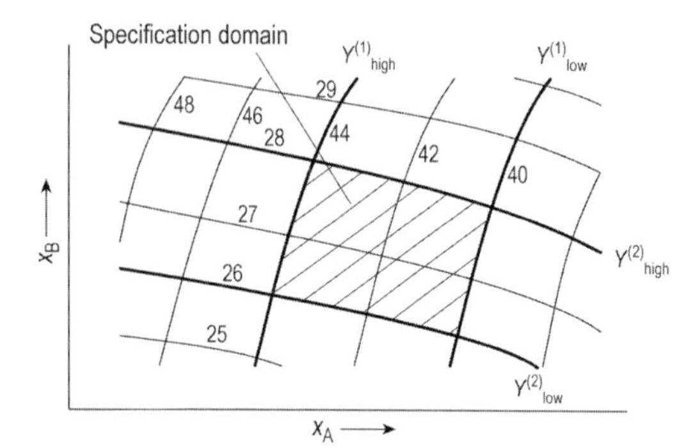

Figure 21. Superimposed contour plot for two responses used for finding specification domains for the two factors x_A and x_B, given the specifications for the responses $Y^{(1)}$ and $Y^{(2)}$

6. Optimization Methods

A major purpose of using response surface modeling techniques is optimization. There are several approaches to optimization depending on the complexity of the situation. Single-response optimizations are easier to handle and are treated in Sections 6.1 and 6.2. Multi-response optimization requires weighting the different responses according to their importance. This can be quite difficult, and the relevant techniques are treated in Section 6.3.

6.1. Basic EVOP Approach Using Factorial Designs

A very basic approach to optimization was proposed in the 1960s by G. E. P. Box and N. R. Draper [15]. It is known as Evolutionary Operation, or EVOP (and has nothing to do with evolution strategy or genetic optimization algorithms). When performing optimization experiments in a running production unit, it is not possible to vary factors over very large domains. Hence, effects will be small and very hard to detect against an underlying natural variability of the response. The idea is to simply repeat a small factorial design for, for instance, two factors, as often as is necessary so that the random noise can be averaged out and the effects become apparent.

The direction of maximal improvement of a response can be deduced from these effects and can subsequently be used to find a new position for a second factorial design. At this new position the second factorial is repeated until significant effects or a satisfactory result is obtained. This procedure can be repeated until a stable optimum is found (Fig. 22).

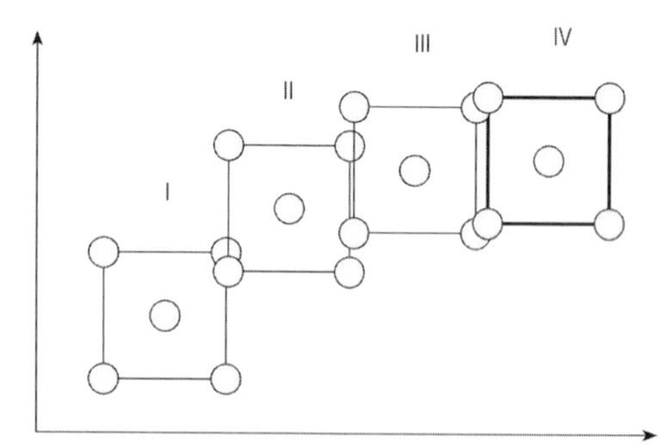

Figure 22. Example of a moving EVOP design (2^2 factorial design with a center point) that has found a stable optimum on the right

6.2. Model-Based Approach

When it is less difficult to establish the significance of a model, as is usually the case on a laboratory scale or in a pilot plant, the model can directly be used for optimization. Gradients of the polynomial models are calculated from the model coefficients because they point to the directions of maximum change in the responses. These directions are then used in an iterative search for a maximum or minimum, as was proposed by BOX and WILSON [18].

For quadratic and interaction models, it makes sense to use conjugate gradients, which correct the direction of maximum change by the curvature of the model function (Fig. 23). In this way the search for an optimum can often be accelerated. When a predicted optimum lies within the experimental domain, it is usually quite a good estimate of the real optimum, particularly when a response surface model has been used. When it is outside of the domain, where, by construction, the model is most probably not adequate in describing the process or system under study, it should be validated in the fashion described in Section 6.4.

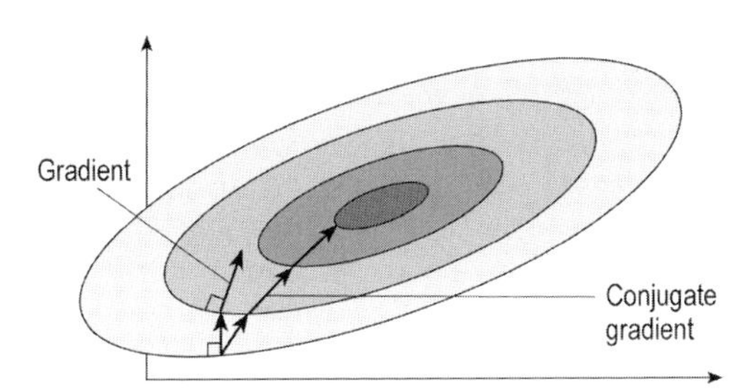

Figure 23. How the conjugate gradient compares to the gradient

6.3. Multi-Response Optimization with Desirability Functions

In an industrial environment, it is usually not sufficient to maximize or minimize one response only. A typical goal is to maximize yields, minimize costs, and reach specification intervals or target values for quality characteristics. But how should one proceed when these goals are contradictory? This question arises quite frequently in practice.

For example, yield, cost, and the quality characteristics are used as responses in an experimental design. The design should be chosen to allow adequate models to be fitted to all responses. For each response, a model will be fitted to its corresponding measured data. Now optimization can be started. Before being able to apply any mathematical optimization algorithm, the goal of the optimization has to be translated from a multidimensional target to a simple maximization or minimization problem. This can effectively be done by using desirability functions. A desirability function $d_j(y^{(j)})$ for one response $y^{(j)}$, measures how desirable each value of this response is. A high value of $d_j(y^{(j)})$ indicates a high desirability of $y^{(j)}$. Many types of desirability functions have been proposed in literature (e.g., [35]). Two-sided desirability functions are used if a target value or a value within specification limits is to be reached (Fig. 24, left), whereas one-sided desirability functions are used if the corresponding response is to be maximized (Fig. 24, right) or minimized.

An example of a two-sided desirability function is:

$$d_j\left(y^{(j)}\right)=0 \text{ for } y^{(j)}>y^{(j)}_{\max} \; or \; y^{(j)}<y^{(j)}_{\min},$$

$$d_j\left(y^{(j)}\right)=\left(y^{(j)}-y^{(j)}_{\min}\right)^s/\left(y^{(j)}_{\text{target}}-y^{(j)}_{\min}\right)^s$$
$$for\; y^{(j)}_{\text{target}}>y^{(j)}>y^{(j)}_{min},$$

$$d_j\left(y^{(j)}\right)=\left(y^{(j)}_{\max}-y^{(j)}\right)^t/\left(y^{(j)}_{\max}-y^{(j)}_{\text{target}}\right)^t$$
$$for\; y^{(j)}_{\max}>y^{(j)}>y^{(j)}_{target}.$$

This function depends on a target value $y^{(j)}_{\text{target}}$, a minimum acceptable value $y^{(j)}_{\min}$, a maximum acceptable value $y^{(j)}_{\max}$, and two exponents $s,t>0$ (Fig. 24, left). High values of s and t emphasize the importance of reaching the target; smaller values leave more room to move within $y^{(j)}_{\min}$ and $y^{(j)}_{\max}$.

To build one desirability function for all responses, the geometric mean is usually taken:

$$D(y)=\left[d_1\left(y^{(1)}\right)\cdot d_2\left(y^{(2)}\right)\cdot\ldots\cdot d_k\left(y^{(k)}\right)\right]^{1/k}$$

By substituting the y by the predicted values of the model functions, desirability D becomes a function of the factor variables, $D = D(\boldsymbol{x})$. To find the best possible factor settings, this func-

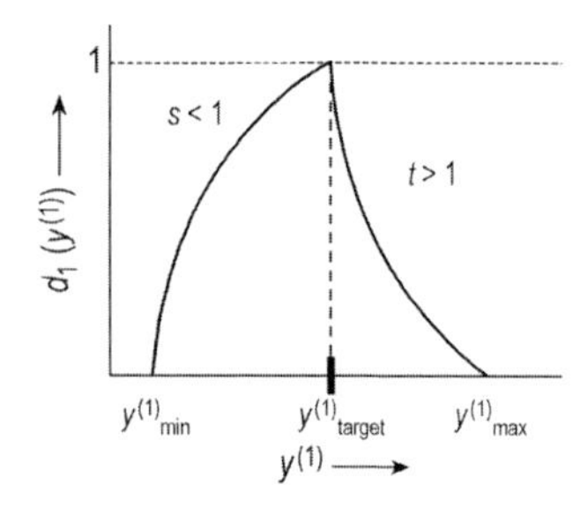

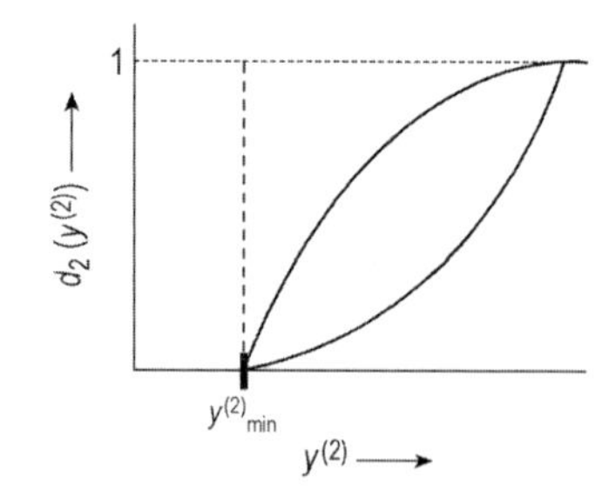

Figure 24. Two responses with corresponding desirability functions: a two-sided desirability function for $y^{(1)}$ (target value $y^{(1)}_{\text{target}}$ desired) and two different one-sided desirability functions for $y^{(2)}$ (maximization of $y^{(2)}$ desired)

tion D should be maximized by using any efficient mathematical optimization procedure.

In many typical applications, production cost and yield tend to increase simultaneously, and the optimizer, in trying to raise yield and lower cost, reaches an optimum somewhere within the experimental domain. Then it is essential to have good information about the quality of predictions in the interior of the domain. It may be necessary to fit a quadratic model to obtain such good predictions.

In a similar fashion, quality characteristics are often contradictory, e.g., more textile strength means less skin and body comfort; again, the optimizer will find an optimal compromise, which means suboptimal settings for the individual responses within the experimental domain.

Often, the predicted optimum is at the border of the experimental region. It is in fact quite typical that, in cases where linear or interaction models are used and no compromises as described above are necessary, two or more corners of the experimental domain are found to be locally optimal. From a mathematical point of view this seems to present a problem, but in fact it is a very promising situation for further improvement of the product or the process under study. Although the models were made to work inside the experimental region, they often also give very useful information just outside the experimental domain.

To be able to make a compromise between different y, it is sometimes more practical to use desirability functions d_j that are not exactly 0 for $y^{(j)} > y^{(j)}_{\text{max}}$ or $y^{(j)} < y^{(j)}_{\text{min}}$, but only close to 0. This allows the comparison of factor settings for which desirability would otherwise be zero, because some of the y are outside of specification. In any case, constructing desirability functions is quite a delicate task, and it is useful to play around with the desirability function when looking for an optimum.

When extrapolating outside the domain of x, remember that predictions using the model deteriorate when the experimental region is left. This means that predicted optima outside the region must always be validated by further experiments.

6.4. Validation of Predicted Optima

There are several reasons why predicted optima from empirical models should always be validated by further experiments:

- Errors in observed responses propagate into predictions of the model and hence into the position of the predicted optima
- The model may be insufficient to describe the relevant behavior of the experimental system under study and lead to bad predictions of optima
- The desirability function used in multiresponse optimization may not correctly depict all aspects of the real goal
- Predicted optima may lie outside the initial experimental domain, where the quality of the model is doubtful
- Predicted optima may be unsatisfying in that not all responses lie near the desired values or within specified ranges.

It is possible within the scope of linear modeling to calculate confidence intervals for all responses, and hence to get a good idea of the quality of prediction. This calculation of confidence intervals is only correct for qualified models as explained in Chapter 5.

Confidence intervals of this type must be mistrusted if there is a significant lack of fit or if Q^2 is very low. They should also be mistrusted if residuals of observed responses in the vicinity of the predicted optima are very large.

In this case and when predicted optima are outside the experimental domain, further experiments are inevitable. A good strategy for this is to find a sequence of experimental settings for which the predictions gradually improve (according to the model). This sequence will start within the experimental domain and will typically leave it at some point (see Fig. 25). Experiments in this sequence should be carefully performed and checked against predicted results. This procedure will usually lead to very good results. It should be emphasized, however, that these experiments are purely confirmatory in nature, they may lead to better products and better processes but they will not lead to better (fitted) models.

7. Designs for Special Purposes

The design methods that have been discussed until now cover many situations in which the cause and effect relationship between several factors and one or more responses of an experimental system are investigated. However there are cases where further methods are necessary to solve modeling problems. Four characteristic situations are treated in this chapter:

- Designs for modeling and optimizing mixtures are used in product optimization
- Designs for categorical factors may be used for raw product screening and for implementing blocking (as described in Section 2.3)
- So-called optimal or D-optimal designs which are implemented for advanced modeling when nonstandard models or irregular experimental domains are to be investigated
- Robust design techniques have the aim of not only optimizing response values but also response variability.

Further interesting topics, such as

- Nested designs, where the levels of factor B depend on the setting of factor A [16, 24]
- The field of QSAR (quantitative structure activity relationships), which is becoming more and more interesting in conjunction with the screening of active ingredients in medicines
- The whole field of DoE for nonlinear dynamic models involving differential algebraic equations

and others, cannot be treated here. The advanced reader is invited to research on his own, particularly on the latter two themes, which are still in constant motion (both of them incidently make extensive use of D-optimal designs to be shortly discussed in the following).

7.1. Mixture Designs

All designs described above assume that all factor variables can be set and varied independently of each other in an arbitrary manner. This is the typical situation in process optimizations, where factors are technical parameters such as temperatures, pressures, flow rates, and concentrations. However, when the goal is to model the properties of a mixture in order to find optimal ratios for its components, then this factor independence may no longer be presumed. There are many branches of the chemical industry and rubber industries where mixtures are investigated, and there is a whole range of characteristic problems:

- Optimizing the melting temperature of a metal alloy
- Reducing cost in producing paints while keeping quality at least constant
- Optimizing the taste of a fruit punch or the consistency of a yogurt
- Increasing adhesion of an adhesive
- Finding the right consistency of a rubber mixture for car tires

These are cases where experimental design methods must be modified to cope with the fact that all components of the mixture must add up to unity, i.e., for three components, $x_A + x_B + x_C = 1$. Nevertheless there are useful models and designs for mixtures. They are based on the so-called mixture triangle or the mixture simplex when four or more components are involved (Fig. 26).

All simplices have the following properties, which facilitate analysis and the visualization of results:

- Response plots can be generated based on the mixture triangle (as contour plots).

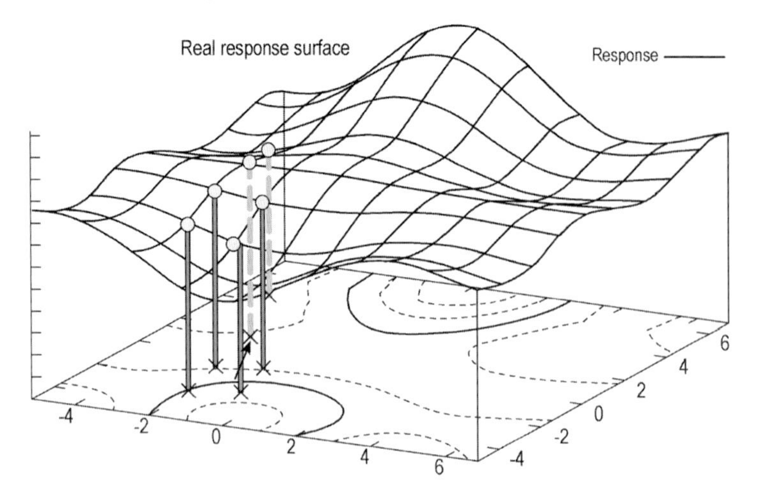

Figure 25. Predicting optima based on the model for increasing domains: A strategy in optimization consists of predicting a series of optima while slowly increasing the domain. Experiments should be performed at these predicted optima for validation purposes. In this way either a satisfactory optimum is reached or a further design is employed

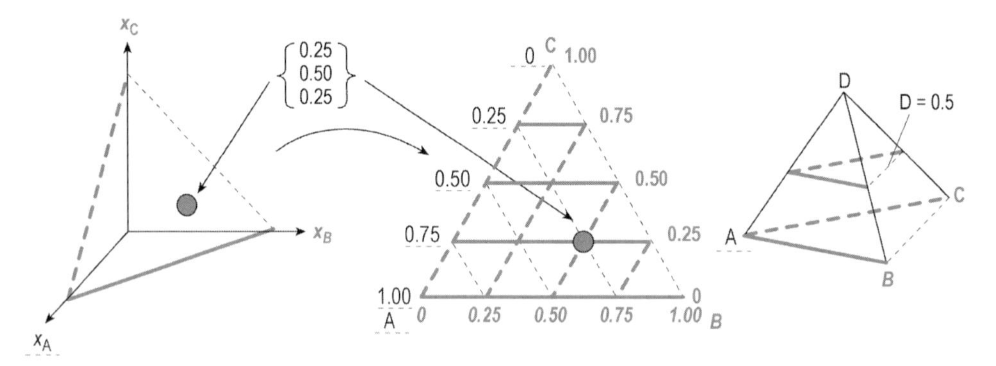

Figure 26. For three-component mixtures the cube is no longer the appropriate geometrical object for modeling the experimental domain. The equation $x_A + x_B + x_C = 1$ defines a triangle. Unfolding this triangle leads to the mixture triangle for three factors (left).
Mixtures with four components can be modeled by a simplex; setting one of the factors constant again leads to a triangle for the others, e.g., $x_D = 0.5$, i.e., $x_A + x_B + x_C = 0.5$ (right)

- A simplex is very similar to the cube in that its boundaries and also its sections parallel to its boundaries are again simplices in a lower dimension. This allows contour plots to be used in the mixture triangle even in situations where more than four components are involved; just set the surplus factors constant and put the three important ones into the mixture triangle.
- Lower and upper levels of the mixture factors can be visualized geometrically as cutting off parts of the simplex. An active upper level means cutting off a corner, an active lower level means cutting off the base (or side) of a simplex. When upper levels are active, the experimental domain is no longer a regular simplex, but only a subset thereof, a so-called irregular mixture region.

The best experimental designs for regular mixture regions are so-called simplex lattice designs. Simplex lattice designs place experiments at the corners and along edges and axes of a simplex. Depending on the complexity of the model that is to be fitted, different strategies in picking out the points are used. The most typical design for a linear model is the axial simplex design which consists of experiments at the corners, at the centroid (equal amounts of all components) and midpoints of the axes from the corners to the centroid (Fig. 27, Table 6). Details can be found in [36].

Table 6. Design matrix for an axial simplex design for three mixture-components

Run no.	Comp. A	Comp. B	Comp. C
1	1	0	0
2	0	1	0
3	0	0	1
4	0.6666	0.1667	0.1667
5	0.1667	0.6666	0.1667
6	0.1667	0.1667	0.6666
7	0.3333	0.3333	0.3334

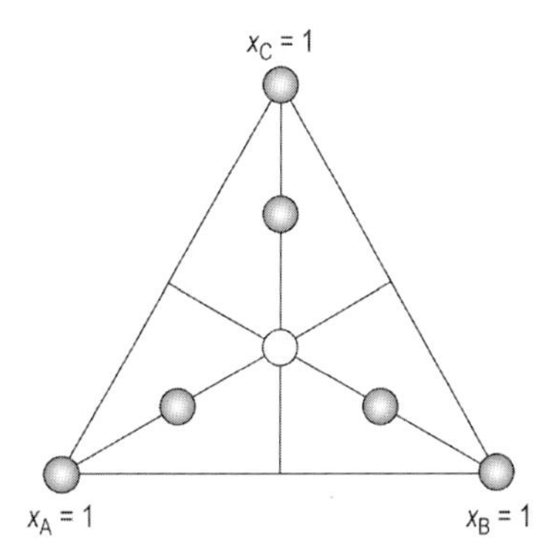

Figure 27. Geometrical view of an axial simplex design for three-component mixtures

When investigating mixtures, it is important to keep in mind that, although it is possible to correlate changes in the responses to changes in the factor settings, it is principally impossible to tell which factors have caused these changes in the responses, because whenever one factor has changed, at least one other factor has also changed. So which one should be made responsible? Example 1: A fruit punch containing pineapple, grapefruit, and orange juice seems to become tastier when pineapple juice is added. Example 2: A long drink containing pineapple and orange juice and vodka seems to become less bitter when pineapple juice is added. Is it possible, in the two examples, to attribute the change in taste to the change in the amount of pineapple?

When mixtures are modeled and optimized, it is not always necessary to use simplex designs and corresponding models. There are principally three ways to proceed, only one of which involves using simplex designs:

- Define component ratios as factors and use the classical factorial design or a CCD (instead of ratios, any other transformation leading to independent pseudocomponents can be used).
- Use a simplex design, as described above, or D-optimal design, as described in Section 7.3.
- Identify a filler component, for example, x_C, which does not have an effect on any of the responses, use a factorial design, a CCD, or a D-optimal design for the remaining components (together with the constraint $x_A + x_B \leq 1$, if necessary), and regression analysis for a model in which the filler variable does not appear.

Analyzing mixture designs requires some care, because due to the dependencies amongst the mixture factors the models must be adapted to the situation. H. SCHEFFÉ and D. R. COX were the pioneers who developed ways to use models correctly for the mixture problem. Details cannot be included here. The reader is referred to the literature [25, 36 – 38].

7.2. Designs for Categorical Factors

Not all factors that influence a product or a process can be quantified. Examples of nonquantifiable or so-called categorical (or discrete) factors are:

- The supplier of a raw material, who may influence some of the quality characteristics of an end product
- The date or the time of the year may influence how well a process will perform
- Different persons doing the same experiments
- The type of catalyst or solvent
- Mutants of a strain of bacteria in a fermentation process

All these are possible influencing factors that cannot be quantified in a satisfactory manner. (In the case of investigating solvents it would be a very good idea to consider polarity as a quantifiable factor instead of just the type of solvent.) Hence, particularly when more than two instances of a categorical variable must be considered, new design techniques are necessary.

Typical questions that arise in conjunction with categorical or discrete factors are:

- How large is the impact of a qualitative factor? Is it significant?
- If so, which instance yields the best results?
- Does the categorical factor interact with other factors? Are there interactions between different categorical factors?
- Is a possible optimum for the other (continuous) factors robust with respect to varying the categorical factor? If so, where does it lie?

If there are just two instances or categories of a qualitative factor, it is easy to encode them as "–" and "+" and to use factorial designs. The categorical or qualitative factor can then be used in calculations like a continuous quantitative factor. Effects and coefficients can be calculated in the usual way, and they measure how large the influence of changing categories is. The same is true for interactions with other factors; they measure how the influence of these factors on the responses changes when changing categories.

Treating categorical factors with three or more instances is much more delicate. Within generalized linear modeling, it is possible to treat factors with three or more instances by increasing the number of dimensions of the problem. Dummy variables are used to differentiate between instances: Let p_1, p_2 be two dummy variables and encode instances i_1, i_2, i_3 in the fashion shown in Table 7.

Table 7. How to encode three instances of a categorical factor by using two dummy variables

Coding for	p_1	p_2
Instance i_1	1	0
Instance i_2	0	1
Instance i_3	– 1	– 1

Geometrically the three instances become three points in the two-dimensional coordinate system of the dummy variables. The coded design with example values for a response y as well as the corresponding geometrical visualization is shown in Figure 28.

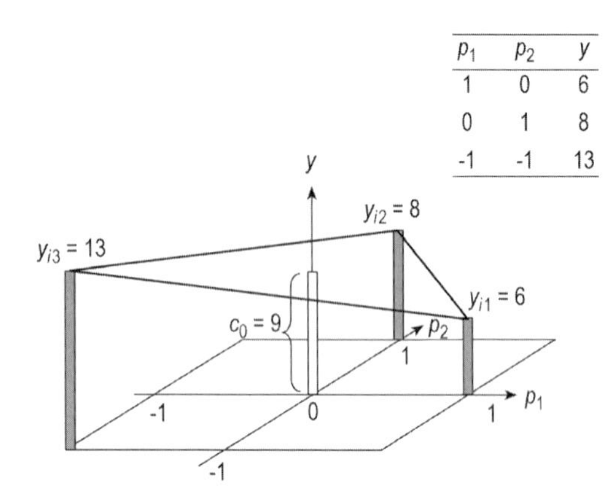

Figure 28. Geometrical and tabular representation of the design and the response values

Now the advantage of using dummy variables becomes apparent: Analysis of data can be done using the same regression methods as described in Section 4.4, and the interpretation of results, although being somewhat different, is still relatively straightforward:

The model has the form

$$y = c_0 + c_1 p_1 + c_2 p_2$$

where c_1 and c_2 are coefficients pertaining to the dummy variables p_1 and p_2.

Model predictions for the instances i_1, i_2, i_3 become:

$$\hat{y}(i_1) = c_0 + c_1$$
$$\hat{y}(i_2) = c_0 + c_2$$
$$\hat{y}(i_3) = c_0 - c_1 - c_2$$

So the coefficients c_1, c_2 are actually a measure of the extent to which instances i_1 and i_2 differ from the mean c_0.

By additionally setting $c_3 = -c_1 - c_2$ the same measure has been found for instance i_3.

For the example shown in Figure 28 the coefficients are $c_0 = 9$, $c_1 = -3$, $c_2 = -1$, $c_3 = 4$.

Designs for categorical variables will normally be adapted to the problem at hand and generated by a computer algorithm. In fact optimal designs, as described in Section 7.3, should be used.

It is interesting to note that dummy variables can be used to model interactions: Just use these variables in interaction terms of a model. The coefficients then measure by how much the effect of another factor varies from a given instance to its mean effect taken over all instances (i.e., its

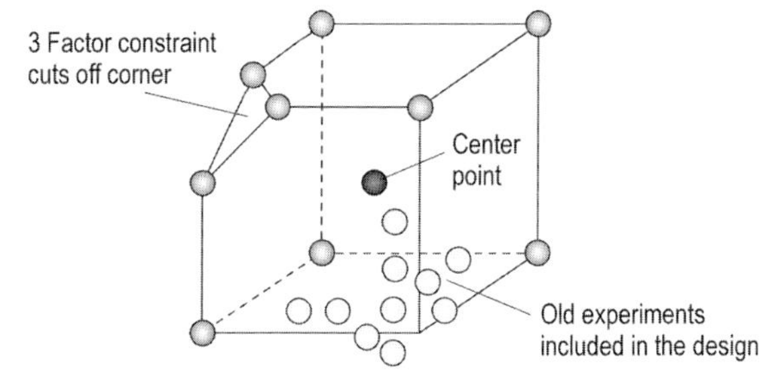

Figure 29. An optimal design with constraints and inclusion of some old experiments

main effect). It would go beyond the scope of this article to go into details about this, in particular about calculation of degrees of freedom. A short digression on dummy variables and their use for modeling blocking can be found in [32].

7.3. Optimal Designs

The usual procedure in planning an experimental design is to identify factors and responses, to determine ranges for the factors, and then to choose a standard design amongst those that have been discussed. This design may be factorial, Plackett – Burman, CCD, Box – Behnken, Simplex, or the like. However, in some situations none of the standard designs are suitable. Such situations typically are:

- Use of nonstandard models
- Use of nonstandard experimental domains (i.e., constraints involving more than one factor)
- Special restrictions on the number of runs or tests that can be performed
- Use of mixture designs when either upper factor levels are active (this means that the regular simplex structure is corrupted) or when normal continuous factors are to be investigated in the same design
- One or more categorical factors with three or more discrete settings are present.

In these cases it is common practice to generate a design by using a mathematical algorithm that maximizes the information that the design will contain. Such designs depend on the model to be used and on the experimental domain to be covered.

Criteria involved in optimizing designs are based on the extended design matrix, usually denoted by $\boldsymbol{X}$, which is built up of the design matrix and extended by a column for each term in the model. The basic idea behind optimal designs is to ensure that there are no correlations between the columns of this $\boldsymbol{X}$ matrix. For if there are correlations here, then the influence of the terms involved cannot be resolved (this situation is similar to that described in Section 7.1 in conjunction with mixture components, which are always correlated).

Working with optimal designs (Fig. 29) involves:

- Defining the experimental domain (including possible constraints)
- Choosing the appropriate model
- Specifying experiments that the experimenter explicitly wants to perform or has already performed
- Selecting the optimization criteria for the design
- Specifying approximately how many experimental runs can be performed.

Typically not just one design is generated, but a whole number of designs of differing size (test or run number). This makes it possible to evaluate designs by comparing the optimality criteria for different designs. Typical criteria are D-optimality, G-optimality, A-optimality, and E-optimality. They are essentially variance-minimizing design criteria in the sense that the variances of model predictions or model coefficients are minimized. The most commonly used criterion is D-optimality, which leads to a maximized determinant of the squared $\boldsymbol{X}$ matrix. For more details, in particular, for further optimality criteria, the reader is referred to special literature on optimal designs [39, 40].

Optimal designs will have a high quality if a sufficient number of experimental runs or tests has been allowed for. They are analyzed by regression analysis like other designs, and the corresponding models can be used for prediction and optimization purposes.

7.4. Robust Design as a Tool for Quality Engineering

Sometimes the goal of investigation goes beyond estimating the effects of influencing factors and predicting process behavior or quantifiable product characteristics. A typical application of DoE techniques is toward finding conditions, i.e., factor settings, for which not only quantifiable responses are optimal (usually in a multivariate way as described in Section 6.3) but also the variability of response values is minimal with respect to disturbing or environmental factors that act during production or field use of the product. Typically, these influences can not all be controlled or are too expensive to be controlled even during the experimentation phase.

Doing DoE with the goal of reducing variability is known as robust design. In a robust design, control factors or design factors are varied in a design as described above, typically in a screening design, and experiments are repeated at each trial run so as to estimate the standard deviation (or variance) at each trial. This dispersion measure is employed as a new response value. Statistical analysis and optimization tools such as those described above are then used to quantify and finally to minimize variation, while at the same time improving product characteristics.

TAGUCHI [1, 19, 20] introduced an additional idea into robust design, namely, the concept of so-called noise factors or environmental factors [1, 11]. These are introduced as perturbing factors in a designed experiment with the idea that they simulate possible external influences which may effect the quality of products after they have left the plant. These factors, which will in fact increase the standard deviation, are varied in a second experimental design which is performed for each of the trial runs in the design for the design factors. To distinguish between the two designs, the design for the environmental factors is sometimes called outer array and that for the design factors is called inner array. The use of outer arrays is recommended, when noise factors can rather cheaply be varied independently of the controlled factors.

Example: During the production of CDs a transparent coating is applied to protect the optical layer. CDs are placed onto a turntable, a droplet of lacquer is applied, and the CD is spun to spread the lacquer. Parameters to be varied are the size of the droplet, the speed and the acceleration of the turntable and the temperature in the room. CDs are to be subjected to different extreme climatic conditions in order to simulate their performance in the field. This is to be done in a climate chamber. Factors to be varied in the corresponding outer array are humidity and temperature in the climate chamber. For the inner array a 2^{4-1} factorial design with 3 realizations at the center point is chosen, for the outer array a simple 2^2 full factorial (Table 8). Results from the four experiments in the outer

Table 8. A design for a quality engineering problem, using a 2^{4-1} factorial with center point as inner array and a 2^2 factorial as an outer array.

Run	Design factors; A, B, C, D				Environmental factors: H, T				Mean	Variance
					H = −	H = −	H = +	H = +		
					T = −	T = +	T = −	T = +		
	A	B	C	D	y1	y2	y3	y4	y-mean	y-var
1	−	−	−	−						
2	+	−	−	+						
3	−	+	−	+						
4	+	+	−	−						
5	−	−	+	+						
6	+	−	+	−						
7	−	+	+	−						
8	+	+	+	+						
9	0	0	0	0						
10	0	0	0	0						
11	0	0	0	0						

array are taken together, and the mean and the variance are calculated and subjected to effect calculation or regression analysis.

8. Software

Selected DoE software and suppliers are listed in Table 9.

Table 9. A selection of DoE software tools and providers (as of Nov. 2005).

Tool	Company	WEB link
Design-Expert	Stat-Ease Inc.	http://www.statease.com
D.o.E. FUSION	S-Matrix Corp.	http://www.s-matrix-corp.com
ECHIP	ECHIP Inc.	http://www.echip.com
JMP, SAS/QC	SAS Institute Inc.	http://www.sas.com
MINITAB	Minitab Inc.	http://www.minitab.com
MODDE	Umetrics	http://www.umetrics.com
Starfire, RS/Series	Brooks Automation Inc.	http://www.brooks.com
STATGRAPHICS Plus	Manugistics Inc.	http://www.statgraphics.com
STATISTICA	StatSoft Inc.	http://www.statsoft.com
STAVEX	AICOS Technologies AG	http://www.aicos.com

9. References

1. R. N. Kacker: "Off-Line Quality Control, Parameter Design, and the Taguchi Method (with Discussion)", *J. Quality Technol.* **17** (October 1985) no. 4, 176 – 209.
2. M. S. Phadke: *Robuste Prozesse durch Quality Engineering (Quality Engineering Using Robust Design)*, gfmt, München, 1990 (Prentice Hall, London, 1989).
3. T. Pfeifer: *Qualitätsmanagement: Strategien, Methoden, Techniken*, Hanser, München 1993.
4. J. Wallacher: *Einsatz von Methoden der statistischen Versuchsplanung zur Bestimmung von robusten Faktorkombinationen in der präventiven Qualitätssicherung*, Fortschr.-Ber. VDI Reihe 16 Nr. 70, VDI-Verlag, Düsseldorf 1994.
5. G. E. P. Box, S. Bisgaard: The Scientific Context of Quality Improvement, *Quality Progr.* **June** (1987) 54 – 61.
6. ISO 3534–3: 1999 (E/F): Statistics — Vocabulary and symbols — Part 3: Design of experiments.
7. A. Orth, M. Schottler, O. Wabersky: Statistische Versuchsplanung, Serie: Qualität bei Hoechst, Hoechst AG 1993.
8. S. Soravia: "Quality Engineering mit statistischer Versuchsmethodik", *Chem.-Ing.-Tech.* **68** (1996) no. 1 + 2, 71 – 82.
9. DuPont Quality Management & Technology: *Design of Experiments — A Competitive Advantage*, E. I. du Pont de Nemours and Company 1993.
10. R. A. Fisher: *The Design of Experiments*, 8th ed., Oliver & Boyd, London 1966.
11. S. Bisgaard: Industrial Use of Statistically Designed Experiments: Case Study References and Some Historical Anecdotes, *Quality Eng.* **4** (1992) no. 4, 547 – 562.
12. O. L. Davies (ed.): *The Design and Analysis of Industrial Experiments*, 2nd ed., Oliver & Boyd, London 1956.
13. K. H. Simmrock: "Beispiele für das Auswerten und Planen von Versuchen", *Chem.-Ing.-Tech.* **40** (1968) no. 18, 875 – 883.
14. G. E. P. Box, N. R. Draper: *Empirical Model-Building and Response Surfaces*, John Wiley & Sons, New York 1987.
15. G. E. P. Box, N. R. Draper: *Evolutionary Operation — A Statistical Method for Process Improvement*, John Wiley & Sons, New York 1998.
16. G. E. P. Box, W. G. Hunter, J. S. Hunter: *Statistics for Experimenters: Design, Innovation and Discovery*, 2nd ed; John Wiley & Sons, New York 2005.
17. G. E. P. Box, H. L. Lucas: Design of Experiments in Nonlinear Situations, *Biometrika* **46** (1959) 77 – 90.
18. G. E. P. Box, K. B. Wilson: On the Experimental Attainment of Optimum Conditions, *J. Roy. Statist. Soc.* **B 13** (1951) 1 – 45.
19. G. Taguchi: *System of Experimental Design*, **vols. I and II**, Kraus International Publications, New York 1987.
20. B. Gunter: "A Perspective on the Taguchi Methods", *Quality Progr.* (June 1987) 44 – 52.
21. J. S. Hunter: "Statistical Design Applied to Product Design", *J. Quality Technol.* **17** (October 1985) no. 4, 210 – 221.
22. K. R. Bhote: *Qualität — Der Weg zur Weltspitze (World Class Quality)*, IQM, Großbottwar 1990 (American Management Association, New York 1988).
23. B. Mittmann: "Qualitätsplanung mit den Methoden von Shainin", *Qualität und Zuverlässigkeit (QZ)* **35** (1990) no. 4, 209 – 212.

24. D. C. Montgomery: *Design and Analysis of Experiments*, 6th ed., John Wiley & Sons, New York 2005.
25. E. Scheffler: *Statistische Versuchsplanung und -auswertung — Eine Einführung für Praktiker,* 3., neu bearbeitete und erweiterte Auflage von "Einführung in die Praxis der statistischen Versuchsplanung", Deutscher Verlag für Grundstoffindustrie, Stuttgart 1997.
26. D. E. Coleman, D. C. Montgomery: "A Systematic Approach to Planning for a Designed Industrial Experiment (with Discussion)", *Technometrics* **35** (1993) no. 1, 1 – 27.
27. G. J. Hahn: "Some Things Engineers Should Know About Experimental Design", *J. Quality Technol.* **9** (January 1977) no. 1, 13 – 20.
28. C. D. Hendrix: "What Every Technologist Should Know about Experimental Design", *CHEMTECH* **March** (1979) 167 – 174.
29. R. L. Mason, R. F. Gunst, J. L. Hess: *Statistical Design and Analysis of Experiments with Applications to Engineering and Science*, John Wiley & Sons, New York 2003.
30. E. Spenhoff: *Prozeßsicherheit durch statistische Versuchsplanung in Forschung, Entwicklung und Produktion*, gfmt, München 1991.
31. R. H. Myers, D. C. Montgomery: *Response Surface Methodology: Process and Product Optimization Using Designed Experiments*, John Wiley & Sons, New York 2002.
32. N. R. Draper, H. Smith: *Applied Regression Analysis*, 3rd ed., John Wiley & Sons, New York 1998.
33. H. O. Hartley: "Smallest Composite Designs for Quadratic Response Surfaces", *Biometrics* **15** (1959) 611 – 624.
34. I. E. Frank, J. H. Friedman: "A Statistical View of Some Chemometrics Regression Tools", *Technometrics* **35** (May 1993) no. 2, 109 – 148.
35. G. Derringer, R. Suich: "Simultaneous Optimization of Several Response Variables", *J. Quality Technol.* **12** (Oktober 1980) no. 4, 214 – 219.
36. J. A. Cornell: *Experiments with Mixtures — Designs, Models, and the Analysis of Mixture Data*, 3rd ed., John Wiley & Sons, New York 2002.
37. J. Bracht, E. Spenhoff: "Mischungsexperimente in Theorie und Praxis (Teil 1 und 2)" *Qualität und Zuverlässigkeit (QZ)* **39** (1994) no. 12, 1352 – 1360, *Qualität und Zuverlässigkeit (QZ)* **40** (1995) no. 1, 86 – 90.
38. R. D. Snee: "Experimenting with Mixtures", *CHEMTECH* (November 1979) 702 – 710.
39. A. C. Atkinson, A. N. Donev: *Optimum Experimental Designs*, Oxford University Press, Oxford 1992.
40. F. Pukelsheim: *Optimal Design of Experiments*, John Wiley & Sons, New York 1993.

Microreactors – Modeling and Simulation

STEFFEN HARDT, TU Darmstadt, Darmstadt, Germany

See also: Computational Fluid Dynamics, Molecular Dynamics Simulations

1. **Introduction** 401
2. **Flow Distributions** 405
2.1. **Straight Microchannels** 405
2.2. **Periodic and Curved Channel Geometries** 406
2.3. **Multichannel Flow Domains** 407
3. **Heat Transfer** 411
3.1. **Straight Microchannels** 411
3.2. **Periodic and Curved Channel Geometries** 412
3.3. **Multichannel Flow Domains** 413
3.4. **Micro Heat Exchangers** 415
4. **Mass Transfer and Mixing** 418
4.1. **Simple Mixing Channels** 418
4.2. **Chaotic Micromixers** 420
4.3. **Multilamination Micromixers** 425
4.4. **Hydrodynamic Dispersion** 427
5. **Chemical Kinetics** 429
5.1. **Numerical Methods for Reacting Flows** 430
5.2. **Reacting Channel Flows** 431
5.3. **Heat-Exchanger Reactors** 433
6. **Conclusion and Outlook** 434
7. **References** 436

1. Introduction

In the field of conventional, macroscopic process technology modeling and simulation approaches are by now used on a routine basis to design and optimize processes and equipment. Many of the models employed have been developed for and carefully adjusted to specific processes and reactors and allow flow and heat and mass transfer to be predicted, often with a high degree of accuracy. In comparison, modeling and simulation approaches for microreactors are more immature, but have great potential for even more reliable computer-based process engineering. In general, the purposes of computer simulations are manifold, such as feasibility studies, optimization of process equipment, failure modeling, and modeling of process data. For each of these tasks within the field of chemical engineering simulation methods have been applied successfully.

Microreactors are developed for a variety of different purposes, specifically for applications which require high heat- and mass-transfer coefficients and well-defined flow patterns. The spectrum of applications includes gas and liquid flow as well as gas – liquid or liquid – liquid multiphase flow. The variety and complexity of flow phenomena clearly pose major challenges to the modeling approaches, especially when additional effects such as mass transfer and chemical kinetics must be taken into account. However, there is one aspect which makes the modeling of microreactors in some sense much simpler than that of macroscopic equipment: the laminarity of the flow. Typically, in macroscopic reactors the conditions are such that a turbulent flow pattern develops, which makes the use of turbulence models [1] necessary. With turbulence models the stochastic velocity fluctuations below the scale of grid resolution are accounted for in an effective manner, without the need to explicitly model the time evolution of these fine details of the flow field. Heat- and mass-transfer processes strongly depend on the turbulent velocity fluctuations, and for this reason the accuracy of the turbulence model is of paramount importance for a reliable prediction of reactor performance. However, currently (2006) there is no model available which is capable of describing turbulent flow phenomena in a universal manner and is computationally inexpensive at the same time. For this reason, simulation approaches for microreactors, which usually do not require turbulence models, offer some potential to make predictions with a degree of accuracy unparalleled by models of macroscopic reactors.

When comparing processes in microreactors with those in conventional systems, a few general differences can be identified:

Ullmann's Modeling and Simulation

ISBN: 978-3-527-31605-2

- Flow in microstructures is usually laminar, in contrast to the turbulent flow patterns on the macroscale.
- The diffusion paths for heat and mass transfer are very short, and this makes microreactors ideal candidates for heat- or mass-transfer-limited reactions.
- The surface-to-volume ratio of microstructures is very high. Thus, surface effects are likely to dominate over volumetric effects.
- The fraction of solid wall material is typically much higher than in macroscopic equipment. Thus, solid heat-transfer plays an important role and must be accounted for when designing microreactors.

While the absence of turbulence simplifies many modeling tasks, the predominance of surface effects introduces additional complications, especially in the case of multiphase flow. Some of the fundamental mechanisms, for example, those of dynamic wetting and spreading phenomena, are not yet well understood, and this adds some degree of uncertainty to the modeling of these processes. As more and more practical applications of microfluidic systems emerge, research in the field of fluidic surface and interfacial phenomena gains additional impetus. It is thus hoped that in the following years refined models for microfluidic multiphase systems will be formulated and will add an additional degree of predictability to flow phenomena in microreactors.

When facing the task of formulating a model of a specific microfluidic system, the question arises whether the conventional macroscopic equations describing fluid flow and heat and mass transfer are still valid on the microscale. Systems for chemical processing rarely contain structures with dimensions smaller than 10 μm; the relevant length scale is often in the range of 100 μm. The standard approach to modeling transport processes in microreactors relies on a continuum description, i.e., continuum-field quantities such as pressure and velocity are introduced to represent the mean dynamics of the fluid molecules. The corresponding mathematical formulation is based on the Navier – Stokes equation and convection – diffusion equations for heat and mass transfer. The question to answer is whether or not this mathematical framework is suitable to describe transport processes in microreactors on the relevant length scales. Besides the continuum assumption there is one additional assumption on which the Navier – Stokes and related equations rest. This is related to the local statistical distribution of the particles in phase space: The standard convection – diffusion equations rely on the assumption of local thermal equilibrium. For gas flow, this means that a Maxwell – Boltzmann distribution is assumed for the velocity of the particles in the frame of reference co-moving with the fluid. When gas flow in microreactors at high temperature or low pressure is considered, this assumption may break down. The principle quantity determining the flow regime of gases and deviations from the standard continuum description is the Knudsen number Kn (Eq. 1).

$$\mathrm{Kn} = \frac{\lambda}{L} \tag{1}$$

The Knudsen number is the ratio of two length scales: the mean free path of the gas molecules λ and a characteristic length scale of the flow domain L, for example, the channel diameter. When Kn is larger than unity, a gas molecule is more likely to collide with the channel wall than with another molecule. As the transport of momentum or enthalpy is to a large extent governed by the collisions between molecules, major changes in the flow behavior are expected when the Knudsen number is on the order of unity. This may happen when gas flow through narrow channels is considered, but also when the temperature is high and/or the pressure is low.

Based on the Knudsen number, four different flow regimes can be distinguished [2]:

- Continuum flow with no-slip boundary conditions ($Kn \leq 10^{-2}$)
- Continuum flow with slip boundary conditions ($10^{-2} < Kn \leq 10^{-1}$)
- Transition flow ($10^{-1} < Kn \leq 10$)
- Free molecular flow ($Kn > 10$)

In the first two cases the Navier – Stokes equation can be applied, with modified boundary conditions in the second case. The computationally most difficult case is the transition flow regime, which, however, may be encountered in microreactor systems. Clearly, the defined ranges of Knudsen numbers are not rigid; they instead vary from case to case. However, the numbers given above are guidelines applicable to many situations encountered in practice.

For applications in the field of microreaction engineering, the conclusion can be drawn that the Navier – Stokes equation and other continuum models are valid in many cases, as Knudsen numbers greater than 10^{-1} are rarely obtained. This observation is supported by the fact that a typical mean free path of nitrogen at 1 bar is about 70 nm. However, it may be necessary to use slip boundary conditions to describe the flow in microchannels. The first theoretical investigations on slip flow of gases were carried out already in the 1800s by MAXWELL and VON SMOLUCHOWSKI. The basic concept relies on a so-called slip length L_s, which relates the local shear strain to the relative flow velocity at the wall (Eq. 2)

$$u_{gas} - u_{wall} = L_s \frac{\partial u_{gas}}{\partial y}\Big|_{Wall} \tag{2}$$

where u_{gas}, u_{wall} refer to the gas and wall velocities parallel to the wall and y is the coordinate normal to the wall. The slip length L_s is depends on wall roughness and can be determined experimentally. Similarly, there is a relationship for the temperature jump across a solid surface in the slip-flow regime [2].

In the transition flow regime, which is rarely encountered in microreactors, the usual convection – diffusion equations cease to be valid. The general formulation to be applied in this regime is given by the Boltzmann equation, from which less involved model descriptions such as the Burnett equation can be derived via a low-*Kn* expansion [3]. However, rather than by a continuum model, transition flows are often computed based on the Direct Simulation Monte Carlo (DSMC) method [4]. Instead of modeling gas flow by a set of differential or integro-differential equations, this method is based on tracking the trajectories and interactions of gas molecules directly. DSMC is a time marching approach based on a time discretization with steps smaller than the collision time of the molecules. The computational demand for DSMC is much higher than for continuum approaches.

Similar to the case of gas flows, the question of the range of applicability of continuum models for liquid flow arises. While the kinetic theory of gases provides clear indications of the limits of continuum models and of the onset of rarefaction effects on the microscale, there is no general framework explaining possible deviations of liquid flow phenomena from their macroscopic behavior. The concept of mean free path ceases to be useful for liquids, as the molecules interact with their neighbors in a permanent manner (in a sense, the molecular mean free path is zero for liquids). Some groups have conducted experiments on liquid flows in microstructures and measured pressure drop and heat-transfer coefficients, with largely contradictory results [5 – 11]. A compilation of the published results does not seem to leave room for any simple conclusions of general importance. Furthermore, the physics of Newtonian liquids does not suggest any universal mechanisms by which the flow behavior in channels of several hundred micrometers in width could differ considerably from macroscopic behavior at the same Reynolds number. However, in some specific situations liquid microflows exhibit a behavior which does not occur on the macroscale, but which can be reproduced experimentally and explained by theoretical models.

Boundary slip of liquids is a first example of such effects. Since ca. 1995 it was confirmed that not only gases, but also liquids can exhibit boundary slip, with corresponding experiments indicating a slip length of a few tens of nanometers [12 – 14]. The phenomenon has also been studied by numerical simulations, especially the molecular dynamics (MD) method (→ Molecular Dynamics Simulations), which is based on solving Newton's equations of motion for an ensemble of mutually interacting molecules (see, e.g., [15]). MD simulation results confirm the existence of boundary slip for liquids and indicate how the slip length depends on molecular parameters and shear rate [16].

While slip-flow effects are mostly expected to have an impact on flow behavior on the submicrometer scale, there is another phenomenon which may lead to deviations from the usual model predictions on a larger length scale. On a solid surface exposed to an ionic liquid, charges may accumulate to form a Debye layer above the surface, to which an additional layer of immobilized charges is attached [17]. This so-called electric double layer (EDL) is related to accumulation of charges inside the liquid which act as sources of an electric field. In turn, this field is sensitive to changes in the EDL structure which may be induced by the flow in such a manner that inside a channel a field component in streamwise direction builds up. When ionic charges start to migrate in the electric field, liquid molecules are dragged along with them and the apparent viscosity for channel flow changes. The phenomenon is known as electroviscous effect [18] and may cause a considerable modification of the friction factor in comparatively large channels. The electroviscous effect was studied theoretically for rectangular microchannels [19]. The channel width and height considered were in the range between 20 and 40 μm. Depending on the solute concentration, the friction factor was increased by up to 30 %, which shows that the electroviscous effect manifests itself on length scales much larger than the EDL thickness, typically on the order of a few nanometers. Thus, even in typical microchannel geometries ionic liquids may exhibit some "unexpected" behavior.

Many of the flow phenomena discussed here can be incorporated into continuum models such as the Navier – Stokes equation, an exception being gas flows in the transition and free molecular regimes. Only a modification of the boundary conditions and not of the model equations themselves is usually necessary, i.e., the non-standard physics characteristic for the micro- and nanoscale mainly appears as a boundary effect. Thus, when attempting to model transport processes in microreactors, the continuum hypothesis is a far-reaching concept which covers most of the scenarios encountered in practice. For a more detailed discussion of the special aspects of micro- and nanoflows, see [20]. Along this line of arguments it becomes clear that to a large extent the problem of modeling transport phenomena in microreactors is a scaling problem: The familiar macroscopic model equations may be used, but they must be applied in an unfamiliar and possibly unexplored parameter range as far as the dimensionless groups characterizing the problem are concerned [21]. In that context, care must be taken to pay attention to effects which are usually neglected macroscopically, such as surface tension and viscous heating.

Most of the simulation techniques and results presented in the following chapters are based on a numerical solution of the macroscopic transport equations. As already mentioned above, owing to the laminar flow conditions, many of these studies can be regarded as first-principles calculations which are free of adjustable model parameters. Consequently, simulation results for microreactors are expected to have a high predictive power. The major limitation for the predictive power of the models is the large number of degrees of freedom, which often excludes the simulation of complete reactors and only allows focusing on components such as single channels. Thus, the true art of reactor modeling consists of selecting a suitable subsystem and representing the portions of the reactor neglected by appropriate boundary conditions. For many of the results reported in the following sections such a strategy was employed. When it is not clear how to select an appropriate subsystem or genuine system-level behavior to be studied, a macromodel approach may be chosen. One strategy of doing so is based on partitioning of the reactor domain into certain sections, each of which is characterized by a specific input – output behavior. As input to the model sections, thermodynamic forces such as pressure or temperature gradients are applied by which thermodynamic fluxes such as mass or heat flux are induced. An alternative strategy of macromodel generation is based on an averaging approach. By averaging over a large number of microscopic structures (e.g., channels) effective equations for volume-averaged field quantities can be derived, thus allowing the number of degrees of freedom for system-level modeling to be reduced considerably. In the following chapters, some examples for both of these modeling strategies which allow transport phenomena to be studied on the level of complete systems or at least larger subsystems will be given.

With continuum models being the basis for the simulation of microreactors, computational fluid dynamics (CFD) is the numerical method to be applied. By now, CFD techniques are routinely used in conventional chemical process engineering. In principle the very same methods can be employed for microreactors, with the possible drawback that the numerical algorithms have been optimized mainly for the parameter regime characteristic for macroscopic flows. The fundamentals of CFD are discussed in → Computational Fluid Dynamics; suitable introductions can be found in, e.g., [22].

2. Flow Distributions

Due to the specific microstructuring technology employed to build up microreactors, the geometric shape of the flow domain is often different from that in macroscopic equipment. While usually the elements for fluid transport such as pipes are of circular cross section, the channels in microreactors have a rectangular or trapezoidal cross section. Depending on the specific microstructuring technology, also tub-like grooves or close-to-triangular shapes may be obtained. Furthermore, a characteristic feature common to many microreactors is a comparatively wide flow distributor followed by a large number of parallel microchannels. In the following, the flow distributions in characteristic geometries and methods to obtain approximate flow distributions in highly parallel flow domains are discussed.

2.1. Straight Microchannels

For laminar flow in channels of rectangular cross section, the velocity profile can be determined analytically. For this purpose, incompressible flow is assumed. The flow profile can be expressed in form of a series expansion (see [23] and references therein) which, however, is not always useful for practical applications, where often only a fair approximation of the velocity field over the channel cross section is needed. An approximate solution of the form [24]

$$u(x,y)=u_{\max}\left[1-\left(\frac{x}{a}\right)^{s}\right]\left[1-\left(\frac{y}{b}\right)^{r}\right] \tag{3}$$

was suggested for a rectangular channel oriented along the z-axis with a width of $2a$ in the x-direction and a depth of $2b$ in the y-direction, where u is the local flow velocity and $u_{\max}$ the maximum velocity. The exponents s and r depend on the aspect ratio b/a of the channel. The most common correlations can be found in [23]. Typically, Equation 3 approximates the exact velocity profile with an accuracy of a few percent. Often, not the detailed velocity profile is of interest, but only the friction factor f, which determines the pressure drop over a channel of a given length. The fanning friction factor is defined as

$$f=-\frac{\mathrm{d}p}{\mathrm{d}z}\frac{D_{\mathrm{h}}}{2\rho U^{2}} \tag{4}$$

where $\mathrm{d}p/\mathrm{d}z$ is the pressure gradient along the channel, ρ the density, D_{h} the hydraulic diameter and U the mean flow velocity. The hydraulic diameter is given by $4A/P$, where A is the cross-sectional area of the channel and P its perimeter. A comparatively simple expression for the friction factor in rectangular channels [25] deviates from the analytical solution by less than 0.05 % (Eq. 4)

$$f=\frac{24}{\mathrm{Re}}\left(1-1.3553\alpha+1.9467\alpha^{2}-1.7012\alpha^{3}+0.9564\alpha^{4}-0.2537\alpha^{5}\right) \tag{5}$$

where α is the aspect ratio of the channel ($\alpha \leq 1$ by definition) and the length scale entering the Reynolds number Re is the hydraulic diameter.

Most expressions for the flow profile in rectangular channels assume that the flow is fully developed, i.e., that the flow velocity is oriented along the z-axis and does not change in streamwise direction. Close to the entrance of a channel this assumption is not valid, and the flow undergoes a development from an entrance distribution to a fully developed profile. Correspondingly, the pressure distribution deviates from that observed in a fully developed flow, and expressions for the friction factor such as Equation 5 are not valid in the entrance region. To determine the developing velocity distribution and friction factor, various approaches have been employed, such as analytical calculations based on a linearized inertia term, numerical solutions of the Navier – Stokes equation, and experimental velocity and pressure measurements. An overview of the results is given in [23]. In general, the pressure drop per unit length of an entrance flow is higher than that of a fully developed flow. To compare entrance flow effects, a hydrodynamic entrance length L_{hy} can be defined which is the length necessary to achieve a centerline velocity equal to 99 % of the fully developed value. Usually, a nondimensional quantity is used (Eq. 6).

$$L_{\mathrm{hy}}^{+}=\frac{L_{\mathrm{hy}}}{D_{\mathrm{h}}\mathrm{Re}} \tag{6}$$

Depending on the aspect ratio of the channel, values between 0.01 and 0.1 are found for the nondimensional entrance length [23]. From Equation 6 it can be deduced — given that L_{hy}^{+} is Re-independent — that L_{hy} increases linearly with the hydraulic diameter and the Reynolds number.

A number of authors have considered channel cross sections other than rectangular [25 – 27]. In general, an analytical solution of the Navier – Stokes and enthalpy equations in such channel geometries would be quite involved due to the implementation of the wall boundary condition. For this reason, usually numerical methods are employed to study laminar flow and heat transfer in channels with arbitrary cross-sectional geometry.

2.2. Periodic and Curved Channel Geometries

A key advantage of microreactors is their potential for rapid heat and mass transfer. Naturally, the acceleration of transport processes is related to the decrease of diffusion paths on the microscale. In addition, special channel geometries are explored for which a further speed up of heat and mass transfer is found. Three different channel shapes that have been studied are shown in Figure 1: zigzag sinusoidally curved, and converging-diverging. Velocity and temperature fields in zigzag channels were computed based on the finite-volume method for Reynolds numbers up to 1500 [28]. Above a specific Reynolds number depending on the geometry parameters, flow separation occurs, i.e., in the corners of the channel recirculation zones are formed. While the friction factor for straight channels displays a linear decrease as a function of Reynolds number, as is apparent from Equation 5, the friction factor in zigzag channels becomes nearly independent of the Reynolds number for $Re > 1000$. Data obtained for a zigzag channel with rounded corners show a similar trend [29].

Flow in sinusoidally curved channels for Reynolds numbers up to 500 was also studied with a finite-volume discretization [30]. For most of the parameter space explored, flow separation was not observed. However, for $Re = 500$ and a large enough sine-wave amplitude as compared to the period, there were some indications of recirculation zones forming in the recesses of the channel. Unfortunately, no attempt was made to compute friction factors.

Flow in converging-diverging channels was studied with special attention to the transition between laminar and turbulent flow [31] by using a spectral-element method in which the flow domain is divided into macroelements over which a set of polynomial test functions is defined. This method allowed damping of small-scale fluctuations due to numerical viscosity to be suppressed and is thus well suited to studying the transition from stationary to oscillatory and chaotic flow. The computed streamline patterns indicate a transition from unseparated flow at low Reynolds numbers to a flow with recirculation zones within the recesses of the channel, occurring at a comparatively small Reynolds number between 10 and 20. Streamline patterns for various Reynolds numbers are depicted in Figure 2. The phenomenology of flow distributions in such channel domains was found to be quite diverse. At low Reynolds numbers the flow is stationary, but at a Reynolds number of 150 oscillations begin to develop, with the vortices still being confined in the recesses. At Re = 400 the viscous forces are no longer strong enough to confine the vortices in the recesses and vortex ejection is observed. At even higher Reynolds numbers the flow becomes aperiodic and chaotic.

In some cases, for example, when long residence times are desired, comparatively long microchannels must be integrated into a compact micro reaction device, a task which can only be achieved with curved channels. In addition, straight channels are often not suited for connecting a microflow domain to the external world. One of the first theoretical studies on flow in curved channels [32, 33] investigated the secondary flow perpendicular to the main flow direction induced by inertial forces by means of a perturbative analysis. The dimensionless group characterizing such transverse flows is the Dean number.

$$K = \mathrm{Re}\sqrt{\frac{D_\mathrm{h}}{R}} \tag{7}$$

where D_h is the hydraulic diameter and R the mean radius of curvature of the channel. The typical secondary flow pattern in curved channels is given by two counterrotating vortices separated by the plane of curvature. The strength of these vortices increases with increasing Dean number.

Originally, the studies of secondary flow in curved channels were performed for pipes of circular cross section. Of much greater relevance to

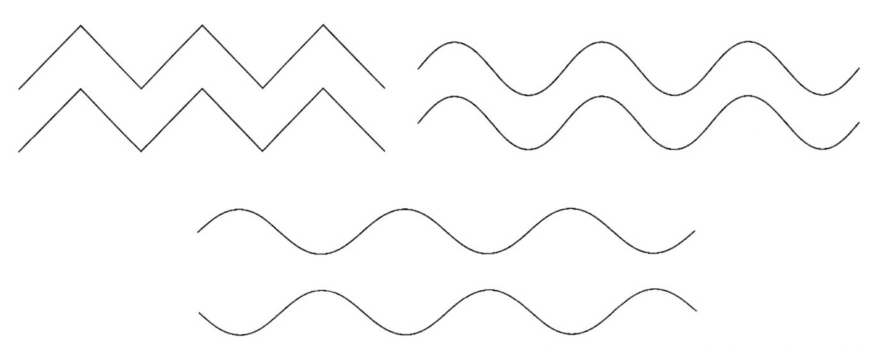

Figure 1. Different channel shapes for which flow distributions have been computed: zigzag (upper left), sinusoidally curved (upper right), and converging-diverging (bottom)

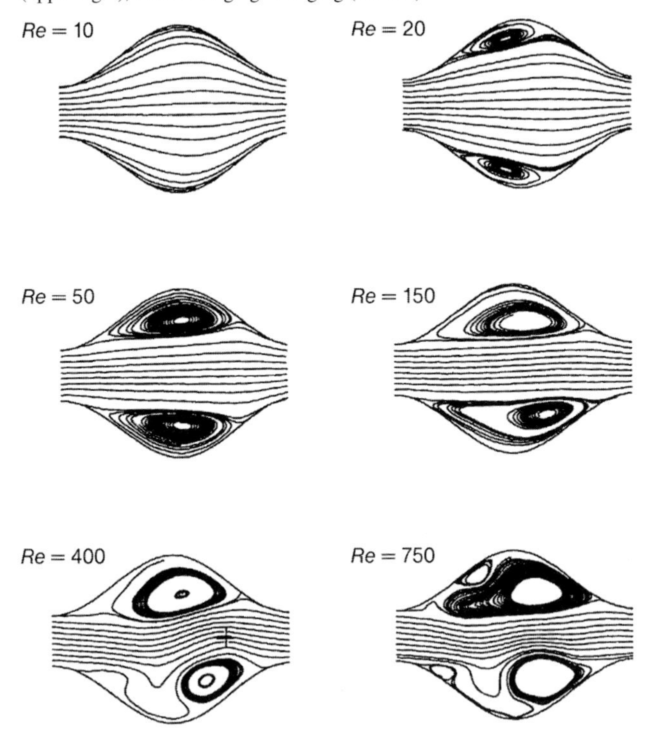

Figure 2. Streamline patterns in a converging-diverging channel for various Reynolds numbers, taken from [31]

microreactors are rectangular channels, which were the subject of detailed numerical investigations [34] into the formation of vortices in channels of square cross section, for which a complex branch structure was found describing the transition between flow patterns with two and four counterrotating vortices. Up to a Dean number of about 113, a two-vortex solution is found, followed by a transition regime between Dean numbers of 113 and 130 with coexisting two- and four-vortex configurations. For $K > 130$ the flow exhibits a four-vortex pattern which, however, coexists with a two-vortex pattern of broken symmetry. The friction factor in curved channels was found to be larger than in straight channels for most of the Dean number range considered, and not surprisingly different branches of the flow map are associated with different friction values.

2.3. Multichannel Flow Domains

In chemical microprocess technology it is important to guarantee well-defined and reproducible reaction conditions in microchannels, but a high throughput should also be achieved. For this purpose, the process fluid is guided through a large number of parallel microchannels, where heat exchange and/or chemical reactions occur. One of the characteristic problems of microprocess technology is to equally distribute the incoming fluid over the microchannels. Fluid maldistribution would induce unequal residence times in different channels, with undesired consequences for the product distribution of a chemical reaction being conducted inside the reactor. When a process with various competing side reactions and byproducts is considered, the contact time of the process fluid in the reaction region should be as well defined as possible to maximize the selectivity of the process. However, not only the maldistribution of the fluid over a multitude of microchannels induces variations of the contact time, but also hydrodynamic dispersion of concentration tracers in the channels themselves.

Various concepts for the equipartition of fluid over a multitude of microchannels have been developed. One concept relies on guiding the incoming fluid through a flow splitter with subsequent bifurcations [35]. The most widely used design is based on a comparatively wide inlet region leading into a multitude of narrow microchannels. A corresponding geometry, developed for methanol reforming, is shown in Figure 3. The figure shows the computational domain of the CFD model of the reactor together with the computed streamlines, where use of reflection symmetry was made and only one quarter of the portion of the reactor to be considered was modeled. An inlet pipe leads to a flow distribution chamber connected with a multitude of microchannels. The essence of such designs is to be seen in the pressure barrier of the microchannels. The narrower the microchannels are, the higher will be the pressure drop in the channels themselves as compared to the pressure drop in the flow distribution chamber, and the more uniform the flow distribution will be.

During the development of the microreformer of Figure 3 one of the goals was to design the flow manifold in such a way that the volume flows in the different reaction channels are approximately the same [36]. In spite of the recirculation zones found, for the chosen design a flow variation of about 2 % between different channels was predicted from CFD simulations. In the application under study, a washcoat catalyst layer is applied to the microchannels. Thickness variations of the catalyst layer are likely to play the leading role in the nonuniformities in flow distribution to be expected. Hence, a flow manifold with intrinsic volume flow variations of a few percent over different reaction channels is usually satisfactory.

For the computation of the flow distribution in the methanol-reforming reactor, a reduced-order flow model for the microchannels was used. In such a model a fully developed flow profile as given by Equation 3 is assumed, which means that entrance flow effects are neglected. This approximation is justified when the entrance length is small compared to the total length of the channel. A major advantage of a reduced-order flow model is the significant reduction of the degrees of freedom entering the simulation. Each microchannel is then only represented by a single degree of freedom, that is, the total volume flow, and resolution of the corresponding flow domain by a computational grid is no longer required. Thus the simulation of microfluidic devices comprising a large number of channels becomes possible at moderate computational cost.

Finite-volume grids in combination with reduced-order flow models for microchannels allow flow distribution problems to be solved in 3D, for which the standard approach would be computationally too expensive. However, when the goal is to find designs with optimum flow equipartition by tuning specific geometric parameters, it is advisable to set up models with significantly fewer degrees of freedom. The flow distribution in multichannel microreactors was studied with the help of macromodels [37]. The geometry considered was a microstructured plate of a heat-exchanger stack developed at the Institute of Microtechnology Mainz. The plate together with arrows indicating the flow direction is shown on the left side of Figure 4; on the right side the model is displayed together with the geometric parameters. The model is based on the idea of subdividing the flow domain into a number of virtual channel segments with rectangular cross section over which the flow is distributed. Through the channel segments in the inlet zone the volume flows $V_1, V_2, \ldots, V_{\mathrm{Nc}}$ are transported, where a part of the flow branches off into the actual microchannels of width W_c

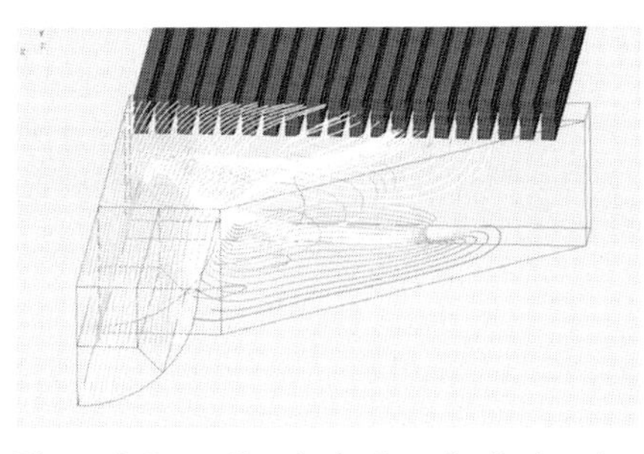
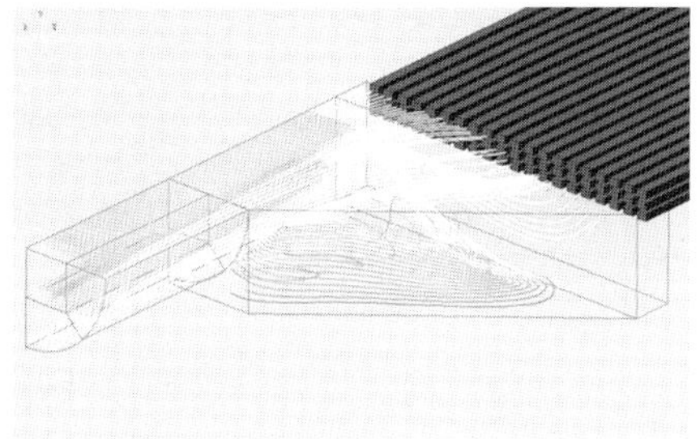

Figure 3. Streamlines in the flow distribution chamber of a multichannel reactor

and depth e. For each of the channel segments a relationship between the pressure drop Δp and the average flow velocity u of the form

$$\Delta p = 32\lambda_{\mathrm{Nc}} \frac{\mu L u}{D_{\mathrm{h}}^2} \tag{8}$$

was used, where the hydraulic diameter is defined as

$$D_{\mathrm{h}} = \frac{2we}{w+e} \tag{9}$$

In these expressions, μ is the dynamic viscosity, L the length of the channel segment, w and e are its width and depth, and λ_{Nc} is a correction factor accounting for the noncircularity of the channels. Clearly, the above formulas rely on the assumption of a hydrodynamically developed flow.

By means of the model depicted in Figure 4, Equation 8, and Equation 9 it is possible to compute the flow distribution just by solving a comparatively low-dimensional system of linear algebraic equations. The problem resembles the task of computing the current distribution of an electric circuit. A priori it is not clear whether the approximation to subdivide the flow domain in the described way is justified. To assess the quality of the chosen approximation, the flow field was computed by means of the finite-volume method [37]. The results obtained suggest that, owing to the orientation of the isobars of the flow and the absence of recirculation loops, the chosen subdivision into channel segments is a reasonable simplification, at least for the geometry and flow regime considered. The model allows the flow distribution to be studied for a variety of different geometries with a minimum computational effort. As a result of simulations, optimized flow chamber geometries leading to a maximum of flow equipartition were determined.

While the strategy described above is a way to obtain effective, low-dimensional models for multichannel reactors in various cases, it clearly has its limitations. In geometries with hundreds of microchannels, another method can be utilized to compute the flow field. In a situation where the flow is distributed over a large number of parallel microchannels, the multitude of channels can be regarded as a porous medium. For porous media effective, volume-averaged transport equations have been known for a long time (for an overview, see [38]). To solve a flow distribution problem for a multichannel geometry, the flow domain can be split into several regions, one of which is the flow distribution chamber, and another the region comprising the multitude of channels. In the flow distribution chamber, the ordinary transport equations are solved, whereas in the multichannel domain, the effective, volume-averaged description of the transport processes is used. A corresponding 2D geometry is shown in Figure 5. The incoming flow is distributed over a large number of channels of width of w_{c} separated by walls with a width of $w - w_{\mathrm{c}}$.

The most straightforward porous-medium model which can be used to describe the flow in the multichannel domain is the Darcy equation [38], which represents a simple model used to relate the pressure drop and the flow velocity inside a porous medium. Applied to the geometry of Figure 5 it is written as

$$\frac{\mathrm{d}}{\mathrm{d}y}\langle p\rangle_{\mathrm{f}} + \varepsilon \frac{\mu}{K}\langle u\rangle_{\mathrm{f}} = 0 \tag{10}$$

with the porosity ε and permeability K given as

$$\varepsilon = \frac{w_{\mathrm{c}}}{w}, \quad K = \frac{\varepsilon\, w_{\mathrm{c}}^2}{12} \tag{11}$$

The flow velocity, pressure and dynamic viscosity are respectively denoted as u, p, and μ and

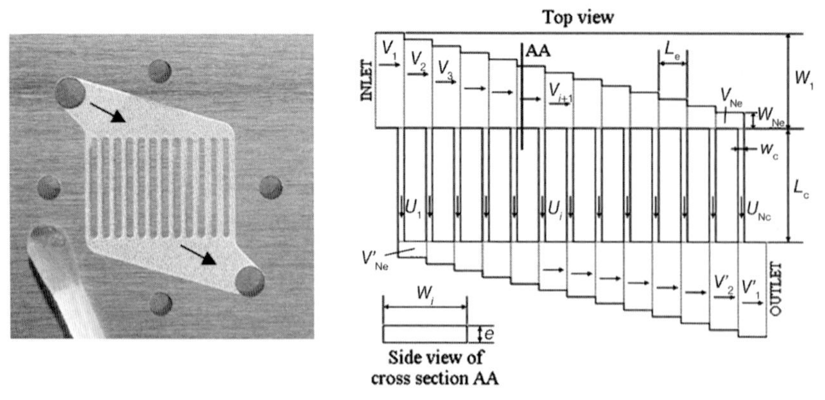

Figure 4. Geometry of the microstructured plates (left) and subdivision of the flow domain into channel segments (right), as considered in [37]

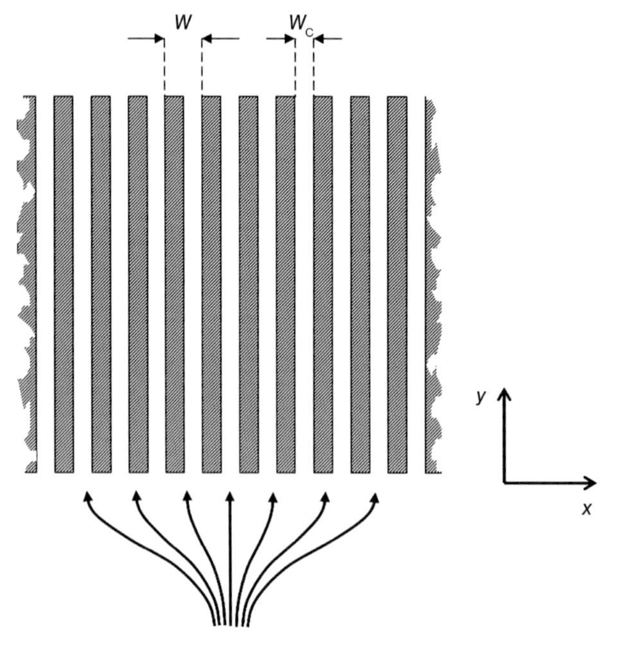

Figure 5. Multichannel geometry with channels and separating walls of uniform width

the symbol $<\ldots>_f$ represents an average over the fluid phase. [39]. An extended Darcy equation was used to model the flow distribution in a microchannel cooling device. In general, the permeability K must be regarded as a tensor quantity accounting for the anisotropy of the medium. Furthermore, the description can be generalized to include heat-transfer effects in porous media.

Compared to the use of reduced-order flow models, the porous-medium approach allows an even larger multitude of microchannels to be dealt with. Furthermore, for comparatively simple geometries with only a limited number of channels, it represents a simple way to provide qualitative estimates of the flow distribution. However, as a course-grained description it does not reach the same level of accuracy as reduced-order models. Compared to the macromodel approach [37], the porous-medium approach has a broader scope of applicability and can also be applied when recirculation zones appear in the flow distribution chamber. However, the macromodel approach is computationally less expensive and can ideally be used for optimization studies.

3. Heat Transfer

Heat-transfer phenomena belong to the key issues to be studied in microreactors. Due to the small thermal diffusion paths, microreactors have the potential to enable fast heat transfer and to control temperature distributions with very high accuracy. Correspondingly, modeling and simulation of heat-transfer phenomena and their reliable prediction is of paramount importance for process design. The simulation of temperature distributions in microreactors often requires the solution of a conjugate heat transfer problem, i.e., the temperature fields in the fluid phase and in the solid wall material must be computed. Owing to the microstructuring technologies used to fabricate microreactors, the share of wall material in the total reactor volume is often higher than in conventional equipment. Hence, solid heat conduction effects become important and usually have to be taken into account when the temperature field inside a reactor is to be computed.

3.1. Straight Microchannels

A standard geometry of channels in microfluidic systems is a rectangular or close-to-rectangular cross section. With a given velocity profile, the temperature field inside rectangular channels can often be determined analytically. However, the problem of determining a temperature profile in a channel geometry is much more multifaceted than the computation of a flow distribution. While in most cases a zero-velocity boundary condition at the channel walls is prescribed for the flow, the wall-boundary conditions for the temperature field can be diverse. On the one hand, either a heat flux or a temperature can be prescribed. On the other hand, the thermal boundary conditions on the four walls of a rectangular channel may be different. Due to this complexity, the solution for temperature field itself is usually not reported. Rather, the Nusselt number, defined as

$$\mathrm{Nu} = \frac{\alpha D_{\mathrm{h}}}{k} \tag{12}$$

is determined. In this expression, D_{h} is the hydraulic diameter, k the thermal conductivity of the fluid, and α the heat-transfer coefficient measuring the transmitted thermal power per unit area divided by a characteristic temperature difference. The Nusselt number is a dimensionless quantity characterizing the efficiency of heat transfer. Similar to the velocity field, the temperature field assumes an invariant profile far enough downstream from the channel entrance. However, due to the continuous heat transfer from or to the channel walls, only the shape of the temperature distribution stays invariant, but the normalization changes. Close to the entrance of the channel a thermally developing flow may be observed. A thermal entrance flow is not a priori related to a hydrodynamic entrance flow, i.e., a thermally developing flow may be observed even in regions where the flow is hydrodynamically developed and vice versa.

An overview of the Nusselt numbers obtained for rectangular channels with various aspect ratios and various thermal boundary conditions has been given [23]. Depending on the thermal boundary conditions, the Nusselt number for thermally fully developed flow either increases or decreases when the aspect ratio is increased. When the Nusselt number for thermally developing and hydrodynamically developed flow is plotted as a function of position along the channel, a divergence is observed when the channel entrance is approached. This means that, with decreasing distance to the channel entrance, increasing heat transfer efficiency is found. The same observation is made when a simultaneously (i.e. hydrodynamically and thermally) developing flow is considered. Analogously to hydrodynamically developing flows, a thermal entrance length L_{th} can be defined. It is given as the duct length required for the Nusselt number to fall within a 5 % interval of the fully developed value. Again, a dimensionless quantity

$$L_{\mathrm{th}}^{+} = \frac{L_{\mathrm{th}}}{D_{\mathrm{h}} \mathrm{Re\,Pr}} \tag{13}$$

is used, where the Prandtl number Pr is the ratio of the momentum diffusivity (i.e. the kinematic viscosity) and the thermal diffusivity. The dimensionless thermal entrance length is a quantity only depending on the aspect ratio of the channel and the thermal boundary conditions. Hence, L_{th} is a linear function of the hydraulic diameter, the Reynolds number, and the Prandtl number.

Apart from rectangular channels, heat transfer has also been studied in channels with different cross-sectional geometries, and Nusselt numbers for wall boundary conditions of fixed temperature and fixed heat flux are tabulated for different geometric parameters [25, 26].

3.2. Periodic and Curved Channel Geometries

Similar to the computation of friction factors, heat transfer has not only been studied in straight channels, but also in channels with specific periodic shapes. Nusselt numbers in the zigzag channels shown in Figure 1 were computed by using a finite-volume approach in two dimensions [28]. With increasing Reynolds number the Nusselt number increases significantly as compared to a straight channel, i.e., a parallel-plate arrangement. For suitable geometric parameters, a Reynolds number in the range of 1000 and a Prandtl number of 8, a heat-transfer enhancement factor of about 13 was computed. For a Prandtl number of 0.7 (corresponding to air) the ratio of the heat fluxes achieved with the zigzag and the straight channels at identical values of pumping power and pressure drop were compared. The zigzag channel outperforms the straight channel by up to a factor of 6, i.e., for the same pumping power or pressure drop a considerably higher heat flux is achieved. This is an interesting result in view of the design of micro heat exchangers, in which the goal is often to increase the heat flux while limiting the pressure drop.

A zigzag geometry with rounded corners was studied in 2D by the finite-volume method [40]. The Prandtl number was kept fixed at 0.7 and the Reynolds number was varied between 100 and 1000. For the range of geometric parameters considered, a heat-transfer enhancement factor of about 3 was found at $Re = 1000$, and this confirms the potential of corrugated channels for heat-transfer enhancement.

Heat transfer in sinusoidally curved channels was studied for a Prandtl number of 1.0. [30] Surprisingly, at a Reynolds number of 100 the Nusselt number is lower than the corresponding value in a parallel-plate channel. Such a suppression of heat transfer stands in contradiction to results for sine-wave channel walls of comparatively large amplitude [41], for which heat-transfer enhancement increased with increasing Reynolds number, with an enhancement factor of about 4 for $Re = 800$. Both of these studies were restricted to steady-state flows, and steady-state solutions could be determined for the range of Reynolds numbers considered. Experimental work on flow and heat transfer in sinusoidally curved channels [42] indicate heat-transfer enhancement and do not show evidence of decreasing Nusselt number in any range of Reynolds numbers. However, the flow patterns begin to exhibit some unsteady behavior for Reynolds numbers greater than about 200 in the geometry considered. Interestingly, oscillations were predominantly observed close to the exit of the channel and were found to move upstream when the Reynolds number was increased. Much of the heat-transfer enhancement was attributed to the unsteady character of the flow [42].

The flow and temperature field in an elementary cell of the converging-diverging channel geometry for which flow fields were calculated in [31] was computed on the basis of the finite-volume method using periodic boundary conditions [43]. As far as the transition between a steady-state and an unsteady flow is concerned, similar results were found. At Reynolds numbers beyond the transition to the unsteady flow regime, the temperature field evaluated for a Prandtl number of 0.7 becomes quite complex. This is indicated in Figure 6, which shows the evolution of (normalized) temperature contour lines at a Reynolds number of 328. To quantify heat-transfer enhancement, the time-averaged Nusselt number was computed. The increase in Nusselt number compared to a straight channel was found to be significant and reached values as high as 7.54 at a Reynolds number of 520. Flow and heat transfer in converging-diverging channels were analyzed based on a perturbation analysis derived from lubrication theory and on the finite-volume method [44]. Most of these studies were limited to Reynolds numbers up to 100. In this regime the increase in Nusselt number compared to a parallel-plate geometry was moderate, amounting to a factor of about 1.3 at $Re = 100$ and $Pr = 1$. The results also showed an approximately linear increase of heat-transfer enhancement as a function of Prandtl number. The studies of [45] indicated that heat-transfer enhancement in converging-diverging channels is significantly reduced if the separation between the wavy channel walls is bigger than the wavelength of the undulations. Here a cubic

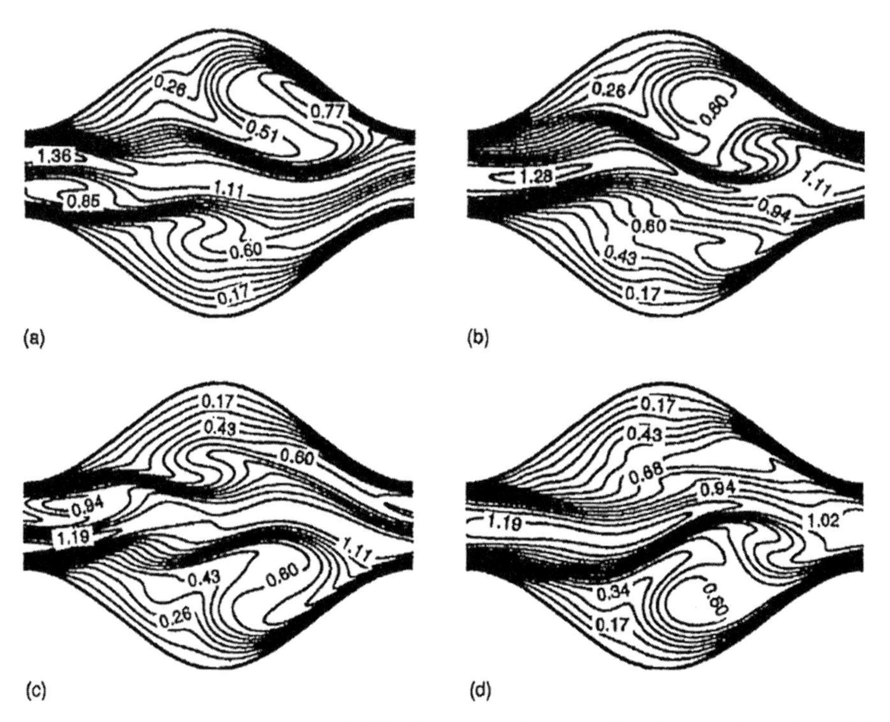

Figure 6. Evolution of the temperature field in the recesses of a converging-diverging channel, as obtained in [43]

spline collocation method was used to numerically solve the transport equations and, in contrast to the other studies, a cyclic structure of the flow and dimensionless temperature fields with the period of the wall structures was not assumed. The results obtained at $Re = 500$ and $Pr = 6.93$ show a rather moderate heat-transfer enhancement factor of about 1.8 compared to a parallel-plate geometry, which is further reduced for smaller Prandtl numbers.

Similar to the structure of the flow field, heat transfer has also been studied in curved channel geometries. The complicated branch structure with competing patterns of two and four counterrotating vortices in channels of square cross section is reflected in the Nusselt number [34]. When plotting the Nusselt number as a function of Dean number, different branches are found corresponding to symmetric and asymmetric secondary flow patterns with two and four vortices. However, the relative difference between the different branches is not very pronounced and should be hard to measure experimentally. For a Dean number of 210 and a Prandtl number of 0.7 a heat-transfer enhancement factor of about 2.8 was determined, that is, curved channels as well as other channels with specific periodically varying cross sections may be used for applications where rapid heat transfer is desired.

3.3. Multichannel Flow Domains

When a large number of parallel microchannels is considered, the problem of computing the temperature distribution inside the channels and the channel walls becomes quite involved. In such a case the use of the method of reduced-order flow models discussed in Section 2.3 is not as straightforward as in the case of a pure flow distribution problem, due to thermal cross-talk between the different channels. In principle, heat transfer in a multichannel domain can be described on the basis of a porous-medium model, in which an average is taken over an ensemble of channels and continuous field quantities are introduced. However, in the standard models for porous media, a thermal equilibrium between the fluid and the solid phase is assumed, i.e., only a single temperature field is used. Especially in the regime of large Péclet numbers the thermal-equilibrium assumption may break down, and it may become necessary to define separate temperature fields for the fluid and the solid phase. It is thus advisable to take one step back and to derive the mean-field model equations that allow heat transfer in multichannel domains to be described on the basis of a volume-averaging approach.

A generic geometry of a multichannel stack to which the mean-field equations should be applied is shown in Figure 7. Many microchannels, usually several hundred up to several thousand, are arranged in layers stacked onto each other. As a whole the multichannel stack forms a cuboid of extension L_x, L_y, L_z in x-, y-, and z-direction, respectively. The microchannels of width w_f and depth h_f are assumed to be parallel to the z-axis and are separated by solid walls of thickness w_s and height h_s, respectively.

The need for modeling single channels and their specific geometry is eliminated when regarding the reaction fluid and wall material as interpenetrating continua. Hence, in the following it is assumed that the reactor stack is filled with interpenetrating fluid and solid phases which interact via the exchange of heat. Momentum exchange between the two phases is not accounted for explicitly and it is assumed that the velocity distribution over the cross section of the reactor is known a priori. The phase volume fractions can be derived from the geometric parameters h_f, h_s, w_f, and w_s and are given as

$$\Phi_\mathrm{f}=\frac{w_\mathrm{f}h_\mathrm{f}}{(w_\mathrm{f}+w_\mathrm{s})(h_\mathrm{f}+h_\mathrm{s})},\quad \Phi_\mathrm{s}=1-\Phi_\mathrm{f} \tag{14}$$

The temperature field of the solid phase can change due to heat conduction or due to heat transfer from the fluid phase. For the fluid, heat conduction does not have to be taken into account since the flow domain consists of disconnected channels, but convective heat transfer due to the flow velocity as well as heat transfer from the solid is included. In the xy plane no conduction within the fluid without intermediate transfer to the solid walls can occur. The reason for not including heat conduction in the z-direction is the fact that in most cases of practical relevance it is negligible when compared to convective transport. Following the derivation of [46], the fluid and solid temperature fields T_f and T_s are obtained as solutions of the equations

$$\begin{aligned}&\rho_\mathrm{f}c_\mathrm{f}\left(\frac{\partial}{\partial t}+u_i\frac{\partial}{\partial x_i}\right)T_\mathrm{f}=-\alpha a_\mathrm{V}\left(T_\mathrm{f}-T_\mathrm{s}\right)+S_\mathrm{c}\\&\rho_\mathrm{s}c_\mathrm{s}\frac{\partial}{\partial t}T_\mathrm{s}=k_{ij}\frac{\partial}{\partial x_i}\frac{\partial}{\partial x_j}T_\mathrm{s}-\alpha a_\mathrm{V}\left(T_\mathrm{s}-T_\mathrm{f}\right)\end{aligned} \tag{15}$$

where c_f, c_s are the fluid and solid specific heats, u_i is the average velocity of the fluid (averaged over a channel cross section), α the heat transfer coefficient from the channel walls to the fluid, a_V the specific surface area (wall surface/reactor volume), and the Einstein convention of summation over repeated indices has been used. This convention will be used throughout this chapter when vector or tensor quantities are involved. The fluid and solid densities are volume-averaged quantities and given as

$$\rho_\mathrm{f}=\Phi_\mathrm{f}\rho_\mathrm{f}^{(0)},\quad \rho_\mathrm{s}=\Phi_\mathrm{s}\rho_\mathrm{s}^{(0)}, \tag{16}$$

where the superscript (0) indicates the corresponding material densities. The equation for the fluid temperature contains a source term S_c which could, e.g., represent the input of thermal energy from a chemical reaction. Owing to the anisotropy of the solid matrix the thermal conductivity is direction-dependent. For this reason a thermal conductivity tensor k_ij appears in Equation 15. The heat-transfer coefficient α can be obtained from well-known correlations for rectangular channels. The model of Equation 15 can be regarded as a mean-field model expected to describe the average temperatures of the fluid and the solid phase without incorporating the local fluctuations which are due to temperature gradients within single channels or channel walls.

To compute the thermal conductivity tensor k_ij, the multitude of walls shown in Figure 7 can be regarded as a network of thermal resistors. On this basis, the components of the thermal conductivity tensor are derived as

$$\begin{aligned}&k_{xx}\approx(w_\mathrm{f}+w_\mathrm{s})\left(\frac{h_\mathrm{f}+h_\mathrm{s}}{h_\mathrm{s}}w_\mathrm{f}+w_\mathrm{s}\right)^{-1}k_\mathrm{s}\\&k_{yy}\approx(h_\mathrm{f}+h_\mathrm{s})\left(\frac{w_\mathrm{f}+w_\mathrm{s}}{w_\mathrm{s}}h_\mathrm{f}+h_\mathrm{s}\right)^{-1}k_\mathrm{s}\\&k_{zz}\approx\Phi_\mathrm{s}k_\mathrm{s}\end{aligned} \tag{17}$$

where k_s represents the thermal conductivity of the solid material. All other components of the thermal conductivity tensor vanish.

The mean-field model described here was developed to compute temperature fields in microreactors with low computational effort. Previously, a 2D version of such a model was used to determine the temperature distribution in a

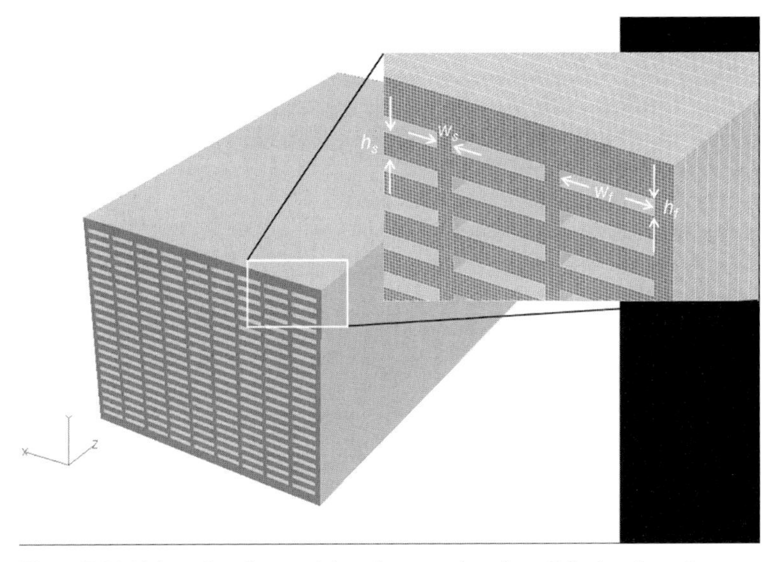

Figure 7. Multichannel stack comprising a large number of parallel microchannels

microchannel heat sink [39]. To assess the quality of the 3D mean-field model, the resulting temperature fields were compared with results from a full conjugate heat-transfer model explicitly accounting for all of the geometrical details of the multichannel stack. The results of such a comparison are displayed in Figure 8. The temperatures were evaluated along the centerline of the reactor pointing in the z-direction. The heated region with nonzero source term extends up to 1 mm downstream from the inlet plane. The temperature curves show that in the inlet region the fluid and solid temperatures are not equilibrated, and this underlines the necessity of a two-temperature model. The same conclusion was drawn on the basis of the 2D model [39]. Overall, a good agreement between the mean-field and the full model is found, with maximum local differences in temperature of 2 – 3 K and a total temperature range of about 35 K.

The times needed to numerically solve the equations of the two models were found to differ by orders of magnitude. This makes the mean-field description an especially promising method for fast evaluation of reactor designs. When developing a micro reaction system it is often unclear how exactly the properties of the channel walls (geometry, thermal conductivity) influence the reaction performance. Usually, the goal is to achieve a temperature distribution as uniform as possible, in order to suppress unwanted side reactions and to increase the selectivity of the process. Based on the volume-averaging approach presented here it should be easy to assess and compare the thermal performance of different reactors and to identify favorable designs.

3.4. Micro Heat Exchangers

Micro heat exchangers are used for rapid heat transfer between a hot and a cold fluid; owing to the small thermal diffusion paths, the size of the system can be reduced compared to conventional devices. Among the different flow schemes (cocurrent, countercurrent, and cross-current) especially the countercurrent scheme has been studied theoretically, because of the practical relevance of countercurrent heat exchangers on the one hand, and the fairly simple model structure compared to cross-current heat exchangers on the other. A schematic drawing of a countercurrent micro heat exchanger is shown in Figure 9. A characteristic feature of such devices is the fact that the thickness of the wall material separating the channels is of the same order of magnitude as the channel depth itself. This distinguishes micro heat exchangers from their macroscopic counterparts, in which the width of the flow passage is usually well above the wall thickness.

Owing to the comparatively large share of wall material, longitudinal heat conduction in the channel walls must be taken into account

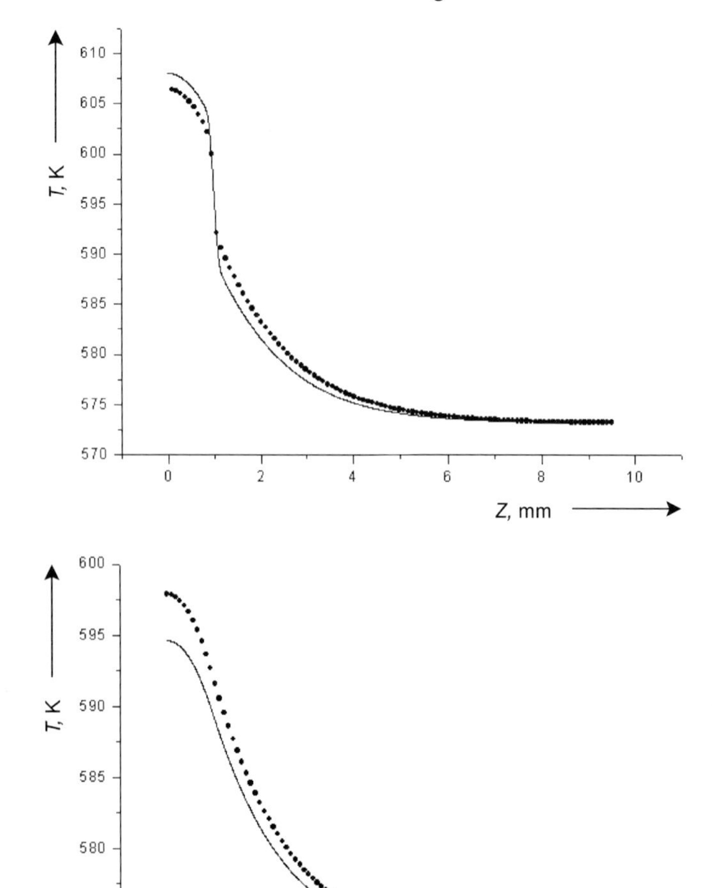

Figure 8. Fluid (left) and solid (right) temperatures in a multichannel stack as obtained from the full (dots) and the mean-field model (lines)

when attempting to formulate a heat-exchanger model. In contrast, the models for conventional heat exchangers usually account only for transverse heat transfer from one fluid to another.

The performance of countercurrent micro heat exchangers was investigated, especially with respect to wall conduction effects and the choice of wall material [47], for a flow of nitrogen in channels with a length of 10 mm, a depth of 50 μm, and separating walls with a thickness between 125 and 500 μm. To reduce the computational effort, the heat exchanger geometry was only discretized in the axial direction, i.e., in the direction of the flow. In the direction perpendicular to the flow a fixed heat-transfer coefficient was used to describe the exchange of heat between the fluid and the channel walls.

When the thermal conductivity of the wall material is varied, characteristic temperature profiles for the channels with the hot and cold fluids and the channel wall are obtained. For very small values of the thermal conductivity, no heat is exchanged and the temperatures stay approximately constant. At intermediate wall thermal conductivities, almost linear temperature profiles are obtained, while at very high thermal conductivities the wall assumes a constant temperature and the temperatures of the

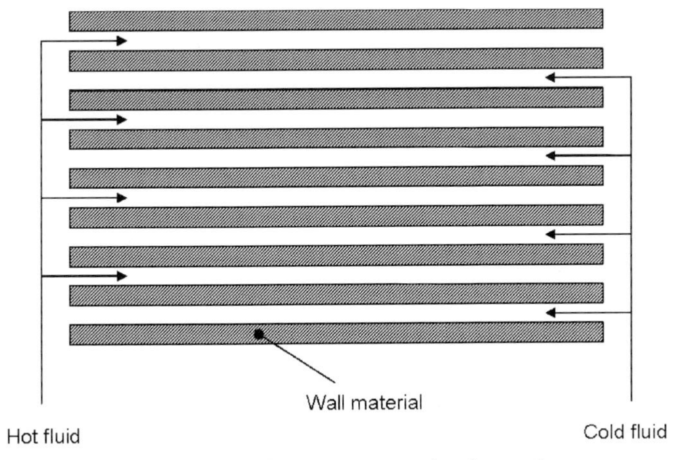

Figure 9. Schematic drawing of a countercurrent micro heat exchanger

fluids change rapidly in the entrance regions of the channels. A quantity characterizing the performance of a heat exchanger is the efficiency, which is the ratio of the transmitted heat and the maximum transmittable heat, given as

$$\varepsilon = \frac{T_{\text{in}}^{\text{hot}} - T_{out}^{hot}}{T_{\text{in}}^{\text{hot}} - T_{\text{in}}^{\text{cold}}} \tag{18}$$

in the case of equal volume flows of two identical incompressible fluids; T_{in} and T_{out} denote the temperatures at the channel in- and outlet, respectively. The heat-exchanger efficiency was calculated for a couple of characteristic geometries as a function of the wall thermal conductivity [47]. A typical result for equal volume flows is displayed in Figure 10. At low conductivities almost no heat is transferred, as expected. At high conductivities the wall assumes a uniform temperature due to axial heat conduction and the device becomes indistinguishable from a cocurrent heat exchanger, which is limited to an efficiency of 50 % at equal volume flows. At values around 1 W m^{-1} K^{-1} the curve assumes a maximum, as axial heat conduction inside the walls is suppressed to a sufficient degree and transverse conduction to the neighboring channels is still efficient. Hence, in typical micro heat exchangers the use of low-conductivity materials such as glass or polymers is preferable, and the common stainless steel materials are expected to reduce the efficiency.

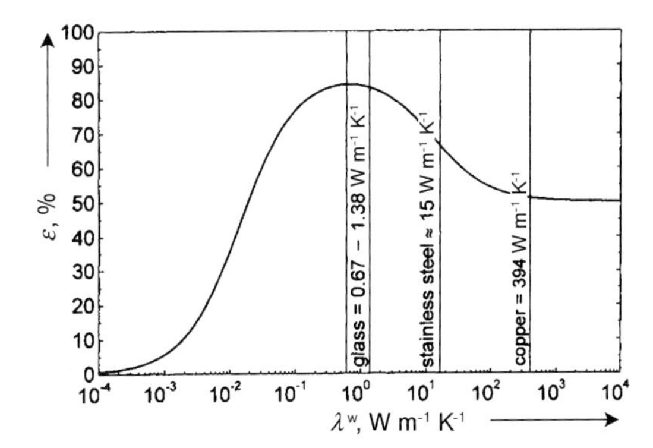

Figure 10. Heat-exchanger efficiency as a function of the thermal conductivity of the wall, taken from [47]

Heat losses to the exterior by conduction through the walls belong to the aspects not considered in the analysis in [47]. The vertical edges of the rectangular walls shown in Figure 9 are in thermal contact with the housing or the flow distribution manifold of the heat exchanger, and by these means heat is transferred to the surroundings. This effect was studied [48] by using a similar pseudo-1D model as [47], discretized only in the flow direction. Equal mass flow rates and identical incompressible fluids were assumed. The results were expressed as functions of the two independent dimensionless parameters of the problem, the conduction parameter

$$\lambda = \frac{kA}{\dot{m}\, c_p L} \tag{19}$$

and the number of transfer units

$$\text{NTU} = \frac{\alpha A}{\dot{m}\, c_p} \tag{20}$$

where k is the thermal conductivity of the wall material, A the cross-sectional area of a wall, $\dot{m}$ the mass flow, c_p the specific heat of the fluid, L the channel length, and α the heat-transfer coefficient between the fluid and the wall. As a result of the computational model, the heat-exchanger efficiency (or effectiveness) and the ratio of the conduction to the flow losses was obtained (Fig. 11). The conduction losses represent the heat flux dissipated to the surroundings via conduction through the walls, while the flow losses account for the fact that the cold fluid is not heated to the inlet temperature of the hot fluid (the thermodynamic limit) and vice versa. Figure 11 shows that the conduction losses are usually larger than the flow losses. Especially for a highly conductive wall material they exceed the flow losses by one order of magnitude or more. Thus, in combination with the results of Figure 10 the need for a low-thermal-conductivity construction material for countercurrent micro heat exchangers is underpinned in a most distinct way.

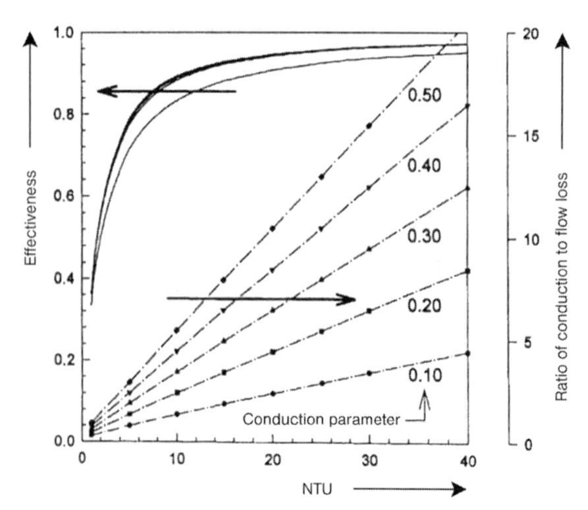

Figure 11. Effectiveness and ratio of conduction to flow loss as a function of NTU and conduction parameter for a countercurrent micro heat exchanger, taken from [48]

The studies of [47, 48] are both based on a description which assumes a vanishing transverse heat-transfer resistance within the channel walls, i.e., temperature gradients within the walls pointing in streamwise direction. A more elaborate model allowing for computation of the complete 2D temperature field in the wall material based on a Fourier expansion [49] confirmed that, with the exception of the close vicinity of the channel inlets and within the parameter range studied, transverse gradients are negligible, and thus corroborates the usefulness of pseudo-1D models for micro heat exchangers.

4. Mass Transfer and Mixing

Similar to heat transfer, fast mass transfer is one of the key aspects of microreactors. Again, due to the short diffusion paths, microreactors enable rapid mass transfer and a uniform solute concentration within the flow domain. Good control of reactant concentration throughout the whole reactor volume is a prerequisite for highly selective chemical reactions and helps to avoid hazardous operation regimes. In addition, overcoming mass-transfer limitations by rapid mixing allows the rapid intrinsic kinetics of chemical reactions to be exploited and enables higher yield and conversion. However, when dealing with liquid-phase reactions, fast mixing remains a challenge even at length scales of 100 μm or less due to the small diffusion constants in liquids.

Micromixing has been a very active field of research [50 – 52]. Not all of the micromixers reported in the literature are suitable candidates for microprocess engineering. On the one hand, the Reynolds numbers in microreactors are usually higher than those in Lab-on-a-chip devices which are often of the order of unity. On the other hand, microreactor processes are typically continuous flow processes for which batch-type micromixers are of little value. Owing to these differentiations, this section will mainly concentrate on mixing concepts and designs suitable for microprocess engineering and will highlight some specific simulation methods for micromixers.

4.1. Simple Mixing Channels

The simplest micromixer is the so-called mixing tee (Fig. 12, left). Two inlet channels merge into a common mixing channel where mixing of the two co-flowing fluid streams occurs. The mixing characteristics of T-type mixers were investigated by using CFD methods [53]. Mixing of gases in a channel of 500 μm width was considered and certain geometric parameters such

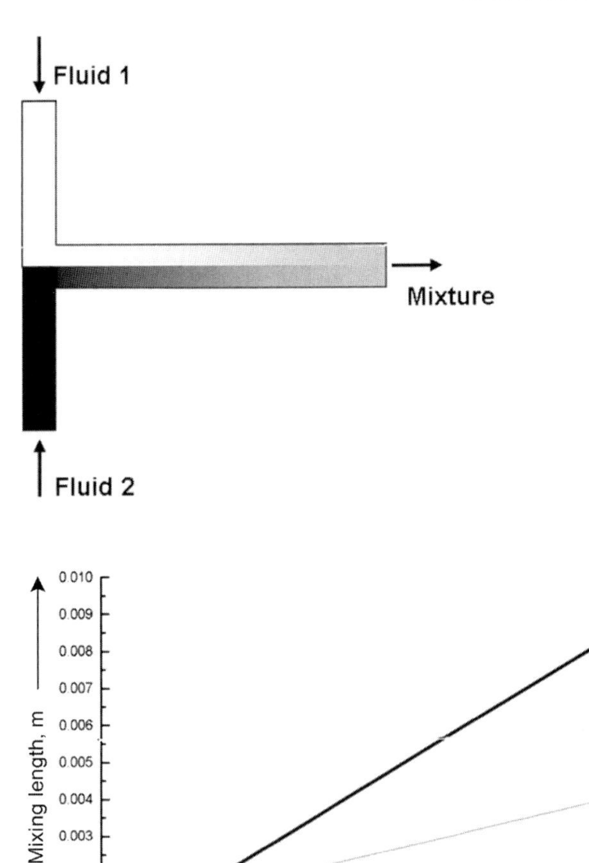

Figure 12. Schematic design of a mixing tee (left) and CFD results for mixing of gases in a channel of 500 μm width and 300 μm depth (right), taken from [53]

as the aspect ratio of the mixing channel or the angle at which the two inlet channels meet were varied. To quantify mixing, a mixing length was defined as the length in flow direction after which the gas composition over all positions of a channel cross section deviates by no more than 1 % of the equilibrium composition. The CFD results for the mixing length as a function of Péclet number $Pe = \bar{u}\ d_{\mathrm{h}}/D$ are displayed on the right side of Figure 12, where $\bar{u}$, d_{h}, D are the average flow velocity, hydraulic channel diameter, and molecular diffusivity, respectively. The maximum Reynolds number considered was about 17.

In addition to the CFD results, estimates of the mixing length based on the Fourier number

$$\mathrm{Fo} = \frac{Dt}{l^2} \tag{21}$$

are shown in the figure. The Fourier number relates a residence time t in the mixing channel to the molecular diffusion constant D and a characteristic length scale l, which is the width of the channel. For a given value of Fo and a given flow rate the length along the mixing channel necessary to achieve the corresponding Fourier number was determined. As is apparent from Figure 12, the Fourier number is a reasonable indicator for mixing, which occurs for Fourier numbers between 0.1 and 1.0. Consequently, l^2/D is an order-of-magnitude representation of the mixing timescale. The linear increase of the CFD-based mixing length as a function of Péclet number points to a very simple mixing mechanism by diffusion between co-flowing fluid lamellae. Obviously, complex convection-dominated mixing mechanisms (e.g. driven by swirls or recirculating flows) are absent in the simple mixing-tee configuration for the range of Reynolds numbers studied.

When the Reynolds number is increased, a regime with a flow pattern of broken symmetry is reached [54]. At a Reynolds number of about 140 the symmetric flow pattern becomes unstable and an engulfment flow with an increased area of material interface between the two fluids is found. This is illustrated in Figure 13, which shows maps of the computed concentration field over cross sections of the mixing channel for Reynolds numbers ranging from 119 (a) to 239 (f). The increase in material interface is related to an increase in mixing efficiency, and thus a mixing tee is more efficient than predicted by simple models accounting for mass transfer through an interface located in the symmetry plane of the mixer.

In cases where symmetry breaking and the corresponding engulfment flows are absent (i.e. at sufficiently small Reynolds numbers) the steady-state concentration field in a mixing tee can be computed based on an analytical approach. The underlying assumption is that the flow moves with uniform velocity. Ref. [51] gives the following Fourier series expansion for the concentration field in a mixing channel of width w with two streams of equal viscosity and a volume flow ratio of $r/(1-r)$

$$\frac{c(x,y)}{c_0} = r + \frac{2}{\pi} \sum_{n=1}^{\infty} \frac{\sin n r \pi}{n} \cos \frac{n \pi y}{w} \exp\left(- \frac{2n^2\pi^2}{Pe + \sqrt{Pe^2 + 4n^2\pi^2}} \frac{x}{w} \right) \tag{22}$$

where the concentrations in the two streams at the channel entrance are $c = c_0$ and $c = 0$ and Pe is the Péclet number based on the channel width. The coordinates in length and in width direction of the microchannel are denoted as x and y, respectively. With such an analytical expression the evaluation of mixing performance in a simple microchannel is very time efficient. However, an even simpler analysis based on the Fourier number already gives the correct order of magnitude for the mixing length, as indicated in Figure 12.

4.2. Chaotic Micromixers

The performance of mixing tees is usually not sufficient for liquid-mixing applications, since corresponding diffusion constants are of the order of 10^{-9} m^2/s or smaller, and mixing times far below one second may be required. As described above, mixing in straight channels occurs mainly by diffusion between co-flowing streams with a characteristic time constant of l^2/D. A reduction of the mixing timescale can be achieved if diffusion is superposed by convective transport. On the macroscale, convective transport is often due to turbulent eddies, and only in the very final stages of the mixing process does diffusion play a role. Unfortunately, in microscopic geometries flow is usually laminar, and a different convection mechanism must be devised to speed up mixing.

Much attention has been devoted to creating flow schemes in micromixers which enhance the mixing performance by convective transport. In that context it was found that chaotic flows bear a great potential for speeding up mixing. The corresponding principle of chaotic advection has been studied for quite some time [55], but it has only more recently been implemented in microchannel flows. Chaotic advection is synonymous with complex flow patterns in which extreme stretching and backfolding of fluid volumes occurs. It is very difficult to numerically study chaotic micromixing of liquids. The reason lies in the discretization artifacts from which most CFD schemes suffer, specifically the phenomenon of numerical diffusion [56, 57]. Owing to discretization errors artificial fluxes are induced which, for specific differencing schemes, take the form of diffusive fluxes and thus artificially increase the diffusion constant. The magnitude of numerical diffusion depends on the cell-based Péclet number (evaluated with the extension of a grid cell as length scale) and the relative orientation of the grid cells and the flow velocity. Specifically, if the flow is parallel to the grid lines, the diffusivity perpendicular to the flow direction is zero. In the case of a simple, regular flow pattern it is often possible to align the computational grid with the velocity

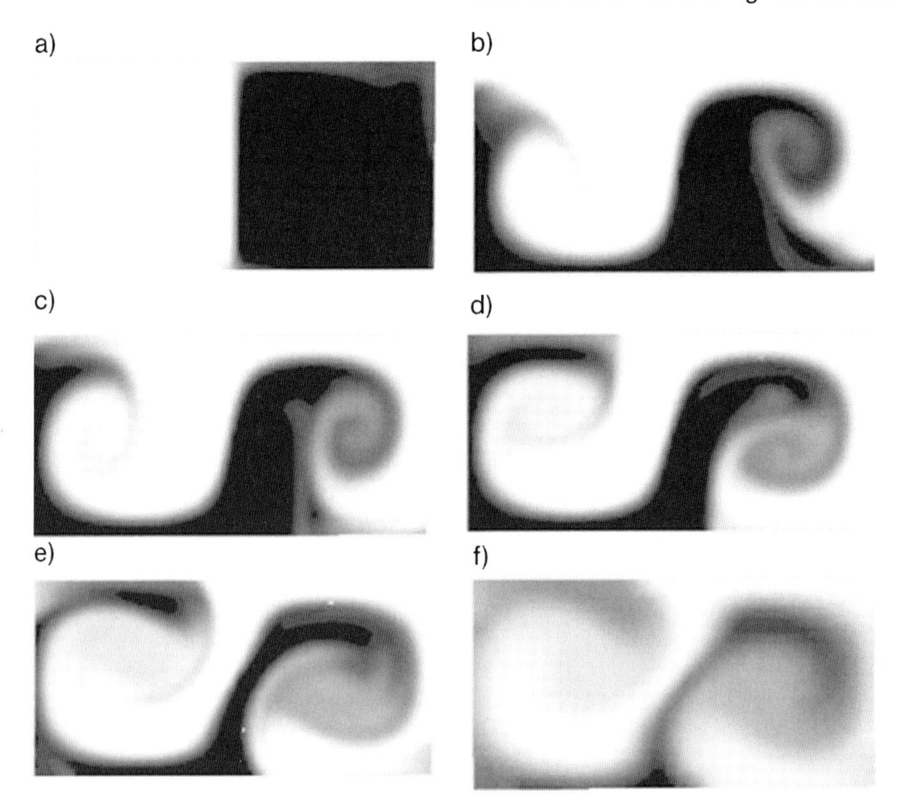

Figure 13. Concentration field over the channel cross section of a mixing tee for Reynolds numbers of 119 (a), 139 (b), 146 (c), 159 (d), 186 (e), and 239 (f), reproduced from [53]

field and thereby suppress numerical diffusion without having to prohibitively increase grid resolution. However, in the case of a chaotic flow this is no longer possible and numerical diffusion often severely falsifies the computational results.

To evaluate the performance of chaotic micromixers, the standard CFD methods for solving convection – diffusion equations must be abandoned and replaced by a method which is largely free of discretization artifacts. Such a method is Lagrangian particle tracking, which is based on tracking the paths of massless particles in a given flow field. Assuming that the flow velocity u_i is given, the particle trajectories are obtained as solution of the equation

$$\frac{\mathrm{d}x_i}{\mathrm{d}t} = u_i \qquad (23)$$

where x_i denotes the position vector of a particle. The mixing process is then studied by seeding the inlet of the mixer with a multitude of particles and recording how these particles have been redistributed by the flow at a downstream position. The fact that Equation 23 does not contain any contribution of Brownian particle motion is often not a real drawback. If the flow itself displays a large degree of chaoticity, much of the redistribution of the initial concentration field is done solely by convective transport, and diffusion only completes the mixing process. In a coarse-grained description in which concentrations are evaluated by computing the particle numbers in specific grid cells, the inclusion of Brownian motion would not modify the results significantly in many cases.

The computational effort for Lagrangian particle tracking can be reduced if the mixing section exhibits some periodicity. The structures inducing the chaotic flow are often arranged in a certain spatial period for which the flow field may be computed using periodic boundary conditions. Particles are then seeded onto the entrance plane of this section of the mixing channel and tracked up to its exit plane. The fluid streamlines map each point of a channel cross section (x_k, y_k) to another point (x_{k+1}, y_{k+1}) of the corresponding cross section after one period of the fluidic structures, as shown in Figure 14. Such a so-called Poincaré map can be written formally as

$$(x_{k+1}, y_{k+1}) = P\left[(x_k, y_k)\right]. \tag{24}$$

The function P can be computed either from an analytical or a numerical representation of the flow field, making use of Equation 23. In such a way, a 3D convection problem is essentially reduced to a mapping between two-dimensional Poincaré sections. The complete mixing process can be analyzed by successive application of the Poincaré map.

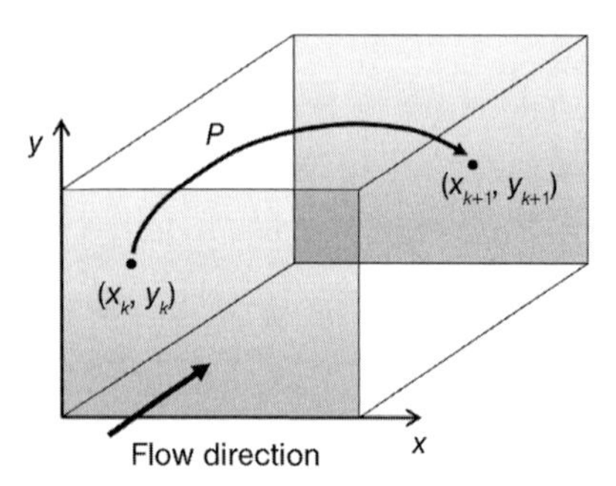

Figure 14. Section of a mixing channel with a map P connecting the points of two cross-sectional planes

Lagrangian particle tracking techniques have been used to evaluate the performance of several chaotic micromixers. One of the first micromixers for which convincing evidence for chaotic mixing was given is based on a design known as staggered herringbone mixer (SHM) [58]. The corresponding mixer geometry is shown in Figure 15. The bottom of the channel of height h and width w contains a staggered arrangement of grooves, whereby a fraction p of the bottom channel wall has grooves inducing a helical flow with a left-handed recirculation, and the remaining bottom wall fraction induces a right-handed recirculation. Schematic views of the channel cross section including projections of the streamlines are shown at the top of Figure 15. If $p \neq 1/2$, the vortex pattern is asymmetric, and a superposition of two patterns exhibiting the larger of the two vortices on the left and on the right side, respectively, could result in chaotic flow. In a generalized sense, this is an implementation of the "blinking-vortex principle" [59]: two vortex structures can be superposed in an alternating fashion to create a chaotic flow pattern. In the SHM the superposition is achieved by the alternating, staggered groove patterns shown in Figure 15. The lower part of the figure shows channel cross sections with two streams of fluorescent and clear solutions after 0, 0.5, and 1 cycle. The images were recorded using a confocal microscope. The two different vortices are clearly visible, and the third frame shows first indications of a chaotic disturbance of the flow.

Mixing in the SHM has been analyzed using Lagrangian particle tracking [60]. By comparison with a mixer containing only simple groove structures (as opposed to staggered herringbone structures) evidence was given for the superior performance of chaotic micromixers. The Poincaré sections obtained from particle tracking were compared to the confocal micrographs of Figure 15 [58]. The computed particle distributions (Figure 16) show strong similarities to the distributions of fluorescence intensity, and thus prove that mixing in complex flows can realistically be described by Lagrangian particle tracking. With standard CFD techniques resting on the solution of a convection – diffusion equation for the concentration field this is often not possible, since discretization errors would have blurred the thin lamellae visible in Figure 16.

Lagrangian particle tracking has also been applied to analyze chaotic mixing in a number of other channel designs. Simulation studies [61] on a twisted microchannel [62] showed that for Reynolds numbers above 10 mixing is mostly chaotic, apart from a few unmixed, regular islands. Cross-channel micromixers have been studied by a number of authors [63 – 65]. In such devices multiple side channels intersect with the main mixing channel. An oscillating pressure is imposed on the side channels such that the time-averaged flow rate entering the main channel is zero. The flow entering from the side channels distorts the material interface of the two fluids to be mixed, thereby increasing the interfacial area and promoting mixing. The analysis of mass transfer based on Lagrangian particle tracking has shown that, for suitable values of the pressure excitation amplitude and frequency, chaotic mixing is induced [63].

Many of the studies on micromixing reported in this section are more or less geared towards Lab-on-a-chip applications, i.e., the tar-

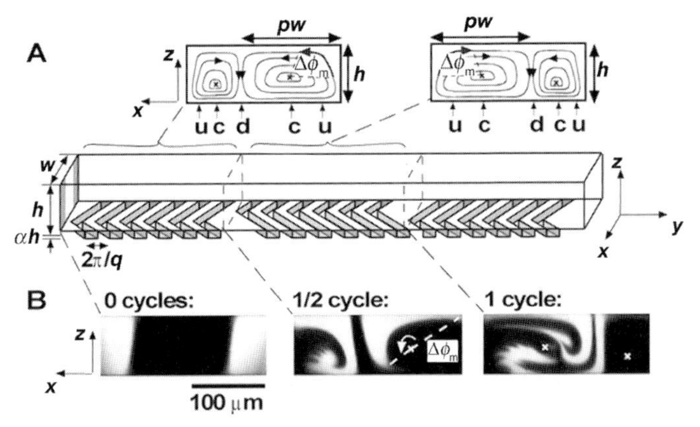

Figure 15. Micromixer geometry with staggered groove structures on the bottom wall, as considered in [58] (reprinted with permission, © 2002, AAAS). The top of the figure shows a schematic view of the channel cross section with the vortices induced by the grooves. At the bottom, confocal micrographs show the distribution of two liquids over the cross section. Flow is from left to right.

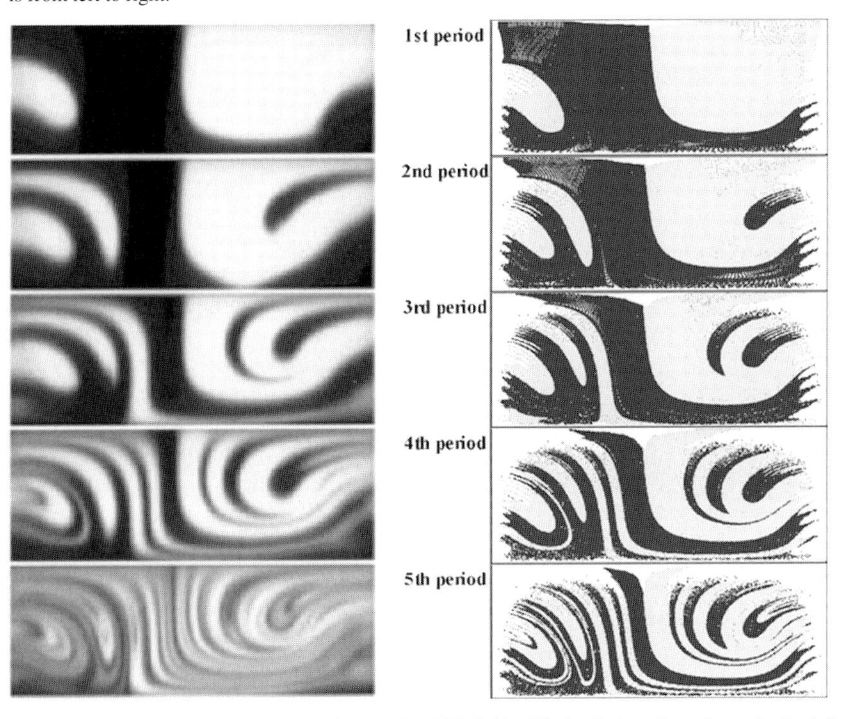

Figure 16. Comparison of confocal micrographs [58] (left) with the Poincaré sections obtained from Lagrangian particle tracking [60] (right).

get Reynolds number is about 10 or smaller. In microreactors throughput is often a more important issue than in Lab chips and, correspondingly, Reynolds numbers are higher and may easily reach values of several hundreds. A quite simple micromixer which is well suited for this Reynolds number regime was analyzed theoretically and experimentally [66]. The design of the mixer is shown at the bottom of Figure 17. The mixing channel consists of a number of arclike segments of alternating curvature, two of which are shown in the figure. As discussed in Section 2.2, a helical flow develops in such curved channels, with vortices appearing in the projection of the velocity field onto the channel cross section. For Dean numbers larger than 140 the flow ex-

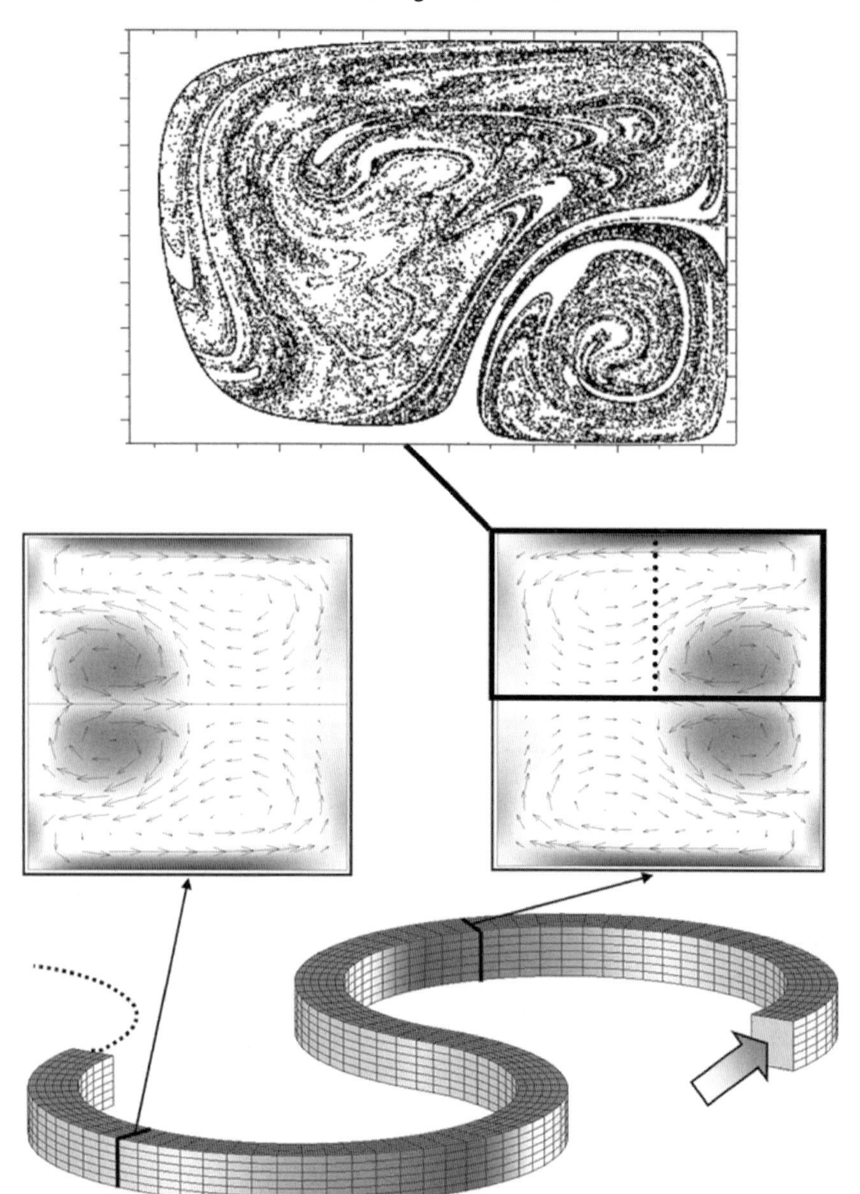

Figure 17. Two segments of a meandering mixing channel [66] (bottom) and CFD results showing the projected velocity fields on different cross-sectional planes (middle). The upper part of the figure shows the positions of massless particles initially seeded along the material interface after six channel segments.

hibits four counterrotating vortices, as indicated by the CFD results shown in the middle of Figure 17. In the mixer proposed by [66] a chaotic flow is created by alternating between the flow patterns in different channel segments, with the smaller pair of vortices either appearing in the right or the left half of the cross-sectional plane.

To study the mixing performance, a modified version of the usual Lagrangian particle tracking was developed. Usually, two types of tracer particles are distributed over the entire entrance plane and the mixing quality is computed by counting the number of particles in grid cells on a plane located at a downstream position. Based on this method it is difficult to obtain reliable results for states close to complete mixing. In an alternative method [66] only the material interface is seeded with particles (indicated as the dotted line in the upper half of one channel cross section in Fig. 17) and the length of the intersection of this interface with cross-sectional planes at various downstream positions is computed. This length scale should provide a good measure for mixing, since all the mass transfer proceeds through the material interface, and can be computed with high accuracy, since the same number of particles usually distributed over a plane can now be distributed along a line.

The upper part of Figure 17 shows the distribution of particles along the material interface after six channel segments for a Dean number of 200. The particles are distributed over almost the entire cross-sectional plane and thus document the pronounced and very complex stretching and folding of the interface. For Dean numbers above 100 an exponential growth of interfacial length scale as a function of residence time in the channel could clearly be proven. From these results it is apparent that mixing proceeds in a chaotic manner. This statement was underpinned by experimental results which showed that the position of close-to-complete mixing along the channel is virtually independent of the flow rate.

4.3. Multilamination Micromixers

The chaotic micromixers described in Section 4.1 offer a quite elegant solution to the micromixing problem, as complicated manifolds for distributing the fluid streams are no longer required and mixing is achieved by the intrinsic structure of the flow field. However, if the flow is chaotic in one part of the channel domain and regular in the other, unmixed islands, which usually deteriorate process performance, will remain. Thus, before using a chaotic micromixer in practice its mixing performance should be carefully evaluated over the whole range of Reynolds numbers considered. In comparison, multilamination mixers often require much more complex distribution manifolds and channel structures, but their mixing performance can usually be better controlled and predicted over the operation range envisioned.

The principle of mixing by multilamination relies on the creation of a multitude of thin fluid lamellae between which mixing occurs by diffusion. Thus, the mixing times achievable with multilamination mixers are directly related to the diffusion timescales between neighboring fluid lamellae. Figure 18 exemplifies the basic flow structure in a multilamination mixer: A multitude of fluid streams is guided through a flow chamber of varying width. For the mixer in the figure the streams are focused into a narrow channel by which the characteristic diffusion path is drastically reduced and mixing is enhanced.

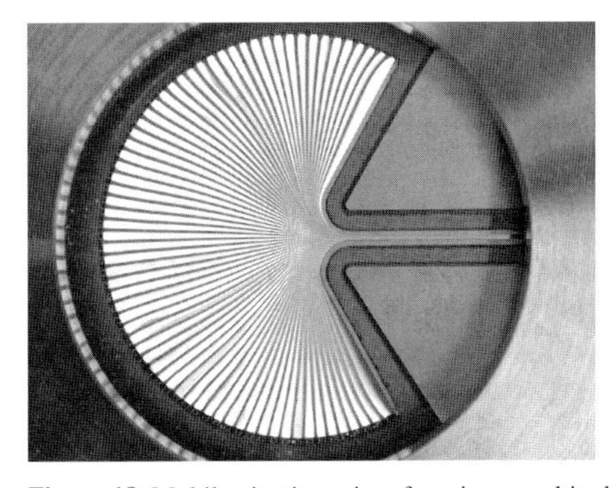

Figure 18. Multilamination mixer focusing a multitude of fluid streams through an inlet chamber into a narrow mixing channel. Alternating fluid inlets are distributed over a circular arc-shaped manifold.

In a similar way as in Section 4.1 an analytical formula can be derived which allows for a fast computation of the concentration field in a multilamination mixer. By transforming the spatial coordinate in streamwise direction (x in Eq. 22) into a time coordinate by dividing the spatial differerential by the average flow velocity and integrating from the inlet to the outlet of the mixer, a 2D spatial problem is reformulated as a problem with one space and one time coordinate. Thus, in a frame of reference co-moving with the flow and assuming a plug-flow profile, the convection – diffusion problem is transformed into a pure diffusion problem. The resulting diffusion equation can be solved by the usual Fourier-transform techniques. A corresponding analytical model for multilamination mixing has been developed [67] which takes into account changes of the channel width along the flow path, i.e., it incorporates effects of focusing or defocusing, with the former often being used to speed up mixing. Specifically, a coordinate s orthogonal to the fluid lamellae is introduced which may be a curvilinear coordinate in regions where focusing or defocusing occurs. It is assumed that s lies between $-w/2$ and $w/2$, where w is a measure for the channel width. The concentration field is then expanded in harmonics

$$\frac{c(s,t)}{c_0}=\frac{1}{2}+\sum_{n=0}^{\infty}a_n(t)\sin\frac{k_n s}{w},\quad k_n=(2n+1)\pi \tag{25}$$

where only the sine terms contribute if the total number of fluid lamellae is even and initial concentrations of c_0 and 0 are considered. The time dependence of the concentration field arises from the fact that a frame of reference co-moving with the flow is chosen, as described above. Inserting this expansion into the diffusion equation in the co-moving frame of reference allows $a_n(t)$ to be determined as

$$a_n(t)=a_n(0)\exp\left(-Dk_n^2\int_0^t\frac{dt'}{w^2(t')}\right), \tag{26}$$

where D is the molecular diffusivity. The function $w(t)$ incorporates the changes in channel width. In contrast to a channel of constant width an integral must be evaluated to determine the Fourier coefficients. This integral represents the history of the mixing process and accounts for the changes in mixing speed due to width variations.

Clearly, the model of Equations 25 and 26 should also be applicable to a simple mixing channel with only two fluid streams as considered in Section 4.1, i.e., Equation 22 should be recovered when transforming to a formulation with two spatial coordinates instead of a space – time description. However, the difference between the two formulations is that Equation 22 also takes into account diffusion in streamwise direction, an effect that becomes important at small Péclet numbers. Nevertheless the model for multilamination mixing is useful in most practical cases, since diffusion in streamwise direction is usually negligible. This model is free of numerical diffusion and thus not only allows faster, but also more reliable prediction of mixing performance in specific cases. The fast evaluation of mixing performance allows parameter studies to conducted which would have been extremely time consuming when using standard CFD techniques [68].

Reducing the solution of the convection – diffusion equation for the concentration field to a 1D problem in the co-moving frame-of-reference clearly reaches its limits if the fluid lamellae become deformed and are no longer arranged in parallel. Such a situation may occur if the channel depth is not small compared to its width and the design comprises a sudden expansion or contraction of the flow passage. Multilamination mixers with cornerflow geometries have been analyzed by CFD techniques, and situations where the multilamination pattern suffers strong deformations identified [69]. The complex flow patterns found in such mixers may also mislead experimentalists who attempt to evaluate the mixing performance by photometric techniques based on the diffusion of a solute [69].

Creating a multitude of lamellae such as in Figure 18 usually requires a quite complex flow distribution manifold. There is, however, an alternative multilamination principle which enables multiplication of the number of lamellae along the mixing channel: split-and-recombine (SAR) mixing. An SAR mixer consists of successive mixing elements each of which causes a doubling of the number of lamellae. An example of such a mixing element is shown in Figure 19. The flow entering the element in positive z-direction is split into two substreams which are then recombined. Thereby the number of fluid lamellae is increased from two to four. For n successive mixing elements, the number of lamellae is increased to 2^{n+1} and the diffusion paths are reduced correspondingly. The exponential reduction of diffusion paths resembles mass transfer in chaotic flows, where fluid volumes are stretched and folded such that the width of fluid filaments decreases exponentially. For this reason SAR mixers are usually very efficient, without the need for incorporating very small channels. Furthermore, in contrast to some chaotic mixers SAR mixers guarantee a largely uniform mixing quality over the entire channel cross section.

For the channel design shown in Figure 19 it could be shown that lamellae multiplication works in a close-to-ideal manner for Reynolds numbers below about 30 [70]. For higher Reynolds numbers inertial forces become important and cause deformation and finally merging of lamellae of the same type. Altogether, experiments indicate that at comparatively small Reynolds numbers the design shown

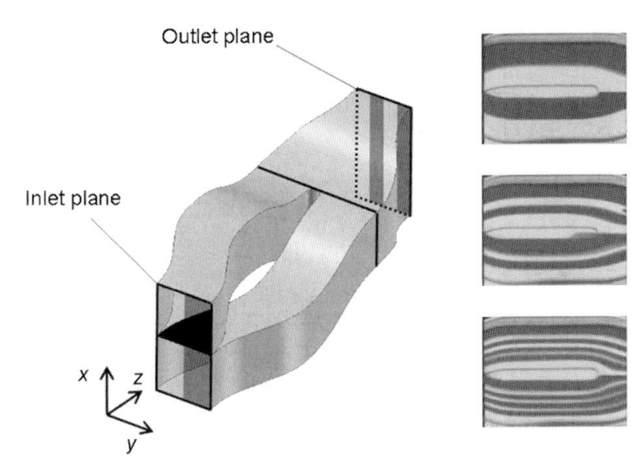

Figure 19. Mixing element of an SAR mixer (left) and optical micrographs showing lamellae multiplication in the first three mixing elements (right)

enables an efficient liquid mixing at low pressure drops, and thus the SAR mixer is especially qualified for the processing of highly viscous liquids [70].

As far as the computation of concentration fields in SAR mixers is concerned, the challenges are greater than in ordinary multilamination mixers. The reason is that the number of lamellae does not remain constant; rather, there are discontinuous transitions between patterns with n and $2n$ lamellae. A strategy to deal with this problem has been formulated [67]: based on solving a 1D diffusion equation in a co-moving frame-of-reference (Eqs. 25 and 26), the mixing element shown in Figure 19 is subdivided in two sections. In the first section, the flow passes through two separate branches, and in the second section it only passes through one branch. In each of these sections the time evolution of the Fourier coefficients is computed by Equation 26. At the interface, the downstream Fourier coefficients are obtained from a sum involving the upstream Fourier coefficients. In addition, the evolution of the concentration field in a complete SAR mixer comprising several successive mixing elements is obtained in an iterative manner from the analysis of a single mixing element, i.e., the Fourier coefficients on the outlet plane shown in Figure 19 are identified with the coefficients on the inlet plane of the next element. Thus the degree of mixing was computed as a function of position along the channel, and reasonable agreement with experimental data was found [67]. Thus, the usefulness of the 1D model for multilamination mixing is corroborated by experimental data, and reasonable results can be obtained even for such complex situations involving flow passages of varying width as well as flow splitting and recombination.

There are alternative flow topologies for SAR mixers which have been proven to cause a multiplication of fluid lamellae. The design of mixing element shown in Figure 19 was especially optimized with regard to extending the Reynolds number range in which the SAR principle is implemented, consistent with the throughput requirements typical for microreactors. The disadvantage is that this mixing element poses major challenges for microfabrication. When only very small Reynolds numbers must be considered the design constraints are less severe, and a design which is easier to fabricate has been proposed [71]. The flow topology is such that a doubling of the number of fluid lamellae is achieved after two SAR steps, and the functionality with respect to lamellae multiplication has been proven by using dyed aqueous solutions. In the same spirit as above, the model of Equation 25 and 26 can be used to compute the evolution of concentration fields in such a mixer if the upstream and downstream Fourier coefficients at the branching points are matched with each other.

4.4. Hydrodynamic Dispersion

Mass transfer of a solute dissolved in a fluid is not only the fundamental mechanism of mixing processes, it also determines the residence-time distribution in microfluidic systems. As mentioned in Section 2.3, in many applications it is desirable to have a narrow residence-time distribution of concentration tracers being transported through a microfluidic system. An initially narrow concentration of tracer will suffer broadening (i.e., dispersion) due to two different effects. First, in some regions of the flow domain of a system the fluid velocity will be smaller than in others, and this leads to a longer residence time of molecules being transported preferably through these parts of the domain. However, due to Brownian motion the molecules will also sample some of the other regions with higher flow velocity. Hence, molecular diffusion may reduce the dispersion of a concentration tracer. On the other hand, by diffusion an initially localized concentration tracer in a fluid at rest will be dispersed. From these arguments it becomes clear that hydrodynamic dispersion depends on a quite subtle interplay of convective and diffusive mass transfer, and the evolution of a concentration tracer as it is transported through the flow domain depends on various factors such as the flow profile, the magnitudes of the flow velocity, and the diffusion constant.

The key analysis of hydrodynamic dispersion of a solute flowing through a tube [72, 73] assumed a Poiseuille flow profile in a tube of circular cross section and showed that for large enough times the dispersion of a solute is governed by a one-dimensional convection – diffusion equation

$$\frac{\partial \bar{c}}{\partial t}+\bar{u}\,\frac{\partial \bar{c}}{\partial x}=D_e\frac{\partial^2 \bar{c}}{\partial x^2} \tag{27}$$

where $\bar{c}$ is the concentration averaged over the cross section of the tube, $\bar{u}$ the average velocity, and D_{e} an effective diffusivity, also denoted dispersion coefficient, which is given by

$$D_e=D+\frac{\bar{u}^2 R^2}{48D} \tag{28}$$

where D is the molecular diffusivity, and R the radius of the tube. The factor 1/48 multiplying the velocity-dependent term is generic for tubes of circular cross section and is modified when other geometries are considered. In many cases the second term, which can be rewritten as $DPe^2/48$, dominates over the first, which is a purely diffusive contribution. Hence, due to convection a concentration tracer is usually dispersed much more strongly than it would have been by diffusion alone. A notable feature of Equation 27 and 28 is their independence of any initial condition. Independent of how the solute is distributed over the channel cross section and along the channel initially, the Taylor – Aris [72, 73] description is valid in the limit of long times (t · ·). When exactly this limit is reached with a given accuracy depends on the initialization of the concentration field. A rough guideline is provided by the Fourier number of Equation 21, evaluated with the tube radius as length scale. The Fourier number can be regarded as a dimensionless time coordinate which compares the actual time to the time a molecule needs to sample the cross-sectional area of the tube. The validity of the description should be related to the condition that the Fourier number assumes values on the order of 1 or larger.

The above analysis was extended to arbitrary time values for the dispersion of an initially plug-like profile [74], i.e.

$$c\,(x,r,0)=\{\begin{matrix} c_0\;(|x|\leq l/2)\\ 0\;(|x|>l/2)\end{matrix} \tag{29}$$

where r is the radial coordinate of the tubular geometry, and l the length of the plug. A generalized evolution equation was derived for the area-averaged concentration of the form

$$\frac{\partial \bar{c}}{\partial t}=\sum_{n=1}^{\infty}k_n\,(t)\,\frac{\partial^n \bar{c}}{\partial x^n} \tag{30}$$

which is valid without any restriction on t. The derivatives in the infinite series appearing on the right-hand side are multiplied by time-dependent dispersion coefficients k_{n}. In the Taylor – Aris limit, all of the dispersion coefficients except k_1, which describes the convection of the tracer with the flow, and k_2, which determines the spreading of the tracer, are negligible. When moving to smaller times, the time-dependence of k_2 needs to be taken into account, while all higher dispersion coefficients are still negligible [74]. Only at very small times do the higher dispersion coefficients become important. For the case considered, k_2 can be regarded as time-independent for Fourier numbers greater than about 0.5 [74].

Even if the results discussed above highlight some of the most important aspects of hydrodynamic dispersion, they were based on cylindrical ducts, which are not the generic geometry used in the field of microfluidics. In chemical microprocess technology, tubular sections are used to connect different units, but the channels contained in microreactors typically have a rectangular or tublike cross section. Dispersion in rectangular channels was studied in detail [75]. The evolution equation (Eq. 27) is still valid in this case, but the expression for the dispersion coefficient (Eq. 28) needs to be modified. While a simple analytical expression for the dispersion coefficient related to flow between parallel plates could still be obtained [76], the corresponding expression for rectangular channels is a complicated series expansion. This is not surprising, since the exact form of the flow profile in a rectangular channel is given in form of an infinite series as well. A simpler relationship approximates the exact expression for the dispersion coefficient in rectangular channels within an error of 10 % [77]. In addition, the tub-like channel cross sections which are typically obtained by isotropic etching processes were considered by employing a numerical scheme that allowed computation of the dispersion coefficient. On this basis, different channel geometries were compared, and favorable and less favorable designs identified.

In cases where hydrodynamic dispersion and the corresponding broadening of residence-time distributions deteriorate the performance of a process, the question arises which channel design minimizes dispersion. Already from the Taylor – Aris analysis it becomes clear that enhanced mass transfer perpendicular to the main flow direction reduces the broadening of concentration tracers. Such a mass transfer enhancement can be achieved by the secondary flow occurring in a curved channel. This aspect was investigated for ducts of circular cross section [78] under the assumption that the diameter of the duct is small compared to the radius of curvature, and the convection – diffusion equation for the concentration field was solved numerically. More specifically, a two-dimensional problem defined on the cross-sectional plane of the duct was solved based on a combination of a Fourier series expansion and an expansion in Chebyshev polynomials. The solution is of the general form

$$D_{\mathrm{e}}^{\mathrm{cur}}=D\left[1+\mathrm{Pe}^{2}f\left(K,\mathrm{Sc}\right)\right], \tag{31}$$

where $D_{\mathrm{e}}^{\mathrm{cur}}$ is the dispersion coefficient in curved ducts, D the molecular diffusivity, Pe the Péclet number of the flow, and the function f depends on the Dean number K defined in Equation 7 and on the Schmidt number Sc, which is the ratio of the kinematic viscosity and the diffusivity. The following asymptotic behavior for the ratio of the dispersion coefficients in curved and straight ducts was found [78]:

$$\frac{D_{\mathrm{e}}^{\mathrm{cur}}}{D_{\mathrm{e}}}\begin{cases}=1 & K\to 0\\ \propto K^{-1} & K\to\infty\end{cases}. \tag{32}$$

In curved channels secondary flow patterns of two or four counterrotating vortices are formed. These vortices redistribute fluid volumes in a plane perpendicular to the main flow direction. Such a transverse mass transfer reduces dispersion, a fact reflected in the K^{-1} dependence in Equation 32 at large Dean numbers. For small Dean numbers, the secondary flow is negligible, and the dispersion in curved ducts equals the Taylor – Aris dispersion of straight ducts.

Similar to the case of mixing, the transverse redistribution of matter in a curved channel can be improved in a sequence of channel segments of alternating curvature. By superposition of different vortex patterns a chaotic flow is induced [66]. By Lagrangian particle tracking in combination with a CFD computation of the flow field it can be shown that the residence-time distribution of tracers in the channel built from segments as shown in Figure 17 becomes very narrow and approaches a deltalike distribution of zero width. In practice, however, the width of the residence-time distribution is limited by the molecular diffusivity of the tracers, which was neglected in the simulations [66]. Nevertheless, a substantial reduction of hydrodynamic dispersion can be achieved in a meandering channel when compared to a straight channel.

5. Chemical Kinetics

Most plants or reactors in chemical microprocess technology inevitably contain a unit in which chemical conversion takes place. The goal might be to produce a fine chemical with a high yield and selectivity or to screen a large number of reactions in parallel. Hence, a thorough understanding of chemical kinetics is a key requirement for the successful design and optimization of micro reaction devices. For this purpose, reliable models of reaction kinetics coupled to the transport equations of momentum, heat, and matter are needed.

The type of kinetic model to be used depends on the type of reaction considered. For a homogeneous reaction occurring in the bulk of the fluid, a power-law kinetic model is often appropriate (see, e.g., [79]). In such models the rate of a certain reaction depends on a product of powers of the species concentration. On the other hand, heterogeneously catalyzed reactions are often conducted in microreactors. In a strict sense, power-law kinetics do not capture the dynamics of such processes over the full range of pressure, temperature, and concentrations. Instead, a more complicated kinetic model of, e.g., the Langmuir – Hinshelwood type [80] would have to be used. Nevertheless, power-law kinetics is frequently applied to heterogeneously catalyzed processes in a limited parameter range to simplify the description.

Independent of the specific modeling strategy, the kinetic equations often exhibit a nonlinear dependence on the species concentrations and a reaction rate rapidly increasing with temperature. In combination with the transport equations for mass, momentum, and heat, the resulting numerical problem is usually challenging due to the nonlinearities and the multitude of timescales involved. For this reason, methods are needed to eliminate some of these difficulties and to simplify the numerical structure.

5.1. Numerical Methods for Reacting Flows

The solution of the species concentration equations in combination with the momentum and the enthalpy equation generally requires an iterative procedure. A rough sketch of the numerical structure of a stationary reacting-flow problem is given as

$$\begin{bmatrix} A_{cc} & A_{cu} & A_{cT} \\ A_{uc} & A_{uu} & A_{uT} \\ A_{Tc} & A_{Tu} & A_{TT} \end{bmatrix} \begin{bmatrix} c \\ u \\ T \end{bmatrix} = \begin{bmatrix} b_c \\ b_u \\ b_T \end{bmatrix}, \qquad (33)$$

where c, u, and T denote the vectors of concentration, velocity, and temperature fields, respectively. Owing to the nonlinear nature of the problem, the coefficients of the different matrices $A_{\alpha\beta}$ still depend on the unknowns c, u, and T. The cross-coupling between different field quantities is provided by those matrices $A_{\alpha\beta}$ with $\alpha \neq \beta$. The set of nonlinear algebraic equations is solved iteratively, i.e., starting with an initial guess the approximation is successively improved until convergence is reached. Depending on the nature of the chemical reaction term entering the species-concentration equation, different strategies may be applied to solve Equation 33. For intrinsic kinetics characterized by a much shorter timescale than transport of momentum, heat, and matter, it is often preferable to set up an iteration scheme in which a number of iterations of the species-concentration equation are performed during one iteration cycle of the remaining equations. However, for a fast reaction which is limited by heat and mass transfer (e.g. in a situation where the reactants are not premixed), comparable iteration cycles of the species-concentration equation and the remaining equations might be sufficient.

Apart from the coupling of chemical kinetics to the transport equations, the chemical reaction dynamics themselves may pose numerical challenges when a number of different reactions are superposed. In such a case the rate of disappearance of a chemical species i can be written as

$$-\frac{\dot{n}_i}{V} = R_{ij} r_j, \qquad (34)$$

where r_j is the rate of the jth reaction, and R_{ij} a matrix defining how a specific reaction contributes to a change in concentration of the chemical species involved. For brevity the Einstein convention of summation over repeated indices was used. Quite frequently it occurs that the timescales characterizing the different reactions vary by orders of magnitude, such that the fast reactions are already completed while the slow reactions have not yet progressed to any appreciable degree. The corresponding stiff differential equations are usually solved by an implicit time-integration scheme which allows comparatively large time steps without suffering from numerical instabilities or predicting unrealistic asymptotic states [81]. However, implicit time integrators involve the solution of a (generally nonlinear) algebraic system of equations for each time step, which is done by some iterative scheme such as Newton's method. For reaction systems with a broad spectrum of timescales these iteration schemes can fail to converge, with the consequence that very small time steps must be chosen. Such a situation is related to high computational costs, and methods are needed to simulate extremely stiff reaction systems more efficiently.

Methods based on the partitioning of a reaction system into fast and slow components were proposed [82 – 84]. A key assumption made in this context is the separation of the space of concentration variables into two orthogonal subspaces Q_{s} and Q_{f} spanned by the slow and fast reactions. With this assumption the time variation of the species concentrations is given as

$$-\frac{\dot{n}_i}{V}=(Q_{s,j})_i\left(\dot{y}_s\right)_j+(Q_{f,j})_i\left(\dot{y}_f\right)_j. \quad (35)$$

The notation is such that $(Q_{\mathrm{s},j})_i$, $(Q_{\mathrm{f},j})_i$ denote the ith component of the jth basis vector in the subspace of slow and fast reactions, respectively. The corresponding expansion coefficients are $\left(\dot{y}_s\right)_j$ and $\left(\dot{y}_f\right)_j$, respectively, and are expressed by the reaction rates via

$$\left(\dot{y}_s\right)_i=\left(Q^T_{s,j}\right)_i R_{jk} r_k, \quad (36)$$

$$\left(\dot{y}_f\right)_i=\left(Q^T_{f,j}\right)_i R_{jk} r_k. \quad (37)$$

If the timescale of the fast reactions is much shorter than that of the slow reactions, it can be assumed that the former are completed at an initial stage of the latter. Mathematically, this assumption reads

$$\left(Q^T_{f,j}\right)_i R_{jk} r_k=0. \quad (38)$$

Equation 38 represents a set of algebraic constraints for the vector of species concentrations expressing the fact that the fast reactions are in equilibrium. The introduction of constraints reduces the number of degrees of freedom of the problem, which now exclusively lie in the subspace of slow reactions. In this a way the fast degrees of freedom have been eliminated, and the problem is now much better suited for numerical solution methods. It has been shown that, depending on the specific problem to be solved, the use of simplified kinetic models allows the computational time to be reduced by two to three orders of magnitude [85].

5.2. Reacting Channel Flows

In chemical microprocess technology one class of reactor designs deserves the term "generic", since many of the microreactors reported in the literature are based on this design concept. The design comprises at least one rectangular microchannel, often a multitude thereof, with a solid catalyst attached to the channel walls. The reacting fluid flows through the channel, while the reagents diffuse to the channel wall where they undergo chemical reactions. There are two versions of this design concept (Fig. 20). Either a smooth surface, often a metal layer, acts as catalyst, or the reaction occurs in a catalytically active porous medium. Clearly, the advantage of the porous catalyst layer is the higher specific surface area offering more reaction sites to the reagents. However, fully coupled simulations of reaction – convection – diffusion processes in catalyst-coated channels are quite rare, so most of the studies reported in this section are based on the concept of wall-catalyzed reactions.

For the reasons described above, reaction – convection – diffusion problems tend to be difficult to solve numerically. Hence, the simulation of reacting flows in three dimensions or parameter studies of micro reaction devices may be very time consuming. To enable rapid prototyping of microreactors, efficient modeling strategies with a minimum expenditure of computational resources are needed. The modeling approach developed in [86] allows a limited class of reacting microchannel flows to be assessed very quickly. It assumes a microchannel of length L and depth h with a first-order reaction occurring at one of the channel walls (Fig. 21). In cases where the flow profile is independent of the axial position in the channel and the problem can be approximated by a two-dimensional model, the mass transport equation for a chemical species a can be written in dimensionless form as

$$\frac{\mathrm{Pe}\,h}{L}u(\eta)\frac{\partial c_{\mathrm{a}}}{\partial\zeta}=\frac{\partial^2 c_{\mathrm{a}}}{\partial\eta^2} \quad (39)$$

where the axial and transverse coordinates ζ and η were nondimensionalized by the channel length L and the channel depth h. The reactant concentration is denoted by c_{a}, the velocity by u, and the Péclet number is expressed by the average velocity $\bar{u}$ and the diffusion constant D as $\mathrm{Pe}=\bar{u}\,h/D$. Equation 39 is solved subject to the boundary conditions of an impermeable upper channel wall and a first-order reaction with rate constant k occurring at the lower channel wall.

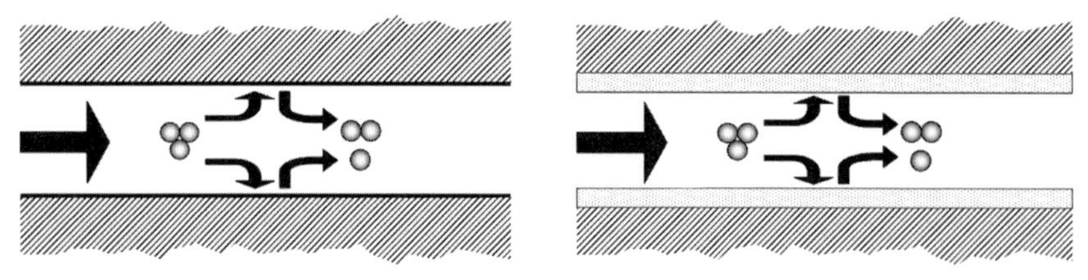

Figure 20. Reaction channels with a smooth surface (left) and a porous medium (right) as catalyst

Such a first-order reaction term surely does not adequately capture the mechanism of heterogeneous catalysis, but it may be a reasonable approximation of the kinetics in a limited parameter or operation range. An important dimensionless group characterising the reactive flow is the Damköhler number, defined as

$$\mathrm{Da}=\frac{kh}{D} \tag{40}$$

which characterizes the ratio of the diffusive and reactive timescales. The mass-transport equation has a separable solution of the form

$$c_{\mathrm{a}}(\zeta,\eta)=\bar{c}_{\mathrm{a}}(\zeta)\, f_{\mathrm{a}}(\eta), \tag{41}$$

where

$$\bar{c}_{\mathrm{a}}(\zeta)=\int_0^1 c_{\mathrm{a}}(\zeta,\eta)\,\mathrm{d}\eta. \tag{42}$$

By inserting this ansatz into Equation 39, the solution can be determined in form of an eigenfunction expansion [87]. The parameter controlling the number of terms of this expansion having to be taken into account is $Pe \cdot h/L$, which is usually of the order $0.01 - 1$ in microreactors. For this reason, often only the first term contributes. With the entrance condition $\bar{c}_{\mathrm{a}}(\zeta)=1$ the axial dependence can then be written as

$$\bar{c}_{\mathrm{a}}(\zeta)=\exp(-\lambda_{\mathrm{a}}\zeta), \tag{43}$$

where the eigenvalue λ_{a} is given as the solution of a nonlinear algebraic equation. Comparing the analytical results to full numerical simulations showed good agreement [86]. In addition to isothermal flows, analytical solutions were also determined for nonisothermal reacting flows and the model extended to second-order kinetics. Thus, a class of models was developed which may provide a simple characterization of reacting flows in microchannels without the need to do a full numerical simulation.

A similar analytical model for reacting flows in microchannels was used to assess the quality of simple plug-flow models which may be used to estimate reaction rate constants [88]. Microreactors lend themselves to measuring intrinsic rate constants of chemical reactions as, due to the short diffusion paths, heat and mass transfer limitations can be eliminated. The simplest way to deduce the rate constant k of a first-order heterogeneously catalyzed reaction at the walls of a tube is by assuming the reaction to occur in the volume of a plug-flow reactor. In this way the wall reaction is replaced by a pseudohomogeneous reaction and the velocity profile of the flow is ignored, which means that effectively a one-dimensional model is used. By measuring the inlet and outlet concentration of the reacting component, the rate constant is then obtained as

$$k=\frac{\bar{u}\,R}{2L}\ln\left(\frac{c_{\mathrm{a}}(\zeta=1)}{c_{\mathrm{a}}(\zeta=0)}\right). \tag{44}$$

where the notation is chosen similar to the previous paragraph, and L and R are the length and radius of the tube, respectively. Two effects are not taken into account by this expression. First, radial concentration gradients are ignored. Second, dispersion in the tube as discussed in Section 4.4 is neglected.

The one-dimensional model of reacting flows was extended to include Taylor – Aris dispersion [88] by considering an equation of the form

$$\frac{\mathrm{d}^2 c_{\mathrm{a}}}{\mathrm{d}\zeta^2}-\frac{\mathrm{Pe}^*}{2}\frac{\mathrm{d}c_{\mathrm{a}}}{\mathrm{d}\zeta}-\beta c_{\mathrm{a}}=0, \tag{45}$$

where Pe^* is a modified Péclet number containing the Taylor – Aris dispersion constant instead of the diffusivity, and β a dimensionless parameter representing the pseudohomogeneous reaction. To study the influence of dispersion on chemical conversion, the solution of Equation 45 was compared to the solution of the corresponding two-dimensional problem, obtained in a similar way as sketched in the previous paragraph. It turned out that for a

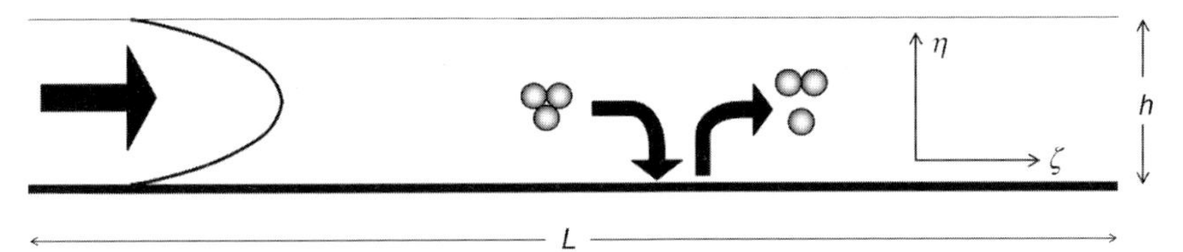

Figure 21. Two-dimensional model geometry of a microchannel with a reaction occurring at the lower channel wall

Damköhler number of 1, no satisfactory agreement between the one- and the two-dimensional models was achieved. The inclusion of Taylor – Aris dispersion improved the concentration profiles to a certain degree with respect to a plug-flow model, but the main reason for the deviations are the radial concentration gradients, which are not accounted for in the one-dimensional models. Hence, when attempting to extract intrinsic reaction-rate constants from comparisons of experimental results with results of one-dimensional reactor models, care should be taken to work in a regime of Damköhler numbers significantly smaller than 1.

5.3. Heat-Exchanger Reactors

The design of multichannel microreactors for gas-phase reactions is typically based on a stack of microstructured platelets. For strongly endothermic or exothermic reactions, it lends itself to alternate between layers of reaction channels and heating or cooling gas channels which supply energy to or withdraw it from the reaction. Such a set up is similar to the heat exchanger design depicted in Figure 9. Within this class of microreactor designs a choice can be made between different flow schemes of the gas streams in adjacent layers (co-, counter- or cross-current). The countercurrent coupling of an endothermic reaction to a heating gas stream in a multilayer architecture was studied [41] for a 2D model (Figure 22).

The dynamics of a heterogeneously catalyzed gas-phase reaction occurring in a nanoporous medium in combination with heat and mass transfer was simulated using a finite-volume approach. In contrast to other studies of similar type, heat and mass transfer in the nanoporous medium were explicitly accounted for by solving volume-averaged transport equations in the porous medium. Such an approach made it possible to compare the transport resistances in the gas phase and in the porous medium and to study the trade-off between maximization of catalyst mass and minimization of mass-transfer resistance due to pore diffusion. A typical concentration profile of a reacting chemical species which is converted by the catalyst is displayed in Figure 23. Due to the small pore size with an average diameter of 40 nm, the effective diffusivity in the porous medium is quite small and large concentration gradients build up, whereas in the microchannel the gradients are negligible. Typical catalyst effectiveness factors for a 100 μm catalyst layer were found to be of the order of 0.4. One of the outstanding potentials of microreactors is efficient utilization of the catalyst material. In conventional fixed-bed technology, catalyst pellets for liquid reactions are usually of a size of 2 – 5 mm [79]. Due to diffusive limitations in such comparatively large pellets, reactions often occur in a region close to the surface.

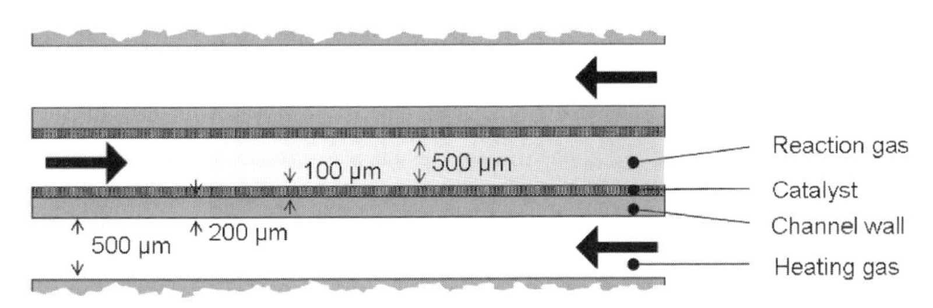

Figure 22. 2D model of a countercurrent heat-exchanger reactor with a nanoporous catalyst layer deposited on the channel wall

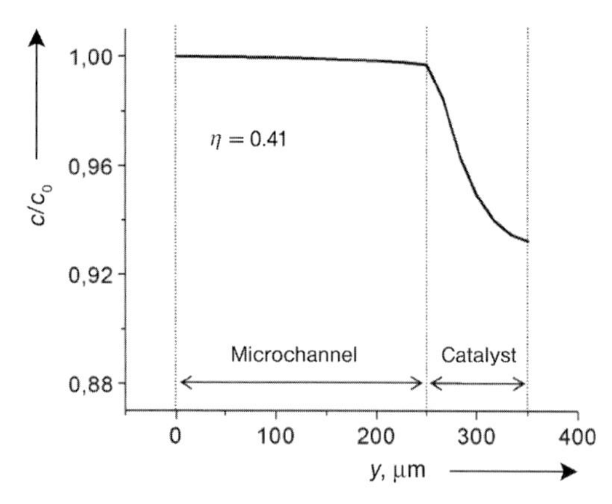

Figure 23. Normalized concentration profile of a reacting species across a microchannel of 500 μm width with a 100 μm catalyst layer deposited on the wall

A main objective of [41] was to study the influence of heat transfer on the achievable molar flux per unit reactor volume of the product species. Unstructured channels were compared to channels containing microfins. Heat-transfer enhancement due to microfins resulted in a different axial temperature profile with a higher outlet temperature in the reaction gas channel. Due to this effect and by virtue of the temperature dependence of the reaction rate, an improvement of heat transfer resulted in a significantly higher specific product molar flux. For the system under study, the heat-transfer enhancement achievable with microfins was found to increase the specific molar flux by about a factor of two. Such model studies show that a complex interplay between flow, heat transfer, and mass transfer may occur in microreactors and underline the need for fully coupled simulations incorporating conjugate heat transfer and transport in porous media.

The optimization of heat transfer in a heat-exchanger reactor was also the objective of [89]. Specifically, the exothermic water gas shift (WGS) reaction

$$CO+H_2O \leftrightarrow H_2+CO_2 \qquad (46)$$

which is utilized in fuel reformers to reduce the level of carbon monoxide, was considered. When the temperature of an exothermic, reversible reaction such as the WGS reaction increases, the kinetics are accelerated but the equilibrium is shifted more towards the feed components. Due to this fact, neither very low nor very high temperatures are optimal when the goal is to maximize the space – time yield for a given conversion. Rather, there is a specific temperature trajectory, i.e., a specific functional dependence of the reaction temperature on time, which allows maximization of the space – time yield. Due to their short thermal diffusion paths, microreactors allow the temperature profile in a reaction channel to be controlled much better than conventional equipment.

A counter-current heat-exchanger reactor for the WGS reaction with integrated cooling gas channels for removal of the reaction heat was analyzed [89]. The computational domain of the 2D model on the basis of the finite-element method is sketched in Figure 24. The reactor design does not allow for detailed adjustment of the temperature profile in the reaction gas channel, but by varying the cooling gas inlet temperature and the ratio of cooling gas and reaction gas flow rate, different temperature profiles can be imposed.

The reaction dynamics studied in the simulations is such that by release of reaction heat close to the inlet, the reaction gas temperature rises considerably. Owing to the fast kinetics at high temperatures, much of the CO conversion is already achieved in the inlet region. The remaining conversion is then mainly due to a reduction of the temperature and the corresponding shift in chemical equilibrium. This part of the process requires a large channel length. Thus, even minor differences in conversion achievable by improved temperature control may result in a considerable reduction of reactor size and the required amount of catalyst. When considering more advanced reactor designs allowing for fine-tuning of the temperature trajectory, CFD simulations are indispensable for performance optimization.

6. Conclusion and Outlook

The discussion of modeling and simulation techniques for microreactors shows that the toolbox available at present is quite diverse and goes well beyond the standard capabilities of CFD methods available in commercial solvers. In microreactors special methods needed for the modeling of noncontinuum physics play only a minor role, and most of the effects are described by the

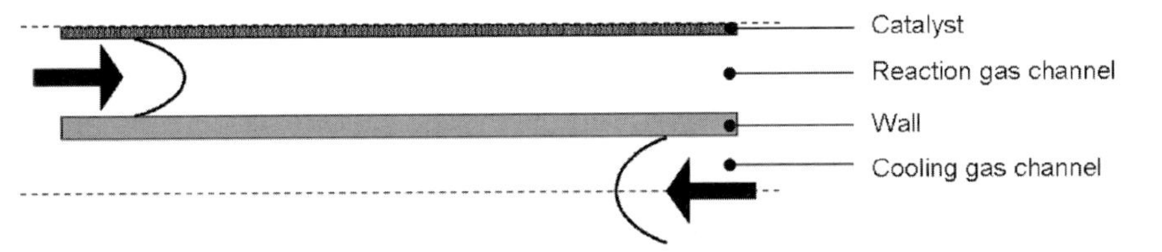

Figure 24. Model of a counter-current heat-exchanger reactor for exothermic reactions. The dashed lines indicate symmetry planes

standard continuum equations. However, even if the laminar nature of the flow somehow reduces the difficulty of simulation problems compared to macroscopic flows, there are a number of problems which are extremely difficult and require very fine computational grids. Among these problems is the numerical study of mixing in liquids, which often severely suffers from discretization artefacts.

A subject of growing importance which has not been discussed in this overview due to space limitations is modeling and simulation of multiphase microflows. Macroscopic multiphase flows are often modeled by using an Euler – Euler formulation of interpenetrating continua, that is, a volume-averaged description where phase boundaries are no longer explicitly resolved. When the spatial extension of bubbles or droplets becomes comparable to the channel dimensions, the gas – liquid or liquid – liquid interfaces must be modeled explicitly. There are methods available for such kinds of problems (e.g. the volume-of-fluid method) which are, however, computationally much more expensive than the Euler – Euler formulations employed for macroscopic flows. It is not only for the computational requirements that the predictive power of multiphase microflow computations is much lower than that of their counterparts for single-phase flow. As an example, moving contact lines are relevant for gas – liquid flows, and in some cases flow is accompanied by phase-change phenomena. Both the motion of a gas – liquid – solid contact line and the evaporation of liquid heavily depend on complex transport phenomena in the close vicinity of the contact-line region which are only partially understood until now. Consequently, the simulation of such phenomena as flow boiling in microchannels is in its infancy and its predictive power is very limited. Such examples underline the necessity of further research on multiphase microflows. Currently, corresponding numerical schemes are available, but the models do not adequately account for the boundary conditions at the contact line.

A further topic which has not been discussed in this overview and which is only beginning to evolve refers to system-level simulations of microprocess devices. In conventional process technology there is often an interplay between detailed CFD and system-level simulations of physicochemical processes. Owing to their large computational requirements, CFD methods can usually not be employed for the simulation of chemical reactors or plants consisting of a number of different subunits. Rather, CFD results enter into system models which treat the subunits in a simplified manner by just referring to characteristic curves specifying input – output behavior. In such a way, the flow in a pipe, for example, is characterized by a flow rate as a function of applied pressure difference. In conventional process technology a library of component models already exists which may be further extended and refined by incorporating the results of CFD simulations. However, in microprocess technology such system-level approaches do not yet seem to have been developed to an appreciable degree. A natural step for the further development of modeling and simulation techniques for microreactors would be to formulate system-level models which allow coupling to CFD simulations. In such a way also the commercial application of microreactors could gain additional impetus, as system-level models would allow fast evaluation of processes unparalleled by the comparatively time-consuming approaches presently available.

7. References

1. S. B. Pope, *Turbulent Flows*, Cambridge University Press, Cambridge 2000.
2. M. Gad-el-Hak, "The Fluid Mechanics of Microdevices", *J. Fluids Eng.* **121** (1999) 5 – 33.
3. C. Cercignani, *The Boltzmann Equation and its Applications*, Springer-Verlag, New York 1988.
4. G. Bird, *Molecular Gas Dynamics and the Direct Simulation of Gas Flows*, Oxford University Press, New York 1994.
5. J. Pfahler, J. Harley, H. Bau, H. J. Zemel: "Liquid Transport in Micron and Submicron Channels", *Sensors Actuators A* **21 – 23** (1990) 431 – 434.
6. X. F. Peng, G. P. Peterson, B. X. Wang: "Frictional Flow Characteristics of Water Flowing through Rectangular Microchannels", *Experimental Heat Transfer* **7** (1994) 249 – 264.
7. X. F. Peng, G. P. Peterson: "The Effect of Thermofluid and Geometrical Parameters on Convection of Liquids through Rectangular Microchannels", *Int. J. Heat Mass Transfer* **38** (1995) 755 – 758.
8. S. M. Flockhart, R. S. Dhariwal: "Experimental and Numerical Investigation into the Flow Characteristics of Channels Etched in (100) Silicon", *J. Fluids Eng.* **120** (1998) 291 – 295.
9. T. M. Harms, M. J. Kazmierczak, F. M. Gerner: "Developing Convective Heat Transfer in Deep Rectangular Microchannels", *Int. J. Heat Fluid Flow* **20** (1999) 149 – 157.
10. G. M. Mala, D. Li: "Flow Characteristics of Water in Microtubes", *Int. J. Heat Fluid Flow* **20** (1999) 142 – 148.
11. W. Qu, G. M. Mala, D. Li: "Pressure-Driven Water Flows in Trapezoidal Silicon Microchannels", *Int. J. Heat Mass Transfer* **43** (2000) 353 – 364.
12. J. Baudry, E. Charlaix, A. Tonck, D. Mazuyer: "Experimental Evidence for a Large Slip Effect at a Nonwetting Fluid-Solid Interface", *Langmuir* **17** (2001) 5232 – 5236.
13. V. S. J. Craig, C. Neto, D. R. M. Williams: "Shear-Dependent Boundary Slip in an Aqueous Newtonian Liquid", *Phys. Rev. Lett.* **87** (2001) 54504 – 54507.
14. E. Bonaccurso, M. Kappl, H. J. Butt: "Hydrodynamic Force Measurements: Boundary Slip of Water on Hydrophilic Surfaces and Electrokinetic Effects", *Phys. Rev. Lett.* **88** (2002) 76103 – 76106.
15. M. Rieth, W. Schommers (eds.): *Theoretical and Computational Nanotechnology*, American Scientific Publishers, Stevenson Ranch 2005.
16. P. A. Thompson, S. M. Troian: "A General Boundary Condition for Liquid Flow at Solid Surfaces", *Nature* **389** (1997) 360 – 362.
17. R. F. Probstein, *Physicochemical Hydrodynamics*, Wiley, New York 1994.
18. R. J. Hunter, *Zeta Potential in Colloid Science: Principles and Applications*, Academic Press, New York 1981.
19. C. Yang, D. Li, J. H. Masliyah: "Modeling Forced Liquid Convection in Rectangular Microchannels with Electrokinetic Effects", *Int. J. Heat Mass Transfer* **41** (1998) 4229 – 4249.
20. G. E. Karniadakis, A. Beskok: *Micro Flows — Fundamentals and Simulation*, Springer-Verlag, New York 2002.
21. V. Hessel, S. Hardt, H. Löwe: *Chemical Micro Process Engineering*, Wiley-VCH, Weinheim 2004.
22. J. H. Ferziger, M. Peric: *Computational Methods for Fluid Dynamics*, 3rd ed., Springer-Verlag, Berlin 2002.
23. J. P. Hartnett, M. Kostic: "Heat Transfer to Newtonian and non-Newtonian Fluids in Rectangular Ducts", *Adv. Heat Transfer* **19** (1989) 247 – 356.
24. H. F. P. Purday: *An Introduction to the Mechanics of Viscous Flow*, Dover, New York 1949
25. R. K. Shah, A. L. London: "Laminar Flow Forced Convection in Ducts", *Adv. Heat Transfer* Suppl. 1, 1978.
26. R. K. Shah: "Laminar Flow Friction and Forced Convection Heat Transfer in Ducts of Arbitrary Geometry", *Int. J. Heat Mass Transfer* **18** (1975) 849 – 842.
27. D. H. Richardson, D. P. Sekulic, A. Campo: "Low Reynolds Number Flow Inside Straight Microchannels with Irregular Cross Sections", *Heat Mass Transfer* **36** (2000) 187 – 193.
28. Y. Asako, M. Faghri: "Finite-Volume Solutions for Laminar Flow and Heat Transfer in a Corrugated Duct", *J. Heat Transfer* **109** (1987) 627 – 634.
29. R. C. Xin, W. Q. Tao: "Numerical Prediction of Laminar Flow and Heat Transfer in Wavy Channels of Uniform Cross-Sectional Area", *Numerical Heat Transfer* **14** (1988) 465 – 481.

30. V. K. Garg, P. K. Maji: "Flow and Heat Transfer in a Sinusoidally Curved Channel", *Int. J. Eng. Fluid Mech.* **1** (1988) 293 – 319.
31. A. M. Guzmán, C. H. Amon: "Dynamical Flow Characterization of Transitional and Chaotic Regimes in Converging-Diverging Channels", *Int. J. Eng. Fluid Mech.* **321** (1996) 25 – 57.
32. W. R. Dean: "Note on the Motion of a Fluid in a Curved Pipe", *Philos. Mag.* **4** (1927) 208 – 223.
33. W. R. Dean: "The Stream-line Motion of Fluid in a Curved Pipe", *Philos. Mag.* **5** (1928) 673 – 695.
34. L. Wang, T. Yang: "Multiplicity and Stability of Convection in Curved Ducts: Review and Progress", *Adv. Heat Transfer* **38** (2004) 203 – 254.
35. Y. Chen, P. Chen: "Heat Transfer and Pressure Drop in Fractal Tree-Like Microchannel Nets", *Int. J. Heat Mass Transfer* **45** (2002) 2643 – 2648.
36. V. Cominos et al.: "A Methanol Steam Micro-Reformer for Low Power Fuel Cell Applications", *Chem. Eng. Comm.* **192** (2005) 685 – 698.
37. J. M. Commenge, L. Falk, J. P. Corriou, M. Matlosz: "Optimal Design for Flow Uniformity in Microchannel Reactors", *AIChE J.* **48** (2000) 345 – 358.
38. M. Kaviany: *Principles of Heat Transfer in Porous Media*, Springer-Verlag, New York 1995.
39. S. J. Kim, D. Kim, D. Y. Lee: "On the Local Thermal Equilibrium in Microchannel Heat Sinks", *Int. J. Heat Mass Transfer* **43** (2000) 1735 – 1748.
40. R. C. Xin, W. Q. Tao: "Numerical Predictions of Laminar Flows and Heat Transfer in Wavy Channels of Uniform Cross Section", *Numerical Heat Transfer* **14** (1988) 465 – 481.
41. S. Hardt, W. Ehrfeld, V. Hessel, K. M. van den Bussche: "Strategies for Size Reduction of Microreactors by Heat Transfer Enhancement Effects", *Chem. Eng. Comm.* **190** (2003) 540 – 559.
42. T. A. Rush, T. A. Newell, A. M. Jacobi: "An Experimental Study of Flow and Heat Transfer in Sinusoidal Wavy Passages", *Int. J. Heat Mass Transfer* **42** (1999) 1541 – 1553.
43. G. Wang, S. P. Wanka: "Convective Heat Transfer in Periodic Wavy Passages", *Int. J. Heat Mass Transfer* **38** (1995) 3219 – 3230.
44. D. R. Sawyers, M. Sen, H. C. Chang: "Heat Transfer Enhancement in Three-Dimensional Corrugated Channel Flow", *Int. J. Heat Mass Transfer* **41** (1998) 3559 – 3573.
45. C. C. Wang, C. K. Chen: "Forced Convection in a Wavy Wall Channel", *Int. J. Heat Mass Transfer* **45** (2002) 2587 – 2595.
46. S. Hardt, T. Baier: "A Computational Model for Heat Transfer in Multichannel Microreactors", Proc. of ICMM05 - 3rd Int. Conf. on Microchannels and Minichannels, June 13–15, 2005, Toronto, Canada, paper ICMM 2005-75079.
47. T. Stief, O.-U. Langer, K. Schubert: "Numerical Investigations of Opimal Heat Conductivity in Micro Heat Exchangers", *Chem. Eng. Technol.* **22** (1999) 297 – 302.
48. R. B. Peterson: "Numerical Modeling of Conduction Effects in Microscale Counterflow Heat Exchangers", *Microscale Thermophys. Eng.* **3** (1999) 17 – 30.
49. G. Maranzana, I. Perry, D. Maillet: "Mini- and Microchannels: Influence of Axial Conduction in the Walls", *Int. J. Heat Mass Transfer* **47** (2004) 3993 – 4004.
50. V. Hessel, H. Löwe, F. Schönfeld: "Micromixers — A Review on Passive and Active Mixing Principles", *Chem. Eng. Sci.* **60** (2005) 2479 – 2501.
51. N. T. Nguyen, Z. Wu, "Micromixers — A Review", *J. Micromech. Microeng.* **15** (2005) R1 – R16.
52. S. Hardt, K. S. Drese, V. Hessel, F. Schönfeld: "Passive Micromixers for Applications in the Microreactor and μTAS Fields", *Microfluidics Nanofluidics* **1** (2005) 108 – 118.
53. D. Gobby, P. Angeli, A. Gavriilidis: "Mixing Characteristics of T-Type Microfluidic Mixers", *J. Micromech. Microeng.* **11** (2001) 126 – 132.
54. D. Bothe, C. Stemich, H. J. Warnecke: "Theoretische und experimentelle Untersuchungen der Mischvorgänge in T-förmigen Mikroreaktoren · Teil 1: Numerische Simulation und Beurteilung des Strömungsmischens", *Chem. Ing. Tech.* **76** (2004) 1480 – 1484.
55. J. M. Ottino: *The Kinematics of Mixing: Stretching, Chaos and Transport*, Cambridge University Press, Cambridge 1989.
56. C. A. J. Fletcher: *Computational Techniques for Fluid Dynamics*, 2nd ed. Springer-Verlag, Berlin 1991.
57. B. Noll: *Numerische Strömungsmechanik*, Springer-Verlag, Berlin 1993.

58. A. D. Stroock et al.: "Chaotic Mixer for Microchannels", *Science* **295** (2002) 647 – 651.
59. H. Aref: "Stirring by Chaotic Advection", *Int. J. Eng. Fluid Mech.* **143** (1984) 1 – 21.
60. T. G. Kang, T. H. Kwon: "Colored Particle Tracking Method for Mixing Analysis of Chaotic Micromixers", *J. Micromech. Microeng.* **14** (2004) 891 – 899.
61. M. A. Stremler, F. R. Haselton, H. Aref: "Designing for Chaos: Applications for Chaotic Advection at the Microscale", *Phil. Trans. R. Soc. Lond. A* **362** (2004) 1019 – 1036.
62. R. H. Liu et al.: "Passive Mixing in a Three-Dimensional Serpentine Microchannel", *J. Microelectromech. Syst.* **9** (2000) 190 – 197.
63. X. Niu, Y. K. Lee: "Efficient Spatial-Temporal Chaotic Mixing in Microchannels", *J. Micromech. Microeng.* **13** (2003) 454 – 462.
64. P. Tabeling, M. Chabert, A. Dodge, C. Jullien, F. Okkels: "Chaotic Mixing in Cross-channel Micromixers", *Phil. Trans. R. Soc. Lond. A* **362** (2004) 987 – 1000.
65. F. Bottausci, I. Mezic, C. D. Meinhart, C. Cardonne: "Mixing in the Shear Superposition Micromixer: Three-Dimensional Analysis", *Phil. Trans. R. Soc. London A* **362** (2004) 1001 – 1018.
66. F. Jiang, K. S. Drese, S. Hardt, M. Küpper, F. Schönfeld: "Helical Flows and Chaotic Mixing in Curved Microchannels", *AIChE J.* **50** (2004) 2297 – 2305.
67. S. Hardt, H. Pennemann, F. Schönfeld: "Theoretical and Experimental Characterization of a Low Reynolds Number Split-And-Recombine Mixer", *Microfluidics Nanofluidics* **2** (2006) 237 – 248.
68. K. S. Drese: "Optimization of Interdigital Micromixers via Analytical Modeling — Exemplified with the SuperFocus Mixer", *Chem. Eng. J.* **101** (2004) 403 – 407.
69. S. Hardt, F. Schönfeld: "Laminar Mixing in Interdigital Micromixers with Different Mixing Chambers — Part 2: Numerical Simulations", *AIChE J.* **49** (2003) 578 – 584.
70. F. Schönfeld, V. Hessel, C. Hofmann: "An Optimised Split-and-Recombine Micro-Mixer with Uniform "Chaotic" Mixing", *Lab on a Chip* **4** (2004) 65 – 69.
71. H. Chen, J. C. Meiners: "Topologic Mixing on a Microfluidic Chip", *Appl. Phys. Lett.* **84** (2004) 2193 – 2195.
72. G. I. Taylor: "Dispersion of Soluble Matter in Solvent Flowing Slowly through a Tube", *Proc. Roy. Soc. A* **219** (1953) 186 – 203.
73. R. Aris: "On the Dispersion of a Solute in a Fluid Flowing through a Tube", *Proc. Roy. Soc. A* **235** (1956) 67 – 77.
74. W. N. Gill, R. Sankarasubramanian: "Exact Analysis of Unsteady Convective Diffusion", *Proc. Roy. Soc. Lond. Ser. A* **316** (1970) 341 – 350.
75. M. R. Doshi, P. M. Daiya, W. N. Gill: "Three Dimensional Laminar Dispersion in Open and Closed Rectangular Conduits", *Chem. Engn. Sci.* **33** (1978) 795 – 804.
76. R. Aris: "On the Dispersion of a Solute by Diffusion, Convection and Exchange between Phases", *Proc. R. Soc. London Ser. A* **252** (1959) 538 – 550.
77. D. Dutta, D. T. Leighton: "Dispersion Reduction in Pressure-Driven Flow through Microetched Channels", *Anal. Chem.* **75** (2001) 57 – 70.
78. P. Daskopoulos, A. M. Lenhoff: "Dispersion Coefficient for Laminar Flow in Curved Tubes", *AIChE J.* **34** (1988) 2052 – 2058.
79. R. H. Perry, D. W. Green: *Perry's Chemical Engineers' Handbook*, 7th ed. McGraw-Hill, New York 1997.
80. J. M. Thomas, W. J. Thomas: *Principles and Practice of Heterogeneous Catalysis*, VCH, Weinheim 1997.
81. W. H. Press, S. A. Teukolsky, W. T. Vetterling, B. P. Flannery: *Numerical Recipes in Fortran 77*, Cambridge University Press, Cambridge 1992.
82. U. Maas: "Efficient Calculation of Intrinsic Low-Dimensional Manifolds for the Simplification of Chemical Kinetics", *Comput. Visualization Sci.* **1** (1998) 69 – 82.
83. U. Maas, S. B. Pope: "Simplifying Chemical Kinetics: Intrinsic Low-Dimensional Manifolds in Composition Space", *Combustion Flame* **88** (1992) 239 – 264.
84. M. Kiehl: "Partitioning Methods for the Simulation of Fast Reactions", *Zentralblatt Angew. Math. Mech.* **78** (1998) 967 – 970.
85. S. B. Pope, "Computationally Efficient Implementation of Combustion Chemistry using In Situ Adaptive Tabulation", *Combustion Theory Modeling* **1** (1997) 41 – 63.
86. D. Gobby, I. Eames, A. Gavriilidis: "A Vertically-Averaged Formulation of Catalytic Reactions in Microchannel Flows", Proc. of IMRET 5: 5th Int. Conf. on Microreaction

Technology, Springer-Verlag, Berlin 2001, pp. 141 – 149.

87. R. E. Walker: "Chemical Reaction and Diffusion in a Catalytic Tubular Reactor", *Phys. Fluids* **4** (1961) 1211 – 1216.
88. J. M. Commenge, L. Falk, J. P. Corriou, M. Matlosz: "Microchannel Reactors for Kinetic Measurement: Influence of Diffusion and Dispersion on Experimental Accuracy", Proc. of IMRET 5: 5th Int. Conf. on Microreaction Technology, Springer-Verlag, Berlin 2001, pp. 131 – 140.
89. W. E. TeGrotenhuis, D. L. King, K. P. J. Brooks, G. B. R. S. Wegeng: "Optimizing Microchannel Reactors by Trading-Off Equilibrium and Reaction Kinetics through Temperature Management", Proc. of IMRET6: 6th Int. Conf. on Microreaction Technology, AIChE Pub. No. 164 (2002) 18 – 28.

Author Index

Author Index

Biegler, Lorenz T., Carnegie Mellon University, Pittsburgh, Pennsylvania, United States (Chap. 10) *Mathematics in Chemical Engineering*, **3**

Bockhorn, Henning, Technische Hochschule, Darmstadt, Federal Republic of Germany, *Mathematical Modeling*, **203**

Bopp, Philippe A., Université Bordeaux I, France *Molecular Dynamics Simulation*, **323**

Boyd, Donald B., Indiana University-Purdue University at Indianapolis (IUPUI), Indianapolis, IN, United States *Molecular Modeling*, **307**

Buhn, Jörn B., Technische Universität Darmstadt, Darmstadt, Germany *Molecular Dynamics Simulation*, **323**

Finlayson, Bruce A., Department of Chemical Engineering, University of Washington, Seattle, Washington, United States (Chap. 1, 2, 3, 4, 5, 6, 7, 8, 9, 11 and 12) *Mathematics in Chemical Engineering*, **3**

Gatica, Jorge E., Department of Chemical and Biomedical Engineering, Cleveland State University, Cleveland, OH, USA *Model Reactors and Their Design Equations*, **151**

Grossmann, Ignacio E., Carnegie Mellon University, Pittsburgh, Pennsylvania, United States (Chap. 10) *Mathematics in Chemical Engineering*, **3**

Hampe, Manfred J., Technische Universität Darmstadt, Darmstadt, Germany *Molecular Dynamics Simulation*, **323**

Hardt, Steffen, TU Darmstadt, Darmstadt, Germany *Microreactors – Modeling and Simulation*, **401**

Hlavacek, Vladimir, Laboratory for Ceramic and Reaction Engineering, Department of Chemical Engineering, University of Buffalo, Buffalo, NY, USA *Model Reactors and Their Design Equations*, **151**

Orth, Andreas, University of Applied Sciences, Frankfurt am Main, Germany (Chap. 4, 5, 6, 7 and 8) *Design of Experiments*, **363**

Paschedag, Anja R., University of Technology Berlin, Berlin, Germany (Chap. 1) *Computational Fluid Dynamics*, **341**

Puszynski, Jan A., Chemical and Biological Engineering Department, South Dakota School of Mines and Technology, Rapid City, SD, USA *Model Reactors and Their Design Equations*, **151**

Soravia, Sergio, Process Technology, Degussa AG, Hanau, Germany (Chap. 1, 2, 3 and 8) *Design of Experiments*, **363**

Viljoen, Hendrik J., Department of Chemical and Biomolecular Engineering, University of Nebraska, Lincoln, NE, USA *Model Reactors and Their Design Equations*, **151**

Ullmann's Modeling and Simulation

ISBN: 978-3-527-31605-2

Subject Index

Subject Index

A
Adams integration method 54
Adams–Bashforth integration methods 51
Adams–Moulton integration method 54
Adiabatic reactor
 heat exchange in 195
 temperature profile in 195
Algebraic equations
 solution of sets of 5
Analysis of variance (ANOVA)
 in design of experiments 386
ANSYS 89
Approximation gap 108
Approximations
 piecewise 15
Axilrod–Teller interaction 328

B
Batch reactor
 for nonhomogeneous, gas–liquid systems 158
 for nonhomogeneous, solid–solid systems 157, 159
 homogeneous, isothermal 154
 nonisothermal 155
Beek and Singer model 179
Bernoulli distribution 127
Bifurcation theory 58
Binomial distribution function 128
Bodenstein number 177
Bodenstein number 246
Born–Oppenheimer approximation 325
Boundary element method 94
Boundary finite element method 94
Boundary integral equation 94
Boussinesq approximation 183
Box–Behnken designs 383
Box–Wilson method 193
Branch
 of a multiple-valued function 28
Branch-and-bound enumeration 112
Branch-and-bound method 113, 114
Branch-and-bound node 108
Brent's method 10
Brown's method 11
Brownian dynamics 338
Broyden's method 11
Buckingham potential 328
Bulirsch–Stoer recursion algorithm 15
Burgers viscosity equation 77
Burke–Plummer equation 183

C
Cambridge Structural Database 310
Carman–Kozeny equation 183
Car–Parrinello MD 325
Catalyst pellet
 one-phase model 174
 two-phase model 173
Cauchy condition
 in computational fluid dynamics 345
Cauchy's integral 27
Cauchy's theorem 26
Cauchy–Riemann equation 25
Central composite designs (CCD) 383
Central differencing scheme (CD) 356
Chebyshev polynomial 14
 for spectral method 83
Chemical Abstracts Service's Registry database 310
CHIP 121
Chi-square distribution 131
Cholesky decomposition 8
Collocation method
 for ordinary differential equations 65
Collocation methods 123
Complex number 22
Complex plane 22
 integration in 26
Complex variable 22
 analytic functions 25
 elementary functions 23
Computational fluid dynamics (CFD)
 in microreactor simulation 404
 micromixing models 349
 programs for 88
Comsol Multiphysics 88
Confidence level 131
Confidence interval 218
Conformational modeling 311
CONOPT 102, 104, 121
Constraint programming (CP) 120
Continuous stirred-ank reactor (CSTR)
 isothermal heterogeneous system 166
Continuous stirred-tank reactor (CSTR)
 cascade of 170
 kinetics 165
 nonisothermal, mathematical treatment of 167
 residence-time distribution 164
Continuous, ideally mixed, stirred-tank reactor
 concentration ratio of different reaction order 163
 kinetics 165
Continuous, ideally mixed, stirred-tank reactor 162
Convective diffusive equation 77
Convex hull relaxation 120
Coordinate system
 cylindrical 48
 spherical 48
Cost estimate
 for a chemical plant accuracy of estimation 130
Coulomb potential 328
Counterflow separation column
 mathematical modeling 272
Courant number 78
Courant–Friedrich–Lewy criterion 356
Covariance 129
CPLEX 112
Crank–Nicolson method 54, 357
Crystallizer
 continuously operating, size distribution 246
Cubic B-splines 73
Curvilinear coordinate 48
Cylindrical coordinate system 48

D
D-optimal designs 383
DAKOTA 106
Darcy–Oberbeck–Boussinesq model 183
Darcy's law 183
 modified 183, 184
Del operator 44
Density-functional theory 325
Derivative-free optimization 106

Ullmann's Modeling and Simulation

ISBN: 978-3-527-31605-2

Design of experiments
  class of models used in 381
  experimental investigations 365
  experimental strategies 368
  for special purposes 393
  optimization methods 390
  regression analysis 384
  standard 382
  visualizing models 384
Deterministic mathematical model 210
Deterministic system
  probability density functions 242
DICOPT 119
Difference equations, linear
  arising in staged operations, such as distillation ect. 12
Differential equation, ordinary
  sensitivity of the solution to the value of parameters 61
Differential–algebraic system 56
Differential equation
  mathematical models, based on 231
Diffusion equation 37, 79
Dirac delta function 235
Direct search method
  for solving nonlinear equations 223
Dirichlet condition
  in computational fluid dynamics 345
Dissipative particle dynamics 338
Distillation
  column, mathematics for 56
  solution of difference equations, arising in 12
Drug design modeling 314
Dyadics 41
  divergence of 45
  divergence theorem 47

E

ECLiPSe 121
Eddy break up model 349
Eigenvalue
  eigenvalue problem, numerical methods for 92
  of matrices 12
Electrostatic (Coulomb) interaction 327
Elliptic partial differential equation 75
Engulfment model 349
Enthalpy 136
Entropy 136
Equations
  linear algebraic 7
  linear difference equations 12
Equations
  nonlinear, direct search methods 223
  systems of nonlinear, gradient methods for solving 225
Ergun equation 184
Error analysis
  in experiments 132
Ethylene
  vapor-phase catalytic oxidation of 185
Euler integration method
  for ordinary differential equations 51
Euler–Euler method
  in computational fluid dynamics 351
Euler–Lagrange method
  in computational fluid dynamics 351
Euler–Newton continuation 60
Evaporation cooler
  simultaneous mass and heat transfer, dynamic model 269
Evolutionary operation (EVOP)
  in design of experiments 390
Ewald summation 327
Experiments
  error analysis 132
  factorial design of, and analysis of variance 133
Extended cutting plane method 113, 116
Extraction
  solution of difference equations, arising in 12

F

Factorial design
  2:2 factorial design 371
  2:3 factorial design 374
  fractional 377
Factorial design of experiments 133
Factorial designs 370
Finite difference method 81
  for ordinary differential equations 63
Finite element method 15, 71, 82
Flamelet model 349
Flow
  in packed-bed reactors 183
Flow phenomenon
  applications of the law of conservation of mass 267
  mathematical models, application of the principle of
    conservation of momentum 254
  mathematical models for 254
  reactors with turbulent flow, mathematical model for 247
Fluent 89
Fluid mechanics
  mathematics 86
  vector formulas useful in 45
Force field modeling 312
Fourier series 29
Fourier transform 30
  of discrete data 21
Fredholm equation of the first and second kind 90
  numerical methods for 91
Fredholm integral equation 89

G

Galerkin finite element method 71
GAMS 119
Gauss–Hermite polynomials
  quadrature formula for 19
Gaussian quadrature points and weights 19
Gauss–Newton approximation 104
Gauss–Ostrogradski integration principle 259
Gauss–Seidel method 9
Gay Berne model 328
Gear's backward difference formula 55
Generalized Benders decomposition 113
Generalized Benders decomposition 115
Genetic algorithms 105
Geometric mean 125
Glauber distribution 105
Global polynomial 13
Gradient-based nonlinear programming 96
Gradient method
  for solving nonlinear systems of equations 225
Green's theorem 47
Green's function 92
GRG2 104

H

Hagen–Poiseuille flow 255
Hamilton–Cayley theorem 13
Hammerstein equation 90
  numerical methods for 91
Heat conduction
  equation 37
  equation, unsteady 86
Heat conduction
  mathematical model for 262
Heat-transfer coefficient
  for hot air at the wall of a pipe with turbulent flow, model
    for 217

Heat transfer
    mathematical modeling 263
    simultaneous mass and heat transfer, dynamic model 269
Heaviside expansion 35
Hermite cubic polynomial 16
Hermite polynomial 14
Homology modeling 314
Homotopy methods 11
Hopf bifurcation
    analysis 285, 286
Hyperbolic partial differential equation 75
Hypergeometric distribution function 128

I
ILOG Solver 121
Impeller
    marine-type 164
    turbine 164
Integer programming (IP) 96
Integral equation 89
Integral transform 29
Intermolecular potential
    three-body and higher order 328
Internal energy 136
Interpolation
    two-dimensional, and quadrature 22
IPOPT 101, 104, 121
Iterative methods
    for solution of sets of linear algebraic equations 9

J
Jacobi method 84
Jacobi polynomial 14
Jacobian matrix 11
Jet stirred reactor
    mathematical model for 256

K
Karush–Kuhn–Tucker condition 97, 115
Kinetics
    of continuous stirred-tank reactors 165
    of continuous, ideally mixed stirred tank reactors 165
Kinetics
    modeling of the mean reaction rates 299
KNITRO 104
Kolmogorov length 256
Kreisselmeier–Steinhauser function 122
Kronecker delta 42
Kronecker delta 254

L
Lack-of-fit test
    in design of experiments 386
Lagrangian particle tracking 421
Laguerre polynomial 14
Laminar flow
    mathematical model for 255
LANCELOT 102, 104
Langmuir–Hinshelwood theory
    oxidation rate of NO 192
Laplace equation 26
Laplace transform 33
Laplacian operator 45
Large-eddy simulation (LES)
    in computational fluid dynamics 348
Lattice Boltzmann method 87
Lattice Boltzmann method (LBM) 357
Laurent series 27
Legendre polynomial 15
Leibniz formula 47
Lennard–Jones ansatz 328
Lennard–Jones potential 327
Level set method 87
Likelihood function 229
Linear independence constraint qualification (LICQ) 98
Linear programming (LP) 95
    branching strategies 112
    depth-first strategies 112
Local optimization method 97
Logic-based optimization 119
Loop reactor
    mathematical modeling 279, 280
LOQO 104
Lorentz–Berthelot rule 328
LP/NLP-based branch-and-bound method 116

M
MacCormack method 78
MacLaurin series 27
Mangasarian–Fromovitz constraint qualification 98
Maple 88
Mass-expansion coefficient 184
Mass transfer
    simultaneous mass and heat transfer, dynamic model 269
    with chemical reactions, mathematical modeling 274
    without chemical reaction, mathematical modelling 267
Mathcad 88
Mathematica 88
Mathematical model
    for a set of experimental data 13
Mathematical model
    based on differential equations 231
    based on transport equations for probability density functions 212
    empirical 213
Mathematical modeling 203
Mathematics
    computer software for 88
    full discretization method 123
    optimization problems 95
    partial discretization method 122
    variational methods 121
Matlab 88
Matrix
    eigenvalues of 12
    for solution of sets of linear algebraic equations 7
    properties 5
    QR factorization of 6
    singular value decomposition 6
Maximum likelihood method 229
McCabe–Thiele diagram 273
MDL Information Systems' Available Chemicals Directory 310
Mean deviation 125
Metropolis distribution 105
Micromixing 418
Microreactors
    3D solution to flow distribution problems 408
    chemical kinetics in 429
    flow distribution 405
    heat transfer in 411
    reduced-order flow model 408
Microreators
    mass transfer 418
    mixing 418
Middle value 125
MINLP_BB 119
MINOPT 119
MINOS 103, 104
Mixed integer programming 110
    linear 110
    nonlinear 112
Mixed-integer linear programming (MILP) 96, 110

Mixed-integer linear programming (MINLP) 112
  extensions of 117
Mixed-integer nonlinear programming (MINLP) 96
Mixing
  in microreactors 418
  split-and-recombine (SAR) mixing 426
Model predictive control (MPC) 102
Molecular dynamics simulation
  data for 330
  intermolecular interactions 327
  intramolecular interactions 326
  simple average 335
  time-correlation function 335
Monte Carlo method
  in molecular modelling 311
Monte-Carlo simulation 324

N
Navier–Stokes equation 86
Navier–Stokes equations 344
Nelder–Mead algorithm 106
NEOS 104
Neumann condition
  in computational fluid dynamics 345
Neville's algorithm 14
Newton–Raphson method 100, 190
Nonequilibrium molecular dynamics 338
Nonlinear programming (NLP) 95
  algorithmic details 103
  convex 97, 99
  global optimization 106
  linear programs 100
  local optimality conditions 97
  nonconvex 97
  nonconvex, with bilinear, linear fractional, and concave separable terms 107
  quadartic programs 100
  quauad 102
  solvers 104
Normal probability distribution 126
NPSOL 104
Nucleic Acid Database 310
Null space algorithm 100
NVE simulation 332
Nyquist critical frequency 21

O
One-factor-at-a-time method 369
Optimization
  classes 95
  dynamic 121
  global 106
  problems 95
Ordinary differential equation
  as initial value problems 49
  computer software for solving 57
Orthogonal collocation method
  for ordinary differential equations 65
  on finite elements 65
OSL 112
ougen–Watson–Langmuir–Hinshelwood kinetics 223
Outer approximation method 114
Outer-approximation algorithm 107
Overall upper bound 108

P
Péclet number 177
Packed-bed reactor
  bed porosity of 176
  mathematical treatment of 178
Packed–bed reactor
  pressure-drop effects 184
Packed-bed reactor 171
  energy balance, mathematical treatment of 172
  mass balance, mathematical treatment of 172
  mass transport in 176
  mathematical treatment of 55
  radial flow 185
  reactor performance 185
  single tube 185
  spherical flow 185
Pair potential 327
Parabolic partial differential equation 75
Parseval equation 31
Partial differential equation 75
  solution by using transforms 37
Partial differential equation 258
Plackett–Burman designs 383
Poisson distribution 128
Poisson's equation 95
Polymath 88
Pontryagin's maximum principle 121
Potential of mean forces 335
Power-law kinetics 185
Probability
  of outcomes, and statistics 125
Probability density function 127
Probability density function approach (PDF) 349
Probability density function 212
  examples of calculating 242
  mathematical models, based on transport equations for 232
  single-point, transport equations for 237
Protein Data Bank 310

Q
Quadratic programming (QP) 95
Quadratic programs (QP)
  convex 100
Quadratic upstream interpolation for convective kinetics (QUICK) 356
Quadrature formulas
  for calculation of integrals 18
Quantum mechanical modeling 311
Quasi-Newtonian approximation 101, 103

R
Radial pair distribution function 335
Range space method 100
Reaction field technique 327
Reactor
  model for a chemical reactor with axial diffusion 63
  optimization of 189
  stirred tank, mathematics for 57
Reactor
  cascade of ideally mixed, mathematical models 278
  heterogeneous catalytic reactions in, mathematical modeling 287
  homogeneous phase, mathematical modeling 276
  ideally mixed nonisothermal flow reactor, stability analysis 294
  ideally mixed stirred-tank reactor, residence time distribution function 242

residence time distribution function 244
isothermal, mathematical modeling 277
isothermal, sensitivity analysis 286
isothermal, stability analysis 280
jet-stirred fluid-phase with turbulent flow, mathematical model 232
jet stirred, mathematical model for 256
loop reactor, mathematical modeling 279
nonisotherm nonisothermal, stability analysis 294
residence time distribution, functions for 242
with finite mixing, residence time distribution function 245
with turbulent flow, mathematical model for 232
Rectification column
mathematical modeling 273
Reduced gradient methods 102
Regression measure
in design of experiments 385
Regression, linear
used in design of experiments 384
Regression, linear 214
standardized computer programs 219
Regression, nonlinear 222
Residence time distribution function
for chemical reactors 242
Response surface designs 380
Reynolds averaging
in computational fluid dynamics 347
Reynolds-averaged Navier–Stokes (RANS) 347
Reynolds flux
mathematical modeling 299
Reynolds stress
mathematical modeling 298
mathematical models for 249, 256
Richardson extrapolation 53
Riesz–Fischer theorem 31
Romberg's method
for integrals 20
Root-mean-square 125
Runge–Kutta method 51
Runge–Kutta–Feldberg method 52
Runge–Kutta–Gill method 52

S
Schmidt number 258
Schur complement method 100
Secant method 10
Second derivative (Hessian) matrix 97
Sensitivity analysis
of a model 60
Sensitivity analysis
of a model 216
Shooting method
for ordinary differential equations 62
Simplex method 224
Simpson's rule
for integrals 18
Simulated annealing 105
SNOPT 104
SOCS 104, 121
SOLVER 102, 104
Spatial branch-and-bound algorithm 107, 109
Spherical coordinate system 48
Splines 16
Split-and-recombine (SAR) mixing 426
SQP algorithm 119
SRQP 104
Standard deviation 125
Statistical modeling 313
Statistical mathematical model 211
Statistical system
probability density functions 247
Statistical variable 210
Statistics
probability of outcomes, use of statical methods 125
Stiffness
of a system of differential equations 55
Stirred-tank reactor
mathematics for 57
Stokes theorem 47
Student's t-distribution 126
Styrene
by catalytic dehydration of ethylbenzene, mathematical modeling 287
Subgrid-scale models (SGS models)
of Smagorinsky type 348
Successive quadratic programming (SQP) 100
Sutton–Chen model 328

T
Taylor–Aris dispersion 428, 432
Taylor–Galerkin method 78
Thermal-expansion coefficient 184
Thermodynamics
differentials of thermodynamic functions 135
partial derivatives of thermodynamic functions 137
state functions 135
Thiele modulus 74
Time-correlation function
in molecular dynamics simulation 335
Transport equation
for single-point probability density functions 237
Trapezoid rule
for integrals 18
Trust-region method 104
Turbulent diffusion flame
combustion of hydrogen in, mathematical modeling 301
combustion of propane, model for 252
Turbulent flow
nonreactive, free jets, mathematical model for 255

U
Upwind differencing scheme (UD) 355
Urysohn equation 90
Urysohn equation of the second kind
numerical methods for 91

V
Variance 125
Variance analysis
in design of experiments 386
Vector 40
curl of 44
divergence of 44
divergence theorem 46
Vector analysis 40
Vector differential operator 44
Vector differentiation 43
Vector integration 46
Vector operation 42
Vectorial search method 224
Velocity autocorrelation function 337
Volterra equation of the second kind 89
numerical methods for 91
Volterra integral equation 89

W
Wall heat-transfer coefficient
for packed-bed reactors 178

X
XPRESS 112

ULLMANN'S

Modeling and Simulation

Related Titles

Puigjaner, L., Heyen, G. (Eds.)
Computer Aided Process and Product Engineering
2 Volumes
2006
ISBN 978-3-527-30804-0

Keil, F. J. (Ed.)
Modeling of Process Intensification
2007
ISBN 978-3-527-31143-9

Paschedag, A. R.
CFD in der Verfahrenstechnik Allgemeine Grundlagen und mehrphasige Anwendungen
2004
ISBN 978-3-527-30994-8

Wiley-VCH (Ed.)
Ullmann's Chemical Engineering and Plant Design
2 Volumes
2004
ISBN 978-3-527-31111-8

Wiley-VCH (Ed.)
Ullmann's Electronic Release 2007
Online plus CD-ROM
2007
ISBN 978-3-527-31602-1

Wiley-VCH (Ed.)
Ullmann's Encyclopedia of Industrial Chemistry
Sixth Edition, 40 Volumes
2003
ISBN 978-3-527-31096-8